Energy Supply within Sustainable Agricultural Production: Challenges, Policies and Mechanisms

Editors
Piotr Gradziuk
Polish Academy of Sciences, Institute of Rural and Agricultural Development, Economic Modelling Department
Poland

Bogdan Klepacki
Institute of Economics and Finance of the Warsaw University of Life Sciences (SGGW) in Warsaw
Poland

Mariusz J. Stolarski
University of Warmia and Mazury in Olsztyn
Poland

Editorial Office
MDPI
St. Alban-Anlage 66
4052 Basel, Switzerland

This is a reprint of articles from the Special Issue published online in the open access journal *Energies* (ISSN 1996-1073) (available at: https://www.mdpi.com/journal/energies/special_issues/energy_supply_within_sustainable_agricultural_production_challenges_policies_mechanisms).

For citation purposes, cite each article independently as indicated on the article page online and as indicated below:

LastName, A.A.; LastName, B.B.; LastName, C.C. Article Title. *Journal Name* **Year**, *Volume Number*, Page Range.

ISBN 978-3-0365-5175-3 (Hbk)
ISBN 978-3-0365-5176-0 (PDF)

Contents

About the Editors

Piotr Gradziuk

MSc in Agricultural Economics from the Faculty of Agriculture, Agricultural Academy in Lublin, Poland (currently under new name University of Life Sciences in Lublin). PhD in Economics, Faculty of Economics, Maria Curie-Skłodowska University in Lublin and D.Habil. in Economics, Faculty of Economics, Warsaw University of Life Sciences. From 1982 to 2015, he worked in the Faculty of Agricultural Sciences, University of Life Sciences in Lublin. He is currently employed as Associate Professor, Polish Academy of Sciences, Institute of Rural and Agricultural Development, Warszawa. Fields of scientific specialization: effectiveness of renewable energy sources utilization; socio-economic and institutional changes in rural areas; efficiency and productivity.

2018–2022, "LIFT: Low-Input Farming and Territories - Integrating knowledge for improving ecosystem-based farming" in the EU Horizon 2020 programme (ID: 770747), http://lift-h2020.eu, co-author.

2017–2021, "SURE-Farm: Towards SUstainable and REsilient EU FARMing systems" in the EU Horizon 2020 programme (ID: 727520), https://surefarmproject.eu, co-author.

2018–2021, "POWER4BIO - emPOWERing regional stakeholders for realising the full potential of european BIOeconomy" in the EU Horizon 2020 programme (ID: 818351), https://power4bio.eu, co-author.

https://www.irwirpan.waw.pl/12/team/employees/2/team/employees/2/wpis/62

https://scholar.google.com/citations?user=dcj0h7gAAAAJ&hl=en

https://www.researchgate.net/profile/Piotr-Gradziuk

Bogdan Klepacki

A graduate of the Faculty of Economics and Agriculture of the Warsaw University of Life Sciences, and professor of economic sciences. Author of about 600 publications, including about 360 original papers and scientific monographs. Supervisor of 29 positively completed doctoral dissertations, including 9 habilitated doctors and 1 professor among the doctoral students. His main research areas are: economic and organizational effects of expanding the area of farms, possibilities of improving the efficiency of farming, adaptation processes in enterprises, assessment of the effectiveness of technological progress in agriculture, economic importance of logistics, with particular emphasis on agrology (logistics in the food sector).

Dr. h.c. University of Agriculture in Krakow, Honorary Professor of the Universities of Life Sciences in Kiev and Dublany (Lviv). Third term member of the Economic Sciences Committee (KNE) of the Polish Academy of Sciences, two-term chairman of the former Committee of Agricultural Economics and Rural Development of the Polish Academy of Sciences, former president of the Association of Agricultural and Agribusiness Economists (SERiA), former long-term chairman of the Scientific Council of the Institute of Rural and Agricultural Sciences (IRWiR) of the Polish Academy of Sciences.

Mariusz J. Stolarski

Professor of agricultural sciences, working at the University of Warmia and Mazury in Olsztyn (UWM), conducting research into the following: bioresources, bioeconomy, renewable energy sources (RES), biomass growing, productivity, quality, the use of perennial and annual alternative crops for energy and industrial purposes, production technology and logistics of biofeedstock acquisition and processing to obtain higher added value bioproducts and conversion to solid, liquid, and gaseous biofuels, the cascaded use of various agricultural- and forest-origin biofeedstock types, the bioconversion of various types of biomass residues by insects, the assessment of the economic, energetic, and environmental effectiveness of cultivation, production, logistics, and processing of biomaterials, and the characterization and assessment of the usability of biomass and other RES on individual, local, regional, national, and international levels. He is the co-author of over 80 original publications with the IF, and the co-author of 11 willow varieties and 1 patent. https://orcid.org/0000-0003-0712-9678;

https://www.researchgate.net/profile/Mariusz-Stolarski

Preface to "Energy Supply within Sustainable Agricultural Production: Challenges, Policies and Mechanisms"

The inspiration for accepting the proposal to join in the editing process of this Special Issue came from a statement in the preamble to Directive (EU) 2018/2001 on the promotion of the use of energy from renewable sources that increasing the use of RES "has a fundamental part to play in promoting the security of energy supply, sustainable energy at affordable prices, technological development and innovation as well as technological and industrial leadership while providing environmental, social and health benefits as well as major opportunities for employment and regional development, especially in rural and isolated areas, in regions or territories with low population density or undergoing partial deindustrialisation". Agriculture and rural areas may play a very important role in solving these problems to a greater extent than before. The Issue intends to link together research on agricultural sustainability and its generation of energy. It also aims to contribute to policy debate on supporting agricultural development locally, regionally or globally to assure its input into sustainability through increasing delivery of renewable energy. Hence, the relevant questions were what enhances the sustainability of agricultural production and how that can be practically achieved through the increase of efficiency in use of energy and other related resources. This Special Issue aims to be of service to policy makers dealing with sustainability, energy, and agricultural policies. The papers also offer insights into solutions for working in times of crises, such as global economic crises (e.g., 2008/2009), world embargoes (e.g., Russian embargo of 2014 and 2022, associated with the war in Ukraine) or COVID-19, as both supply and demand fluctuate in times of extraordinary conditions arising from economic, social, environmental, and other challenges.

As the guest editors of this Special Issue, we would like to kindly thank MDPI and the *Energies* team for providing this extraordinary opportunity of learning and growth as well as the editorial staff, especially Ms. Reka Kovacs, for their continuous support and consideration. We must acknowledge the fact that such interactions are an excellent platform especially for young researchers for their scientific growth and we hope the readers enjoy this research.

Piotr Gradziuk, Bogdan Klepacki, and Mariusz J. Stolarski
Editors

energies

MDPI

Article

The Potential of Agricultural Biogas Production in Ukraine—Impact on GHG Emissions and Energy Production

Adam Wąs [1,*], Piotr Sulewski [1], Vitaliy Krupin [2], Nazariy Popadynets [3], Agata Malak-Rawlikowska [1], Magdalena Szymańska [4], Iryna Skorokhod [5] and Marcin Wysokiński [1]

1 Institute of Economics and Finance, Warsaw University of Life Sciences—SGGW, Nowoursynowska 166, 02-787 Warsaw, Poland; piotr_sulewski@sggw.edu.pl (P.S.); agata_malak_rawlikowska@sggw.edu.pl (A.M.-R.); marcin_wysokinski@sggw.edu.pl (M.W.)
2 Institute of Rural and Agricultural Development, Polish Academy of Sciences, Nowy Świat 72, 00-330 Warsaw, Poland; vkrupin@irwirpan.waw.pl
3 M. I. Dolishniy Institute of Regional Research of NAS of Ukraine, Kozelnytska 4, 79026 Lviv, Ukraine; popadynets.n@gmail.com
4 Agricultural Institute, Warsaw University of Life Sciences—SGGW, Nowoursynowska 159, 02-776 Warsaw, Poland; magdalena_szymanska@sggw.edu.pl
5 Lesya Ukrainka Eastern European National University, Volia Ave. 13, 43025 Lutsk, Ukraine; skorokhodiryna1@gmail.com
* Correspondence: adam_was@sggw.edu.pl; Tel.: +48-22-59-342-18

Received: 4 October 2020; Accepted: 30 October 2020; Published: 3 November 2020

Abstract: Renewable energy production is gaining importance in the context of global climate changes. However, in some countries other aspects increasing the role of renewable energy production are also present. Such a country is Ukraine, which is not self-sufficient in energy supply and whose dependency on poorly diversified import of energy carriers regularly leads to political tensions and has socio-economic implications. Production of agricultural biogas seems to be a way to both slow down climatic changes and increase energy self-sufficiency by replacing or complementing conventional sources of energy. One of the most substantial barriers to agricultural biogas production is the low level of agricultural concentration and significant economies of scale in constructing biogas plants. The aim of the paper was thus to assess the potential of agricultural biogas production in Ukraine, including its impact on energy self-sufficiency, mitigation of greenhouse gas (GHG) emissions and the economic performance of biogas plants. The results show that due to the prevailing fragmentation of farms, most manure cannot be processed in an economically viable way. However, in some regions utilization of technically available manure for agricultural biogas production could cover up to 11% of natural gas or up to 19% of electricity demand. While the theoretical potential for reducing greenhouse gas emissions could reach 5% to 6.14%, the achievable technical potential varies between 2.3% and 2.8% of total emissions. The economic performance of agricultural biogas plants correlates closely with their size and bioenergy generation potential.

Keywords: agricultural biogas; bioenergy; biomethane; GHG emission; economic performance; regional analysis; Ukraine

1. Introduction

Production of agricultural biogas can be analyzed and assessed from three essential perspectives, i.e., the ecological, economic and social. In particular the environmental benefits of agricultural biogas production are often emphasized [1–3]. This is important, since one of the goals of the UN's Sustainable

The growth of the total renewable energy generation is planned to gradually increase and reach 8% in 2020, rising to 25% by 2035, reaching 23 Mtoe (or 4.4 times the actual value for 2018).

The strategy [31] stresses that in order to achieve the goals for renewable energy sources it is crucial to increase the use of biomass in the generation of electricity and heat by: (1) stimulating biomass use as a fuel in enterprises that produce biomass as a byproduct, (2) informing about the possibilities of biomass use in individual heating, and (3) supporting the creation of competitive biomass markets. The creation of proper logistics system and infrastructure aimed to collect and transport the biological raw material is necessary to ensure the achievement of these goals.

Despite the government and business efforts, the growth both of total renewable energy generation and the energy from biomass alone, biofuels and wastes are falling behind their expected growth rates defined in the above strategy, which underlines the need to intensify efforts toward structural transformations in energy generation. Biogas generation plants, in this case, are among the crucial drivers of change that can help achieve the targeted transformation values, serving both the country's energy independence and working towards a nationwide switch to renewable energy.

Biogas plants in Ukraine are a relatively new form of energy generation. Even though the first such plant was built in 1993 on a pig farm in the Zaporizka region, until 2012, only four biogas plants were functioning in agricultural enterprises [34,60]. There may be a slight confusion, since in 2013 [61] Ukraine's State Agency on Energy Efficiency and Energy Saving reported the first biogas plant as operating only in that year. However, the reason for this is that the agency monitors only the biogas plants supplying the energy utilizing the feed-in tariff, which was introduced for biogas plants in 2013. Since then, their numbers have been steadily growing, reaching 21 units by the end of 2019 with an overall generation capacity of 59 MWe [62].

Overall, state support for renewable energy generation in Ukraine intensified in 2008 with the introduction of feed-in tariffs [63]. Nevertheless, for several years there has been a visible imbalance in the development of particular types of renewable energy generation plants, as the feed-in tariff was not available to some types of plant. This was the case of the biogas plants, as the only bioenergy generation supported by the feed-in tariff was based on crop biomass. In April 2013 the legislation was changed, and all types of generation plants based on biogas and biomass were covered. Nevertheless, the tariff levels were highly differentiated between the renewable energy types, thus giving most preferences to solar and hydro energy. The rate for solar energy was set at €0.3393 to €0.3586/kWh, hydro energy at €0.116316 to €0.19386/kWh, while the tariff for generation based on biomass and biogas was €0.1239/kWh [64] (pp. 19–24) and was expected to gradually decrease until 2030. It was only in 2017 that the tariff for biomass and biogas was set at a constant €0.1239/kWh with 2030 as the cut off year [65], which created additional security and potential viability for current and future investments. One of the key advantages of the feed-in tariff that it is set in euro, since the value of Ukrainian currency (UAH) has been highly volatile in the past decade. There are also numerous preferences for investors, such as preferential import tariffs for equipment bought for the construction of power plants based on renewable energy technologies [66].

Experts emphasize several issues with the Ukrainian feed-in tariff, all of which are connected with institutional aspects of the Ukrainian economy and legislation [67]: (1) the tariff cannot be applied for mixed energy generation, (2) while each investor needs to know if it will be possible to sell the energy at the feed-in tariff, this is not possible until the investment is fully operational and production is permitted, (3) in order to receive the permit for the feed-in tariff, 50% of the energy generation plant construction/equipment costs need to be of Ukrainian origin.

Due to particular economic and institutional aspects, so far the main investors in Ukrainian biogas generation plants have been the agro holdings [68], with the largest agricultural enterprises being involved in primary agricultural production itself. However, as the construction of biogas plants gains in intensity due to recently fixed feed-in tariff, it is expected that more entities will use this opportunity to expand their potential income.

2.3. Ukrainian Agriculture as a Feedstock Supplier for Biogas Generation

Despite the country's long-term political, financial and economic instability, agriculture in Ukraine is one of the few sectors managing to increase production and steadily expand on foreign markets. The 10% share of the agricultural sector in real Ukrainian GDP and 39% share in total export value reveal one of the current essential specializations of the Ukrainian economy [69].

The good conditions for agriculture in Ukraine have long been known [70], as country's geographic position and its climate and soil quality are elements of a highly beneficial environment both for crop and livestock production. In 1991 (the last year when the share of livestock production exceeded the value of crops) the relationship between the value of the agricultural subsectors (crops to livestock) was 49.4% to 50.6% [71]. Adaptation of the Ukrainian agricultural sector to market conditions and its structural transformation since the beginning of the 1990s have changed these proportions, gradually shifting the focus towards crops and showing a gradually intensifying decline in livestock. As of the 2018 [72] (p. 287) value-wise the proportion between the two subsectors was 73.7% for crops to 26.3% for livestock. The livestock inventory at the end of 2018 [72] (p. 287) included 3.3 million head of cattle (including 1.9 million dairy cows), 6.0 million pigs, 1.3 million sheep and goats, and 211.7 million head of poultry.

Currently, livestock production in Ukraine and its development trend varies greatly depending on the subsectors. Thus, the beef and dairy products are in continuing decline, together with the production of pigs. The poultry subsector stabilized in the early 2000s and since then by 2019 had almost regained its initial level (falling back by only 14.0% compared to the 1991 figures).

Key factors influencing the decline of the livestock sector were the issues with adaptation to the market conditions in the 1990s, a shift the focus of crops from domestic to foreign markets (with the domestic market shrinking from ca. 52 million people in 1991 to almost 42 million by 2019) with a simultaneous decrease in areas under fodder crops (over six-fold, from 12 million ha in 1991 to 1.8 million ha in 2018) leading to a price increase, an overall profitability decrease in livestock production, as well as the inability of farms to comply with the changing requirements regarding production processes, quality and safety of products. Due to these factors, livestock production is either concentrated in medium and large agricultural enterprises (those producing marketable products) or small family farms for self-sustenance (mostly producing non-marketable commodities).

Despite the current difficult situation in the livestock sector, there is a potential for positive change. According to economic forecasts [73] the sector's physical output is to increase by 2030, even though the declining trend in livestock numbers will remain. The driving force is the domestic consumption of meat and dairy products as a result of growing consumer income.

The key messages in Ukrainian scientific publications [74–76] stress the need to support the transformation of the livestock sector. The sector should be reformed towards innovative and cost-effective production technologies and processes, as well as to ensure and diversify its income-generation abilities to maintain its resilience. It is crucial to maintain country's food security [77]; measures aimed to support the development of livestock sector were therefore also included in the 2020 national budget [78]. The search for reserves, including cutting costs and expanding revenues, is one of the ways to improve the economic viability of livestock farms, in which the generation of biogas, as a by-product of their primary economic activity, could aid farms of various sizes and organizational forms in additional income generation. This would also contribute to building the capacity for national energy independence and help to transform the energy sector into more environmentally and climate-friendly renewable technologies.

3. Materials and Methods

The production of agricultural biogas depends mainly on the availability of a suitable substrate. Data on the number of main farm animals (cattle, pigs, poultry) were therefore collected in order to determine the possibility of producing agricultural biogas. The polarized structure of Ukrainian agriculture, resulting in the presence both of large commercial entities and numerous small individual farms, made it necessary to collect data on the number of animals regarding the legal form of the farm, distinguishing commercial farms and small family farms. The basic source of data used in the study was that published by the State Statistics Service of Ukraine (Ukrstat), as well additionally obtained from Ukrstat's detailed unpublished data.

Determining the number of animals made it possible to estimate the quantity of natural fertilizer (animal excrement) available for biogas production. The estimates were based on the research and legislation sources providing amounts of manure obtained per group of animals [79,80]. The theoretical potential of biogas production was established based on the amount of substrate determined [81]. The assumed manure production and biogas yield coefficients are presented in Table 1.

Table 1. Manure production and biogas yield coefficients.

	Cattle	Pigs	Poultry
Manure production (/head/year)	16.8	2.92	0.1
Biogas yield (m^3/t of fresh matter)	25	24	51.3

Source: own elaboration.

The volume of available substrates and potential biogas was determined for regions of Ukraine based on data from agricultural enterprises but excluding small family farms. This is due to the significant fragmentation of family farms, resulting in a very high number of units (4.6 million in 2018 [82], a relatively small number of animals per farm, as well as the limited investment opportunities of these farms, it was decided to omit the theoretical potential of biogas for these farms in further analyses.

The technical potential of agricultural biogas production was determined assuming that it is technically possible to create a biogas plant with a minimum power of the CHP unit exceeding 10 kWe, which if feeding only with slurry and manure requires a stocking level of about 30 livestock units (LU), which is an equivalent of 30 adult cows (for details see: https://ec.europa.eu/eurostat/statistics-explained/index.php/Glossary:Livestock_unit_(LSU)). Based on the Ukrstat data on herd size, the commercial farms were divided into categories regarding potential biogas generation: below a technical threshold, small, medium and large.

In the small farms, with herds from 30 to 100 LU, a micro-biogas installation ensuring an appropriate amount of biogas for a 25 kWe cogeneration unit was analyzed as a representative example. For medium-sized farms, biogas plants with a scale corresponding to the aggregate power of 100 kWe were assumed, with 750 kWe for the largest farms. The number of potential biogas plants required to achieve the technical potential of biogas production on a national scale was determined on the basis of assumptions about the amount of manure available. The structure of farms in terms of the size of animal herds was analyzed at national level. Due to the insufficient number of objects, mostly the largest farms, Ukrstat does not provide complete data on the herds' structure in the regions.

The technical potential of biogas production on farms was calculated on the basis of the amount of manure available that could be used for it. In order to compare the amount of energy produced in biogas plants to the current demand, the amount of methane that could be produced was determined (assuming that biogas contains 55% methane) as was the amount of electricity that could be produced from it. The comparison of the amount of biomethane produced with natural gas consumption can be considered to a limited extent. The existing technologies for purifying biogas to grid parameters are still challenging to obtain for small-scale installations and relatively expensive [83]. The purpose of this comparison is only to determine the possible scale of natural gas substitution by biomethane from a biogas plant.

Comparing the potential electricity generation with existing demand assumes the use of existing technologies. For the calculations, it was assumed that 1 m^3 of biogas has an energy value of 20 MJ [84] and the CHP aggregate efficiency for electricity production is 40% [84,85], which finally gives 2.2 kWh electricity per cubic meter of biogas. Due to transport difficulties and thus low chances of commercial use the heat generated during cogeneration is omitted from the bill.

Under the above assumptions, economic analyses were carried out for the three sizes of biogas plants. The essential technical parameters of the biogas plant considered were determined using the tools provided as a result of the work on the Bio Energy Farm 2 (BEF2) project [86]. The analyses took the operating conditions of biogas plants similar to those in south-east Poland (near the Ukrainian border). For economic analyses, the Ukrainian feed-in tariff for biogas generated electricity (0.1239/kWh—see Section 2.2 for details) was applied. The assumptions regarding the investment costs were based on the results of the BEF2 project and similar studies [87]. A 15-year operation of the installation without general repairs was assumed.

The environmental effects of using the technical potential of biogas production were then assessed. It was assumed that the utilization of a significant part of manure would reduce emissions due to manure management in proportion to the amount of manure used. In addition, it was assumed that the potential use of biogas as a substitute for natural gas would reduce GHG emissions by 1023 kg CO_2e per 1 m^3 [88] while substituting 1 kWh of electricity by electricity generated in biogas plants would reduce CO_2e emissions by 660 g [89]. The profitability of the biogas plants was analyzed, taking into account revenues from electricity sold and the operating and maintenance costs. The substrate cost was not taken into account, as using manure available on-farm was assumed. The economic viability criterion was an internal rate of return of 0% (IRR > 0%), which means that investment outlays will be recovered after 15 years of operation. The fulfilment of such an assumption does not mean that a given investment is attractive from an economic point of view, but only that it does not generate losses.

4. Results

The parameter that directly determines the amount of biogas produced from manure is the number of animals kept. Theoretically, manure from any animals can be converted into biogas. However, taking into account the technical requirements and financial expenditures, only some of the manure can be managed effectively. Due to the dual structure of farms, it can be assumed that in Ukraine only farms listed as enterprises meet conditions to operate a biogas plant. Thus only approximately 1.14 million head of cattle, 3.4 million head of pigs and 118.9 million head of poultry may be the assumed as suppliers of the substrate for the biogas plants (Table 2).

Table 2. Farm animal stocks in Ukraine (2018) in thousand head.

	Cattle		Pigs		Poultry		Sheep and Goats	
	All Farms	**of Which Enterprises**	**All Farms**	**of Which Enterprises**	**All Farms**	**of Which Enterprises**	**All Farms**	**of Which Enterprises**
Ukraine total	3339.3	1138.2	6024.8	3395.6	211,614.7	118,812.9	1269.9	182.3
Vinnytska	239.4	81.4	251.4	91.4	32,588.6	24,107.1	33	3.5
Volynska	130.3	44.7	285.9	81.8	7560.1	4634.9	16.3	1.2
Dnipropetrovska	122.7	31.9	362.4	280.1	19,521.9	15,325.4	57.6	10.5
Donetska	60.3	27.6	455.7	423.3	5146.7	3181.3	41.6	6.8
Zhytomyrska	189.4	55.4	146.6	40.3	7491.7	583.1	27.5	5.1
Zakarpatska	122.9	1.9	242.7	19.7	3240.7	161.3	153.8	8.4
Zaporizka	91.5	19.3	217.5	145.1	4784.6	2527	63.6	21
Ivano-Frankivska	136.2	12.2	310.7	214.7	4812.7	1772.7	28.1	3.8
Kyivska	117.1	82.9	480.7	394.5	28,389.2	19,913.4	31.2	9.5
Kirovohradska	89.7	25.8	220.5	133.8	4996.7	157.4	36.6	4.3
Luhanska	54.1	16.8	43.1	26.1	996.8	49.6	25.2	2.4
Lvivska	170.9	17.6	417.3	263.2	9114.4	3615.7	31.6	4.6
Mykolayivska	98.5	17.2	83.1	41.2	2554.2	739.1	49.5	9.7
Odeska	154.9	22	173	64.1	3173.5	135.9	319.1	44.1
Poltavska	231.3	142.8	322.2	229.3	5650.3	2692.2	47.6	7.7
Rivnenska	118.6	29.5	243.7	34.1	7332.4	2213.2	15.4	0.5
Sumska	146.3	74.9	114.9	51.3	4892.8	1259.1	38.4	5.1

Table 2. *Cont.*

	Cattle		Pigs		Poultry		Sheep and Goats	
	All Farms	of Which Enterprises	All Farms	of Which Enterprises	All Farms	of Which Enterprises	All Farms	of Which Enterprises
Ternopilska	138.7	30.8	339.3	163.7	5241.8	2043.4	14.4	1.2
Kharkivska	180.8	88.7	194.8	99.1	8021.9	3147.4	71	6.8
Khersonska	96	15.4	111.6	63	5828.9	3557.5	41.6	13.5
Khmelnytska	230.2	67.5	325.9	163.2	7091.6	4519.2	27.3	2.3
Cherkaska	161	117.5	358.8	221.8	26,032.7	21,200.1	28.4	3.2
Chernivetska	81.5	8	141.8	52.1	3531.2	1036.4	43.8	4.6
Chernihivska	177	106.4	181.2	98.7	3619.3	240.5	27.3	2.5

Source: own elaboration based on data from the Ukrstat.

This number of animals can provide about 40.9 million tonnes of manure, which, however, makes less than half of the total amount of manure produced (Table 3). When assessing the substrate resources, it is also worth paying attention to the regional diversification of manure production, which determines the biogas potential. In some regions, the amount of available manure produced in enterprises is meagre (the lowest in Zakarpatska—0.11 million tonnes), while in others it is many times higher (e.g., Cherkaska 4.75 million tonnes; Kyivska 4.54 million tonnes). It is worth noting that in regions where enterprises produce little manure, much more is usually produced on individual farms at the same time.

Table 3. Manure production by Ukrainian regions (million tonnes/year).

	Manure from Enterprises (Million t/year)				Manure from Individual Farm (Million t/year)				Ukraine Total
	Cattle	Pig	Poultry	Total	Cattle	Pig	Poultry	Total	
Ukraine total	19.11	9.92	11.92	40.94	36.96	7.68	9.31	53.94	94.88
Vinnytska	1.37	0.27	2.42	4.05	2.65	0.47	0.85	3.97	8.02
Volynska	0.75	0.24	0.46	1.45	1.44	0.60	0.29	2.33	3.78
Dnipropetrovska	0.54	0.82	1.54	2.89	1.52	0.24	0.42	2.19	5.08
Donetska	0.46	1.24	0.32	2.02	0.55	0.09	0.20	0.84	2.86
Zhytomyrska	0.93	0.12	0.06	1.11	2.25	0.31	0.69	3.25	4.36
Zakarpatska	0.03	0.06	0.02	0.11	2.03	0.65	0.31	2.99	3.10
Zaporizka	0.32	0.42	0.25	1.00	1.21	0.21	0.23	1.65	2.65
Ivano-Frankivska	0.20	0.63	0.18	1.01	2.08	0.28	0.30	2.67	3.68
Kyivska	1.39	1.15	2.00	4.54	0.57	0.25	0.85	1.68	6.22
Kirovohradska	0.43	0.39	0.02	0.84	1.07	0.25	0.49	1.81	2.65
Luhanska	0.28	0.08	0.00	0.36	0.63	0.05	0.10	0.77	1.13
Lvivska	0.30	0.77	0.36	1.43	2.57	0.45	0.55	3.58	5.00
Mykolayivska	0.29	0.12	0.07	0.48	1.37	0.12	0.18	1.67	2.15
Odeska	0.37	0.19	0.01	0.57	2.23	0.32	0.30	2.85	3.42
Poltavska	2.40	0.67	0.27	3.34	1.49	0.27	0.30	2.05	5.39
Rivnenska	0.50	0.10	0.22	0.82	1.50	0.61	0.51	2.62	3.44
Sumska	1.26	0.15	0.13	1.53	1.20	0.19	0.36	1.75	3.28
Ternopilska	0.52	0.48	0.20	1.20	1.81	0.51	0.32	2.65	3.85
Kharkivska	1.49	0.29	0.32	2.09	1.55	0.28	0.49	2.31	4.41
Khersonska	0.26	0.18	0.36	0.80	1.35	0.14	0.23	1.72	2.52
Khmelnytska	1.13	0.48	0.45	2.06	2.73	0.48	0.26	3.46	5.53
Cherkaska	1.97	0.65	2.13	4.75	0.73	0.40	0.48	1.62	6.36
Chernivetska	0.13	0.15	0.10	0.39	1.23	0.26	0.25	1.75	2.14
Chernihivska	1.79	0.29	0.02	2.10	1.19	0.24	0.34	1.77	3.86

Source: own elaboration based on data from the Ukrstat.

The calculations showed that the total potential of agricultural biogas production from manure could be estimated at 2.9 billion m^3. However, only about 1.3 m^3 can be produced from manure produced in enterprises (Table 4). Due to the fragmentation of the sector, most of the potential is thus hardly useable in practice. The very numerous (4.6 million) small farms have a low number of animals (on average 10 tonnes of manure per farm per year), which means that it would be impossible to ensure substrate supply even for a small, 10 kWe micro biogas plant. There are large differences in potential between particular regions. More than 50% of the total potential (enterprises) is located in five regions (Cherkaska, Kyivska, Vinnytska, Dnipropetrovska, Poltavska). The most considerable contribution to the generation of biogas potential from enterprises would be poultry, whose share in biogas production would amount to 46.2% (611.3 million m^3), then cattle (36%—477.8 million m^3) and the least pig production (17.8%). A slightly different contribution to the generation of biogas potential can be observed in the case of individual farms—over 58% (923 million m^3) would be generated by cattle.

Table 4. Theoretical biogas yield in manure-based biogas plant in Ukraine.

	Biogas from Enterprises (Million m^3/year)				Biogas from Individual Farms (Million m^3/year)				Ukraine Total
	Cattle	Pig	Poultry	Total	Cattle	Pig	Poultry	Total	
Ukraine total	477.8	238.0	611.3	1327.1	923.9	184.3	477.5	1585.7	2912.7
Vinnytska	34.2	6.4	124.0	164.6	66.3	11.2	43.6	121.2	285.8
Volynska	18.8	5.7	23.8	48.3	35.9	14.3	15.1	65.3	113.6
Dnipropetrovska	13.4	19.6	78.9	111.9	38.1	5.8	21.6	65.5	177.3
Donetska	11.6	29.7	16.4	57.6	13.7	2.3	10.1	26.1	83.7
Zhytomyrska	23.3	2.8	3.0	29.1	56.2	7.4	35.5	99.2	128.3
Zakarpatska	0.8	1.4	0.8	3.0	50.8	15.6	15.8	82.3	85.3
Zaporizka	8.1	10.2	13.0	31.3	30.3	5.1	11.6	47.0	78.3
Ivano-Frankivska	5.1	15.0	9.1	29.3	52.0	6.7	15.6	74.4	103.7
Kyivska	34.8	27.6	102.5	164.9	14.4	6.0	43.6	64.0	228.9
Kirovohradska	10.8	9.4	0.8	21.0	26.8	6.1	24.9	57.8	78.8
Luhanska	7.1	1.8	0.3	9.1	15.7	1.2	4.9	21.7	30.9
Lvivska	7.4	18.4	18.6	44.4	64.3	10.8	28.3	103.4	147.9
Mykolayivska	7.2	2.9	3.8	13.9	34.1	2.9	9.3	46.4	60.3
Odeska	9.2	4.5	0.7	14.4	55.8	7.6	15.6	79.0	93.5
Poltavska	59.9	16.1	13.9	89.9	37.1	6.5	15.2	58.9	148.7
Rivnenska	12.4	2.4	11.4	26.2	37.4	14.7	26.3	78.4	104.6
Sumska	31.4	3.6	6.5	41.5	30.0	4.5	18.7	53.1	94.6
Ternopilska	12.9	11.5	10.5	34.9	45.3	12.3	16.5	74.1	109.0
Kharkivska	37.2	6.9	16.2	60.4	38.7	6.7	25.1	70.4	130.8
Khersonska	6.5	4.4	18.3	29.2	33.8	3.4	11.7	48.9	78.1
Khmelnytska	28.3	11.4	23.3	63.0	68.3	11.4	13.2	92.9	156.0
Cherkaska	49.3	15.5	109.1	173.9	18.3	9.6	24.9	52.7	226.7
Chernivetska	3.4	3.7	5.3	12.3	30.9	6.3	12.8	50.0	62.3
Chernihivska	44.7	6.9	1.2	52.8	29.6	5.8	17.4	52.8	105.6

Source: own elaboration based on data from the Ukrstat.

As mentioned above, the fragmentation of small-scale individual farms limits the practical possibilities for developing agricultural biogas plants, which can only be considered within the category of agricultural enterprises. However, it should be noted that enterprise farms are not uniform in terms of the scale of livestock production either. Estimates show that there are 1947 enterprises below the assumed minimum threshold of livestock production for biogas plant installation (Table 5), which in total would be able to generate 17.2 million m^3 of biogas, which gives 37.7 GWh of electricity, using an average CHP of 2.4 kWe. Such small installations are not used in practice. Hence the possible use of the potential of the smallest enterprises would require cooperation between them (which, however, would generate a problem with substrate transport between farms).

On the other hand, the potential of farms with a breeding scale allowing for the installation of a CHP unit with a capacity of at least 25 kWe can be assumed as technically and organizationally realistic [90]. There are 2527 such enterprises in Ukraine, of which almost half (1259 units) have a substrate to power a CHP unit with an average power of about 24.8 kW. These enterprises can generate over 113 million m^3 of biogas and 250 GWh of electricity. Almost 40% of these (983 units) are enterprises with a substrate that makes it possible to power a 105.5 kWe aggregate. In total, they would be able to generate nearly 830 GWh of electricity. With an installed capacity of 750 kWe (11.3%, i.e., 285 plants), the largest enterprises, which, however, have the most potential for biogas and electricity

It is also understandable that the energy-generation potential varies greatly depending on the local peculiarities: landscape and climatic predispositions to either crop or livestock farming (and their particular types), availability of technical, economic and financial potential. For these reasons, the large agro-holdings have concentrated land and allocated their production capacities in the regions with the most beneficial conditions for their activities. As is known, siting of biogas generation plants is most efficient near or even on a local livestock farm (as bioenergy inputs are not transportable over long distances).

Most research, however, does not approach the issue of biogas potential from the regional perspective, especially for all the Ukrainian regions (with one of the exceptions being Kudria [98]). Our analyses indicate that the highest degree of self-sufficiency in meeting energy needs would be observed in central and western Ukraine. In some regions, the biogas produced would make it possible to cover nearly 11% of natural gas consumption or almost 20% of electricity (Vinnytska region). One of the studies [42] on regional capacities for renewable energy generation states that, based on its biomass energy potential (including biogas, biodiesel and bioethanol), the western region of Ukraine (combining Volynska, Zakarpatska, Ivano-Frankivska, Lvivska, Rivnenska, Ternopilska, Chernivetska) could fully cover its natural gas needs. Other research [44] focusing solely on the Lvivska region specified that only biogas plants based on agricultural residues (both crop and livestock) could generate enough to replace 163 million m^3 of natural gas, or 22.9% of its regional consumption as of 2016. Our estimates prove the substantial capacity for biogas generation in the particular regions, yet need to mention the differences in values which come from different methodical approaches: both Bashynska [42] and Yankovska [44] based their evaluations on both the crop and livestock production values.

As mentioned in the introduction, one of the fundamental prerequisites for biogas production (as well as the use of other alternative energy sources) is seeking environmental benefits. The production of biogas, including agricultural biogas, is indicated as an important way to reduce GHG emissions [20,99,100]. According to the World Biogas Association calculations, "biogas and biomethane industries have the potential to reduce global GHG emissions by 10–13%" [101]. The analyses show that, as a result of using the potential, the total level of greenhouse gas emissions would decrease by 5% to 6.14%. In comparison, for the technical potential, it would be 2.3% to 2.8%. The scope for reduction results both from the reduction of emissions due to management and the replacement of conventional energy by renewable energy. The influence of the second factor is particularly important in countries where the energy mix is dominated by high-emission sources such as coal (a situation that also partially applies to Ukraine). However, it is worth bearing in mind that the environmental impacts of biogas generation from manure also depend on many other factors, such as substrate, technology and operating practices [100].

The economic dimension is the third key aspect of biogas production. The economic viability assessment of investments in biogas plants in Ukraine presented in the literature provides similar outcomes. In 2017 Shanda Consult [96] published report which concludes that "small" (<300 kWe) biogas plants are not economically attractive. Slightly bigger plants could be justified only if using heat for one's own purposes, while a chance of fast payback in four to five years is possible for units larger than 1 MWe. This observation is related to the phenomenon of economies of scale, reflecting the degression of unit costs as the scale of production increases. Our analyses show that the IRR of medium (100 kWe) and large (750 kWe) plants analyzed are positive, although close to the average inflation rate in Ukraine. It might thus be concluded that, even if neither investment generated losses in a 15-year period, they would be one of the last priorities on farmers' investment "wish list". Generally, our results confirm Shanda Consult outcome on the economic viability of biogas plants. However, we assumed that plants which do not generate losses are economically viable, while the [96] experts set higher requirements to consider the investment economically attractive.

Financial support is a decisive factor for the profitability of many renewable energy sources, especially in the early stages of development [8,59,93,102,103]. There are many different support mechanisms [104] in Ukraine, one of the key being the subsidized feed-in tariffs.

6. Conclusions

The analyses indicate that in absolute terms Ukraine has a significant potential for the production of agricultural biogas from animal manure, reaching nearly 3 billion m^3. However, the practical possibilities of using this potential are severely limited by the dual structure of agriculture. More than half of the available manure is produced on small livestock farms that are too small-scale to consider investing in biogas plants. Our analyses show that under the current conditions, only biogas plants with a CHP aggregate capacity of around 100 kW could provide a positive return on investment. In practice, therefore, the economically justified production of agricultural biogas can only be carried out by agricultural enterprises, which allows for the satisfaction of over 3% of the country's electricity demand. However, policymakers might consider the possibility of creating programs to support cooperation between small farmers to create collective biogas initiatives (e.g., biogas cooperatives). Increasing the real possibilities of using the agricultural potential in the production of biogas is important both for increasing energy self-sufficiency and reducing GHG emissions from agriculture. The production of biogas from manure will not completely solve these problems, however, as shown by the analyses; in some regions of Ukraine it may make a noticeable contribution to meeting energy needs and reducing GHG emissions from the agricultural sector. In the context of global efforts to replace non-renewable energy sources with renewable sources and increase climate neutrality, the increased importance of agricultural biogas in the energy mix should be one of the goals of state energy policy. However, the potential growth of the agricultural biogas production in electricity generation in Ukraine requires growing interest in this issue from agricultural producers themselves. It is necessary to make them aware that converting manure into biogas can bring economic benefits for farmers and also environmental and social benefits for society as a whole. However, the development of agricultural biogas production also requires further in-depth scientific analyses that would enable the adjustment of academic knowledge from other countries to Ukrainian conditions.

It should be noted, however, that when interpreting this paper's results, one should remember that these are only estimates based on the available statistical data. The substantial obstacle to more precise estimates was the lack of detailed accessible data regarding the particular elements of agricultural activity in Ukraine, including detailed data on the livestock population (divided by the type, age, weight), production technologies (extensive, intensive) and breeding systems. While the agricultural enterprises report their operational data to the Ukrstat, there is still a substantial gap in the data about individual (household) farms, which were not well captured in the analysis. However, having regard to the available literature, the work presented substantially fills the knowledge gap about the problem discussed, and the research results may be the basis for further investigations based on more detailed data when available.

Author Contributions: Conceptualization, A.W., P.S. and V.K.; methodology, A.W. and P.S.; validation, V.K., N.P. and M.S.; formal analysis, A.W. and P.S.; investigation, P.S. and A.W.; resources, V.K. and P.S.; data curation, V.K., P.S., I.S. and N.P.; writing—original draft preparation, P.S., A.W., V.K. and A.M.-R.; writing—review and editing, A.W., P.S., A.M.-R., N.P., M.W., I.S. and M.S.; visualization, A.W. and V.K.; supervision, A.W. and P.S.; funding acquisition, M.W. All authors have read and agreed to the published version of the manuscript.

Funding: This research received no external funding.

Conflicts of Interest: The authors declare no conflict of interest.

References

1. Ferdes, M.; Zabava, B.; Dinca, M.; Paraschiv, G.; Toma, L. Environmental impact associated with biogas production—A short review. In Proceedings of the International Symposium ISB-INMA TEH Agricultural and Mechanical Engineering Conference, Bucharest, Romania, 31 October–1 November 2019.
2. Bijnagte, J.W. *Public Final Report BioEnergy Farm II*. BioEnergy Farm II publication, Cornelissen Consulting Services B.V. 2017. Available online: https://www.bioenergyfarm.eu/wp-content/uploads/2015/05/D1.1_Public-Report-BioEnergy-Farm-II.pdf (accessed on 21 September 2020).

3. Michel, J.; Weiske, A.; Möller, K. The effect of biogas digestion on the environmental impact and energy balances in organic cropping systems using the life-cycle assessment methodology. *Renew. Agric. Food Syst.* **2010**, *25*, 204–218. [CrossRef]
4. *Transforming Our World: The 2030 Agenda for Sustainable Development Resolution Adopted by the General Assembly on 25 September 2015*; Seventieth Session, Agenda Items; United Nations: New York, NY, USA, 2015.
5. C2ES Global Emissions. Available online: https://www.c2es.org/content/international-emissions/ (accessed on 20 August 2020).
6. FAO. *The Contribution of Agriculture to Greenhouse Gas Emissions*; FAO: Rome, Italy, 2020.
7. IRENA. Global Energy Transformation. Available online: https://irena.org/publications/2019/Apr/Global-energy-transformation-A-roadmap-to-2050-2019Edition (accessed on 12 August 2020).
8. Van de Graaf, T.; Bond, K.; Overland, I. *A New World: The Geopolitics of the Energy Transformation*; Global Commission on the Geopolitics of Energy Transformation: Abu Dhabi, UAE, 2019.
9. UN. *World Economic and Social Survey 2013: Sustainable Development Challenges*; United Nations: New York, NY, USA, 2013.
10. Bank, W. Sustainable Energy for All (SE4ALL) Database from the SE4ALL Global Tracking Framework Led Jointly by the World Bank, International Energy Agency, and the Energy Sector Management Assistance Program. Available online: https://data.worldbank.org/indicator/EG.FEC.RNEW.ZS (accessed on 12 September 2020).
11. Ministry of Energy and Environmental Protection of Ukraine. Ukraine's Greenhouse Gas Inventory 1990–2018 (Draft); Kyiv, Ukraine, 2020. Available online: https://mepr.gov.ua/files/docs/Zmina_klimaty/2020/Ukraine_NIR_2020%20draft.pdf (accessed on 12 September 2020).
12. USAID. Greenhouse Gas Emissions Factsheet: Ukraine|Global Climate Change. Available online: https://www.climatelinks.org/resources/greenhouse-gas-emissions-factsheet-ukraine (accessed on 18 October 2020).
13. Nielsen-Holm, J.B.; Oleskowicz-Popiel, A.; Seadi, T. Energy Crop Potentials for Bioenergy in EU-27. In Proceedings of the 15th European Biomass Conference and Exhibition, Berlin, Germany, 7–11 May 2007.
14. Lukehurst, C.; Bywater, A. Exploring the Viability of Small Scale Anaerobic Digesters in Livestock Farming. Available online: https://www.ieabioenergy.com/publications/exploring-the-viability-of-small-scale-anaerobic-digesters-in-livestock-farming/ (accessed on 18 October 2020).
15. Shih, J.S.; Burtraw, D.; Palmer, K.; Siikamäki, J. Air emissions of ammonia and methane from livestock operations: Valuation and policy options. *J. Air Waste Manag. Assoc.* **2008**, *58*, 1117–1129. [CrossRef]
16. Lusk, P. *Methane Recovery from Animal Manures the Current Opportunities Casebook*; National Renewable Energy Lab.: Golden, CO, USA, 1998.
17. Bentley, C.; Gooch, C.A.; Pronto, J.; Scott, N. Greenhouse gas emissions from a community anaerobic digester with mixed organic wastes. In Proceedings of the An ASABE Meeting Presentation, Pittsburgh, PA, USA, 20–23 June 2010; p. 34.
18. IPCC. *2006 IPCC Guidelines for National Greenhouse Gas Inventories*; Institute for Global Environmental Strategies: Hayama, Japan, 2006.
19. Oenema, O.; Oudendag, D.; Velthof, G.L. Nutrient losses from manure management in the European Union. *Livest. Sci.* **2007**, *112*, 261–272. [CrossRef]
20. Mirosz, L.; Amrozy, M.; Trząski, A.; Wiszniewska, A. What Policymakers Should Know about Micro-Scale Digestion. Available online: https://www.bioenergyfarm.eu/wp-content/uploads/2015/05/D3.1_EN-Guide-for-Policymakers-BEF2template.docx.pdf (accessed on 3 September 2020).
21. Majewski, E.; Sulewski, P.; Wąs, A. *Ewolucja Wspólnej Polityki Rolnej Unii Europejskiej w kontekście wyzwań Trwałego Rozwoju*; Wydawnictwo SGGW: Warszawa, Poland, 2018.
22. Lauer, M.; Hansen, J.K.; Lamers, P.; Thrän, D. Making money from waste: The economic viability of producing biogas and biomethane in the Idaho dairy industry. *Appl. Energy* **2018**, *222*, 621–636. [CrossRef]
23. O'Connor, S.; Ehimen, E.; Pillai, S.C.; Lyons, G.; Bartlett, J. Economic and environmental analysis of small-scale anaerobic digestion plants on Irish dairy farms. *Energies* **2020**, *13*, 637. [CrossRef]
24. Koryś, K.A.; Latawiec, A.E.; Grotkiewicz, K.; Kuboń, M. The review of biomass potential for agricultural biogas production in Poland. *Sustainability* **2019**, *11*, 6515. [CrossRef]
25. Szymańska, M.; Sosulski, T.; Szara, E.; Wąs, A.; Sulewski, P.; Van Pruissen, G.W.; Cornelissen, R.L. Ammonium sulphate from a bio-refinery system as a fertilizer—Agronomic and economic effectiveness on the farm scale. *Energies* **2019**, *12*, 4721. [CrossRef]

26. Frost, P.; Lukehurst, C.T.; Seadi, T. Utilisation of Digestate from Biogas Plants as Biofer/Tiliser. Available online: https://www.researchgate.net/publication/228367700_Utilisation_of_digestate_from_biogas_plants_as_biofertiliser (accessed on 17 September 2020).
27. Szymańska, M.; Szara, E.; Sosulski, T.; Stępień, W.; Pilarski, K.; Pilarska, A. Chemical properties and fertilizer value of ten different anaerobic digestates. *Fresenius Environ. Bull.* **2018**, *27*, 3425–3432.
28. Szymańska, M.; Nowaczewska, D.; Świerżewska, E.; Wrzosek-Jakubowska, J.; Gworek, B. An attempt to assess physicochemical properties of soil fertilized with fresh and treated digestate from biogas plant. *Przem. Chem.* **2016**, *95*, 572–576. [CrossRef]
29. USDA. Biogas Opportunities Roadmap. Voluntary Actions to Reduce Methane Emissions and Increase Energy Independence. Available online: https://www.epa.gov/sites/production/files/2015-12/documents/biogas-roadmap.pdf (accessed on 2 September 2020).
30. European Commission. Optimal Use of Biogas from Waste Streams an Assessment of the Potential of Biogas from Digestion in the EU Beyond. 2020. Available online: https://ec.europa.eu/energy/sites/ener/files/documents/ce_delft_3g84_biogas_beyond_2020_final_report.pdf (accessed on 21 September 2020).
31. Cabinet of Ministers of Ukraine. Energy Strategy of Ukraine until 2035: Security, Energy Efficiency, Competitiveness. Available online: https://zakon.rada.gov.ua/laws/file/text/58/f469391n10.pdf (accessed on 15 August 2020).
32. National Investment Council. Agricultural Sector of Ukraine. Securing the Global Food Supply. Available online: https://www.agroberichtenbuitenland.nl/documenten/rapporten/2018/07/04/ua-report-investment-council-ua-agriculture (accessed on 11 September 2020).
33. Pryshliak, N.V.; Tokarchuk, D.M.; Palamarenko, Y.V. *Ensuring National Energetic and Ecological Security Based on Biofuels Obtained from Bioenergy Crops and Wastes*; Vinnytsia National Agrarian University: Vinnytsia, Ukraine, 2019.
34. Geletuha, G.G.; Kucheruk, P.P.; Yu, B.M. *Prospects of Biogas Generation and Use in Ukraine: Analytical Report No 4*. Available online: https://saf.org.ua/wp-content/uploads/2019/04/position-paper-uabio-4-ua.pdf (accessed on 17 August 2020).
35. Kutsenko, Y.M. Substantiatin of biogas production in agricultural enterprises. *Pr. Tavriyskogo Agrotechnologichnogo Universytetu* **2014**, *14*, 100–107.
36. Kaletnik, H.; Zdyrko, N.; Fabiyanska, V. Biogas in Households as the Base for Energy Independence of the Rural Areas in Ukraine. Available online: http://nbuv.gov.ua/UJRN/efmapnp_2018_8_3 (accessed on 15 August 2020).
37. Sakun, L.; Riznichenko, L.; Vielkin, B. Prospects of biogas market development in Ukraine and abroad. *Ekon. Organ. Upr.* **2020**, *37*, 160–170. [CrossRef]
38. Kurbatova, T.; Hyrchenko, Y. Economic prospects of biogas sector development based on using agriculture's organic waste. *Mod. Econ.* **2019**, *14*, 121–129. [CrossRef]
39. Panchuk, M.V.; Shlapak, L.S. Analysis of prospects for development of production and use of biogas in Ukraine. *Rozvìdka Rozrobka Naft. Gazov. Rodovyshch.* **2016**, *60*, 26–33.
40. Gradovyi, V. Ecological-Economic Evaluation and Means to Ensure Biogas Generation. Available online: http://dspace.tneu.edu.ua/jspui/bitstream/316497/17253/1/Градовий.pdf (accessed on 11 August 2020).
41. Garmash, S.; Mitina, N.; Zubareva, I. *Prosects Obtainement of Biogas from Organic Wastes in Ukraine*; Zaporizka State Engineering Academy: Zaporizhia, Ukraine, 2016; p. 3.
42. Bashynska, Y.I. Organizational-Economic Bases of Using of the Renewable Energy Potential in the Western Region of Ukraine. Available online: https://ird.gov.ua/irdd/d20170529_a805_BashynskaYI.pdf (accessed on 23 August 2020).
43. Ukrainian League of Industrialists and Entrepreneurs. Prospects of Biomass Market Development in the EU and Ukraine. Available online: https://uspp.ua/assets/doc/uspp-biomass.pdf (accessed on 15 August 2020).
44. Yankovska, K. Economic Efficiency of Biomass Use for Energy Supply of Agricultural Enterprises. Available online: http://www.lnau.edu.ua/lnau/attachments/4772_Dis.pdf (accessed on 29 August 2020).
45. Eurostat. *Agri-Environmental Indicator-Greenhouse Gas Emissions Statistics Explained*; Eurostat: Luxembourg, 2017.
46. Tubiello, F.N.; Salvatore, M.; Ferrara, A.F.; House, J.; Federici, S.; Rossi, S.; Biancalani, R.; Golec, R.D.C.; Jacobs, H.; Flammini, A.; et al. The contribution of agriculture, forestry and other land use activities to global warming, 1990–2012. *Glob. Chang. Biol.* **2015**, *21*, 2655–2660. [CrossRef]

47. Olesen, J.; Trnka, M.; Kersebaum, K.; Skjelvåg, A.; Seguin, B.; Peltonensainio, P.; Rossi, F.; Kozyra, J.; Micale, F. Impacts and adaptation of European crop production systems to climate change. *Eur. J. Agron.* **2011**, *34*, 96–112. [CrossRef]
48. Del Prado, A.; Crosson, P.; Olesen, J.E.; Rotz, C.A. Whole-farm models to quantify greenhouse gas emissions and their potential use for linking climate change mitigation and adaptation in temperate grassland ruminant-based farming systems. *Animal* **2013**, *7*, 373–385. [CrossRef] [PubMed]
49. Appels, L.; Lauwers, J.; Degreve, J.; Helsen, L.; Lievens, B.; Willems, K.A.; Van Impe, J.; Dewil, R. Anaerobic digestion in global bio-energy production: Potential and research challenges. *Renew. Sustain. Energy Rev.* **2011**, *15*, 4295–4301. [CrossRef]
50. Khalid, A.; Arshad, M.; Anjum, M.; Mahmood, T.; Dawson, L. The anaerobic digestion of solid organic waste. *Waste Manag.* **2011**, *31*, 1737–1744. [CrossRef]
51. Braun, R.; Wellinger, A. Potential of Co-Digestion. Available online: http://task37.ieabioenergy.com/files/daten-redaktion/download/publi-task37/Potential%20of%20Codigestion%20short%20Brosch221203.pdf (accessed on 2 September 2020).
52. Murphy, J.; Braun, R.; Weiland, P.; Wellinger, A. Biogas from Crop Digestion. Available online: https://www.ieabioenergy.com/wp-content/uploads/2011/10/Update_Energy_crop_2011.pdf (accessed on 15 September 2020).
53. Petersen, B.; Snapp, S. What is sustainable intensification? Views from experts. *Land Use Policy* **2015**, *46*, 1–10. [CrossRef]
54. Persson, M.; Jönsson, O.; Wellinger, A. Biogas Upgrading to Vehicle Fuel Standards and Grid Injection. Available online: https://www.researchgate.net/publication/285649841_Task_37_-_Biogas_upgrading_to_vehicle_fuel_standards_and_grid_injection (accessed on 7 September 2020).
55. Fischer, K.; Fischer, T.; Krieg, A. Farm-Scale Biogas Plants. Available online: https://www.kriegfischer.de/fileadmin/public/docs/texte/farm_scale_Biogas_Plants.pdf (accessed on 15 August 2020).
56. Bruins, M.E.; Sanders, J.P. Small-scale processing of biomass for biorefinery. *Biofuel Bioprod. Biorefin.* **2012**, *6*, 135–145. [CrossRef]
57. De Dobbelaere, A.; De Keulenaere, B.; De Mey, J.; Lebuf, V.; Meers, E.; Ryckaert, B.; Schollier, C.; van Driessche, D. *Small-Scale Anaerobic Digestion; Case studies in Western Europe*; Mia Demeulemeester; Inagro Vzw: Rumbeke-Beitem, Belgium, 2015. [CrossRef]
58. Kost, C.; Shammugam, S.; Jülch, V.; Nguyen, H.T.; Schlegl, T. *Levelized Cost of Electricity Renewable Energy Technologies*; Fraunhofer Institute for Solar Energy Systems: Freiburg, Germany, 2018.
59. Zillman, D.N.; Redgwell, C.; Omorogbe, Y.O.; Barrera-Hernández, L.K.; Barton, B. *Beyond the Carbon Economy: Energy Law in Transition*; Oxford Scholarship Online: New York, NY, USA, 2012; ISBN 9780191701054. [CrossRef]
60. Sokurenko, N. Biogas Will Warm Ukraine Up. Available online: https://a7d.com.ua/agropoltika/ekologja/4713-bogaz-zgrvatime-ukrayinu.html (accessed on 4 September 2020).
61. SAEE. Generation of Thermal Energy from Renewable Sources, Current State. Available online: https://saee.gov.ua/en/activity/vidnovlyuvana-enerhetyka/suchasny-stan (accessed on 3 October 2020).
62. Mostova, M. Biogas Sector in Ukraine: Substantial Prospects and Reality. Available online: https://energytransition.in.ua/sfera-biohazu-v-ukraini-velyki-perspektyvy-ta-real-nist (accessed on 13 August 2020).
63. Parliament of Ukraine Law of Ukraine. On Changes Implemented to the Law of Ukraine 'On Power Engineering' Regarding the Stimulation of Use of Alternative Energy Sources. Available online: https://zakon.rada.gov.ua/laws/show/601-17#Text (accessed on 10 September 2020).
64. International Finance Corporation Stimulation of Renewable Energy in Ukraine Based on Feed-in Tariff. Available online: https://saee.gov.ua/documents/green-tariff.pdf (accessed on 23 August 2020).
65. Parliament of Ukraine Law of Ukraine. On Changes to Several Laws of Ukraine Regarding Ensuring Competitive Conditions for Generation of Electric Energy from Alternative Energy Sources. Available online: https://zakon.rada.gov.ua/laws/show/2712-19#Text (accessed on 21 August 2020).
66. State Agency on Energy Efficiency and Energy Saving of Ukraine Tax Exemptions and Support for Energy-Efficient Equipment. Available online: https://saee.gov.ua/uk/business/preferentsii/derzh-pidtrymka/podatkovi-ta-mytni-pilgy (accessed on 20 September 2020).
67. State Agency on Energy Efficiency and Energy Saving of Ukraine. Available online: https://saee.gov.ua/documents/Ukraine_EE_2013_ENG.pdf (accessed on 30 August 2020).

68. Bioenergy Association of Ukraine Biogas Projects in Ukraine. Available online: https://uabio.org/en (accessed on 25 August 2020).
69. Cabinet of Ministers of Ukraine. Economic Renewal: Program for Stimulation of the Economy to Overcome the Consequences of COVID-19. Available online: https://www.kmu.gov.ua/storage/app/sites/1/18-Department/Prezentacii/ProgramaEkonomichnestymyluvannia/analitichni-materiali-do-programi-stimulyuvannya-1.pdf (accessed on 3 September 2020).
70. Food and Agriculture Organization of the United Nations. Reviving Ukraine's Breadbasket. Available online: http://www.fao.org/support-to-investment/news/reviving-the-breadbasket/en (accessed on 16 August 2020).
71. Garkavyi, V. Prospect for Development of Agricultural Production in Ukraine. Available online: http://nbuv.gov.ua/j-pdf/ape_2015_10_17.pdf (accessed on 2 September 2020).
72. *State Statistics Service of Ukraine Statistical Yearbook of Ukraine for 2018*; Kyiv, Ukraine, 2019. Available online: http://www.ukrstat.gov.ua/druk/publicat/kat_u/2019/zb/11/zb_yearbook_2018_e.pdf (accessed on 16 August 2020).
73. Bogonos, M.; Stepaniuk, O. Agrarian Perspectives of Ukraine 2017–2030. Available online: https://apd-ukraine.de/images/APD_APR_06-2017_AGMEMOD_Baseline_ukr.pdf (accessed on 31 August 2020).
74. Parhomets, M.; Uniyat, L. Income of Livestock Sectors and Directions for Their Increase in Agrarian Companies of a Region. Available online: http://dspace.tneu.edu.ua/bitstream/316497/4469/1/Пархомець М.pdf (accessed on 18 August 2020).
75. Diyesperov, V.S. Livestock Breeding as the Most Problematic Sector. Available online: http://eapk.org.ua/sites/default/files/eapk/2016/2/5.pdf (accessed on 7 September 2020).
76. Shepel, T. Methodic Approaches to Assessment of Efficiency of Livestock Production in Individual Household Farms. Available online: http://ed.pdatu.edu.ua/article/view/127612/122356 (accessed on 2 September 2020).
77. Vasylieva, N.; Kruse, J.R. Models on providing food security: Case of Ukraine. *Probl. Perspect. Manag.* **2018**, *16*, 344–352. [CrossRef]
78. Ministry for Development of Economy, Trade and Agriculture of Ukraine. The Government has Defined Six Main Directions of Support for Agricultural Producers' Support for 2020. Available online: https://www.rv.gov.ua/news/uryad-viznachiv-6-osnovnih-napryamiv-pidtrimki-agrariyiv-na-2020-rik (accessed on 27 August 2020).
79. Grobenko, O.; Lyubvytsky, S.; Gurieva, T. *Guidelines in the Discipline "Machine Use in Animal Husbandry"*; Ministry of Agrarian Policy of Ukraine: Mykolaiv, Ukraine, 2005.
80. State Sanitary Doctor of The USSR Sanitary Rules and Regulations for Placement, Arrangement and Operation of Small Farms for Keeping Animals (Livestock, Poultry, Animals) in Settlements of the Ukrainian SSR. Available online: https://zakon.rada.gov.ua/laws/show/n0002400-89#Text (accessed on 3 October 2020).
81. World Biogas Assciation Gloobal. Potential of Biogas. Available online: https://www.worldbiogasassociation.org/global-potential-of-biogas (accessed on 18 September 2020).
82. UkrStat. *Main Agricultural Characteristics of Households in Rural Area in 2018*; State Statistics Service of Ukraine: Kiev, Ukraine, 2019.
83. Elizabeth, K.; Warren, H. A Techno-Economic Comparison of Biogas Upgrading Technologies in Europe. Available online: https://citeseerx.ist.psu.edu/viewdoc/download?doi=10.1.1.456.1353&rep=rep1&type=pdf (accessed on 3 September 2020).
84. Nsair, A.; Cinar, S.O.; Alassali, A.; Abu Qdais, H.; Kuchta, K. Operational parameters of biogas plants: A review and evaluation study. *Energies* **2020**, *13*, 3761. [CrossRef]
85. Barragán-Escandón, E.A.; Ruiz, J.M.O.; Tigre, J.D.C.; Zalamea-León, E.F. Assessment of power generation using biogas from landfills in an equatorial tropical context. *Sustainability* **2020**, *12*, 2669. [CrossRef]
86. CCS BioEnergyFarm 2. Available online: https://www.bioenergyfarm.eu/en/ (accessed on 1 October 2020).
87. Breure, M.S.; Langeveld, J.W.; Pombo, J. A generalised model for the calculation of renewable energy production costs with special emphasis on co-digestion biogas chains in Spain. In Proceedings of the Poster Presented at the European Biomass Conference and Exhibition, Stockholm, Sweden, 12–16 June 2017.
88. Wartości Opałowe i Wskaźniki Emisji w roku 2012 do Raportowania w Ramach Wspólnotowego Systemu Handlu Uprawnieniami do Emisji za rok 2015 (Eng. Calorific Values and Emission Indices for Year 2012 for Reporting in European Union Emission Trading System Sys). Available online: https://www.kobize.pl/uploads/materialy/download/2014/WO_i_WE_do_stosowania_w_SHE_2015.pdf (accessed on 3 September 2020).

89. Koffi, B.; Cerutti, A.; Kona, A. *CoM Default Emission Factors for the Eastern Partner countries Dataset Version 2017*; JRC: Rome, Italy, 2017.
90. Paterson, M. Implementation Guide for Small-Scale Biogas Plants. Available online: https://www.bioenergyfarm.eu/wp-content/uploads/2015/05/D3.2_EN-Implementation-guideline.pdf (accessed on 24 August 2020).
91. Havrysh, V.; Kalinichenko, A.; Mentel, G.; Olejarz, T. Commercial Biogas Plants: Lessons for Ukraine. *Energies* **2020**, *13*, 2668. [CrossRef]
92. Lebuhn, M.; Munk, B.; Effenberger, M. Agricultural biogas production in Germany—From practice to microbiology basics. *Energy Sustain. Soc.* **2014**, *4*, 1–21. [CrossRef]
93. REMAP 2030: Renewable Energy Prospects for Ukraine (Background Paper). Available online: https://www.irena.org/-/media/Files/IRENA/Agency/Publication/2015/Apr/IRENA_REmap_Ukraine_paper_2015.pdf (accessed on 8 August 2020).
94. Yevdokimov, Y.; Chygryn, O.; Pimonenko, T.; Lyulyov, O. Biogas as an alternative energy resource for Ukrainian companies: EU experience. *Innov. Mark.* **2018**, *14*, 7–15. [CrossRef]
95. Majewski, E.; Sulewski, P.; Wąs, A. *Potencjał i Uwarunkowania Produkcji Biogazu Rolniczego w Polsce*; Wydawnictwo SGGW: Warszawa, Poland, 2016.
96. Shanda Consult. The Biogas Market of Ukraine. Available online: https://shandaconsult.com/biogas-market-ukraine (accessed on 22 September 2020).
97. SAEE. In-Depth Review of the Energy Efficiency Policy of The Republic of Belarus. Available online: https://www.energycharter.org/fileadmin/DocumentsMedia/IDEER/IDEER-Belarus_2013_en.pdf (accessed on 2 September 2020).
98. Kudria, S.O. Potential of Renewable Sources of Energy in Ukraine, Kyivska Oblast and Chornobyl Zone. Available online: http://www.ive.org.ua/wp-content/uploads/tpp-may-2017.pdf (accessed on 5 August 2020).
99. Budzianowski, W.M.; Postawa, K. Renewable energy from biogas with reduced carbon dioxide footprint: Implications of applying different plant configurations and operating pressures. *Renew. Sustain. Energy Rev.* **2017**, *68*, 852–868. [CrossRef]
100. Agostini, A.; Battini, F.; Padella, M.; Giuntoli, J.; Baxter, D.; Marelli, L.; Amaducci, S. Economics of GHG emissions mitigation via biogas production from Sorghum, maize and dairy farm manure digestion in the Po valley. *Biomass Bioenergy* **2016**, *89*, 58–66. [CrossRef]
101. EBA. *The Contribution of the Biogas and Biomethane Industries to Medium-Term Greenhouse Gas Reduction Targets and Climate- Neutrality by 2050*; EBA: Brussels, Belgium, 2020.
102. Wiesenthal, T.; Dowling, P.; Morbee, J.; Thiel, C.; Schade, B.; Russ, P.; Simoes, S.; Peteves, S.; Schoots, K.; Londo, M. Technology Learning Curves for Energy Policy Support. *JRC Sci. Policy Rep.* **2012**, *332*. [CrossRef]
103. Renewable Energy Technologies: Cost Analysis Series Biomass for Power Generation Acknowledgement. Available online: https://www.irena.org/documentdownloads/publications/re_technologies_cost_analysis-biomass.pdf (accessed on 24 August 2020).
104. Status Review of Renewable Support Schemes in Europe for 2016 and 2017 Public Report. Available online: https://www.ceer.eu/documents/104400/-/-/80ff3127-8328-52c3-4d01-0acbdb2d3bed (accessed on 13 August 2020).

Publisher's Note: MDPI stays neutral with regard to jurisdictional claims in published maps and institutional affiliations.

energies

MDPI

Article

Biogas Plant Exploitation in a Middle-Sized Dairy Farm in Poland: Energetic and Economic Aspects

Patrycja Pochwatka [1], Alina Kowalczyk-Juśko [1,*], Piotr Sołowiej [2], Agnieszka Wawrzyniak [3] and Jacek Dach [4]

1 Department of Environmental Engineering and Geodesy, University of Life Sciences in Lublin, Leszczyńskiego 7, 20-069 Lublin, Poland; patrycja.pochwatka@up.lublin.pl
2 Department of the Electrotechnics Energetics Electronics and Automatics, University of Warmia and Mazury in Olsztyn, Oczapowskiego 11, 10-736 Olsztyn, Poland; pit@uwm.edu.pl
3 Institute of Technology and Life Sciences, Falenty, Hrabska 3, 05-090 Raszyn, Poland; a.wawrzyniak@itp.edu.pl
4 Department of Biosystems Engineering, Poznan University of Life Sciences, Wojska Polskiego 50, 60-637 Poznan, Poland; jacek.dach@up.poznan.pl
* Correspondence: alina.jusko@up.lublin.pl

Received: 3 November 2020; Accepted: 17 November 2020; Published: 19 November 2020

Abstract: Although cow manure is a valuable natural fertilizer, it is also a source of extreme greenhouse gas emissions, mainly methane. For this reason, this study aims to determine the impact of investments in a biogas plant on the energy and economic aspects of the operation of a dairy farm. A farm with a breeding size of 600 livestock units (LSU) was adopted for the analysis. In order to reach the paper's aim, the analysis of two different scenarios of dairy farm functioning (conventional–only milk production, and modern–with biogas plant exploitation) was conducted. The analysis showed that the investment in biogas plant operations at a dairy farm and in using cow manure as one of the main substrates is a more profitable scenario compared to traditional dairy farming. Taking into account the actual Polish subsidies for electricity produced by small biogas plants, the scenario with a functioning biogas plant with a capacity of 500 kW brings €332,000/a more profit compared to the conventional scenario, even when taking into account additional costs, including the purchase of straw to ensure a continuous operation of the installation. Besides, in the traditional scenario, building a biogas plant allows for an almost complete reduction of greenhouse gas emissions during manure storage.

Keywords: biogas plant; energetic optimization; substrates; manure; wheat straw

1. Introduction

1.1. Methane as a Greenhouse Gas

Methane, right after carbon dioxide, is considered to be the greenhouse gas that is the most harmful to the climate [1]. Almost 59% of the world's emissions of this gas are of anthropogenic origin, of which the largest share (40–53%) is agriculture, especially intensive production [2]. In the EU, the share of agriculture in anthropogenic methane emissions is 53%, 26% is methane from waste, and 19% comes from energy production [3]. The reduction of methane emissions slows down negative climate changes and improves air quality [4]. Most of the legal acts concerning climate policy pay mostly attention to limiting carbon dioxide emissions. However, more and more attention is paid to the reduction of methane emissions [5]. The European Green Deal Communication emphasizes that reducing methane emissions related to the energy sector is one of the goals in achieving climate neutrality by 2050 [6]. In turn, the EU strategy to reduce methane emissions emphasizes that biogas

can also be potentially used as an energetic substrate for biogas plants; however, in the described situation, the distance from the dairy factory was too far for the economic transport of waste for the planned biogas plant.

2. Materials and Methods

In order to reach the paper's aim, the analysis of two different scenarios of dairy farm functioning (conventional—only milk production, and modern—with biogas plant exploitation) were conducted. The main schema of the analytical and calculation procedures are shown in Figure 1.

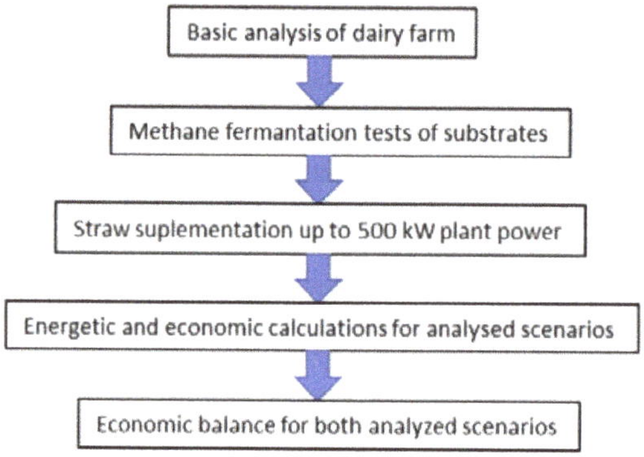

Figure 1. Methodological procedures for the energetic and economic analysis of both scenarios.

The paper's idea is based on a real investment planned to be realized in one of the Poznan University of Life Sciences (PULS) experimental farms. In this farm, dairy production from 600 cows is realized, and the investment for building a biogas plant with 500 kW of electric power is planned for 2021. At the end of 2019, PULS already finished a similar biogas plant at Przybroda experimental university farm (25 km west of Poznan) (Figure 2). However, the Przybroda biogas plant mainly uses biowaste (including food waste and waste from the agro-food industry) as its main substrates.

Figure 2. 500 kW biogas plant at Przybroda PULS experimental farm.

The investment procedure made in the case of the biogas plant building should be based on strong economic and energetic arguments. It has to be underlined that the decision about the construction of biogas plants is usually related to a cost of several million euros. That is why this decision should be taken while considering several steps that let one reach the proper choice. The good practice is to follow the way that is shown in Figure 3.

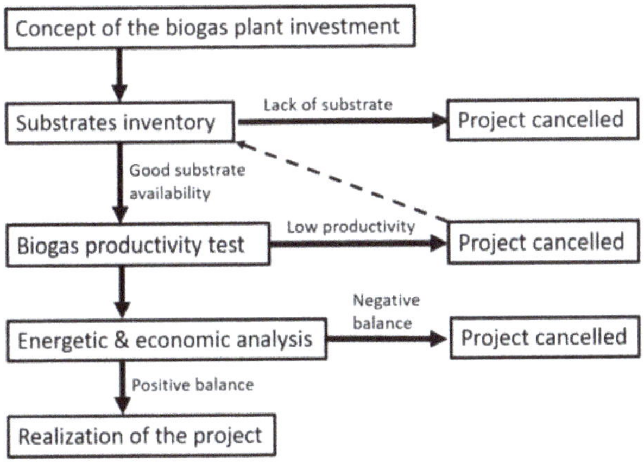

Figure 3. The flowchart of the proper way for making decisions for biogas plant investment.

2.1. *Biogas Production Efficiency from Substrates*

The biogas production efficiency analysis from substrates (manure and straw) was made in the Ecotechnologies Laboratory at Poznan University of Life Sciences, the biggest Polish biogas laboratory. The test methodology was done within German norms DIN 38414/S8 [92] and VDI 4630 [93], the most popular analytical procedure used in the European biogas laboratories. The Ecotechnologies Laboratory was the first Polish biogas laboratory that passed the quality proficiency test organized by the German organizations Verband deutscher landwirtschaftlicher Untersuchungs- und Forschungsanstalten (VDLUFA) and Kuratorium für Technik und Bauwesen in der Landwirtschaft e.V. (KTBL).

The fermentation process runs in the glass reactors (volume 2 dm^3), well-closed with a weck system. The reactors are maintained at a stable temperature (39 °C, ±1 °C) by their placement inside the aquarium with heat water. The samples (about 50 g of organic dry matter) are put inside reactors (always in three replications) and then filled with inoculum. The blank samples (in three replications as well) are only filled with inoculum. Inoculum is produced from the liquid part of the digestate obtained from a typical biogas plant. The produced biogas goes to the plexiglass cylinder, where its volume is measured every 24 h. The gas composition (CH_4 0–100%, CO_2 0–100%, O_2 0–25%, H_2S 0–10,000 ppm, and NH_3 0–1000 ppm) is measured using the special portable tool GA5000. This instrument is calibrated each week using the gases to calibration (CH_4 65%, CO_2 35%, H_2S 500 ppm, and NH_3 100 ppm) supplied by Linde Gas company. The obtained results for each material are reduced by the production from the blank samples and then recalculated to be expressed as biogas and methane production from 1 Mg of FM (fresh matter) (as well as of TS and VS) of the analyzed sample.

The final results of the methane production efficiency from the substrates analyzed in laboratory tests sent from the Ecotechnologies Laboratory to the contractors are always expressed in CH_4 m^3 from 1 Mg of FM. This value is indispensable for further energetic calculations of planned biogas plant exploitation.

2.2. *Energetic and Economic Calculations*

The investment plan is to build the biogas plant with 500 kW of electric power in Dynamic Biogas technology. The reason for this size is related to Polish law regulations. Whole administrative procedures are simplified in building a biogas plant with a maximum electric power of 0.5 MW. Furthermore, this size is enough to reach a reasonable profit from electric energy production.

The electric energy (E_e) production from methane produced yearly from used substrates was calculated as follow:

$$Ee = VCH_4 \cdot WCH_4 \cdot \eta e \quad [MWh] \tag{1}$$

where:

VCH_4—volume of methane produced from digested substrates [m^3];
WCH_4—energetic value of 1 cubic meter of methane [9.968 kWh m^{-3}];
ηe—electric efficiency of CHP unit [0.4].

A similar equation should be used to calculate the heat (thermal energy, E_t) amount emitted from the cogeneration unit (CHP):

$$Et = VCH_4 \cdot WCH_4 \cdot \eta t \quad [MWh] \tag{2}$$

where:

VCH_4—volume of methane produced from digested substrates [m^3];
WCH_4—energetic value of 1 cubic meter of methane [9.968 kWh m^{-3}];
ηt—thermal efficiency of CHP unit [0.45].

The calculation of biogas plant electric power (P_e) is based on the electric energy amount produced during the whole year and the working time of CHP. This equation is presented below:

$$Pe = Ee/t \quad [MW] \tag{3}$$

where:

Ee—amount of electric energy produced yearly [MWh];
T—working time of CHP during the whole year of exploitation [8400 h].

The amount of heat can also be expressed in GJ, which is the most common unit that is used in reality by enterprises and people. This amount can be easily calculated because 1 MWh is equal to 3.6 GJ. As in the case of electric power (P_e), a similar equation was used for thermal power (P_t) calculations:

$$P_t = E_t/t \quad [MW] \tag{4}$$

where:

Et–amount of heat produced yearly [MWh];
T–working time of CHP during the whole year of exploitation [8400 h].

It should be underlined that in a typical biogas plant, the real working time (7200–8100 h/a) is firmly lower than the 8400 h/year value fixed before. This is related to many exploitation problems and breaks of biogas production by, i.e., mixing system failures and other mechanical damages. However, the biogas plants working in Dynamic Biogas technology have a unique construction of steel fermenters (with a central, vertical mixing system working with only one mixer that is easily exchanged) and can reach up to 8500 working hours per year in reality.

The investment cost for building the 0.5 MW biogas plant in DB technology is 9 million PLN (€1.970 million with a currency rate of €1 = 4.56 PLN). This means that the depreciation cost (for a 15-year period of exploitation) is €131,580/a.

3. Results

3.1. Basic Characterization of Analyzed Dairy Farm

The analyzed farm has over 1000 ha of planted area and a herd of 600 dairy cows (breeding size of 600 livestock units (LSU)), which produce milk at a high level of 11,000 L/LSU/a. The milk is delivered to the dairy for 0.29 €/L. The sale of milk is one of the primary income sources for the farm, as it amounts to €1.925 million (Table 1).

Table 1. Characterization of the analyzed dairy farm.

Parameter	Unit	Value
Price of milk	€/L	0.29
Milk production	L/LSU/a	11,000
Number of cows	LSU	600
Income from the sale of milk	€1000/a	1925
Manure weight	Mg/a	9855

After being removed from breeding buildings, cow manure is stored in piles on concrete platforms and spread into the field three times a year for fertilization. The total mass of produced farmyard manure is 9855 Mg per year. It should be emphasized that in the current situation, manure stored for several months a year is a source of uncontrolled methane emissions, which in the future may become a financial problem for the farm if the European Commission pushes the taxation of CH_4 emissions, on an equal basis as CO_2 currently, through.

3.2. Methane Productivity Analysis

The basic parameters, such as dry matter content (TS), organic dry matter (VS), methane content, biogas, and methane yield of substrates, used for the fermentation tests are presented in Table 2.

Table 2. The basic parameters and biogas yield of the tested materials.

Substrate	TS	VS	C:N Ratio	Biogas	CH_4 Content	CH_4
	[% FM]	[% TS]		[m^3/Mg FM]	[%]	[m^3/Mg FM]
Cow manure	21.56	85.62	21:1	79.90	56.37	45.04
Wheat straw	92.67	96.96	95:1	468.49	56.72	265.72

The results of TS show the big difference between both materials. Farmyard manure contains mostly water (over 78%) and more than 15% ash, so the methane productivity from one ton of fresh matter (45.04 m^3 /Mg FM) is almost six times lower when compared to the wheat straw result (265.72 m^3/Mg FM). That is why, from a biogas plant exploitation point of view, the usage of straw for installation feeding has a much higher profitability when compared to manure. The biogas obtained from the fermentation of wheat straw and manure had a similar methane content: 56.7 and 56.4% methane, respectively.

The fermentation charts for manure and wheat straw are presented in Figures 4 and 5. It has to be underlined that the samples of manure and straw had a different initial mass, so the results presented in the figures cannot be directly compared. However, for comparison, the biogas and methane productivities calculated in m^3 from 1 Mg of both materials are presented in Table 2. Moreover, Figures 4 and 5 are important for presenting the dynamic of the gas production and fermentation time.

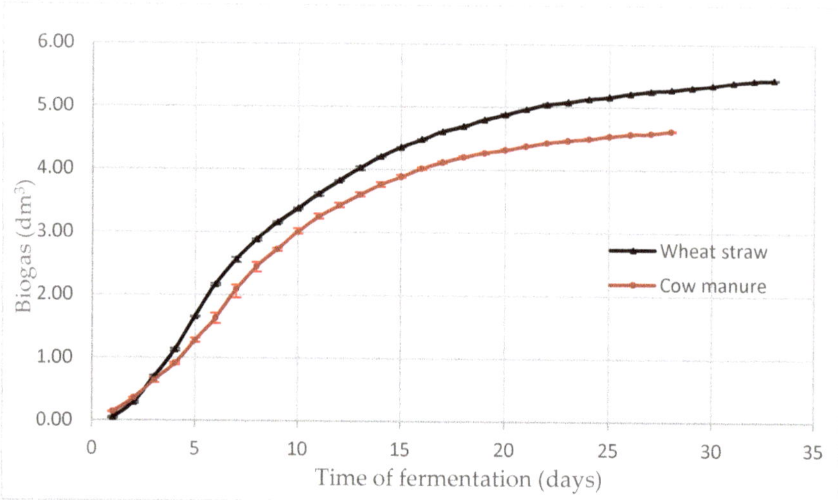

Figure 4. Production of biogas during the samples' fermentation process (daily measurements) for both materials, with error bars (the initial sample masses were different).

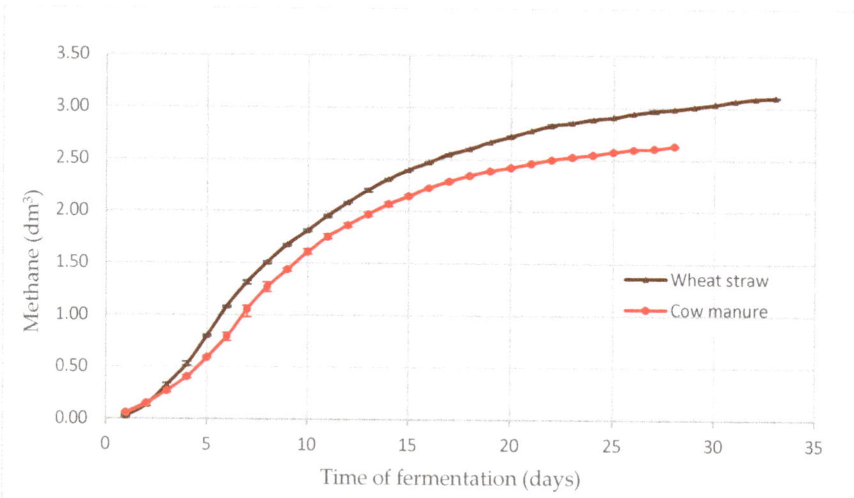

Figure 5. Methane production during the samples' fermentation process (daily measurements) for both materials, with error bars (the initial sample masses were different).

The runs of fermentation show that cow manure is digested in a shorter time (HRT (Hydraulic Retention Time) = 28 days) than wheat straw (33 days). This phenomena is also visible in the periods for 80% and 90% of the total methane production (Figure 6).

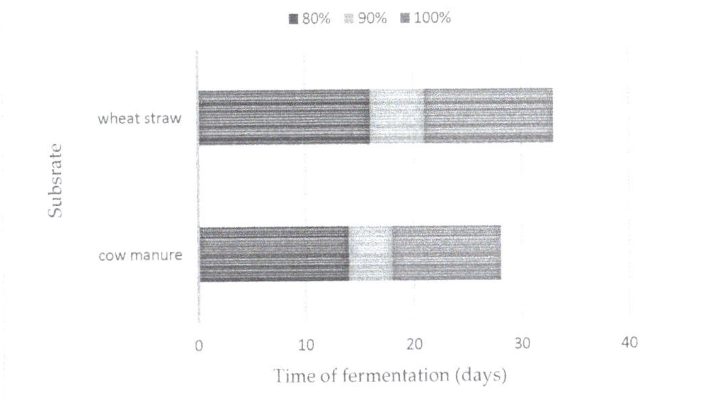

Figure 6. Time required for 80, 90, and 100% of total methane production for the analyzed materials.

80% of total methane production was reached by cow manure on the 14th day; however, wheat straw needed 16 days to reach the same level (two days later). Those differences between manure and straw are bigger for 90% of the total production (three days), and the biggest difference is for the end of fermentation (five days).

3.3. Energetic and Economic Calculations

In the energy calculations, the first stage included the calculation of the electric power of the biogas plant, while only assuming work that used manure as a substrate. On this basis, it has been calculated that the production of methane from manure will amount to over 443,000 m^3 annually, which will allow for the production of 1770 MWh of electricity by a 0.211 MW installation (Table 3).

As this capacity is less than half of the planned installation (500 kW), it was decided to obtain the missing capacity (0.289 MW) by using 2300 Mg of wheat straw (Table 3).

Table 3. Energetic aspects of the biogas plant in the analyzed dairy farm.

Parameter	Unit	Value
CH_4 yield of manure	m^3/Mg	45.04
Amount of methane	m^3/a	443,869
Electric energy	MWh/a	1770
Electric power of the installation	MW_e	0.211
Additional substrate: wheat straw		
The mass of straw	Mg/a	2300
CH_4 yield of straw	m^3/Mg	265.72
Amount of methane	m^3/a	611,156
Electric energy	MWh/a	2437
Electric power of the installation	MW_e	0.289
Total amount of methane	m^3/a	1,055,025
Total electric energy	MWh/a	4207
Total electric power	MW_e	0.500
Electricity price	EUR/MWh	158.3
Electricity value	EUR/a	666,044
Amount of heat	MWh/a	4732
Amount of heat	GJ/a	17,037
Heat power	MW	0.563
Price for heat	EUR/GJ	8.77
Heat value	EUR/a	134,500

Ultimately, the designed installation will generate 1,055,025 m^3 of methane, which, with a CHP unit electrical efficiency of 40%, allows for the production of 4207 MWh of electricity. When related to the 8400 h of operation of the cogeneration unit, this amount of energy allows 500 kW of electric power to be reached for the planned installation.

Since the current price (including subsidies from the state) for electricity for a 50–500-kW biogas plant in Poland is 158.3 €/MWh, the generated energy will bring revenues of over €666,000 per year.

The amount of heat, calculated according to a methodology similar to that of electricity, is over 17 TJ. When calculating the heat price, the price of heat produced from Poland's most popular energy carrier, i.e., hard coal, was used. In Poland, the value of heat generated from coal is €8.77/GJ. On this basis, the revenue from the sale of heat can be calculated, amounting to €134,500 per year (Table 4).

Table 4. Economic balance of the analyzed dairy farm in a conventional scenario (milk production) and in a scenario with milk production and biogas running.

Parameter	Standard Scenario	Biogas Plant Scenario
	Profits [kEUR/a]	
Sold milk	1925	1925
Manure as fertilizer	259	
Digestate as fertilizer		74
Electricity production		666
Heat		135
	Costs [kEUR/a]	
Straw cost		−76
Depreciation		−132
Service costs and others		−77
Balance	**2184**	**2516**

The total (simplified) economic balance of the dairy farm operation in two scenarios (traditional and in the version with a biogas plant processing cow manure and straw) shows a significant advantage of the second variant. The total profit in scenario II (€2.516 million) is €332,000 higher than in the traditional scenario. Considering the cost of building a biogas plant (€1.970 million), this means that this installation will be paid back after six years of operation.

The described situation may change even more favorably for scenario II if the European Commission introduces taxes for methane emissions, similar to the current situation in the case of carbon dioxide. It should be taken into account that agriculture is a source of high methane emissions, and especially large amounts of methane are produced from heaps of cattle manure. The necessity to pay for emissions of methane—a greenhouse gas that is 21 times more powerful than CO_2—will significantly worsen the profitability of scenario I (traditional). On this basis, it can be assumed that in the future, the construction of biogas plants at medium and large dairy farms will be necessary due to the reduction of uncontrolled methane emissions from stored manure.

4. Discussion

The paper describes an energetic and economic analysis of a dairy farm and small biogas plant coupling in order to increase farm profitability and decrease the uncontrolled methane emission from manure stored in heaps. However, some doubts may relate to the term "small" biogas plant, as in the described farm the planned installation has 500 kW of electric power and cannot be sufficiently fed with manure produced from 600 cows. In fact, when compared to the most popular biogas plant size in India and Southeast Asia (fermenters with a volume of only several cubic meters even, often located underground), this is a huge difference [73,94]. The main reason for this considerable difference in the definition of a small biogas plant is related to climate conditions. In Poland, temperatures reaching even −20 °C in wintertime make it almost impossible to build small biogas plants like those in Asia because all constructed installations should be equipped with a heating system that allows the temperature inside fermenters to be kept at the level of 38–42 °C [95,96]. Some trials in Poland, involving the construction of small biogas plants with a fermenter volume of 20–60 m^3 and an installed power of 8–12 kW, showed that those installations were not able to produce enough biogas during wintertime to heat themselves and stabilize the temperature inside the fermenters. Furthermore, this is due to the high investment costs of operating such micro biogas plants, and, as a solution to this situation, Polish legislation defines a "small" biogas plant as an installation with installed electric power between 50 to 500 kW [97,98].

The usage of manure as a substrate for biogas plants is mentioned by many scientists as the best solution from an energetic, economic, and environmental point of view [33,99,100]. This can provide additional profit to animal farms (as was calculated in this paper) and can also reduce uncontrolled methane emissions to the atmosphere from manure stored in piles. What is especially interesting for human health is that anaerobic digestion can also destroy the antibiotics that are present in manure [101]. The massive usage of antibiotics in animal production and the subsequent uncontrolled stream of solved medicaments to the environment via animal (and human) excrements is the main reason for the creation of "super-bacteria"—resistant to all known antibiotics.

Some studies underline that (in general) animal manures should be treated by anaerobic digestion; however, these materials have a relatively weak biogas potential in order to guarantee a high productivity of biogas plants [102,103]. This was also the case in our study because whole manure production could cover less than half of the power in the planned installation. That is why the usage of additional substrates is required [80,104]. One of the most popular substrates is straw, which is easy to collect and has a high productivity. Many researchers have underlined the high value of different straws used for biogas feeding, like maize straw [77,105] or cereals [106]. In the described study, we need an additional 2300 Mg/a of wheat straw to keep the highest productivity of the planned biogas plant. It has to be underlined that straw usage for biogas plant feeding does not generate a conflict between food and biofuels production as is the case for maize silage usage. This is because straw is treated as an agricultural byproduct and not as the main yield.

The concept described in this paper is based on using only manure and straw for the feeding fermentation process. The reason for this concept is that both materials are produced on farms. However, it has to be underlined that the codigestion of different biowastes can generate a synergy effect and increase the total biogas production on a level that is higher than the sum of biogas generated separately by each substrate [107,108]. An excellent way to increase biogas productivity is the usage of dairy waste (i.e., cheese waste), brewery waste, and other materials (sometimes specific materials like biochar) [109,110]. In the described case, however, the dairy factory's distance was too big for the economic transport of dairy waste to the planned biogas plant.

5. Conclusions

Based on the research and results obtained, the following conclusions were formed:

1. Investment in a biogas plant operating at a dairy farm and using cow manure as one of the main substrates is a more profitable scenario than traditional dairy farming.
2. Currently, in Polish conditions, a scenario with a biogas plant operating with a capacity of 500 kW brings 332,000 €/a more profit than a conventional scenario (only milk production), even when taking into account additional costs, including the purchase of straw to ensure the continuous operation of the installation (manure from 600 cows will not provide substrate coverage).
3. In the analyzed case, the cost of building a biogas plant will pay off in less than six years. However, it is important to underline that this concerns the biogas plant with a power of 500 kW working with cow manure and wheat straw. Smaller-scale installations, as well as other substrates, can change this revenue time.
4. The economic advantage of the dairy farm scenario with a biogas plant over the conventional variant will be even more significant if the European Commission introduces charges for methane emissions from agriculture (which will affect particular dairy farms that store manure in piles).

5. It should be necessary to develop and conduct energetic and economic analyses for scenarios based on smaller size installations and alternative substrates (dairy waste, food-industry waste, etc.).

Author Contributions: Conceptualization, P.P., A.K.-J., P.S., A.W. and J.D.; methodology, P.S. and J.D.; software, P.P.; validation, A.K.-J., P.S. and J.D.; formal analysis, P.P. and A.W.; investigation, P.P. and J.D.; resources, A.K.-J.; data curation, P.P. and A.W.; writing—original draft preparation, P.P. and A.K.-J.; writing—review and editing, P.P., A.K.-J., P.S., A.W. and J.D.; visualization, P.P.; supervision, J.D.; project administration, A.K.-J. and J.D.; funding acquisition, P.P. and A.K.-J. All authors have read and agreed to the published version of the manuscript.

Funding: This research was funded from the 'Excellent science' program of the Ministry of Science and Higher Education as a part of the contract no. DNK/SP/465641/2020 "The role of the agricultural engineering and environmental engineering in the sustainable agriculture development".

Acknowledgments: This paper was created in the framework of the project "Decision support system for sustainable and GHG optimised milk production in key European areas" No. ICT-AGRI-3 ID 39288, Acronym MilKey, realized at Poznan University of Life Sciences.

Conflicts of Interest: The authors declare no conflict of interest.

References

1. IPCC. *Climate Change 2013: The Physical Science Basis. Contribution of Working Group I to the Fifth Assessment Report of the Intergovernmental Panel on Climate Change*; Cambridge University Press: Cambridge, UK; New York, NY, USA, 2013; p. 1535. Available online: https://www.ipcc.ch/report/ar5/wg1/ (accessed on 1 October 2020).
2. IEA. World Energy Outlook. 2013. Available online: https://edgar.jrc.ec.europa.eu/overview.php?v=50_GHG (accessed on 1 October 2020).
3. EEA. EEA Greenhouse Gas—Data Viewer. 2013. Available online: https://www.eea.europa.eu/ds_resolveuid/f4269fac-662f-4ba0-a416-c25373823292 (accessed on 1 October 2020).
4. Cole, N.A.; Meyer, B.E.; Parker, D.B.; Neel, J.; Turner, K.E.; Northup, B.K.; Jennings, T.; Jennings, J.S. Effects of diet quality on energy metabolism and methane production by beef steers fed a warm-season grass-based hay diet. *Appl. Anim. Sci.* **2020**, *36*, 652–667. [CrossRef]
5. EU 2030 Climate Target Plan Impact Assessment, Commission Staff Working Document Impact Assessment: Accompanying the Document Communication from the Commission to the European Parliament, the Council, the European Economic and Social Committee and the Committee of the Regions Stepping up Europe's 2030 Climate Ambition Investing in a Climate-neutral Future for the Benefit of our People. Available online: https://eur-lex.europa.eu/legal-content/EN/TXT/?uri=CELEX:52020SC0176 (accessed on 2 October 2020).
6. EGD. Communication on The European Green Deal. 2016. Available online: https://ec.europa.eu/info/publications/communication-european-green-deal_en (accessed on 2 October 2020).
7. Matthews, C.; Crispie, F.; Lewis, E.; Reid, M.; O'Toole, P.W.; Cotter, P.D. The rumen microbiome: A crucial consideration when optimising milk and meat production and nitrogen utilisation efficiency. *Gut Microbes* **2019**, *10*, 115–132. [CrossRef] [PubMed]
8. Rojas-Downing, M.M.; Nejadhashemi, A.P.; Harrigan, T.; Woznicki, S.A. Climate change and livestock: Impacts, adaptation, and mitigation. *Clim. Risk Manag.* **2017**, *16*, 145–163. [CrossRef]
9. Leytem, A.B.; Dungan, R.S.; Bjorneberg, D.L.; Koehn, A.C. Emissions of Ammonia, Methane, Carbon Dioxide, and Nitrous Oxide from Dairy Cattle Housing and Manure Management Systems. *J. Environ. Qual.* **2011**, *40*, 1383–1394. [CrossRef]
10. Bryszak, M.; Szumacher-Strabel, M.; Huang, H.; Pawlak, P.; Lechniak, D.; Kołodziejski, P.; Yanza, Y.R.; Patra, A.K.; Váradyová, Z.; Cieslak, A. Lupinus angustifolius seed meal supplemented to dairy cow diet improves fatty acid composition in milk and mitigates methane production. *Anim. Feed Sci. Technol.* **2020**, *267*, 114590. [CrossRef]
11. Benchaar, C.; Hassanat, F. Methane emissions of stored manure from dairy cows fed conventional or brown midrib corn silage. *J. Dairy Sci.* **2019**, *102*, 10632–10638. [CrossRef]
12. Kolling, G.J.; Stivanin, S.C.B.; Gabbi, A.M.; Machado, F.S.; Ferreira, A.L.; Campos, M.M.; Tomich, T.R.; Cunha, C.S.; Dill, S.W.; Pereira, L.G.R.; et al. Performance and methane emissions in dairy cows fed oregano and green tea extracts as feed additives. *J. Dairy Sci.* **2018**, *101*, 4221–4234. [CrossRef]

13. Lee, M.A.; Todd, A.; Sutton, M.A.; Chagunda, M.G.G.; Roberts, D.J.; Rees, R.M. A time-series of methane and carbon dioxide production from dairy cows during a period of dietary transition. *Cogent Environ. Sci.* **2017**, *3*, 1385693. [CrossRef]
14. Aemiro, A.; Hanada, M.; Umetsu, K.; Nishida, T. The effect of Sunphenon 30S-O on methane emission, nutrient intake, digestibility and rumen fermentation. *Anim. Feed Sci. Technol.* **2016**, *214*, 34–43. [CrossRef]
15. Gislon, G.; Colombini, S.; Borreani, G.; Crovetto, G.M.; Sandrucci, A.; Galassi, G.; Tabacco, E.; Rapetti, L. Milk production, methane emissions, nitrogen, and energy balance of cows fed diets based on different forage systems. *J. Dairy Sci.* **2020**, *103*, 8048–8061. [CrossRef]
16. Wanapat, M.; Cherdthong, A.; Phesatcha, K.; Kang, S. Dietary sources and their effects on animal production and environmental sustainability. *Anim. Nutr.* **2015**, *1*, 96–103. [CrossRef] [PubMed]
17. Knapp, J.R.; Laur, G.L.; Vadas, P.A.; Weiss, W.P.; Tricarico, J.M. Invited review: Enteric methane in dairy cattle production: Quantifying the opportunities and impact of reducing emissions. *J. Dairy Sci.* **2014**, *97*, 3231–3261. [CrossRef]
18. Benchaar, C.; Hassanat, F. Frequency of diet delivery to dairy cows: Effect on nutrient digestion, rumen fermentation, methane production, nitrogen utilization, and milk production. *J. Dairy Sci.* **2020**, *103*, 7094–7109. [CrossRef] [PubMed]
19. Bell, M.J.; Saunders, N.; Wilcox, R.H.; Homer, E.M.; Goodman, J.R.; Craigon, J.; Garnsworthy, P.C. Methane emissions among individual dairy cows during milking quantified by eructation peaks or ratio with carbon dioxide. *J. Dairy Sci.* **2014**, *97*, 6536–6546. [CrossRef] [PubMed]
20. Yan, T.; Mayne, C.S.; Gordon, F.G.; Porter, M.G.; Agnew, R.E.; Patterson, D.C.; Ferris, C.P.; Kilpatrick, D.J. Mitigation of enteric methane emissions through improving efficiency of energy utilization and productivity in lactating dairy cows. *J. Dairy Sci.* **2010**, *93*, 2630–2638. [CrossRef] [PubMed]
21. Uddin, M.E.; Larson, R.A.; Wattiaux, M.A. Effects of dairy cow breed and dietary forage on greenhouse gas emissions from manure during storage and after field application. *J. Clean. Prod.* **2020**, *270*, 122461. [CrossRef]
22. Wattiaux, M.A.; Uddin, M.E.; Letelier, P.; Jackson, R.D.; Larson, R.A. INVITED REVIEW: Emission and mitigation of greenhouse gases from dairy farms: The cow, the manure, and the field. *Appl. Anim. Sci.* **2019**, *35*, 238–254. [CrossRef]
23. Xue, B.; Wang, L.Z.; Yan, T. Methane emission inventories for enteric fermentation and manure management of yak, buffalo and dairy and beef cattle in China from 1988 to 2009. *Agric. Ecosyst. Environ.* **2014**, *195*, 202–210. [CrossRef]
24. Kozłowski, K.; Dach, J.; Lewicki, A.; Malińska, K.; Do Carmo, I.E.P.; Czekała, W. Potential of biogas production from animal manure in Poland. *Arch. Environ. Prot.* **2019**, *45*, 99–108. [CrossRef]
25. Thiruvenkatachari, R.R.; Carranza, V.; Ahangar, F.; Marklein, A.; Hopkins, F.; Venkatram, A. Uncertainty in using dispersion models to estimate methane emissions from manure lagoons in dairies. *Agric. For. Meteorol.* **2020**, *290*, 108011. [CrossRef]
26. VanderZaag, A.C.; Flesch, T.K.; Desjardins, R.L.; Baldé, H.; Wright, T. Measuring methane emissions from two dairy farms: Seasonal and manure-management effects. *Agric. For. Meteorol.* **2014**, *194*, 259–267. [CrossRef]
27. Ma, G.; Ndegwa, P.; Harrison, J.H.; Chen, Y. Methane yields during anaerobic co-digestion of animal manure with other feedstocks: A meta-analysis. *Sci. Total Environ.* **2020**, *728*, 138224. [CrossRef] [PubMed]
28. Jafari-Sejahrood, A.; Najafi, B.; Faizollahzadeh Ardabili, S.; Shamshirband, S.; Mosavi, A.; Chau, K. wing Limiting factors for biogas production from cow manure: Energo-environmental approach. *Eng. Appl. Comput. Fluid Mech.* **2019**, *13*, 954–966. [CrossRef]
29. Ding, L.; Lu, Q.; Xie, L.; Liu, J.; Cao, W.; Shi, Z.; Li, B.; Wang, C.; Zhang, G.; Ren, S. Greenhouse gas emissions from dairy open lot and manure stockpile in northern China: A case study. *J. Air Waste Manag. Assoc.* **2016**, *66*, 267–279. [CrossRef] [PubMed]
30. Wang, J.; Duan, C.; Ji, Y.; Sun, Y. Methane emissions during storage of different treatments from cattle manure in Tianjin. *J. Environ. Sci.* **2010**, *22*, 1564–1569. [CrossRef]
31. Zhang, N.; Bai, Z.; Ledgard, S.; Luo, J.; Ma, L. Ammonia mitigation effects from the cow housing and manure storage chain on the nitrogen and carbon footprints of a typical dairy farm system on the North China Plain. *J. Clean. Prod.* **2021**, *280*, 124465. [CrossRef]
32. Flesch, T.K.; Desjardins, R.L.; Worth, D. Fugitive methane emissions from an agricultural biodigester. *Biomass Bioenergy* **2011**, *35*, 3927–3935. [CrossRef]

33. Díaz-Vázquez, D.; Alvarado-Cummings, S.C.; Meza-Rodríguez, D.; Senés-Guerrero, C.; de Anda, J.; Gradilla-Hernández, M.S. Evaluation of biogas potential from livestock manures and multicriteria site selection for centralized anaerobic digester systems: The case of Jalisco, Mexico. *Sustainability* **2020**, *12*, 3527. [CrossRef]
34. Feng, L.; Ward, A.J.; Moset, V.; Møller, H.B. Methane emission during on-site pre-storage of animal manure prior to anaerobic digestion at biogas plant: Effect of storage temperature and addition of food waste. *J. Environ. Manag.* **2018**, *225*, 272–279. [CrossRef]
35. Wang, Q.; Thompson, E.; Parsons, R.; Rogers, G.; Dunn, D. Economic feasibility of converting cow manure to electricity: A case study of the CVPS Cow Power program in Vermont. *J. Dairy Sci.* **2011**, *94*, 4937–4949. [CrossRef]
36. Massé, D.I.; Jarret, G.; Hassanat, F.; Benchaar, C.; Saady, N.M.C. Effect of increasing levels of corn silage in an alfalfa-based dairy cow diet and of manure management practices on manure fugitive methane emissions. *Agric. Ecosyst. Environ.* **2016**, *221*, 109–114. [CrossRef]
37. Sommer, S.G.; Petersen, S.O.; Søgaard, H.T. Greenhouse Gas Emission from Stored Livestock Slurry. *J. Environ. Qual.* **2000**, *29*, 744–751. [CrossRef]
38. VanderZaag, A.; Amon, B.; Bittman, S.; Kuczyński, T. Ammonia Abatement with Manure Storage and Processing Techniques. In *Costs of Ammonia Abatement and the Climate Co-Benefits*; Springer: Berlin/Heidelberg, Germany; Dordrecht, The Netherlands, 2015; pp. 75–112. [CrossRef]
39. Hindrichsen, I.K.; Wettstein, H.R.; Machmüller, A.; Kreuzer, M. Methane emission, nutrient degradation and nitrogen turnover in dairy cows and their slurry at different milk production scenarios with and without concentrate supplementation. *Agric. Ecosyst. Environ.* **2006**, *113*, 150–161. [CrossRef]
40. Kupper, T.; Häni, C.; Neftel, A.; Kincaid, C.; Bühler, M.; Amon, B.; VanderZaag, A. Ammonia and greenhouse gas emissions from slurry storage—A review. *Agric. Ecosyst. Environ.* **2020**, *300*, 106963. [CrossRef]
41. Browne, J.D.; Gilkinson, S.R.; Frost, J.P. The effects of storage time and temperature on biogas production from dairy cow slurry. *Biosyst. Eng.* **2015**, *129*, 48–56. [CrossRef]
42. Luo, T.; Khoshnevisan, B.; Huang, R.; Chen, Q.; Mei, Z.; Pan, J.; Liu, H. Analysis of revolution in decentralized biogas facilities caused by transition in Chinese rural areas. *Renew. Sustain. Energy Rev.* **2020**, *133*, 110133. [CrossRef]
43. Garcia, N.H.; Mattioli, A.; Gil, A.; Frison, N.; Battista, F.; Bolzonella, D. Evaluation of the methane potential of different agricultural and food processing substrates for improved biogas production in rural areas. *Renew. Sustain. Energy Rev.* **2019**, *112*, 1–10. [CrossRef]
44. Zemo, K.H.; Panduro, T.E.; Termansen, M. Impact of biogas plants on rural residential property values and implications for local acceptance. *Energy Policy* **2019**, *129*, 1121–1131. [CrossRef]
45. Korberg, A.D.; Skov, I.R.; Mathiesen, B.V. The role of biogas and biogas-derived fuels in a 100% renewable energy system in Denmark. *Energy* **2020**, *199*, 117426. [CrossRef]
46. Rosa, A.P.; Chernicharo, C.A.L.; Lobato, L.C.S.; Silva, R.V.; Padilha, R.F.; Borges, J.M. Assessing the potential of renewable energy sources (biogas and sludge) in a full-scale UASB-based treatment plant. *Renew. Energy* **2018**, *124*, 21–26. [CrossRef]
47. RED I. Renewable Energy—Recast to 2030 (RED II) EU Science Hub. 2013. Available online: https://ec.europa.eu/jrc/en/jec/renewable-energy-recast-2030-red-ii (accessed on 10 October 2020).
48. Kozłowski, K.; Pietrzykowski, M.; Czekała, W.; Dach, J.; Kowalczyk-Juśko, A.; Jóźwiakowski, K.; Brzoski, M. Energetic and economic analysis of biogas plant with using the dairy industry waste. *Energy* **2019**, *183*, 1023–1031. [CrossRef]
49. Lauer, M.; Hansen, J.K.; Lamers, P.; Thrän, D. Making money from waste: The economic viability of producing biogas and biomethane in the Idaho dairy industry. *Appl. Energy* **2018**, *222*, 621–636. [CrossRef]
50. McVoitte, W.P.A.; Clark, O.G. The effects of temperature and duration of thermal pretreatment on the solid-state anaerobic digestion of dairy cow manure. *Heliyon* **2019**, *5*, e02140. [CrossRef] [PubMed]
51. Klavon, K.H.; Lansing, S.A.; Mulbry, W.; Moss, A.R.; Felton, G. Economic analysis of small-scale agricultural digesters in the United States. *Biomass Bioenergy* **2013**, *54*, 36–45. [CrossRef]
52. Klootwijk, C.W.; Van Middelaar, C.E.; Berentsen, P.B.M.; de Boer, I.J.M. Dutch dairy farms after milk quota abolition: Economic and environmental consequences of a new manure policy. *J. Dairy Sci.* **2016**, *99*, 8384–8396. [CrossRef]

53. Vergote, T.L.I.; Bodé, S.; De Dobbelaere, A.E.J.; Buysse, J.; Meers, E.; Volcke, E.I.P. Monitoring methane and nitrous oxide emissions from digestate storage following manure mono-digestion. *Biosyst. Eng.* **2020**, *196*, 159–171. [CrossRef]
54. Khayum, N.; Anbarasu, S.; Murugan, S. Biogas potential from spent tea waste: A laboratory scale investigation of co-digestion with cow manure. *Energy* **2018**, *165*, 760–768. [CrossRef]
55. Vanden Nest, T.; Amery, F.; Fryda, L.; Boogaerts, C.; Bilbao, J.; Vandecasteele, B. Renewable P sources: P use efficiency of digestate, processed animal manure, compost, biochar and struvite. *Sci. Total Environ.* **2021**, *750*, 141699. [CrossRef]
56. Iocoli, G.A.; Zabaloy, M.C.; Pasdevicelli, G.; Gómez, M.A. Use of biogas digestates obtained by anaerobic digestion and co-digestion as fertilizers: Characterization, soil biological activity and growth dynamic of *Lactuca sativa* L. *Sci. Total Environ.* **2019**, *647*, 11–19. [CrossRef] [PubMed]
57. Ehmann, A.; Bach, I.M.; Bilbao, J.; Lewandowski, I.; Müller, T. Phosphates recycled from semi-liquid manure and digestate are suitable alternative fertilizers for ornamentals. *Sci. Hortic.* **2019**, *243*, 440–450. [CrossRef]
58. Czekała, W.; Bartnikowska, S.; Dach, J.; Janczak, D.; Smurzyńska, A.; Kozłowski, K.; Bugała, A.; Lewicki, A.; Cieślik, M.; Typańska, D.; et al. The energy value and economic efficiency of solid biofuels produced from digestate and sawdust. *Energy* **2018**, *159*, 1118–1122. [CrossRef]
59. Czekała, W.; Lewicki, A.; Pochwatka, P.; Czekała, A.; Wojcieszak, D.; Jóźwiakowski, K.; Waliszewska, H. Digestate management in polish farms as an element of the nutrient cycle. *J. Clean. Prod.* **2020**, *242*, 118454. [CrossRef]
60. Hung, C.Y.; Tsai, W.T.; Chen, J.W.; Lin, Y.Q.; Chang, Y.M. Characterization of biochar prepared from biogas digestate. *Waste Manag.* **2017**, *66*, 53–60. [CrossRef] [PubMed]
61. Törnwall, E.; Pettersson, H.; Thorin, E.; Schwede, S. Post-treatment of biogas digestate-An evaluation of ammonium recovery, energy use and sanitation. In *Proceedings of the Energy Procedia*; Elsevier Ltd.: Amsterdam, The Netherlands, 2017; Volume 142, pp. 957–963. [CrossRef]
62. Baral, K.R.; Labouriau, R.; Olesen, J.E.; Petersen, S.O. Nitrous oxide emissions and nitrogen use efficiency of manure and digestates applied to spring barley. *Agric. Ecosyst. Environ.* **2017**, *239*, 188–198. [CrossRef]
63. Castro, L.; Escalante, H.; Jaimes-Estévez, J.; Díaz, L.J.; Vecino, K.; Rojas, G.; Mantilla, L. Low cost digester monitoring under realistic conditions: Rural use of biogas and digestate quality. *Bioresour. Technol.* **2017**, *239*, 311–317. [CrossRef] [PubMed]
64. Maldaner, L.; Wagner-Riddle, C.; VanderZaag, A.C.; Gordon, R.; Duke, C. Methane emissions from storage of digestate at a dairy manure biogas facility. *Agric. For. Meteorol.* **2018**, *258*, 96–107. [CrossRef]
65. Yan, Y.; Zhang, L.; Feng, L.; Sun, D.; Dang, Y. Comparison of varying operating parameters on heavy metals ecological risk during anaerobic co-digestion of chicken manure and corn stover. *Bioresour. Technol.* **2018**, *247*, 660–668. [CrossRef]
66. Dahlin, J.; Herbes, C.; Nelles, M. Biogas digestate marketing: Qualitative insights into the supply side. *Resour. Conserv. Recycl.* **2015**, *104*, 152–161. [CrossRef]
67. Marin-Batista, J.D.; Villamil, J.A.; Qaramaleki, S.V.; Coronella, C.J.; Mohedano, A.F.; de la Rubia, M.A. Energy valorization of cow manure by hydrothermal carbonization and anaerobic digestion. *Renew. Energy* **2020**, *160*, 623–632. [CrossRef]
68. Møller, H.B.; Moset, V.; Brask, M.; Weisbjerg, M.R.; Lund, P. Feces composition and manure derived methane yield from dairy cows: Influence of diet with focus on fat supplement and roughage type. *Atmos. Environ.* **2014**, *94*, 36–43. [CrossRef]
69. Li, Y.; Achinas, S.; Zhao, J.; Geurkink, B.; Krooneman, J.; Willem Euverink, G.J. Co-digestion of cow and sheep manure: Performance evaluation and relative microbial activity. *Renew. Energy* **2020**, *153*, 553–563. [CrossRef]
70. Duan, N.; Khoshnevisan, B.; Lin, C.; Liu, Z.; Liu, H. Life cycle assessment of anaerobic digestion of pig manure coupled with different digestate treatment technologies. *Environ. Int.* **2020**, *137*, 105522. [CrossRef] [PubMed]
71. Valentinuzzi, F.; Cavani, L.; Porfido, C.; Terzano, R.; Pii, Y.; Cesco, S.; Marzadori, C.; Mimmo, T. The fertilising potential of manure-based biogas fermentation residues: Pelleted vs. liquid digestate. *Heliyon* **2020**, *6*, e03325. [CrossRef] [PubMed]

72. Igliński, B.; Buczkowski, R.; Iglińska, A.; Cichosz, M.; Piechota, G.; Kujawski, W. Agricultural biogas plants in Poland: Investment process, economical and environmental aspects, biogas potential. *Renew. Sustain. Energy Rev.* **2012**, *16*, 4890–4900. [CrossRef]
73. Singh, P.; Singh, N. Political economy of bioenergy transitions in developing countries: A case study of Punjab, India. *World Dev.* **2019**, *124*, 104630. [CrossRef]
74. Stürmer, B.; Leiers, D.; Anspach, V.; Brügging, E.; Scharfy, D.; Wissel, T. Agricultural biogas production: A regional comparison of technical parameters. *Renew. Energy* **2021**, *164*, 171–182. [CrossRef]
75. Piwowar, A.; Dzikuć, M.; Adamczyk, J. Agricultural biogas plants in Poland-Selected technological, market and environmental aspects. *Renew. Sustain. Energy Rev.* **2016**, *58*, 69–74. [CrossRef]
76. Gieseke, D.; Lambertz, C.; Gauly, M. Relationship between herd size and measures of animal welfare on dairy cattle farms with freestall housing in Germany. *J. Dairy Sci.* **2018**, *101*, 7397–7411. [CrossRef]
77. Mazurkiewicz, J.; Marczuk, A.; Pochwatka, P.; Kujawa, S. Maize Straw as a Valuable Energetic Material for Biogas Plant Feeding. *Materials* **2019**, *12*, 3848. [CrossRef]
78. Kowalczyk-Juśko, A.; Pochwatka, P.; Zaborowicz, M.; Czekała, W.; Mazurkiewicz, J.; Mazur, A.; Janczak, D.; Marczuk, A.; Dach, J. Energy value estimation of silages for substrate in biogas plants using an artificial neural network. *Energy* **2020**, *202*, 117729. [CrossRef]
79. Cieślik, M.; Dach, J.; Lewicki, A.; Smurzyńska, A.; Janczak, D.; Pawlicka-Kaczorowska, J.; Boniecki, P.; Cyplik, P.; Czekała, W.; Jóźwiakowski, K. Methane fermentation of the maize straw silage under meso- and thermophilic conditions. *Energy* **2016**, *115*, 1495–1502. [CrossRef]
80. Dach, J.; Boniecki, P.; Przybył, J.; Janczak, D.; Lewicki, A.; Czekała, W.; Witaszek, K.; Rodríguez Carmona, P.C.; Cieślik, M. Energetic efficiency analysis of the agricultural biogas plant in 250 kWe experimental installation. *Energy* **2014**, *69*, 34–38. [CrossRef]
81. Wandera, S.M.; Qiao, W.; Algapani, D.E.; Bi, S.; Yin, D.; Qi, X.; Liu, Y.; Dach, J.; Dong, R. Searching for possibilities to improve the performance of full scale agricultural biogas plants. *Renew. Energy* **2018**, *116*, 720–727. [CrossRef]
82. Abraham, A.; Mathew, A.K.; Park, H.; Choi, O.; Sindhu, R.; Parameswaran, B.; Pandey, A.; Park, J.H.; Sang, B.I. Pretreatment strategies for enhanced biogas production from lignocellulosic biomass. *Bioresour. Technol.* **2020**, *301*, 122725. [CrossRef] [PubMed]
83. Ramos-Suárez, J.L.; Gómez, D.; Regueiro, L.; Baeza, A.; Hansen, F. Alkaline and oxidative pretreatments for the anaerobic digestion of cow manure and maize straw: Factors influencing the process and preliminary economic viability of an industrial application. *Bioresour. Technol.* **2017**, *241*, 10–20. [CrossRef]
84. Li, P.; He, C.; Li, G.; Ding, P.; Lan, M.; Gao, Z.; Jiao, Y. Biological pretreatment of corn straw for enhancing degradation efficiency and biogas production. *Bioengineered* **2020**, *11*, 251–260. [CrossRef] [PubMed]
85. Victorin, M.; Davidsson, Å.; Wallberg, O. Characterization of Mechanically Pretreated Wheat Straw for Biogas Production. *Bioenergy Res.* **2020**, *13*, 833–844. [CrossRef]
86. Yu, Q.; Liu, R.; Li, K.; Ma, R. A review of crop straw pretreatment methods for biogas production by anaerobic digestion in China. *Renew. Sustain. Energy Rev.* **2019**, *107*, 51–58. [CrossRef]
87. Gallegos, D.; Wedwitschka, H.; Moeller, L.; Zehnsdorf, A.; Stinner, W. Effect of particle size reduction and ensiling fermentation on biogas formation and silage quality of wheat straw. *Bioresour. Technol.* **2017**, *245*, 216–224. [CrossRef]
88. Feng, Y.; Zhao, X.; Guo, Y.; Yang, G.; Xi, J.; Ren, G. Changes in the material characteristics of maize straw during the pretreatment process of methanation. *J. Biomed. Biotechnol.* **2012**, *2012*. [CrossRef]
89. Khoshnevisan, B.; Shafiei, M.; Rajaeifar, M.A.; Tabatabaei, M. Biogas and bioethanol production from pinewood pre-treated with steam explosion and N-methylmorpholine-N-oxide (NMMO): A comparative life cycle assessment approach. *Energy* **2016**, *114*, 935–950. [CrossRef]
90. Marks, S.; Dach, J.; Morales, F.J.F.; Mazurkiewicz, J.; Pochwatka, P.; Gierz, Ł. New trends in substrates and biogas systems in Poland. *J. Ecol. Eng.* **2020**, *21*, 19–25. [CrossRef]
91. Mainardis, M.; Flaibani, S.; Trigatti, M.; Goi, D. Techno-economic feasibility of anaerobic digestion of cheese whey in small Italian dairies and effect of ultrasound pre-treatment on methane yield. *J. Environ. Manag.* **2019**, *246*, 557–563. [CrossRef] [PubMed]
92. DIN 38 414/S8. *German Standard Methods for the Examination of Water, Waste Water and Sludge; Sludge and Sediments (Group S); Determination of the Amenability to Anaerobic Digestion (S 8)*; DIN Deutches Institut für Normung e. V.: Berlin, Germany, 2012.

93. VDI 4630. *Fermentation of Organic Materials. Characterization of the Substrate, Sampling, Collection of Material Data, Fermentation Tests*; Verein Deutscher Ingenieure e.V.: Düsseldorf, Germany, 2016.
94. Malik, W.; Mohan, C.; Annachhatre, A.P. Community based biogas plant utilizing food waste and cow dung. *Mater. Today-Proc.* **2020**, *28*, 1910–1915. [CrossRef]
95. Fjørtoft, K.; Morken, J.; Hanssen, J.F.; Briseid, T. Methane production and energy evaluation of a farm scaled biogas plant in cold climate area. *Bioresour. Technol.* **2014**, *169*, 72–79. [CrossRef] [PubMed]
96. Pham, C.H.; Vu, C.C.; Sommer, S.G.; Bruun, S. Factors affecting process temperature and biogas production in small-scale rural biogas digesters in winter in northern Vietnam. *Asian-Australas. J. Anim. Sci.* **2014**, *27*, 1050–1056. [CrossRef] [PubMed]
97. Igliński, B.; Piechota, G.; Iwański, P.; Skarzatek, M.; Pilarski, G. 15 Years of the Polish agricultural biogas plants: Their history, current status, biogas potential and perspectives. *Clean Technol. Environ. Policy* **2020**, *22*, 281–307. [CrossRef]
98. Chasnyk, O.; Sołowski, G.; Shkarupa, O. Historical, technical and economic aspects of biogas development: Case of Poland and Ukraine. *Renew. Sustain. Energy Rev.* **2015**, *52*, 227–239. [CrossRef]
99. Achinas, S.; Willem Euverink, G.J. Rambling facets of manure-based biogas production in Europe: A briefing. *Renew. Sustain. Energy Rev.* **2020**, *119*, 109566. [CrossRef]
100. Zareei, S. Evaluation of biogas potential from livestock manures and rural wastes using GIS in Iran. *Renew. Energy* **2018**, *118*, 351–356. [CrossRef]
101. Hosseini Taleghani, A.; Lim, T.-T.; Lin, C.-H.; Ericsson, A.C.; Vo, P.H. Degradation of Veterinary Antibiotics in Swine Manure via Anaerobic Digestion. *Bioengineering* **2020**, *7*, 123. [CrossRef]
102. Li, Y.; Zhao, J.; Achinas, S.; Zhang, Z.; Krooneman, J.; Euverink, G.J.W. The biomethanation of cow manure in a continuous anaerobic digester can be boosted via a bioaugmentation culture containing Bathyarchaeota. *Sci. Total Environ.* **2020**, *745*, 141042. [CrossRef]
103. Ahmed, S.; Einfalt, D.; Kazda, M. Co-Digestion of Sugar Beet Silage Increases Biogas Yield from Fibrous Substrates. *Biomed Res. Int.* **2016**, *2016*. [CrossRef]
104. Koryś, K.A.; Latawiec, A.E.; Grotkiewicz, K.; Kuboń, M. The Review of Biomass Potential for Agricultural Biogas Production in Poland. *Sustainability* **2019**, *11*, 6515. [CrossRef]
105. Zbytek, Z.; Dach, J.; Pawłowski, T.; Smurzyńska, A.; Czekała, W.; Janczak, D. Energy and economic potential of maize straw used for biofuels production. In *MATEC Web of Conferences*; EDP Sciences: Ullis, France, 2016; Volume 60, p. 4008. [CrossRef]
106. Sieborg, M.U.; Jønson, B.D.; Larsen, S.U.; Vazifehkhoran, A.H.; Triolo, J.M. Co-Ensiling of Wheat Straw as an Alternative Pre-Treatment to Chemical, Hydrothermal and Mechanical Methods for Methane Production. *Energies* **2020**, *13*, 4047. [CrossRef]
107. Ai, P.; Chen, M.; Ran, Y.; Jin, K.; Peng, J.; Abomohra, A.E.F. Digestate recirculation through co-digestion with rice straw: Towards high biogas production and efficient waste recycling. *J. Clean. Prod.* **2020**, *263*, 121441. [CrossRef]
108. Siddique, M.N.I.; Wahid, Z.A. Achievements and perspectives of anaerobic co-digestion: A review. *J. Clean. Prod.* **2018**, *194*, 359–371. [CrossRef]
109. Misson, G.; Mainardis, M.; Incerti, G.; Goi, D.; Peressotti, A. Preliminary evaluation of potential methane production from anaerobic digestion of beach-cast seagrass wrack: The case study of high-adriatic coast. *J. Clean. Prod.* **2020**, *254*, 120131. [CrossRef]
110. Mainardis, M.; Flaibani, S.; Mazzolini, F.; Peressotti, A.; Goi, D. Techno-economic analysis of anaerobic digestion implementation in small Italian breweries and evaluation of biochar and granular activated carbon addition effect on methane yield. *J. Environ. Chem. Eng.* **2019**, *7*, 103184. [CrossRef]

Publisher's Note: MDPI stays neutral with regard to jurisdictional claims in published maps and institutional affiliations.

energies

MDPI

Article

The Economic Effects of New Patterns of Energy Efficiency and Heat Sources in Rural Single-Family Houses in Poland

Krzysztof Księżopolski [1,*], Mirosław Drygas [2], Kamila Pronińska [3] and Iwona Nurzyńska [4]

1 Warsaw School of Economics, Collegium of Socio-Economics, 02-554 Warsaw, Poland
2 Department of European Integration, Institute of Rural and Agriculture Development, Polish Academy of Sciences, 00-330 Warsaw, Poland; mdrygas@irwirpan.waw.pl
3 Faculty of Political Science and International Studies, University of Warsaw, 00-927 Warsaw, Poland; k.proninska@uw.edu.pl
4 Independent Researcher, 00-330 Warsaw, Poland; iwona.nurzynska@gmail.com
* Correspondence: kksiez@sgh.waw.pl

Received: 3 November 2020; Accepted: 30 November 2020; Published: 2 December 2020

Abstract: In the face of severe air pollution and implementation of energy and climate policy, it remains a challenge to develop effective strategies addressing the problem of solid fuels use in single-family houses (SFH) in rural areas in Poland. This study investigated the correlations between thermal modernization of SFH, the changes of heat sources from coal to clean energy, including heat pumps driven by prosumers' photovoltaic (PV) installation, and the disposable income of households in Polish rural areas. It also provided an analysis of the current support mechanisms promoting energy efficiency and PV development. The application of simulation modelling of energy consumption and costs in the research has proved that comprehensive thermal modernization of rural SFH constructed in the period of 1945–1970 and investments supporting PV/heat pump systems would enable the most cost-effective way of heating to be implemented. Considering that, today, spending on energy for heat puts a burden on the budget of rural households, especially those living in the SFH aged 50 years and more that dominate rural areas in Poland, the changes in energy supply–demand patterns would be an enhancement of their economic, energy and environmental security. The research argued that, in the wider process of energy transformation and solving air pollution problems, the role of rural households should not be neglected in public policy.

Keywords: rural areas; energy efficiency; photovoltaic systems; energy security; support mechanisms; public policy; energy policy; prosumer energy; single-family houses

1. Introduction

Energy efficiency and development of renewable energy sources have been identified as key areas of actions aimed at mitigating climate change. A need to improve energy efficiency has been highlighted by many international bodies, including the Intergovernmental Panel on Climate Changes (IPCC), United Nations (UN), and European Union (EU) [1,2]. In this context, the role of the residential buildings has been growing in importance across the EU. The residential sector, or households, constituted 26.1% of the EU's final energy consumption in 2018 [3]. Although the EU as a whole improved energy efficiency in this sector by around 29% (2.1%/year) over the period 2000–2017 [4], considerable differences in energy efficiency between the individual countries and regions were established. In many studies, introduced political measures, financial incentives and energy performance standards have been found a key factor in successful promotion of energy efficiency in the residential sector of individual EU countries [5,6]. Given the huge variations between the

Member States in both adapted policy measures and the existing energy efficiency potential in urban and rural areas, a case study-based approach of improving bottom-up data as well as understanding of the individual countries' needs and specificity has been postulated [5,7,8]

In terms of energy efficiency progress in household sector over the period 2008–2017, Poland was ranked 21st among the EU. With 1.13%/year rate of energy efficiency improvements, Polish households were below the EU's average and far below the improvements rate achieved by Poland's industry. Their role in meeting energy efficiency target and pursuing wider energy transformation can be substantial not only due to 28% share in final energy consumption [9] but also the fact that the most commonly used energy carrier in Polish households is coal. At the same time, there are considerable differences between urban and rural areas in Poland regarding income structure, the main source of heat, buildings' quality and their respective thermal modernisation needs and levels of pollution. Single-family houses aged 50 years and more are the vast majority of buildings located in Polish rural areas. Most of them are characterized by poor thermal insulation performance and the use of traditional fuels—coal and biomass—for heating. For this reason, they are a major source of air pollution in rural areas.

The scientific interest in energy efficiency and renewable energy development in rural areas in Poland has been growing. Studies on sectoral polices and potential for development of individual RES have been popular among Polish researchers [10–12]. They have often emphasized the importance of development of renewable energy production in rural areas based on local sources. Thus the role of biomass in both the context of sustainable development, achieving energy security (including greater self-sufficiency) and providing an alternative source of income for rural communities has been highlighted [10,13–15]. Problem of technologically obsolete heat sources used by Polish rural households has been analysed from the different perspectives. One of them is air pollution and ecological soundness perspective. Authors have focused on the emissions of atmospheric pollutants generated from heat sources in rural areas and displayed an environmental effect of the replacement of old heat sources [15,16]. The low use of renewable energy sources (RES) for energy purposes in rural areas has been presented as a reason for exceeding the air pollution standards [16], threatening not only environmental but also health security. Thus, air pollution as a result of the contemporary structure of energy for heat use in rural households has been discussed in a wider context of sustainable development and social security issues. Problems related to old and energy-intensive rural residential buildings and the domination of ineffective solid fuels and heat equipment used by rural households have been also analysed in relation to economic disparities between the urban and rural areas. In this approach, authors have concentrated on the energy poverty of rural households, indicating the need to broaden knowledge of rural communities on energy savings, develop RES and improve energy efficiency [17–19]. From a political and administrative perspective, there is a common view that both the development of local RES as well as the reduction of energy use per rural household requires changes in regulations, better support systems as well as environmental education among residents of rural areas [12,16,20,21].

So far, Poland's energy and climate security policy has been focused largely on the changes in the centralized large-scale energy system. Recently, the interest has been shifted to the role of households in the process of improving energy efficiency and transition from fossil fuels to renewable energy sources (RES). In terms of both economic and energy security, the rural areas present worse indexes than the urban ones. Among others, they depend more on solid fuels for heat, use more unit of energy per 1 inhabitant, and pay higher bills for electricity [22]. Nevertheless, there is a gap in scientific studies regarding exploration of how the economic situation, including an analysis of how the disposable income of a Polish rural household would be affected by the improvements in energy efficiency and changes to zero-emission sources. In this research, the main focus has been put on rural households due to the substantial income disparities between urban and rural areas, respectively, higher energy poverty and the fact that old single-family houses (SFH) dominate among the residential buildings in rural areas in Poland. Today, spending on energy for heat in 50-year-old (or older) SFH puts a burden on disposable income of rural households. As a result, their economic and energy security largely depends on the ability to change the present patterns of energy for heat supply and demand. Thus, it is important to fill the identified literature gap and address the question of links between improvements in heat consumption/production and an economic situation of rural households in Poland.

The aim of this article is to find the correlations between energy efficiency improvements of single-family buildings, the change of heat source from coal boilers to more environmentally sound sources, including zero-emission ones, and the disposable income and expenditure of rural households. In order to meet the energy and ecological security needs of rural areas, a special focus has been put on the economic analysis of a combined effect of comprehensive thermal modernization and installation of heat pump driven by prosumers' PV system. We attempt to find how the heating costs of rural SFH are affected by thermal modernization and change of heat source and if this translates to any considerable changes in disposable income of rural households. The study focuses on the most common type of rural residential buildings, i.e., single-family houses (SFH) and specifies the following tasks: (1) assessment of energy efficiency of SFH in rural areas and their thermal modernization potential; (2) assessment of heating costs with the use of different heat sources; (3) assessment of heating costs with the use of heat pump driven by PV prosumer installation (under Prosumer public support mechanism); (4) estimation of disposable income of rural households; (5) estimation of heating costs reduction impact on disposable income; (6) analysis of support mechanisms regarding energy efficiency and PV installations dedicated to rural areas.

Improvements in energy efficiency of SFH are perceived as one of the basic methods in enhancing energy security in terms of both availability and affordability of energy for heat. In this research, we focus on an affordability side of a household energy security, i.e., the heating costs and estimated financial savings as an effect of the changes in energy efficiency and heat source patterns in rural SFH. Thermal modernization, exchange of heat sources and development of PV installations, are three potential areas of actions, which shall improve an energy balance and economic situation in rural areas. To verify these links, the research deals with official statistical data for the single-family houses, final energy consumption and income structure in Polish rural areas and elaborates a simplified model for assessment of the energy demand for space heating in the SFH built in different periods. Correlations are searched between the age of the SFH building, its usable area and the final energy consumption.

In international studies, issues of improving energy efficiency in the SFH sector are widely discussed and presented from different perspectives. From a technical perspective, studies often focus on barriers of SFH thermal modernization (e.g., studies on Nordic countries [23,24]) Others concentrate on the impact of the renovation of SFH on energy consumption and other patterns like indoor climate [25], indoor air quality or thermal comfort [26]. Recent studies discuss many innovative clean energy technologies [27], which could be used in renovation of residential houses in order to improve their energy efficiency. Research on the use of different systems based on renewable energy, including the experimental investigations on heat pumps driven by PV systems in SFH have been rising in importance and popularity [28,29]. Such studies regarding SFH in Poland's climatic conditions show the applications of this renewable energy technology after thermal modernization of SFH can be cost effective and recommended [28]. The heat pump–PV system has been also considered one of solutions for rural SFH in this study. Taking the economic approach, researches focus on the financial processes of investments [30] and the cost effectiveness of SFH renovations, including renovating to Passive House level [31]. In this study, we focus solely on the reduction of heating costs (i.e., not the investments costs) after thermal modernization and change of heat source in model SFH in rural areas in Poland, and estimate its impact on the disposable income of a household. From the sustainable development perspective, some empirical studies show how high energy efficient cities may influence and guide energy efficiency in surrounding areas [32]. Interestingly, across the scientific disciplines, studies prove the need of public policy support in encouraging investments in energy efficiency especially in poorer regions [33,34]. In Poland, the scale of the investments in the area of thermal modernization of single-family buildings and fighting low emission also strongly requires dedicated public aid programs which can be financed both from EU and domestic funds. Such investments contribute to the realization of the EU objectives to fight climate changes enhanced by the European Green Deal announced in 2019 [35]. However, some public support measures ("Clean Air Program" in particular) have been controversial, since they encourage rural households to exchange old coal boilers to new more efficient ones instead of stimulating the change to zero-emission sources. Therefore, the theoretical deliberations and economic calculations discussed in this article have been compiled with analysis of existing support schemes.

2. Research Process and Methods

For the purpose of this study, the focus has been put on the residential buildings dominating rural areas in Poland, i.e., single-family houses (SFH). In order to estimate energy and financial savings for rural household resulting from the improvements in the thermal modernization of SFH and changes of heat sources from coal boilers to gas, electricity, heat pump/PV installations, the research process has been divided into the following steps.

During the first phase of the research process, on the basis of available data on the age structure of rural residential buildings and calculations of their final energy consumption [kWh/m^2/year], the two models of semi-family buildings for rural areas were defined (see Table 1). Secondly, energy and cost reduction coming from thermo-modernization of model SFH buildings and/or the change of heat source with special regard to PV/heat pump system was estimated. In this simulation energy savings were estimated first at the micro-level, i.e., of individual rural household, and later at the country level. In order to assess the nationwide potential and feasibility in improving energy efficiency and formulate recommendations, at the final stage of research it was important to review current support mechanism and discuss prospects for comprehensive thermal modernization and development of prosumers' PV in SFH in rural areas in Poland. At the final stage of the research process, we analysed the results.

In the simulation of energy and cost savings, certain assumptions were made regarding the characteristics of SFH buildings, the used thermal-modernization and heat technologies. Firstly, a comprehensive thermal modernization involves investments in windows and modernization and optimization of heating system as specified in Table 1.

Secondly, in methodology of costs estimations the following assumptions were made for the two model SFH buildings:

- Number of floors, 2;
- Geometric, 330 m^3;
- Heating space, 130 m^2;
- Usable area, 136.7 m^2.

Two variants after thermal-modernization were applied: (a) The central heating system will not change—efficiency of the heating system is 50%; (b) Replacement of the heat source with one of the following sources:

- 5th generation boiler—efficiency of the central heating system amounts to 80%;
- Installation powered by natural gas—efficiency of the central heating system amounts to 90%;
- Heat pump installation. The COP index was assumed to be 2.

Table 1. Model single-family houses (SFH) configuration parameters.

SFH	Before Modernization	After Thermal Modernization
construction of external walls	29 cm thick brick wall, air gap, facade brick. Heat transfer coefficient U = 0.65 [W/(m^{2}K)]. The area of external walls is 201 m^2	Thermal insulation of external walls (area of 201 m^2) with 20 cm thick mineral wool with a thermal conductivity coefficient of lambda = 0.036 W/(m^{2}K). Heat transfer coefficient of external walls U = 0.15 [W/(m^{2}K)]
roof	86 m2 roof area. Heat transfer coefficient U = 0.6 [W/(m^{2}K)]	86 m^2 roof area. Thermal insulation of external walls with 20 cm thick mineral wool with a thermal conductivity coefficient of lambda = 0.036 W/(m^{2}K). Roof heat transfer coefficient U = 0.14 [W/(m^{2}K)]
floor	86 m2. Heat transfer coefficient U = 0.6 [W/(m^{2}K)]	86 m^2. Floor insulation with 15 cm thick mineral wool with a thermal conductivity coefficient of lambda = 0.036 W/(m^{2}K). The floor's heat transfer coefficient is U = 0.2 W/(m^{2}K).
windows	total area of windows in the external walls: 21.3 m^2. Four windows, 1.435 m × 1.635 m each, on the southern and northern elevations. Two windows with dimensions 1.135 m × 0.565 m on the E and W elevations. The heat transfer coefficient U = 2.6 W/(m^{2}K) was assumed for all windows.	total area of windows in the external walls: 21.3 m^2. Four windows, 1.435 m × 1.635 m each, on the southern and northern elevations. Two windows with dimensions 1.135 m × 0.565 m on the E and W elevations. The heat transfer coefficient U = 1.1 W/(m^{2}K) was assumed for all windows.

Assumptions regarding fuel prices and PV prosumer energy price settlement mechanism in Poland:

- Hard coal, i.e., eco-pea coal with a calorific value of 28 GJ/tonne and a price of PLN 900/tonne;
- Coal had a calorific value of 19 GJ/ton and a price of PLN 450/ton;
- Price of natural gas with a calorific value of 34 MJ/m^3 at the price of PLN 2.2/m^3 (PLN 1.1/m^3 of gas + PLN 1.1/m^3 for gas transmission);
- Electricity price: PLN 0.55/kWh;
- PV prosumer mechanism: on grid installations of 10 kW receiving 80% of the deposited energy within a fixed period of 365 days.

The separate primary data sets were used in the research to extract the data for rural households and SFH. They came from Statistics Poland (GUS), Eurostat, Ministry of Infrastructure and Construction (now Ministry of Infrastructure), Ministry of Agriculture, and National Census of Population and Housing of 2011.

3. Results of Research

3.1. Characteristics of SFH and Energy for Heat Balance in Rural Areas in Poland

3.1.1. Buildings Characteristics

According to the National Census of Population and Housing conducted in 2011 (NSP 2011), there were 5.56 million residential buildings in Poland [36]. Single-family houses (SFH) constituted a significant share of housing mainly in rural areas. In 2011, SFH accounted for 58.72% (3.3 million) of all housing in the rural areas while in the urban areas its share was 31.22% (1.7 million). At the same time, there is a significant difference between the urban and rural residential housing with regard to the average usable area. The useable area is on average larger in the non-urban housing (96.1 m^2) than in the city (62.7 m^2). As far as the age of the buildings is concerned, the SFH are very heterogeneous. Buildings from 1918–1944 accounted for almost 15% of the SFH located in Poland. However, the peak in construction of SFH characterized years after II World War. Buildings from 1945–1970 account for 25% of all SFH. The next two decades added a significant number of SFH, thus the buildings constructed in 1971–1978 and the 1980s account for, respectively, 12% and 13.6% of all SFH in Poland (see Figure 1).

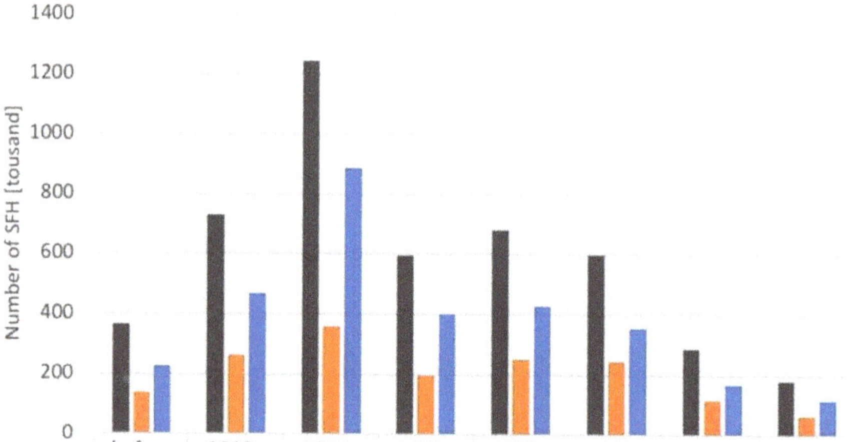

	before 1918	1918–1944	1945–1970	1971–1978	1979–1988	1989–2002	2003–2007	2008–2011
total	365	730	1246	595	680	598	286	182
urban	139	262	356	196	252	245	116	64
rural	226	468	889	399	428	353	170	118

total urban rural

Figure 1. Single-family houses in Poland by construction period.

The age of SFH, or in other words construction technology, impacts an energy balance of a household. It is scientifically proven that investments in more efficient heating systems and thermo modernization of residential buildings, can significantly lower a demand for energy per single-family building and improve environmental standards of these buildings [6,37,38]. Although, studies show that the real size of savings depends also on changes in residents' behaviours [39] as well as climatic zones of selected SFH [37]. In a study regarding SFH in Poland, the authors emphasize that, due to the lower greenhouse gases emissions and lower life cycle costs, wooden SFH can be perceived as more attractive for investments than brick ones [38].

Across the EU, we observe decreasing consumption of heat per m^2 or per dwelling as a result of investments in energy efficient technologies. Climatic conditions and age of residential buildings rank Poland sixth and eleventh among EU countries with the highest heat consumption per m^2 and the highest heat consumption per dwelling, respectively [40].

In Poland the newest SFH buildings, i.e., constructed after 2008, use two times less energy than the same type buildings from 1918–1944. Considering buildings responsible for 25% of all SFH in rural areas, i.e., from 1945–1970, their energy use is 1.7 times higher than of the contemporary ones. At the country level, overall demand for energy for heat is a result of both the age of buildings (which strongly impacts on energy intensity), the total number of buildings constructed in the different time periods and their usable area. Not surprisingly, the highest demand for energy and at the same time, the greatest thermal modernization potential can be found in SFH from 1918–1970. Yet, in this group of buildings, the largest heat consumers are SFH built between 1945–1970 (see Figure 2). More than 71% of them are located in rural areas.

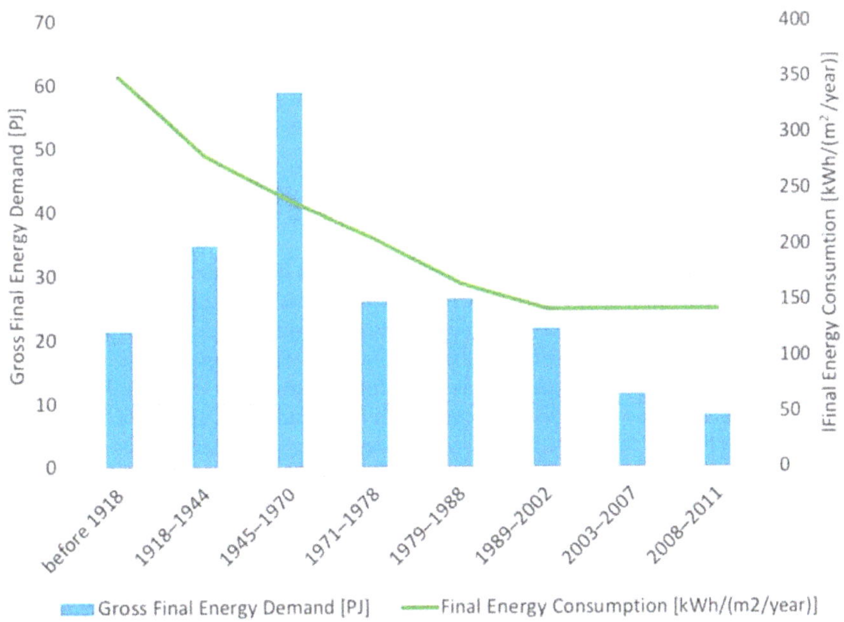

Figure 2. The relation between age and energy demand of SFH (rural) in Poland.

3.1.2. Energy for Heat Consumption Balance in Rural Areas

Regardless the efforts for a deeper and wider diversification of primary energy sources, Poland's energy security still largely depends on coal. In 2018 and 2019 the share of coal (hard coal and lignite) in gross inland primary energy consumption was 48% and 42%, respectively [41,42]. For Polish households' solid fuels, coal in particular, are basic source of energy. In 2018 more than 45% of households in Poland used solid fuels for space heating. These fuels were also used for heating water in 25.6% of households and for cooking in 3.2% of them [43]. Hard coal with 32.1% share in energy consumption per 1 inhabitant, was single most important source of energy for heat. To compare, only 19.6% of energy consumption came from the district heat [43] (p.84).

However, there are significant differences between urban and rural households in Poland in relation to heat sources. According to data for 2012 urban territories were dependent on district heat supply in 59.9%. The second and third most important sources of heat for Polish cities were solid fuels (28.6%) and natural gas (11.5%), respectively. In contrast in the rural areas, households' energy security relied almost entirely on solid fuels, which represented 89.8% of heat consumption. Only 6.2% of heat came from natural gas, whereas the centralized district heat was responsible for around 4% of heat used by the rural households [44] (p. 64). In recent years, the implementation of energy-climate policy in Poland has focused on a large-scale energy system, which are under ETS. In result although some changes could be seen in final energy consumption structure at the country level, from the perspective of rural households not much has changed. As the latest data provided by Statistics Poland show, in 2018 the district heat was commonly used in urban areas with 58.3% of households relying on this source of heat. At the same time, only 3.5% of rural households used district heat. For rural areas, solid fuels remain the most important, basic sources of energy for heat. According to official data, 88.4% of rural households were dependent on solid fuels [43] (p. 59).

It is estimated that coal boilers are a dominant source of heat for SFH buildings constructed from 1945 to 1988. In 2017, boilers and stoves based on solid fuels were the main source for heating in 70% of SFH. In a further 14% of SFH, a wood or other type of biomass boiler was the single most important

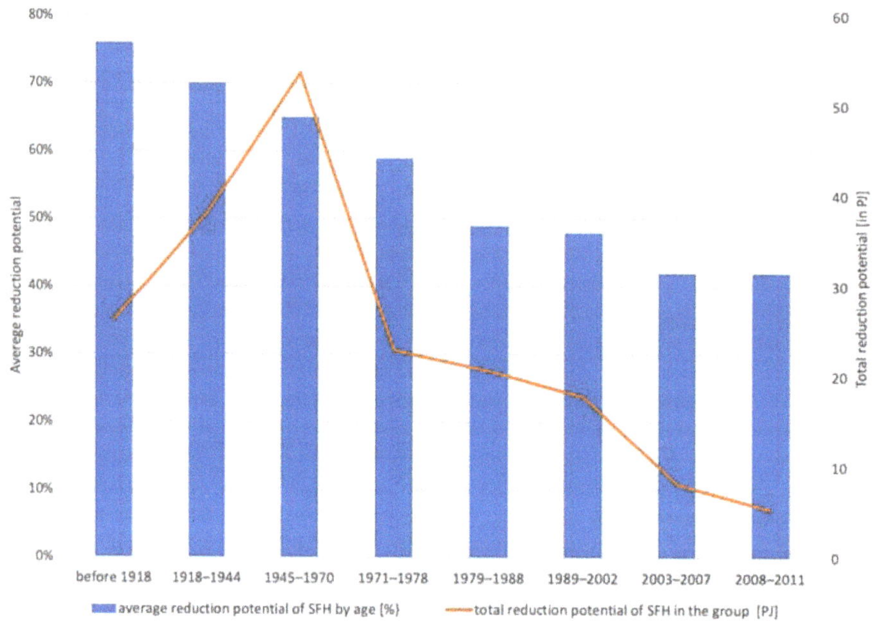

Figure 3. Potential of domestic final energy consumption reduction in SFH.

3.2.2. Changes in the Disposable Income Structure of a Household in Rural Areas

In Poland, some 40% of the population (15.4 million) live in rural areas and for the last 20 years the number of rural inhabitants has been on a steady upward trend. Furthermore, living conditions in rural areas do not diverge considerably from the quality of life of residents of smaller towns (up to 20,000 residents) [49]. At the same time, despite the fact that a share of agriculture in the socio-economic structure of rural areas is on a downward trend an income earned from agriculture activity plays important but diminishing role for rural households. Nowadays, 80% of the rural employed (6.6 million) make on living outside agriculture [50]. After accession to the EU, a process of convergence of income between rural population and non-rural citizens has been observed in Poland. This phenomenon contributed to the fact that a share of people at risk of poverty declined. Nevertheless, there is still income disparity between farming households and non-farming ones. In 2019, a place of residence still strongly differentiates the dynamics and structure of household income and expenditure.

The average monthly disposable income per person in households living in cities was 28.4% higher than in the countryside (by 29.9% in 2018). These differences were due to the level of income received by households, as well as to a higher number of people in rural households. The situation was similar for household expenditure. Spending per person in households living in cities was 34.8% higher than in the countryside (up 34.5% in 2018) [51].

In general, in the period of 2018–2019 annual per capita income growth in Polish households was on upward trend. The average monthly disposable income amounted to PLN 1819 and was 5.0% higher in real terms than in 2018. In turn, the average expenditure per person amounted to PLN 1252 and was higher by 3.1% [51]. It means that the share of expenses in disposable income amounted to 68.8%. Based on available statistical data, it can show that economic situation of Polish farmers in terms of both the average monthly disposable income (PLN 1667) and level of expenditures (PLN 914) was below country's averages The average expenditure was in these households was 27% below the average household expenditure in Poland [51].

The percentage share of expenditure on the use of a flat or house and energy carriers for farmers' households was 13.9%. The costs of energy carriers alone were 9.3% and they included the use of electricity and energy carriers for heat generation and water heating. Wherein the average consumption of electricity consumption per year was 3797 kWh with a cost of PLN 1488 [44]. Among all the expenditure on energy carriers the highest burden is associated with energy for heat which constituted 65.1% of all energy used by household [52] (p. 1).

Average monthly energy (electricity, gas and other fuels) expenditures per capita in households of farmers was PLN 81.28 while the average number of residents in rural households was 3.4 [51]. On the basis of modelling of energy for heat use in SFH constructed between 1945 and 1970 (0.889 million of 3.27 million of residential buildings in rural areas) [36], it is possible to estimate an impact of comprehensive thermal modernization and the change of heat source to PV/heat pump system on the disposable income of rural households. A representative rural household energy and housing spending is PLN 276 per month and PLN 3316 per year. Subtracting the cost of average use of electricity in rural household (PLN 1488) shows that yearly cost of heating is PLN 1828.

Simulation modelling for SFH constructed in years 1945–1970 (see Table 1) shows that heating costs of such building in case of low rank coal use is PLN 3404 while the use of natural gas increases costs to PLN 9958 per year. Compilation of official Statistics Poland data and the result of the simulation modelling exhibit that numerous SFH in rural areas are under heated. Taking into account climate conditions in Poland, especially low temperatures during wintertime, heating comfort is one of key elements of energy security at a household level.

The compilation of the average cost of heating calculated by Statistics Poland, i.e., PLN 1828, and costs of heating in SFH after thermal modernization with integrated PV/heat pump installation, i.e., PLN 705 gives financial savings of PLN 1123 per farmers' household. It means than a representative farmer's family saves PLN 330 per capita per year or only PLN 28 per capita per month. The level of monthly expenditure decrease is 3%, yet the thermal comfort of living increases.

3.3. Current Support Mechanisms for the Thermal Modernization of Buildings and the Development of Prosumerism in Poland—The Role of the CAP Funds

The deep transformation of the economy and clean energy transition must be fair and socially acceptable. It is estimated that more than 50 million households in the European Union suffer from energy poverty [53] due to a combination of high energy expenditure and low household incomes. To address the issue, in 2016 the EU launched its flagship legislative proposal "Clean Energy for All Europeans" under which " ... the consumer will find it easier to invest in renewable energy, most obviously in solar panels, and then consume, store or sell the energy they produce" [54]. Estimates suggest that by 2030, energy communities could own some 17% of installed wind capacity and 21% of solar. By 2050, almost half of EU households are expected to be producing renewable energy [35]. In 2018, energy poverty affected approximately 12.2% of the Polish population, i.e., approximately 4.6 million people [12]. This means that these people do not have sufficient own funds to carry out the insulation of single-family houses or apartments.

Both the EU and Poland offer a number of various publicly financed programs which address the problems of energy efficiency and climate change mitigation. Rising interest in prosumerism has been lately supported by the Polish government via a number of new regulations and aid instruments.

Currently, there are several instruments and domestic programs in Poland to encourage and financially support investors/natural persons interested in increasing the level of energy efficiency in their households and, as a result, reducing expenditure on the purchase of electricity. In general, the central programs offered by the Polish government focus on two areas of support. The first is to support the renovation of existing and under construction single-family residential buildings in order to improve their thermal insulation, the second is to support the development of renewable energy sources in the prosumer system.

In August 2019, the so-called Prosumer Package was introduced in Poland A change was introduced consisting in increasing the permissible capacity of micro-installations from 40 to 50 kW. Definition of a prosumer was extended to include apart from single-family buildings small and medium-sized enterprises, provided that the production of green energy is not their main form of gainful activity. In villages and in urban-rural communes, it was possible to establish energy cooperatives, also with a number of privileges due to prosumers.

The most important privilege of a prosumer is the possibility of settling energy in a cashless, so-called discount system. This system is very beneficial. Cashless billing of the amount of energy under the discount system takes place annually. The generated green energy can be used in real time, covering the current needs of a household or company. In a situation where the amount of energy produced from RES exceeds the current possibilities of its use, the surplus is discharged to the power grid operator.

The bidirectional prosumer counter shows the amount of the receivables, expressed in kWh, at the end of the billing period. Prosumers with installations up to 10 kW and up to 50 kW may receive, respectively, a max. of 80% and 70% of the deposited energy.

Considering the main research objective of this paper it is important to emphasize that from a technical standpoint of view, PV installations can meet the electricity demand only provided they operate within the PV prosumer system based on the grid, which functions as an electricity backup and storage. An additional charge of the "PV backup" solution (20% of the produced energy by the prosumer) was fully taken into consideration in the study and cost calculations.

3.3.1. Clean Air Program

The Clean Air Program [55] is the main program planned for 2018–2029, aiming to contribute to the reduction of emissions to the atmosphere of harmful substances resulting from the combustion of low-quality fuel and the use of obsolete installations in households.

In the planned budget of EUR 23.1 billion, EUR 14.24 billion was allocated for non-returnable subsidies, and EUR 8.9 billion for loans.

For the first time in history, a program could be used to finance single-family housing investments, both already built single-family residential buildings or those under construction. The program covered a wide range of activities, in particular:

- Disassembly and replacement of heat sources;
- Installation of modern devices and installations;
- Installation of renewable energy sources: solar collectors, photovoltaic micro installations;
- Thermal modernization of single-family buildings.

In the first period of operation, a significant limitation was the introduction of the regulation stating that the costs of a photovoltaic micro-installation and solar collectors could be co-financed in 100% only in the form of a loan.

The maximum value of eligible costs, on which the amount of the subsidy is calculated, was set at EUR 12,000. The loans bear a variable interest rate not less than 2% per annum, for up to 15 years. It is possible to combine two sources of financial support, both loans and subsidies [56]

3.3.2. Clean Air Program 2.0

The second edition of the modified Clean Air Program 2.0 [55] began on 21 October 2020. It was aimed at Poles with lower incomes the program offers a higher level of support, extends the implementation period of projects by an additional six months and extends the list of equipment and materials qualified for use. An important element is also the tightening of cooperation between the National Fund for Environmental Protection and Water Management and communes.

According to the new rules, in force from 15 May 2020, natural persons have a chance to receive a subsidy of up to EUR 8500 for the implementation of eco-investments.

Program beneficiaries can apply for support up to EUR 5600 replacing a heat source and installing a photovoltaic installation. If the investment includes a heat pump and a PV installation, it can be even EUR 6750. Moreover, for all beneficiaries, a thermal insulation tax relief of up to EUR 12,000 was introduced.

3.3.3. My Electricity Program

The My Electricity Program [57] results so far include 73,000 applications for subsidies and 408 MW of installed capacity. Thus, as much as 1/3 of the power of prosumer PV sources comes from installations co-financed by the program. The program specifically supporting the segment of photovoltaic (PV) micro-installations.

The budget of the program is EUR 225 million and is intended for non-returnable forms of financing up to 50% of eligible costs of micro-installations, not more than EUR 1125 for one project.

The beneficiaries of the program are natural persons producing electricity for their own needs.

3.3.4. Stop SMOG Program

The applicant in the Stop SMOG Program [58] is a commune which obtains up to 70% of the subsidy for investment costs from the state budget. The program is intended for energy poor people who own or co-own single-family residential buildings.

Scope of the Program:

- Replacement of high-emission heat sources with low-emission ones;
- Thermos-modernization of single-family residential buildings;
- Connection to the heating or gas network.

3.3.5. Agro Energy Program

The Agro Energy Program [59] is addressed towards the agricultural sector the program will run until 2025. The beneficiaries of the program are a natural or legal person who is the owner or leaseholder of agricultural real estate, the total area of agricultural land is in the range from 1 ha to 300 ha and at least one year before submitting the application runs the farm personally or, respectively, conducting agricultural activity or economic activity in the field of agricultural services.

Projects involving the purchase and installation of photovoltaic or wind installations with an installed electrical power of more than 10 kW and not more than 50 kW.

The budget planned for the implementation of the program is approximately EUR 45 million, including up to EUR 38 million subsidies and up to EUR 7 million as a loan.

The amount of support in the form of a subsidy is up to 20% of eligible costs for energy generating installations with a capacity of

- $10 < \text{kW} \leq 30$ up to EUR 3000;
- $30 < \text{kW} \leq 50$ up to EUR 5600.

3.3.6. Increased Use of Photovoltaic Installations in Electricity Generation in Poland in 2018–2019

Over the last two years, Poland has made great progress in popularizing micro-photovoltaic energy sources in the prosumer system (See Table 4). An even faster growth rate is recorded after the first half of 2020. On this basis, it can be concluded that the growing ecological awareness of the inhabitants of Poland, including rural areas, will be the driving force behind the further rapid development of the use of renewable energy sources to satisfy the energy needs of households.

Table 4. Development of electricity production in the prosumer system using photovoltaic micro-installations in 2018–2019 [60].

Year	Number of Prosumers	Installed Capacity in MW	Energy Fed into the Grid in MWh
2018	51,000	344	130,200
2019	149,000	900	325,280
2019/2018 in %	292	262	250

Poland also benefits from EU-funded programs which finance green investments, also in rural areas [61]. In the period of 2014–2020, rural communities in Poland can apply for support under the Rural Development Program 2014–2020 financed from European Fund for Rural Development (EFFRD) of the CAP. RDP in Poland provides for financing of the measure titled Basic services and village renewal in rural areas [62]. Support under this measure covers also investments in renewable energy and energy saving. In Poland, this measure consumed over 1.5 billion euro.

The capacity to adequately reflect the Green Deal Strategy in the EU and finance climate friendly investments depends on a share of the EU budget devoted to its objectives [63]. On 27 May 2020, the Commission put forward two key financial instruments, the Next Generation EU Fund and Multiannual Financial Framework 2021–2027 which is further discussed in Section 4.

3.4. *The Perspectives for the Support Mechanisms Development*

The European Green Deal (EGD) is a new EU sustainable growth strategy announced in 2019 [35]. The fundamental objective of EGD is to stimulate actions contributing to mitigate negative consequences of climate changes and protect and preserve natural resources for future generations. Efforts need to be taken across the entire economy including industry production, transportation infrastructure and agri-food sector. Changes shall also occur in the area of every-day consumption patterns including diet change and reducing food waste. Furthermore, the delivery of EGD objectives requires reshaping of various public policies to address the climate and environmental challenges and degradation of biodiversity. Both experts and policymakers urge for accelerated public and private investments enabling just and inclusive energy transition leading to climate neutrality objective by 2050. The design of policies and programs, including those dedicated to rural areas, which are to be financed under the new EU budget for 2021-2027 and stimulated by the innovative instrument of NGEU, shall consider the just energy transition and climate neutrality objective of the EGD.

Both the Cohesion Policy of the EU and the Common Agriculture Policy will play a fundamental role to meet the objective of EU climate neutrality by 2050 and contribute to achieving the Union's new 2030 climate targets [64]. As a general principle, all EU expenditure should be consistent with the Paris Agreement objectives [65].

For the last 2–3 years, Poland has been experiencing a strong increase in prosumerism accompanied by the rising environmental awareness of the Polish society. The increase in a number of green investments was possible mainly because Poland has launched a couple of public aid programs promoting and supporting prosumerism. This support shall be continued.

Poland will become of one of the largest beneficiaries of the new Just Transition Fund of the EU to be launched after 2021. It is expected that Poland will receive some 20% of the total allocation of that Fund with a key objective to help Member States to depart from fossil fuels and promote green investments also in rural areas.

4. Discussion

Improvements in thermal modernization and the change of heat source increase energy efficiency and thermal comfort in SFH. This has been proved in many studies regarding different EU countries [6,66,67]. Most of SFH located in Polish rural areas were built between 1945 and 1970. They are characterized by energy inefficiency—their average final energy consumption accounts 240 kWh/m^2/year—and high reliance on solid fuel for heating, including low-ranked coal. The comparison of official data and the results of simulation of heat costs in model SFH buildings showed that single-family houses in Polish rural areas are under-heated. It means that economic surplus is not the only reason behind the decisions to thermo-modernize SFH and change its heat source. However, the price disparities between different primary energy carriers and the financial situation of rural households lead to observation that the changes in disposable income are not significant enough to encourage thermal-modernization and/or the changes of heat source to renewable energy. Moreover, this tendency has been accelerated by the attachment of rural communities to traditional heat sources as well as public policy which has supported replacement of an old coal boiler with a new one (i.e., 5. generation coal boiler). Some studies show the potential role of environmental impulses (including the rising ecological awareness in rural areas), yet in reference to Polish rural areas, and farmers' households in particular, further research in this field should be carried out.

An overview of the existing public policy tools and the support mechanisms proves that the number of incentives has been growing. Yet, the key question should be weather these measures are adequate to meet the ambitions relating to energy and climate policy. An enhancement of economic impulses for Polish rural households to change a heat source to zero-emission one (such as indicated in this study PV/heat pump system) should be more prioritized by decision makers. Considering the development so far of environmental and energy policy regulations and support mechanisms in Poland, one should expect that actions taken at the EU level will be crucial.

The COVID19 pandemic, which hit economies of all EU member states, made the European Commission to propose new ambitious financial instruments. After difficult negotiations of the European Council [68] in July 2020, the final financial proposal was approved. The key instrument to be launched to address economic crisis caused by COVID-19 is the NextGenerationEU Fund (NGEU) designed as a one-off emergency instrument amounting to EUR 750 billion to be put in place for a temporary period and used exclusively for crisis response and recovery. It allows the Commission to raise new financing on the financial markets for 2021–2024. Based on the European Council conclusions in July 2020, the funds borrowed may be used for loans up to EUR 360 billion and for grants up to EUR 390 billion (all in 2018 prices). The Commission will borrow money on behalf of the Union to be repaid after 2027 and by 2058 at the latest. MS will access the Fund via grants and loans. The key instrument under NGEU is the Recovery and Resilience Facility of EUR 672.5 billion of which loans represents EUR 360 billion. The share of the NGEU dedicated to rural development accounts for 1%. All the amounts indicated are to be considered as exceptional budgetary allocations and will not be part of future MFF proposals.

The key instrument to finance the Green Deal will be the EU budget for 2021–2027 (Multiannual Financial Framework) worth EUR 1.074 billion (originally EUR 1.100 billion). Additional funding of EUR 10 billion (originally EUR 30 billion) from the Just Transition Fund of NGEU will be allocated for Climate Action Plan dedicated to green investments.

The CAP with CAP budget (2018 prices) amounting to EUR 356 billion consisting of agriculture and maritime policy is expected to provide answers to a number of rising challenges including climate change and collapse of biodiversity as well as environment and climate action. The CAP aims at ensuring a sustainable agriculture with respect to economic, social and environmental aspects.

The CAP post 2020 must continue offering a number of ways to contribute to climate and environmental objectives. The Member States and stakeholders will have to ensure that the national strategic plans for agriculture and rural development policy financed by the CAP shall fully reflect the goals of the Green Deal. Especially, rural development tools can support the transition via investing in green infrastructure, in knowledge transfer and innovation. The financial proposals provide for a minimum of 30% of rural development funds dedicated towards interventions that address specific environmental and climate-related objectives. To facilitate this process on 20 May 2020, the European Commission announced two strategic documents, i.e., the Farm to Fork Strategy (F2F) [69] and the Biodiversity Strategy [70].

5. Conclusions

Improvements in energy efficiency in the EU's residential sector have already been analysed extensively and presented from different scientific perspectives. Nevertheless, we see the room for studies taking into consideration the specificity of rural areas in different EU countries. From energy and climate security perspectives, investments in energy efficiency and clean energy sources, including zero-emission heating systems, in rural areas can contribute to meeting EU's energy and climate targets. From sustainable development perspective, they can also improve the quality of living of the villagers. Our study contributes to this discussion and shows that single-family houses located in Polish rural areas have a large potential for improving energy efficiency. This is an effect of both the age structure of SFH buildings and the fact that prior to 2017 there were no public policy intervention instruments in this field.

The simulation results for two models of SFH buildings prevailing in rural areas and discussed in this research paper univocally show the existing potential for improving energy efficiency and introducing the new heat sources, including the zero-emission ones (PV/heat pump installation) in rural Poland. The process of changes in supply and consumption patterns would enhance energy and environmental security in rural areas. As far as economic security of rural household is concerned, it has been proved that thermal modernization and change of heat source provide for considerable percentage reduction of heating costs yet, it has a limited impact on disposable income. Analysis of available data for farmer's living in SFH (built in over 1945–1970) also showed only a 3% reduction in total household expenditure. This low impact on disposable income and expenditure of rural household will not encourage improvements in energy efficiency and changes of heat sources in rural SFH. Thus, new public policy measures dedicated to rural areas will be required.

Despite the results, which proved that heat pump driven by PV installation operating in the framework of Polish Prosumer support mechanism is the most cost-effective way of heating in both models of SFH buildings, a wider application of this technological solution will depend on additional factors. Among them, the existing and future public policy support mechanisms will play a crucial role. It is particularly important, since significant reductions in heat costs can also be achieved with thermal modernization alone, i.e., without changing the heat source so far.

Public aid, financed both by the EU and domestic budget, shall be continued in order to strengthen and encourage further changes in energy supply–demand patterns in rural areas. The development of prosumers contributes to mitigating negative implications of climate changes but also improves the economic situation of households in Poland. Reduction of SFH heating costs is extremely important in case of poorer rural citizens. Together with the wide range of aid instruments launched by the Polish government and those supported by EU-financed programs, it shall lead to further development and use of renewable energy sources.

Author Contributions: Conceptualization, K.K.; methodology, K.K.; formal analysis K.K. and K.P. (Sections 1, 2, 3.1 and 3.2) and M.D. and I.N. (Sections 3.3 and 3.4); investigation, K.K.; writing—original draft preparation K.K. and K.P. (Sections 1, 2, 3.1, 3.2 and 4, Section 5); writing—original draft preparation M.D. and I.N. (Sections 3.3, 3.4 and 4, Section 5); editing, K.P.; visualization, K.K. All authors have read and agreed to the published version of the manuscript.

Funding: This research received no external funding.

Conflicts of Interest: The authors declare no conflict of interest.

References

1. Economidou, M.; Todeschi, V.; Bertoldi, P.; D'Agostino, D.; Zangheri, P.; Castellazzi, L. Review of 50 years of EU energy efficiency policies for buildings. *Energy Build.* **2020**, *225*, 110322. [CrossRef]
2. Ürge-Vorsatz, D.; Metz, B. Energy efficiency: How far does it get us in controlling climate change? *Energy Effic.* **2009**, *2*, 87–94. [CrossRef]
3. Eurostat. *Energy Consumption in Households*; Eurostat: Brussels, Belgium, 2020.

4. ODYSSEE-MURE ODYSSEE-MURE. Available online: https://www.odyssee-mure.eu/publications/efficiency-by-sector/households/ (accessed on 22 October 2020).
5. Filippini, M.; Hunt, L.C.; Zorić, J. Impact of energy policy instruments on the estimated level of underlying energy efficiency in the EU residential sector. *Energy Policy* **2014**, *69*, 73–81. [CrossRef]
6. Pardo, N.; Thiel, C. Evaluation of several measures to improve the energy efficiency and CO2 emission in the European single-family houses. *Energy Build.* **2012**, *49*, 619–630. [CrossRef]
7. Knoop, K.; Lechtenböhmer, S. The potential for energy efficiency in the EU Member States–A comparison of studies. *Renew. Sustain. Energy Rev.* **2017**, *68*, 1097–1105. [CrossRef]
8. Heinonen, J.; Junnila, S. Residential energy consumption patterns and the overall housing energy requirements of urban and rural households in Finland. *Energy Build.* **2014**, *76*, 295–303. [CrossRef]
9. Statistics Poland. *Energy Efficiency in Poland in Years 2008–2018*; Statistics Poland: Warsaw, Poland, 2019.
10. Koryś, K.A.; Latawiec, A.E.; Grotkiewicz, K.; Kuboń, M. The Review of Biomass Potential for Agricultural Biogas Production in Poland. *Sustainability* **2019**, *11*, 6515. [CrossRef]
11. Klepacka, A.M.; Florkowski, W.J.; Meng, T. Clean, accessible, and cost-saving: Reasons for rural household investment in solar panels in Poland. *Resour. Conserv. Recycl.* **2018**, *139*, 338–350. [CrossRef]
12. Pająk, K.; Kvilinskyi, O.; Fasiecka, O.; Miśkiewicz, R. Energy security in regional policy in Wielkopolska region of Poland. *Ekon. Śr.* **2017**, *2*, 122–138.
13. Szymańska, D.; Chodkowska-Miszczuk, J. Endogenous resources utilization of rural areas in shaping sustainable development in Poland. *Renew. Sustain. Energy Rev.* **2011**, *15*, 1497–1501. [CrossRef]
14. Dudek, T. The Impacts of the Energy Potential of Forest Biomass on the Local Market: An Example of South-Eastern Poland. *Energies* **2020**, *13*, 4985. [CrossRef]
15. Michał, K.; Sowiżdżał, A.; Tomaszewska, B. Energetic and Environmental Aspects of Individual Heat Generation for Sustainable Development at a Local Scale—A Case Study from Poland. *Energies* **2020**, *13*, 454. [CrossRef]
16. Piwowar, A.; Dzikuć, M. Development of Renewable Energy Sources in the Context of Threats Resulting from Low-Altitude Emissions in Rural Areas in Poland: A Review. *Energies* **2019**, *12*, 3558. [CrossRef]
17. Świerszcz, K.; Szczurek, T.; Mitkow, S.; Zalewski, J.; Ćwik, B. Knowledge of the Problem of Fuel Poverty Among Local Government Authorities: In The Aspect of Local Energy Security. *J. East. Eur. Res. Bus. Econ.* **2019**, *2019*, 1–12. [CrossRef]
18. Bouzarovski, S.; Herrero, S.T. Geographies of injustice: The socio-spatial determinants of energy poverty in Poland, the Czech Republic and Hungary. *Post Communist Econ.* **2017**, *29*, 27–50. [CrossRef]
19. Sałach, K.; Lewandowski, P. *Energy Poverty in POLAND, 2012 Description and Changes over Time*; IBS: Warsaw, Poland, 2018.
20. Sobczyk, W. *Sustainable Development of Rural Areas*; Social Science Research Network: Rochester, NY, USA, 2014.
21. Księżopolski, K.; Pronińska, K. *Studium Barier Administracyjnych i Proceduralnych w Rozwoju OZE na Obszarach Wiejskich/Study of Administrative and Procedural Barriers in the Development of Renewable Energy in Rural Areas*; Fundacja Europejski Fundusz Rozwoju Wsi Polskiej/Foundation of the European Fund for the Development of Polish Villages: Warsaw, Poland, 2017; p. 68.
22. Statistics Poland. *Energia 2020/Energy 2020*; Główny Urząd Statystyczny: Warsaw, Poland, 2020.
23. Azizi, S.; Nair, G.; Olofsson, T. Analysing the house-owners' perceptions on benefits and barriers of energy renovation in Swedish single-family houses. *Energy Build.* **2019**, *198*, 187–196. [CrossRef]
24. Bjørneboe, M.G.; Svendsen, S.; Heller, A. Initiatives for the energy renovation of single-family houses in Denmark evaluated on the basis of barriers and motivators. *Energy Build.* **2018**, *167*, 347–358. [CrossRef]
25. Bjørneboe, M.G.; Svendsen, S.; Heller, A. Evaluation of the renovation of a Danish single-family house based on measurements. *Energy Build.* **2017**, *150*, 189–199. [CrossRef]
26. Ekström, T.; Bernardo, R.; Blomsterberg, Å. Cost-effective passive house renovation packages for Swedish single-family houses from the 1960s and 1970s. *Energy Build.* **2018**, *161*, 89–102. [CrossRef]
27. de Azevedo, A.R.G.; Marvila, M.T.; Tayeh, B.A.; Cecchin, D.; Pereira, A.C.; Monteiro, S.N. Technological performance of açaí natural fibre reinforced cement-based mortars. *J. Build. Eng.* **2020**, *33*, 101675. [CrossRef]
28. Chwieduk, B.; Chwieduk, D. Analysis of operation and energy performance of a heat pump driven by a PV system for space heating of a single family house in Polish conditions. *Renew. Energy* **2020**. [CrossRef]

29. Kuboth, S.; Weith, T.; Heberle, F.; Welzl, M.; Brüggemann, D. Experimental Long-Term Investigation of Model Predictive Heat Pump Control in Residential Buildings with Photovoltaic Power Generation. *Energies* **2020**, *13*, 6016. [CrossRef]
30. Kragh, J.; Rose, J. Energy renovation of single-family houses in Denmark utilising long-term financing based on equity. *Appl. Energy* **2011**, *88*, 2245–2253. [CrossRef]
31. Ekström, T.; Blomsterberg, Å. Renovation of Swedish Single-family Houses to Passive House Standard-Analyses of Energy Savings Potential. *Energy Procedia.* **2016**, *96*, 134–145. [CrossRef]
32. Zhang, H.; Fan, L.W.; Zhou, P. Handling heterogeneity in frontier modeling of city-level energy efficiency: The case of China. *Appl. Energy* **2020**, *279*, 115846. [CrossRef]
33. Zhang, L.; Guo, S.; Wu, Z.; Alsaedi, A.; Hayat, T. SWOT Analysis for the Promotion of Energy Efficiency in Rural Buildings: A Case Study of China. *Energies* **2018**, *11*, 851. [CrossRef]
34. Azizi, S.; Nair, G.; Olofsson, T. Adoption of Energy Efficiency Measures in Renovation of Single-Family Houses: A Comparative Approach. *Energies* **2020**, *13*, 6042. [CrossRef]
35. European Commission. *Communication on The European Green Deal*; European Commission: Brussels, Belgium, 2019.
36. National Census of Population and Housing. 2011. Available online: https://stat.gov.pl/en/national-census/national-census-of-population-and-housing-2011/ (accessed on 24 October 2020).
37. Dascalaki, E.G.; Argiropoulou, P.A.; Balaras, C.A.; Droutsa, K.G.; Kontoyiannidis, S. Benchmarks for Embodied and Operational Energy Assessment of Hellenic Single-Family Houses. *Energies* **2020**, *13*, 4384. [CrossRef]
38. Grygierek, K.; Ferdyn-Grygierek, J.; Gumińska, A.; Baran, Ł.; Barwa, M.; Czerw, K.; Gowik, P.; Makselan, K.; Potyka, K.; Psikuta, A. Energy and Environmental Analysis of Single-Family Houses Located in Poland. *Energies* **2020**, *13*, 2740. [CrossRef]
39. Lopes, M.A.R.; Antunes, C.H.; Martins, N. Energy behaviours as promoters of energy efficiency: A 21st century review. *Renew. Sustain. Energy Rev.* **2012**, *16*, 4095–4104. [CrossRef]
40. ODYSSEE-MURE Sectoral Profile-Households. Available online: https://www.odyssee-mure.eu/publications/efficiency-by-sector/households/household-eu.pdf (accessed on 31 October 2020).
41. BP. *Statistical Review of World Energy 2019*, 68th ed.; BP: London, UK, 2019.
42. BP. *Statistical Review of World Energy 2020*, 69th ed.; BP: London, UK, 2020.
43. Statistics Poland. *Zużycie Energii w Gospodarstwach Domowych w 2018 r. Energy Consumption in Households in 2018*; Statistics Poland: Warsaw, Poland, 2019; p. 206.
44. Statistics Poland. *Zużycie Energii w Gospodarstwach Domowych w 2012 Roku. Energy Consumption in Households in 2012*; Statistics Poland: Warsaw, Poland, 2014.
45. *Efektywność Energetyczna w Polsce. Przegląd 2017. Domy Jednorodzinne. Smog./Energy Efficiency in Poland. Review 2017. Single-Family Houses. Smog.*; Instytut Ekonomii Środowiska, European Climate Founda on: Cracow, Poland, 2018.
46. Register of Agricultural Biogas Producers. Available online: https://www.kowr.gov.pl/uploads/pliki/oze/biogaz/Rejestr%20wytw%C3%B3rc%C3%B3w%20biogazu%20rolniczego%20z%20dnia%2030.10.2020%20r.pdf (accessed on 30 October 2020).
47. Igliński, B.; Piechota, G.; Iwański, P.; Skarzatek, M.; Pilarski, G. 15 Years of the Polish agricultural biogas plants: Their history, current status, biogas potential and perspectives. *Clean Technol. Environ. Policy* **2020**, *22*, 281–307. [CrossRef]
48. Eurostat Share of Fuels in the Final Energy Consumption in the Residential Sector for Space Heating, 2018 (%). Available online: https://ec.europa.eu/eurostat/statistics-explained/index.php?title=File:Share_of_fuels_in_the_final_energy_consumption_in_the_residential_sector_for_space_heating,_2018_(%25).png (accessed on 24 October 2020).
49. Nurzyńska, I.; Wilkin, J. *Rural Poland 2018. The Report on the State of Rural Areas*; Scholar Publishing House: Warsaw, Poland, 2018.
50. Statistics Poland. *Agriculture in 2019*; Statistics Poland: Warsaw, Poland, 2020; p. 101.
51. Statistics Poland. *Sytuacja Gospodarstw Domowych w 2019 r. w Świetle Wyników Badania Budżetów Gospodarstw Domowych/The Situation of Households i, n 2019 in the Light of the Results of the Household Budget Survey*; Statistics Poland: Warsaw, Poland, 2020.

52. Statistics Poland. *Skrót. Zużycie Energii w Gospodarstwach Domowych w 2018 r. Summary. Energy Consumption in Households in 2018*; Statistics Poland: Warsaw, Poland, 2019; p. 3.
53. EU. Energy Poverty Observatory. Available online: https://www.energypoverty.eu/about/about-observatory (accessed on 23 October 2020).
54. European Comission. *Clean Energy for All Europeans*; European Union: Brussels, Belgium, 2019; p. 24.
55. Program Czyste Powietrze. Available online: www.czystepowietrze.gov.pl (accessed on 23 October 2020).
56. National Fund for Environmental Protection and Water Management. Available online: https://www.nfosigw.gov.pl/en/ (accessed on 24 October 2020).
57. Narodowy Fundusz Ochrony Środowiska i Gospodarki Wodnej/National Fund for Environmental Protection and Water Management Program Mój Prąd/My Electricity Program. Available online: https://appwwwmojprad.blob.core.windows.net/mojprad/documents/Program_priorytetowy_M%C3%B3j_Pr%C4%85d.pdf (accessed on 22 October 2020).
58. Narodowy Fundusz Ochrony Środowiska i Gospodarki Wodnej/National Fund for Environmental Protection and Water Management Program Czyste Powietrz Stop Smog/Clean Air Program Stop Smog. Available online: http://www.czystepowietrze.gov.pl/stop-smog/ (accessed on 22 October 2020).
59. Narodowy Fundusz Ochrony Środowiska i Gospodarki Wodnej/National Fund for Environmental Protection and Water Management Program Priorytetowy "Agroenergia"–Aktualizacja 2020/ Priority Program Agro Energy-update. 2020. Available online: http://www.nfosigw.gov.pl/oferta-finansowania/srodki-krajowe/programy-priorytetowe/agroenergia/agroenergia-2020/ (accessed on 24 October 2020).
60. Energy Regulation Office. Available online: https://www.ure.gov.pl/pl/urzad/informacje-ogolne/aktualnosci/8771,Rekordowy-rok-dla-fotowoltaiki.html (accessed on 24 October 2020).
61. Mosconi, E.M.; Colantoni, A.; Gambella, F.; Cudlinová, E.; Salvati, L.; Rodrigo-Comino, J. Revisiting the Environmental Kuznets Curve: The Spatial Interaction between Economy and Territory. *Economies* **2020**, *8*, 74. [CrossRef]
62. *Regulation (EU) No 1305/2013 of the European Parliament and of the Council of 17 December 2013 on Support for Rural Development by the European Agricultural Fund for Rural Development (EAFRD) and Repealing Council Regulation (EC) No 1698/2005*; European Union: Brussels, Belgium, 2013.
63. White, D. Just Transitions/Design for Transitions: Preliminary Notes on a Design Politics for a Green New Deal. *Capital. Nat. Social.* **2020**, *31*, 20–39. [CrossRef]
64. Fanelli, R.M. The Spatial and Temporal Variability of the Effects of Agricultural Practices on the Environment. *Environments* **2020**, *7*, 33. [CrossRef]
65. *Paris Agreement*; United Nations: New York, NY, USA, 2015.
66. Hurnik, M.; Specjal, A.; Popiolek, Z.; Kierat, W. Assessment of single-family house thermal renovation based on comprehensive on-site diagnostics. *Energy Build.* **2018**, *158*, 162–171. [CrossRef]
67. Specjał, A.; Lipczyńska, A.; Hurnik, M.; Król, M.; Palmowska, A.; Popiołek, Z. Case Study of Thermal Diagnostics of Single-Family House in Temperate Climate. *Energies* **2019**, *12*, 4549. [CrossRef]
68. European Council. *Special Meeting of the European Council*; Conclusions; European Council: Brussels, Belgium, 2020.
69. European Commission. *A Farm to Fork Strategy for a Fair, Healthy and Environmentally-Friendly Food System*; European Commission: Brussels, Belgium, 2020.
70. European Commission. *EU Biodiversity Strategy for 2030 Bringing Nature Back Into Our Lives*; European Commission: Belgium, Brussels, 2020.

Publisher's Note: MDPI stays neutral with regard to jurisdictional claims in published maps and institutional affiliations.

energies

MDPI

Article

The Linkages between Crude Oil and Food Prices

Monika Roman, Aleksandra Górecka * and Joanna Domagała

Institute of Economics and Finance, Warsaw University of Life Sciences, 166 Nowoursynowska St., 02-787 Warsaw, Poland; monika_roman@sggw.edu.pl (M.R.); joanna_domagala@sggw.edu.pl (J.D.)
* Correspondence: aleksandra_gorecka@sggw.edu.pl

Received: 16 November 2020; Accepted: 9 December 2020; Published: 11 December 2020

Abstract: This paper aims to indicate the linkages between crude oil prices and selected food price indexes (dairy, meat, oils, cereals, and sugar) and provide an empirical specification of the direction of the impact. This paper reviews the fuel–food price linkage models with consideration to the time series literature. This study adopts several methods, namely the Augmented Dickey–Fuller test, Granger causality test, the cointegration test, the vector autoregression model, and the vector error correction model, for studying the price transmission among the crude oil and five selected food groups. The data series covers the period between January 1990 and September 2020. The empirical results from the paper indicate that there are long-term relationships between crude oil and meat prices. The linkage of crude oil prices occurred with food, cereal, and oil prices in the short term. Furthermore, the linkages between the analyzed variables increased in 2006–2020.

Keywords: food prices; crude oil prices; cointegration; vector autoregressive model; Granger causality

1. Introduction

The role of crude oil in the worldwide economy [1–3] is considered essential as it is one of the most crucial sources of energy, which, in turn, constitutes an essence of the modern global economy. In the past, oil occupied most of the energy area [4], and Rahmas [5] suggested that the oil dominion would extend over the twenty-first century, too.

There were some significant spikes in the price of oil. The first was noticed in May 1974, followed by the Yom-Kippur War in 1973, when imported crude oil's actual price per barrel jumped to 69.64 USD, followed by January 1981, just after the Iranian Revolution in 1979, with the price per barrel hitting 115.81 USD. The central peak during the considered period coincided with the time of the world's financial crisis in 2007 and 2008. Further, in March 2012, another peak was observed when real prices increased up to 126.10 USD per barrel (Figure 1).

Crude oil is a critical input for most services and goods and has a massive impact on people's lives. It has a broad scope of applications, supplying various sectors of economies including agriculture, transportation, and industry, as well as households, because it serves for the production of fuel. Therefore, people's quality of life shifts up and down when the price of crude oil is unbalanced and irrational [6]. Therefore, oil price fluctuation may influence the prices of other products [7].

Recent studies on crude oil price influence mostly influence the stock market. For example, Xu et al. [8] argue the heterogeneous nature of the correlation between the stock market of different countries and the crude oil market. Moreover, the work indicates that, when compared, the short-run correlation between fuel and the stock market is lower than the long-term correlation.

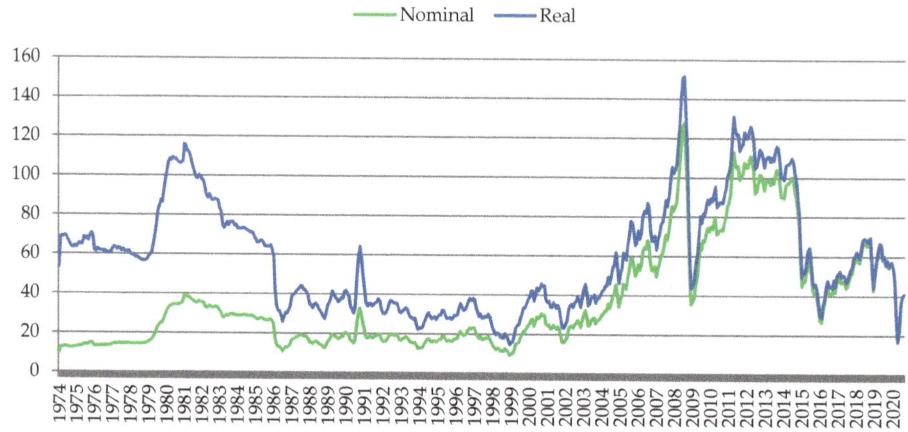

Figure 1. Monthly imported crude oil prices from January 1974 until August 2020 (USD per barrel). Source: own elaboration based on [9].

Research presenting crude oil prices and their relation to GDP level and economic growth constitutes another line of consideration, the results of which indicate the existence of a significant impact of crude oil price on economic growth [6,10–13].

Not only is oil the primary source of energy but it also serves in the production of other forms of energy such as electricity or refinery products, which, in turn, serve for manufacturing various goods or impact transportation processes. Hence, the third field of research is illustrated by the number of studies presenting crude oil price levels with the prices of multiple goods classified as food and nonfood products. The study of Sarwar and Tivari [14] regarding Pakistan demonstrates the nonlinearity of the relationship between the nonfood Consumer Price Index and oil prices. Increases in oil prices lead to increases in prices, whereas there is no such phenomenon in the opposite direction.

Food products are investigated separately as they constitute essential living costs, and numerous studies have been conducted that present food product prices in terms of crude oil prices. There are no studies, however, based on groups of agricultural commodities. Therefore, this paper aims to identify and describe the relations between selected groups of agricultural commodities, such as dairy, meat, oil, cereal, sugar product, and crude oil prices. To aid this process, we have established the short- and long-term linkages between variables and determined the directions of their mutual influence.

The article is structured as follows: Section 2 indicates the literature review, and Section 3 discusses the materials and methods used. Section 4 lays out the outcome of the empirical analysis, and, finally, Section 5 presents the discussion, closing with conclusions.

2. Literature Review

The recent years saw a rise in the number of papers published on the relationship between fuel and food prices (Table 1). Out of these, it is possible to identify three groups of studies. In the first, the researchers were unable to find evidence for the relationship between analyzing data. On the other hand, several studies indicate that there are linkages when investigating this relationship. The final group focuses on discovering studies that point to neutrality between variables in one period but find evidence in the second period. These instances correspond to food (2006) or the financial crisis (2008).

Table 1. Overview of previous studies and results.

Authors, Year	Methods	Data (Source)	Time/Geographical Coverage	Results	
				Neutrality Hypothesis	Crude Oil/Energy Prices Driving Prices of Agricultural/Food Goods
Ding, Zhang (2020) [15]	Spread CRB Index, Dickey–Fuller test	Crude oil, corn, cattle gold, and copper prices daily data (Thomson Datastream)	2005–2018	+	
Hau, Zhu, Huang, Ma (2020) [16]	Model TVP-SVM, Model MCMC estimation	Corn, soybean, bean, strong wheat, cotton, pulp, natural rubber; weekly data	2003–2004, 2007–2011, China	+	
Fowowe (2016) [17]	ECM, Nonlinear causality tests	Maize, sunflower, and soybeans; weekly data (the EIA, the Johannesburg Stock Exchange)	2001–2014, South Africa	+	
Ibrahim (2015) [18]	NARDL model	Food and oil prices annual data	1971–2012, Malaysia	+	
Nazlioglu, Soytas (2011) [19]	Toda and Yamamoto causality test	Monthly data	1994–2010, Turkey	+	
Gilbert (2010) [20]	Granger causality test, 2SLS, 3SLS OLS,	Quarterly data	1971–2008	+	
Zhang, Lohr, Escalante, Wetzstein (2010) [21]	VECM	Crude oil, soybean, corn, wheat prices, monthly data	1989–2008	+	
Vo, Vu, Vo (2019) [22]	SVAR model IRF model, variance decomposition echnique	Crude oil prices, corn, wheat, sugarcane, soybeans, coconut, soybean and palm oil, palm kernel oil, barley, coffee, cocoa, rice, tea, cotton prices; monthly data; WB	January 2000–July 2018, 2000–2006, 2006–2013, 2013–2018		+
Taghizadeh-Hesary, Rasoulinezhad, Yoshino (2019) [7]	Panel-VAR model	Food prices, crude oil and biofuel price, inflation and real interest rate, agricultural land, employment in the agriculture sector, GDP (World Development Indicators, the FAO, the BP, the EIA, Statistical Review of World Energy)	2000–2016, 8 Asian countries		+

Table 1. *Cont.*

Authors, Year	Methods	Data (Source)	Time/Geographical Coverage	Results	
				Neutrality Hypothesis	Crude Oil/Energy Prices Driving Prices of Agricultural/Food Goods
Su, Wang, Tao, Oana-Ramona (2019) [23]	Vertical market integration model Ciaian and Kancs, bootstrap full-sample causality test, Granger causality test, bivariate VAR models	Crude oil spot price Worldwide; maize and soybeans, tea and cocoa beans, monthly data, (WTI)	1990–2017		+
Pal, Mitra (2019) [24]	DCC model, Pearson correlations	Crude oil, corn, soybeans, wheat, and oat prices, daily spot closing prices, (WTI)	2000–2018; U.S.		+
Pasrun, Rosnawintang, La Ode, La, La Ode (2018) [25]	VAR model, Granger causality test	Crude oil price, rice price, monthly data	January 2000–September 2017		+
Ji, Bouri, Roubaud, Shahzad (2018) [26]	Copula model	Daily data,	2000–2017		+
Al-Maadid, Caporale, Spagnolo, Spagnolo (2017) [27]	Bivariate VAR-GARCH(1,1) model	Crude oil and ethanol prices and coffee, cacao, corn, sugar, soybeans, and wheat prices, daily data (Bloomberg)	January 1st, 2003 to June 6th, 2015		+
Bergmann, O'Connor, Thummel (2016) [28]	VAR model, multivariate GARCH model	Palm oil, butter, and crude oil prices	January 1995–December 2005; EU and World		+
Hamulczuk (2016) [29]	Correlation coefficient	Energy prices and agrifood prices	1995–2015,		+

Table 1. *Cont.*

Authors, Year	Methods	Data (Source)	Time/Geographical Coverage	Results	
				Neutrality Hypothesis	Crude Oil/Energy Prices Driving Prices of Agricultural/Food Goods
Mawejje (2016) [30]	Cointegration techniques	Energy, meat, dairy, cereal, edible oil, sugar prices, monthly data; the Uganda Bureau of Statistics, Bank of Uganda, FAO	2000–2011		+
Fernandez-Perez, Frijns, Tourani-Rad (2016) [31]	SVAR	Daily data	2006–2016		+
McFarlane (2016) [32]	Dickey–Fuller test, Johansen tests, VAR	Corn, sugar, wheat, and crude oil prices, weekly data	1999–2005, 2006–2012, The U.S		+
Cabrera, Schulz (2016) [33]	Correlation GARCH model, multivariate multiplicative volatility model	Energy, agricultural product prices	2003–2012, Germany		+
Nwoko, Aye, Asogwa (2016) [34]	GARCH (1, 1) model, Dickey–Fuller test, Phillip–Perron test, Granger causality test, VAR model	Oil price (food crop prices (US EIA, Federal Ministry of Agriculture), annual data	2000–2013, Nigeria		+
Zhang, Qu (2015) [35]	ARMA-GARCH	Daily data	2004–2014		+
Koirala, Mishra, D'Antoni, Mehlhorn (2015) [36]	Copula model	Daily data	2011–2012		+

Table 1. *Cont.*

Authors, Year	Methods	Data (Source)	Time/Geographical Coverage	Results	
				Neutrality Hypothesis	Crude Oil/Energy Prices Driving Prices of Agricultural/Food Goods
Rezitis (2015) [37]	Panel-VAR model, Granger causality tests	US dollar exchange rates, crude oil prices, 5 fertilizer prices, 30 selected agricultural prices, monthly data	June 1983–June 2013		+
Natanelov, Alam, McKenzie, Huylenbroeck (2011) [38]	VECM, TVECM	Monthly data	1989–2010		+
Chang, Su (2010) [39]	EGARCH	Daily data	2004–2008		+
Balcombe, Rapsomanikis (2008) [40]	VECM, AVECM, TVECM	Weekly data	2000–2006		+

Abbreviations: SVAR—structural vector autoregressive model; DCC—dynamic conditional correlation model; VAR—vector autoregression; (V)ECM -(vector) error-correction model; (T)VECM—(threshold) VECM, (A)VECM—(asymmetric) vector error-correction model; ARMA-GARCH—autoregressive moving average with generalized autoregressive conditional heteroskedasticity, EGARCH—exponential GARCH, (N)ARDL—(nonlinear) autoregressive distributed lag model; OLS—ordinary least squares, 2SLS—two-stage least squares, 3SLS—three-stage least squares, ECM—error-correction model, WTI—The West Texas Intermediate; FAO—Food and Agriculture Organization; EIA—the Energy Information Administration; BP—British Petroleum, WB—World Bank.

Some studies show no straight influence of crude oil on groups of food prices. In one example, Ding and Zhang [15] used copper, cattle, oil, corn, and gold data captured between 2005 and 2018. The authors demonstrated the long-term connection between crude oil and industrial metal markets; however, they did not confirm fuel–food linkages. Hau et al. [16] investigated the heterogeneous nature of the relationship between crude oil and China's agricultural futures. The work of Fowowe [17] featured a cointegration test with nonlinear Granger causality tests. His findings pointed to a lack of a short- or long-term price link between crude oil and food product prices in South Africa. Ibrahim [18] analyzed the case of Malaysia through a nonlinear autoregressive distributed lag model (NARDL) model. No long-term relationship between investigated variables was found as a result of this work. However, he found that, in the short term, agricultural product price inflation is caused by fluctuations in the oil price. In their work, Nazlioglu and Soytas [19] tested for causality between agricultural commodity and crude oil price and the exchange rate with the Toda–Yamamoto procedure but failed to discover any linkages formed between fuel and food prices. The researchers found neither direct nor exchange-rate-driven transmission. Gilbert [20] concluded that the significant correlation between analyzing prices is due to monetary and financial developments and rising demand. The limitations of the usage of agricultural products for biofuel production were not supported by his findings. Zhang et al. [21] insisted that the rising prices of fuel do not directly affect food product prices.

In contrast, numerous researchers point to increasing crude oil prices as the main cause of significant shocks the agricultural markets experienced. The 2007/2008 food crisis was mainly driven by the sharp increase in the prices of agricultural goods as well as crude oil and biofuels. The interaction between agricultural commodities and biofuels was extensively studied. The rising price of energy encouraged policy changes aimed to produce biofuels from corn and soybean. An increase in the prices of agricultural commodities with energy-producing capabilities could be caused by the biofuels segment expansion, resulting in high food prices. Several studies [23,41,42] pointed to the bidirectional causal link between crude oil and food prices. Contrastingly, Vo et al. [22] emphasized the fact that the contribution of individual oil shocks to agricultural price fluctuations is not uniform, and the same is true for the aggregate demand shocks and their effects on the food prices. Their findings present the significance of the fuel market in clarifying variabilities in the prices and related agricultural goods changeability. Taghizadeh-Hesary et al. [7] pointed to a connection between the security of energy and food through price volatility. Because oil price growths have an adverse effect on food security, diversification of the energy usage appears to be a necessity, relinquishing the reliance on fossil fuels in favor of an optimal relationship between energy resources (renewable and nonrenewable). Such a solution will be of great benefit not just for the security of energy but food security as well. Pal and Mitra [24], using three generalized autoregressive conditional heteroskedasticity (GARCH) models, discovered a relatively strong relationship between crude oil and energy crops; however, the value of this index for food crops was relatively low. Su et al. [23] submitted evidence of bidirectional causality existing among oil and food prices over selected subperiods. Ji et al. [26] discovered the tail dependence among food products and energy. Pasrun et al. [25] indicated a lack of long-term connections between the exchange rates and the prices of crude oil and rice. Only a short-term relationship based on the causality test transpired. Al-Maadid et al. [27] studied the nature of relationships between food and energy prices. Their results indicate the existence of outstanding linkages between the prices of agricultural commodities and petroleum products. Bergmann et al. [28] studied the transmission of volatility in the prices of palm oil, butter, and fuel markets with the application of the vector autoregression (VAR) model. The results indicate the spillover of oil prices into butter prices. Mawejje [30] found long-term linkages between agricultural commodities and energy prices in Uganda. McFarlane [32] explored the relationship in the US market between the prices of agricultural goods and oil. He found significant cointegration between the variables in 1999 and 2005 and the second between 2006 and 2012. Cabrera and Schulz [33] showed that prices move together and maintain a long-run balance despite the fact that market shocks appear. However, no evidence was discovered pointing to relations from rapeseed to crude oil in either the long-run or short-run. The study of Fernandez-Perez et al. [31] puts forward a conclusion that oil prices affect corn, soybeans, and wheat, whereas soybeans and wheat have an effect on ethanol. Hamulczuk [29] confirmed an increasing connection between Brent crude oil and food index prices. There are numerous roots of increase in price relationships, among them a policy of developed economies, the main focus of which is biofuels and their promotion and consumption. Nwoko et al. [34] mainly focused on the effect that oil prices apply on the relation of food prices in Nigeria in 2000–2013. The results obtained revealed a consequential short-term relationship between the volatility of variables. Other authors, based on the research from China [35], pointed to an irregularity in oil price shocks and food products. Koirala et al. [36] found a significant correlation between agricultural commodities and future energy prices. Rezitis [37] concluded that the prices of international agricultural commodities are influenced by crude oil prices and US dollar exchange rates. Chang and Su [39], using a bivariate exponential GARCH (EGARCH) model, pointed to crude oil and its relationship with corn prices. Natanelov et al. [38] presented that biofuels policy mitigates joint oil and corn price developments until a certain price threshold is exceeded. Balcombe and Rapsomanikis [40] used Bayesian techniques to investigate long-run relations, and their study resulted in a long-term balance between ethanol, crude oil, and sugar prices.

Researchers apply various methods to conduct their investigations of agricultural and energy commodities. They are mentioned in Table 1.

3. Materials and Methods

To identify the linkages between crude oil and food prices, we selected 5 groups of food commodities: dairy, meat, oils, cereals, and sugar products. The statistical variables were monthly real crude oil prices [9] and real food price indexes [43]. The time series are shown in Figure 2. Putting the issue into a time perspective, the research covered the period from January 1990 until September 2020. Following Al-Maadid et al. [27] and Vo et al. [22], we divided the full period into two subperiods: (i) 1990–2005 (before the 2006 food crisis) and (ii) 2006–2020 (after the crisis). Table 2 presents the results of Pearson's correlation with division into subperiods. It should be noted that, in the second period, a moderate correlation occurred between crude oil and food, cereal, and oil prices (logarithms of prices). In the analysis of the first price differences, the prices of food and oils had the highest correlation coefficients.

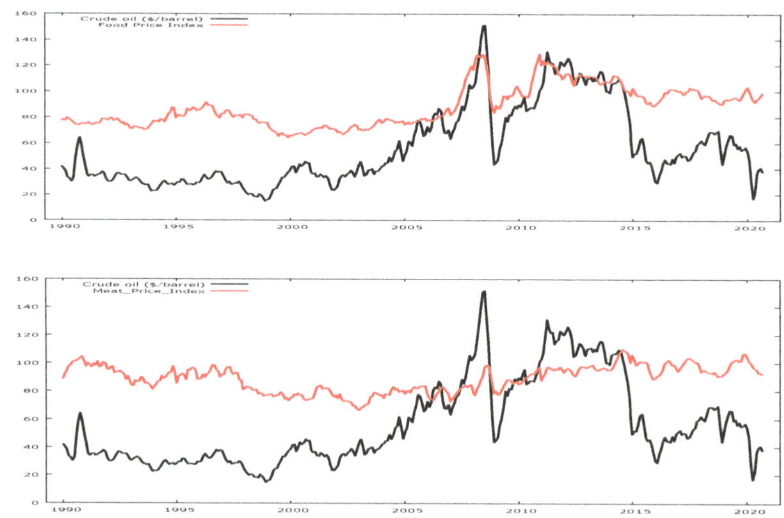

Figure 2. *Cont.*

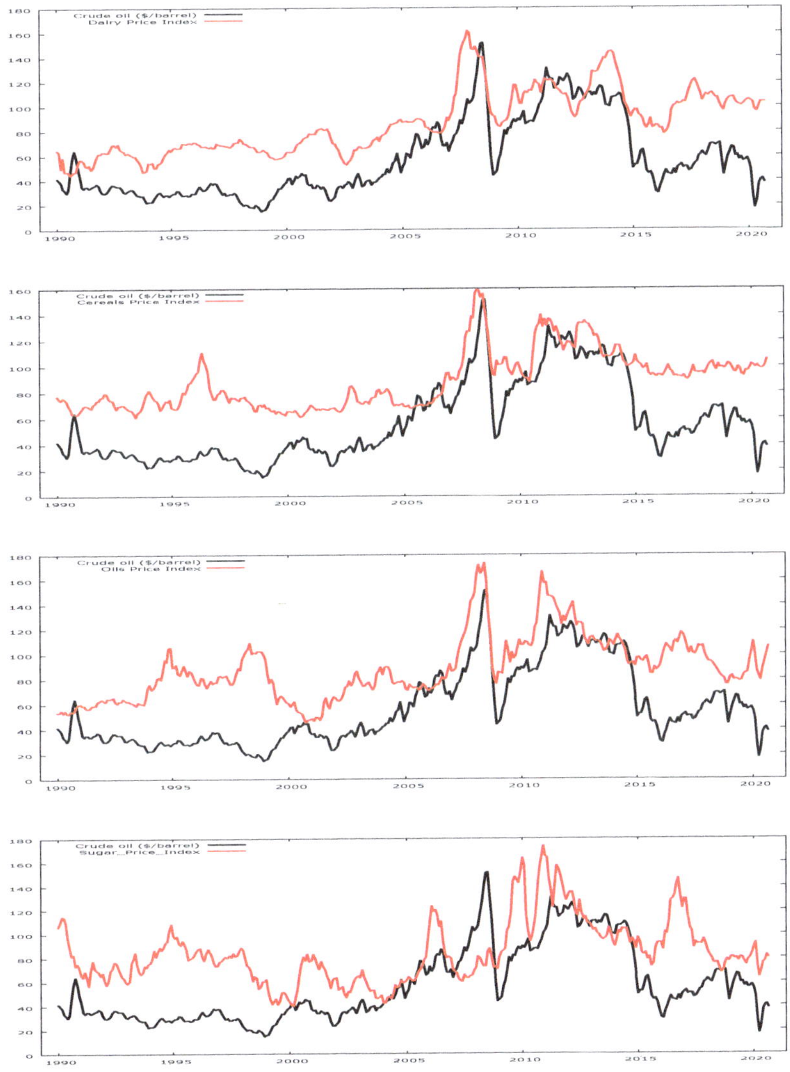

Figure 2. Monthly food price indices. Source: based on [9,43].

Table 2. Correlation coefficients between crude oil and food price indices.

	L_Crude Oil				Dl_Crude Oil		
	1999–2020	1990–2005	2006–2020		1999–2020	1990–2005	2006–2020
l_Food	0.742	−0.125	0.603	dl_Food	0.195	−0.174	0.393
l_Meat	0.275	−0.025	−0.085	dl_Meat	0.159	0.007	0.282
l_Dairy	0.783	0.341	0.557	dl_Dairy	0.124	−0.095	0.293
l_Cereals	0.732	−0.121	0.595	dl_Cereals	−0.001	−0.223	0.137
l_Oils	0.592	−0.353	0.603	dl_Oils	0.202	−0.141	0.436
l_Sugar	0.509	−0.072	0.321	dl_Sugar	0.169	−0.011	0.308

5% critical value: 0.1021 for 1999–2020, 0.142 for 1990–2005 and 2006–2020. Source: own calculations.

Firstly, to select the appropriate research methodology, we used the Augmented Dickey–Fuller test (ADF) [44]. Based on the test results, we chose the methods for analysis (Table 3). The optimal lag for the tests was selected with the Akaike Information Criterion (AIC). The variables are integrated I(1), except for the sugar in the full period.

Table 3. Unit root testing results.

	1990–2020	1990–2005	2006–2020
l_Food	−2.769	−1.768	−2.873
dl_Food	−14.449 ***	−12.984 ***	−8.507 ***
l_Meat	−2.141	−2.525	−2.947
dl_Meat	−10.460 ***	−4.214 ***	−10.082 ***
l_Dairy	−3.021	−2.694	−3.048
dl_Dairy	−14.612 ***	−13.606 ***	−7.020 ***
l_Cereals	−3.260	−3.044	−3.166
dl_Cereals	−13.326 ***	−7.387 ***	−8.957 ***
l_Oils	−3.120	−2.267	−2.989
dl_Oils	−8.043 ***	−5.743 ***	−5.948 ***
l_Sugar	−3.568 **	−2.676	−2.477
dl_Sugar	−12.747 ***	−9.406 ***	−9.494 ***
l_Crude oil	−2.430	−2.335	−2.556
dl_Crude oil	−13.048 ***	−8.923 ***	−9.247 ***

** $p < 0.05$, *** $p < 0.01$. Source: own calculations.

The first step was testing for linear cointegration. According to Engle and Granger [45], *"two-time series are cointegrated if their linear combination is stationary series, I(0)."* The analysis of cointegration allows to state the existence of a long-run connection between analyzed variables. In order to test the long-run relationship, a Johansen cointegration test was used. The test is based on the VAR [46,47]:

$$X_t = C + \sum_{i=1}^{p} A_i X_{t-i} + e_t, \tag{1}$$

where: X_t—endogenous variable vector, C—constant vector, A_i—coefficient matrix, e_t—white noise vector which is independently and identically distributed with $e_t \sim \text{IID}(0,\Sigma)$ where Σ is a positive definitive matrix.

If the endogenous variables are cointegrated, then Equation (1) can be shown in a vector error-correction model (VECM) $(p - 1)$ as follows [46,47]:

$$\Delta(X_t) = C + \Pi X_{t-1} + \sum_{i=1}^{p-1} \tau_i \Delta(X_{t-1}) + e_t, \tag{2}$$

where: $\Pi = \sum_{i=1}^{p-1} A_i - I$, I—identity matrix; $\Gamma_i = -\sum_{j=i+1}^{p-1} A_i$; Π—called a long-run matrix coefficient, Γ_i—short-run matrix coefficient. To study cointegration in the Johansen procedure, the order of the Π matrix is used, this corresponds to the sum of independent cointegration vectors. The Johansen test is based on the trace or maximum eigenvalue test:

$$LR_{trace}(r) = -(T - p) \sum_{i=r+1}^{k} n(1 - \lambda_i) \tag{3}$$

$$LR_{max}(r) = -(T - p) ln(1 - \lambda_{r+1}), \tag{4}$$

where: T—sample size, λ_i—i-th greatest canonical correlation (eigenvalues of matrix Π). The LR_{trace} tested the H_0: the number of vectors is equal to r against the H_1: the number of vectors is equal to n.

The LR_{max} tested the H_0: the number of vectors is equal to r against the H_1: the number of vectors is equal to $r + 1$.

As a result of using Johansen's procedure, three options may appear [48]: *"(i) the rank of the Π matrix is equal to 0—then model (2) is a VAR model for the increments of variables, in which there is no long-run dependence; (ii) the rank of the matrix Π is bigger than 0 and less than r, then the number of cointegration vectors is equal to this rank; (iii) the matrix Π is of a full order, then the series of variables are stationary and model (2) is a VAR model for the levels of the variables."*

In the next stage, the VAR and VECM models were estimated. If there are relationships in this assessment, then the last step will be the Granger causality test and Impulse Response Function test (IRF). The causality test is used to determine the cause-effect relations, where *"variable x is the Granger cause of variable y when the values of variable y can be more accurately foreseen, considering the future value of variable x than when disregarding those values. In the Granger causality test, H_0 is tested: all β_k coefficients equal zero, which is interpreted as a lack of causality."* The Granger causality test can be shown as follows [49]:

$$Y_t = \beta_0 + \sum_{j=1}^{m} \beta_j Y_{t-j} + \sum_{k=l}^{n} \beta_k X_{t-k} + u_t, \tag{5}$$

$$X_t = \beta_0 + \sum_{j=1}^{m} \beta_j X_{t-j} + \sum_{k=l}^{n} \beta_k Y_{t-k} + u_t, \tag{6}$$

where: Y_t—values of variable Y; X_t—values of variable X; β—structural model parameters; t—time variable; u_t—random model element.

4. Results

4.1. Long-Run Analysis

As the variables are I(1), the Johansen cointegration procedure was performed first. The test was used to verify the long-run relationship between crude oil and food index prices. The results of the cointegration test in two subperiods are presented in Table 4. It should be noted that the statistical values of the two tests are smaller than the required critical values (at $p = 0.05$). The exceptions are the results for meat and crude oil in 2006–2020. This demonstrates the long-run connections between these variables. However, in most cases, the long-term relationship between other variables does not exist. This means that the prices of these variables do not follow each other in the long-term. It should be noted that short-term linkages may exist. For that reason, the next step is to identify the short-term link between analyzed variables in the next part. For this purpose, the VAR model (for food, dairy, cereals, oils, and sugar) and the VECM model (for meat) were used.

Table 4. Cointegration results for oil and food price indices.

	Rank	1990–2005				2006–2020			
		LR_{trace}		LR_{max}		LR_{trace}		LR_{max}	
		Stat.	***p*-Value**	**Stat.**	***p*-Value**	**Stat.**	***p*-Value**	**Stat.**	***p*-Value**
l_Food	0	8.990	0.171	8.900	0.126	22.300	0.131	11.285	0.497
	1	0.091	0.828	0.091	0.819	11.014	0.088	11.014	0.088
l_Meat	0	20.513	0.204	15.769	0.159	32.632	0.005	21.617	0.020
	1	4.744	0.639	4.744	0.640	11.015	0.088	11.015	0.088
l_Dairy	0	23.120	0.106	15.788	0.158	22.398	0.128	13.481	0.301
	1	7.332	0.321	7.332	0.321	8.916	0.190	8.916	0.190

Table 4. *Cont.*

Rank		1990–2005				2006–2020			
		LR_{trace}		LR_{max}		LR_{trace}		LR_{max}	
		Stat.	*p*-Value	Stat.	*p*-Value	Stat.	*p*-Value	Stat.	*p*-Value
l_Cereals	0	21.006	0.182	14.256	0.246	24.618	0.070	14.345	0.240
	1	6.749	0.382	6.749	0.383	10.273	0.117	10.273	0.117
l_Oils	0	9.395	0.941	5.186	0.973	22.431	0.127	14.123	0.255
	1	4.209	0.713	4.209	0.715	8.308	0.234	8.308	0.234
l_Sugar	0	23.177	0.104	13.759	0.280	19.150	0.278	11.936	0.434
	1	9.417	0.160	9.417	0.160	7.214	0.332	7.214	0.333

Source: own calculations.

4.2. Short-Run Analysis

Due to the lack of cointegration among analyzed variables (except meat), the VAR model was estimated. Using the AIC criterion, it was found that the appropriate lag length was $p = 3$. Since the variables are I(0), the VAR ($p - 1$) model is estimated, and the lag length is two ($p - 1 = 2$). The effects of VAR (2) in the first difference are shown in the Appendix A (Table A1). On the basis of R^2, it should be concluded that the quality of the model fit is not satisfactory. For example, about 30% of the variation in crude oil is explained by crude oil prices and only 2% by food price.

The results indicate that crude oil price does not have a short-term impact on food, dairy, cereal, oil, and sugar price volatility in the second subperiod; however, food, cereal, and oil prices have a favorable short-term impact on crude oil price volatility. Further proof of the short-run connection between analyzed prices can be inferred from the IRF test. Figure 3 presents only statistically significant impulse responses. The reaction of the price of crude oil to the food, cereal, and oil price is positive. The change in impact occurs after the fourth month for oils and the fifth month for food and cereals.

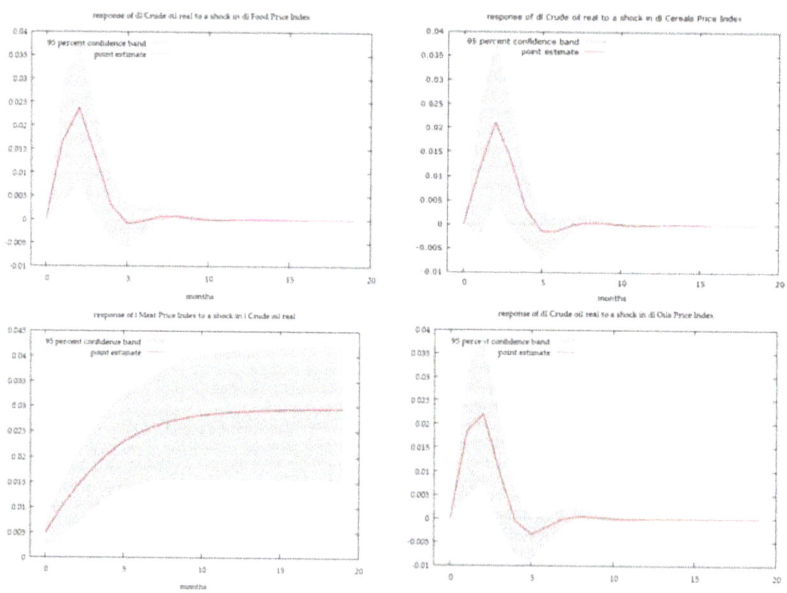

Figure 3. Impulse response function between crude oil price and selected food price volatility. Source: own calculations.

Because there is one cointegrating rank in the relationship between meat and crude oil prices in the second subperiod, the VECM model was used. The result estimation is presented in Table A2 in the Appendix A. The coefficient in the long-term linkage in the VECM model (with the restricted trend and unrestricted constant) is 6.16, which means that a 1% increase or decrease in crude oil price is a response to a 6.16% increase or decrease in meat prices. Therefore, the price of crude oil is an exogenous variable for meat prices. This is evidenced by the significant EC coefficient in the meat price equation. The coefficient for the error correction term in the meat price equation is 0.027, whereas for the crude oil price, the equation is −0.008. Therefore, the disequilibrium in the price system is revised in one month by 2.7% via the reaction of the meat and by 0.8% via the reaction of crude oil. Moreover, the meat price response is positive and persistent throughout the entire period (Figure 3).

4.3. *Causal Relationship*

After estimating the VAR and VECM models, the next step is a determination of causality. For this purpose, the Granger causality test was implemented to investigate the mutual influence of the researched prices. (Table 5). The direction of the relationship in the second subperiod can be deduced from the results in Tables A1 and A2. There is only a one-way (← or →) connection between crude oil and food prices in two subperiods in the short-run. In the first subperiod, in the performed tests approach, we can determine that the food, cereals, and dairy ex-work prices were a Granger cause for crude oil future prices. The food, cereal, and oil future prices were a cause for the crude oil ex-work prices in the second subperiod. It should be noted that crude oil prices are the cause of Granger for meat prices in two subperiods.

Table 5. Granger causality test.

	1990–2005			2006–2020		
	Stat.	*p*-Value		Stat.	*p*-Value	
dl_Crude oil ≠ > dl_Food	1.151	0.333	Crude	0.020	0.980	Crude
dl_Food ≠ > dl_Crude oil	3.359	0.010	oil←Food	5.277	0.006	oil←Food
dl_ Crude oil ≠ > dl_Meat	3.344	0.011	Crude	5.185	0.007	Crude
dl_Meat ≠ > dl_ Crude oil	1.628	0.167	oil→Meat	1.020	0.363	oil→Meat
dl_ Crude oil ≠ > dl_Dairy	1.866	0.173	Crude	0.732	0.482	Crude oil x
dl_Dairy ≠ > dl_ Crude oil	4.697	0.031	oil←Dairy	2.140	0.121	Dairy
dl_ Crude oil ≠ > dl_Cereals	1.176	0.322	Crude	0.497	0.609	Crude
dl_Cereals ≠ > dl_ Crude oil	2.533	0.041	oil←Cereals	3.704	0.027	oil←Cereals
dl_ Crude oil ≠ > dl_Oils	2.137	0.061	Crude oil x	0.571	0.566	Crude
dl_Oils ≠ > dl_ Crude oil	1.237	0.292	Oils	4.853	0.009	oil←Oils
dl_ Crude oil ≠ > dl_Sugar	0.250	0.617	Crude oil x	0.142	0.867	Crude oil x
dl_Sugar ≠ > dl_ Crude oil	0.751	0.387	Sugar	0.548	0.579	Sugar

←/→ the direction of causality, x—no causality Source: own calculations.

5. Discussion and Conclusions

The paper showed an empirical examination into the linkages between the prices of crude oil and selected groups of agricultural commodities. We used monthly data from January 1990 until September 2020. The food prices are for the meat, dairy, cereal, oil, and sugar product groups. Except for meat price, the results indicate no evidence of long-term linkages between the prices of crude oil and food products, whereas the Granger causality tests confirmed that the global oil price reacts to the prices of food products (dairy, oil, cereal) in the short term.

Each farm, especially focused on mass animal husbandry, needs specialized machines that make the work faster and more efficient. Animal husbandry with machine utilization is an example of extensive farming implemented in developing and highly developed countries. Regardless of the specific breeding directions, the significant aspect prompting the efficiency of breeding is the available farm infrastructure with its equipment. For example, feeding cards increase food quality. Likewise, without the support of modernly equipped buildings and safety standards meeting hygiene standards and cleaning machines, it is impossible to keep the costs of breeding at a level that guarantees sufficient income. Additionally, in the case of poultry farming, proper temperature in the boiler is essential. The very strong mechanization of animal farming has a strong relationship with energy use. Vehicles, machines, and heating systems are in use thanks to diesel, which increases the production costs, which, in turn, affects the meat price. There might be enlightenment for the strong correlation between the prices of petroleum and meat products. The results suggest that the development in the mechanization process in the agriculture sector may lead to a situation in which an increase in demand for agricultural commodities will be accompanied by a growth in demand for crude oil [50]. Hence, the surge in food consumption roots a rise in the demand for food and thus may affect the volatility of crude oil prices. Apart from agricultural machinery, crude oil is used for the production of fertilizers, plant protection products, and costs of transport, which can additionally be translated into food prices.

The second explanation may be related to the usage of some groups of agricultural commodities for biofuel production. There has been an important rate change between fuel and food when the Renewable Fuel Standard was enacted in 2005 in the US. Hence, there are noticeably stronger linkages between crude oil prices and volatility in food commodities, which are closely related to biofuel production. These relationships were confirmed by researchers in former studies (e.g., Coronado et al. [51]; Vacha et al. [52]). In March 2020, the price index was lower than in February. However, the decline was not the result, as might be expected, of the fall in demand due to the coronavirus lockdown but the oil price slump. A significant part of the world's crops, e.g., sugar cane in Brazil, maize in the USA, or rape in Poland, is intended for the production of biofuel as an alternative energy source. Therefore, when the crude oil prices fell sharply in the world's markets, biofuel producers also had to adjust their prices. Our results confirm the short-run relationship between the price of crude oil and the prices of cereals and oils.

The lack of long-term dependencies among crude oil and most of the analyzed groups of the prices of agricultural goods was also set, inter alia, by Fowowe et al. [17], Zhang et al. [21], and Pasrun et al. [25]. Furthermore, some authors confirm the long-term relationship between the prices of crude oil and agricultural commodities [30,33]. These studies were based on the analysis of individual products (e.g., wheat, maize, butter, etc.) whereas our analyses concern groups of agricultural commodities; therefore, they are not directly comparable. Mawejje [30], analyzing the groups, agreed that energy prices have long-term cointegrating linkages with food prices. However, it should be noted that the results related to a different research period, 2000–2011, than the adopted research period seems to have a significant impact on the result of this type of calculation and their comparison.

The results obtained have an important practical global context. Firstly, the results should apply to investors involved in hedging prospects between petroleum and food markets. The outcomes can inform them that risk in food markets is not dependent on hazards in the oil market [53]. Moreover, the lack of effect of crude oil price level on the fluctuation in the prices of agricultural goods indicates that agricultural policy relating to mitigating volatility of food prices should be based on other issues rather than fluctuations in crude oil markets [19,54]. Furthermore, there are continuous developments in the biofuel market, hence the production of agricultural commodities for energy purposes is increasing. Therefore, we believe that the analysis of their issues should be continued. To obtain more accurate results, we recommend the analyses of individual agricultural commodities, not their groups.

Author Contributions: Conceptualization, M.R., A.G. and J.D.; methodology, M.R.; software, M.R.; validation, M.R., A.G. and J.D.; formal analysis, M.R., A.G. and J.D.; investigation, M.R., A.G. and J.D.; resources, J.D. and A.G.; data curation, M.R.; writing—original draft preparation, M.R., A.G. and J.D.; writing—review and editing, M.R., A.G. and J.D.; visualization, M.R., A.G. and J.D.; supervision, M.R., A.G. and J.D.; project administration, M.R., A.G. and J.D.; funding acquisition, M.R., A.G. and J.D. All authors have read and agreed to the published version of the manuscript.

Funding: This research received no external funding.

Conflicts of Interest: The authors declare no conflict of interest.

Appendix A

Table A1. Selected VARs statistics.

VAR(2)	Coef.		Coef.
	dl_Crude Oil		**dl_Food**
dl_Crude oil (−1)	0.499 ***	dl_Crude oil (−1)	0.001
dl_Crude oil (−2)	−0.328 ***	dl_Crude oil (−2)	−0.004
dl_Food (−1)	0.6444 ***	dl_Food (−1)	0.342 ***
dl_Food (−2)	0.38	dl_Food (−2)	0.105
Constant	−0.004	Constant	0.001
R^2	0.320	R^2	0.153
	dl_Crude oil		**dl_Dairy**
dl_Crude oil (−1)	0.560 ***	dl_Crude oil (−1)	0.020
dl_Crude oil (−2)	−0.323 ***	dl_Crude oil (−2)	−0.034
dl_Dairy (−1)	0.213	dl_Dairy (−1)	0.521 ***
dl_Dairy (−2)	0.209	dl_Dairy (−2)	−0.002
Constant	−0.003	Constant	0.000
R^2	0.296	R^2	0.279
	dl_Crude oil		**dl_Cereals**
dl_Crude oil (−1)	0.551 ***	dl_Crude oil (−1)	−0.031
dl_Crude oil (−2)	−0.298 ***	dl_Crude oil (−2)	0.024
dl_Cereals (−1)	0.283 *	dl_Cereals (−1)	0.334 ***
dl_Cereals (−2)	0.261	dl_Cereals (−2)	0.036
Constant	−0.004	Constant	0.002
R^2	0.308	R^2	0.118
	dl_Crude oil		**dl_Oils**
dl_Crude oil (−1)	0.488 ***	dl_Crude oil (−1)	0.0156
dl_Crude oil (−2)	−0.311 ***	dl_Crude oil (−2)	−0.046
dl_Oils (−1)	0.378 ***	dl_Oils (−1)	0.423 ***
dl_Oils (−2)	0.111	dl_Oils (−2)	−0.041
Constant	−0.004	Constant	0.001
R^2	0.317	R^2	0.174
	dl_Crude oil		**dl_Sugar**
dl_Crude oil (−1)	0.603 ***	dl_Crude oil (−1)	0.013
dl_Crude oil (−2)	−0.288 ***	dl_Crude oil (−2)	−0.031
dl_Sugar (−1)	−0.052	dl_Sugar (−1)	0.313 ***
dl_Sugar (−2)	−0.074	dl_Sugar (−2)	−0.077
Constant	−0.003	Constant	−0.001
R^2	0.283	R^2	0.097

* $p < 0.1$, *** $p < 0.01$. Source: own calculations.

Table A2. Selected VECMs statistics.

Selected Statistic	Stat.
AIC	−6.421
BIC	−6.242
Long-run relationship	1 * ln_Crude oil − 6.164 × ln_Meat + 0.012 × time
EC (ln_Crude oil)	−0.008
EC (ln_Meat)	0.027 ***

*** $p < 0.01$. Source: own calculations.

References

1. Hamilton, J.D. Oil and the macroeconomy since World War II. *J. Polit. Econ.* **1983**, *91*, 228–248. [CrossRef]

2. Kilian, L.; Vigfusson, R.J. Do oil prices help forecast us real GDP? The role of nonlinearities and asymmetries. *J. Bus. Econ. Stat.* **2013**, *31*, 78–93. [CrossRef]
3. Wang, J.; Huang, Y.; Ma, F.; Chevallier, J. Does high-frequency crude oil futures data contain useful information for predicting volatility in the US stock market? New evidence. *Energy Econ.* **2020**, *91*, 1–13. [CrossRef]
4. Trench, C.J.; Miesner, T.O. The role of energy pipelines and research in the United States. In *Pipeline Research Council*; International Inc.: Bothell, WA, USA, 2006.
5. Rahman, M. Oil and Gas: The Engine of the World Economy. In Proceedings of the Presentation as OPEC Secretary General at the Tenth International Financial and Economic Forum, Vienna, Austria, 10–11 November 2004.
6. Jahangir, S.R.; Dural, B.Y. Crude oil, natural gas, and economic growth: Impact and causality analysis in Caspian Sea region. *Int. J. Manag. Econ.* **2018**, *54*, 169–184. [CrossRef]
7. Taghizadeh-Hesary, F.; Rasoulinezhad, E.; Yoshino, N. Energy and Food Security: Linkages through Price Volatility. *Energy Policy* **2019**, *128*, 796–806. [CrossRef]
8. Xu, S.; Du, Z.; Zhang, H. Can Crude Oil Serve as a Hedging Asset for Underlying Securities? Research on the Heterogenous Correlation between Crude Oil and Stock Index. *Energies* **2020**, *13*, 3139. [CrossRef]
9. EIA. Short-Term Energy Outlook. Available online: https://www.eia.gov/outlooks/steo/marketreview/crude.php (accessed on 10 October 2020).
10. Umar, G.; Kilishi, A.A. Oil price shocks and the Nigeria economy: A variance autoregressive (VAR) model. *Int. J. Bus. Manag.* **2010**, *5*, 39. [CrossRef]
11. Benedictow, A.; Fjærtoft, D.; Løfsnæs, O. Oil dependency of the Russian economy: An econometric analysis. *Econ. Model.* **2013**, *32*, 400–428. [CrossRef]
12. Bayraktar, Y.; Taha, E.; Yildiz, F. A causal relationship between oil prices current account deficit, and economic growth: An empirical analysis from fragile five countries. *Ecoforum. J.* **2016**, *5*, 3.
13. Ftiti, Z.; Guesmi, K.; Teulon, F.; Chouachi, S. Relationship between crude oil prices and economic growth in selected OPEC countries. *J. Appl. Bus. Res.* **2016**, *32*, 11. [CrossRef]
14. Sarwar, S.; Tivari, A. Analyzing volatility spillovers between oil market and Asian stock markets. *Resour. Policy* **2020**, *66*, 101608. [CrossRef]
15. Ding, S.; Zhang, Y. Cross market predictions for commodity prices. *Econ. Model.* **2020**, *91*, 455–462. [CrossRef]
16. Hau, L.; Zhu, H.; Huang, R.; Ma, X. Heterogeneous dependence between crude oil price volatility and China's agriculture commodity futures: Evidence from quantile-on-quantile regression. *Energy* **2020**, 213. [CrossRef]
17. Fowowe, B. Do oil prices drive agricultural commodity prices? Evidence from South Africa. *Energy* **2016**, *104*, 149–157. [CrossRef]
18. Ibrahim, M.H. Oil and food prices in Malaysia: A nonlinear ARDL analysis. *Agric. Food Econ.* **2015**, *3*, 1–14. [CrossRef]
19. Nazlioglu, S.; Soytas, U. World oil prices and agricultural commodity prices: Evidence from an emerging market. *Energy Econ.* **2011**, *33*, 488–496. [CrossRef]
20. Gilbert, C.L. How to understand high food prices. *J. Agric. Econ.* **2010**, *61*, 398–425. [CrossRef]
21. Zhang, Z.; Lohr, L.; Escalante, C.; Wetzstein, M. Food versus fuel: What do prices tell us? *Energy Policy* **2010**, *38*, 445–451. [CrossRef]
22. Vo, D.H.; Vu, T.N.; Vo, A.T.; McAleer, M. Modeling the Relationship between Crude Oil and Agricultural Commodity Prices. *Energies* **2019**, *12*, 1344. [CrossRef]
23. Su, C.W.; Wang, X.Q.; Tao, R.; Oana-Ramona, L. Do oil prices drive agricultural commodity prices? Further evidence in a global bio-energy context. *Energy* **2019**, *172*, 691–701. [CrossRef]
24. Pal, D.; Mitra, S.K. Correlation dynamics of crude oil with agricultural commodities: A comparison between energy and food crops. *Econ. Model.* **2019**, *82*, 453–466. [CrossRef]
25. Pasrun, A.; Rosnawintang, R.; Saidi, L.D.; Tondi, L.; Sani, L.O.A. The Causal Relationship between Crude Oil Price, Exchange Rate and Rice Price. *Int. J. Energy Econ. Policy* **2018**, *8*, 90–94.
26. Ji, Q.; Bouri, E.; Roubaud, D.; Shahzad, S.J.H. Risk spillover between energy and agricultural commodity markets: A dependence-switching CoVaR-copula model. *Energy Econ.* **2018**, *75*, 14–27. [CrossRef]
27. Al-Maadid, A.; Caporale, G.M.; Spagnolo, F.; Spagnolo, N. Spillovers between food and energy prices and structural breaks. *Int. Econ.* **2017**, *150*, 1–18. [CrossRef]

28. Bergmann, D.; O'Connor, D.; Thummel, A. An analysis of price and volatility transmission in butter, palm oil and crude oil markets. *Agric. Food Econ.* **2016**, *4*, 1–23. [CrossRef]
29. Hamulczuk, M. Energy prices vs agri-food prices – energy security and food security. *Humanit. Soc. Sci.* **2016**, *23*, 37–51.
30. Mawejje, J. Food prices, energy and climate shocks in Uganda. *Agric. Food Econ.* **2016**, *4*, 1–18. [CrossRef]
31. Fernandez-Perez, A.; Frijns, B.; Tourani-Rad, A. Contemporaneous interactions among fuel, biofuel and agricultural commodities. *Energy Econ.* **2016**, *58*, 1–10. [CrossRef]
32. McFarlane, L. Agricultural commodity prices and oil prices: Mutual causation. *Outlook Agric.* **2016**, *45*, 87–93. [CrossRef]
33. Cabrera, B.L.; Schulz, F. Volatility linkages between energy and agricultural commodity prices. *Energy Econ.* **2016**, *54*, 190–203. [CrossRef]
34. Nwoko, I.C.; Aye, G.C.; Asogwa, B.C. Effect of oil price on Nigeria's food price volatility. *Cogent Food Agric.* **2016**, *2*, 1–14. [CrossRef]
35. Zhang, C.; Qu, X. The effect of global oil price shocks on China's agricultural commodities. *Energy Econ.* **2015**, *51*, 354–364. [CrossRef]
36. Koirala, K.H.; Mishra, A.K.; D'Antoni, J.M.; Mehlhorn, J.E. Energy prices and agricultural commodity prices: Testing correlation using copulas method. *Energy* **2015**, *81*, 430–436. [CrossRef]
37. Rezitis, A.N. The relationship between agricultural commodity prices, crude oil prices and US dollar exchange rates: A panel VAR approach and causality analysis. *Int. Rev. Appl. Econ.* **2015**, *29*, 403–434. [CrossRef]
38. Natanelov, V.; Alam, M.J.; McKenzie, A.M.; Huylenbroeck, G.V. Is there co-movement of agricultural commodities futures prices and crude oil? *Energy Policy* **2011**, *39*, 4971–4984. [CrossRef]
39. Chang, T.-H.; Su, H.-M. The substitutive effect of biofuels on fossil fuels in the lower and higher crude oil price periods. *Energy* **2010**, *35*, 2807–2813. [CrossRef]
40. Balcombe, K.; Rapsomanikis, G. Bayesian estimation and selection of nonlinear vector error correction models: The case of the sugar-ethanol-oil nexus in Brazil. *Am. J. Agric. Econ.* **2008**, *90*, 658–668. [CrossRef]
41. Nazlioglu, S. World oil and agricultural commodity prices: Evidence from nonlinear causality. *Energy Policy* **2011**, *39*, 2935–2943. [CrossRef]
42. Rosa, F.; Vasciaveo, M. Agri-Commodity Price Dynamics: The Relationship between Oil and Agricultural Market. In Proceedings of the International Association of Agricultural Economists (IAAE) Triennial Conference, Foz do Iguaçu, Brazil, 18–24 August 2012; pp. 18–24.
43. FAO Food Price Index. Available online: http://www.fao.org/fileadmin/templates/worldfood/Reports_and_docs/Food_price_indices_data.xls (accessed on 10 October 2020).
44. Dickey, D.A.; Fuller, W.A. Distribution of the estimators for autoregressive time series with a unit root. *J. Am. Stat. Assoc.* **1979**, *74*, 427–431.
45. Engle, R.F.; Granger, C.W.J. Co-integration and error correction: Representation, estimation and testing. *Econometrica* **1987**, *55*, 251–276. [CrossRef]
46. Lutkepohl, H. Vector Autoregressive and Vector Error Correction Models. In *Book Applied Time Series Econometrics*; Lutkepohl, H., Kratzig, M., Eds.; Cambridge University Press: New York, NY, USA, 2004; pp. 86–108.
47. Neusser, K. *Time Series Econometrics*; Springer International Publishing: Cham, Switzerland, 2016; pp. 305–310.
48. Kusideł, E. *Modelowanie Wektorowo-Autoregresyjne VAR. Metodologia i Zastosowanie w Badaniach Ekonomicznych*; Absolwent: Łódź, Poland, 2000; pp. 40–55.
49. Granger, C.W.J. Investigating Causal Relations by Econometric Models and Cross-spectral Methods. *Econometrica* **1969**, *37*, 424–438. [CrossRef]
50. Baumeister, C.; Kilian, L. Do oil price increases cause higher food prices? *Econ. Policy* **2014**, *29*, 691–747. Available online: https://www.jstor.org/stable/24030050 (accessed on 11 December 2020). [CrossRef]
51. Coronado, S.; Rojas, O.; Romero-Meza, R.; Serletis, A.; Chiu, L.V. Crude oil and Biofuel Agricultural Commodity Prices. In *Expectations, Uncertainty and Asset Price Dynamics*; Fredj, J., Ed.; Springer: Berlin/Heidelberg, Germany, 2018.
52. Vacha, L.; Janda, K.; Kristoufek, L.; Zilberman, D. Time-frequency dynamics of biofuel-fuel-food system. *Energy Econ.* **2013**, *40*, 233–241. [CrossRef]
53. Nazlioglu, S.; Erdem, C.; Soytas, U. Volatility spillover between oil and agricultural commodity markets. *Energy Econ.* **2013**, *36*, 658–665. [CrossRef]

54. Reboredo, J.C. Do food and oil prices co-move? *Energy Policy* **2012**, *49*, 456–467. [CrossRef]

energies

MDPI

Article

Construction and Empirical Verification of the Agri-Environmental Index (AEI) as a Tool for Assessing the Green Performance of Agriculture

Armand Kasztelan and Anna Nowak *

Department of Economics and Agribusiness, University of Life Sciences in Lublin, Akademicka 13, 20-950 Lublin, Poland; armand.kasztelan@up.lublin.pl
* Correspondence: anna.nowak@up.lublin.pl; Tel.: +48-513-131-856

Abstract: In this paper, the aggregate index of green performance of agriculture (Agri-Environmental Index (*AEI*)) was proposed and empirically verified. For this purpose, a taxonomic method was used, i.e., the linear ordering method, which allows for the construction of a synthetic metric for the assessment of performance. Based on 16 agri-environmental indicators from the Organization for Economic Cooperation and Development (OECD) and Eurostat database, green performance indexes were constructed for 20 European countries. The constructed indexes are based on the multi-line impact of agriculture on the environment, with a particular focus on energy issues. During the analyses, answers to the following research questions were sought: Is the *AEI* an appropriate tool for evaluating the green performance of agriculture? What is the overall situation in this matter in EU countries? Which areas in terms of the impact of agriculture on the environment require remedial actions? The results of surveys show that the level of green performance in countries is still low (an average of 0.3069). The article indicates the areas that require special attention in the context of continuation of greening processes in the agricultural sector.

Keywords: green agriculture; agri-environmental indicators; energy efficiency; green performance index; taxonomic methods; zero unitarization method; comparative analysis

Citation: Kasztelan, A.; Nowak, A. Construction and Empirical Verification of the Agri-Environmental Index (AEI) as a Tool for Assessing the Green Performance of Agriculture. *Energies* **2021**, *14*, 45. https://dx.doi.org/10.3390/en14010045

Received: 25 November 2020
Accepted: 21 December 2020
Published: 23 December 2020

1. Introduction

Agriculture is a sector of the economy that has special links with the natural environment. These links are complex and reflect biological processes, changes in the natural environment conditions, socio-economic factors, and agricultural and environmental policy [1,2]. They are additionally compounded by a spatial variation of the impact of agriculture on the environment in different countries [3,4]. On the one hand, the performance of agriculture depends on land and water resources, and on the other hand, agricultural production often takes place at the expense of the environment (e.g., resulting in soil degradation, deteriorated quality of water, reduced biological diversity, increased greenhouse gas emissions), which undermines environmental sustainability [1,5–8]. Industrial agriculture was highly efficient but generated environmental and social consequences of global significance [9]. Due to the need for altering the paradigm of European agriculture after the period of industrialization, the concept of sustainable agriculture was recognized as a priority direction of development reflected in the common agricultural policy of the EU [10]. This gives rise to specific prospects for the development of this sector but also to several challenges. In fulfilling its environmental sustainability mission, present-day agriculture should integrate a wide spectrum of objectives connected with the demand for food and agricultural commodities with the environmental challenges [8,11]. Challenges to be faced by agriculture also include increased competition around alternative uses of the natural resources, maintaining biological diversity, food safety, and climate change mitigation [12,13]. This also means a need for constructing an adequate policy of energy efficiency in agriculture in which the effects of production are considered in terms of energy expenditure and minimization of environmental impacts.

Recently, in connection with a growing interest in environmental issues, many studies have been undertaken investigating the relationship between the economy as a whole or its respective sectors and the natural environment. One of the research lines is environmental (green) performance. Mutingi et al. [14] used the term 'green performance' with reference to supply chains. In turn, Li and Lin [15] extended this reference to include the economy as a whole. Gallego-Álvarez et al. [16] use 'environmental performance' and 'environmental sustainability' as alternative terms. Several papers have analyzed environmental performance in the EU from different angles. In macroeconomic terms, the environmental performance of countries is defined as a country's ability to produce environmental public goods [17]. In turn, Beltrán-Esteve and Picazo-Tadeo [18] evaluated environmental performance in the European Union using Luenberger productivity indicators and DEA techniques. The environmental performance of specific countries is also assessed based on EPI (Environmental Performance Index), which is an integrated measure of how well they can handle environmental issues from the perspective of human health and the ecosystem [19]. Schultze and Trommer [20] emphasize that empirical studies make use of multiple environmental performance measures selected mainly based on practicability.

A significant research gap exists in a comprehensive assessment of this phenomenon at the level of sectors of the economy with a key impact on the environment. Without any doubt, agriculture can be deemed such a sector. A significant role of agriculture in implementing the concept of sustainable development prompted the authors of this paper to investigate the methods for evaluating the green performance of agriculture. In this paper, this term is defined as all positive effects of the environmental impact of the agricultural sector connected with preventing and reducing the emissions of pollutants, a sustainable use of resources and measures aimed at greening agriculture (including the development of the renewable energy sector and organic agriculture), to ensure sustainability of the whole agroecosystem.

The relationships between agriculture and the environment are reflected by agri-environmental indicators (*AEI*). They have become increasingly important for measuring the environmental consequences of agricultural practices and monitoring of progress towards sustainable development [1,4,21,22]. The characteristic of *AEI* methods is that they provide a conceptual framework to define and bring together a set of agri-environmental indicators [23]. The methods of evaluation based on a set of indicators were elaborated at international [24] and domestic [25,26] level but also at regional [23] and farm [27–29] level as well as with reference to agricultural systems [28,30]. The general framework of and approach to the set of agri-environmental indicators were designed by the Organization for Economic Co-operation and Development [31]. They were assigned to six groups of indicators: soil, water, air, biodiversity, farm management, and agricultural inputs. The indicators were used in many scientific studies for evaluating a relationship between agriculture and the environment in selected countries [25]. Also, the European Commission, as a result of the IRENA (International Renewable Energy Agency) operation, identified 28 agri-environmental indicators [32,33].

Considering the large diversity of agri-environmental factors, analyzing one or more specific indicators separately does not have major advantages [11]. Although studies concerning the evaluation of a relationship between agriculture and the natural environment do exist, they are usually limited to analyzing selected aspects only such as pesticides [34], nitrogen [35] and biodiversity [36]. On the other hand, there are no studies comprehensively approaching the issues of measuring the green performance of agriculture using an advanced set of indicators. The need for such research was also mentioned by Czyżewski et al. [37]. In view of the above-presented arguments, the purpose of this paper is the evaluation of the green performance of agriculture in 20 member states of the European Union. An aggregate index using multiple variables expressed as agricultural and environmental indicators was constructed. Therefore, this work is a genuine contribution to research concerning the evaluation of the environmental impact of the agricultural sector through designing a synthetic measure—the *AEI*. When constructing the synthetic index (*AEI*), the linear ordering method with the median and standard deviation was applied. Thanks to this, this method is characterized by a high resistance to the occurrence of extreme observations, which is particularly important from the point of view of comparative analysis of the EU countries [38–40].

The authors try to answer the following questions: Is the Agri-Environmental Index an appropriate tool for evaluating the green performance of agriculture? What is the overall situation in this matter in EU countries? Which areas in terms of the impact of agriculture on the environment require remedial actions? The following structure was adopted in the paper. The next chapter presents the construction method of the *AEI* index. In the third part, based on the values of the indexes, a comparative analysis of selected EU countries was carried out, the results obtained were discussed and the directions for further research were indicated. The last part contains conclusions drawn from the analyses.

2. Materials and Methods

The main aim of the research was to construct a synthetic index of green performance of agriculture (*AEI*). This measure takes into account the multi-line impact of agriculture on the natural environment, with a particular focus on energy issues. The index was constructed using a taxonomic linear ordering method based on median and standard deviation [41,42].

Methods of constructing synthetic measures are the subject of numerous publications [41,43–45]. Based on these foundations, the *AEI* index was designed by using the following procedure [40,44]:

1. The selection as well as construction of the partial indicators describing an agri-environmental performance from the OECD and Eurostat database;
2. The standardization of the indicators according to their impact (stimulants/destimulants) on the phenomenon studied (green performance of agriculture);
3. The construction of the synthetic measure; *AEI* indexes for respective countries;
4. The linear hierarchization of selected EU countries, based on the *AEI*.

The problems that arise when selecting the indicators are primarily the difficulty in their proper defining and the lack of available data. For this reason, the agri-environmental indicators from the OECD and Eurostat databases were used. Ultimately, according to the data availability, 16 indicators (Table 1) and 20 countries were selected for the *AEI* calculation. The average values for the reference years, 2008–2017 were chosen for the analysis. Since some agri-environmental indicators are given in absolute values, to ensure their comparability, they were relativized, e.g., by converting them to a unit of agricultural land area in a given country. The indicators used for designing the synthetic measure were selected to reflect the multi-directional relations between this sector and the natural resources (earth, water, and air).

A significant role in the adopted set of partial indicators is ascribed to energy efficiency indicators: Total final energy consumption in agriculture and production of renewable energy from agriculture. Taking the first of the above-mentioned indicators into account relates to the fact that agriculture, as an energy user, contributes to the depletion of non-renewable energy resources and to global warming through energy-related emissions [46]. In this context, the need for minimizing the expenditure of energy in the agricultural sector has been identified [47]. Biological, technical and technological progress in agricultural production contributes to increasing efficiency of production but at the same time leads to increasing energy expenditure connected primarily with consumption of energy accumulated in the means of production. The adopted indicator refers to the direct use of energy by agriculture. It comprises all energy carriers used directly in the process of agricultural production, including electricity, refined oil products, fuels derived from natural gas, and renewable fuels [48]. The second indicator considered in the studies shows that the agricultural sector both emits greenhouse gas and consumes energy but at the same time has a potential to generate renewable energy. Due to the wide variety of renewable energy resources that can be processed in agriculture, this sector can play a significant role both in generating energy and implementing the objectives of the climate policy [49].

Table 1. Indicators selected for the analysis.

Indicator Symbol	Indicator Name (Unit of Measure)	Stimulant/ Destimulant	Characteristic/Impact on the Environment
x_1	Nitrogen balance (inputs—outputs) (kg/ha)	D	a positive nitrogen balance increases the risk of soil, water, and air pollution
x_2	Phosphorus balance (inputs—outputs) (kg/ha)	D	a positive phosphorus nitrogen balance increases the risk of soil, water, and air pollution
x_3	Total sales of agricultural pesticides (kg/ha)	D	the greater the use of pesticides, the greater the risk of environmental pollution
x_4	Agriculture freshwater abstraction (m^3/ha)	D	the greater the abstraction, the greater the pressure on the environment
x_5	Irrigation area (% total agriculture land area)	S	areas actually irrigated; irrigation infrastructure reduces water abstraction+
x_6	Irrigable area (% total agriculture land area)	S	areas with irrigation infrastructure, but not always irrigated; irrigation infrastructure reduces water abstraction
x_7	Permanent pasture (% total agriculture land area)	S	promote biodiversity, regulate biochemical cycles, and limit the transfer of nitrogen to waters
x_8	Organic farming (% total agriculture land area)	S	processes related to organic farming favour the minimization of pollution and waste
x_9	Total final energy consumption in agriculture (kg of oil equivalent (toe)/ ha)	D	the less energy consumption, the less pressure on the environment
x_{10}	Agricultural ammonia (NH_3) (% of total ammonia emissions)	D	ammonia emissions cause air pollution, negatively affecting the quality of soil and water as well as biodiversity
x_{11}	Total greenhouse gas emissions from agriculture (% of total emissions)	D	increase in greenhouse gas emissions contributes to the global warming
x_{12}	Farmland Birds Index (index)	S	a higher index favours biodiversity
x_{13}	Agricultural land classified as having low wind erosion risk (%)	S	wind erosion destroys fertile topsoil and organic matter, deposits unwanted nutrients and salt, threatening plants and animals
x_{14}	Agricultural land classified as having moderate water erosion risk (%)	S	water erosion negatively affects the soil, plants, and wildlife, as well as the water quality itself
x_{15}	Renewable energy production from agriculture (% of total production)	S	the higher the share of energy production from renewable sources, the lower the pressure on the environment (use of non-renewable resources, environmental pollution, climate change)
x_{16}	Organic carbon content in arable land (tonnes/ha)	D	high carbon deposits in the soil increase the risk of greenhouse gas emissions

Source: own elaboration based on the OECD and Eurostat database.

Based on the characteristics of the agri-environmental indicators in the OECD and Eurostat database [50,51], eight were considered to be larger-the-better characteristics (stimulants) with a positive influence on the synthetic measure, and eight were regarded as smaller-the-better characteristics (de-stimulants) which reduced the *AEI* [52]. The values of the variables describing respective countries are presented as a matrix of observations:

$$X = \begin{bmatrix} x_{11} & \cdots & x_{1m} \\ \vdots & \ddots & \vdots \\ x_{n1} & \cdots & x_{nm} \end{bmatrix} \quad (1)$$

As the diagnostic data set contained indicators that could not be directly aggregated, they were standardized using the zero unitarization method [53]:

For stimulants:

$$z_{ij} = \frac{x_{ij} - \min(x_{ij})_i}{\max(x_{ij})_i - \min(x_{ij})_i} \quad (2)$$

For de-stimulants:

$$z_{ij} = \frac{\max(x_{ij})_i - x_{ij}}{\max(x_{ij})_i - \min(x_{ij})_i} \quad (3)$$

where:

z_{ij} is the normalized value of the *j*-th variable in the *i*-th country;

x_{ij} is the initial value of the *j*-th variable in the *i*-th country.

This method was chosen because it was the only one that met all seven postulates formulated with regard to the use of standardization Equations, i.e., (1) elimination of labels describing the features; (2) order of magnitude of variables allowing comparison; (3) equal length of the variation interval for all standardized features (constant range) and equal lower and upper limit of their variation interval; (4) possibility of standardizing features that are both positive and negative and features that are only negative; (5) possibility of standardizing features with values close to zero; (6) non-negativity of values of the standardized features; (7) existence of simple formulas normalizing the nature of features [54]. Diagnostic features standardized as described above get values from 0 to 1. The closer to 1, the better the situation in terms of the analyzed feature, and the closer to 0, the worse the situation. Standardization results for individual indicators and countries can be found in Supplementary Materials, Annex S1.

The normalized values of agri-environmental indicators were the basis for calculating the median and the standard deviation for each of the selected EU countries. The median values were determined using the formula [38,39]:

$$Me_i = \frac{z_{(\frac{m}{2})i} + z_{(\frac{m}{2}+1)i}}{2} \quad (4)$$

For an even number of observations, or:

$$Me_i = z_{(\frac{m}{2}+1)i} \quad (5)$$

For an odd number of observations, where:

$z_{i(j)}$ is the *j*-th statistical ordinal for the vector $(z_{i1}, z_{i2}, \ldots, z_{im})$, $i = 1, 2, \ldots, n$; $j = 1, 2, \ldots, m$.

In turn, the standard deviation values were calculated according to the following formula:

$$Se_i = \sqrt{\frac{1}{m}\sum_{j=1}^{m}(z_{ij} - \overline{z})} \quad (6)$$

In the last step, the *AEI* indexes for each country were developed (Supplementary Materials, Annex S1):

$$AEI_i = Me_i(1 - Se_i)AEI_i < 1 \quad (7)$$

Values of the *AEI* closer to 1 indicate a higher level of green performance of agriculture for the specific country, resulting in a higher rank. This allowed the comparison of the selected EU countries and classifying them into uniform groups according to their level of green performance:

$$\text{group I}: \ AEI_i \geq \overline{AEI} + S \text{ high level}$$

$$\text{group II}: \ \overline{AEI} + S > AEI_i \geq \overline{AEI} \text{ medium–high level}$$

$$\text{group III}: \ \overline{AEI} > AEI_i \geq \overline{AEI} - S \text{ medium–low level}$$

$$\text{group IV}: \ AEI_i < \overline{AEI} - S \text{ low level}$$

where $\overline{AEI}$ is the mean value of the synthetic measure, and S is the standard deviation of the synthetic measure.

3. Results and Discussion

The synthetic measure describing the level of green efficiency of agriculture using the presented method was calculated for 20 member states of the European Union. Table 2 presents a division of the member states into four groups depending on the adopted value of the measure. On the other hand, Figure 1 presents a ranking of the analyzed member states according to the value of the aggregate measure. The studies show that the mean *AEI* for the analyzed EU member states was 0.3069. It can therefore be concluded that the level of green efficiency of agriculture in the analyzed group of EU countries is very low. Simultaneously, a high variability in the analyzed phenomenon is noticeable between countries covered by the study. This is demonstrated by the fact that the values of *AEI* deviated from the mean value by 0.0880. The analysis of the four identified groups of countries leads to the conclusion that group I, characterized by the highest level of eco-efficiency of the analyzed sector, is represented by three countries, i.e., Portugal (0.4931), Austria (0.4144) and Greece (0.4139). An identical number of countries was recorded in the group in which the level of the analyzed phenomenon was the lowest. That group comprised Hungary, Lithuania, and Belgium. In Belgium, as the country with the lowest level of green efficiency of agriculture, *AEI* was only 0.1358. Eight countries presented a medium–high level of agriculture eco-efficiency, including four member states that joined the EU in 2004 or later (Slovak Republic, the Czech Republic, Slovenia, and Latvia). A medium–low level of the analyzed phenomenon was observed for six countries. A deeper analysis of agri-environmental indicators shows that the main problems in this regard relate to:

- The high share of agricultural lands classified as having high water and wind erosion risk (average z_i for 20 countries: respectively 0.1314 and 0.1383),
- The irrigation (0.1890) and irrigable (0.2465) rates of agricultural land areas,
- The low share of the renewable energy from agriculture (0.3059) as well as organic farming (0.3081),
- The high rates of agriculture sector in ammonia (NH_3) emissions.

Table 2. Classification of 20 EU member states according to the value of the *AEI*.

Group Number	Green Performance Level	*AEI* Range	Countries
I	high	0.3950–	Portugal, Austria, Greece
II	medium–high	0.3069–0.3950	Slovak Republic, Czech Republic, Slovenia, Germany, Spain, Latvia, Netherlands, Sweden
III	medium–low	0.2189–0.3069	United Kingdom, Denmark, France, Estonia, Luxembourg, Poland
IV	low	−0.2189	Hungary, Lithuania, Belgium

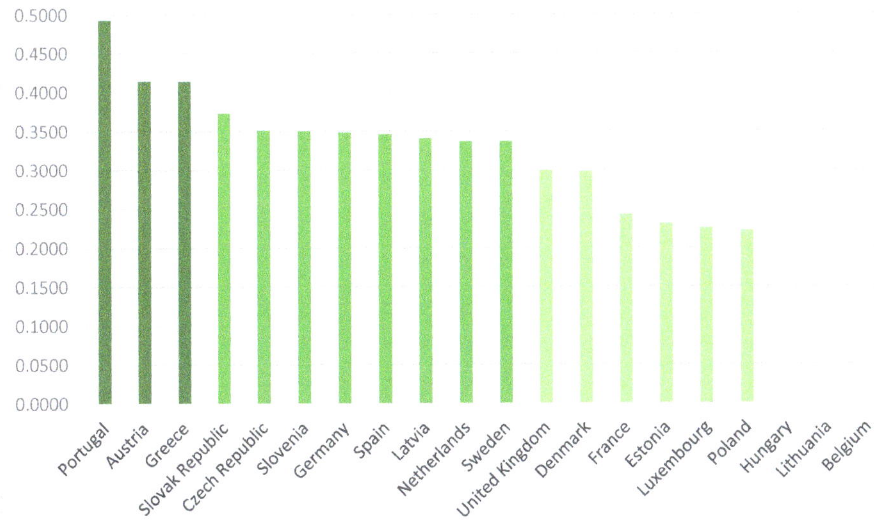

Figure 1. Ranking of selected EU countries based on *AEI*.

The indicated areas require undertaking intensive corrective measures, and hence should be especially monitored by international and national institutions to outline action strategies for the agricultural sector in the following years.

On the other hand, areas can be identified in which very good results have been already achieved, i.e., regarding the volume of water used for agricultural purposes (0.8693), regarding the use of pesticides (0.8446) or phosphorus balance indicator (0.7942). Another characteristic of agriculture in EU countries is a low consumption of final energy (0.7476) and a low emission of greenhouse gases (0.7155).

Considering the shape of energy efficiency indicators, on the one hand a very low share of agriculture in renewable energy production is still observed, so this indicator has a negative impact on the overall green performance of the agricultural sector (*AEI*), and on the other hand, agriculture in the analyzed EU countries features a low level of final energy consumption, which in turn improves green performance.

Looking closer at the countries, their weaknesses and strengths regarding green performance can be identified. Portugal, scoring best out of 20 member states of the EU, predominantly owes its success to high scores (0.8000–1.0000) earned for 5 indicators, including: the share of agriculture in total ammonia emissions, biodiversity of agricultural areas, relatively low consumption of final energy in agriculture and low content of organic carbon in arable land. On the other hand, indicators of wind and water erosion and activities to increase the production of energy from renewable sources need to be improved. In turn, Belgium was the country with the worst results (0.0000–0.1000) in seven areas, i.e., water and wind erosion risk, irrigation, and irrigable areas, farmland bird index, total final energy consumption in agriculture and ammonia emissions from agriculture. Moreover, the average values (0.5000) were not exceeded for nine indicators (56.3%).

The Benelux countries are characterized by a low balance of nitrogen in soil. The Netherlands and Belgium also had the highest pesticide use indicators. In contrast, in the south of Europe, i.e., in Greece, Spain, and Portugal, considerable amounts of water are used for agricultural needs, which should be certainly associated with the climate of these countries. As many as 15 out of 20 analyzed states (except Denmark, Portugal, Spain, the Netherlands and Greece) have a low share of areas subject to irrigation processes. Considering the structure of use of arable land, Hungary, the Netherlands, Luxembourg, the United Kingdom, France, and Poland were characterized by the lowest percentage of land for organic farming. As regards the share of agriculture in final energy consumption, countries with the worst score are the Netherlands, Belgium, and Luxembourg. In turn, Denmark, France, and Lithuania have the highest indicators of greenhouse gas and ammonia emissions.

Soil in EU member states contains about 79 billion tons of coal. The CO_2 storage capacity is sensitive to climatic conditions, so there is a high risk that soils will become the main source of greenhouse gas emissions due to global warming [55]. For this reason, the *AEI* was designed using the organic carbon content in arable land. The least coal in soil is found in Hungary, and the United Kingdom performs the worst in this respect. In contrast, the risk of erosion by wind and/or water is a problem that bothers all the analyzed countries to a greater or lesser degree. Taking into account the share of agriculture in renewable energy production the leaders are the Netherlands, Belgium, and Germany, while at the lower end of the ranking are Estonia, Sweden, and Slovenia.

When reviewing the existing literature, it is difficult to indicate studies that address the issue of a comprehensive evaluation of the green performance of agriculture, based on the construction of synthetic measures. The research conducted so far has focused mainly on the selection of one or several indicators. For instance, Szuba-Barańska [56] and Mrówczyńska-Kamińska investigated the impact of agriculture on the environment in Central and Eastern Europe. To this end, they used selected agri-environmental indicators only, including greenhouse gas emissions and general pollution of the environment (ammonia, methane, nitrous oxide, and carbon dioxide). Their studies revealed that Lithuania (alongside Romania and Latvia) was a country with the strongest negative impact of agriculture on the natural environment. The evidence is the highest level of emissions of greenhouse gases and air pollutants into the environment. The results of studies presented in this paper partly corroborate these observations as Lithuania ranks second to last in terms of green performance of agriculture. Ilić et al. [57] evaluated environmental performance of agriculture in EU countries based on the EPI for the problem area—Agriculture. They divided EU countries into three groups according to the value of EPI, taking nine problem areas into account, and two groups according to the area of agriculture. However, the index designed for agriculture consisted of two indicators only, i.e., nitrogen use efficiency and nitrogen balance. Due to the narrower range of indicators involved, this study only partly coincides with the study presented in this paper. However, the unfavorable position of Belgium and Luxembourg is confirmed. It is worth emphasizing that Belgium is a country with a high level of intensity of using technology in agriculture, which gives rise to several negative consequences in the form of adverse changes in the natural environment [58]. The agri-environmental indicators were also used by Turčeková et al. [59] for evaluating agri-environmental performance in 27 member states of the EU. The authors applied a radial output-oriented DEA model in their study, taking into account the following variables: greenhouse gas emissions (as output), arable land, labor force, fertilizers consumption, and agricultural subsidies (as inputs). It was demonstrated that Slovakia, the Czech Republic, and Poland were countries with the highest environmental performance. One of the reasons for this fact was the low level of technology involved in agriculture. The studies focused solely on environmental performance from the point of view of a relationship between greenhouse gas emissions and production outlays. However, they did not provide a full image of the level of environmental performance of agriculture. This testifies to the reasonableness of seeking more comprehensive measures of the impact of agriculture on the natural environment.

The proposed synthetic index method allowed for a comprehensive and unambiguous assessment of the level of green performance of agriculture sectors. By analyzing individual indicators in the studied countries, the authors obtained information about the strengths and weaknesses of activities in this area in individual countries, while synthetic measures (*AEI*) allowed for a comparison and general assessment of the level of this phenomenon in the EU countries.

The conducted analysis has indicated potential directions for further research. Due to the large information gap, only 20 EU countries were included. Individual countries show shortcomings in the reporting of individual agri-environmental indicators. Therefore, an important area of research should be the development of an effective information gathering system and the development of the existing set of indicators for an even more comprehensive assessment of green performance of agriculture. Moreover, it would be worth extending the existing analysis to include countries from other regions of the world (e.g., USA, China, Japan, Australia, South American countries).

4. Conclusions

New barriers, primarily deemed equivalent to exhaustion of natural resources, necessitated the verification of economic growth paradigms towards sustainable growth. This concept is essential for the agricultural sector that on the one hand relies on natural resources in the production process, and on the other hand has a significant impact on the environment. In the agricultural sector the production effects are associated with the emergence of various environmental hazards. Environmental protection is currently one of the priorities in EU policy and at the same time one of the major challenges for agriculture. The environmental effects of agricultural activity have been made a part of economic and agricultural research. An area that needs further investigation and deeper analysis is green performance of agriculture.

The evaluation of green performance of the agricultural sector is a complex issue, which makes it a difficult subject for analysis. As a multi-criterion concept, it requires aggregate measures based on integration of various impacts of agriculture on the environment. The added value of the research carried out lies in the development of a comprehensive method of evaluation the green performance of agriculture sector by constructing a synthetic index (*AEI*). The information value of this index will allow for better integration and improvement of activities around monitoring, planning, and implementation of agricultural policy assumptions. At the same time, the analysis of synthetic measures of agri-environmental indicators made it possible to identify the strengths and weaknesses of green performance of agriculture both at the European and national levels.

When answering the research questions posed in the paper, it should be stated that the proposed method allows for an unambiguous assessment of the studied phenomenon, enabling a comparative analysis between individual countries. In response to the second question, it can be stated that the general level of green performance of agriculture is low, as demonstrated by the average *AEI* value for all analyzed countries (0.3069). On the other hand, the analysis of the average standardized values of agri-environmental indicators shows that many aspects of agricultural activity require decisive intervention both at the national and international level. In the upcoming years, one of the main challenges should be increasing the share of agriculture in the production of energy from renewable sources as well as increasing the share of organic agriculture in the crop structure. Furthermore, dynamic action requires a high indicator of agriculture sector in ammonia (NH_3) emissions.

The limitations at the research stage, including the absence of data for certain agri-environmental indicators and/or countries, point to lines of further actions and research. In the first place, it should be considered how the existing system for collecting data from the member states could be improved. This will allow comprehensive analyses of *AEI*. In addition, dynamic changes in the conditions of functioning of the agricultural sector will certainly necessitate developing the existing set of indicators. Further lines of research should take into account a wider range of variables to allow more accurate assessment of the effect of agriculture on the natural environment, especially in the context of the implemented energy policy.

Supplementary Materials: The following are available online at https://www.mdpi.com/1996-1073/14/1/45/s1, Annex S1: Standardization of the agri-environmental indicators and the *AEI* calculation.

Author Contributions: Conceptualization, A.K. and A.N.; methodology, A.K. and A.N.; formal analysis, A.K. and A.N.; resources, A.N. and A.K.; data curation, A.K.; writing—original draft preparation, A.N. and A.K.; writing—review and editing, A.N. and A.K.; visualization, A.K.; funding acquisition, A.N. and A.K. All authors have read and agreed to the published version of the manuscript.

Funding: The APC was funded by the University of Life Sciences, Faculty of Agrobioengineering, grant number RKA/S/8/2020 (EFI) and the vouchers.

Conflicts of Interest: The author declares no conflict of interest.

References

1. Piorr, H.P. Environmental policy, agri-environmental indicators and landscape indicators. *Agric. Ecosyst. Environ.* **2003**, *98*, 17–33. [CrossRef]
2. Pawłowska-Tyszko, J.; Pawłowski, M.; Konat, G. *Environmental Protection Instruments in Polish Agriculture in the Context of Its Sustainable Development*; Institute of Agricultural and Food Economics National Research Institute: Warsaw, Poland, 2018; pp. 36–39.
3. The Organisation for Economic Co-operation and Development (OECD). *A Green Growth Strategy for Food and Agriculture*; Preliminary Report; OECD Publishing: Paris, France, 2011.
4. United Nations Economic Commission for Europe (UNECE). Agri-Environmental Indicators. In Proceedings of the Committee on Environmental Policy Conference of European Statisticians, Geneva, Switzerland, 30 October–1 November 2012.
5. Nigel, J. Concepts of efficiency in ecological economics: Sisyphus and the decision maker. *Ecol. Econ.* **2006**, *56*, 359–372.
6. Tanentzap, A.J.; Lamb, A.; Walker, S.; Farmer, A. Resolving conflicts between agriculture and the natural environment. *PLoS Biol.* **2015**, *13*, e1002242. [CrossRef] [PubMed]
7. Klepacki, B.; Gołasa, P.; Wysokiński, M. Efficiency of greenhouse gas emissions in European Union agriculture. *Village Agric.* **2016**, *3*, 129–144.
8. Femia, A.; Hass, J.; Lumicisi, A.; Romeiro, A. *Agri-Environmental Statistics and Indicators: A Literature Review and Key Agri/Environmental Indicators*; Global Strategy Technical Report, No. 27; Food and Agriculture Organization of the United Nations: Rome, Italy, 2017.
9. Kremen, C.; Miles, A. Ecosystem services in biologically diversified versus conventional farming systems: Benefits, externalities, and trade-offs. *Ecol. Soc.* **2012**, *17*, 40. [CrossRef]
10. Pe'er, G.; Bonn, A.; Bruelheide, H.; Dieker, P.; Eisenhauer, N.; Feindt, P.H.; Hagedorn, G.; Hansjürgens, B.; Herzon, I.; Lomba, L.; et al. Action needed for the EU Common Agricultural Policy to address sustainability challenges. *People Nat.* **2020**, *2*, 305–316. [CrossRef]
11. Zhang, T.Q.; Zheng, Z.M.; Lal, R.; Lin, Z.Q.; Sharpley, A.N.; Shober, A.L.; Smith, D.; Tan, C.S.; Van Cappellen, P. Environmental indicator principium with case references to agricultural soil, water, and air quality and model-derived indicators. *J. Environ. Qual.* **2018**, *47*, 191–202. [CrossRef] [PubMed]
12. Ahtiainen, H.; Pouta, E.; Liski, E.; Myyrä, S.; Assmuth, A. Importance of economic, social, and environmental objectives of agriculture for stakeholders: A Meta-Analysis. *Agroecol. Sustain. Food* **2015**, *39*, 1047–1068. [CrossRef]
13. Musvoto, C.; Nortje, K.; De Wet, B.; Mahumani, B.K.; Nahman, A. Imperatives for an agricultural green economy in South Africa. *S. Afr. J. Sci.* **2015**, *111*, 1–8. [CrossRef]
14. Mutingi, M.; Mapfaira, H.; Monageng, R. Developing performance management systems for the green supply chain. *J. Remanuf.* **2014**, *4*, 6. [CrossRef]
15. Li, J.; Lin, B. Green Economy Performance and Green Productivity Growth in China's Cities: Measures and Policy Implication. *Sustainability* **2016**, *8*, 947. [CrossRef]
16. Gallego-Álvarez, I.; Vicente-Galindo, M.P.; Galindo-Villardón, M.P.; Rodríguez-Rosa, M. Environmental Performance in Countries Worldwide: Determinant Factors and Multivariate Analysis. *Sustainability* **2014**, *6*, 7807–7832. [CrossRef]
17. Duit, A. *Understanding Environmental Performance of States: An Institution Centered Approach and Some Difficulties*; Working Paper Serie 7; Göteborg University: Göteborg, Sweden, 2005.
18. Beltrán-Estevea, M.; Picazo-Tadeob, A.J. Assessing environmental performance in the European Union: Ecoinnovation versus catching-up. *Energy Policy* **2017**, *104*, 240–252. [CrossRef]
19. Emerson, J.W.; Hsu, A.; Levy, M.A.; de Sherbinin, A.; Mara, V.; Esty, D.C.; Jaiteh, M. *Environmental Performance Index and Pilot Trend Environmental Performance Index*; Yale Center for Environmental Law and Policy: New Haven, CT, USA, 2012.
20. Schultze, W.; Trommer, R. The concept of environmental performance and its measurement in empirical studies. *J. Manag. Control* **2011**, *22*, 375–412. [CrossRef]
21. Riley, J. The indicator explosion: Local need and international challenges. *Agric. Ecosyst. Environ.* **2001**, *87*, 119–120. [CrossRef]
22. Bockstaller, C.; Guichard, L.; Makowski, D.; Aveline, A.; Girardin, P.; Plantureux, S. Agri-environmental indicators to assess cropping and farming systems. A review. *Agron. Sustain. Dev.* **2008**, *28*, 139–149. [CrossRef]
23. Payraudeau, S.; Van der Werf, H.M.G. Environmental impact assessment for a farming region: A review of methods. *Agric. Ecosyst. Environ.* **2005**, *107*, 1–19. [CrossRef]
24. Delbaere, B.; Serradilla, A.N.E. *Environmental Risks from Agriculture in Europe: Locating Environmental Risk Zones in Europe Using Agri-Environmental Indicators*; EC NC-European Centre for Nature Conservation: Tilburg, The Netherlands, 2004.

25. Rigby, D.; Woodhouse, P.; Young, T.; Burton, M. Constructing a farm level indicator of sustainable agriculture agriculture practice. *Ecol. Econ.* **2001**, *39*, 463–478. [CrossRef]
26. Saunders, C.M.; Kaye-Blake, W.; Campbell, R.; Benge, J. Agricultural Environmental Indicators: Using OECD Agri-Environmental Indicators to Assess New Zealand Kiwifruit Orchards. In Proceedings of the 15th New Zealand Agricultural and Resource Economics Society Conference, Nelson, New Zealand, 27–28 August 2009.
27. Van der Werf, H.G.M.; Petit, J. Evaluation of environmental impact of agroculture at the farm level: A comparison and analysis of 12 indicator-based methods. *Agric. Ecosyst. Environ.* **2002**, *93*, 131–145. [CrossRef]
28. Halberg, N.; Verschuur, G.; Goodlass, G. Farm level environmental indicators; are they useful? An overview of green accounting systems for European farms. *Agric. Ecosyst. Environ.* **2005**, *105*, 195–212. [CrossRef]
29. Meyer-Aurich, A. Economic and environmental analysis of sustainable farming practices—A Bavarian case study. *Agric. Syst.* **2005**, *86*, 190–206. [CrossRef]
30. Clark, M.; Tilman, D. Comparative analysis of environmental impacts of agricultural production systems, agricultural input efficiency, and food choice. *Environ. Res. Lett.* **2017**, *12*, 064016. [CrossRef]
31. The Organisation for Economic Co-operation and Development (OECD). *Environmental Performance of Agriculture in OECD Countries Since 1990*; OECD Publications: Paris, France, 2008.
32. European Commission. *Development of Agri-Environmental Indicators for Monitoring the Integration of Environmental Concerns into the Common Agricultural Policy*; Commission Communication COM; European Commission: Brussels, Belgium, 2006.
33. European Environment Agency (EEA). *Agriculture and Environment in EU-15—The IRENA Indicator Report*; European Environment Agency Report, No. 6; Office for Official Publications of the European: Copenhagen, Denmark, 2005.
34. Reus, J.; Leenderste, P.; Bockstaller, C.; Fomsgaard, I.; Gutsche, V.; Lewis, K.; Nilsson, C.; Pussemier, L.; Trevisan, M.; van der Werf, H.; et al. Comparing and evaluating eight pesticide environmental risk indicators developed in Europe and recommendations for future use. *Agric. Ecosyst. Environ.* **2002**, *90*, 177–187. [CrossRef]
35. Buczko, U.; Kuchenbuch, R.O. Environmental indicators to assess the risk of diffuse nitrogen losses from agriculture. *Environ. Manag.* **2010**, *45*, 1201–1222. [CrossRef] [PubMed]
36. Bockstaller, C.; Lassere-Joulin, F.; Slezack-Deschaumes, S.; Piutti, S.; Villerd, J.; Amiaud, B.; Plantureux, S. Assessing biodiversity in arable farmland by means of indicators: An overview. *OCL* **2011**, *18*, 137–144. [CrossRef]
37. Czyżewski, B.; Matuszczak, A.; Muntean, A. Influence of agricultural policy on the environmental sustainability of European farming. *J. Environ. Prot. Ecol.* **2018**, *19*, 426–434.
38. Strahl, D. *Methods for Evaluation of Regional Development (Metody Oceny Rozwoju Regionalnego)*; AE Publishing House: Wrocław, Poland, 2006.
39. Grzebyk, M.; Stec, M. Sustainable Development in EU Countries: Concept and Rating of Levels of Development. *Sustain. Dev.* **2015**, *23*, 110–123. [CrossRef]
40. Kasztelan, A. How Circular Are the European Economies? A Taxonomic Analysis Based on the INEC (Index of National Economies' Circularity). *Sustainability* **2020**, *12*, 7613. [CrossRef]
41. Hellwig, Z. The Application of the Taxonomic Method to the Typological Division of Countries by Level of their Development as Well as Resources and Structure of Qualified Staff. (Zastosowanie metody taksonomicznej do typologicznego podziału krajów ze względu na poziom ich rozwoju oraz zasoby i strukturę wykwalifikowanych kadr). *Przegląd Stat.* **1968**, *4*, 307–326.
42. Łogwiniuk, K. The Use of Taxonomic Methods in the Comparative Analysis of the Access to the ICT Infrastructure by Schoolchildren in Poland. *Econ. Manag.* **2011**, *3*, 7–23.
43. The Organisation for Economic Co-operation and Development (OECD). *Handbook on Constructing Composite Indicators*; Methodology and User Guide; OECD Publishing: Paris, France, 2008.
44. Piotrowska, E.; Roszkowska, E. Analysis of the Diversity of Polish Voivodeships in Terms of the Level of Innovation (Analiza zróżnicowania województw Polski pod względem poziomu innowacyjności). *Optim. Studia Ekon.* **2011**, *2*, 65–85.
45. Godlewska, J.; Sidorczuk-Pietraszko, E. Taxonomic Assessment of Transition to the Green Economy in Polish Regions. *Sustainability* **2019**, *11*, 5098. [CrossRef]
46. Eurostat. Agri-Environmental Indicator—Energy Use. Available online: https://ec.europa.eu/eurostat/statistics-explained/index.php/Agri-environmental_indicator_-_energy_use (accessed on 10 April 2020).
47. Harchaoui, S.; Chatzimpiros, P. Can Agriculture Balance Its Energy Consumption and Continue to Produce Food? A Framework for Assessing Energy Neutrality Applied to French Agriculture. *Sustainability* **2018**, *10*, 4624. [CrossRef]
48. Gołaszewski, J. (Ed.) *Efektywność Energetyczna w Rolnictwie Europejskim—Studium Przypadków Energy Efficiency in European Agriculture—Case Studies*; UWM Publishing House: Olsztyn, Poland, 2013.
49. Sulewski, P.; Majewski, E.; Wąs, A. The importance of agriculture in the renewable energy production in Poland and the EU. *Probl. Agric. Econ.* **2017**, *1*, 50–74.
50. OECD. Available online: https://stats.oecd.org/ (accessed on 18 October 2020).
51. Eurostat. Available online: https://ec.europa.eu/eurostat/data/database (accessed on 18 April 2020).
52. Kasztelan, A.; Nowak, A.; Hawlena, J. Green Growth in Agriculture in the European Union: Myth or Reality. *Eur. Res. Stud. J.* **2019**, *22*, 35–48. [CrossRef]
53. Kijek, A. Sector Risk of the Processing Industry. In *Modeling and Assessment (Ryzyko Sektorowe Przemysłu Przetwórczego: Modelowanie i Ocena)*; UMCS Publishing House: Lublin, Poland, 2013.

54. Jarocka, M. The Choice of a Formula of the Data Normalization in the Comparative Analysis of Multivariate Objects (Wybór formuły normalizacyjnej w analizie porównawczej obiektów wielocechowych). *Econ. Manag.* **2015**, *1*, 113–126.
55. Jones, A. *The European Environment State and Outlook 2010*; Publications Office of the European Union: Luxembourg, 2010.
56. Szuba-Barańska, E.; Mrówczyńska-Kamińska, A. An attempt to assess the impact of agriculture on the environment in the countries of Central-Eastern Europe. *J. Agribus. Rural Dev.* **2016**, *3*, 401–412. [CrossRef]
57. Ilić, I.; Krstić, B.; Jovanović, S. Environmental performances of agriculture in the European Union countries. *Econ. Agric.* **2017**, *64*, 41–55.
58. Nowak, A.; Krukowski, A.; Różańska-Boczula, M. Assessment of sustainability in agriculture of the European Union countries. *Agronomy* **2019**, *9*, 890. [CrossRef]
59. Turčeková, N.; Svetlanská, T.; Kollár, B.; Záhorský, T. Agri-Environmental Performance of EU Member states. *Agris On-Line Pap. Econ. Inform.* **2015**, *7*, 199–208. [CrossRef]

energies

Article

The Importance of Local Investments Co-Financed by the European Union in the Field of Renewable Energy Sources in Rural Areas of Poland

Aldona Standar [1,*], Agnieszka Kozera [2] and Łukasz Satoła [3]

[1] Department of Economics and Economic Policy in Agribusiness, Poznań University of Life Sciences, Wojska Polskiego 28, 60-637 Poznań, Poland
[2] Department of Finance and Accounting, Poznań University of Life Sciences, Wojska Polskiego 28, 60-637 Poznań, Poland; agnieszka.kozera@up.poznan.pl
[3] Department of Management and Economics of Enterprises, University of Agriculture in Krakow, al. Mickiewicza 21, 31-120 Kraków, Poland; lukasz.satola@urk.edu.pl
* Correspondence: standar@up.poznan.pl

Citation: Standar, A.; Kozera, A.; Satoła, Ł. The Importance of Local Investments Co-Financed by the European Union in the Field of Renewable Energy Sources in Rural Areas of Poland. *Energies* **2021**, *14*, 450. https://doi.org/10.3390/en14020450

Received: 22 December 2020
Accepted: 12 January 2021
Published: 15 January 2021

Abstract: Local investments for the development of renewable energy sources (RESs) constitute an important element of sustainable rural development. They are conducive to the social and economic development of the said areas, and improve the environmental values and living conditions of their inhabitants. However, such advancement in rural areas is not possible without adequate financial support, including the funds from the EU budget. Therefore, the main objective of the research is to assess the scale, scope and importance of local investments in renewable energy sources in rural areas of Poland in 2014–2020, cofinanced from EU funds. The study covered 1117 projects, whose beneficiaries were rural and urban–rural municipalities. Evaluation of the municipal investment activities in acquiring EU subsidies in the area of environmentally friendly energy was conducted using selected methods of descriptive statistics and the analysis of variance. Subsequently, with the use of logistic regression, the study identified the main socioeconomic, financial and environmental conditions of the investment activities of the local government entities in RES in rural areas. Empirical studies allowed for the positive verification of the research hypothesis, which assumed that "The highest investment activity in the field of local projects co-financed from EU funds, related to the development of RES in rural areas, may be attributed to municipalities performing primarily agricultural functions, located in Eastern Poland". The municipalities' own income potential and investment activity are of major importance for the acquisition of EU funds used in RES financing. Municipalities at a lower development level demonstrated a greater activity in accessing these funds. They view the development of RES as an opportunity for accelerated growth.

Keywords: local investments; renewable energy sources; municipal economy; EU funds; rural areas; Poland

1. Introduction

In recent years, particularly after Poland's accession to the European Union (EU), a considerable amount of attention has been given to perspectives and threats of the rural development, especially in the context of their sustainable development [1]. In Poland, rural areas cover more than 90% of the country's area [2] and are characterized by a significant degree of developmental diversity, particularly at the municipal level [3–6]. The widely discussed issues of sustainable rural development cover many problems, such as activities aimed at rationalizing the management of natural resources [7,8], local development planning and protection of the environmental values of these areas [9]. The use of renewable energy sources (RESs) may improve the quality of the natural environment in rural areas [10]. According to the Energy Law Act, RES are "the sources which, in the course of energy processing, use wind power, solar power, geothermal energy, sea waves, sea currents and tidal energy, or energy acquired from the fall of rivers, and biomass energy, energy from landfill biogas and biogas produced in the process of sewage disposal and treatment, or decomposition of plant and animal remains" [11]. Renewable sources of energy constitute an alternative to fossil fuels and contribute to the reduction of greenhouse gas emissions [12], the diversification of energy supplies and decrease of dependence on uncertain and unstable fossil fuel markets, particularly oil and gas [13]. However, it is worth noting that the development of RES has to be coordinated with other actions that improve the energy efficiency of the economy. Without taking such actions, a phenomenon known in economic literature as the Jevons paradox may occur. In his research, Jevons has shown that technical change related to increased efficiency of use of a given resource usually leads to increased resource consumption [14–16]. The economic development level of a country is also a vital aspect. In the early stage of development, economic growth processes are accelerated at the expense of high resource consumption and, consequently, environmental pollution. After a certain economic development level is exceeded, the so-called turning point occurs and the environmental protection expenditures start to grow. This dependence in relation to wealth and income inequality was spotted by Kuznets. Next, it was adapted to environmental and natural resource economics [17,18]. The use of renewable energy resources without taking any other actions aimed at social and economic development may turn out to be another unsustainable strategy [19].

The policy of the European Union with regard to RES is unambiguous—European economy is to strive to reduce carbon dioxide emissions, which is evidently related to the development of RES [20,21]. Efficient use of energy, and human, economic and natural resources, constitute a fundamental principle of sustainable energy development (SED) [22]. The Green Deal initiative, too, includes measures taken to ensure a better resource efficiency through the transition to a clean circular economy and the reduction of pollution levels. These goals are supposed to be attained through measures, which include investments in state-of-the-art environmentally friendly technologies. Hence, this paper contributes to research on the allocation of EU funds to RES. Already in early 2021, the Commission will adopt a new, more ambitious EU strategy for accommodating climate change in order to intensify its security measures against climate change. Fighting against climate change and attaining climate neutrality will require considerable investments and collective measures. Hence, it is important to investigate on financial resources allocated to these goals [23].

In Poland, rural areas exhibit significant potential for RES development. This is indicated by considerable biomass resources, roof surfaces, watercourses and areas unsuitable for agricultural activity. In rural areas, the investments in RES may have an important economic and social function [10]. RESs exhibit potential for the improvement of living conditions, environmental quality and conducting various types of activities in peripheral areas (e.g., in rural areas of Eastern Poland) and constitute an additional source of income of the population [24]. However, in order to develop RES in rural areas, it is necessary to acquire appropriate financial support. An important role in the construction of green energy installations is given to local investments of basic territorial units (local self-government units) in Poland, i.e., municipalities that use means from various EU funds for this purpose [25].

The main objective of the research is to assess the scale, scope and importance of local investments in renewable energy sources in rural areas of Poland in 2014–2020, cofinanced from EU funds. The empirical studies aimed at verification of the research hypothesis, which assumed that "The highest investment activity in the field of local projects co-financed from EU funds, related to the development of RES in rural areas, may be attributed to municipalities performing primarily agricultural functions, located in Eastern Poland".

2. Literature Review

2.1. Financing Investments in Energy from RES—Strategic Assumptions of the European Union and Poland

The energy sector constitutes one of the most relevant economic sectors in every country [26]. Its key importance results from the role that energy has in the functioning of modern society and economies. The ability to produce energy in sufficient volume to satisfy the current and future demand of domestic recipients becomes a fundamental dimension of energy security. Energy security is currently considered one of the main areas of the state, and the national economy security, which to a certain extent proves the country's independence [27].

Local self-government units conduct a number of tasks aimed at improving the living standard of their inhabitants, and ensuring the dynamics of local development. Within the responsibilities of municipal management, apart from the basic tasks (related to, e.g., the supply of water and sewage disposal) recently, there is also increasing emphasis on the supply of electricity, heat and gas [28,29]. For many years, the organizational and legal solutions adopted in Poland have consolidated the significant role of large concerns from the energy and fuel sectors in providing energy carriers (electricity and natural gas).

The currently arising development challenges, and the necessity to undertake actions involving the requirements of the environmental protection and sustainable development result, among other things, in a significantly increased interest in unconventional, renewable energy sources. Due to a much greater dispersion of RES compared to non-renewable sources, primarily from fossil fuels, the problem of their decentralized production is becoming a topical issue [30]. Combined with the decentralized energy demand represented by households and business entities, the role of local self-government units is substantially rising and will continue to do so in the future [31]. They are the ones who, having the best recognition of local conditions and needs, are the most inclined to become an important part of the energy market. According to Charles Tiebout's theory (decentralization hypothesis), local goods and services should be delivered in a decentralized manner, at the lowest level of territorial division, which ensures the economic efficiency of this process [32].

The presented conditions prove that the relevance of municipalities (the basic entities of the local self-government sector in Poland) as entities participating in the energy distribution is increasing. Moreover, their role may be considered in a twofold manner. Firstly, as entities capable of participating in the generation and distribution of energy obtained from renewable sources by themselves or through their subsidiaries, e.g., municipal enterprises. In the latter case, municipalities may participate, as a consultative body, in the efforts of their inhabitants or entrepreneurs in acquiring funds for the implementation of investments related to the use of renewable energy sources.

The processes of dynamic economic development, occurring in most countries in the world, results in a significant increase in energy demand. This increase, and the consequent rise in the consumption of energy substrates, is also associated with a rising number of population and improved standard of living [33,34]. The model of economic growth that was present for many years, resulted in overexploitation of natural resources and environmental pollution. It was noted that, due to the depletion of resources, it is not possible to proceed with this model. Currently, many countries, especially in the EU, are taking actions aiming at ensuring environmental protection and more sustainable development, which refer to the environmental Kuznets curve [35,36]. The increase in economic energy consumption may result in the insufficiency of conventional energy sources in meeting global energy demand [37]. However, it is worth noting that the exploitation of renewable energy sources alone, without taking any other actions aimed at improving energy efficiency will be insufficient. As, according to Jevons paradox, the emergence of new energy sources leads to increased exploitation of these resources. Therefore, it is necessary to take measures to reduce energy consumption and increase energy efficiency in the economies of many countries around the world [38,39].

As, according to Jevons paradox, the emergence of new energy sources leads to increased exploitation of these resources. Therefore, it is necessary to take measures to reduce energy consumption and increase energy efficiency in the economies of many countries around the world [38,39]. In Poland, most of the energy is produced as a result of the coal combustion, which, apart from the natural issues related to the depletion of coal resources, generates significant problems in terms of air pollution and environmental protection. There is also the problem of the country fulfilling its commitments, related to the agreements with the European Union, which Poland ratified. The absence or insufficient reduction of pollutants generated in the production of electricity may result in the imposition of high penalties on entities that do not meet the environmental requirements. An excessive greenhouse gas emission constitutes a continuous problem for the Polish energy sector [40].

One of the methods that reduce the emission of harmful substances into the atmosphere is to implement the idea of a low emission economy (cf. [41]). It involves all actions aimed at reducing greenhouse gas emissions while respecting the principles of sustainable development that consider competitiveness and innovativeness in the global market [42]. Documents of the European Commission state that the transition to a competitive, low-carbon economy would indicate that by 2050, the European Union should reduce emissions by 80% compared to their 1990 levels [21]. The implementation of the low-carbon economy is thus one of the key problems that the economies of the EU Member States encounter. The transition of the EU economy, particularly the energy supply sector, to one that would fulfill the low-carbon economy postulates is an important and a topical challenge. In order to address this challenge, high costs would have to be borne by both the private and public sectors [43]. Its importance is further increased in conditions of rising air pollution and consequent climate changes [44].

The process of energy production from conventional sources results in a high amount of pollution being introduced into the water and atmosphere. This is one of the reasons why many countries around the world and within the EU are undertaking steps to minimize pollution and promote alternative solutions for unconventional energy sources. Due to the importance of climate change and the depletion of conventional sources, such actions are supported institutionally. Support programs for various undertakings fulfilling the postulates of a low-carbon economy are implemented at the level of the European Union, individual Member States and local self-government units. The basic EU document that determines the direction of necessary actions related to changes in energy production is the climate and energy package. Until 2030, the Climate and Energy Policy includes common, EU assumptions and objectives for 2021–2030. In the 2030 perspective, the most important ones include [45,46]:

- Reducing a minimum of 40% of greenhouse gas emissions (compared to 1990 levels),
- Increasing the share of energy from renewable sources in the total energy consumption to a minimum of 32%,
- Increasing energy efficiency by a minimum of 32.5%.

The basic assumptions of the Climate and Energy Policy were adopted by the European Council in 2014, while the RES and energy efficiency objectives were increased in 2018. The revision of the objectives, including the increase of RES requirements, clearly indicates that, to a large extent, the EU associated their development with an opportunity to reduce the emission of harmful substances into the atmosphere and the negative impact on climate change.

2.2. Opportunities and Threats for the Development of RES Energy Related to the Use of Resources in Rural Areas

In Poland, the production of energy from renewable sources is based primarily on biofuels, which account for its 80%. Wind power also has a significant share (12%), while much less energy is generated from biogas (2.9%), water (2%), and solar energy (0.7%).

Insignificant amounts of energy are produced from geothermal sources, heat pumps and municipal waste incineration [47,48].

Decisions made at the EU level result in an increased interest in energy from renewable sources, i.e., energy generated from water, biomass, wind, solar radiation and geothermal sources. The advantage of renewable energy sources over conventional ones consists in their availability in unlimited quantities, the fact that they do not, or emit a small amount of greenhouse gases, and their higher level of affordability. The rising awareness of the population in the field of environmental protection, and the sources and scale of its pollution from the process of energy production in a conventional way also lead to social pressure on public authorities. As a result of this pressure, among other things, some public institutions are dynamizing activities increasing the share of energy produced from renewable sources in the national energy demand. At the national level, it is crucial to ensure energy security, understood as the ability to deliver a sufficient amount of energy at a price that consumers can pay while respecting the rules of environmental protection [26,49,50]. Recently, the concept of energy security has also been emphasized in dimensions such as sustainability and energy efficiency [51]. In this context, an important yet ambiguous role of RES is also observed [52].

The production of energy from renewable sources is more decentralized than in the case of conventional ones, where frequently the production sites are located in the vicinity of energy sources extraction sites (e.g., lignite mines). The dispersion of renewable energy production sites, which indicates a decentralization of the whole system, is conducive to increasing the role of local government units, particularly municipalities, within the said system.

In Poland, the basic territorial self-government units, i.e., municipalities, may constitute an entrepreneurial entity provided that they conduct a communal economy in the public utility sphere. The municipalities' own tasks in the field of, among others, water supply system and provision, sewage system, disposal and treatment of municipal sewage, and electricity, heat and gas supply, may be conducted by the local, self-government budget companies [53]. Therefore, it can be concluded that municipalities may be active in the field of renewable energy production. However, it does not exhaust the whole range of possibilities for municipal activities in the scope of RES development. They may also conduct informative and promoting activities on their territories, to advocate the use of RES by residents and business entities.

Rural areas, which account for more than 90% of Poland's territory [2], exhibit great potential for RES development [54–56]. They may become a significant supplier of renewable energy, obtained from solar, wind, water and biomass processing [57]. The RES sector development offers a great opportunity for agricultural diversification and multifunctional rural development as well [58,59]. Another significant problem consists of the outdated national electricity grid. As a result, there are considerable difficulties with connecting power generators to the main power supply point, and the distribution of electricity. For this reason, the emergence of dispersed and small energy producers (the development of prosumer energy) constitutes one of the most appropriate solutions for the RES sector development in rural areas of Poland [7].

In Poland, there are two main areas for the implementation of investments in RES by the local self-government units. The first one is the south-eastern part of the country, specifically Małopolskie and Lubelskie voivodeships, where over 25% of the investments are located (and the use of solar energy prevails). The second area is western and central Poland (Łódzkie, Wielkopolskie and Pomorskie voivodeships), where 30% of all RES projects were implemented, primarily wind power plants [60].

Rural areas are currently becoming an important part of the debate on the long-term development of renewable energy. This indicates a new, intelligent specialization of rural areas as a provider of green energy, whose actions thus positively contribute to the prevention of climate change [61]. Since renewable energy sources constitute a part of the European Union's energy and climate policy, they are supported by structural funds. In the previous programming period (2007–2013), EU funds allowed for acquiring funding for numerous investments in RES. The total value of projects cofinanced from these funds amounted to 2 billion euro [62]. In Poland, local self-government units are included in the primary beneficiaries of EU cohesion policy, also in terms of investments in RES (cf. [63–65]).

3. Materials and Methods

Member States are obliged to spend EU funds transparently, which involves publishing the appropriate data. In Poland, this responsibility applies to the Ministry of Investment and Economic Development [66]. From the base of nearly 75 thousand projects implemented within the 2014–2020 EU Perspective, those fulfilling the 04 objective of supporting the transition to a low-carbon economy in all sectors were identified for the study purposes. Projects involving RES in the field of electricity, wind, solar, biomass and other types of energy were subsequently identified. Three of the most expensive projects were excluded in order to eliminate the impact of outliers on the results of empirical research. Ultimately, the study covered 1117 projects, whose beneficiaries were rural and urban–rural municipalities. Therefore, 1548 rural and 628 urban–rural municipalities (as of 31 December 2018) constituted the subject of the research. The remaining empirical data on the investment activity and the socioeconomic position of the examined territorial units were obtained from the Local Data Bank of Central Statistical Office of Poland [2]. The results are presented in Polish currency (the key data were converted to euro as per the weighted average exchange rate of the National Bank of Poland [67].

Research on the evaluation of the investment activity of territorial units in RES in rural areas of Poland between 2014 and 2020 consisted of three stages. The first one presents the theoretical basis for the implementation of RES investments in Poland and subsequently evaluated the investment activity of municipalities in acquiring EU subsidies in the area of environmentally friendly energy. Then, the results of the modeling of socioeconomic, financial and environmental conditions of the investment activity in RES in rural areas, cofinanced from EU funds, were presented.

To test the significance of differences between the average values of implemented RES projects, an analysis of variance was applied. Due to the lack of normality in the distribution of the analyzed variables in the distinguished groups, a non-parametric ANOVA analysis of Kruskal–Wallis rank was performed. Moreover, selected methods of descriptive statistics were used as well. Tabular and graphical methods of data presentation were applied.

The process of acquiring EU projects related to renewable energy sources by local self-government units is conditioned by many factors, including the socioeconomic, financial and environmental ones. The empirical research was also aimed at determining the relations between the probability of occurrence of EU projects related to RES in a given territorial unit and a group of independent variables constituting the socioeconomic, financial and environmental conditions, presented in Table 1. The explanatory variables were selected based on substantive grounds (i.e., the authors' knowledge of and several years of experience in that domain, and other authors' research). Statistical features were also taken into consideration. Strongly intercorrelated variables were removed from the set of explanatory variables. To do that, an analysis was carried out of correlation coefficients between the explanatory variables covered by this study.

When a dependent variable is a categorical variable assuming two values (usually the occurrence and absence of a phenomenon), suitable methods of modeling the analyzed phenomenon for a dichotomous variable should be applied (e.g., logistic regression or discriminative model). Logistic regression was used for this purpose. This method is applied when the dependent variable is dichotomous—assumes two values—0 and 1, where: value 1 indicates the possession of a given attribute while 0—the lack thereof [68]. The usefulness of such a method may be attributed to the fact that the explanatory variables (i.e., predictors) can be measured on a metric and non-metric scale. The most important advantage of modeling data using this method is that the result is a single mathematical formula. It allows determining to what extent and in which direction individual variables influence the modeled phenomenon.

The logistic function, which the logistic regression model is based on, assumes the following form [69]:

$$f(z) = \frac{e^z}{1 + e^z} \tag{1}$$

It adopts values from 0 to 1. The value of the logistic function approaches 0 when z aims at minus infinity. However, when z aims at plus infinity, the logistic function approaches 1. The surveyed rural and urban–rural municipalities in Poland were divided into two separate classes:

$$y_i = \begin{cases} 0 \text{ territorial units that did not implement EU projects related to RES} \\ 1 \text{ territorial units that implemented EU projects related to RES} \end{cases}$$

The logit regression method is most frequently used to determine the probability of occurrence of a certain Y event, provided that the events $X_1, X_2, \ldots, X_K$ transpire as well. This method is applied, among others, in modeling the probability of the examined unit being in a certain state ($Y = 1$), and enables the identification of statistically significant factors influencing this probability. Such probability is determined by the following logit regression model [70]:

$$P(y_i = 1/X) = \frac{e^{\beta_0+\beta_1 x_1+\ldots+\beta_K x_K}}{1 + e^{\beta_0+\beta_1 x_1+\ldots+\beta_K x_K}}, \tag{2}$$

where:

y_i—ith ($i = 1, \ldots, n$) observation on a dichotomous explanatory variable that assumes a value of 1 or 0;
$k = 1, 2, \ldots, K$;
$X_{i1}, \ldots, X_{iK}$—explanatory variables (socio-economic, financial and environmental characteristics);
$P(y_i = 1/X)$—the probability that the variable Y would assume a value equal to 1 for the values of the explanatory variables $X = [X_{ik}, \ldots, X_{iK}]$, $i = 1, \ldots, n$; $k = 1, \ldots, K$;
$\beta_0, \beta_1, \ldots, \beta_K$—structural parameters of the model.

The parameters of the $\beta_0, \beta_1, \ldots, \beta_K$ model are most frequently determined by the highest reliability method, maximizing the logarithm of the reliability function in relation to the model parameters through iterative mathematical procedures. In the logit regression models, the structural parameters of the β_k model are not directly interpretable, similarly to the linear model. They should not be interpreted in terms of changes in the value of X_k, only in terms of the direction of the relationship between X_k and Y. The symbol of the parameter X_k defines the direction of the influence of X_k on Y. With a positive β_k, along with the increase of X_k rises the probability that $Y = 1$, while negative values of β_k are associated with a decrease in the probability that $Y = 1$. Possibilities of interpretation are provided by the transformation of the estimated equation into the so-called quotient of probability [71]:

$$\Psi = \frac{P_i}{1 - P_i} \tag{3}$$

The quotient determines the relative probability of occurrence of a given event. In the logit regression model, which shapes the probability values, it is possible to estimate the level of probability as a function of independent variables. This quotient is simplified to the form of:

$$\psi = e^{\beta_0+\beta_1 x_1+\ldots+\beta_K X_K}. \tag{4}$$

Expression e^{β_k} constitutes a relative change in the probability of an event occurring as a result of a factor described by X_k, assuming the stability of the remaining variables included in the model. This value is interpreted by comparing it with value 1 and expressing the difference as a percentage. If [72]:

- $e^{\beta_k} > 1$, then the factor described by X_k variable is considered to stimulate the probability of occurrence of a given phenomenon, with a stable influence of the other variables included in the model,
- $e^{\beta_k} < 1$, then the factor described by the X_k variable is considered to reduce the probability of occurrence of a given phenomenon, with a stable influence of the other variables included in the model,
- $e^{\beta_k} = 1$, then the factor described by X_k variable is considered to have no effect on the probability of occurrence of a given phenomenon, with a stable influence of the other variables included in the model.

The evaluation of the logit model accuracy is conducted by means of *chi*-squared statistics and the so-called *pseudo*-R^2 measures [73]. The value of *chi*-squared statistics verifies the hypothesis that the explanatory variables introduced into the model do not introduce any significant information beyond that provided by the constant term. Verification of the statistical significance of variables is conducted based on the Wald test. In addition to the accuracy, the model predictive ability is also evaluated using the measure of predictive ability of a model determined based on the accuracy table—the so-called count (overall model accuracy) [74].

Table 1. Definition of explanatory variables adopted in the logit regression models.

Variable Designation	Variable Name
	Socioeconomic Conditions
x_{11}	Population density (persons per km^2)
x_{12}	Old-age dependency ratio
x_{13}	Natural increase per 10 thousand population
x_{14}	Percentage of councillors with higher education (%)
x_{15}	Cumulative net migration rate per 1000 persons from 2016–2018
x_{16}	Entities entered into the REGON (National Business Register) per 10 thousand population
x_{17}	Entities of the national economy employing more than 49 people per 10 thousand inhabitants
x_{18}	Natural persons conducting business activity per 10 thousand inhabitants
x_{19}	Foundations, associations and social organisations per 10 thousand inhabitants
x_{10}	Percentage of persons using the water supply system (%)
x_{11}	Percentage of persons using the sewage system (%)
x_{12}	Percentage of people using the gas network (%)
x_{13}	Number of dwellings per 1000 inhabitants
x_{14}	Average floor space per capita (in m^2)
x_{15}	Beneficiaries of social care per 10 thousand population
x_{16}	Unemployment rate (%)
x_{17}	Accommodations per 10 thousand inhabitants
x_{18}	Share of the agricultural holdings of 15 ha and more (%)
x_{19}	Number of persons employed in individual agricultural holdings per 100 population of working age
	Financial Conditions
x_{20}	Total income per capita in PLN
x_{21}	Level of personal income per capita in PLN
x_{22}	Shares in taxes constituting the state budget revenue (PIT and CIT) per capita in PLN
x_{23}	General subsidy level excluding the educational part per capita in PLN
x_{24}	Current transfers per capita in PLN
x_{25}	Share of the current income in total income (%)
x_{26}	Share of the personal income in total income (%)
x_{27}	Level of property expenditures per capita in PLN
x_{28}	European Union funds for financing EU programmes and projects in 2014–2019 per capita in PLN
x_{29}	Share of the operating surplus in total income (%)
x_{30}	Share of property expenditures in total expenditures (%)
x_{31}	Burdening current expenditures with remuneration and related expenses

Table 1. *Cont.*

Variable Designation	Variable Name
	Financial Conditions
x_{32}	Share of the operating surplus and income from the sale of assets
x_{33}	Operating surplus per capita in PLN
x_{34}	Total liabilities per capita in PLN
x_{35}	Share of total liabilities in total income (%)
x_{36}	Burdening total income with debt service expenditures (%)
x_{37}	Burdening personal income with debt service expenditures (%)
x_{38}	Share of due liabilities in total liabilities
	Environmental Conditions
x_{39}	Afforestation (%)
x_{40}	Underwater land (%)
x_{41}	Protected area in total surface area (%)
x_{42}	Agricultural land (%)
x_{43}	Built-up and urbanised areas (%)
x_{44}	Ecological land (%)
x_{45}	Wastelands (%)
x_{46}	Share of the industry in water consumption (%)
x_{47}	Water consumption per capita (in m^3)

Source: Own study based on [75–79].

4. Results of Empirical Studies

4.1. Local Investments in Environmentally Friendly Energy, Cofinanced by the European Union in Poland

Objectives of the EU's cohesion policy indicate support for investment in energy infrastructure while the adopted system of thematic targets of the said policy includes the term "low-carbon economy", referring to the reduction of carbon emissions. Poland belongs to the group of EU countries where the share of energy obtained from renewable sources constitutes less than 15% while the average for the Member States amounted to as high as 30% in 2018 [80]. In 2018, EU leaders determined an objective, according to which, by 2030, 32% of the energy consumption within the Union is to be derived from renewable sources [81]. Achieving such a formulated strategic target requires the involvement of significant funds to finance investments in this area. The primary sources of funding for energy projects are EU funds, both from the general budget and the structural funds.

Poland has favorable geographical and environmental conditions for the production of energy from renewable sources, particularly considering the potential for the production of environmentally friendly energy through agriculture (cultivation of energy crops and production of energy from biogas) and rural areas [24,82]. The largest resources of renewable energy sources occur in rural areas, yet the same areas simultaneously have the greatest problems with ensuring continuity and quality of the supplied energy [83]. Power cuts hinder agricultural activity and restrict opportunities for the development of entrepreneurship in rural areas. Difficulties in ensuring a stable energy supply of satisfactory quality result from several reasons, particularly outdated or underdeveloped infrastructure of energy distribution. Energy balancing may be supported by local renewable energy sources. Both the RES potential of rural areas and the dispersion of buildings speak in favor of such a solution. The employment of locally available energy sources, particularly those directly related to agricultural production, may improve local energy security and facilitate farmers' fulfillment of environmental protection requirements [9]. In rural areas, the support of RES development through EU funds is consistent with the concept of multifunctional development of agriculture, rural areas and, as Klepacka [22] notes, the idea of sustainable development. It constitutes a certain alternative to the dominant role of agricultural activity—the production of food raw materials. The second pillar of the Common Agricultural Policy, which funds the Rural Development Program for 2014–2020 [9], stipulates multiannual support for rural development aimed at increasing competitiveness of farms, and sustainable development of the natural resources management, including the use of energy from renewable sources.

The empirical studies aimed at assessing the scale and importance of local investments in the field of RES in rural areas of Poland, cofinanced from the EU funds. To do that, the energy projects, cofinanced by the EU budget and implemented by the basic, territorial, self-government units, i.e., municipalities in 2014–2020 were characterized. The attention was focused primarily on projects implemented by rural and urban–rural municipalities. The research revealed that 2777 projects in the field of energy have been conducted within the 2014–2020 Perspective, amounting to PLN 12.99 billion (EUR 2.92 billion), of which nearly 75.8% has been financed from the EU budget. Local self-governments constitute their most active beneficiaries. They have been implementing 1153 projects (i.e., almost 42% of all energy-related projects), amounting to PLN 4.32 billion (EUR 0.97 billion). Private entities are party to a much higher number of implemented projects. However, they are characterized by a significantly higher capital intensity [66].

Within the local self-government sector, the most active local investments in the field of RES are realized by its basic entities, i.e., municipalities. Projects implemented by municipalities constitute as much as 95.1% of the total number and over 97% of the total value of projects realized by the local self-government sector. Therefore, municipalities are the most active group of beneficiaries. In the examined period, an average municipality implemented activities in the field of RES whose value amounted to PLN 3.67 million (EUR 0.83 million). Considering their administrative type, the diversity of implemented projects in terms of an average level is high, which was confirmed by the Kruskal–Wallis test ($KW = 4.97$ with $p = 0.002$). On average, the largest activities of this type were undertaken by cities, smaller by urban–rural self-governments, while the smallest occurred in cities with poviat rights and rural municipalities. Analyzing the percentage of entities benefiting from this form of support, the highest was a characteristic of cities with poviat rights—nearly every second city used such support. In the case of other types of municipalities, the percentage of those that acquire EU funds is significantly lower. In total, rural and urban–rural municipalities are characterized by the highest number of implemented projects—67% and 20% respectively. The projects accounted respectively for 60% and 23% of the total value of undertaken activities in the field of RES (Table 2).

Table 2. Characteristics of the renewable energy source (RES) projects, cofinanced by the European Union and implemented under the 2014–2020 Financial Perspective by municipalities in Poland.

Type of Municipality	Number of Projects	Average Value of Projects (PLN Million)	Total Value of Projects (PLN Million)	Number of Municipalities Acquiring Resources from the EU Funds	Percentage of Municipalities Acquiring Resources from the EU Funds (%)
urban–rural	224	4.14	928	168	26.75
rural	745	3.32	2476	539	34.82
urban	92	5.44	500	69	29.24
cities with poviat rights	59	3.47	204	30	45.45
Total	1120	3.67	4108	806	32.53

Source: Own study based on [66].

Interestingly, it is possible to distinguish a significant group of municipalities whose success in acquiring the first EU subsidy in the field of energy has translated into further successes. The leaders in acquiring subsidies are primarily rural and urban–rural municipalities: Stanin, Ryki, Rajgród, Psary, Inowłódź, Drohiczyn and Daszyna (each of them is implementing four projects). In total, as many as 210 local self-governments (26%) implemented more than one project in the field of RES. The support obtained consists of a non-refundable grant. Half of the projects have been completed, the remaining ones are to be finished in 2021 at the latest.

All projects are implemented by municipalities within the Regional Operational Programs, i.e., programs dedicated to the needs of beneficiaries from a particular region. In the case of support obtained by the surveyed municipalities, their spatial diversity may be observed. In this respect, statistically significant differences were confirmed by the Kruskal–Wallis test ($KW = 11.04$ with $p = 0.000$). Considering the level of investments in the field of RES, by far, the highest amount of funds was spent by the municipalities from Lubelskie, Podkarpackie and Śląskie voivodeships, which in total constituted more than a half of all projects. There is also a significant yet smaller concentration of the implemented projects in RES. In this case, Lubelskie, Podlaskie and Śląskie voivodeships were the leading municipalities. The aforementioned results have translated into a large variation in values of an average project (Table 3).

Table 3. Characteristics of the RES projects, cofinanced by the European Union and implemented by urban–rural and rural municipalities under the 2014–2020 Financial Perspective in Poland, divided into Regional Operational Programs (ROP).

Programmes	Total Municipalities			Urban-Rural Municipalities			Rural Municipalities		
	AVP * (PLN Million)	VP ** (PLN Million)	NP ***	AVP * (PLN Million)	VP ** (PLN Million)	NP ***	AVP * (PLN Million)	VP ** (PLN Million)	NP ***
Dolnośląskie ROP	3.36	30.22	9	4.21	25.24	6	2.10	4.21	2
Kujawsko-Pomorskie ROP	1.07	76.19	71	1.21	12.09	10	1.10	54.89	50
Lubelskie ROP	3.82	1079.77	283	4.47	165.34	37	3.66	830.55	227
Lubuskie ROP	0.44	0.88	2	0.50	0.50	1	0.38	0.38	1
Łódzkie ROP	3.51	417.71	119	6.79	115.46	17	2.85	262.19	92
Małopolskie ROP	3.28	374.01	114	1.97	72.81	37	3.34	220.48	66
Mazowieckie ROP	8.82	273.36	31	7.71	46.25	6	8.97	215.25	24
Opolskie ROP	9.59	38.34	4	12.10	36.30	3	0.00	0.00	0
Podkarpackie ROP	7.32	629.56	86	7.38	169.83	23	7.07	360.34	51
Podlaskie ROP	1.45	205.46	142	1.34	55.11	41	1.01	82.53	82
Pomorskie ROP	7.19	122.25	17	6.83	13.65	2	7.74	100.66	13
Śląskie ROP	3.61	508.38	141	7.96	95.46	12	1.97	149.63	76
Świętokrzyskie ROP	4.61	124.48	27	4.90	53.89	11	4.41	66.22	15
Warmińsko-Mazurskie ROP	0.71	26.82	38	1.10	7.71	7	0.62	16.01	26
Wielkopolskie ROP	7.34	168.76	23	7.74	54.18	7	7.25	94.29	13
Zachodniopomorskie ROP	1.86	18.57	10	1.07	4.26	4	1.13	4.52	4
Total	3.67	4094.77	1117	4.14	928.08	224	3.32	2462.17	742

* AVP—the average value of projects, ** VP—the value of projects, *** NP—the number of projects. Source: Own study based on [66].

The concentration of RES investment activity in the area of several voivodeships occurred also in the case of projects implemented by urban–rural and rural municipalities. Particularly great interest in this type of project has been observed in municipalities of Eastern Poland, especially Lubelskie Voivodeship, whose local self-governments lead both in terms of value and number of implemented projects. Such a result was influenced by the exceptional investment activity of rural municipalities. Surprising may be the fact that, in the Opolskie Voivodeship, only three municipalities decided to conduct the surveyed investments in their areas. Additionally, low interest in such undertakings was observed in Zachodniopomorskie and Pomorskie voivodeships, which are predisposed to implement wind energy (Figure 1).

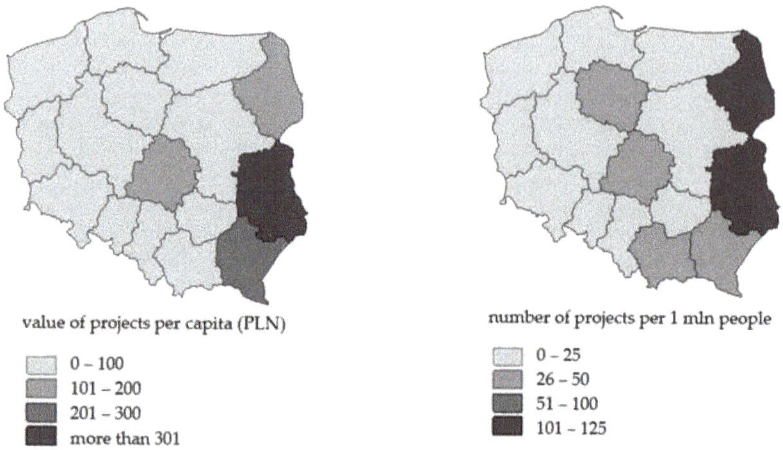

Figure 1. Projects in RES, implemented by urban–rural and rural municipalities, cofinanced by the EU under the 2014–2020 Financial Perspective in Poland by regions. Source: Own study based on [66].

The majority of Polish municipalities acquired subsidies within the 010 Renewable Energy: solar area (879 projects, i.e., 78% of their total number), absorbing more than PLN 3.40 billion (EUR 0.77 billion), which constitutes 83% of their total amount (Table 4). It is worth noting that activities in the scope of solar energy were the most capital-intensive. The second area of interest for Polish municipalities is the 083 Air Quality Measures. According to the design nomenclature, more than half of these projects involved 22 activities concerning the environment and climate change (projects involved: reducing low-stack emission by promoting and funding the replacement of residential wood heaters with new heating devices using gas or biomass), every third one involved 18 Public Administration (including, among others, public utility buildings thermal efficiency improvement), and the remaining ones involved 10 grid energy, gas fuels, steam, hot water and air for air conditioning systems (included the replacement of heat sources with those indicated in the title). Within the remaining areas, only around a few to a dozen projects were implemented. Such a structure of implemented RES projects was characteristic for urban–rural and rural municipalities, while rural self-governments also implemented individual projects within the remaining support areas, i.e., hydroelectric, geothermal, marine, biomass, wind or electrical energy. The Kruskal–Wallis test (2.97 with $p = 0.002$) confirmed the statistically significant allocation of funds within the areas of the supporting transition towards a low CO_2 emissions economy (Table 4).

Implementation of the said projects was aimed at reducing the negative impact on the environment and decreasing the energy consumption. Municipalities received subsidies primarily for measures improving the energy efficiency of facilities from the public and residential sectors, frequently by increasing the efficiency of existing installations and equipment. These projects involved the reconstruction of heating installations, ventilation and air-conditioning systems, insulation of facilities, replacement of windows, external doors and low-efficiency boilers, change of water and sewage systems and lighting systems into energy-efficient ones, installation of RES in electrically modernized facilities, installation of cooling systems, including those from RES, and installation or reconstruction of electric energy and heat-generating units in high-efficiency combined heat and power (CHP). The subsidy could be granted for an economically justified construction with the lowest possible emission of CO_2 and other air pollutants. In turn, any reconstruction of the existing installations to high-efficiency CHP was bound to result in a reduction of at least 30% of CO_2 emission (the 2014–2020 WROP was used as an example [84]).

Table 4. Characteristics of the RES projects, cofinanced by the European Union and implemented by urban–rural and rural municipalities under the 2014–2020 Financial Perspective, divided into areas of support.

Area of Project Intervention	Total Municipalities			Urban-Rural Municipalities			Rural Municipalities		
	AVP * (PLN Million)	VP ** (PLN Million)	NP ***	AVP * (PLN Million)	VP ** (PLN Million)	NP ***	AVP * (PLN Million)	VP ** (PLN Million)	NP ***
005 Electricity (storage and transmission)	1.63	1.63	2	0.00	0.00	0	0.00	0.00	0
010 Renewable Energy: Solar	3.89	3422.63	879	4.45	708.22	159	3.57	2175.31	610
011 Renewable Energy: from Biomass	1.88	20.71	11	4.22	8.43	2	1.36	12.28	9
012 Other Renewable Energy [(a)]	2.09	43.93	21	2.14	6.42	3	2.60	28.65	11
083 Air Quality Measures	2.97	605.87	204	3.42	205.01	60	2.20	245.93	112
Total	3.67	4094.77	1 117	4.14	928.08	224	3.32	2462.17	742

* AVP—the average value of projects, ** VP—the value of projects, *** NP—the number of projects. [(a)] Including hydroelectric, geothermal and marine) and integration of renewable energy (including storage, conversion of electricity to gas and infrastructure). Source: Own study based on [66].

4.2. The Shaping of Investment Activity of Rural and Urban–Rural Municipalities in the Field of Environmentally Friendly Energy Projects Cofinanced by the European Union

In rural areas, the investment activity of territorial units in environmentally friendly energy is highly diversified regionally, as shown in Figure 1. In order to identify the main determinants of investment activity of rural and urban–rural municipalities in RES, a logit model was applied. It was used to determine the direction and strength of influence of individual factors on the activity of territorial units in acquiring projects in the area of environmentally friendly energy, cofinanced from the EU budget. This method is applied, among others, in modeling the probability of the examined unit being in a certain state ($Y = 1$), and enables the identification of statistically significant factors influencing this probability. Logit regression models are used in the case of the dichotomous explained variable (0—the municipality did not acquire any project, 1—the municipality acquired at least one project in the area of RES, cofinanced by the EU funds under the 2014–2020 Financial Perspective). Three logit models were estimated, demonstrating respectively the influence of socioeconomic, financial and environmental factors on the investment activity of the surveyed territorial units in acquiring projects cofinanced by the EU in the field of environmentally friendly energy. The assessment of the constructed models was conducted based on the analysis of classification tables (assessment of predictive ability), chi-squared statistics (assessment of model significance) and the Wald test (determination of the significance of explanatory variables). The results of the estimated logit models were presented in Table 5. The estimated models were characterized by high accuracy with respect to the empirical data, and high statistical significance ($p < 0.05$) of parameters occurring with explanatory variables.

The empirical studies revealed that, among the socioeconomic factors, the municipal investment activity in acquiring RES projects cofinanced from EU funds is significantly influenced by the demographic situation, level of entrepreneurship development, number of persons employed on agricultural holdings per 100 population of working age and the social situation. Empirical studies proved that the rising value of explanatory variables, such as population density and net migration rate per 1000 persons, has increased the probability of municipalities acquiring a project cofinanced by the EU in the field of RES. High demographic potential is associated with higher income tax revenues of basic territorial units, which translates into higher potential and investment capacity (Table 5).

Table 5. Results of the estimated logit models parameters of investment activity in RES of rural and urban–rural municipalities in Poland in 2014–2020 [a].

Explanatory (Independent) Variables		Coefficient	Standard Error	Probability Ratio	p-Values	Relevance [b]
Socio-economic [c]	const	0.3130	0.4651	1.3676	0.5009	
	Population density (persons per km^2)	0.0053	0.0010	1.0053	<0.0001	***
	Cumulative net migration rate per 1000 persons from 2016–2018	0.0173	0.0041	1.0174	<0.0001	***
	Entities entered into the REGON (National Business Register) per 10 thousand population	−0.0022	0.0004	0.9978	<0.0001	***
	Share of the agricultural holdings of 15 ha and more (%)	−0.0228	0.0053	0.9774	<0.0001	***
	Number of persons employed in individual agricultural holdings per 100 population of working age	0.0177	0.0042	1.0179	<0.0001	***
	Beneficiaries of social community care per 10 thousand population	0.0006	0.0002	1.0006	0.0009	***
	Percentage of councillors with higher education (%)	0.0067	0.0041	1.0067	0.0986	*
Financial [d]	const	0.4104	0.1778	1.5074	0.021	**
	Share of the personal income in total income (%)	−0.0237	0.0047	0.9766	<0.0001	***
	European Union funds for financing EU programmes and projects in 2014–2019 per capita in PLN	0.0006	0.0002	1.0006	0.011	**
	Level of property expenditures per capita (in PLN)	0.0003	0.0001	1.0003	<0.0001	***
Environmental [e]	const	0.3423	0.12307	1.4082	0.0054	***
	Protected area in total surface area (%)	−0.0030	0.0012	0.9970	0.0102	**
	Underwater land (%)	−0.0832	0.0252	0.9202	0.0010	***
	Built-up and urbanised areas (%)	−0.0355	0.0179	0.9651	0.0469	**
	Wastelands (%)	0.0715	0.0401	1.0741	0.0749	*

[a] The models were construed on the basis of balanced samples (approximately 700 municipalities), which acquired at least 1 project in RES, cofinanced by the EU (1) and over 700 municipalities, which did not exhibit any investment activity in this area (0). [b] If p-value < 0.001 it is given three stars (***), 0.001 < p-value < 0.05—two stars (**) and 0.05 < p-value < 0.1—one star (*). [c] Collective test of model coefficients: $\chi^2 = 180.7$, $p = 0.000$ and number of cases of correct prediction = 64.1%. [d] Collective test of model coefficients: $\chi^2 = 48.1$, $p = 0.000$ and number of cases of correct prediction = 60.2%. [e] Collective test of model coefficients: $\chi^2 = 28.0$, $p = 0.000$ and number of cases of correct prediction = 52.8%. Source: Own calculations using the *Gretl* program, based on [2,66,85].

Projects in the area of environmentally friendly energy, cofinanced by EU, are more frequently implemented by municipalities with a lower level of socioeconomic development, including territorial units characterized by a lower level of entrepreneurship development and typical agricultural functions. The results of the estimated logit model indicate that the higher the number of entities registered in the REGON (National Business Register) per 10 thousand population, the lower the probability of implementation of RES projects by the municipality. On the other hand, the higher the number of persons employed on agricultural holdings per 100 population of working age, the higher the probability of implementation of the discussed projects (Table 5).

Rural areas are facing serious problems such as depopulation and limited potential to develop non-agricultural economic functions. Moreover, the local self-governments focus their attention on the social benefits resulting from investments in RES, which are expressed primarily in the ability to create additional, stable jobs for less-skilled employees and economic activation of rural areas. Municipalities with a higher number of beneficiaries of environmental social assistance per 10 thousand population, and those with a high number of persons working on agricultural holdings per 100 population of working age, were characterized by an increased probability of acquiring a project in RES, cofinanced by the EU (Table 5).

The percentage of councilors with higher education was also a statistically significant variable, demonstrating the level of human capital in territorial self-government units. Many definitions reduce human capital to the issue of education. The human capital consists of all the predispositions, knowledge and skills, together with the competencies that allow them to be applied in specific actions [86]. The research determined that a rise in the percentage of councilors with a higher level of education increases the probability of a given territorial unit to acquire a project cofinanced from the EU in the area of RES (Table 5).

The study determined that, among the financial factors, the level of municipality's own income potential, investment activity and the activity in the field of acquiring EU funds are important for their investment activity in obtaining RES-related projects and projects cofinanced from the EU funds. As indicated earlier, projects in the area of environmentally friendly energy, which are cofinanced by the EU, are more frequently implemented by municipalities with a lower level of development and typically agricultural functions. The results of the estimated logit model for financial conditions determined that the higher the own income potential (quantified by the amount of own income per capita), the lower the probability of acquiring EU funds to cofinance the projects in RES. This confirms that entities interested in the implementation of such projects consist of agricultural municipalities, for which even limited financial independence [87] is not a significant barrier for applying for EU funds in RES. At the same time, the research revealed that municipalities with higher investment activity are more inclined to implement the discussed, projects cofinanced by the EU (Table 5). Such a result indicates a significant role of the beneficiary's experience in the process of applying for an EU subsidy. In a municipality, in which the employees have already conducted such projects and learned the procedures, the language of documentation and the method of completing application forms, it is easier to achieve success in the subsequent competition [88].

The research determined that among the last, i.e., the third group of environmental factors, the investment activity of municipalities in acquiring projects related to RES and cofinanced by the EU is significantly influenced by the percentage of protected areas, the underwater land, built-up and urbanized areas (%) and wastelands (%). However, among the aforementioned environmental factors, only a higher share of wastelands in the municipality's surface area translated into a higher probability of acquiring a project cofinanced by the EU related to RES. For instance, apart from roofs, photovoltaic panels may be installed on wastelands, by placing the installations required to produce electricity in their area. On the other hand, a higher share of the underwater land does not translate into a higher probability of acquiring the discussed projects by local self-governments (Table 5). Considering that Polish area, for the most part, consists of the lowlands (with no big natural slopes), it does not create favorable conditions for the construction of large hydroelectric power plants. In Poland, the majority of hydroelectric power plants are built on rivers.

5. Political Implications

All energy-related projects, cofinanced by the EU and implemented by the surveyed municipalities involved the 04 objective of supporting the transition to a low carbon economy in all sectors (Table 6). The said objective is consistent with the direction of the EU Climate Policy of reducing greenhouse gas emissions and targets of the EU Strategy for the Baltic Sea Region [89] in the field of climate change adaptation projects. Directions of the interventions undertaken within the framework of the 04 objective will enable approaching the achievement of the determined objectives of the Europe 2020 Strategy [90], involving primary energy consumption, the share of energy from renewable sources in the gross final energy consumption and the reduction of greenhouse gas emissions. Under the national law, the implementation of these activities is included in the Energy Policy of Poland until 2030 [91], the Energy Security and Environment Strategy [92], the Transport Development Strategy until 2020 (with a perspective until 2030) [93] and program documents in this area, primarily in the National Energy Efficiency Action Plan for Poland [94] and the National Renewable Energy Action Plan [95].

Table 6. Intervention priorities under the 04 objective of supporting the transition to a low-carbon economy in all sectors.

Increasing Energy Efficiency of the Economy	Reducing Emissions Generated by Transport in Urban Agglomerations	Increasing Energy Production from Renewable Sources
increasing the energy efficiency of public facilities and multifamily residential buildings; increasing energy efficiency in enterprises; improving heating and cooling systems, as well as supporting low-carbon strategies; reducing the energy consumption by constructing intelligent, medium and/or low voltage distribution networks; increasing energy production in highly efficient installations (supporting highly efficient CHP)	developing low-emission public transport and other environmentally friendly forms of urban mobility	increasing energy production from renewable sources and the development of networks for RES; increasing the efficiency of system operation by constructing intelligent, medium and/or low voltage distribution networks; supporting the national industry providing the equipment necessary for the production of energy from renewable sources, as an industry with significant development potential in the light of the increasing share of RES in the energy mix. Supporting this type of projects will also be consistent with the 03 thematic objective (CT3), serving the development of enterprises.

Source: Own study based on [96].

As emphasized in the Partnership Agreement—Programming of the 2014–2020 Financial Perspective [90], considering the dependence of the Polish economy on coal as a primary energy source, the process of developing a low-carbon economy will be more time-consuming and costly than in the case of many other EU countries. These measures aim not only at reducing primary energy consumption and CO_2 emissions but also increasing the competitiveness of the economy.

In their new Multiannual Financial Framework for 2021–2027, the EU specified the priority projects to be cofinanced with structural funds. In 2021–2027, it is planned to allocate EUR 379 billion to goal 3 (natural resources and environment) and EUR 442 billion to goal 2 (cohesion and values). The plans provide for an allocation of EUR 114 billion more to climatic and environmental measures than in 2014–2020. Such an important change means giving priority to environmental issues, which, as a consequence, will translate into a greater use of renewable energies. The financial resources are supposed to support the implementation of the Green Deal. The environmental standards applicable in the EU are among the world's most stringent ones. Green growth is a key component in the EU policy designed to guarantee that Europe will follow an organic path of economic growth. The EU also plays a key role in promoting sustainable development at a global level [23].

The studies and findings derived from it could provide a basis for the creation of a new regional-level RES policy in Poland. Research findings confirm that considerable differences exist between the geographies in the implementation of RES investments cofinanced by the EU. Municipalities located in Eastern Poland proved to be the most efficient beneficiaries of that support despite being at lower levels of development and investment potential. So far, municipalities at higher development levels, located in territories affected by greater environmental pollution, have been passive in their quest for funds from the EU budget. They, too, should intensify the measures taken to develop RES in their territories. To do that, it is necessary to design a regionally diversified government policy for supporting RES investments implemented by LGUs.

6. Discussion

The interest in RES results, among other things, from the increase in energy demand, the ongoing climate change, the use of the surplus of agricultural raw materials and the maintenance of energy security in the EU [97]. Research carried out by, among others, Ossowska and the research team [98] indicates that in recent years, one might spot positive changes concerning the consumption of energy from renewable sources in the European Economic Community but these changes vary depending on a country. The type of renewable energy used depends on the geographical location and the economic and financial efficiency of each source. For example, Northern Europe has a more environmentally friendly energy policy. On the other hand, Fisher [99] and Ringel and Knodt [100] indicate that Central and Eastern Europe countries rely on their own energy. In countries with large coal reserves, such as Poland, it is more difficult to stop the exploitation of this energy source in favor of more environmentally friendly resources. Research carried by Ossowska and the research team [98] also shows that less developed countries are distinguished by lower greenhouse gas emissions (including Central and Eastern Europe countries), while GDP growth results in increased greenhouse gas emissions.

Beneficiaries may also obtain support for RES projects from the European Union, whose main objective consists in the transition to a low-carbon economy [101,102]. Rural areas exhibit significant potential for RES development, while municipalities, as the basic territorial self-government units in Poland, are the main creators of local development. The supply of electricity constitutes one of the basic tasks of the municipal economy. Due to the rising importance of this issue, it is becoming a cause for reflection around the world (e.g., [103,104]), while the degree of preparation of local authorities may constitute a key determinant of the effectiveness of RES projects [54].

The key factor for the development of RES seems to be an appropriate energy policy and financial support. Poland, due to the number of rural areas, has favorable conditions for RES development. However, there is a regional disparity, which is influenced by many factors. Particularly great interest in this type of project was observed among rural entities in Eastern Poland, especially in the Lubelskie Voivodeship. Significant interest in RES activities in this voivodeship was also confirmed by Gradziuk and Gradziuk [97]. As Kazak et al. [54] indicate, EU funds in the field of RES have different success rates in Poland and other Member States [105], e.g., the Baltic States [106], Romania [107] and Italy [108].

Investment projects concerning renewable energy sources are an opportunity to stimulate local development in peripheral rural areas (especially in Eastern Poland) affected by population ageing and depopulation cf. [109,110]. The renewable energy sector creates a variety of jobs in manufacturing, services and construction requiring a variety of qualifications and skills. Its development not only increases but also improves the quality of jobs in the industry. The research conducted by Wasiuta [111] shows that the development of RES is not only an opportunity for local communities to create jobs but also an opportunity for local government units to introduce various types of taxes. The existing diversity concerning the implementation of investment projects involving environmentally friendly forms of energy, cofinanced by the EU, is influenced by socioeconomic, financial and environmental conditions.

The projects in RES, cofinanced by the EU, are more frequently implemented by municipalities with a lower level of development, characterized by a lower level of entrepreneurship development and typical agricultural functions. On the one hand, it results from the potential for the development of RES in the area of the said municipalities (the availability of land) while on the other, it may be considered an attempt to specialize the area in activities that provide the possibility to dynamize the economic development of the municipality. Moreover, the research conducted by Gradziuk and Gradziuk [55] proved a great diversity in acquiring EU funds by municipalities on the example of Lubuskie Voivodeship. It was furthermore found that the value of investments in RES was significantly and negatively correlated with the level of GDP per capita. Kazak et al. [54] added that the allocation of EU funds was not directed at the most profitable parts of Poland in terms of renewable energy production. The importance of experience in acquiring EU support was further proved by Standar and Puslecki [88]. Furthermore it is easier to conduct proenvironmental activities in places where the area is still undeveloped. As Kazak et al. [54] note the density of development and the degree of dispersion of potential customers affect the size and economic impact of RES implementation.

7. Conclusions

The research revealed that 2777 projects in the field of energy have been conducted within the 2014–2020 Perspective, amounting to nearly PLN 13 billion (EUR 2.93 billion), of which approximately 75.8% has been financed from the EU budget. The most active beneficiaries are local self-government units and among them—municipalities, whose projects account for 95.1% of the total number and over 97% of the total value of projects implemented by the local self-government sector. In total, rural and urban–rural municipalities have implemented the highest number of RES projects. Interestingly, every fourth municipality conducts a minimum of two such projects, noticing high needs in this area. Statistically, solar energy-related activities were significantly more often undertaken, as confirmed by the Kruskal–Wallis test. These activities were characterized by high capital intensity. Up to date, half of them have been completed.

All projects are implemented by municipalities within the Regional Operational Programs, i.e., programs dedicated to the needs of beneficiaries from a particular region. Although each regional program includes actions aimed at the implementation of RES, the Kruskal–Wallis test proved the existence of a statistically significant spatial concentration of investment activity. Particularly great interest in this type of projects was observed among rural entities in Eastern Poland, especially in the Lubelskie Voivodeship.

In order to identify the main determinants of investment activity of rural and urban-rural municipalities in RES, a logit model was applied. It was used to determine the direction and strength of influence of individual factors on the activity of territorial units in acquiring projects in the area of environmentally friendly energy, cofinanced from the EU budget. The projects in RES, cofinanced by the EU, are more frequently implemented by municipalities with a lower level of development, characterized by a lower level of entrepreneurship development and typical agricultural functions. On the one hand, it results from the potential for the development of RES in the area of the said municipalities (the availability of land) while on the other, it may be considered an attempt to specialize the area in activities that provide the possibility to dynamize the economic development of the municipality.

The analyses determined that, among the financial factors, the level of municipality's own income potential, investment activity and the activity in the field of acquiring EU funds are important for their investment activity in obtaining RES projects and projects cofinanced from the EU funds. The lower own income potential of the RES project beneficiaries indicates a typically agricultural character of these municipalities (see [81]) and is associated with a lower level of development, as indicated above.

Additionally, the analyses proved a significant influence of the selected environmental factors on increasing the investment activity of the territorial units in the field of RES. The only positive impact on the investment activity of municipalities in such an area exhibited the share of wastelands in the total area of agricultural land. It indicates that it is easier to conduct proenvironmental activities in places where the area is still undeveloped.

Finally, it should be stated that empirical studies allowed for the positive verification of the research hypothesis, which assumed that "The highest investment activity in the field of local projects co-financed from EU funds, related to the development of RES in rural areas, may be attributed to municipalities performing primarily agricultural functions, located in Eastern Poland". The occurrence of significant disparities in the implementation of projects is not beneficial from the perspective of economic, social and territorial cohesion, and the achievement of sustainable development objectives. It should be emphasized that, for instance, the negative consequences on climate, do not have a territorial range of a particular municipality but rather involve them all, and only joint actions may stop them. Therefore, actions should be undertaken to persuade the remaining local self-government units about the importance and rightness of investments in RES.

Selection of the topic is appropriate due to the topicality of the issues and the great importance of renewable energy sources in the energy transformation. Moreover, the discussed topic is important from the economic point of view, based on the EU energy and climate policy and the resulting requirements to limit the use of conventional energy sources. Even during the last summit of EU leaders, the goals of energy policy were changed, which additionally proves the topicality of the article, even if it concerns one country (and 16 regions in the EU). However, empirical research covered by this paper does not exhaust all topics involved in the local government's investment activity related to RES in rural areas. Nevertheless, this paper brought more value to this research topic because such studies have not been conducted in other countries. The research methods used, and the analysis carried out on that basis, required the use of microdata for projects implemented by municipalities, Polish lowest-level LGUs. Often, it is impossible to access such data for other EU countries. Furthermore, the countries differ in the scope of public tasks carried out by local government units; this would make it difficult to directly compare the findings. However, the findings from this research are of particular importance to the objectives of the Polish regional-level energy policy. They also may provide an incentive and set a standard for scientists from other countries in order for them to conduct similar studies. The potential lines of future research could be indicated by an analysis of how advanced is the local government in using RES, what is the share of renewable energies in the local energy consumption and what is the impact of RES investments on local rural development.

Author Contributions: Conceptualization, A.K., A.S. and Ł.S.; data curation, A.K. and A.S.; formal analysis, A.K., A.S. and Ł.S.; methodology, A.K. and A.S.; project administration, A.K., A.S. and Ł.S.; resources, A.K., A.S. and Ł.S.; software, A.K. and A.S.; visualization, A.K. and A.S.; writing—original draft preparation, Ł.S., A.K. and A.S.; writing—review and editing, A.K., A.S. and Ł.S.; funding acquisition, A.K., A.S. and Ł.S.; investigation, A.K., A.S. and Ł.S.; supervision, A.K., A.S. and Ł.S.; validation, A.K., A.S. and Ł.S. All authors have read and agreed to the published version of the manuscript.

Funding: The research was cofinanced from resources of the Faculty of Economics Poznań University of Life Sciences and Faculty of Agriculture and Economics University of Agriculture in Krakow.

Institutional Review Board Statement: Not applicable.

Informed Consent Statement: Not applicable.

Data Availability Statement: Not applicable.

Conflicts of Interest: The authors declare no conflict of interest.

References

1. Haider, L.J.; Boonstra, W.J.; Peterson, G.D.; Schlüter, M. Traps and sustainable development in rural areas: A review. *World Dev.* **2018**, *101*, 311–321. [CrossRef]
2. Local Data Bank, Central Statistical Office, Warsaw, Poland. Available online: https://bdl.stat.gov.pl/BDL/dane/podgrup/temat (accessed on 1 September 2020).
3. Rosner, A.; Stanny, M. *Socio-Economic Development of Rural Areas in Poland. The European Fund for the Development of Polish Villages Foundation*; Institute of Rural and Agricultural Development, Polish Academy of Sciences: Warsaw, Poland, 2017.
4. Kozera, A.; Głowicka-Wołoszyn, R. Identification of functional types of rural communes in Poland. In Proceedings of the 2018 International Scientific Conference Economic Sciences for Agribusiness and Rural Economy, Warsaw, Poland, 7–8 June 2018; Gołębiewski, J., Ed.; Warsaw University of Life Sciences Press: Warsaw, Poland, 2018; pp. 109–115. [CrossRef]
5. Bański, J.; Mazur, M. Classification of rural areas in Poland as an instrument of territorial policy. *Land Use Policy* **2016**, *54*, 1–17. [CrossRef]
6. Bartkowiak, N. The diversity of socioeconomic development of rural areas in Poland in The Western Borderland and the problem of post-state farm localities. *Oeconomia Copernic.* **2017**, *8*, 417–431. [CrossRef]
7. Szymańska, D.; Chodkowska-Miszczuk, J. Endogenous resources utilization of rural areas in shaping sustainable development in Poland. *Renew. Sustain. Energy Rev.* **2011**, *15*, 1497–1501. [CrossRef]
8. Drejerska, N.; Gołębiewski, J. The role of Poland's primary sector in the development of the country's bioeconomy. *Ital. Rev. Agric. Econ.* **2017**, *72*, 311–326. [CrossRef]
9. *Rural Development Programme for 2014–2020 (Program Rozwoju Obszarów Wiejskich na Lata 2014–2020)*; Ministerstwo Rozwoju Wsi i Rolnictwa: Warsaw, Poland. Available online: https://www.gov.pl/web/rolnictwo/-program-rozwoju-obszarow-wiejskich-2014-2020-prow-2014-2020 (accessed on 10 December 2020).
10. Akella, A.K.; Saini, R.P.; Sharma, M.P. Social, economical and environmental impacts of renewable energy systems. *Renew. Energy* **2009**, *34*, 390–396. [CrossRef]
11. Energy Law Act of 10 April 1997, Dz. U./Journal of Laws/of 2006, No. 89, Item. 625 as Amended (Ustawa z Dnia 10 Kwietnia 1997 Roku Prawo Energetyczne, Dz. U. z 2006 r., Nr 89, poz. 625 z późn. zm.). Available online: https://isap.sejm.gov.pl/isap.nsf/download.xsp/WDU19970540348/U/D19970348Lj.pdf (accessed on 10 December 2020).

12. Yang, J.; Li, X.; Peng, W.; Wagner, F.; Mauzerall, D.L. Climate, air quality and human health benefits of various solar photovoltaic deployment scenarios in China in 2030. *Environ. Res. Lett.* **2018**, *13*, 064002. [CrossRef]
13. Panwar, N.L.; Kaushik, S.C.; Kothari, S. Role of renewable energy sources in environmental protection: A review. *Renew. Sustain. Energy Rev.* **2011**, *15*, 1513–1524. [CrossRef]
14. York, R.; McGee, J.A. Understanding the Jevons paradox. *Environ. Sociol.* **2016**, *2*, 77–87. [CrossRef]
15. Alcott, B. Jevons' paradox. *Ecol. Econ.* **2005**, *54*, 9–21. [CrossRef]
16. Sorrell, S. Jevons' Paradox revisited: The evidence for backfire from improved energy efficiency. *Energy Policy* **2009**, *37*, 1456–1469. [CrossRef]
17. Dinda, S. Environmental Kuznets curve hypothesis: A survey. *Ecol. Econ.* **2004**, *49*, 431–455. [CrossRef]
18. Jebli, M.B.; Youssef, S.B. The environmental Kuznets curve, economic growth, renewable and non-renewable energy, and trade in Tunisia. *Renew. Sustain. Energy Rev.* **2015**, *47*, 173–185. [CrossRef]
19. Pieńkowski, D. Paradoks Jevons' a a konsumpcja energii w Unii Europejskiej (The Jevons Effect and the Consumption of Energy in the European Union). *Probl. Ekorozw. Probl. Sustain. Dev.* **2012**, *7*, 105–116.
20. Vavrek, R.; Chovancová, J. Energy Performance of the European Union Countries in Terms of Reaching the European Energy Union Objectives. *Energies* **2020**, *13*, 5317. [CrossRef]
21. EU COM. *Communication from the Commission to the European Parliament, the Council, the European Economic and Social Committee and the Committee of the Regions A Roadmap for Moving to a Competitive Low Carbon Economy in 2050*; European Commission (EU COM): Brussels, Belgium, 2011.
22. Klepacka, A. Significance of renewable energy sources in sustainable development. *Ann. Pol. Assoc. Agric. Agribus. Econ.* **2019**, *XXI*, 55–64. [CrossRef]
23. An Official Website of the European Union. Available online: https://europa.eu/european-union/topics/environment_pl (accessed on 7 January 2021).
24. Piwowar, A.; Dzikuć, M. Development of renewable energy sources in the context of threats resulting from low-altitude emissions in rural areas in Poland: A review. *Energies* **2019**, *12*, 3558. [CrossRef]
25. Banaszewska, M.; Bischoff, I. The Political Economy of EU-funds: Evidence from Poland. *Jahrb. Natl. Stat.* **2017**, *237*, 191–224. [CrossRef]
26. Winzer, C. Conceptualizing energy security. *Energy Policy* **2012**, *46*, 36–48. [CrossRef]
27. Kruyt, B.; van Vuuren, D.P.; de Vries, H.J.; Groenenberg, H. Indicators for energy security. *Energy Policy* **2009**, *37*, 2166–2181. [CrossRef]
28. Cullmann, A.; Nieswand, M.; Seifert, S.; Stiel, C. A (re) municipalization trend among energy utilities: Truth or myth? *DIW Econ. Bull.* **2016**, *6*, 227–232.
29. Berlo, K.; Wagner, O.; Heenen, M. The incumbents' conservation strategies in the German energy regime as an impediment to re-municipalization—An analysis guided by the multi-level perspective. *Sustainability* **2017**, *9*, 53. [CrossRef]
30. Brinker, L.; Satchwell, A.J. A comparative review of municipal energy business models in Germany, California, and Great Britain: Institutional context and forms of energy decentralization. *Renew. Sustain. Energy Rev.* **2020**, *119*, 109521. [CrossRef]
31. Voorn, B.; Van Genugten, M.L.; Van Thiel, S. Re-interpreting re-municipalization: Finding equilibrium. *J. Econ. Policy Reform.* **2020**, *3*, 1–14. [CrossRef]
32. Eberts, R.W.; Gronberg, T.J. Jurisdictional homogeneity and the Tiebout hypothesis. *J. Urban Econ.* **1981**, *10*, 227–239. [CrossRef]
33. Bölük, G.; Mert, M. Fossil & renewable energy consumption, GHGs (greenhouse gases) and economic growth: Evidence from a panel of EU (European Union) countries. *Energy* **2014**, *74*, 439–446. [CrossRef]
34. Mardani, A.; Streimikiene, D.; Nilashi, M.; Arias Aranda, D.; Loganathan, N.; Jusoh, A. Energy consumption, economic growth, and CO_2 emissions in G20 countries: Application of adaptive neuro-fuzzy inference system. *Energies* **2018**, *11*, 2771. [CrossRef]
35. Farhani, S.; Mrizak, S.; Chaibi, A.; Rault, C. The environmental Kuznets curve and sustainability: A panel data analysis. *Energy Policy* **2014**, *71*, 189–198. [CrossRef]
36. López-Menéndez, A.J.; Pérez, R.; Moreno, B. Environmental costs and renewable energy: Re-visiting the Environmental Kuznets Curve. *J. Environ. Manag.* **2014**, *145*, 368–373. [CrossRef]
37. Helm, D. The future of fossil fuels—Is it the end? *Oxf. Rev. Econ. Policy* **2016**, *32*, 191–205. [CrossRef]
38. Freire-González, J.; Puig-Ventosa, I. Energy efficiency policies and the Jevons paradox. *Int. J. Energy Econ. Policy* **2015**, *5*, 69.
39. Copiello, S. Building energy efficiency: A research branch made of paradoxes. *Renew. Sustain. Energy Rev.* **2017**, *69*, 1064–1076. [CrossRef]
40. Dzikuć, M.; Kuryło, P.; Dudziak, R.; Szufa, S.; Dzikuć, M.; Godzisz, K. Selected Aspects of Combustion Optimization of Coal in Power Plants. *Energies* **2020**, *13*, 2208. [CrossRef]
41. Foxon, T.J. A coevolutionary framework for analysing a transition to a sustainable low carbon economy. *Ecol. Econ.* **2011**, *70*, 2258–2267. [CrossRef]
42. Gradziuk, P.; Gradziuk, B. Gospodarka niskoemisyjna–nowe wyzwanie dla gmin wiejskich. *Wieś Rol.* **2016**, *1*, 105–126. [CrossRef]
43. Owen, R.; Brennan, G.; Lyon, F. Enabling investment for the transition to a low carbon economy: Government policy to finance early stage green innovation. *Curr. Opin. Environ. Sustain.* **2018**, *31*, 137–145. [CrossRef]
44. Winkler, H.; Marquand, A. Changing development paths: From an energy-intensive to low-carbon economy in South Africa. *Clim. Dev.* **2009**, *1*, 47–65. [CrossRef]

45. European Commission. Proposal for a Regulation of the European Parliament and of the Council on the Governance of the Energy Union, Amending Directive 94/22/EC, Directive 98/70/EC, Directive 2009/31/EC, Regulation (EC) No 663/2009, Regulation (EC) No 715/2009, Directive 2009/73/EC, Council Directive 2009/119/EC, Directive 2010/31/EU, Directive 2012/27/EU, Directive 2013/30/EU and Council Directive (EU) 2015/652 and Repealing Regulation (EU) No 525/2013: COM (2016) 759. Available online: https://eur-lex.europa.eu/legal-content/EN/TXT/?uri=CELEX%3A52016PC0759 (accessed on 1 January 2021).
46. Veum, K.; Bauknecht, D. How to reach the EU renewables target by 2030? An analysis of the governance framework. *Energy Policy* **2019**, *127*, 299–307. [CrossRef]
47. Marks-Bielska, R.; Bielski, S.; Pik, K.; Kurowska, K. The Importance of Renewable Energy Sources in Poland's Energy Mix. *Energies* **2020**, *13*, 4624. [CrossRef]
48. Ogarek, P. Przyszłość odnawialnych źródeł energii w Polsce w oparciu o Politykę Energetyczną Polski do 2040 roku (The future of renewable energy sources in Poland based on Poland's Energy Policy until 2040). In *Bezpieczeństwo Energetyczne Polski i Europy Uwarunkowania–Wyzwania–Innowacje*; Ruszel, M., Podmiotko, S., Eds.; Instytut Polityki Energetycznej im. I. Łukasiewicza: Rzeszów, Poland, 2019; pp. 98–119. Available online: https://www.instytutpe.pl/wp-content/uploads/2019/10/Bezpieczeństwo-energetyczne-Polski-i-Europy-.pdf (accessed on 18 November 2020).
49. Ren, J.; Sovacool, B.K. Quantifying, measuring, and strategizing energy security: Determining the most meaningful dimensions and metrics. *Energy* **2014**, *76*, 838–849. [CrossRef]
50. Ang, B.W.; Choong, W.L.; Ng, T.S. Energy security: Definitions, dimensions and indexes. *Renew. Sustain. Energy Rev.* **2015**, *42*, 1077–1093. [CrossRef]
51. Azzuni, A.; Breyer, C. Definitions and dimensions of energy security: A literature review. *Wiley Interdiscip. Rev. Energy Environ.* **2018**, *7*, e268. [CrossRef]
52. Lucas, J.N.V.; Francés, G.E.; González, E.S.M. Energy security and renewable energy deployment in the EU: Liaisons Dangereuses or Virtuous Circle? *Renew. Sustain. Energy Rev.* **2016**, *62*, 1032–1046. [CrossRef]
53. Kosiński, E.; Trupkiewicz, M. Gmina jako podmiot systemu wspierania wytwarzania energii elektrycznej z odnawialnych źródeł energii (A municipality as a part of the suport system for generetion of electricity from renewable Energy sources). *Ruch Praw. Ekon. Socjol.* **2016**, *78*, 93–107.
54. Kazak, J.K.; Kamińska, J.A.; Madej, R.; Bochenkiewicz, M. Where Renewable Energy Sources Funds are Invested? Spatial Analysis of Energy Production Potential and Public Support. *Energies* **2020**, *13*, 5551. [CrossRef]
55. Gradziuk, P.; Gradziuk, B. Renewable energy sources as a development opportunity for peripheral areas/Odnawialne źródła energii szansą rozwojową obszarów peryferyjnych. *Econ. Reg. Stud.* **2020**, *13*, 184–198. [CrossRef]
56. Gradziuk, P.; Gradziuk, B.; Trocewicz, A.; Jendrzejewski, B. Potential of straw for energy purposes in Poland—Forecasts based on trend and causal models. *Energies* **2020**, *13*, 5054. [CrossRef]
57. Benedek, J.; Sebestyén, T.T.; Bartók, B. Evaluation of renewable energy sources in peripheral areas and renewable energy-based rural development. *Renew. Sustain. Energy Rev.* **2018**, *90*, 516–535. [CrossRef]
58. Bergmann, A.; Hanley, N.; Wright, R. Valuing the attributes of renewable energy investments. *Energy Policy* **2006**, *34*, 1004–1014. [CrossRef]
59. Waldenström, C.; Ferguson, R.; Sundberg, C.; Tidåker, P.; Westholm, E.; Åkerskog, A. Bioenergy from agriculture: Challenges for the rural development program in Sweden. *Soc. Nat. Resour.* **2016**, *29*, 1467–1482. [CrossRef]
60. Chodkowska-Miszczuk, J. Odnawialne źródła energii i ich wykorzystanie jako nowe trendy na obszarach wiejskich w Polsce. *Studia Obsz. Wiej.* **2014**, *35*, 227–241.
61. Poggi, F.; Firmino, A.; Amado, M. Planning renewable energy in rural areas: Impacts on occupation and land use. *Energy* **2018**, *155*, 630–640. [CrossRef]
62. Chodkowska-Miszczuk, J.; Biegańska, J.; Środa-Murawska, S.; Grzelak-Kostulska, E.; Rogatka, K. European Union funds in the development of renewable energy sources in Poland in the context of the cohesion policy. *Energy Environ.* **2016**, *27*, 713–725. [CrossRef]
63. Kaya, O.; Klepacka, A.M.; Florkowski, W.J. Achieving renewable energy, climate, and air quality policy goals: Rural residential investment in solar panel. *J. Environ. Manag.* **2019**, *248*, 109309. [CrossRef] [PubMed]
64. Hernik, J.; Noszczyk, T.; Rutkowska, A. Towards a better understanding of the variables that influence renewable energy sources in eastern Poland. *J. Clean. Prod.* **2019**, *241*, 118075. [CrossRef]
65. Štreimikiene, D. Review of financial support from EU Structural Funds to sustainable energy in Baltic States. *Renew. Sustain. Energy Rev.* **2016**, *58*, 1027–1038. [CrossRef]
66. *Database of Implemented EU Projects*; Ministry of Development Funds and Regional Policy: Warsaw, Poland. Available online: https://www.funduszeeuropejskie.gov.pl/strony/o-funduszach/projekty/lista-projektow/lista-projektow-realizowanych-z-funduszy-europejskich-w-polsce-w-latach-2014-2020/ (accessed on 22 September 2020).
67. *Archive of Medium Courses*; National Bank of Poland: Warsaw, Poland; Available online: https://www.nbp.pl/home.aspx?f=/kursy/arch_a.html (accessed on 19 December 2020).
68. Hosmer, D.W.; Lemeshow, S. *Applied Logistic Regression*; Wiley & Sons: New York, NY, USA, 2000.
69. Stanisz, A. *Przystępny Kurs Statystyki z Zastosowaniem Statistica PL na Przykładach z Medycyny, tom 2, Modele Liniowe i Nieliniowe*; Wydawnictwo StatSoft Polska: Kraków, Poland, 2007.
70. Hilbe, J.M. *Logistic Regression Models*; Chapman & Hall/CRC Press: Boca Raton, FL, USA, 2009.

71. Kleinbaum, D.G.; Klein, M. *Logistic Regression*; Springer: New York, NY, USA, 2002.
72. Cramer, J.S. *Logit Models from Economics and Other Fields*; Cambridge University Press: Cambridge, UK, 2011.
73. Maddala, G.S. *Introduction to Econometrics*, 3rd ed.; Wiley and Sons Ltd.: Chichester, UK; New York, NY, USA, 2001.
74. Bewick, V.; Cheek, L.; Ball, J. Statistics review 14: Logistic regression. *Crit. Care* **2005**, *9*, 112–118. [CrossRef] [PubMed]
75. Fierek, D.; Janiszewska, D. Lokalne uwarunkowania województwa pomorskiego dla produkcji biogazu (Local Determinants of Pomorskie Voivodeship for Biogas Production). *Zesz. Nauk. Wydziału Nauk Ekon.* **2016**, *20*, 69–81.
76. Bartkowiak-Bakun, N.; Standar, A. *Ocena Oddziaływania Wybranych Działań PROW 2007–2013 na Zrównoważony Rozwój Obszarów Wiejskich Polski: (Analiza Regionalna) (Assessment of the Impact of Selected RDP 2007–2013 on the Sustainable Development of Polish Rural Areas: (Regional Analysis)*; Wydawnictwo Uniwersytetu Przyrodniczego w Poznaniu: Poznań, Poland, 2015.
77. Bartkowiak, N.; Poczta, W. *Czynniki Rozwoju Obszarów Wiejskich Wielkopolski (Development Factors of Rural Areas of Wielkopolska)*; Wydawnictwo Uniwersytetu Przyrodniczego w Poznaniu: Poznań, Poland, 2015.
78. Standar, A. Assessing the Scale of and Financial Reasons behind Differences in the Local Government Units' Investment Expenditures in the Context of Reducing Disparities in Socio-Economic Development. In *International Scientific Days 2018. "Towards Productive, Sustainable and Resilient Global Agriculture and Food Systems": Conference Proceedings*; Horská, E., Kapsdorferová, Z., Hallová, M., Eds.; Wolters Kluwer ČR: Prague, Czech Republic, 2018; pp. 1462–1475.
79. Głowicka-Wołoszyn, R.; Kozera, A.; Wysocki, F. Identyfikacja wewnętrznych uwarunkowań samodzielności i atrakcyjności finansowej gmin wiejskich województwa wielkopolskiego z wykorzystaniem biplotu (Internal Determinants Identification of the Autonomy and Financial Attractiveness of Rural Gminas in Wielkopolskie Voivodship Using Biplot). *Wiadomości Stat.* **2017**, *8*, 74–84.
80. *Energy from Renewable Sources in 2018*; Central Statistical Office: Warsaw, Poland, 2019.
81. Fact Sheets on the European Union, European Parliament. Available online: https://www.europarl.europa.eu/factsheets/pl/sheet/68/polityka-energetyczna-zasady-ogolne (accessed on 5 December 2020).
82. Marks-Bielska, R.; Bielski, S.; Novikova, A.; Romaneckas, K. Straw stocks as a source of renewable energy. A case study of a district in Poland. *Sustainability* **2019**, *11*, 4714. [CrossRef]
83. Robles, R.; Dalia, P. Interactions between agriculture and energy policies: The case of Castile and Leon (Spain). *Rural Dev.* **2011**, *5*, 222–228.
84. *Wielkopolski Regionalny Program Operacyjny 2014–2020*; Zarząd Województwa Wielkopolskiego: Poznań, Poland, 2015.
85. *Indicators for the Evaluation of the Financial Position of Territorial Self-Government Units in 2016–2018 (Wskaźniki do Oceny Sytuacji Finansowej JST w Latach 2016–2018)*; Ministry of Finance: Warsaw, Poland, 2019. Available online: https://www.gov.pl/web/finanse/wskazniki-do-oceny-sytuacji-finansowej-jednostek-samorzadu-terytorialnego-w-latach-2016-2018 (accessed on 22 October 2020).
86. Satola, L. Kapitał ludzki źródłem przewagi konkurencyjnej gmin w aspekcie rozwoju lokalnego (Human capital as a source of competitive advantage of communes in the context of local government). *Rocz. Nauk. Stowarzyszenia Ekon. Rol. Agrobiz.* **2009**, *11*, 293–298.
87. Satoła, Ł.; Standar, A.; Kozera, A. Financial autonomy of local government units: Evidence from Polish rural municipalities. *Lex Localis Local Self-Gov.* **2019**, *17*, 321–342. [CrossRef]
88. Standar, A.; Puślecki, Z.W. *Ocena Zastosowania Środków Polityki Regionalnej Unii Europejskiej Przez Samorządy Gminne Województwa Wielkopolskiego*; Wydawnictwo Uniwersytetu Przyrodniczego w Poznaniu: Poznań, Poland, 2011.
89. EU Strategy for the Baltic Sea Region (Strategia UE dla regionu Morza Bałtyckiego), Komisja Wspólnot Europejskich, Bruksela. 2009. Available online: https://www.gov.pl/web/gospodarkamorska/strategia-ue-dla-regionu-morza-baltyckiego (accessed on 7 December 2020).
90. *Europe 2020 A Strategy for Smart, Sustainable and Inclusive Growth*; European Commission: Brussels, Belgium, 2010. Available online: https://eur-lex.europa.eu/legal-content/en/ALL/?uri=CELEX%3A52010DC2020 (accessed on 7 December 2020).
91. *The Energy Policy of Poland Until 2030 (Polityka Energetyczna Polski do 2030 Roku)*; Ministerstwo Gospodarki: Warsaw, Poland, 2009. Available online: https://www.gov.pl/web/klimat/polityka-energetyczna-polski-do-2030-roku (accessed on 7 December 2020).
92. The Energy Security and Environment Strategy (Strategia Bezpieczeństwo Energetyczne i Środowisko z Perspektywa do 2020). Uchwała Rady Ministrów z Dnia 15 Kwietnia 2014 r. (M.P. z 2014, poz. 469). Available online: http://isap.sejm.gov.pl/isap.nsf/DocDetails.xsp?id=wmp20140000469 (accessed on 7 December 2020).
93. The Transport Development Strategy Until 2020 (with a Perspective Until 2030) (Strategia Rozwoju Transportu do 2020 roku (z Perspektywą do 2030 Roku)). Załącznik do Uchwały nr 58 Rady Ministrów z Dnia 15 Kwietnia 2014 r. Available online: https://www.gov.pl/documents/905843/1047987/Strategia_Rozwoju_Transportu_do_2020_roku.pdf/ead3114a-aac7-3cdd-c71d-7f88267ce596 (accessed on 7 December 2020).
94. *National Energy Efficiency Action Plan for Poland (Krajowy Plan Działań Dotyczący Efektywności Energetycznej dla Polski)*; Ministerstwo Energii: Warsaw, Poland, 2017. Available online: https://www.gov.pl/web/klimat/krajowy-plan-dzialan-dotyczacy-efektywnosci-energetycznej (accessed on 7 December 2020).
95. *The National Renewable Energy Action Plan (Krajowy Plan Działań w Zakresie Odnawialnych Źródeł Energii)*; Ministerstwo Gospodarki: Warsaw, Poland, 2010. Available online: https://www.senat.gov.pl/gfx/senat/userfiles/_public/k7/komisje/2011/kgn/288/288-1.pdf (accessed on 7 December 2020).

96. *Partnership Agreement–Programming of the 2014–2020 Financial Perspective (Umowa Partnerstwa. Programowanie Perspektywy Finansowej 2014–2020)*; Ministerstwo Rozwoju i Inwestycji: Warsaw, Poland, 2014. Available online: https://www.funduszeeuropejskie.gov.pl/strony/o-funduszach/dokumenty/umowa-partnerstwa/ (accessed on 7 December 2020).
97. Gradziuk, P.; Gradziuk, B. Próba oceny efektów absorpcji środków z funduszy europejskich na rozwój wykorzystania odnawialnych źródeł energii w województwie lubelskim. *Rocz. Nauk. Ekon. Rol. I Rozw. Obsz. Wiej.* **2017**, *104*, 95–105. [CrossRef]
98. Ossowska, L.; Janiszewska, D.; Bartkowiak-Bakun, N.; Kwiatkowski, G. Energy Consumption Versus Greenhouse Gas Emissions in EU. *Eur. Res. Stud. J.* **2020**, *XXIII*, 185–198. [CrossRef]
99. Fischer, S. The EU's new energy and climate: Policy framework for 2030. *SWP Comments* **2014**, *55*, 1–8.
100. Ringel, M.; Knodt, M. The governance of the European Energy Union: Efficiency, effectiveness and acceptance of the Winter Package 2016. *Energy Policy* **2018**, *112*, 209–220. [CrossRef]
101. Martí-Ballester, C.P. Do European renewable energy mutual funds foster the transition to a low-carbon economy? *Renew. Energy* **2019**, *143*, 1299–1309. [CrossRef]
102. Kachniarz, M.; Szewranski, S.; Kazak, J. The Use of European Funds in Polish and Czech Municipalities. A Study of the Lower Silesia Voivodship and Hradec Kralove Region. In *IOP Conference Series: Materials Science and Engineering*; IOP Publishing: Bristol, UK, 2019; Volume 471, p. 112047.
103. Ghiani, E.; Giordano, A.; Nieddu, A.; Rosetti, L.; Pilo, F. Planning of a Smart Local Energy Community: The Case of Berchidda Municipality (Italy). *Energies* **2019**, *12*, 4629. [CrossRef]
104. Malinauskaite, J.; Jouhara, H.; Egilegor, B.; Al-Mansour, F.; Ahmad, L.; Pusnik, M. Energy efficiency in the industrial sector in the EU, Slovenia, and Spain. *Energy* **2020**, *208*, 118398. [CrossRef]
105. Bointner, R.; Pezzutto, S.; Grilli, G.; Sparber, W. Financing innovations for the renewable energy transition in Europe. *Energies* **2016**, *9*, 990. [CrossRef]
106. Streimikiene, D.; Klevas, V.; Bubeliene, J. Use of EU structural funds for sustainable energy development in new EU member states. *Renew. Sustain. Energy Rev.* **2007**, *11*, 1167–1187. [CrossRef]
107. Bostan, I.; Lazar, C.M.; Asalos, N.; Munteanu, I.; Horga, G.M. The three-dimensional impact of the absorption effects of European funds on the competitiveness of the SMEs from the Danube Delta. *Ind. Crops Prod.* **2019**, *132*, 460–467. [CrossRef]
108. Carfora, A.; Romano, A.A.; Ronghi, M.; Scandurra, G. Renewable generation across Italian regions: Spillover effects and effectiveness of European Regional Fund. *Energy Policy* **2017**, *102*. [CrossRef]
109. Żmija, K.; Fortes, A.; Tia, M.N.; Šūmane, S.; Ayambila, S.N.; Żmija, D.; Satoła, Ł.; Sutherland, L.A. Small farming and generational renewal in the context of food security challenges. *Glob. Food Secur.* **2020**, *26*, 100412. [CrossRef]
110. Satoła, Ł. Problems of the aging of the farmers' population in small farms in Poland. Scientific Papers Series-Management. *Econ. Eng. Agric. Rural Dev.* **2019**, *19*, 305–310.
111. Wasiuta, A. Renewable energy sources and their impact on Polish labor market in the context of global energy problems. In *International Scientific Days 2018. Towards Productive, Sustainable and Resilient Global Agriculture and Food Systems*; Horská, E., Kapsdorferová, Z., Hallová, M., Eds.; Wolters Kluwer ČR: Prague, Czech Republic, 2018; pp. 2726–2737. Available online: http://spu.fem.uniag.sk/mvd2018/isd2018_proceedings/isd_conference_proceedings.pdf (accessed on 5 January 2021).

energies

MDPI

Article

Regional Diversification of Potential, Production and Efficiency of Use of Biogas and Biomass in Poland

Grzegorz Ślusarz [1], Barbara Gołębiewska [2,*], Marek Cierpiał-Wolan [1], Jarosław Gołębiewski [2], Dariusz Twaróg [3] and Sebastian Wójcik [4]

1 Institute of Economics and Finance, University of Rzeszów, ul. M. Ćwiklińskiej 2, 35-601 Rzeszów, Poland; gslusarz@ur.edu.pl (G.Ś.); M.Cierpial-Wolan@stat.gov.pl (M.C.-W.)
2 Institute of Economics and Finance, Warsaw University of Life Sciences—SGGW, Nowoursynowska 166, 02-787 Warsaw, Poland; jaroslaw_golebiewski@sggw.edu.pl
3 Statistical Office in Rzeszów, Jana III Sobieskiego 10, 35-001 Rzeszów, Poland; D.Twarog@stat.gov.pl
4 Institute of Mathematics, University of Rzeszów, ul. Pigonia 1, 35-959 Rzeszów, Poland; swojcik@ur.edu.pl
* Correspondence: barbara_golebiewska@sggw.edu.pl; Tel.: +48-22-5934228

Abstract: Energy obtained from renewable sources is an important element of the sustainable development strategy of the European Union and its member states. The aim of this research is, therefore, to assess the potential and use of renewable energy sources and their effectiveness from the regional perspective in Poland. The research covered the years 2012 and 2018. The diversification of production and potential of renewable energy sources was defined on the basis of biogas and biomass. Calculations made using the data envelopment analysis (DEA) method showed that, in 2012, only three voivodeships achieved the highest efficiency in terms of the use of biogas and biomass resources; in 2018, this number increased to four. Comparing the effective units in 2012 and 2018, it can be seen that their efficiency frontier moved upwards by 56% in terms of biogas and 21% in terms of to biomass. Despite a large relative increase in the production of heat from biogas by 99% compared to the production of heat from biomass by 38%, the efficiency frontier for biogas did not change considerably. It was found that the resources of solid biomass are used far more intensively than the resources of biogas. However, in the case of biogas, a significant increase in the utilization of the production potential was observed: from 3.3% in 2012 to 6.4% in 2018, whereas in the same years, the utilization of solid biomass production potential remained at the same level (15.3% in 2012, 15.4% in 2018). It was also observed that, at the level of voivodeships, the utilization of biogas and biomass production potential is negatively correlated with the size of this potential. The combined potential of solid biomass and biogas can cover the demand of each of the studied regions in Poland in terms of thermal energy. The coverage ranges from 104% to 1402%. The results show that when comparing biomass and biogas, the production of both electricity and heat was dominated by solid biomass. Its high share occurred especially in voivodeships characterized by a high share of forest area and a low potential for biogas production (Lubuskie Voivodeship, Zachodniopomorskie Voivodeship).

Keywords: biogas; biomass; data envelopment analysis (DEA); efficiency ranking; renewable energy; Poland; regional potential

Citation: Ślusarz, G.; Gołębiewska, B.; Cierpiał-Wolan, M.; Gołębiewski, J.; Twaróg, D.; Wójcik, S. Regional Diversification of Potential, Production and Efficiency of Use of Biogas and Biomass in Poland. *Energies* **2021**, *14*, 742. https://doi.org/10.3390/en14030742

Academic Editor: Alberto-Jesus Perea-Moreno
Received: 23 December 2020
Accepted: 27 January 2021
Published: 31 January 2021

1. Introduction

Due to the development of civilization, more and more energy resources are necessary to satisfy basic social needs as well as production. Lack of integration in resource assessment and policy making leads to inconsistent strategies and inefficient use of resources [1]. Fossil fuels play a dominant role in global energy systems [2], although according to Arıoğlu et al. [3], renewable energy sources are becoming the fastest growing energy source in the world. As pointed out by Moomaw et al. [4], 85% of the primary energy used by global economies comes from fossil fuels. However, the share of energy from renewable sources in the structure of energy consumption is growing rapidly, especially in Europe. In 2018, the share of energy from fossil fuels in the EU decreased to 70.2% of total energy [5].

In recent years, more and more attention has been paid to the development and use of energy from renewable sources [6,7]. Moomaw et al. [4] report that renewable energy sources play a role in the provision of energy services in a sustainable manner, and in particular in mitigating climate change. Gielen and his team [8] also note that renewable energy can meet two-thirds of the total global energy demand and contribute, to a large extent, to the reduction in greenhouse gas emissions responsible for climate change.

Activities in the field of production of energy from renewable sources are undertaken in many countries of the world, not only in EU countries, pointing to their importance for the development of rural areas [9–11]. As Lemaire [12] shows, in rural South Africa, small energy companies can play useful roles in supplementing conventional systems.

Renewable energy and energy efficiency are the two main components of sustainable energy systems. Abolhosseini and his team [13] indicate that electricity consumption will constitute an increasing share of global energy demand over the next two decades. Therefore, the development of renewable energy sources is becoming one of the most important challenges in light of the increasingly energy-consuming socio-economic development and the need to reduce the share of fossil, high-emission sources of energy production. This is due to the so-called climate and energy package implemented in EU countries, which assumed a 20% reduction in greenhouse gas emissions by 2020 compared to 1990 and an increase in the share of renewable energy sources (RESs) in total energy consumption by 20% [14]. Such a limitation is also reflected in the assumptions of the EU climate and energy policy in the 2030 perspective [15]. For Poland, according to this document, this share accounted for at least 15% in 2020 [16]. Due to the slow pace of development of renewable energy sources, this goal will require many years of multidirectional activities. These activities should take into account the spatial differentiation of development conditions, and thus the diversified opportunities for the development of renewable energy sources. This task is difficult because the structure of electricity produced in Poland has been dominated by energy produced from high-emission solid fuels for many years. In 2015, hard coal accounted for 46.5% of these fuels, lignite 32.2%, natural gas 3%, renewable energy 13.7%, and the remaining 4.6% [17]. High-emission fossil fuels are a common global problem. They are the cause of more than 70% of greenhouse gas emissions in the world [18]. Hence, the issue of the development of a low-emission economy and energy generation based on low-emission sources is of particular importance.

Rural areas, along with the agriculture and forestry sectors located within them, have an important role to play in generating renewable energy sources, especially as agricultural land and forests dominate the land use structure in all regions of Poland [19]. This diversification may undoubtedly affect the potential possibilities of producing biomass, which is the largest source of renewable energy, mainly obtained in agriculture [20]. Solid biofuels as well as raw materials for the production of liquid fuels from biomass (biodiesel and bioethanol) and some biogases are produced from biomass [21].

Thus, the agricultural sector may not only be an emitter of greenhouse gases and a consumer of energy, but may also have the potential to generate energy [22,23]. As indicated by Hengeveld et al. [24], an increasing number of local and regional initiatives show a growing interest in decentralized energy production, in which biogas can play a role. Carrosio [25], who studied the evolution of agricultural biogas production in Italy, also believes that in order to achieve a more sustainable development of bioenergy, the existing institutional framework should be reformed by reorganizing subsidies and involving farmers in local projects. It is therefore expected that the amount of biogas produced on agricultural land will increase in the coming years.

This issue receives little recognition in the countries of Central and Eastern Europe and in Poland itself. Such studies have so far been conducted in China [26,27], in Vietnam [28], or in Ukraine [29], but most studies focus mainly on technical or technological issues related to the production of this energy [30–32], while less work is devoted to the spatial differentiation of the efficiency of biomass and biogas potential. The issues of the spatial distribution of infrastructure [33,34], the deployment of the renewable energy industry [35], and spatial planning in terms of resource availability and use [36], were addressed, including waste for biogas production [37,38]. A spatial analysis of the power density of renewable energy was carried out [39]. Research in the field of solar and wind energy was also often conducted [40,41]. Spatial aspects also concern the differences between EU countries [42,43].

Therefore, even if renewables are distributed all over the world, location plays a huge role in deciding which resources to use, not only globally but also locally. In addition, it is important to assess the energy efficiency of different regions, which can help to identify differences in energy efficiency, which can be the basis for improving this efficiency.

The main aim of the research is to look for answers to the fundamental question regarding the potential of and resources required for use of biomass and biogas in Poland and the effectiveness of their use, the calculation of which was carried out using the data envelopment analysis (DEA) method, widely used in the field of environmental and energy economics [44–47].

The analyses highlighted the regions in Poland where the production of biogas and biomass is highly developed and where they constitute the largest share in total energy. Answers were also sought to the question regarding the efficiency of energy production from biomass and biogas in the regions and whether they can achieve energy self-sufficiency based on local resources.

2. Research Methodology

The research was carried out at the regional level in Poland. The spatial scope of detailed research covered all voivodeships (regions) of the country. To compare the efficiency of useful heat production by voivodeships, data on the consumption of biogas and biomass from the report on heat from renewable sources of Statistics Poland were used. The adopted approach takes into account those renewable energy sources that can occur throughout the country without major restrictions, can be produced in each region, and are related to agriculture. The data of Local Data Bank of Statistics Poland were also used.

2.1. Partial Ranking and Hasse Diagram

The ratio of potential of heat production from biogas to biogas consumption was calculated, as well as a similar indicator for biomass for each voivodeship. If one region dominates another with respect to indicators under comparison, it is easy to determine which region is better. In another case, there is a need to establish a trade-off between indicators to make a comparison. The partial ranking shows only unambiguous comparisons and introduces a partial ordering in the region under comparison. The partial ranking can be visualized with a Hasse diagram with takes a form of graph. In the Hasse diagram, if there is an arrow from one object to another, it means that the first of them is better than the second one in terms of each analyzed variable. The situation when one can move from the first to the second object through a sequence of arrows with the same direction is interpreted in a similar way. If there is no sequence of arrows between objects, such objects are incomparable.

2.2. Data Envelopment Analysis

The data envelopment analysis (DEA) method was used to organize the voivodeships in terms of effectiveness, but without indicating the distance between the other voivodeships and the leaders.

The basis of this method is a set of variables that are inputs and a set of variables that are outputs. The DEA method is widely used to study energy efficiency as a total factor energy efficiency evaluation method [48]. DEA belongs to the group of nonparametric methods of linear programming, in which efficiency is defined as the quotient of the weighted sum of effects to the weighted sum of inputs [49]. The DEA allows the effectiveness of each Decision-Making Unit (DMU) to be maximized by selecting weights assigned to inputs and effects [50]. This allows for the identification of relatively effective and ineffective units and the measurement of the ineffectiveness of the latter. This approach also makes it possible to compare DMUs with very diverse structures, natures of effects and inputs, and to estimate unobservable technological elements directly from inputs and effects without applying restrictive assumptions about the parameters in the production process [51]. In this method, at least one object always has 100% efficiency. The efficiency of all the others is compared with the objects with 100% efficiency; hence, the relative efficiency is obtained.

When presenting the idea of the basic DEA model, it should be noted that y_{jk} $(j = 1, \ldots, n)$ stands for effects (outputs) and x_{ik} $(i = 1, \ldots, m)$ stands for inputs of k-th DMU $(k = 1, \ldots, r)$. In DEA, for each fixed DMU (say t), a system of weights for inputs v_{it} $(i = 1, \ldots, m)$ and effects w_{jt} $(j = 1, \ldots, n)$ must be found that maximizes the following expression

$$Ef_t = \frac{\sum_j^n w_{jt} y_{jt}}{\sum_i^m v_{it} x_{it}} \quad (1)$$

and meets the conditions of:

$$w_{jt} > 0 \; (j = 1, \ldots, n), \; v_{it} > 0 \; (i = 1, \ldots, m) \quad (2)$$

$$\sum_i^m v_{it} x_{it} = 1 \quad (3)$$

$$Ef_k = \frac{\sum_j^n w_{jt} y_{jk}}{\sum_i^m v_{it} x_{ik}} \leq 1 \; (k = 1, \ldots, r) \quad (4)$$

In this research, the CCR model described above was used [52,53], with constant returns to scale with two inputs and one effect. Consumption of biogas and biomass consumption were assumed as inputs, while the total production of heat from biogas and biomass was assumed as the effect. Data analysis showed that—in this case—constant returns to scale are more appropriate than the alternative—i.e., variable returns to scale. The calculations were performed using the "deaR" package [54] in the R-Studio program. A sensitivity analysis of DEA results was performed using the all-factors-at-once approach. The influence of changes of all variables simultaneously on the relative efficiency was investigated by performing 10,000 simulations. For each variable, a new value was drawn within ±1% of the original value based on a uniform distribution. The simulation results were presented as an interval for the relative efficiency. This interval was defined by quantiles of the order of 0.025 and 0.975 of simulation results. Thus, 95% of simulation results were covered by the interval. In order to estimate the potential of biogas production, publicly available statistical data of Statistics Poland were used and the methodology applied in the study by Bujakowski et al. [55] additionally took into account losses and damages in agricultural crops. Biogas is produced in the process of anaerobic digestion of organic waste.

2.3. Biomass Potential

In order to estimate the potential of solid biomass, it was assumed that it would come from plant production, including straw surplus, hay surplus, energy crops, orchards, forest production as well as annual felling and care cuts. When calculating the potential offered by the timber management, assumptions from the methodology presented in the work of Bujakowski et al. [55] were included. In order to assess the surplus of straw and hay that can be used for energy purposes, the methodology presented in the study by Ludwicka et al. [56] was used. By calculating the share of the above for special purpose energy crops, it was assumed that the land use factor for growing these plants is 1/10, which is a safe border eliminating competition between the production of raw materials and production for food purposes [57]. When calculating the potential of solid biomass, the biomass that could be obtained from the care and replacement of stands in orchards was also taken into account [58].

3. Results

3.1. Regional Differentiation of Biomass and Biogas Potential in Poland

In Poland, and in particular in voivodeships with a large share of the agricultural economy, an upward trend in electricity consumption has been observed for several years. Electricity consumption is growing faster in rural areas. The share of energy from renewable sources is also increasing, although the growth dynamics is not as high as the EU average [19]. Energy obtained from renewable sources in Poland in 2018 came mainly from solid biofuels (69.3%), wind energy (12.4%), and liquid biofuels (10.2%) [59].

The structure of production of energy from renewable sources for Poland results primarily from the geographic conditions characteristic of our country and possible resources to be managed [60]. The share of energy from renewable sources in the production of primary energy in total increased in 2014–2018 from 12.1% to 14.3% [59]. However, it was significantly differentiated regionally (Table 1), which is justified due to the different potential of the resources used for its production. This applies, for example, to organic waste in landfills, animal and vegetable waste on farms, or to the structures and sizes of farms, as indicated by Wąs et al. [29]. This applies both to the production of biomass and biogas, which are analyzed in this paper.

Table 1. Regional differentiation of potential of biogas and biomass production and use in 2012 and 2018.

Voivodeships	Production Potential and Its Use in 2012				Production Potential and Its Use in 2018			
	Biogas		Biomass		Biogas		Biomass	
	Production Potential (dam³)	Use of Potential (%)	Production Potential (tons)	Use of Potential (%)	Production Potential (dam³)	Use of Potential (%)	Production Potential (tons)	Use of Potential (%)
Dolnośląskie	321,324.6	2.9	4,563,028.1	16.8	314,818.7	7.7	3,953,491.8	5.5
Kujawsko-pomorskie	469,096.0	3.7	2,892,793.3	51.5	481,521.1	5.2	3,101,466.5	72.8
Lubelskie	367,761.7	2.5	3,747,584.7	1.6	335,812.8	7.2	3,873,572.5	1.5
Lubuskie	141,629.0	12.4	4,358,377.3	5.6	140,665.7	3.6	3,969,589.0	9.1
Łódzkie	486,559.7	1.6	2,601,736.8	14.8	476,215.4	2.7	2,451,250.0	9.3
Małopolskie	341,333.1	2.8	2,955,906.9	15.1	355,141.3	4.7	2,589,329.1	5.9
Mazowieckie	987,486.1	2.0	5,764,615.9	13.2	1,094,449.0	8.3	3,890,607.5	31.2
Opolskie	185,369.8	1.9	2,309,445.6	6.3	172,393.0	3.2	2,142,324.1	3.3
Podkarpackie	173,942.8	4.5	4,119,273.1	4.2	167,445.7	8.9	3,273,198.6	7.2
Podlaskie	595,891.8	1.0	4,924,012.2	11.1	632,614.3	2.1	3,174,245.3	7.5
Pomorskie	304,462.6	14.6	4,386,315.3	35.9	305,960.8	23.6	3,932,233.7	37.4
Śląskie	379,862.3	6.4	2,526,229.1	25.5	338,600.2	9.0	2,202,212.8	22.5
Świętokrzyskie	168,925.5	2.2	1,906,426.8	42.2	159,623.7	3.4	1,914,166.3	12.5
Warmińsko-mazurskie	364,333.5	1.9	7,056,542.1	3.7	379,785.0	5.6	4,621,153.2	6.9
Wielkopolskie	945,417.0	1.7	5,798,389.8	13.1	1,022,612.0	4.1	4,771,492.2	8.2
Zachodniopomorskie	233,444.0	4.3	6,011,437.7	17.6	235,268.1	7.1	5,201,121.4	10.5
Poland	6,466,839.5	3.3	65,922,114.7	15.3	6,612,926.7	6.4	55,061,454.0	15.4

Source: own study based on the data of the Local Data Bank of Statistics Poland and the Statistics Poland survey of heat from renewable sources (G-02o).

The data in Table 1 show that Pomorskie Voivodeship is the leader in terms of using the potential of biogas production. In 2012, it achieved its potential in terms of biogas production, 14.6%, while in 2018 the use of the potential for biogas production increased to 23.6%. The opposite of Pomorskie Voivodeship is Podlaskie Voivodeship, which was the last in the country in terms of using the potential of biogas production, both in 2012 and 2018.

Comparing the degree of use of the potential of biogas with the degree of use of the potential of solid biomass, it can be clearly seen that the resources of solid biomass are used far more intensively. In particular, this applies to voivodeships where large power plants and combined heat and power plants operate using the so-called green blocks powered by biomass or cogeneration installations burning solid biomass produced on the site. Such large centers operate in Poland in the following voivodeships: Kujawsko-Pomorskie, Pomorskie, Śląskie, and Świętokrzyskie, which translates into a high degree of use of the biomass potential in these voivodeships.

3.2. The Importance of Biogas and Biomass in Meeting Energy Demand in the Regions

In the further part of the research, the potentials of biogas and biomass were compared, and their total potential in the production of electricity and heat was determined. Then, the data obtained in this way were compared with the statistical data of Statistics Poland on electricity and heat consumption in 2012 and 2018 (Table 2). The equations describing the results of the calculations shown in Table 2 are included in Appendix A.

Table 2. Comparison of the total electricity and heat production potential in 2012 and 2018.

Voivodeships	2012				2018			
	Heat Production Potential/total Consumption (%)	Electricity Production Potential/Total Production (%)	Solid Biomass/Biogas in Heat Production	Solid Biomass/Biogas in Electricity Produc-tion	Heat Production Potential/Total Consumption (%)	Electricity Production Potential/Total Production (%)	Solid Biomass/Biogas in Heat Production	Solid Biomass/Biogas in Electricity Production
Dolnośląskie	489.7	146.2	23.0	28.4	379.5	180.7	20.2	26.1
Kujawsko-pomorskie	411.7	437.1	10.0	13.1	445.7	224.3	10.3	14.1
Lubelskie	455.2	832.5	16.4	21.6	677.4	860.6	18.6	24.2
Lubuskie	1633.7	791.0	49.1	66.2	1402.2	577.6	44.7	63.4
Łódzkie	225.2	37.1	8.9	11.7	242.9	32.4	8.5	11.5
Małopolskie	289.4	214.4	14.1	18.1	249.3	212.4	11.8	15.8
Mazowieckie	222.0	126.6	9.7	12.5	159.0	67.9	6.0	8.0
Opolskie	840.7	115.9	20.2	24.1	764.3	95.0	20.0	25.5
Podkarpackie	812.5	705.9	38.0	50.5	852.4	611.5	31.5	41.8
Podlaskie	1018.6	3091.2	13.6	16.9	729.5	1573.1	8.2	11.5
Pomorskie	440.9	589.4	23.2	30.6	376.9	462.5	20.5	28.6
Śląskie	111.0	39.0	10.7	14.3	103.9	43.7	10.5	14.3
Świętokrzyskie	598.5	110.8	18.4	24.8	742.0	82.9	19.4	26.8
Warmińsko-mazurskie	1298.3	4028.6	31.6	38.3	844.9	1884.6	19.5	26.7
Wielkopolskie	575.4	214.0	10.0	13.1	578.8	252.5	7.6	10.6
Zachodniopomorskie	993.2	312.5	41.2	54.4	736.8	299.8	35.1	48.5
Poland	446.2	187.2	16.6	21.4	392.6	157.3	13.4	18.3

Source: own study based on the data of the Local Data Bank of Statistics Poland.

The data presented in Table 2 show that the combined potential of solid biomass and biogas can significantly cover the demand of each voivodeship in terms of thermal energy. In the case of Warmińsko-Mazurskie Voivodeship, heat production may exceed the region's demand more than 12 times, and in the case of electricity, this amount is more than 40 times higher (to illustrate the results in Table 2 better, heat and electricity consumption are included in Appendix B—Table A1). Even in the case of Ślaskie Voivodeship, the weakest in the ranking, biomass may become an important component of the energy mix that is able to fully cover the demand for heat energy and cover 39.2% of the demand for electricity.

The results also clearly show that the production of both electricity and heat was dominated by solid biomass. Its high share occurred especially in voivodeships characterized by a large share of forest area and a low potential for biogas production—e.g., Lubuskie Voivodeship.

3.3. Spatial Diversification of Biogas and Biomass Efficiency

Individual regions differed in the size and structure of energy sources, as well as the level of efficiency. The data on biogas and biomass consumption and heat production from the unpublished Statistics Poland report on heat from renewable sources being used to compare the effectiveness of useful heat production by voivodeship. The ratio of total heat production from biogas to total biogas consumption was calculated and an analogous index was created for biomass for each of the voivodeships (Table 3).

Table 3. Biogas and biomass efficiency by voivodeship in Poland.

Voivodeships	2012				2018			
	Biogas		Biomass		Biogas		Biomass	
	Average Boiler Efficiency (MJ/m^3)	Weighted Average Heat Density (MJ/m^3)	Average Boiler Efficiency (MJ/kg)	Weighted Average Calorific Value (MJ/kg)	Average Boiler Efficiency (MJ/m^3)	Weighted Average Heat Density (MJ/m^3)	Average Boiler Efficiency (MJ/kg)	Weighted Average Calorific Value (MJ/kg)
Dolnośląskie	14.41	22.77	1.85	11.44	9.15	21.42	4.72	9.86
Kujawsko-pomorskie	6.39	21.60	5.60	11.26	6.44	22.33	8.35	8.72
Lubelskie	11.19	21.72	7.59	15.20	6.91	20.83	8.42	15.19
Lubuskie	4.21	19.97	5.16	9.83	8.51	21.51	5.31	11.77
Łódzkie	10.09	23.10	1.13	11.21	10.48	21.82	1.57	10.88
Małopolskie	9.21	21.95	2.96	15.90	8.95	22.72	6.12	10.42
Mazowieckie	11.90	20.85	5.30	10.34	12.19	20.55	4.12	9.02
Opolskie	10.02	21.22	1.04	12.72	5.59	20.23	3.38	16.01
Podkarpackie	11.54	22.02	4.11	12.80	8.81	21.80	4.64	13.00
Podlaskie	12.87	22.41	3.11	9.48	6.88	21.47	7.01	11.58
Pomorskie	5.62	18.56	3.28	9.02	5.58	19.14	4.02	8.86
Śląskie	9.90	21.31	1.78	15.60	9.42	22.31	2.20	11.91
Świętokrzyskie	8.63	19.90	0.76	12.77	9.56	20.85	2.81	13.02
Warmińsko-mazurskie	12.97	22.02	8.50	11.58	8.84	21.17	8.86	11.84
Wielkopolskie	4.18	21.18	3.48	11.13	6.04	21.51	9.71	12.56
Zachodniopomorskie	8.47	21.88	3.39	13.01	7.25	19.78	4.97	10.60
POLSKA	8.37	20.85	3.48	11.57	8.42	20.91	5.72	10.06

Source: own study based on the data from the Statistics Poland survey of heat from renewable sources (G-02o).

Calorific value is the amount of energy released when 1 kg of fuel burns; energy density is the amount of energy released when 1 m^3 of fuel burns. In the G-02o report, each of the surveyed entities provided calorific values of the fuels, but these values differed even for the same fuel. That is why it was decided to use the concept of average boiler efficiency for biogas and solid biofuel given by instead of calorific value to capture intertemporal technological changes.

$$\text{Efficiency} = \frac{\text{total heat production in voivodeship}}{\text{total fuel consumption in voivodeship}} \left(\frac{\text{MJ}}{\text{m}^3} \text{ or } \frac{\text{MJ}}{\text{kg}} \right) \tag{5}$$

The introduced concept of "efficiency" should not be confused with the concept of energy density and calorific value present in science and technology. In our case, "efficiency" should be treated as an average boiler efficiency for different fuels used in varying proportions and over different periods of time (the installations surveyed were not always in operation throughout the reporting period). In order to illustrate the differences between average boiler efficiency in a given voivodeship and the calorific value of fuel consumed often from different sources, Table 3 also includes the weighted average heat density/calorific value calculated for a given voivodeship. As it is easy to see, the calorific value of biogas is a subject to lower fluctuations than the calorific value of biomass—this is due to the fact that biogases from various sources often have similar calorific values, while in the case of biomass, a much greater variability can be observed depending on the type of fuel. Furthermore, only a part of the biomass obtained is used for heat production; in 2012, in Dolnośląskie Voivodeship, around 24% of the biomass was used for heat production. In the case of Opolskie Voivodeship, it was slightly over 11% and in Łódzkie Voivodeship it was 10.7%. In Śląskie and Świętokrzyskie Voivodeships, it was 16.8% and 11.3%, respectively. Such low biomass consumption for heat generation translated into low average boiler efficiency.

From 2012 to 2018, biogas consumption increased by 98% and production by 99%, which, at the country level, translated into a 0.6% increase in biogas efficiency from 8.37 to 8.42 MJ per kilogram. At the same time, changes in regional terms are very uneven and often inconsistent with the national direction of change. In nine out of sixteen voivodeships, the efficiency of biogas decreased. Undoubtedly, the greatest change took place in Mazowieckie voivodeship, where the production of heat from biogas in 2012–2018 increased ~4.85 times, and the share of production in Poland increased from 13% to 31%. At the same time, biogas consumption increased ~4.73 times, which in total resulted in increase in efficiency from 11.9 to 12.19 MJ per kilogram. This is largely due to changes in the structure of biogas sources. The share of biogas of agricultural origin increased from 29.5% to 45.7% at the expense of biogas from sewage treatment plants. As indicated by Kwaśny et al. [61], biogas of agricultural origin has, on average, a higher methane content than biogas from sewage treatment plants.

On a national scale, during the period under examination, the use of biomass fell by 16%, while production increased by 38%, which together increased the efficiency of biomass by as much as 65%. The increase in efficiency occurred in each of the voivodeships. In 2012, only in Warmińsko-Mazurskie Voivodeship was the efficiency of using one kilogram of biomass over 8 MJ per kilogram. For six years, this level was exceeded by Lubelskie, Kujawsko-Pomorskie, and Wielkopolskie Voivodeships, the latter with the highest change—i.e., from 3.48 to 9.71. In Wielkopolskie Voivodeship, at the same time, the consumption of resources decreased by nearly 50% and production increased by 44%.

Naturally, the question arises as to which voivodeships are the most effective in terms of using biocomponents for heat production. Comparing, for example, Podkarpackie and Opolskie Voivodeships, it is easy to say that Podkarpackie is generally more efficient, both in terms of biogas and biomass use. However, it is not easy to compare Dolnośląskie and Kujawsko-Pomorskie Voivodships. Dolnośląskie Voivodeship makes more efficient use of biogas while Kujawsko-Pomorskie Voivodeship of biomass. Using the partial ranking, the voivodeships can be ranked according to unambiguous comparisons only. The results of the partial ranking are shown in Figures 1 and 2. If there is an arrow from one object to another, it means that the first of them is better than the second one in terms of each analyzed variable. The situation when one can move from the first to the second object through a sequence of arrows with the same direction is interpreted in a similar way. If there is no sequence of arrows between objects, such objects are incomparable. Clusters of voivodeships with similar levels of analysis variables are marked with a gradient color scale, staring from green (the best cluster) and ending on red (the worst cluster).

In 2012, six groups of voivodeships could be distinguished, where a group is shown in the figure as voivodeships at the same height. The best voivodeships are Dolnośląskie and Warmińsko-Mazurskie Voivodeships, which are leaders in efficiency in biogas and biomass production, respectively. Świętokrzyskie Voivodeship, which is the least efficient in terms of biomass use, is at the other end, together with Pomorskie and Wielkopolskie Voivodeships from the fifth group, with a low biogas efficiency. The 2012–2018 period brought big changes in the partial ranking of voivodeships. The number of groups decreased from six to four. Małopolskie, Dolnośląskie, and Mazowieckie Voivodeships joined the group of leaders. Among the weakest, compared to 2012, only Pomorskie Voivodeship remained. Together with Opolskie Voivodeship, they were characterized by the lowest efficiency of biogas. Despite the lowest efficiency of biomass in 2018, Łódzkie Voivodeship achieved the second best result in terms of biogas efficiency, which resulted in its presence in the second highest group of voivodeships.

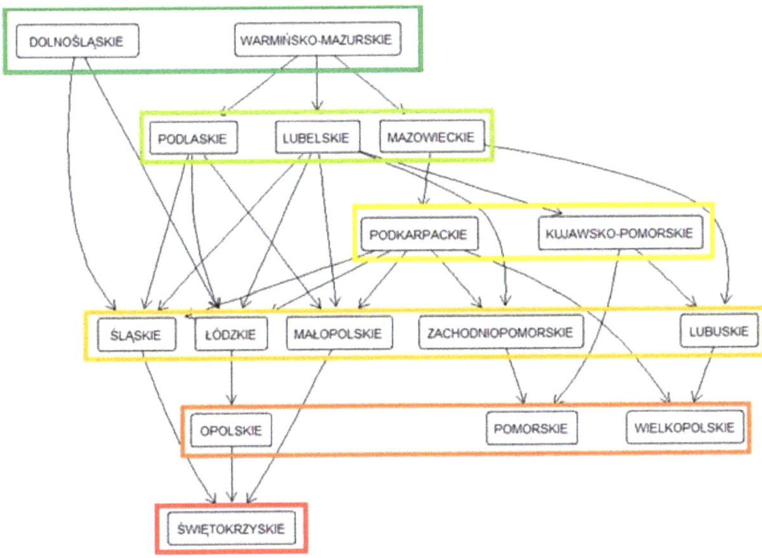

Figure 1. Partial ranking of voivodeships in terms of biogas and biomass efficiency for 2012.

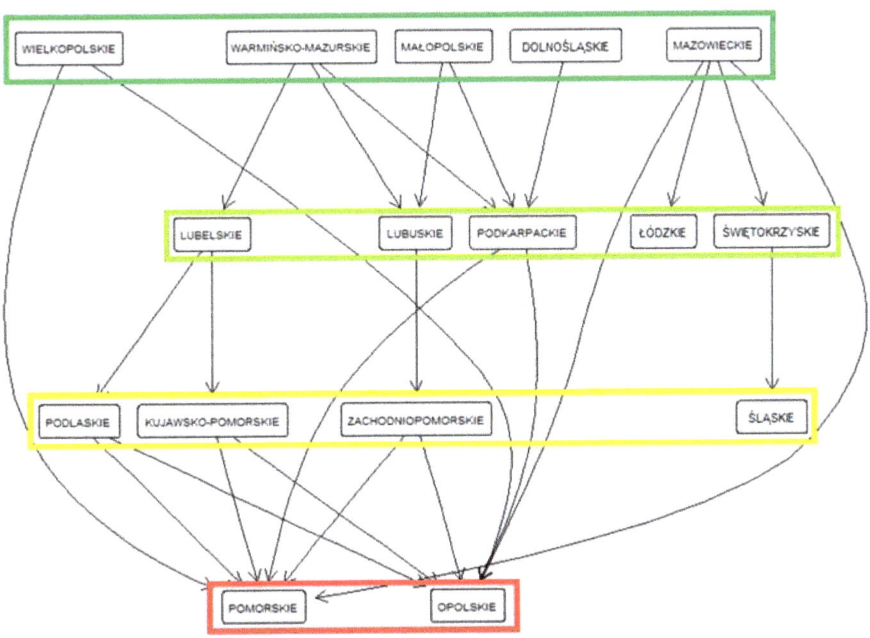

Figure 2. Partial ranking of voivodeships in terms of biogas and biomass efficiency in 2018.

Using the DEA method, the analysis of relative effectiveness of voivodeships in 2012 and 2018 was carried out. The results are presented in Table 4. The DEA analysis showed that in 2012 the following voivodeships were effective: Kujawsko-Pomorskie, Lubelskie, and Warmińsko-Mazurskie Voivodeships. Within six years, the group of effective voivodeships was joined by Wielkopolskie Voivodeship, while Warmińsko-Mazurskie voivodeship were minimally distant from the efficiency frontier. In the period 2012–2018, the disparities in terms of effectiveness decreased. The lowest relative efficiency increased from 17.8% to 23.2%. In 13 out of 16 voivodeships, relative efficiency increased or did not change.

Table 4. Relative efficiency of voivodeships in the production of heat from biogas and biomass including a sensitivity analysis in 2012 and 2018.

Voivodeships	Relative Efficiency		95% interval	
	2012	2018	2012	2018
Dolnośląskie	0.350	0.554	(0.275; 0.677)	(0.074; 0.709)
Kujawsko-Pomorskie	1.000	1.000	(0.526; 1.000)	(0.832; 1.000)
Lubelskie	1.000	1.000	(1.000; 1.000)	(0.189; 1.000)
Lubuskie	0.606	0.64	(0.212; 0.963)	(0.578; 1.000)
Łódzkie	0.178	0.232	(0.116; 0.377)	(0.128; 0.308)
Małopolskie	0.407	0.685	(0.311; 0.655)	(0.233; 0.691)
Mazowieckie	0.648	0.628	(0.788; 1.000)	(0.055; 0.643)
Opolskie	0.152	0.391	(0.128; 0.608)	(0.263; 0.501)
Podkarpackie	0.519	0.548	(0.259; 0.663)	(0.207; 0.807)
Podlaskie	0.610	0.793	(0.359; 0.681)	(0.075; 0.921)
Pomorskie	0.389	0.468	(0.237; 0.854)	(0.441; 1.000)
Śląskie	0.243	0.295	(0.138; 0.345)	(0.259; 0.469)
Świętokrzyskie	0.350	0.350	(0.206; 0.570)	(0.315; 0.969)
Warmińsko-Mazurskie	1.000	0.991	(0.373; 1.000)	(0.214; 1.000)
Wielkopolskie	0.453	1.000	(0.402; 0.998)	(1.000; 1.000)
Zachodniopomorskie	0.746	0.589	(0.620; 1.000)	(0.212, 0.912)

The simulation of the impact of variable changed in the range of ±1% in the DEA results, showing that the relative efficiency of particularly large voivodeships is sensitive to variable changes. For small voivodeships, e.g., Opolskie, the distribution of results is symmetrical, and the range of results is based on quantiles of the order of 0.025 and 0.975 is narrow. For highly efficient voivodeships, the distribution of results was strongly asymmetric and the range was wide (95%) (Figure 3).

The completeness of the G-02o report for 2012 in the case of biogas plants amounted to 99%, while in the case of installations burning solid biomass it was over 98%. In 2018, in the case of biogas plants, the completeness amounted to 92.8%, and in the case of installations burning solid biomass, it was 89%.

In DEA, effective units set the efficiency frontier. Inefficient units lie in the Production Possibility Set. For each of the years, both the efficiency frontier and the Production Possibility Set will be different, so a decrease in the relative efficiency of a voivodeship in subsequent years does not mean that the situation in the voivodeship deteriorated.

The efficiency frontier set by DEA for the CCR model, where the total number of variables (inputs and outputs) does not exceed three in total, can be easily visualized. To this end, indicators were calculated:

$$\text{Ratio1} = \frac{\text{heat production from biogas} + \text{heat production from biomass}}{\text{consumption of biogas}} \left(\frac{\text{MJ}}{\text{m}^3}\right) \quad (6)$$

$$\text{Ratio2} = \frac{\text{heat production from biogas} + \text{heat production from biomass}}{\text{consumption of biomass}} \left(\frac{\text{MJ}}{\text{kg}}\right) \quad (7)$$

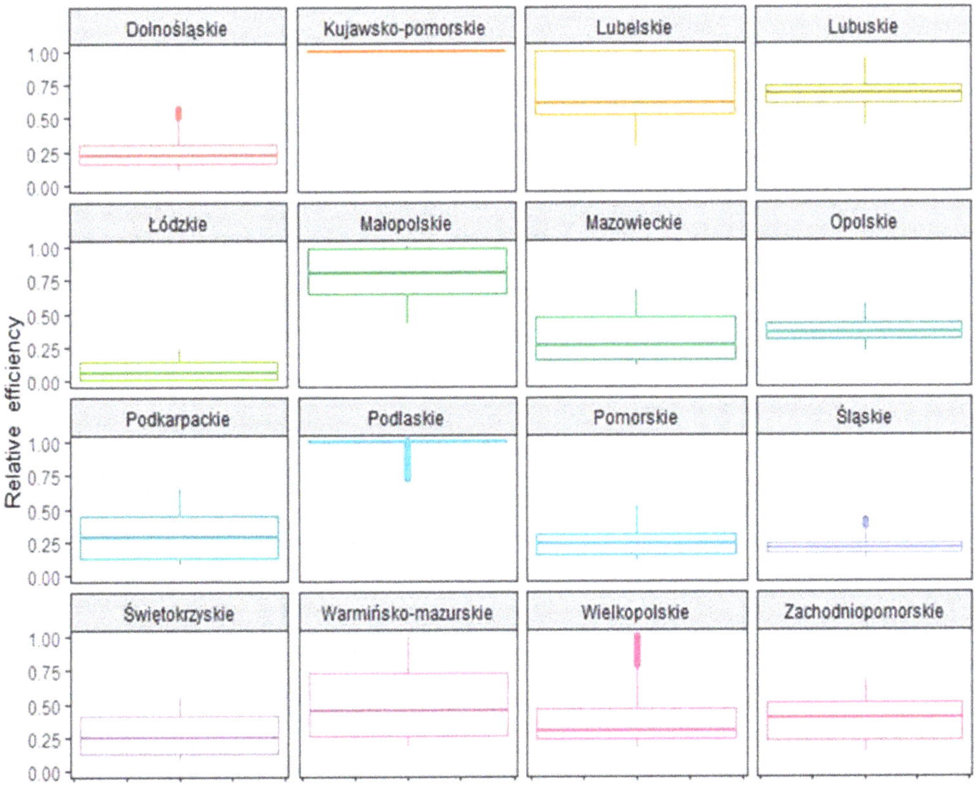

Figure 3. Boxplots of relative efficiency obtained through simulation for voivodeships in 2018.

The numerators of both indicators include heat production from both biogas and biomass, while the denominator refers to the consumption of only one of these components. Thus, these ratios cannot be interpreted in terms of efficiency of heat production from biogas nor biomass, respectively. They are used because they enable the visualization of the efficiency frontier. The results are shown in Figures 4 and 5, covering the years 2012 and 2018, maintaining the same range of scale on both axes. In this way, absolute changes in the years compared can be seen. The pink polygon is the Production Possibility Set and is designated as the convex hull of the effective units, their projections on the axes, and the starting point of the coordinate system. The Production Possibility Set was determined separately for each of the years according to the effective units of that year.

Comparing the effective units in 2012 and 2018, it can be seen that their efficiency frontier moved upwards by 56% due to biogas and 21% due to biomass. In the analyzed period, the change of the efficiency limit due to biomass was actually caused by Kujawsko-Pomorskie Voivodeship, whose share in newly produced heat was 78%. Despite a large relative increase in the production of heat from biogas by 99% compared to the production of heat from biomass by 38%, the efficiency frontier for biogas did not change much. This is due to the fact that the share of biogas heat production in biomass and biogas heat production was 5% in 2012 and 7.3% in 2018.

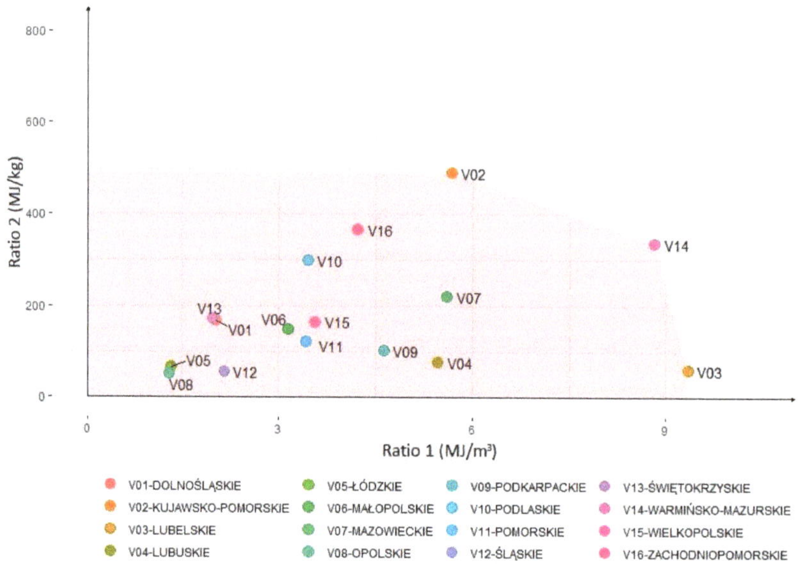

Figure 4. Efficiency of heat production from biogas and biomass in 2012 in Poland.

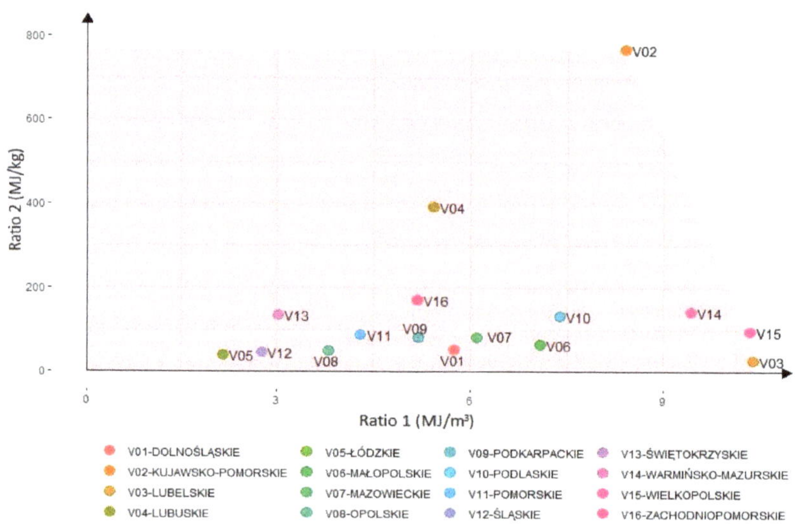

Figure 5. Efficiency of heat production from biogas and biomass in 2018 in Poland.

4. Discussion

The importance of assessing the potential of renewable energy sources and assessing energy efficiency in sustainable development is now growing worldwide [62]. Poschl [63] points to ways to increase the efficiency of biogas production by establishing that energy efficiency could be enhanced depending on the use of raw material resources and the adopted technological process. Similar conclusions are presented by Alluvione et al. [64], who note that energy efficiency of agriculture needs improvement to reduce the dependency on nonrenewable energy sources. Jalali Sepehr et al. [65] indicate that great opportunities in this respect also arise for developing countries with low incomes, which can achieve high energy efficiency results.

Kaygusuz [66] states that, due to high environmental pollution, renewable energy sources appear to be one of the most efficient and effective solutions for developing clean and sustainable energy. This is particularly important due to inappropriate disposal of agricultural residues (e.g., burning of straw), leading to waste of energy potential and atmospheric environmental problems, as shown in a Chinese study [67]. The use of renewable energy sources is also highlighted by Borhanazad et al. [68], in the context of reducing energy poverty in rural areas of Malaysia. The authors point out not only environmental issues, but, as Kumar did [69], also social and economic ones. Similarly, Pakistan has great potential for renewable energy sources, as Raza et al. [70] stress, adding, however, that in rural areas of Pakistan, this potential is not adequately exploited. Sutherland et al. [71] draw attention to rural areas and agriculture's renewable energy potential, pointing out that the agriculture sector plays an important role in renewable energy transitions, owing to its historical involvement in managing key resources, particularly land and biomass.

However, the rapid increase in the use of biomass for energy purposes is causing concern among many experts around the world, especially with regard to potential threats to sustainable development and food security [72]. This is due to the fact that the production of biomass for energy purposes in agriculture may compete with food production due to a reduction in the area under cultivation for food and feed for livestock [73,74]. Then, as Jasiulewicz [75] points out, the production of biomass for solid fuels should mainly use inferior quality soils, including set-aside and fallow soils, as well as soils contaminated with heavy metals, degraded, particularly unsuitable for the production of consumer crops. The problem of such competition is not observed in the case of biogas, since it is mainly produced from agricultural by-products [76], and the most commonly used substrates for biogas production are manure from cattle, pigs, and poultry litter [62]. This means organic waste that is unfit for consumption or not used for other purposes [77]. It seems, therefore, that biogas is a more acceptable option for energy production.

The results of our calculations, based only on what remains of agricultural production, indicated a significant decrease in the technical resources of biomass, which took place between 2012 and 2018. According to our calculations in 2012, the potential for obtaining thermal energy from biomass reached the level of 907 PJ (which fits well the value given by Bartoszewicz-Burczy [78]), of which 856 PJ came from solid biomass (65.9 million tons) and only 51 PJ from biogas (6.46 billion m^3). In 2018, the biomass potential decreased to the level of 752 PJ, where 700 PJ came from solid biomass (55.1 million tons) and 52 PJ from biogas (6.61 billion m^3). The decrease in the potential of solid biomass was caused by a decrease in the amount of wood that could be harvested from pruning (a decrease of about 1.9 million tons) and a decrease in surplus straw (about 2.8 million tons) and hay (about 8.8 million tons) that could be harvested. As can be seen, solid biomass, having a much greater share, is a more unstable energy source than biogas, which, after purification to the biomethane standard, could cover 20%–25% of the demand for natural gas in Poland (according to consumption for 2018). This creates wide opportunities for the development of this energy sector in the areas so far associated in Poland mainly with agricultural activities (e.g., Podlaskie Voivodeship).

In this context, it is also worth paying attention to other European countries that use biomass resources. As can be seen from the example of selected European countries (Table 5), the estimated biomass potential shows a large diversity, which results not only from the methodology of calculation, but also depends on the changes taking place in the biomass sources themselves (development of sewage infrastructure, change in the nature of crops, drought or legal restrictions). The biomass potential in Poland is comparable to countries with a similar area, such as Italy or Germany; however, per capita, it is one of the best among the compared countries in Central Europe (Table 5). Taking into account the consumption of natural gas in 2018, the maximum use of the potential of biogas production in Germany would cover about 10% of the country's demand for natural gas; in the case of Czech Republic, it would be around 12%; for Hungary this value is around 9%, while for Italy it is less than 7% [43].

Table 5. Biomass potential for Central European countries per capita (where: 1 PJ = 10^9 MJ, 1 GJ = 10^3 MJ).

Country	Total Biomass Potential (PJ)	Biomass Potential per Person (GJ/person)
Czech Republic	300	28.16
Germany	560–1050	6.74–12.64
Hungary	153–190	15.60–19.37
Italy	1094–1260	18.08–20.83
Poland	900	23.44
Slovenia	28–53	13.57–25.70
Slovakia	90	16.52

Source: [78].

The importance of renewable energy in the Polish economy is growing, although Poland still has one of the highest carbon dioxide emission figures in Europe in relation to electricity produced [79]. Therefore, the increase in energy efficiency in the regions may contribute to the increase in the national energy supply potential. The use of renewable resources as substitute energy sources is a factor improving the security of energy supply [80]. The increase in energy efficiency of renewable energy sources contributes to the reduction in greenhouse gas emissions [81]. It is, therefore, necessary to increase the efficiency and consumption of renewable energy [82].

The raw material resources and natural conditions, as well as modern technologies, are not always sufficient for the transition to renewable resources. It may turn out that the exploitation of a given energy resource will not be profitable without state support. Currently, such production in many countries is more expensive than energy produced from fossil resources [83]. Similar conclusions were reached by Radziszewska-Zielina and Rumin [84], indicating that unconventional energy sources in Poland are relatively expensive. However, the results of our analyses indicate a very high potential of both biomass and biogas, which could be used more widely, contributing to the reduction in fossil fuel use. A proper state energy policy, which is also indicated by Jedlińska [85] and action at a local level are therefore needed, especially in the era of energy transformation and withdrawal from coal in the energy sector.

5. Conclusions

The research carried out indicates that Poland has significant biomass and biogas potential. However, it is regionally differentiated. At a national scale, Pomorskie Voivodeship is the leader in terms of using the biogas production potential. The opposite is Podlaskie Voivodeship, which both in 2012 and 2018 was lowest ranked in the country. Comparing the years 2012 and 2018, it should be pointed out that, in Poland, the use of the potential for the production of heat from biogas and also from biomass almost tripled, although in the latter case to a lesser extent. Our analyses show that, in the same period, there was an increase in the efficiency of the use of both biogas and biomass, with a higher increase in efficiency for biomass. This is due to structural changes in energy carriers and an increase in the technical efficiency of heat production. The DEA analysis showed that the highest relative technical efficiency in 2012 was achieved by three voivodeships: Kujawsko-Pomorskie, Warmińsko-Mazurskie, and Lubelskie. In 2018, Wielkopolskie Voivodeship joined them. This was associated with the developing agri-food industry producing waste biomass yields, as well as an increase in the market for processed biomass (including an increase in the number of biogas plants from 183 in 2012 to 293 in 2018).

In the case of Podlaskie Voivodeship, which is distinguished in the country by the highest cattle population per 100 ha of farmland, the use of the biogas production potential reached only 2.1% in 2018. However, "activating" the potential to produce heat from biogas would ensure self-sufficiency in this region. The remaining regions, on the other hand, could achieve energy self-sufficiency if the use of solid biomass was increased. At the national scale, however, the potential of biogas may be more important than in the countries of Central Europe, as in the case of Poland, it may cover more than 20% of the demand for natural gas, which, in the case of the upcoming energy transformation (moving away from coal), may prove to be a very important contribution of Polish agriculture. This is confirmed by the results of our research, which indicates that biogas, as opposed to solid biomass, is a more stable energy source (52 PJ in 2018).

It should also be noted that the presented results are based on available statistical data. An obstacle to more accurate estimates was the lack of detailed, available data on elements of the potential of both biomass and biogas. These data, derived from farms and rural areas, may be incomplete, which may affect the accuracy of the calculations. In the case of estimating the potential of biogas production, due to the lack of data, the crop production grown for the input of biogas plants was not taken into account and nor was the share of postproduction waste supplied by the food industry. Moreover, the lower boundaries of the biogas production efficiency from a given raw material were taken into account. In the case of estimating the production potential of solid biomass, due to the lack of data, the share of postproduction waste supplied by the paper and cellulose industry was not taken into account.

However, it should be pointed out that the studies carried out by our team are one of a few in Poland, as well as in Central and Eastern European countries. However, taking into account the available literature, the presented work significantly fills the knowledge gap on the discussed problem.

Author Contributions: Conceptualization, G.Ś. and B.G.; methodology, M.C.-W., D.T. and S.W.; software, S.W. and D.T.; validation, S.W. and D.T.; formal analysis, B.G., D.T., S.W. and G.Ś.; investigation, J.G., M.C.-W., D.T. and S.W.; resources, S.W., M.C.-W. and D.T.; data curation, S.W. and D.T.; writing—original draft preparation, G.Ś., B.G., D.T., S.W. and M.C-W; writing—review and editing, B.G. and J.G.; visualization, B.G. and S.W.; supervision, G.Ś.; project administration, G.Ś., B.G. and J.G.; funding acquisition, G.Ś. and B.G. All authors have read and agreed to the published version of the manuscript.

Funding: This research received no external funding.

Institutional Review Board Statement: Not applicable.

Informed Consent Statement: Not applicable.

Conflicts of Interest: The authors declare no conflict of interest.

Appendix A

The equations describing the calculation of the ratio of heat production potential to consumption in Table 2:

$$\frac{heat}{total} consumption = \frac{heat\ production\ potential * 100\%}{total\ heat\ consumption}, \tag{A1}$$

where:

$$heat\ production\ potential = biogas\ heat\ p.p + straw\ heat\ p.p + hay\ heat\ p.p + wood\ heat\ p.p + energy\ crop\ heat\ p.p, \tag{A2}$$

heat p.p—heat production potential.

$$biogas\ heat\ p.p = biogas\ production\ potential \left[m^3\right] * 23 \left[\frac{MJ}{m^3}\right], \tag{A3}$$

$$biogas\ heat\ p.p = biogas\ production\ potential\ \left[m^3\right] * 23 \left[\frac{MJ}{m^3}\right], \quad (A4)$$

$$straw\ heat\ p.p = excess\ straw\ production\ [kg] * 13.1 \left[\frac{MJ}{kg}\right], \quad (A5)$$

$$hay\ heat\ p.p = excess\ hay\ production\ [kg] * 13.4 \left[\frac{MJ}{kg}\right], \quad (A6)$$

$$wood\ heat\ p.p = wood\ production\ potential\ [kg] * 12.4 \left[\frac{MJ}{kg}\right], \quad (A7)$$

$$energy\ crop\ heat\ p.p = energy\ crop\ production\ potential[kg] * 17 \left[\frac{MJ}{kg}\right]. \quad (A8)$$

The equations describing the calculation of the ratio of electricity production potential to electricity production in the studied years:

$$electricity/total\ production = \frac{electricity\ production\ potential * 100\%}{total\ electricity\ prodution}, \quad (A9)$$

where:

$$electricity\ production\ potential = biogas\ el.\ p.p + straw\ el.p.p + hay\ el.p.p + wood\ el.p.p + energy\ crop\ el.p.p, \quad (A10)$$

el.p.p—electricity production potential.

$$biogas\ el.p.p = biogas\ production\ potential\ \left[m^3\right] * 6.3 \left[\frac{kWh}{m^3}\right], \quad (A11)$$

$$straw\ el.p.p = excess\ straw\ production\ [kg] * 3.46 \left[\frac{kWh}{kg}\right], \quad (A12)$$

$$hay\ el.p.p = excess\ hay\ production\ [kg] * 3.46 \left[\frac{kWh}{kg}\right], \quad (A13)$$

$$wood\ el.p.p = wood\ production\ potential\ [kg] * 4.8 \left[\frac{kWh}{kg}\right], \quad (A14)$$

$$energy\ crop\ el.p.p = energy\ crop\ production\ potential[kg] * 4.1 \left[\frac{kWh}{kg}\right], \quad (A15)$$

Ratio of heat and electricity production potential from solid biomass to biogas production potential:

$$solid\ biomass/biogas\ in\ heat\ production = \frac{solid\ biomass\ heat\ production\ potential}{biogas\ heat\ p.p}, \quad (A16)$$

$$solid\ biomass/biogas\ in\ electricity\ production = \frac{solid\ biomass\ electricity\ production\ potential}{biogas\ el.\ p.p}. \quad (A17)$$

Appendix B

Table A1. Consumption of heat energy and production of electricity.

Voivodeships	2012		2018	
	Thermal Energy Consumption (GJ)	Electricity Production (GWh)	Thermal Energy Consumption (GJ)	Electricity Production (GWh)
Dolnośląskie	12,587,790.0	13,567.7	14,067,201.0	9917.7
Kujawsko-pomorskie	9,989,713.3	3177.6	9,790,920.0	6798.8
Lubelskie	11,249,543.0	2100.7	7,761,207.0	2066.8
Lubuskie	3,476,022.3	2524.7	3,665,305.0	3290.8
Łódzkie	17,100,953.0	34,968.5	14,969,554.0	38,641.0
Małopolskie	14,244,845.0	6384.5	14,587,770.0	5888.7
Mazowieckie	38,045,642.0	22,090.0	38,352,261.0	30,441.0
Opolskie	3,743,473.0	8442.2	3,786,207.0	10,087.2
Podkarpackie	6,682,073.0	2664.7	5,101,140.0	2462.0
Podlaskie	6,840,182.0	723.4	6,362,132.0	1051.2
Pomorskie	13,353,847.0	3426.3	13,953,962.0	4104.9
Śląskie	32,089,653.0	31,249.5	29,878,280.0	24,905.9
Świętokrzyskie	4,371,481.0	8268.3	3,514,462.0	11,213.1
Warmińsko-mazurskie	7,312,422.0	745.6	7,361,499.0	1170.2
Wielkopolskie	14,447,636.5	13,112.6	12,129,926.0	9840.7
Zachodniopomorskie	7,934,129.0	8692.8	9,230,561.0	8159.4
POLSKA	203,469,405.1	162,139.1	194,512,387.0	170,039.5

Source: Local Data Bank of Statistics Poland.

References

1. Howells, M.; Hermann, S.; Welsch, M.; Bazilian, M.; Segerström, R.; Alfstad, T.; Ramma, I. Integrated analysis of climate change, land-use, energy and water strategies. *Nat. Clim. Change* **2013**, *3*, 621–626. [CrossRef]
2. Ritchie, H.; Roser, M. Fossil Fuels. *Our World Data*. 2017. Available online: https://ourworldindata.org/fossil-fuels (accessed on 16 December 2020).
3. Arıoğlu Akan, M.; Selam, A.; Oktay Fırat, S.; Er Kara, M.; Özel, S. A Comparative Analysis of Renewable Energy Use and Policies: Global and Turkish Perspectives. *Sustainability* **2015**, *7*, 16379–16407. [CrossRef]
4. Moomaw, W.; Yamba, F.; Kamimoto, M.; Maurice, L.; Nyboer, J.; Urama, K.; Weir, T. Introduction. In *IPCC Special Report on Renewable Energy Sources and Climate Change Mitigation*; Edenhofer, O., Pichs-Madruga, R., Sokona, Y., Seyboth, K., Matschoss, P., Kadner, S., Zwickel, T., Eickemeier, P., Hansen, G., Schlömer, S., et al., Eds.; Cambridge University Press: Cambridge, UK, 2011.
5. Eurostat. Energy Statistics—An Overview. Available online: https://ec.europa.eu/eurostat/statistics-explained (accessed on 17 December 2020).
6. Chel, A.; Kaushik, G. Renewable energy for sustainable agriculture. *Agron. Sustain. Dev.* **2011**, *31*, 91–118. [CrossRef]
7. Bishoge, O.K.; Kombe, G.G.; Mvile, B.N. Renewable energy for sustainable development in sub-Saharan African countries: Challenges and way forward. *J. Renew. Sustain. Energy* **2020**, *12*, 052702. [CrossRef]
8. Gielen, D.; Boshell, F.; Saygin, D.; Bazilian, M.D.; Wagner, N.; Gorini, R. The role of renewable energy in the global energy transformation. *Energy Strateg. Rev.* **2019**, *24*, 38–50. [CrossRef]
9. Kibria, G. Sustainable Energy for Rural Development in Developing Countries- Economic, Social, and Environmental Benefits of Renewable Energy. A Case Study. *Sci. Technol. Artic.* **2015**, *37*. [CrossRef]
10. Poggi, F.; Firmino, A.; Amado, M. Planning renewable energy in rural areas: Impacts on occupation and land use. *Energy* **2018**, *155*, 630–640. [CrossRef]
11. Gradziuk, P.; Gradziuk, B. Renewable Energy Sources as a Development Opportunity for Peripheral Areas. *Econ. Reg. Stud.* **2020**, *13*, 184–198. [CrossRef]

12. Lemaire, X. Off-Grid electrification with solar home systems: The experience of a fee-for-service concession in South Africa. *Energy Sustain. Dev.* **2011**, *15*, 277–283. [CrossRef]
13. Abolhosseini, S.; Heshmati, A.; Altmann, J. A Review of Renewable Energy Supply and Energy Efficiency Technologies. *IZA Discuss. Pap. Ser.* **2014**, *8145*, 1–35.
14. European Parliament. *Directive 2009/28/EC of the European Parliament and of the Council of 23 April 2009 on the Promotion of the Use of Energy from Renewable Sources and Amending and Subsequently Repealing Directives 2001/77/EC and 2003/30/EC*; European Parliament: Brussels, Belgium, 2009.
15. European Parliament. *Communication from the Commission to the European Parliament and the Council. Energy Efficiency and Its Contribution to Energy Security and the 2030 Framework for Climate and Energy Policy*; European Parliament: Brussels, Belgium, 2014.
16. European Commission. *Rozporządzenie Parlamentu Europejskiego i Rady w Sprawie Rocznych Ograniczeń Emisji Gazów Cieplarnianych Przez Państwa Członkowskie w Latach 2021–2030 na Rzecz Stabilnej Unii Energetycznej i w Celu Wywiązania się ze Zobowiązań Wynikających z Porozumienia Paryskiego, Oraz Zmieniające Rozporządzenie Parlamentu Europejskiego i Rady nr 525/2013 w Sprawie Mechanizmu Monitorowania i Sprawozdawczości w Zakresie Emisji Gazów Cieplarnianych Oraz Zgłaszania Innych Informacji Mających Znaczenie dla Zmiany Klimatu (Regulation of the European Parliament and of the Council on Binding Annual Greenhouse Gas Emission Reductions by Member States from 2021 to 2030 for a Resilient Energy Union and to Meet Commitments under the Paris Agreement and Amending Regulation (EU) No 525/2013 of the European Parliament and the Council on a Mechanism for Monitoring and Reporting Greenhouse Gas Emissions and Other Information Relevant to Climate Change)*; European Commission: Brussels, Belgium, 2016.
17. Government Gazzete of the Republic of Poland. *Strategia na rzecz Odpowiedzialnego Rozwoju (Strategy for Responsible Development) M.P poz. 260 Uchwała NR 8 Rady Ministrów z dnia 14 lutego 2017 roku w Sprawie Przyjęcia Strategii na rzecz Odpowiedzialnego Rozwoju do Roku 2020 (z perspektywą do 2030 roku)*; The Polish Monitor: Warsaw, Poland, 2017.
18. EPA. Greenhouse Gas Emissions. Global Emissions by Economic Sector, Washington: United States Environmental Protection Agency. Available online: https://www.epa.gov/ (accessed on 16 December 2020).
19. GUS. Energy, Central Statistical Officce, Warsaw. Available online: https://stat.gov.pl (accessed on 12 December 2020).
20. Mirowski, T.; Jach-Nocoń, M.; Jelonek, I.; Nocoń, A. The new meaning of solid fuels from lignocellulosic biomass used in low-emission automatic pellet boilers. *Energy Policy J.* **2020**, *23*, 75–86.
21. Pawlak, J. Produkcja i zużycie energii odnawialnej w Polsce ze szczególnym uwzględnieniem rolnictwa (Production and Consumption af Renewable Energy in Poland with a Special Regard to Agriculture). *Probl. Agric. Eng.* **2016**, *4*, 67–76.
22. Sulewski, P.; Majewski, E.; Wąs, A. Miejsce i rola rolnictwa w produkcji energii odnawialnej w Polsce i UE (The importance of agriculture in the renewable energy production in poland and the EU). *Zagadnienia Ekonomiki Rolnej* **2017**, *1*, 50–74. [CrossRef]
23. Woźniak, M. Zrównoważona gospodarka energetyczna na obszarach wiejskich w Polsce (A sustainable energy economy in rural areas in Poland). *Policy J.* **2018**, *21*, 69–84.
24. Hengeveld, E.J.; Bekkering, J.; van Gemert, W.J.T.; Broekhuis, A.A. Biogas infrastructures from farm to regional scale, prospects of biogas transport grids. *Biomass Bioenergy* **2016**, *86*, 43–52. [CrossRef]
25. Carrosio, G. Energy production from biogas in the Italian countryside: Policies and organizational models. *Energy Policy* **2013**, *63*, 3–9. [CrossRef]
26. Yang, Y.; Zhang, P.; Li, G. Regional differentiation of biogas industrial development in China. *Renew. Sustain. Energy Rev.* **2012**, *16*, 6686–6693. [CrossRef]
27. Yin, D.; Liu, W.; Zhai, N.; Wang, Y.; Ren, C.; Yang, G. Regional differentiation of rural household biogas development and related driving factors in China. *Renew. Sustain. Energy Rev.* **2017**, *67*, 1008–1018. [CrossRef]
28. Roubík, H.; Mazancová, J.; Rydval, J.; Kvasnička, R. Uncovering the dynamic complexity of the development of small–scale biogas technology through causal loops. *Renew. Energy* **2019**, 1–21. [CrossRef]
29. Wąs, A.; Sulewski, P.; Krupin, V.; Popadynets, N.; Malak-Rawlikowska, A.; Szymańska, M.; Wysokiński, M. The Potential of Agricultural Biogas Production in Ukraine—Impact on GHG Emissions and Energy Production. *Energies* **2020**, *13*, 5755. [CrossRef]
30. Groscurth, H. Total costs and benefits of biomass in selected regions of the European Union. *Energy* **2000**, *25*, 1081–1095. [CrossRef]
31. Sağlam, U. The Efficiency Assessment of Renewable Energy Sources with Data Envelopment Analysis. *SSRN Electron. J.* **2018**, 1–13. [CrossRef]
32. Sornek, K. Prototypical Biomass-Fired Micro-Cogeneration Systems—Energy and Ecological Analysis. *Energies* **2020**, *13*, 3909. [CrossRef]
33. Chodkowska-Miszczuk, J.; Szymańska, D. Agricultural biogas plants—A chance for diversification of agriculture in Poland. *Renew. Sustain. Energy Rev.* **2013**, *20*, 514–518. [CrossRef]
34. Lombard, A.; Ferreira, S.L.A. The spatial distribution of renewable energy infrastructure in three particular provinces of South Africa. Bulletin of Geography. *Soc. Econ. Ser.* **2015**, *30*, 71–85. [CrossRef]
35. Wang, Q.; Kwan, M.-P.; Fan, J.; Zhou, K.; Wang, Y.-F. A study on the spatial distribution of the renewable energy industries in China and their driving factors. *Renew. Energy* **2019**, *139*, 161–175. [CrossRef]
36. Stoeglehner, G.; Niemetz, N.; Kettl, K.H. Spatial dimensions of sustainable energy systems: New visions for integrated spatial and energy planning. *Energy Sustain. Soc.* **2011**, *1*, 2. [CrossRef]
37. Moreda, I.L. The potential of biogas production in Uruguay. *Renew. Sustain. Energy Rev.* **2016**, *54*, 1580–1591. [CrossRef]

38. Roubík, H.; Mazancova, J. Suitability of small-scale biogas systems based on livestock manure for the rural areas of Sumatra. *Environ. Dev.* **2020**, *33*, 100505. [CrossRef]
39. Van Zalk, J.; Behrens, P. The spatial extent of renewable and non-renewable power generation: A review and meta-analysis of power densities and their application in the U.S. *Energy Policy* **2018**, *123*, 83–91. [CrossRef]
40. Park, B.; Hur, J. Spatial Prediction of Renewable energy resources for reinforcing and expanding power grids. *Energy* **2018**, *164*, 757–772. [CrossRef]
41. Frantál, B.; Nováková, E. En the spatial differentiation of energy transitions: Exploring determinants of uneven wind energy developments in the Czech Republic. *Morav. Geogr. Rep.* **2019**, *27*. [CrossRef]
42. Zawada, M. Spatial differentiation of renewable energy sources use in the European Union countries in the years 2001–2013. Materials, Methods & Technologies. *J. Int. Sci. Publ.* **2015**, *9*, 80–94.
43. Scarlat, N.; Fahl, F.; Dallemand, J.-F.; Monforti, F.; Motola, V. A spatial analysis of biogas potential from manure in Europe. *Renew. Sustain. Energy Rev.* **2018**, *94*, 915–930. [CrossRef]
44. Mardani, A.; Streimikiene, D.; Balezentis, T.; Saman, M.; Nor, K.M.; Khoshnava, S. Data Envelopment Analysis in Energy and Environmental Economics: An Overview of the State-of-the-Art and Recent Development Trends. *Energies* **2018**, *11*, 2002. [CrossRef]
45. Olanrewaju, O.; Jimoh, A.; Kholopane, P. Integrated IDA-ANN-DEA for assessment and optimization of energy consumption in industrial sectors. *Energy* **2012**, *46*, 629–635. [CrossRef]
46. Lee, S.K.; Mogi, G.; Hui, K.S. A fuzzy analytic hierarchy process (AHP)/data envelopment analysis (DEA) hybrid model for efficiently allocating energy R&D resources: In the case of energy technologies against high oil prices. *Renew. Sustain. Energy Rev.* **2013**, *21*, 347–355.
47. Han, Y.; Geng, Z.; Gu, X.; Zhu, Q. Energy efficiency analysis based on DEA integrated ISM: A case study for Chinese ethylene industries. *Eng. Appl. Artif. Intell.* **2015**, *45*, 80–89. [CrossRef]
48. Xu, T.; You, J.; Li, H.; Shao, L. Energy Efficiency Evaluation Based on Data Envelopment Analysis: A Literature Review. *Energies* **2020**, *13*, 3548. [CrossRef]
49. Farrell, M.J. The Measurement of Productive Efficiency. *J. R. Stat. Soc. Ser. A* **1957**, *120*, 253–281. [CrossRef]
50. Mardani, A.; Zavadskas, E.K.; Streimikiene, D.; Jusoh, A.; Khoshnoudi, M. A comprehensive review of data envelopment analysis (DEA) approach in energy efficiency. *Renew. Sustain. Energy Rev.* **2017**, *70*, 1298–1322. [CrossRef]
51. Shabanpour, H.; Yousefi, S.; Saen, R.F. Forecasting efficiency of green suppliers by dynamic data envelopment analysis and artificial neural networks. *J. Clean. Prod.* **2017**, *142*, 1098–1107. [CrossRef]
52. Charnes, A.; Cooper, W.W.; Rhodes, E. Measuring the efficiency of decision making units. *Eur. J. Oper. Res.* **1978**, *2*, 429–444. [CrossRef]
53. Charnes, A.; Cooper, W.W.; Golany, B.; Seiford, L.M.; Stutz, J. Foundations of Data Envelopment. Analysis for Pareto–Koopman's Efficient Empirical Production Functions. *J. Econom.* **1985**, *30*, 91–107. [CrossRef]
54. Coll-Serrano, V.; Bolos, V.; Benitez Suarez, R. deaR: Conventional and Fuzzy Data Envelopment. Analysis. R Package v1.2.1. Available online: https://CRAN.R-project.org/package=deaR (accessed on 12 December 2020).
55. Bujakowski, W.; Barbacki, A.; Grzybek, A.; Hołojuch, G.; Pająk, L.; Skoczek, A.; Skrzypczak, M.; Skrzypczak, S. *Opracowanie Metody Programowania i Modelowania Systemów Wykorzystania Odnawialnych źródeł Energii na Terenach Nieprzemysłowych Województwa Sląskiego, wraz z Programem Wykonawczym dla Wybranych Obszarów Województwa—Część I: Metodyka Opracowania*; Instytut Gospodarki Surowcami Mineralnymi i Energią PAN: Kraków, Poland, 2005; p. 55.
56. Ludwicka, A.; Grzybek, A. Bilans masy rolnej (słomy) na potrzeby energetyki (Balance of the agricultural biomass (straw) for energy purposes). *Problemy Inżynierii Rolniczej* **2010**, *18*, 101–110.
57. Kuś, J.; Faber, A. Alternatywne kierunki produkcji rolniczej. *Studia i Raporty* **2007**, *7*, 138–158.
58. Lach, A.; Dyjakon, A.; Noszczyk, T. Sady jabłoniowe jako lokalne źródło biomasy do ogrzewania gospodarstw domowych w województwie mazowieckim (Apple orchards as a source of local biomass for households heating in the Mazowieckie Voivodeship). In *Rolnictwo XXI Wieku—Problemy i Wyzwania 2018*; Łuczycka, D., Ed.; Idea Knowledge Future: Wroclaw, Poland, 2018; pp. 528–539.
59. GUS. Energy from Renewable Sources in 2018. Statistical Analyses. Warsaw. Available online: http://stat.gov.pl (accessed on 16 December 2020).
60. Tomczyk, M. Gospodarczo-obronne determinanty polityki energetycznej Polski (Poland's present-day energy management). *Zeszyty Naukowe Wydziału Zarządzania i Dowodzenia Akademii Sztuki Wojennej* **2017**, *1*, 181–195.
61. Kwaśny, J.; Banach, M.; Kowalski, Z. Przegląd Technologii Produkcji Biogazu (Technologies of biogas production from different sourcesA review). *Chemia Czasopismo Techniczne Tech. Trans. Chem.* **2012**, *17*, 83–102.
62. Damyanova, S.; Beschkov, V. Biogas as a Source of Energy and Chemicals. Biorefinery Concepts, Energy and Products. In *Biorefinery Concepts*; IntechOpen: London, UK, 2020. [CrossRef]
63. Pöschl, M.; Ward, S.; Owende, P. Evaluation of energy efficiency of various biogas production and utilization pathways. *Appl. Energy* **2010**, *87*, 3305–3321. [CrossRef]
64. Alluvione, F.; Moretti, B.; Sacco, D.; Grignani, C. EUE (energy use efficiency) of cropping systems for a sustainable agriculture. *Energy* **2011**, *36*, 4468–4481. [CrossRef]

65. Jalali Sepehr, M.; Haeri, A.; Ghousi, R. A cross-country evaluation of energy efficiency from the sustainable development perspective. *Int. J. Energy Sect. Manag.* **2019**, *13*, 991–1019. [CrossRef]
66. Kaygusuz, K. Sustainable energy, environmental and agricultural policies in Turkey. *Energy Convers. Manag.* **2010**, *51*, 1075–1084. [CrossRef]
67. Wang, B.; Song, J.; Ren, J.; Li, K.; Duan, H.; Wang, X. Selecting sustainable energy conversion technologies for agricultural residues: A fuzzy AHP-VIKOR based prioritization from life cycle perspective. *Resour. Conserv. Recycl.* **2019**, *142*, 78–87. [CrossRef]
68. Borhanazad, H.; Mekhilef, S.; Saidur, R.; Boroumandjazi, G. Potential application of renewable energy for rural electrification in Malaysia. *Renew. Energy* **2013**, *59*, 210–219. [CrossRef]
69. Kumar, M. Social, Economic, and Environmental Impacts of Renewable Energy Resources. In *Wind Solar Hybrid Renewable Energy System*; Okedu, K.E., Ed.; IntechOpen: London, UK, 2020. [CrossRef]
70. Raza, M.Y.; Wasim, M.; Sarwar, M.S. Development of Renewable Energy Technologies in rural areas of Pakistan. *Energy Sources A Recovery Util. Environ. Eff.* **2019**, 1–21. [CrossRef]
71. Sutherland, L.-A.; Peter, S.; Zagata, L. Conceptualising multi-regime interactions: The role of the agriculture sector in renewable energy transitions. *Res. Policy* **2015**, *44*, 1543–1554. [CrossRef]
72. Popp, J.; Lakner, Z.; Harangi-Rákos, M.; Fári, M. The effect of bioenergy expansion: Food, energy, and environment. *Renew. Sustain. Energy Rev.* **2014**, *32*, 559–578. [CrossRef]
73. Ignaciuk, A.; Vöhringer, F.; Ruijs, A.; van Ierland, E.C. Competition between biomass and food production in the presence of energy policies: A partial equilibrium analysis. *Energy Policy* **2006**, *34*, 1127–1138. [CrossRef]
74. Muscat, A.; de Olde, E.M.; de Boer, I.J.M.; Ripoll-Bosch, R. The battle for biomass: A systematic review of food-feed-fuel competition. *Glob. Food Secur.* **2020**, *25*, 100330. [CrossRef]
75. Jasiulewicz, M. Production of energy from agribiomass in Poland in the background of selected countries of the European Union. *Roczniki Naukowe SERiA* **2015**, *8*, 94–99.
76. Owczuk, M.; Wardzińska, D.; Zamojska-Jaroszewicz, A.; Matuszewska, A. Wykorzystanie odpadów biodegradowalnych do produkcji biogazu jako alternatywnego źródła energii odnawialnej. *Studia Ecologiae et Bioethicae UKSW* **2013**, *11*, 133–144. [CrossRef]
77. Nevzorova, T.; Kutcherov, V. Barriers to the wider implementation of biogas as a source of energy: A state-of-the-art review. *Energy Strategy Rev.* **2019**, *26*, 100414. [CrossRef]
78. Bartoszewicz-Burczy, H. Biomass potential and its energy utilization in the Central European countries. *Energetyka* **2012**, *12*, 860–866.
79. Portal na Rzeczy Czystej Energii i Klimatu. Leonardo Energy. Polska nie Osiągnie celu OZE. Available online: https://leonardo-energy.pl/ (accessed on 18 December 2020).
80. Zhu, D.; Mortazavi, S.M.; Maleki, A.; Aslani, A.; Yousefi, H. Analysis of the robustness of energy supply in Japan: Role of renewable energy. *Energy Rep.* **2020**, *6*, 378–391. [CrossRef]
81. Harmsen, R.; Wesselink, B.; Eichhammer, W.; Worrell, E. The unrecognized contribution of renewable energy to europe's energy savings target. *Energy Policy* **2011**, *39*, 3425–3433. [CrossRef]
82. Garrett-Peltier, H. Green versus brown: Comparing the employment impacts of energy efficiency, renewable energy, and fossil fuels using an input-output model. *Econ. Model.* **2016**, *61*, 439–447. [CrossRef]
83. Janik, W.; Kaproń, H.; Paździor, A. Uwarunkowania rozwoju produkcji energii elektrycznej na bazie źródeł odnawialnych (Conditions for development of electricity production based on renewable energy sources). *Rynek Energii* **2018**, *2*, 1–11.
84. Radziszewska-Zielina, E.; Rumin, R. Analysis of The Profitability of Investment In Renewable Energy Sources On The Example of A Semi-Detached House. In Proceedings of the First International Conference on the Sustainable Energy and Environment Development, Krakow, Poland, 17–19 May 2016; pp. 1–4. [CrossRef]
85. Jedlińska, K.; Olkuski, T. The decision-making process in the EU in the field of energy policy. *Energy Policy J.* **2019**, *22*, 5–18. [CrossRef]

energies

MDPI

Article

Edible Energy Production and Energy Return on Investment—Long-Term Analysis of Global Changes

Bartłomiej Bajan [1,*], Joanna Łukasiewicz [1], Agnieszka Poczta-Wajda [2] and Walenty Poczta [1]

[1] Department of Economics and Economic Policy in Agribusiness, Faculty of Economics, Poznan University of Life Sciences, 60-637 Poznan, Poland; asialukasiewicz1@gmail.com (J.Ł.); poczta@up.poznan.pl (W.P.)
[2] Department of Macroeconomics and Agricultural Economics, Institute of Economics, Poznan University of Economics and Business, 61-875 Poznan, Poland; Agnieszka.Poczta-Wajda@ue.poznan.pl
* Correspondence: bartlomiej.bajan@up.poznan.pl; Tel.: +48-61-846-6379

Abstract: The projected increase in the world's population requires an increase in the production of edible energy that would meet the associated increased demand for food. However, food production is strongly dependent on the use of energy, mainly from fossil fuels, the extraction of which requires increasing input due to the depletion of the most easily accessible deposits. According to numerous estimations, the world's energy production will be dependent on fossil fuels at least to 2050. Therefore, it is vital to increase the energy efficiency of production, including food production. One method to measure energy efficiency is the energy return on investment (EROI), which is the ratio of the amount of energy produced to the amount of energy consumed in the production process. The literature lacks comparable EROI calculations concerning global food production and the existing studies only include crop production. The aim of this study was to calculate the EROI of edible crop and animal production in the long term worldwide and to indicate the relationships resulting from its changes. The research takes into account edible crop and animal production in agriculture and the direct consumption of fossil fuels and electricity. The analysis showed that although the most underdeveloped regions have the highest EROI, the production of edible energy there is usually insufficient to meet the food needs of the population. On the other hand, the lowest EROI was observed in highly developed regions, where production ensures food self-sufficiency. However, the changes that have taken place in Europe since the 1990s indicate an opportunity to simultaneously reduce the direct use of energy in agriculture and increase the production of edible energy, thus improving the EROI.

Keywords: EROI; energy efficiency; edible energy; food production; direct energy use

Citation: Bajan, B.; Łukasiewicz, J.; Poczta-Wajda, A.; Poczta, W. Edible Energy Production and Energy Return on Investment—Long-Term Analysis of Global Changes. *Energies* **2021**, *14*, 1011. https://doi.org/10.3390/en14041011

Academic Editor: Piotr Gradziuk
Received: 18 January 2021
Accepted: 12 February 2021
Published: 15 February 2021

1. Introduction

Since the end of World War II, the world's population has been growing steadily and the projections, by 2100, indicate that it will continue to grow [1]. In this context, the main function of agriculture is to feed the growing world population. In recent decades, the use of fertilizers, pesticides, improved water management and technological innovations have allowed increasing agricultural production [2]. The increase in agricultural productivity is associated with the high use of energy, including fossil fuel energy [3]. Agricultural intensification, starting with the Green Revolution, often has a negative impact on the environment [4,5]. Moreover, access to relatively cheap transport after World War II has contributed to the acceleration of the globalization process [6]. Both the intensification and globalization of production have contributed to the increased consumption of energy, mainly from fossil fuels, with regard to agriculture. From the global perspective, this proved to be a factor that negatively influenced the average energy efficiency of agricultural production [7].

According to many studies, global energy production will remain dependent on fossil fuels at least to 2050 [8]. Despite the constant decrease of net energy production from fossil fuels due to the increasing energy input for its extraction [9], it is still higher than net energy production from renewable sources [10,11]. Brockway et al. [12] claimed that net energy production from fossil fuels is much lower than suggested by previous studies, which may in fact result in a faster-than-expected reduction in the amount of energy available to society. Currently applied agricultural production techniques are still strongly dependent on fossil fuels [13,14], which, concerning the aforementioned problems, increases the need to improve energy efficiency in this sector. Therefore, the need to improve the energy efficiency of agricultural production is related both to the projected population growth, which affects the increase in demand for food and the growing input required to obtain the energy necessary for the food production process.

The literature provides many approaches concerning the measurement of energy efficiency [15] but the most commonly used indicator is the energy return on investment (EROI). EROI can be defined as the relationship between total energy production and the energy used for this production [16]. This concept was first used in research on fish migration [17] and soon afterwards it was widely applied in energy systems analyses [18]. EROI was first used in research on the food production system by Pimentel et al. [19]. Concerning agriculture itself, EROI measures how much edible biomass energy is produced from the invested unit of energy [20]. According to the methodology adopted by the United Nations Food and Agriculture Organization (FAO), edible energy production refers to the energy suitable for human consumption, that is, the energy that can potentially be provided by food produced. A higher EROI value indicates higher energy efficiency in agricultural production. However, the occurrence of high values should not be considered optimal, as exceedingly high EROI might be accompanied by shortages in food production. In general, however, it is desirable for EROI to increase when the demand for food is met.

Concerning the analyses that use EROI to measure agricultural production, there are currently three main directions of research [21]. First, the studies compare the EROI of conventional agriculture to organic farming or other alternative production systems [22–24]. The conclusions of these studies usually indicate that the achievement of higher production in conventional systems is associated with providing higher energy input, as well as putting greater pressure on the environment. The second type of analysis involves focusing on the change of EROI concerning single agricultural products over time [25–28]. This research describes how to optimize energy use for individual agricultural products and thus improve their energy efficiency. The third type of analysis includes calculating the EROI of agriculture of the whole country over a long period [29–32]. These analyses show which areas of production are more energy-efficient and create the basis for optimization of production structures. Some studies have also taken into account the entire food production system [33–35]. They indicate the energy efficiency of individual phases of the food production chain [36].

In most cases, research concerning individual countries proves a decrease in the EROI of agriculture during the early industrialization of a given sector [37]. However, in recent decades, some countries have been able to significantly increase their EROI [3,38]. Research results concerning the EROI of individual countries have one major disadvantage: they are usually incomparable [39]. There are two reasons for this. First, there are methodological differences in determining the system (process) boundaries or conversion factors for individual products, as well as other detailed assumptions about how EROI is calculated. These differences result from specific research investigations and, consequently, different ways of allocating energy inputs and outputs [40]. Despite attempts to create a general computational framework that would allow comparisons to be made [41,42], the final results of EROI depend on the research context. Therefore, it is necessary to indicate precisely what is included in energy consumption and production for each individual analysis.

Secondly, the EROI calculations found in the literature are most often based on data sources of individual countries, which means that there is a lack of international research based on data that would ensure comparability of results between countries. To the best of our knowledge, there are only two publications that analyze this issue on a global scale, which are based on data from the FAO. Conforti and Giampietro [43] investigated the EROI of fossil energy use in crop production between 1990 and 1991 using a sample of 75 countries worldwide. The results indicated that developed countries are characterized by the lowest EROI, amounting from around 1 to 2; the exceptions were Canada and the United States, where EROI was higher than 2 and was similar to the EROI of developing countries. The highest EROI, usually between 15 and 30, was found in African countries. In turn, Pellegrini and Fernández [44] analyzed the EROI of global crop production from 1961 to 2014, based on 58 countries that produce 95% of global crops. The research showed that the highest consumption and production of energy was in the countries with the highest use of irrigation and their EROI decreased significantly during the analyzed period. The authors declared that between 1961 and 2014, the general EROI in the world was in the shape of a U curve, reaching an average of about 3 in the initial years and about 4 in the final years. The highest EROI was characteristic of Africa and the lowest for highly developed regions: Oceania, North America and Europe. However, the studies presented above have one basic disadvantage: they do not take into account animal production, so the analyses presented do not cover the whole area of edible energy production. Thus, in the literature, there is a research gap resulting from the lack of internationally comparable studies on the energy efficiency of agricultural production (crop and animal).

The inclusion of animal production is important due to existence of large differences between energy intensity of food product categories. According to research describing this issue, the energy efficiency of animal production in individual countries is significantly lower than that of crop production [35,45]. It is mainly caused by feed conversion inefficiencies and high energy demands of creating animal fat and muscles. Moreover, the animal production is characterized by much higher range of energy inputs per kilogram of food than crop production [14,46], which can increase the differences in the obtained EROI values depending on the production directions. Furthermore, there are significant differences in structure of food-related energy use around the world. For instance, nearly 50% of total food-related energy use in the United States is associated with animal production [47], while in the Netherlands it is around 35% [48]. Considering above it can be concluded, apart from the differences in the structure, that the energy consumption associated with animal production accounts for a significant share of total food-related energy use, therefore, it should not be omitted in EROI studies.

Hence, the aim of this article is to calculate the EROI of edible crop and animal production over a long period worldwide, as well as indicators characterizing the energy productivity of agriculture and the energy intensity of this production. On the input side, the direct consumption of energy from fossil fuels (such as coal, gasoline and oil) in agriculture and electricity is taken into account. This research is based on uniform, internationally comparable FAO data, which allows for the analysis of the EROI obtained among various regions of the world. The analysis covers all continents but due to the limitations of available data, the research period slightly differs (the data for the years 1970-2018 is provided by the FAO in energy units on a uniform, annual basis but there are some gaps. This is particularly visible in the case of data from the 1970s, which, for some large countries, is missing or some energy sources are not included. For example, the data concerning the direct consumption of energy from coal in Asia covers the period since 1986, although this kind of energy was used earlier in large quantities [49]. In consequence, research periods vary from continent to continent). The period included in the research concerning North America covers 1970–2018, Oceania 1974–2018, Africa 1977–2018, South America 1976–2018, Asia 1986–2018 and Europe 1992–2018.

The main and original contribution of this paper is the inclusion of animal production in the analysis and conducting a dynamic comparative analysis on a global scale. Such research has not been conducted so far, which we consider to be a serious research gap, as the share of animal production in edible energy production in some regions reaches even over 40%. In regions where this share was previously low, its dynamic growth can be observed, which is related to demographic changes and the evolution of consumption patterns.

After the above introduction, the rest of the article is divided according to the following structure: Section 2 includes a description of the data used and the methodology applied. Section 3 presents the research results and discussion. Section 4 contains a summary of the analysis, including suggested directions for further research and policy recommendations.

2. Materials and Methods

The data used for the calculation of production and energy consumption were retrieved from the FAOSTAT database. The number of calories from edible agricultural production was calculated on the basis of FAO food balances using the method proposed by Sadowski and Baer-Nawrocka [50]:

$$EEP = \sum_{i=1}^{n} FS_i * P * SSC_i, \quad (1)$$

where EEP is edible energy production in agriculture (kcal/year), FS_i is the food supply of product i (kcal/person/year), P is the population and SSC_i is the self-sufficiency coefficient of product i. SSC_i is calculated according to the formula:

$$SSC_i = PQ_i/DSQ_i, \quad (2)$$

where PQ_i is the production quantity of product i (tonnes/year), DSQ_i is the domestic supply quantity of product i (tonnes/year).

The aim of introducing SSC to the formula is to include international trade and possible stocks from previous years. Thus, if the value of export of a given product was higher than the value of import, it increased the amount of produced edible energy calculated on the basis of the food supply value; if the import was higher, the amount of edible energy decreased as it did not result from domestic production. Therefore, the SSC balances the equation of production. The advantage of using the presented method based on FAO data is also the fact that the FAO food balances include fodder in the production quantity and subtract it from the domestic supply quantity. The result is that this quantity of energy is subtracted from all the energy obtained in agricultural production, avoiding double counting. Moreover, the above method was chosen because, unlike other methods, it allows the calculation of energy production directly from food balances, without using external datasets, which allows full comparability of the obtained results between continents. The full list of edible products considered in the study with corresponding FAO's item codes is included in the Table A1.

The edible agricultural production energy consumption included the direct consumption of fossil fuels in agriculture (gas-diesel oil, motor gasoline, natural gas, liquefied petroleum gas, fuel oil and coal) and electricity. Fuels used in fishing were also included in the calculation, as edible energy production also covers fishery products. EROI was calculated as the quotient of edible energy production in agriculture and direct energy use in agriculture:

$$Edible\ EROI = EEP/DEU, \quad (3)$$

where DEU is direct energy use in agriculture. As FAO stores the values of energy consumption in terajoules and energy production in kcal, the value of energy consumption has been converted into kcal by multiplying it by the 238,902,957.6.

The EROI values below 1 mean that more energy is consumed than produced in the production process. Values above 1 mean that more energy is produced than is consumed in the production process. In general, as indicated earlier, the higher the value of edible EROI, the better. However, whether a given level of food production is able to meet the needs of a population is important—if it is not and the EROI is high, then this situation should be considered unfavorable.

Since indirect energy consumption in agriculture (e.g., the use of energy for the production of fertilizers, pesticides or agricultural machinery) is an estimate calculated on the basis of many conversion factors that vary in the literature depending on the study and the country, as well as the assumptions made [3], it was not included in this edible EROI calculation. To maintain comparability among the continents, the calculation includes only direct energy consumption, which corresponds to the first level of boundary for energy inputs according to standards proposed by Murphy et al. [39]. The impact of not considering indirect energy consumption is presented in the results and discussion section.

For both edible energy production in agriculture (*EEP*) and direct energy use in agriculture (*DEU*), three indicators were calculated, dividing both values by the number of inhabitants, the area of agricultural land and the value of agricultural edible production. For consistency the population data were retrieved from the FAO's food balances and area of agricultural land from the FAO's land use data. The value of agricultural edible production refers to the gross production value of food at constant 2014–2016 prices, which was also retrieved from FAO database.

The results were presented concerning given decades, taking into account the survey period for each continent. For example, the 2010s are represented by the 2010–2018 period and the 1970s are represented by different periods depending on the continent, which is related to the previously mentioned gaps in FAO data. For each indicator, the average growth rate (AGR) was also calculated as the geometric mean of chain indexes from all the years of the analyzed periods.

3. Results and Discussion

The calculations show that during the researched period, the highest EROI of the production of edible energy was visible in Africa, 24 on average. However, between 1977 and 2018, a decrease was found from about 57 in the first period to about 12 in the last (Table 1). The average growth rate of the indicator was −4%, which indicates similar transformations in energy efficiency in Africa to those that occurred in the rest of the world during agricultural mechanization and industrialization [37]. However, the lowest value of EROI in Africa was obtained in 2007, since than it started growing at a slow pace (Figure 1). The lowest EROI values could be observed in highly developed regions of the world, namely North America, Europe and Oceania, which is similar to the results of the previously discussed studies on the energy efficiency of crop production [43,44]. This may indicate a certain similarity of energy efficiency in case of animal and crop production within continents. The results are similar in the sense that the continents that had the highest or the lowest EROI in the crop production have it also in case of crop and animal production together. However, It does not mean that the values of these EROIs are the same, since they are not comparable. Concerning this study, it is vital to calculate only edible energy production. To do so, one must subtract the production of fodder from the value of agricultural production, as it constitutes a part of crop production but at the same time is also an input in animal production. On the other hand, the conversion factor of energy from crop products being fodder into energy from animal products always exceeds 1, as this is due to, among others, the living needs of animals and energy losses. The organism of an animal is not a perfect machine producing energy without losses from an energy source such as fodder.

North America and Oceania were characterized by EROI fluctuations during the research period, which is partly due to the high impact of individual countries on the final values of indicators. In the case of North America, this country is the United States and for Oceania, it is Australia. The amount of energy used in agriculture and the amount of edible energy produced can be significantly influenced by weather conditions. Such weather conditions are characterized by annual fluctuations in individual countries and can also yield fluctuations. According to the data provided by the FAO, fluctuations in cereal yields in the research period occurred both for the United States and Australia. The average EROI in North America during the research period was about 2.2 and its standard deviation was 0.26. In Oceania, on the other hand, the average EROI was 1.9, the lowest among the continents and its standard deviation was 0.27. For North America, our results differ from those presented by Conforti and Giampietro [43], who concluded that EROI in Canada and the US are closer to Asian countries than to other developed countries. However, as previously indicated, the results of these studies are not fully comparable to ours due to the different methodology.

Table 1. Standard deviation, average growth rate and average Edible energy return on investment (EROI) values for years 1970–2018.

Region	EROI in 1970s	EROI in 1980s	EROI in 1990s	EROI in 2000s	EROI in 2010s	Standard Deviation for Analyzed Period	Average Growth Rate (AGR) for Analyzed Period %
South America (1976–2018)	5.35	4.74	4.14	3.65	3.87	0.63	−0.68
North America (1970–2018)	2.32	1.86	2.30	2.16	2.29	0.26	0.07
Europe (1992–2018)	-	-	1.57	2.12	2.45	0.37	2.21
Asia (1986–2018)	-	5.85	4.86	4.88	4.61	0.41	−0.85
Africa (1977–2018)	57.25	40.49	17.55	12.56	11.78	15.71	−3.97
Oceania (1974–2018)	2.34	1.89	1.96	1.65	1.72	0.27	−1.19

Detailed average results used during the EROI calculations are presented in Table A2.

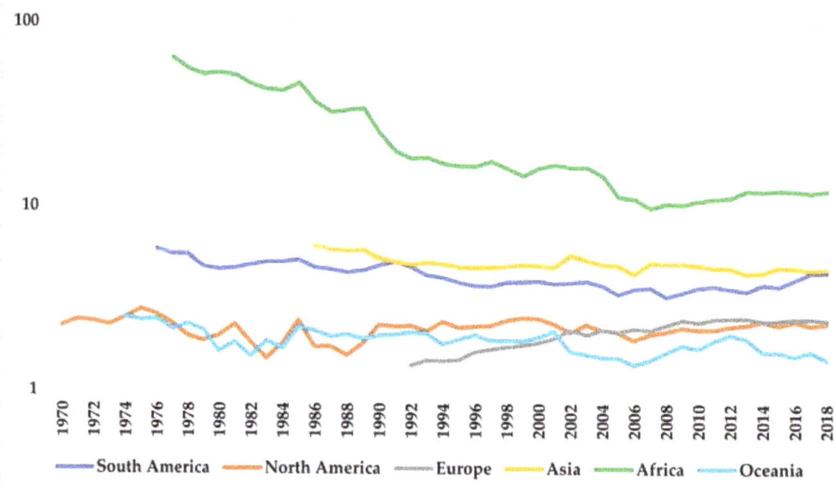

Figure 1. Energy return on investment (EdibleEROI) (kcal/kcal) (note logarithmic scale).

The opposite trend occurred concerning Europe, where at the beginning of the 1990s, the EROI hovered around 1.5; by the end of the research period, it significantly exceeded 2.4. The average growth rate in the years 1992–2018 was 2.2% and was, by far, the highest among the continents surveyed. The change toward higher energy efficiency in Europe was set by the European Union, which in 1993 adopted a law aimed at improving energy efficiency [51] and further reformed it in subsequent years [52,53].

In South America and Asia, the EROIs were on average higher than those in the most developed regions but, during the analyzed period, were characterized by a decrease and slight fluctuations. What is more, the average EROI for South America during the researched period was 4.25, with an average growth rate of −0.68%. In Asia, however, the average EROI was about 4.9 and the average growth rate was −0.85%. These results in general are in line with studies conducted in individual countries [29,30,32]. However, the results are not so unambiguous in the case of individual products. For example, research results by Pracha and Volk [25] confirm the downward trend of the EROI for Pakistan's wheat but for rice production the EROI trend was more volatile. Similarly, results by Infante-Amate and Picado [27] indicate a downward trend in the energy efficiency of coffee production in Costa-Rica.

As it was mentioned before, the animal production is less energy efficient than crop production, thus, the obtained EROI results may be influenced by the share of animal production in edible production on individual continents. It should be assumed that higher share of animal production lowers average EROI values within continent. However, simple Pearson's correlation coefficients for EROI values and share of animal production in edible energy production do not confirm this as their values are relatively low (Table 2). On the other hand, one could argue that the highest share of animal production in edible energy production is observed in Oceania, which also has the lowest values of EROI. Moreover, opposite situation concerns Africa where the EROI values are the highest and the share of animal production in edible energy production is the lowest. However, there are long periods in which the EROI between continents was similar despite the differences in animal production shares in edible energy production, which is especially true for North America and Oceania until the late 1990s and to the lesser extend for Asia and South America during the analyzed period. It can be concluded that, although animal production is less energy-efficient than crop production, it does not mean that its higher shares result in the lack of ability to produce edible energy efficiently. In fact, the obtained results indicate that the development level of continents should be consider as the main driver of edible EROI values regardless of production direction (animal or crop). As it was mention before, the highest EROI values were observed in the least developed regions and the lowest values in the most developed regions in case of crop production alone [43,44] and as indicated by this study, in case of animal and crop production combined. The importance of economic development for the results of EROI is confirmed by Steinhart and Steinhart [54], who, based on the example of the United States, found that the relationship between energy consumption and food production has the shape of a logistic growth curve. Therefore, increases in food production due to increased energy inputs are higher in less developed regions.

Table 2. Share of animal production in edible energy production for years 1970–2018 and its correlation coefficients with EROI values.

Region	1970s	1980s	1990s	2000s	2010s	Pearson's Correlation Coefficients with EROI Values for Analyzed Period
South America (1976–2018)	16.9	17.5	20.1	21.3	25.6	−0.38
North America (1970–2018)	24.3	23.2	22.1	23.2	23.4	−0.14
Europe (1992–2018)	-	-	30.3	28.7	25.8	−0.70
Asia (1986–2018)	-	9.2	12.2	14.2	15.2	−0.64
Africa (1977–2018)	7.3	7.2	7.0	7.7	11.3	−0.28
Oceania (1974–2018)	44.4	42.4	44.0	44.8	36.7	0.20

The literature analyses of the EROI of the agricultural sector showed that the wider the system boundaries, the lower the EROI. This is due to the fact that the increase in the amount of energy consumed resulted in increasing the value of the equation's denominator [55]. This fact should be taken into account when the results are analyzed. For example, the boundaries can be extended to include intermediate energy consumption in agricultural production. It is related to the use of fossil fuels for the production of fertilizers, pesticides or machinery. Many authors also classify other activities as intermediate consumption but the three mentioned in this study are commonly recognized and comprise the largest part of it [56]. If, for example, one were to consider the estimated results for intermediate consumption calculated by Arizpe et al. [57], the results of the Edible EROI would be, on average, lower for North America by 36%, from 2.25 to 1.65 in 1991. As Harchaoui and Chatzimpiros [3] pointed out, extending the system boundaries for the inclusion of other types of production, for example, food processing or household food processing, could result in a decrease of EROI below 1 in some countries. Concerning food, such a low rate may be acceptable because it must be produced regardless of the rationality of the process.

Moving to detailed indicators, the clear differences are apparent between the calculated indicators of edible energy production (Table 3) and energy consumption (Table 4) in particular decades in the researched regions. In the case of production indicators, their increase per capita (per consumer), signifying an increase in food security (food availability), as well as per hectare, indicating higher "energy productivity" of agricultural land, is desirable. In the case of edible energy production, the lower indicators show a higher unit production value. The higher the value of the production, shown in the denominator, the lower the indicator. From the point of view of a producer, it is a favorable situation, as they receive more money per unit of energy produced; such an interpretative approach was adopted when the indicator of edible energy production per value of the production was discussed.

Table 3. Edible energy production in agriculture for years 1970–2018.

Region	Edible Energy Production Indicator	Thousands kcal	1970s	1980s	1990s	2000s	2010s	AGR %
South America (1976–2018)	Per number of citizens	kcal/person	992	1005	1034	1134	1261	0.71
	Per agricultural area	kcal/ha	498	561	694	869	1034	2.00
	Per value of production	kcal/const. Int$	2.08	2.03	1.93	1.72	1.61	−0.55
North America (1970–2018)	Per number of citizens	kcal/person	1280	1379	1417	1435	1459	0.60
	Per agricultural area	kcal/ha	655	801	941	1087	1229	1.84
	Per value of production	kcal/const. Int$	1.62	1.68	1.69	1.64	1.57	−0.02
Europe (1992–2018)	Per number of citizens	kcal/person	-	-	1148	1192	1218	0.20
	Per agricultural area	kcal/ha	-	-	1704	1835	1951	0.61
	Per value of production	kcal/const. Int$	-	-	1.68	1.77	1.70	0.09
Asia (1986–2018)	Per number of citizens	kcal/person	-	844	882	917	977	0.53
	Per agricultural area	kcal/ha	-	1983	1915	2144	2543	1.02
	Per value of production	kcal/const. Int$	-	3.59	3.18	2.74	2.49	−1.21
Africa (1977–2018)	Per number of citizens	kcal/person	764	765	806	845	830	0.13
	Per agricultural area	kcal/ha	302	359	485	622	766	2.53
	Per value of production	kcal/const. Int$	3.17	3.27	3.15	3.00	2.72	−0.35
Oceania (1974–2018)	Per number of citizens	kcal/person	1660	1673	1697	1707	1769	−0.03
	Per agricultural area	kcal/ha	60	69	83	106	142	2.07
	Per value of production	kcal/const. Int$	0.97	1.00	0.97	0.92	0.96	−0.38

Constant Int$ refers to the value of food production at constant 2014–2016 prices.

Table 4. Direct energy consumption in agriculture for years 1970–2018.

Region	Edible Energy Production Indicator	Thousands kcal	1970s	1980s	1990s	2000s	2010s	AGR %
South America (1976–2018)	Per number of citizens	kcal/person	185	212	250	311	326	1.40
	Per agricultural area	kcal/ha	93	118	167	238	267	2.70
	Per value of production	kcal/const. Int$	0.39	0.43	0.46	0.47	0.42	0.13
North America (1970–2018)	Per number of citizens	kcal/person	553	742	616	664	636	0.53
	Per agricultural area	kcal/ha	283	431	409	503	536	1.77
	Per value of production	kcal/const. Int$	0.70	0.91	0.74	0.76	0.68	−0.08
Europe (1992–2018)	Per number of citizens	kcal/person	-	-	732	562	497	−1.96
	Per agricultural area	kcal/ha	-	-	1086	865	796	−1.57
	Per value of production	kcal/const. Int$	-	-	1.07	0.83	0.69	−2.07
Asia (1986–2018)	Per number of citizens	kcal/person	-	144	182	188	212	1.39
	Per agricultural area	kcal/ha	-	339	394	440	552	1.88
	Per value of production	kcal/const. Int$	-	0.61	0.65	0.56	0.54	−0.37
Africa (1977–2018)	Per number of citizens	kcal/person	13	19	46	67	71	4.27
	Per agricultural area	kcal/ha	5	9	28	50	65	6.77
	Per value of production	kcal/const. Int$	0.06	0.08	0.18	0.24	0.23	3.77
Oceania (1974–2018)	Per number of citizens	kcal/person	710	887	867	1034	1026	1.17
	Per agricultural area	kcal/ha	26	37	43	64	82	3.30
	Per value of production	kcal/const. Int$	0.42	0.53	0.50	0.56	0.55	0.82

Constant Int$ refers to the value of food production at constant 2014–2016 prices.

In the case of energy consumption indicators, the situation is slightly more complicated. Low values per hectare usually indicate low production intensity, which on the one hand can cause low productivity of the land but on the other hand, can result in a number of positive effects, for example, lower pressure on the environment. It could be considered from different perspectives, for example, whether the low energy inputs are the choice of a farmer or they resulting from the low level of economic development in the region, which forces farmers to produce with low energy inputs. Another factor influencing the energy input per 1 ha of agricultural land and per capita is population density, which is closely related to the area of agricultural land per 1 inhabitant. What is more, climatic and natural conditions (e.g., the share of permanent grassland, soil quality, length of vegetation period, level temperature, amount and distribution of precipitation) and the associated production structure, including the role of crop and animal production, also have an impact.

It is of key importance to shape the relationship between energy consumption and the production of edible energy as part of agricultural production. These relationships might be analyzed concerning both static and dynamic approaches. In the case of the indicator of energy consumption per unit of production value, lower values that prove better economic efficiency of the energy invested are desirable but the appropriate level of production still needs to be taken into account. Concerning a dynamic approach, it is desirable that energy input consumption grows slower than the value of agricultural production, assuming that production is sufficient to meet the needs of food consumers.

However, although in South America the EROI was decreasing until the last analyzed decade, that is, energy consumption was growing faster than its production, there was a clear increase in productivity per hectare (2% per year on average). Despite the rapid increase in energy production per hectare in South America, it is relatively low compared to the most developed regions (Europe and North America). Indicators per capita grew at a slower rate, which is due to the rapid growth of the South American population over the analyzed period. In contrast, it was shown that the indicator of energy production per capita in the analyzed region has been, in recent years, higher to that achieved in Europe, which illustrates an improvement in food self-sufficiency in South America. The high level of self-sufficiency observed in recent years can also be confirmed by other research [58]. At the same time, the direct energy consumption needed to produce one Int$ remained relatively stable in the analyzed period (AGR was 0.13%); however, the amount of edible energy produced per Int$ (as evidence by a −0.55% decrease in AGR) increased significantly. This might be considered a favorable situation from the point of view of agricultural producers, as the unit value of production increased, while the economic efficiency of the energy invested remained relatively stable.

The observed fluctuations of EROI in North America are reflected in the analyzed direct energy consumption indicators, which generally grew, yet fluctuated. Energy consumption per hectare of agricultural land increased particularly rapidly, indicating a progressive, energy-intensive escalation of production (1.77%) and it increased more moderately per capita (0.53%). Production expressed in energy per hectare and per capita, respectively (1.84% and 0.60%), increased at a slightly faster rate than the increase in energy input, resulting in a visible improvement in productivity. In the 1990s, when EROI was higher than in following decades, there was a decrease in direct energy consumption; this is also reflected by indicators per capita and per hectare of agricultural land. Moreover, the indicators of energy consumption and production per value of production were directly proportional; the exception occurred in the 1990s, when there was an increase in the economic efficiency of energy consumption. At the same time, North America is characterized by a relatively high individual value of edible energy production and low economic efficiency of energy consumption in agriculture. Other research that takes North American countries into account illustrates the occurrence of fluctuations concerning direct fuel consumption in agriculture [59] and points out the increase in energy efficiency in the 1990s, claiming it was due to the change in production direction toward more energy-efficient agriculture [38].

In examining Europe, the increase in EROI is evident, as well as in energy production per hectare and per capita in the analyzed period (with relatively high values of their indicators), in the years 1992–2018. However, the EROI increase was mainly due to a decrease in direct energy consumption in agriculture. Moreover, the energy intensity of production visibly decreased, which is reflected by a decrease in energy consumption per hectare and per capita. Together with increases in production rates, this provided the fastest change toward more energy-efficient edible production compared to other continents. The only indicator that deteriorated slightly was the decrease in the individual value of energy production, reflected by the average increase of production required to obtain the value of one Int$. However, its values were relatively stable in analyzed period. At the same time, in the 1990s, a decrease was found in the direct energy consumption required to produce one Int$ worth of edible energy, which was still the highest among the regions subject to analysis. As indicated by other studies [44], in Europe, before the 1990s, the EROI was decreasing as a result of rapidly increasing energy use in agriculture, even though steady growth in production was maintained. In the 1990s, energy consumption began to decrease while production growth remained steady, resulting in obvious improvements in energy efficiency.

During the analyzed period, direct energy consumption per hectare as well as per capita in Asia grew at faster rate than edible energy production. However, due to high population density and significant population growth in recent decades, Asia has relatively low edible energy production per person, resulting in food availability problems in this region. At the same time, Asia had the highest edible energy production per hectare of agricultural land in the world. This is mainly influenced by the low ratio of area of agricultural land per 1 inhabitant, which makes it necessary to obtain high production from 1 ha of agricultural land and in many regions to yield two or even three crops in one year. High production from 1 ha is also optimal due to the structure of agriculture (many small farms). During the analyzed period, the individual value of edible energy production in Asia increased and the economic efficiency of energy consumption in agriculture improved. This might be reflected by the decrease of the energy use and energy production per value of production (AGR was −0.37% and −1.21%, respectively), which were relatively high at the beginning of the considered period.

In Africa, the rapid EROI decrease until 2007 was mainly due to an increase in direct energy consumption in order to increase production. Energy consumption per person increased sixfold, while per hectare increased more than fourteenfold in the whole analyzed period but the values remained relatively low. What is more, although there was a desirable increase in edible energy production, it was insufficient to guarantee food availability in Africa [60,61]. As mentioned earlier, similar changes to those occurring in Africa were found during the period under consideration, after the industrial revolution on other continents as well; however, now, with the EROI of fossil fuel decreasing, reliance on fossil fuel-based energy production might have a negative impact on its increase in the long term [62]. The rising Africa's edible EROI since 2008 was due to lower increase in energy consumption and simultaneous growth in energy production, however, in the last decade (2010–2018) the decrease in energy production per person can be observed. An increase in EROI in the absence of food self-sufficiency and a decrease in edible energy production per capita in 2010–2018, should be considered as unfavorable. The individual value of edible energy production in the analyzed period also increased, with a simultaneous decrease in economic efficiency from the energy invested. However, it remained high only due to unsatisfactory energy use, which resulted in insufficient production. In this context, the decrease concerning the discussed energy efficiency of production should not be considered something negative.

Unlike the other analyzed regions, Oceania had several times lower energy use and production per hectare of agricultural land than per person. The results concerning Oceania are mainly derived from the results of Australia and, to a lesser extent, New Zealand, countries that are characterized by extensive agricultural production on a large area of agricultural land. The structure of agricultural land is dominated by permanent grasslands (permanent meadows and pastures). During the analyzed period, their share in Australia exceeded 90% and in New Zealand 95%. A large number of areas of agricultural land per 1 inhabitant ensures that Oceania also has the highest rate of edible energy production per person but much of this production is exported. The direct energy consumption per capita is high because an increase in energy consumption ensures an increase in production, the surplus of which is sold abroad. The second reason is the dominance of animal production, as it is less efficient concerning energy use. The individual value of edible production expressed in kcal per Int$ remained more or less constant over the analyzed period, while the economic efficiency of energy use in agriculture declined to some extent, as a result of the increasing size of cattle production in Oceania.

4. Conclusions and Policy Recommendations

The study on edible energy production and its direct consumption in agriculture determined that the highest edible EROI is present in Africa and the lowest in the most developed regions. During the analyzed period, energy consumption in agriculture worldwide increased, contributing to the increase of edible energy production. The only exception was Europe, where, since the 1990s, a decrease was found in direct energy use, mainly concerning energy from fossil fuels, while an increase in energy production was found, resulting in a visible improvement of edible EROI. The changes that took place in Europe, mainly due to the Common Agricultural Policy of the European Union, confirm the possibility of improving the energy efficiency of crop and animal production while reducing the use of fossil fuels, which is particularly important considering the decreasing energy efficiency of their extraction and expectations concerning the reduction of pollution generation while meeting growing food needs.

The analysis broadened the scope of the international comparison of EROI in agriculture present in the literature, including animal production. This proves that in regions with low or high EROI for crop production, correspondingly low or high EROI can be found for animal and crop production combined. This indicates that the regions' ability to effectively convert energy into edible energy does not depend solely on the direction of production (animal or crop). However, it must be remembered that the energy efficiency of animal production is, in fact, significantly lower than that of crop production.

The presented results of EROI should be interpreted only within the scope of the established system boundaries, that is, it is a study on the relationship between edible energy production from agriculture and the direct use of fossil fuels and electricity in agriculture; this constitutes a limitation to the conducted studies. When carrying out the analysis, we tried to avoid using conversion factors for production or energy consumption, which could be based on different sources, so the study included only the factors used by the FAO; however, this limited the scope of the system boundaries. On the one hand, this is the limitation of this article but on the other hand, it also sets out future research directions that could focus on extending the system boundaries using methodologically uniform conversion factors for indirect energy consumption, including animal and crop production in research. Extending the system boundaries in research is undoubtedly a vital issue from the point of view of assessing the energy efficiency of agricultural production, as indirect energy consumption associated with the use of fertilizers, pesticides, irrigation systems and possibly agricultural machinery comprises one of the key inputs; at the same time, significant improvement concerning this matter is possible.

Another direction of further research includes incorporating links of the food supply chain in the analysis. Although agriculture is the most energy-intensive phase of the food chain, other phases such as food processing, logistics, packaging and food waste are also crucial for energy efficiency improvement opportunities.

Finally, a key extension of our research would be the analysis of drivers of global changes in energy efficiency in agricultural production. A review of the literature and research conducted in individual countries suggests that the most important drivers include changes in the structure of food consumption and production, applied technologies and practices in agriculture, climate change, and, perhaps most importantly, policy for energy use in agriculture and food production. However, determining the exact impact of individual factors on the efficiency of global edible energy production requires more detailed analysis and is challenging in terms of obtaining comparable data.

Nevertheless, based on the research results and the literature review, some concluding policy recommendations can be formulated. The example of Europe, in which it was possible to simultaneously improve the energy efficiency of agricultural production and reduce the use of fossil fuels, suggests that the state policy has a key direct impact (instruments supporting the use of alternative energy sources) and indirect impact (instruments supporting the implementation of new technologies and practices in agriculture, stimulating a change in consumption and production structure) on energy efficiency. The EU's Common Agricultural Policy has played a special role by encouraging investment in more sustainable farming methods. The same can be said about rural development programs, which aim to facilitate the supply and use of renewable energy sources. Therefore, drawing on the EU's experience in reducing direct consumption of fossil energy and electricity in agriculture, the following tools should be introduced: (i) measures promoting and supporting the production and use of renewable energy sources such as biofuels, wind energy, solar energy and hydropower systems; (ii) incentives for changing the structure of food consumption and production toward limiting the consumption of meat and switching to the consumption of local and seasonal products; (iii) measures promoting and supporting conservation agriculture and organic farming; and (iv) support for R&D and implementation of innovative farming techniques, such as precision agriculture or irrigation technologies.

It is worth emphasizing that the above policy recommendations, formulated on the basis of experience and success in the field of improving the energy efficiency of agricultural production in European countries, apply mainly to developed countries, although to a certain extent and subject to regional modification, they should also be applied in less developed countries because the separation of agriculture productivity from energy consumption remains a challenge across the globe.

Author Contributions: Conceptualization, B.B.; methodology, B.B.; validation, A.P.-W. and W.P.; formal analysis, B.B. and J.Ł.; investigation, B.B.; data curation, B.B. and J.Ł.; writing—original draft preparation, B.B. and J.Ł.; writing—review and editing, A.P.-W. and W.P.; visualization, B.B. and J.Ł.; supervision, W.P. All authors have read and agreed to the published version of the manuscript.

Funding: This research received no external funding

Conflicts of Interest: The authors declare no conflict of interest.

Appendix A

Table A1. The list of products that have been considered for edible energy production with their corresponding FAO's item code.

FAO's Item Code	Name	FAO's Item Code	Name	FAO's Item Code	Name
2656	Beer	2619	Dates	2645	Spices, Other
2658	Beverages, Alcoholic	2625	Fruits, Other	2532	Cassava and products
2657	Beverages, Fermented	2613	Grapefruit and products	2531	Potatoes and products
2655	Wine	2620	Grapes and products	2534	Roots, Other
2740	Butter, Ghee	2612	Lemons, Limes and products	2533	Sweet potatoes
2743	Cream	2611	Oranges, Mandarines	2535	Yams
2737	Fats, Animals, Raw	2618	Pineapples and products	2633	Cocoa Beans and products
2781	Fish, Body Oil	2616	Plantains	2630	Coffee and products
2782	Fish, Liver Oil	2731	Bovine Meat	2635	Tea (including mate)
2769	Aquatic Animals, Others	2735	Meat, Other	2745	Honey
2775	Aquatic Plants	2732	Mutton & Goat Meat	2542	Sugar (Raw Equivalent)
2768	Meat, Aquatic Mammals	2733	Pigmeat	2541	Sugar non-centrifugal
2513	Barley and products	2734	Poultry Meat	2543	Sweeteners, Other
2520	Cereals, Other	2848	Milk—Excluding Butter	2537	Sugar beet
2514	Maize and products	2680	Infant food	2536	Sugar cane
2517	Millet and products	2899	Miscellaneous	2551	Nuts and products
2516	Oats	2736	Offals, Edible	2578	Coconut Oil
2805	Rice and products	2560	Coconuts—Incl Copra	2575	Cottonseed Oil
2515	Rye and products	2556	Groundnuts (Shelled Eq)	2572	Groundnut Oil
2518	Sorghum and products	2570	Oilcrops, Other	2582	Maize Germ Oil
2511	Wheat and products	2563	Olives (including preserved)	2586	Oilcrops Oil, Other
2744	Eggs	2562	Palm kernels	2580	Olive Oil
2766	Cephalopods	2558	Rape and Mustardseed	2577	Palm Oil
2765	Crustaceans	2561	Sesame seed	2576	Palmkernel Oil
2762	Demersal Fish	2555	Soyabeans	2574	Rape and Mustard Oil
2761	Freshwater Fish	2557	Sunflower seed	2581	Ricebran Oil
2764	Marine Fish, Other	2546	Beans	2579	Sesameseed Oil
2767	Molluscs, Other	2547	Peas	2571	Soyabean Oil
2763	Pelagic Fish	2549	Pulses, Other and products	2573	Sunflowerseed Oil
2617	Apples and products	2642	Cloves	2602	Onions
2615	Bananas	2640	Pepper	2601	Tomatoes and products
2614	Citrus, Other	2641	Pimento	2605	Vegetables, Other

Table A2. Detailed average results used during the EROI calculations for years 1970–2018.

Region	Indicator	Unit	1970s	1980s	1990s	2000s	2010s
South America (1976–2018)	Food supply (*FS*)	Thousands	940.3	949.0	972.2	1033.6	1094.5
	Population (*P*)	Millions	274.2	318.8	382.9	443.3	492.5
	Self-sufficiency coefficient (*SSC*)	-	1.06	1.06	1.06	1.10	1.15
	Edible energy production (*EEP*)	Trillions	272.1	320.5	396.1	502.9	621.2
	Direct energy use (*DEU*)	Trillions	50.8	67.7	95.6	137.9	160.6
North America (1970–2018)	Food supply (*FS*)	Thousands	1095.7	1179.9	1249.7	1301.0	1291.4
	Population (*P*)	Millions	302.5	343.9	390.7	439.3	477.5
	Self-sufficiency coefficient (*SSC*)	-	1.17	1.17	1.13	1.10	1.13
	Edible energy production (*EEP*)	Trillions	387.2	474.2	553.6	630.4	696.7
	Direct energy use (*DEU*)	Trillions	167.2	255.3	240.6	291.7	303.7
Europe (1992–2018)	Food supply (*FS*)	Thousands	-	-	1165.1	1210.2	1230.7
	Population (*P*)	Millions	-	-	728.6	732.8	743.7
	Self-sufficiency coefficient (*SSC*)	-	-	-	0.99	0.99	0.99
	Edible energy production (*EEP*)	Trillions	-	-	836.5	873.8	906.0
	Direct energy use (*DEU*)	Trillions	-	-	533.0	411.9	369.5
Asia (1986–2018)	Food supply (*FS*)	Thousands	-	874.2	911.7	956.1	1019.6
	Population (*P*)	Millions	-	2977.4	3416.3	3891.2	4330.6
	Self-sufficiency coefficient (*SSC*)	-	-	0.96	0.97	0.96	0.96
	Edible energy production (*EEP*)	Trillions	-	2511.6	3013.2	3566.6	4231.0
	Direct energy use (*DEU*)	Trillions	-	429.6	620.4	731.6	918.8
Africa (1977–2018)	Food supply (*FS*)	Thousands	792.6	818.8	864.6	917.3	951.8
	Population (*P*)	Millions	417.4	501.3	646.9	819.7	1031.5
	Self-sufficiency coefficient (*SSC*)	-	0.96	0.93	0.93	0.92	0.97
	Edible energy production (*EEP*)	Trillions	318.7	383.4	521.4	692.6	949.8
	Direct energy use (*DEU*)	Trillions	5.6	9.5	29.7	55.1	72.7
Oceania (1974–2018)	Food supply (*FS*)	Thousands	1104.2	1111.8	1116.0	1129.3	1200.9
	Population (*P*)	Millions	18.6	20.6	23.6	26.8	30.8
	Self-sufficiency coefficient (*SSC*)		1.50	1.51	1.52	1.51	1.47
	Edible energy production (*EEP*)	Trillions	30.9	34.5	40.1	45.8	54.4
	Direct energy use (*DEU*)	Trillions	13.2	18.3	20.5	27.7	31.6

References

1. United Nations (UN). World population prospects 2019. Available online: https://population.un.org/wpp/DataQuery (accessed on 5 November 2020).
2. Mueller, N.D.; Gerber, J.S.; Johnston, M.; Ray, D.K.; Ramankutty, N.; Foley, J.A. Closing yield gaps through nutrient and water management. *Nat. Cell Biol.* **2012**, *490*, 254–257. [CrossRef] [PubMed]
3. Harchaoui, S.; Chatzimpiros, P. Energy, Nitrogen, and Farm Surplus Transitions in Agriculture from Historical Data Modeling. France, 1882–2013. *J. Ind. Ecol.* **2019**, *23*, 412–425. [CrossRef]
4. Pimentel, D. Environmental and Economic Costs of the Application of Pesticides Primarily in the United States. *Environ. Dev. Sustain.* **2005**, *7*, 229–252. [CrossRef]
5. Krausmann, F.; Erb, K.-H.; Gingrich, S.; Haberl, H.; Bondeau, A.; Gaube, V.; Lauk, C.; Plutzar, C.; Searchinger, T.D. Global human appropriation of net primary production doubled in the 20th century. *Proc. Natl. Acad. Sci.* **2013**, *110*, 10324–10329. [CrossRef] [PubMed]
6. Kastner, T.; Kastner, M.; Nonhebel, S. Tracing distant environmental impacts of agricultural products from a consumer perspective. *Ecol. Econ.* **2011**, *70*, 1032–1040. [CrossRef]
7. Parcerisas, L.; Dupras, J. From mixed farming to intensive agriculture: Energy profiles of agriculture in Quebec, Canada, 1871–2011. *Reg. Environ. Chang.* **2018**, *18*, 1047–1057. [CrossRef]

8. Edenhofer, O.; Pichs-Madruga, R.; Sokona, Y.; Minx, J.C.; Farahani, C.; Kadner, S.; Seyboth, K.; Adler, A.; Baum, I.; Brunner, S.; et al. Climate change 2014: Mitigation of Climate Change. In *Contribution of Working Group III to the Fifth Assessment Report of the Intergovernmental Panel on Climate Change*; Cambridge University Press: New York, NY, USA, 2014.
9. Gagnon, N.; Hall, C.A.; Brinker, L. A Preliminary Investigation of Energy Return on Energy Investment for Global Oil and Gas Production. *Energies* **2009**, *2*, 490–503. [CrossRef]
10. Hall, C.A.; Lambert, J.G.; Balogh, S.B. EROI of different fuels and the implications for society. *Energy Policy* **2014**, *64*, 141–152. [CrossRef]
11. King, L.C.; Bergh, J.C.J.M.V.D. Implications of net energy-return-on-investment for a low-carbon energy transition. *Nat. Energy* **2018**, *3*, 334–340. [CrossRef]
12. Brockway, P.E.; Owen, A.; Brand-Correa, L.I.; Hardt, L. Estimation of global final-stage energy-return-on-investment for fossil fuels with comparison to renewable energy sources. *Nat. Energy* **2019**, *4*, 612–621. [CrossRef]
13. Cruse, M.J.; Liebman, M.; Raman, D.R.; Wiedenhoeft, M.H. Fossil Energy Use in Conventional and Low-External-Input Cropping Systems. *Agron. J.* **2010**, *102*, 934–941. [CrossRef]
14. Woods, J.; Williams, A.; Hughes, J.K.; Black, M.; Murphy, R. Energy and the food system. *Philos. Trans. R. Soc. B: Biol. Sci.* **2010**, *365*, 2991–3006. [CrossRef] [PubMed]
15. Mulder, K.; Hagens, N.J. Energy Return on Investment: Toward a Consistent Framework. *Ambio* **2008**, *37*, 74–79. [CrossRef]
16. Kunz, H.; Hagens, N.J.; Balogh, S.B. The Influence of Output Variability from Renewable Electricity Generation on Net Energy Calculations. *Energies* **2014**, *7*, 150–172. [CrossRef]
17. Hall, C.A.S. Migration and Metabolism in a Temperate Stream Ecosystem. *Ecology* **1972**, *53*, 585–604. [CrossRef]
18. Court, V.; Fizaine, F. Long-Term Estimates of the Energy-Return-on-Investment (EROI) of Coal, Oil, and Gas Global Productions. *Ecol. Econ.* **2017**, *138*, 145–159. [CrossRef]
19. Pimentel, D.; Hurd, L.E.; Bellotti, A.C.; Forster, M.J.; Oka, I.N.; Sholes, O.D.; Whitman, R.J. Food Production and the Energy Crisis. *Science* **1973**, *182*, 443–449. [CrossRef]
20. Guzmán, G.I.; De Molina, M.G. Energy Efficiency in Agrarian Systems From an Agroecological Perspective. *Agroecol. Sustain. Food Syst.* **2015**, *39*, 924–952. [CrossRef]
21. Gingrich, S.; Marco, I.; Aguilera, E.; Padró, R.; Cattaneo, C.; Cunfer, G.; Guzmán, G.I.; MacFadyen, J.; Watson, A. Agroecosystem energy transitions in the old and new worlds: Trajectories and determinants at the regional scale. *Reg. Environ. Chang.* **2018**, *18*, 1089–1101. [CrossRef]
22. Atlason, R.S.; Kjaerheim, K.M.; Davidsdottir, B.; Ragnarsdottir, K.V. A Comparative Analysis of the Energy Return on Investment of Organic and Conventional Icelandic Dairy Farms. *Icel. Agric. Sci.* **2015**, *28*, 29–42. [CrossRef]
23. Jónsson, J.; Örvar, G.; Davíðsdóttir, B.; Nikolaidis, N.P.; Giannakis, G.V. Tools for Sustainable Soil Management: Soil Ecosystem Services, EROI and Economic Analysis. *Ecol. Econ.* **2019**, *157*, 109–119. [CrossRef]
24. Pérez-Neira, D.; Schneider, M.; Armengot, L. Crop-diversification and organic management increase the energy efficiency of cacao plantations. *Agric. Syst.* **2020**, *177*, 102711. [CrossRef]
25. Pracha, A.S.; Volk, T.A. An Edible Energy Return on Investment (EEROI) Analysis of Wheat and Rice in Pakistan. *Sustainability* **2011**, *3*, 2358–2391. [CrossRef]
26. Pittelkow, C.M.; Zorrilla, G.; Terra, J.; Riccetto, S.; Macedo, I.; Bonilla, C.; Roel, A. Sustainability of rice intensification in Uruguay from 1993 to 2013. *Glob. Food Secur.* **2016**, *9*, 10–18. [CrossRef]
27. Infante-Amate, J.; Picado, W. Energy flows in the coffee plantations of Costa Rica: From traditional to modern systems (1935–2010). *Reg. Environ. Chang.* **2018**, *18*, 1059–1071. [CrossRef]
28. Macedo, I.; Terra, J.A.; Siri-Prieto, G.; Velazco, J.I.; Carrasco-Letelier, L. Rice-pasture agroecosystem intensification affects energy use efficiency. *J. Clean. Prod.* **2021**, *278*, 123771. [CrossRef]
29. Ozkan, B.; Akcaoz, H.; Fert, C. Energy input–output analysis in Turkish agriculture. *Renew. Energy* **2004**, *29*, 39–51. [CrossRef]
30. Cao, S.; Xie, G.; Zhen, L. Total embodied energy requirements and its decomposition in China's agricultural sector. *Ecol. Econ.* **2010**, *69*, 1396–1404. [CrossRef]
31. Veiga, J.P.S.; Romanelli, T.L.; Gimenez, L.M.; Busato, P.; Milan, M. Energy embodiment in Brazilian agriculture: An overview of 23 crops. *Sci. Agricola* **2015**, *72*, 471–477. [CrossRef]
32. Guzmán, G.I.; De Molina, M.G.; Fernández, D.S.; Infante-Amate, J.; Aguilera, E. Spanish agriculture from 1900 to 2008: A long-term perspective on agroecosystem energy from an agroecological approach. *Reg. Environ. Chang.* **2018**, *18*, 995–1008. [CrossRef]
33. Markussen, M.V.; Østergård, H. Energy Analysis of the Danish Food Production System: Food-EROI and Fossil Fuel Dependency. *Energies* **2013**, *6*, 4170–4186. [CrossRef]
34. Galán, E.; Padró, R.; Marco, I.; Tello, E.; Cunfer, G.; Guzmán, G.; De Molina, M.G.; Krausmann, F.; Gingrich, S.; Sacristán, V.; et al. Widening the analysis of Energy Return on Investment (EROI) in agro-ecosystems: Socio-ecological transitions to industrialized farm systems (the Vallès County, Catalonia, c.1860 and 1999). *Ecol. Model.* **2016**, *336*, 13–25. [CrossRef]
35. Laso, J.; Hoehn, D.; Margallo, M.; García-Herrero, I.; Batlle-Bayer, L.; Bala, A.; Fullana-I-Palmer, P.; Vázquez-Rowe, I.; Irabien, A.; Aldaco, R. Assessing Energy and Environmental Efficiency of the Spanish Agri-Food System Using the LCA/DEA Methodology. *Energies* **2018**, *11*, 3395. [CrossRef]

36. Bajan, B.; Mrówczyńska-Kamińska, A.; Poczta, W. Economic Energy Efficiency of Food Production Systems. *Energies* **2020**, *13*, 5826. [CrossRef]
37. Martinez-Alier, J. The EROI of agriculture and its use by the Via Campesina. *J. Peasant. Stud.* **2011**, *38*, 145–160. [CrossRef]
38. Hamilton, A.; Balogh, S.B.; Maxwell, A.; Hall, C.A.S. Efficiency of Edible Agriculture in Canada and the U.S. Over the Past Three and Four Decades. *Energies* **2013**, *6*, 1764–1793. [CrossRef]
39. Murphy, D.J.; Hall, C.A.; Dale, M.; Cleveland, C.J. Order from Chaos: A Preliminary Protocol for Determining the EROI of Fuels. *Sustainability* **2011**, *3*, 1888–1907. [CrossRef]
40. Hall, C.A.; Dale, B.E.; Pimentel, D. Seeking to Understand the Reasons for Different Energy Return on Investment (EROI) Estimates for Biofuels. *Sustainability* **2011**, *3*, 2413–2432. [CrossRef]
41. Brandt, A.R.; Dale, M. A General Mathematical Framework for Calculating Systems-Scale Efficiency of Energy Extraction and Conversion: Energy Return on Investment (EROI) and Other Energy Return Ratios. *Energies* **2011**, *4*, 1211–1245. [CrossRef]
42. Atlason, R.S.; Unnthorsson, R. Ideal EROI (energy return on investment) deepens the understanding of energy systems. *Energy* **2014**, *67*, 241–245. [CrossRef]
43. Conforti, P.; Giampietro, M. Fossil energy use in agriculture: An international comparison. *Agric. Ecosyst. Environ.* **1997**, *65*, 231–243. [CrossRef]
44. Pellegrini, P.; Fernández, R.J. Crop intensification, land use, and on-farm energy-use efficiency during the worldwide spread of the green revolution. *Proc. Natl. Acad. Sci.* **2018**, *115*, 2335–2340. [CrossRef] [PubMed]
45. Pelletier, N.; Audsley, E.; Brodt, S.; Garnett, T.; Henriksson, P.J.G.; Kendall, A.; Kramer, K.J.; Murphy, D.; Nemecek, T.; Troell, M. Energy Intensity of Agriculture and Food Systems. *Annu. Rev. Environ. Resour.* **2011**, *36*, 223–246. [CrossRef]
46. Carlsson-Kanyama, A.; Ekström, M.P.; Shanahan, H. Food and life cycle energy inputs: Consequences of diet and ways to increase efficiency. *Ecol. Econ.* **2003**, *44*, 293–307. [CrossRef]
47. Pimentel, D.; Williamson, S.; Alexander, C.E.; Gonzalez-Pagan, O.; Kontak, C.; Mulkey, S.E. Reducing Energy Inputs in the US Food System. *Hum. Ecol.* **2008**, *36*, 459–471. [CrossRef]
48. Gerbens-Leenes, P.; Nonhebel, S.; Ivens, W.; Gerbens-Leenes, W. A method to determine land requirements relating to food consumption patterns. *Agric. Ecosyst. Environ.* **2002**, *90*, 47–58. [CrossRef]
49. Siddiqi, T.A. Coal in Asia and the Pacific: A regional overview. *Energy* **1986**, *11*, 1049–1055. [CrossRef]
50. Sadowski, A.; Baer-Nawrocka, A. Food and environmental function in world agriculture—Interdependence or competition? *Land Use Policy* **2018**, *71*, 578–583. [CrossRef]
51. Council Directive 93/76/EEC of 13 September 1993 to Limit Carbon Dioxide Emissions by Improving Energy Efficiency. Available online: https://eur-lex.europa.eu/legal-content/en/ALL/?uri=CELEX%3A31993L0076 (accessed on 10 February 2021).
52. Directive 2006/32/EC of the European Parliament and of the Council of 5 April 2006 on Energy End-Use Efficiency and Energy Services and Repealing Council Directive 93/76/EEC. Available online: https://eur-lex.europa.eu/legal-content/EN/ALL/?uri=CELEX%3A32006L0032 (accessed on 10 February 2021).
53. Directive, E.E. Directive 2012/27/EU of the European Parliament and of the Council of 25 October 2012 on energy efficiency, amending Directives 2009/125/EC and 2010/30/EU and repealing Directives 2004/8/EC and 2006/32. *Off. J.* **2012**, *315*, 1–56.
54. Steinhart, J.S.; Steinhart, C.E. Energy use in the US food system. *Science* **1974**, *184*, 307–316.
55. Pelletier, N. Environmental performance in the US broiler poultry sector: Life cycle energy use and greenhouse gas, ozone depleting, acidifying and eutrophying emissions. *Agric. Syst.* **2008**, *98*, 67–73. [CrossRef]
56. Franzese, P.P.; Rydberg, T.; Russo, G.F.; Ulgiati, S. Sustainable biomass production: A comparison between Gross Energy Requirement and Emergy Synthesis methods. *Ecol. Indic.* **2009**, *9*, 959–970. [CrossRef]
57. Arizpe, N.; Giampietro, M.; Ramos-Martin, J. Food Security and Fossil Energy Dependence: An International Comparison of the Use of Fossil Energy in Agriculture (1991-2003). *Crit. Rev. Plant Sci.* **2011**, *30*, 45–63. [CrossRef]
58. Baer-Nawrocka, A.; Sadowski, A. Food security and food self-sufficiency around the world: A typology of countries. *PLOS ONE* **2019**, *14*, e0213448. [CrossRef] [PubMed]
59. Cleveland, C. The direct and indirect use of fossil fuels and electricity in USA agriculture, 1910–1990. *Agric. Ecosyst. Environ.* **1995**, *55*, 111–121. [CrossRef]
60. Chakona, G.; Shackleton, C.M. Voices of the hungry: A qualitative measure of household food access and food insecurity in South Africa. *Agric. Food Secur.* **2017**, *6*, 1–17. [CrossRef]
61. Fraval, S.; Hammond, J.; Bogard, J.R.; Ng'Endo, M.; Van Etten, J.; Herrero, M.; Oosting, S.J.; De Boer, I.J.M.; Lannerstad, M.; Teufel, N.; et al. Food Access Deficiencies in Sub-saharan Africa: Prevalence and Implications for Agricultural Interventions. *Front. Sustain. Food Syst.* **2019**, *3*, 104. [CrossRef]
62. Court, V.; Lantz, F.; Jouvet, P.-A. Long-term endogenous economic growth and energy transitions. *Energy J.* **2018**, *39*, 29–57. [CrossRef]

energies

Article

Changes in Energy Consumption in Agriculture in the EU Countries

Tomasz Rokicki [1], Aleksandra Perkowska [1,*], Bogdan Klepacki [1], Piotr Bórawski [2], Aneta Bełdycka-Bórawska [2] and Konrad Michalski [3]

[1] Institute of Economics and Finance, Warsaw University of Life Sciences, 02-787 Warsaw, Poland; tomasz_rokicki@sggw.edu.pl (T.R.); bogdan_klepacki@sggw.edu.pl (B.K.)
[2] Department of Agrotechnology and Agribusiness, Faculty of Agriculture and Forestry, University of Warmia and Mazury in Olsztyn, 10-719 Olsztyn, Poland; pboraw@uwm.edu.pl (P.B.); aneta.beldycka-borawska@uwm.edu.pl (A.B.-B.)
[3] Management Institute, Warsaw University of Life Sciences, 02-787 Warsaw, Poland; konrad_michalski@sggw.edu.pl
* Correspondence: aleksandra_perkowska@sggw.edu.pl; Tel.: +48-22-59-342-55

Abstract: The paper's main purpose was to identify and present the current situation and changes in energy consumption in agriculture in the European Union (EU) countries. The specific objectives were the determination of the degree of concentration of energy consumption in agriculture in the EU countries, showing the directions of their changes, types of energy used, and changes in this respect, establishing the correlation between energy consumption and changes in the economic and agricultural situation in the EU countries. All member states of the European Union were deliberately selected for research on 31 December 2018 (28 countries). The research period covered the years 2005–2018. The sources of materials were the literature on the subject, and data from Eurostat. Descriptive, tabular, and graphical methods were used to analyze and present materials, dynamics indicators with a stable base, Gini concentration coefficient, concentration analysis using the Lorenz curve, coefficient of variation, Kendall's tau correlation coefficient, and Spearman's rank correlation coefficient. A high concentration of energy consumption in agriculture was found in several EU countries, the largest in countries with the largest agricultural sector, i.e., France and Poland. There were practically no changes in the concentration level. Only in the case of renewable energy, a gradual decrease in concentration was visible. More and more countries developed technologies that allow the use of this type of energy. However, the EU countries differed in terms of the structure of the energy sources used. The majority of the basis was liquid fuels, while stable and gaseous fuels were abandoned in favor of electricity and renewable sources—according to which, in the EU countries, the research hypothesis was confirmed: a gradual diversification of energy sources used in agriculture, with a systematic increase in the importance of renewable energy sources. The second research hypothesis was also confirmed, according to which the increase in the consumption of renewable energy in agriculture is closely related to the economy's parameters. The use of renewable energy is necessary and results from concern for the natural environment. Therefore, economic factors may have a smaller impact.

Keywords: energy consumption; agriculture; renewable energy sources; development strategies; EU countries

Citation: Rokicki, T.; Perkowska, A.; Klepacki, B.; Bórawski, P.; Bełdycka-Bórawska, A.; Michalski, K. Changes in Energy Consumption in Agriculture in the EU Countries. *Energies* **2021**, *14*, 1570. https://doi.org/10.3390/en14061570

Academic Editor: Piotr Gradziuk

Received: 4 January 2021
Accepted: 6 March 2021
Published: 12 March 2021

1. Introduction

In the European Union (EU), around 50% of agricultural production comes from plant production. Livestock production also generally needs land to feed the animals. About 47% of the land in the EU is used for agriculture. Therefore, it is a sector closely related to land use. Changes in the connection between land and agriculture proceed very slowly, and in the short term, this resource does not change significantly [1–5]. Agriculture in the European Union countries has been and will be diversified. They can be divided into segments.

Countries with a high level of socioeconomic development are most often distinguished, such as France, Germany, the Netherlands, Denmark, and Belgium. The second group includes the remaining EU-15 countries. The other two groups are the countries that joined the EU in 2004 and later [6–13]. In 2011–2013, labor productivity in agriculture in Eastern Europe accounted for only 19% of labor productivity in agriculture in Western Europe. This shows that there are still significant socioeconomic and technological gaps between Western and Eastern Europe, even though Eastern European countries joined the EU as early as 2004 [14–19]. In all Western European countries after 1950, the number of workers in the agricultural sector decreased. At the same time, productivity increased as the operation of machines replaced human labor. This resulted in high energy demand. Simultaneously, the importance of agriculture in generating GDP was gradually diminishing [20,21]. According to Giannakis and Bruggeman [22], the differences between individual countries result from the characteristics of human capital, environmental conditions, and technical efficiency of plant and animal production. By contrast, strong Szabo and Grznár [23] found links between the value of agricultural production and fixed and variable assets, the number of livestock, and the financial support provided. Pietrzak and Walczak [24,25] proved that the agrarian structure is one of the most important agricultural development determinants. Low concentration of land is a significant barrier to agriculture development due to high production costs and generating low income. According to Nowak and Różańska-Boczula [26], such EU member states as the Netherlands, Denmark, Luxembourg, Belgium, Great Britain, and Slovakia have the most significant potential for agricultural production. The first four countries also showed the highest efficiency in the use of production factors, whereas low and average potential and efficiency characterized agriculture in most new member states. Similar dependencies were found by Popescu et al. [27–29], Bularca and Toma [30], Toma [31], and Svoboda et al. [32]. The EU countries also varied in terms of the production profile. There were specializations in agricultural production. Specialization can mean agricultural intensification and concentration. Specialization is related to mechanization, economies of size and scale, technological innovations, comparative advantages, and market forces. Agricultural specialization is also related to farms and agricultural land features, efficiency, and the geographical scale of specialization [33–37].

Energy is one of the basic inputs in agriculture [38]. At the farm level, energy is used directly as well as indirectly. Energy is used directly in plant production, livestock production, and the transport of agricultural products. Indirectly, energy is used outside the farm to produce and transport fertilizers, pesticides, and machines [39–42]. The increase in energy demand in agriculture results from the increase in mechanization. Energy supplies to modern and sustainable agricultural production systems and processing are one of the main factors in the growth of agricultural production [43–47]. In agriculture, various energy sources are used; often, these are hybrid systems that use both traditional and renewable energy sources [48,49].

Karkacier et al. [50] determined a strong relationship between energy use and agricultural productivity. Alipour et al. [51] used rice cultivation to show that water and electricity account for the largest share of total energy inputs in production systems. Chandio et al. [52] indicated that the increase positively influenced agricultural production in gas and electricity consumption. The presented research shows a strong relationship between energy consumption and the value of agricultural production. Energy efficiency in agriculture is one of the primary energy policy goals in countries with a significant agricultural sector [53–55].

The agricultural sector also supplies energy in the form of biomass. Biomass means the biodegradable fraction of products, waste and residues from agricultural production (including substances of plant and animal origin), forestry and related industries, including fisheries and aquaculture, as well as biogas and the biodegradable fraction of industrial and municipal waste [56,57]. Biomass can be used, among others, for the production of biodiesel and bioethanol [58]. In 2010, biomass was the source of 7.5% of the energy generated in the EU, and in 2020 this share amounted to 10%. In the world, energy production from biomass has grown at a rate of 3.3% annually in recent years. The potential of agriculture in this respect is tremendous. It all depends on the progress made in introducing high-efficiency energy crops and on environmental issues [59–65].

The paper's main purpose was to identify and present the current situation and changes in energy consumption in agriculture in the European Union countries. The specific objectives were the determination of the degree of concentration of energy consumption in agriculture in the EU countries, showing the directions of their changes, types of energy used, and changes in this respect, establishing the correlation between energy consumption and changes in the economic and agricultural situation in the EU countries. Two hypotheses were put forward in the study. According to the first, in the EU, there was a gradual diversification of energy sources used in agriculture, with a systematic increase in the importance of renewable energy sources. The second hypothesis assumed that the increase in the consumption of renewable energy in agriculture is closely related to the economy's parameters.

2. Materials and Methods

All member states of the European Union were deliberately selected for research on 31 December 2018 (28 countries). The research period covered the years 2005–2018. The sources of materials were the literature on the subject and data from Eurostat. Descriptive, tabular, and graphical methods, dynamics indicators with a *constant basis*, Gini concentration coefficient, concentration analysis using the Lorenz curve, coefficient of variation, and Pearson's linear correlation coefficient were used for the analysis and presentation of materials.

The first stage of the research presents the energy consumption in agriculture in the EU countries. The primary sources of obtaining energy used for this purpose have been shown. The Gini concentration coefficient (Figure 1) was calculated [66]. It concerned total energy consumption in agriculture. This coefficient was also calculated for energy consumption in agriculture for individual types of energy. The results covered the two years of 2005 and 2018. They were used to determine the degree of concentration of energy consumption in agriculture. It is measured based on energy consumption in agriculture in the EU countries. If such energy consumption occurred in one country, the coefficient would be 1. If it is distributed among more countries, the coefficient becomes lower the closer it is to 0. This proves the even distribution of energy consumption in agriculture among EU countries. The Lorenz curve (Figure 2) is a graphical representation of the concentration of energy consumption in agriculture in EU countries [67].

	Gini coefficient
concept	measure of unevenness (concentration) of distribution of a random variable
formula	$G(y) = \frac{\sum_{i=1}^{n}(2i-n-1)*y_i}{n^2*\bar{y}}$ where: n – number of observations, y_i - value of the "i-th" observation, $\bar{y}$ - the average value of all observations, i.e. $\bar{y} = \frac{1}{n}\sum_{i=1}^{n} y_i$

Figure 1. The Gini coefficient formula.

	Lorenz curve
concept	determines the degree of concentration of a one-dimensional random variable distribution
coordinates	$x_0 = z_0 = 0$, $x_h = \frac{h}{n}$, $z_h = \frac{\sum_{i=1}^{h} y_i}{\sum_{i=1}^{n} y_i}$
	With sorted observations yi., which are non-negative value $0 \le y_1 \le y_2 \le \ldots \le y_n$ is a polyline which apexes , for h = 0,1,…, n

Figure 2. The Lorenz curve formula.

In the second stage of the research, the structure of energy consumption in agriculture was presented. The share of energy sources used in agriculture was shown for sources such as oil and petroleum products, electricity, natural gas, renewables and biofuels, reliable fossil fuels. This division functioned in all EU. Only five countries were selected for analysis. France and Poland used the most energy in agriculture. The first of them belonged to the developed countries, and the second to the developing countries. Romania was in the middle of the EU countries' rate of energy consumption in agriculture. Simultaneously, in the analyzed period, the country recorded the highest increase in energy consumption in agriculture. Greece was also in the middle of the league. In turn, in this country, the largest decrease in energy consumption in agriculture was recorded among all EU countries. Latvia was at the end of the countries' ranking in terms of energy consumption in agriculture, but with relatively high growth dynamics. Apart from France, all analyzed countries were economically developing countries. The countries presented were, therefore, diverse in many aspects.

The third stage presents the share of renewable energy in the total energy consumption in agriculture. The focus was on countries with the highest share of this energy. The use of renewable energy in the economy is significant. There are similar trends in agriculture. At this stage, the differences between individual EU countries were indicated.

The dynamics indicators (Figure 3) for primary groups of energy sources in individual EU countries were calculated in the fourth stage [68]. As a result, knowledge was obtained about the directions and strength of energy consumption changes in agriculture from various sources.

Dynamics indicators	$i = \frac{y_n}{y_0}$ or $i = \frac{y_n}{y_0} \cdot 100\%$ y_n - the level of the phenomenon in a certain period y_0 - the level of the phenomenon during the reference period

Figure 3. Dynamics indicators formula.

In the fifth stage, the coefficients of variation (Figure 4) for individual energy sources in agriculture for 2005–2018 were calculated. As a result, it was possible to determine whether the situation was stable or whether energy consumption was subject to substantial fluctuations [69].

	Coefficient of variation (Cv)
concept	eliminates the unit of measurement from the standard deviation of a series of number by dividing it by the mean of this series of numbers
formula	$C_v = \frac{S}{M}$ where : S - standard deviation from the sample M - arithmetic mean from the sample

Figure 4. The coefficients of variation formula.

In the sixth stage of the research, the relationship between the amount of energy consumption in agriculture in the EU countries and the economy's basic parameters and agriculture was examined. The parameters were selected on purpose based on the literature review. The indicators assessing the economic situation included the value of GDP, final consumption expenditure of households, export, and import of goods and services. The level of economic development is assessed by the parameters of the economy per capita. The following basic parameters were used to assess agricultural production: gross value added of agriculture, forestry, and fishing; area of crops and of grain sowing; and cows' milk production.

At this stage of the research, non-parametric tests were used to establish the correlation between the variables. The first is Kendall's tau correlation coefficient. It is based on the difference between the probability that two variables fall in the same order (for the observed data) and the probability that they are different. This coefficient takes values in the range <–1, 1>. Value 1 means full match, value 0 no match of orderings, and value –1 the complete opposite. The Kendall coefficient indicates not only the strength but also the direction of the relationship. It is a good tool for describing the similarity of the data set orderings. Kendall's tau correlation coefficient is calculated using the formula [70]

$$\tau = P[(x_1 - x_2)(y_1 - y_2) > 0] - P[(x_1 - x_2)(y_1 - y_2) < 0]$$

The given formula estimates Kendall's tau from a statistical sample. All possible pairs of the sample observations are combined, and then the pairs are divided into three possible categories:

P—concordant pairs, when the compared variables within two observations change in the same direction, i.e., either in the first observation both are greater than in the second, or both are less than in the second;

Q—incompatible pairs, when the variables change in the opposite direction, i.e., one of them is more significant for this observation in the pair, for which the other is less than;

T—related pairs when one of the variables has equal values in both observations.

The Kendall tau estimator is then calculated from the formula

$$\tau = \frac{P - Q}{P + Q - T}$$

Additionally, $P + Q + T = \binom{N}{2} = \frac{N(N-1)}{2}$

where

N—sample size

The formula can be represented as

$$\tau = 2\frac{P - Q}{N(N-1)}$$

The second non-parametric test is Spearman's rank correlation coefficient. It is used to describe the strength of the correlation of two features. It is used to study the relationship between quantitative traits for a small number of observations. Spearman's rank correlation coefficient is calculated according to the formula [71]

$$r_S = 1 - \frac{6\sum_{i=1}^{n} d_i^2}{n(n^2-1)}$$

where d_i—differences between the ranks of the corresponding feature xi and feature y_i ($i = 1, 2, \ldots, n$)

The correlation coefficient takes values in the range $-1 \le r_s \le +1$. A positive sign of the correlation coefficient indicates a positive correlation, while a negative sign indicates a negative correlation. The closer the modulus (absolute value) of the correlation coefficient is to 1, the stronger the correlation between the examined variables.

3. Results

In 2005–2018, energy consumption in agriculture in the EU countries decreased by 5.9%. Several sources of energy could be distinguished (Figure 5). The energy consumption from renewable energy sources increased the fastest, as there was an increase in 2005–2018 by 85%. Electricity consumption increased by 25%. There was a decrease in consumption in the remaining cases, i.e., heat by 33%, gas products by 23%, crude oil by 15%, and fossil fuels by 9%. This situation is good for the natural environment.

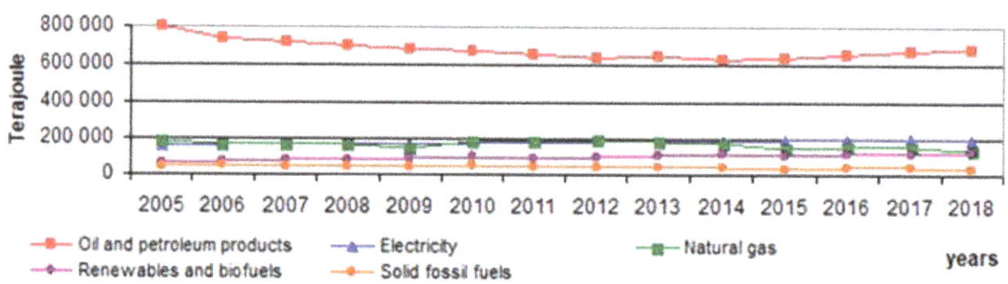

Figure 5. Sources of energy used in agriculture in 2005–2018.

The Gini coefficient was used to determine the concentration of energy consumption in agriculture from its various EU countries' sources. This coefficient is a correct and commonly used measure of inequality. The number of observations was 28 (all EU countries). The results are presented for total energy consumption and five types of energy, i.e., energy from crude oil, electricity, natural gas, renewable sources, and solid fuels. The Gini coefficient for total energy consumption in agriculture in 2005, calculated from the sample, was 0.61, and the estimated coefficient for the population was 0.63. This meant quite a high concentration of energy consumption in agriculture in several EU countries. When the research was repeated in 2018, the results were virtually identical. Therefore, there have been no significant changes in the distribution of energy consumption in agriculture in the EU countries. The Gini coefficients for energy consumption in agriculture were also calculated for individual types of energy. Additionally, the differentiation was presented using the Lorenz concentration curve (Figure 6). In 2018, the concentration of energy consumption was the highest in solid fuels (the coefficient from the sample was 0.95 and the estimated 0.99), and the lowest for electricity (from the sample 0.62, estimated 0.64). In 2018, Poland was responsible for 96% of the solid fuels consumed in agriculture. It was mainly hard coal. In natural gas, the Netherlands accounted for 61% of the consumption of this raw material in agriculture in the EU. The most energy from renewable sources was consumed in Germany (26%) and Poland (17%). It was mainly biodiesel. Overall, there were differences between energy types in the level of consumption concentration. Concentration coefficients were also calculated for the earlier periods, with a frequency every three years. Only between 2014 and 2018, there was a four-year gap. As a result, the results concern the years 2005–2018. Such a summary allows determining the direction and pace of the changes in the concentration of energy consumption in agriculture. Generally, it can be noticed that the concentration of energy consumption in agriculture is maintained in several countries (Table 1). The reasonably stable situation also results from the permanent land stock, which is the primary production factor in agriculture. This limits the possibility of a drastic increase in agricultural production. Another reason may be relatively stable energy consumption and countries using technologies that ensure similar energy efficiency (such as tractors and machines, technologies for fattening animals, milk production, etc.). Only in the case of renewable energy sources, a gradual decrease in the concentration of its consumption in agriculture can be observed. More and more countries are developing technologies that allow the use of this type of energy. Agriculture is a sector that produces more renewable energy than it consumes.

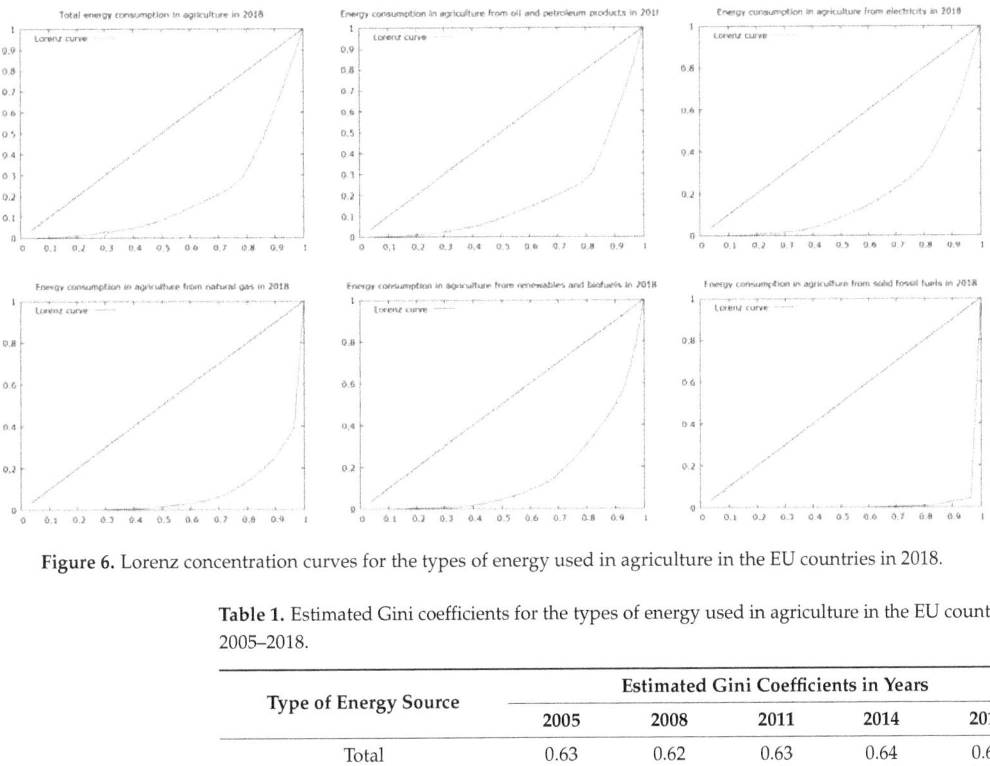

Figure 6. Lorenz concentration curves for the types of energy used in agriculture in the EU countries in 2018.

Table 1. Estimated Gini coefficients for the types of energy used in agriculture in the EU countries in 2005–2018.

Type of Energy Source	Estimated Gini Coefficients in Years				
	2005	2008	2011	2014	2018
Total	0.63	0.62	0.63	0.64	0.64
Oil and petroleum products	0.65	0.65	0.65	0.66	0.66
Electricity	0.66	0.64	0.64	0.64	0.64
Natural gas	0.88	0.86	0.84	0.85	0.87
Renewables and biofuels	0.79	0.76	0.73	0.71	0.71
Solid fossil fuels	0.97	0.98	0.99	0.99	0.99

Each country has a separate history and conditions, also in terms of the energy resources used. Five different countries, i.e., France, Poland, Romania, Greece, and Latvia, were selected for a more detailed analysis. France has the highest energy consumption in agriculture of any EU country. It was also an economically developed country, including in terms of agriculture. In 2005–2018, energy consumption in agriculture in this country decreased by 2.8%. Poland came second in terms of energy consumption. It was a developing country that joined the EU in 2004 and had fairly fragmented agriculture. The total energy consumption in agriculture decreased by 11.7% in the analyzed period, which could be due to agriculture's ongoing transformation. Romania was also one of the countries admitted to the EU in 2004, catching up with Western European countries. In this country, energy consumption in agriculture increased by 163%, by far the highest among all EU countries. Despite this, Romania was in the middle of the EU countries' rate in energy consumption in agriculture. The largest decrease in energy consumption was recorded in Greece, by as much as 77%. As a result, it was ranked 17th in the EU. It was an economically developed country but suffering the effects of the economic crisis and having financial problems. The agricultural sector has also felt the effects of the economic downturn. Latvia was one of the newly admitted countries to the EU, with a small agricultural sector, but with a high energy consumption dynamics, because an increase of 43% was achieved. Despite this, the country ranked 21st in the EU.

In France, almost $\frac{3}{4}$ of the energy used in agriculture came from oil and a dozen or so from electricity (Figure 7). The share of other sources was small. The share of crude oil decreased quite slowly, while electricity and renewable energy sources increased. The situation was, therefore, relatively stable. In Poland's case, crude oil also dominated, but its share systematically decreased from 65% in 2005 to 59% in 2018 (Figure 8). Solid fuels, mainly coal, were also of great importance. A dozen or so percent of energy from renewable sources was also used, mainly it was biodiesel. This share remained at a double-digit level throughout the period considered. The other sources were of little importance.

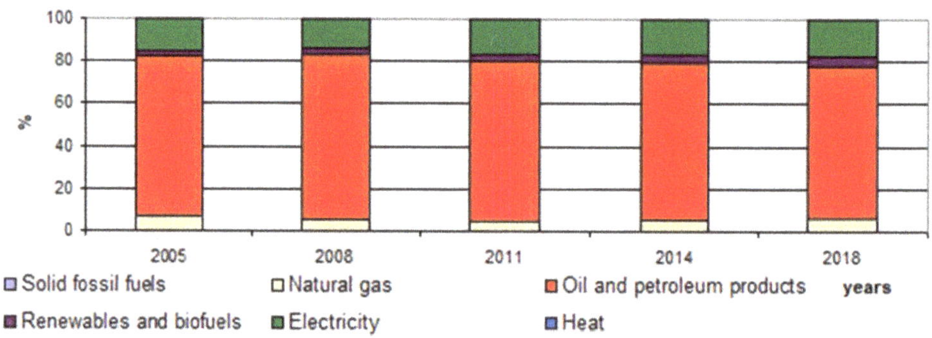

Figure 7. Structure of energy used in agriculture in France in 2005–2018.

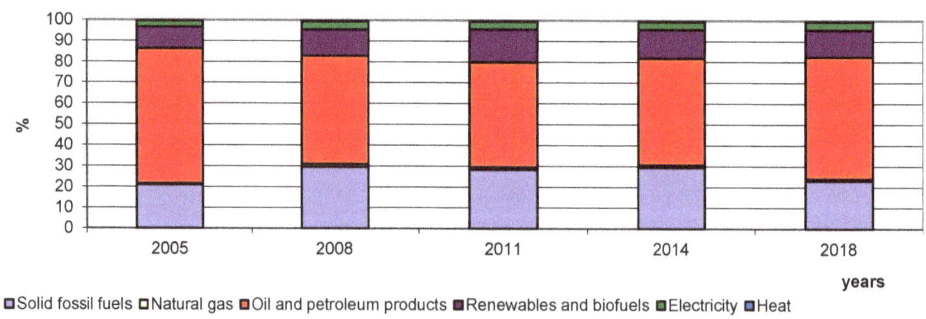

Figure 8. Structure of energy used in agriculture in Poland in 2005–2018.

In Romania, the share of crude oil in the energy used in agriculture was at a similar level as in Poland (Figure 9). In this country, however, the importance of this source grew, as its share increased from 55% in 2005 to 64% in 2018. Almost 20% were used for gas products (also increased in importance), and a dozen for electricity. Noteworthy is the very small share of energy consumed from renewable sources (1–2%). Peat was used for energy purposes at a similar level. It was one of the few countries that used such an energy source. In Greece, there were very large changes in the structure of energy used in agriculture (Figure 10). The reason was a large reduction in energy consumption. Crude oil decreased in importance (a decrease from 76 to 14% in 2005–2018), while electricity gained in importance (an increase from 22 to 74%). A positive aspect was the increase in renewable energy consumption from 1% to 12%. In Latvia, there was the largest consumption of crude oil in agriculture (Figure 11). Its consumption in the analyzed period was around 60–70%. In the case of electricity, it was around 10%. The importance of gaseous products has decreased, while of thermal energy and energy obtained from renewable sources has increased.

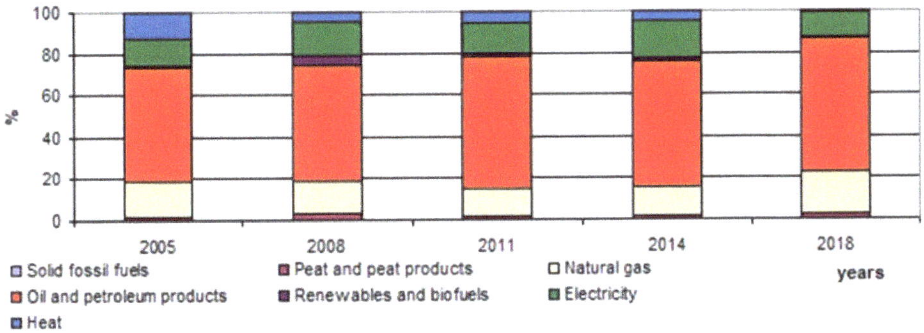

Figure 9. Structure of energy used in agriculture in Romania in 2005–2018.

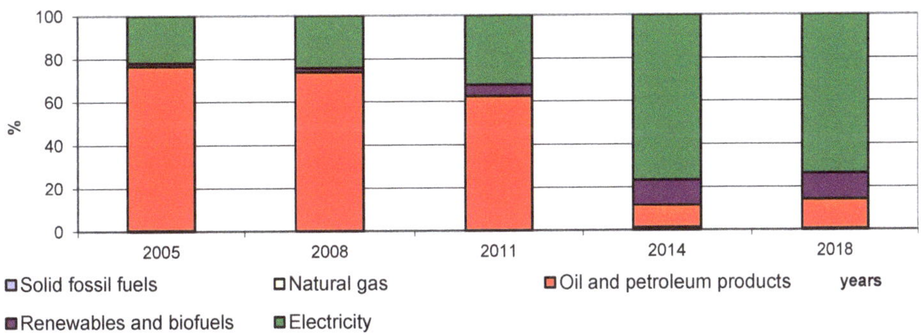

Figure 10. Structure of energy used in agriculture in Greece in 2005–2018.

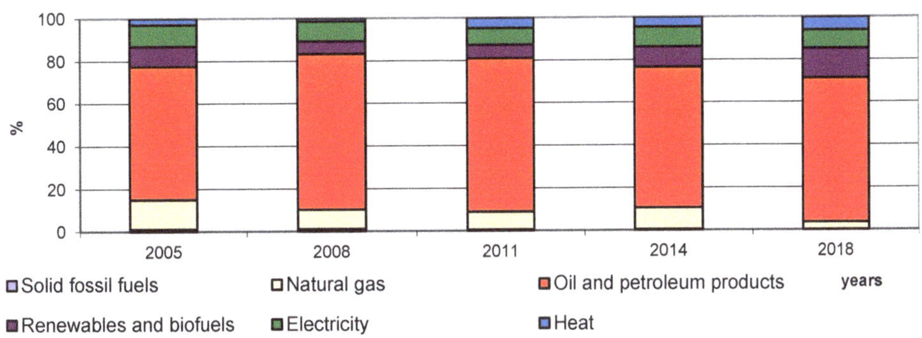

Figure 11. Structure of energy used in agriculture in Latvia in 2005–2018.

In the EU countries, crude oil was the most important, as it satisfied more than half of the needs of the agricultural sector. In 2005, it was 63% and in 2018, 57%. It was followed by electricity (16% in 2018), natural gas (12%), and renewable energy sources (10%). The share of energy from heat and peat combustion was very low. Apart from Greece, all countries presented had a structure similar to the EU average. Liquid fuels dominated, which is obvious, because they were the power source for tractors and agricultural machinery. The share of other sources was smaller and each country had a structure that often corresponded to energy resources found in the country or those that can be easily supplied.

The share of energy from renewable energy sources in total energy consumption in agriculture varied across countries. In 2018, in the top five countries, it was above 20% (Figure 12. Sweden was the clear leader with 35%, followed by Austria (33%), Finland (25%), Germany, and Slovakia (23% each). These were economically developed countries that allocated large resources to the implementation of new technologies ensuring the use of renewable energy. Only Slovakia was a developing country that significantly increased renewable energy use, as in 2005 it was only 1%. As many as 12 countries have achieved or exceeded the 10% share of renewable energy, which is the EU average. There were also economically highly developed countries that had a very small share of renewable energy in energy consumption in agriculture. Examples are Italy (2% in 2018), Spain (3%), and France (5%). Nevertheless, the importance of this energy source was systematically growing.

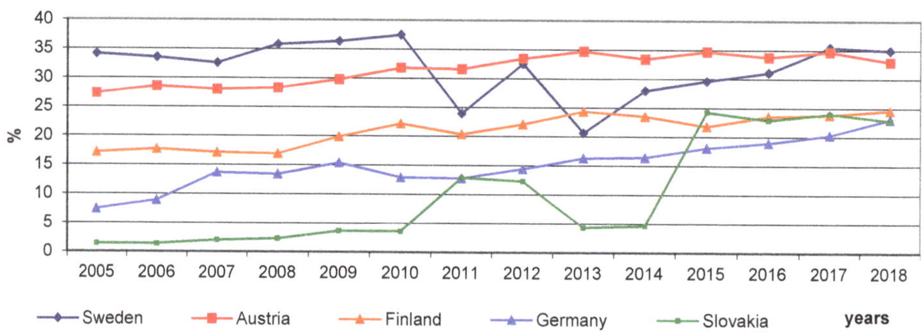

Figure 12. Top 5 EU countries in share of renewable energy in the total energy use in agriculture in the in 2005–2018.

In the next stage, the dynamics indicators for the basic groups of energy sources were calculated. The 2005 level was adopted as the basis (Table 2). Over 14 years, the increase in energy consumption in agriculture was recorded in 11 EU countries, by far the largest in Romania, and significant in Latvia, Germany, and the United Kingdom. In turn, substantial declines occurred in Greece, but also significant in Bulgaria, Sweden, Ireland, and Portugal. The reason for the drops may be the limitation of agricultural production or the use of more effective technology. In turn, when energy consumption increases, the reasons are the opposite. Each country should be analyzed separately due to the natural, economic, and social conditions. In the case of Greece, a substantial reduction in oil consumption can be observed, which may be related to the cessation or reduction of many agricultural production. Increase in this country was recorded in the case of the consumption of renewable energy sources. In Romania, all energy sources' consumption, except for fossil fuels, increased (decrease by 50%). In Latvia, fossil and gas fuels have been abandoned. The consumption of energy from renewable sources in the Netherlands and Belgium proliferated, as it increased by several dozen times. These countries, however, started out from low consumption of this type of energy. In general, countries are moving away from fossil fuels, and mostly from gas and heat. As a rule, oil consumption was reduced. Electricity consumption increased in most countries. In the case of renewable energy, its consumption has been systematically growing in all EU countries. Only Sweden (with a very high share of renewable energy) and Bulgaria recorded declines. Some countries did not use renewable energy in 2005. Therefore, it was not possible to calculate the dynamics index for such countries.

Table 2. Dynamics indicators for energy used in agriculture in EU countries in 2005–2018 (year 2005 = 100).

Countries	Dynamics Indicators for Types of Energy Used in Agriculture in 2005–2018						
	Total	Solid Fossil Fuels	Natural Gas	Oil and Petroleum Products	Renewables and Biofuels	Electricity	Heat
Romania	263.40	49.82	310.99	309.87	198.55	227.27	17.81
Latvia	142.99	0.18	36.10	154.89	211.59	119.67	308.26
Germany	134.80	-	-	93.83	417.30	-	-
United Kingdom	134.02	-	45.92	222.34	195.02	80.95	-
Estonia	118.96	81.01	66.65	137.33	103.81	76.19	70.83
Hungary	114.34	21.00	48.82	169.60	225.12	104.86	150.00
Czechia	113.28	31.94	77.47	98.75	745.13	94.59	55.19
Cyprus	111.99	-	-	86.26	-	147.13	-
Luxembourg	105.71	-	724.74	96.46	160.71	102.11	-
Lithuania	103.22	199.15	73.49	111.29	237.97	110.90	41.34
Austria	100.48	19.45	118.48	80.52	121.42	125.12	159.74
Croatia	99.48	-	102.55	94.28	-	94.48	-
Slovenia	97.30	-	-	93.75	-	-	-
Belgium	97.17	52.00	1 533.80	41.68	2 807.09	457.57	-
France	97.13	-	80.41	91.87	198.37	114.04	-
Finland	96.13	64.38	7.33	74.88	137.67	120.26	123.86
Malta	94.48	-	-	74.33	-	-	-
EU	94.13	90.60	77.36	96.38	185.30	124.82	66.81
Netherlands	94.00	-	74.67	94.45	4 093.37	172.32	39.90
Italy	93.00	-	81.03	90.24	202.12	103.69	879.48
Poland	88.29	97.71	116.89	79.82	110.60	123.50	97.21
Denmark	87.06	21.94	68.03	93.63	108.34	91.36	79.55
Slovakia	80.24	15.49	48.01	73.72	1 280.81	61.22	20.67
Spain	78.84	-	39.84	79.44	389.18	94.71	-
Portugal	73.24	-	101.96	65.91	-	107.53	0.00
Ireland	66.57	-	-	62.59	-	86.74	-
Sweden	62.25	-	26.89	58.02	63.75	78.08	100.00
Bulgaria	60.95	191.79	53.87	50.25	92.53	137.82	2 816.45
Greece	22.92	24.01	-	4.16	186.10	77.22	-

The coefficients of variation for individual energy sources in agriculture were calculated for the years 2005–2018 (Table 3). In the case of total energy, there were no large fluctuations in energy consumption in individual years. The exception was Greece. Energy consumption from crude oil was also relatively stable, apart from Greece, where this demand has decreased very drastically, and Great Britain, where there has been a large increase in oil consumption. There was also little variability in the case of electricity. More significant variability occurred with heat, solid, and gaseous fuels. There was quite a lot of variability in most countries in renewable energy due to the rapidly growing consumption of this energy in almost all EU countries.

To establish the relationship between the amount of energy consumption in agriculture in the EU countries and the basic parameters of the economy and agriculture, Kendall's tau correlation coefficient and Spearman's rank correlation coefficient were calculated (Tables 4 and 5). $p = 0.05$ was adopted as the border value of the significance level. Significant results are marked in bold in the table. Correlation coefficients were calculated for all EU countries for the entire 2005–2018 period. The study tried to check the correlation, which does not indicate that a given factor affects another, but a strong or weak relationship between them. In the case of energy, the total consumption in agriculture and the most critical groups, i.e., crude oil and renewable energy, were used for calculations.

Table 3. Coefficients of variation of energy used in agriculture in EU countries in 2005–2018.

Countries	Coefficients of Variation in the Volume of Energy Used in Agriculture by Types						
	Total	Solid Fossil Fuels	Natural Gas	Oil and Petroleum Products	Renewables and Biofuels	Electricity	Heat
France	0.02	-	0.12	0.03	0.21	0.08	-
Austria	0.02	0.72	0.20	0.06	0.10	0.07	0.20
EU	0.02	0.08	0.09	0.09	0.19	0.07	0.13
Finland	0.03	0.27	0.85	0.07	0.13	0.07	0.08
Croatia	0.03	-	0.12	0.06	-	0.05	-
Slovenia	0.04	-	-	0.05	-	-	-
Lithuania	0.04	0.56	0.22	0.07	0.22	0.07	0.33
Netherlands	0.04	-	0.09	0.04	0.68	0.17	0.34
Italy	0.05	-	0.09	0.06	0.42	0.03	0.72
Denmark	0.06	0.41	0.14	0.07	0.03	0.04	0.11
Slovakia	0.07	0.57	0.20	0.08	0.90	0.15	0.69
Cyprus	0.07	-	-	0.12	-	0.11	-
Poland	0.07	0.08	0.16	0.15	0.07	0.06	0.09
Czechia	0.08	0.34	0.11	0.01	0.66	0.10	0.27
Belgium	0.08	0.56	0.31	0.29	0.49	0.34	-
Luxembourg	0.08	-	1.55	0.10	0.23	0.13	-
Spain	0.10	-	0.67	0.12	0.37	0.14	-
Latvia	0.12	0.75	0.29	0.12	0.38	0.12	0.49
Hungary	0.13	0.99	0.25	0.21	0.34	0.10	0.36
Estonia	0.13	1.79	0.20	0.19	0.40	0.09	0.21
Portugal	0.14	-	0.18	0.17	-	0.09	0.80
Germany	0.15	-	-	0.02	0.36	-	-
Ireland	0.16	-	-	0.19	-	0.04	-
Sweden	0.18	-	0.45	0.14	0.27	0.24	0.00
United Kingdom	0.19	-	0.30	0.51	0.59	0.14	-
Bulgaria	0.20	0.26	0.29	0.30	0.69	0.12	0.77
Romania	0.25	2.20	0.39	0.30	1.01	0.25	0.36
Malta	0.26	-	-	0.36	-	-	-
Greece	0.60	1.38	-	0.98	0.23	0.10	-

Table 4. Kendall's tau correlation coefficients between the volume of use of energy in agriculture in the EU countries and the parameters of the economy and agriculture.

Tested Parameters	Kendall's Tau Correlation Coefficient					
	Total Energy		Oil and Petroleum Products		Renewables and Biofuels	
	τ	*p*-Value	τ	*p*-Value	τ	*p*-Value
Correlation coefficients between energy consumption and						
Value of GDP	**0.978**	**0.001**	−0.341	0.080	**0.868**	**0.001**
Final consumption expenditure of households	**0.978**	**0.001**	−0.341	0.080	**0.868**	**0.001**
Export of goods and services	**0.978**	**0.001**	−0.341	0.080	**0.868**	**0.001**
Import of good and services	**0.978**	**0.001**	−0.341	0.080	**0.824**	**0.001**
GDP per capita	**0.999**	**0.001**	−0.319	0.100	**0.846**	**0.001**
Final consumption expenditure of households per capita	**0.999**	**0.001**	−0.319	0.100	**0.846**	**0.001**
Gross value added of agriculture, forestry, and fishing	**0.934**	**0.001**	−0.341	0.080	**0.780**	**0.001**
Area of agricultural crops	**−0.495**	**0.012**	**0.692**	**0.001**	**−0.604**	**0.002**
Area of grain sowing	**−0.538**	**0.006**	0.077	0.743	**−0.516**	**0.009**
Cows' milk production	**0.495**	**0.016**	−0.165	0.381	**0.560**	**0.006**

Table 5. Spearman's rank correlation coefficients between the volume of use of energy in agriculture in the EU countries and the parameters of the economy and agriculture.

Tested Parameters	Spearman's Rank Correlation Coefficient					
	Total Energy		Oil and Petroleum Products		Renewables and Biofuels	
	τ	*p*-Value	τ	*p*-Value	τ	*p*-Value
Correlation coefficients between energy consumption and						
Value of GDP	**0.996**	**0.010**	−0.442	0.100	**0.947**	**0.010**
Final consumption expenditure of households	**0.996**	**0.010**	−0.442	0.100	**0.947**	**0.010**
Export of goods and services	**0.996**	**0.010**	−0.442	0.100	**0.952**	**0.010**
Import of good and services	**0.996**	**0.010**	−0.437	0.100	**0.930**	**0.010**
GDP per capita	**0.999**	**0.010**	−0.429	0.100	**0.934**	**0.010**
Final consumption expenditure of households per capita	**0.999**	**0.010**	−0.429	0.100	**0.934**	**0.010**
Gross value added of agriculture, forestry, and fishing	**0.982**	**0.010**	−0.442	0.100	**0.903**	**0.010**
Area of agricultural crops	**−0.723**	**0.010**	**0.824**	**0.010**	**−0.780**	**0.010**
Area of grain sowing	**−0.692**	**0.010**	0.169	0.100	**−0.697**	**0.010**
Cows' milk production	**0.618**	**0.050**	−0.178	0.100	**0.675**	**0.010**

In Kendall's tau correlation, significant positive relations were found for all parameters with the total energy supply in agriculture. The strength of the relationship was very significant for the economic parameters. These relationships were solid for both the global performance and per capita performance parameters. The parameters related to agriculture were less related to the energy supply in the agricultural sector. A solid relationship was in the case of gross value added of agriculture, forestry, and fishing. Hostile average relations were in the case of total agricultural area and agricultural area of the grain. In general, these areas slightly decreased, and there was a systematic increase in energy consumption in agriculture. Average positive relations were between cows' milk production and total energy consumption in agriculture. Both parameters tended to increase. Similar relationships were found between renewables and biofuels consumption in agriculture and the studied parameters. Interestingly, in renewable energy, the strength of dependence was lower for the relationship with all the analyzed parameters of the economy and agriculture than in total and crude energy. This may indicate certain independence in the development of these energy sources. It is merely a necessity and additionally contributes to the protection of the natural environment. In this case, social factors are more critical than in the case of other energy sources. Oil and petroleum product consumption relationships were inconsistent with all parameters, except for the total agricultural area. A strong positive correlation was obtained for this parameter. Diesel fuel was mainly used to power tractors and agricultural machines that are used to cultivate the land. So, such dependencies are not strange. The presented correlation results indicate solid relationships between the volume of energy consumption in agriculture and the economic potential and economic development level. The general situation in the economy was more decisive. When favorable, it also fueled agriculture and favored more work. In turn, the economic crisis also affected agriculture and led to a reduction in production. In land-related parameters, these relationships were negative because land resources do not increase but even decrease. In turn, energy consumption in agriculture grew, including as a result of replacing human labor with devices and greater mechanization of labor and the use of crops and agricultural production requiring greater energy consumption per production unit. Milk production increased in animal production, which was positively correlated with energy consumption in agriculture. It must also be said that differences were depending on the type of energy. The lower strength of the relationship was found in the case of renewables and biofuels. The results were generally not significant for oil and petroleum products consumption.

The analysis carried out with the use of Spearman's rank correlation coefficients gave very similar results. The strength of the relationship was slightly different. Both tests confirm the close relationship between total energy consumption in agriculture and economic parameters and smaller ones with agricultural parameters. In the case of renewable energy, the strength of the relationship was also smaller. On the other hand, the consumption of diesel oil was not related to the economic situation or directly to agricultural production parameters. The only exception was the agricultural land area, which strongly influenced diesel fuel consumption for tractors and agricultural machinery.

4. Discussion

Many authors confirm a strong relationship between energy consumption and agricultural productivity [72,73]. The main factor of the increase in productivity was technological progress, taking place precisely due to mechanization and the use of machines requiring energy supply [74–77]. Agriculture's energy use and trends have varied around the world. In the EU, US, and Japan, most energy consumption indicators were decreasing. Only the Netherlands and Spain saw an increase. In developing countries, energy consumption in agriculture has increased [78,79]. There were differences between individual EU countries. Most EU countries could better rationalize their use of inputs (resources), achieving greater production efficiency. Western European countries were more productive than those in Eastern Europe [80,81]. Countries should also find an appropriate compromise between meeting the demand for products through domestic production and import. One cannot forget about environmental protection in these activities [82]. For example, supporting organic farms contributes to the reduced energy consumption in agriculture in the EU [83–86]. On the other hand, saving energy does not only mean saving fuel and electricity. Essential areas of energy saving include reducing the demand for power (machines and devices), reducing the energy consumption of production as a whole, and using alternative energy sources for production. The technical and technological modernization of agriculture directly impacts the energy consumption of production [87–91]. One of the saving methods is the introduction of precision farming [92,93].

Changes in energy consumption from individual sources have been and will be varied. Farajian [94], in his research, predicted that there would be an increase in electricity consumption in agriculture and a decline in diesel fuel consumption. Electricity is consumed in four categories: farm buildings, agricultural land, cultivation procedures, and farms. Electricity is supplied to all machines located in outbuildings, such as milking machines and grain mills. Electrical equipment is also used to harvest and plant irrigation and dry with fans [95–97].

The relationship between economic activity and energy use is fairly well researched. The first results of research on this subject were published in the 1950s [98,99]. The topic was developed in the following decades, with particular emphasis on the US economy, including by Schurr et al. [100], Warren [101] de Janosi and Grayson [102], Solow [103], and Rasche and Tatom [104]. Studies on other countries on the relationship between economic performance and energy consumption were initiated by Kraft and Kraft [105]. The studies covered different periods and different methods were used. One should mention the research by Humphrey and Stanislaw [106] for Great Britain, Zilberfarb and Adams [107] for developing countries, Yu and Choi [108] for international comparison, Adams and Miovic [109] for Western Europe, Abakah [110] for Ghana, Shafik and Bandyopadhyay [111] for the Country Panel, Hawdon and Pearson [112] for the UK, Masih and Masih [113] for the Country Panel, Cheng and Lai [114] for Taiwan, and Naqvi [115] for Pakistan. In the following years, many researchers dealt with the relationship between economic growth and energy consumption. Increasingly longer periods are used from a large number of countries, using increasingly reliable econometric methods. Despite numerous studies, there is still no clear answer to the relationship between economic growth and energy consumption. Some researchers support the hypothesis that energy consumption leads to economic growth, e.g., Apergis and Payne [116], Ozturk et al. [117], Ouedraogo [118], Aslan et al. [119]. Some researchers claim that economic growth affects energy consumption, e.g., Huang and Hwang [120], Narayan et al. [121], Kasman and Duman [122]. Some researchers support the feedback hypothesis that there is a two-way causal relationship between energy use and economic growth, e.g., Constantini and Martini [123], Belke et al. [124], Coers and Sanders [125]. There is also a group of researchers who support the neutrality hypothesis that economic growth and energy consumption are independent, e.g., Wolde-Rufael [126], Kahsai et al. [127], Laughter and Pope [128]. The results and relationships depended on the countries (groups of countries), periods, and methods used.

Different results have been obtained for the relationship between the volume of agricultural production and energy consumption. For example, Dogan et al. [129] studied the electricity consumption in agriculture in Turkey in 1995–2013. They found that the use of electricity in agriculture affected agricultural production in non-coastal regions, while a two-way causal link existed between these variables for coastal regions. Raeeni et al. [130] found, based on Iranian agriculture that a 1% increase in agricultural energy consumption leads to a 1.29% increase in agricultural production in the long run. Similar results were achieved by, among others, Altinay and Karagol [131], Lee and Chang [132], Adebola [133] and Apergis and Payne [134]. Apergis and Payne [135] found a two-way causality between renewable and nonrenewable energy consumption and economic growth in the short and long term. They also found short-term substitutability between the two energy sources. In general, the current trend is to decouple energy consumption from economic growth. According to this assumption, energy consumption should fall and economic growth should follow [136]. The ways to achieve this goal are to reduce the area of crops with the simultaneous advancement of agricultural technology, improving the productivity per unit area. There were, however, significant differences in agricultural technology between countries [137,138]. Developed countries and regions may be more suited to introducing new agricultural technologies to improve productivity than developing countries [139]. In developing countries, compensation for agricultural productivity may come from educating farmers in agricultural knowledge and experience [140,141]. In Western European countries, a situation is observed where agricultural production remains at a similar level, but the consumption of energy allocated for this purpose is falling [142]. There is a significant difference between the old (EU-15) and new EU member states (admitted after 2004) in agricultural productivity. Consequently, there are also differences in the efficiency of energy use [143,144]. It is precisely the increase in energy efficiency in agriculture that should reduce the differences between developed and developing countries [145].

Many studies have confirmed the relationship between energy consumption in agriculture and economic growth [146,147]. These dependencies were analyzed, even taking into account the environmental Kuznets curve. The attention was paid to CO_2 emissions from agriculture-related to energy consumption in agriculture and economic growth. The relationships were one-way [148]. It is precisely the reduction of pollutant emissions into the environment that is the primary goal of agriculture. The emission of pollutants is inextricably linked with energy consumption. Therefore, the aim is to apply energy-efficient technologies and implement innovations in agriculture [149–153].

5. Conclusions

Energy is an indispensable production resource in agriculture. First of all, it is used to power machines and devices operating in this sector. In the EU countries, the total energy consumption has decreased. However, according to its circumstances, there are changes in their origin sources, and each country has its structure in this regard. It was found that the concentration level of energy consumption in agriculture did not change and was relatively high. This situation is influenced by the relative stability of production, which is conditioned, among other things, by the land-owned resources. Another reason may be relatively stable energy consumption and countries with technologies that ensure similar energy efficiency. Only in the case of renewable energy sources, a gradual decrease in the concentration of its consumption in agriculture can be observed. More and more countries are developing technologies that allow the use of this type of energy. Agriculture was a sector that produced more renewable energy than it consumed.

In the EU countries, crude oil was of the most significant importance, as about 60% of the energy used in agriculture came from this source. Electricity and natural gas accounted for a dozen or so percent, and renewable energy accounted for 10%. In most countries, the structure was similar, i.e., with the dominant importance of crude oil. In the case of other energy sources, the proportions were varied. Overall, renewable energies grew in importance in all countries. In the top five countries in 2018, such sources accounted for over 20% of the energy used in agriculture. As a rule, economically developed countries developed this type of technology, but there were also examples of developing countries, such as Slovakia, which dynamically increased the production and energy use from renewable sources. There were also examples of economically developed countries that used very little renewable energy in agriculture, such as Italy and Spain. Overall, the EU is shifting away from fossil and gaseous fuels, the importance of liquid fuels and the growing importance of electricity and renewable energy. The rapid changes created high volatility for fuels that were gaining in importance and those that were losing. Energy sources with stabilized consumption, such as electricity and crude oil, were characterized by low consumption volume variability. There was also a significant stabilization concerning the total energy used in agriculture. The first hypothesis was confirmed. There are processes of diversification of energy sources used in agriculture, but these changes are prolonged. The importance of renewable energy sources is also systematically growing.

A significant influence of the economic situation on energy consumption in agriculture was found. The better it was, the more energy was used in agriculture. The union strength was lower in the case of renewables. Thus, the second hypothesis was confirmed, according to which the increase in the consumption of renewable energy in agriculture is closely related to the economy's parameters. It should be added that this relationship was not very strong. Such energy is a necessity and results from concern for the natural environment. Therefore, economic factors may have a smaller impact.

A close correlation was also established between agricultural parameters concerning land resources and production volume and energy consumption. Thus, the already known regularities were confirmed. In agriculture, energy consumption will increase as a result of increasingly replacing human work with devices. It seems that the increase in mechanization will be faster than the development of energy-consuming technologies. The only chance to achieve progress in mechanization in agriculture without increasing the harmful impact on the environment is by introducing renewable energy sources. The conducted research confirms this trend. These energy sources are also increasingly commonly introduced by all countries, regardless of the level of economic development. In the following years, increasing consumption of energy from renewable sources should be observed. Modern agriculture in the European Union should follow this direction.

Author Contributions: Conceptualization, T.R., A.P., B.K.; data curation, T.R., A.P., P.B., A.B.-B.; formal analysis, T.R., B.K., P.B., A.P.; methodology, T.R., B.K., P.B., A.P.; resources, T.R., A.P., B.K., K.M., A.B.-B.; visualization, T.R., P.B., K.M., A.P.; writing—original draft, T.R., A.P., B.K.; writing—review and editing, T.R., A.P., P.B., A.B.-B., K.M.; supervision, T.R., B.K.; funding acquisition, T.R., B.K., P.B., A.P. All authors have read and agreed to the published version of the manuscript.

Funding: This research received no external funding.

Conflicts of Interest: The authors declare no conflict of interest.

References

1. Giannakis, E.; Bruggeman, A. The highly variable economic performance of European agriculture. *Land Use Policy* **2015**, *45*, 26–35. [CrossRef]
2. Schulp, C.J.; Levers, C.; Kuemmerle, T.; Tieskens, K.F.; Verburg, P.H. Mapping and modelling past and future land use change in Europe's cultural landscapes. *Land Use Policy* **2019**, *80*, 332–344. [CrossRef]

3. Kristensen, S.B.P.; Busck, A.G.; van der Sluis, T.; Gaube, V. Patterns and drivers of farm-level land use change in selected European rural landscapes. *Land Use Policy* **2016**, *57*, 786–799. [CrossRef]
4. Leśniewska-Napierała, K.; Nalej, M.; Napierała, T. The Impact of EU Grants Absorption on Land Cover Changes—The Case of Poland. *Remote Sens.* **2019**, *11*, 2359. [CrossRef]
5. Veronica, P.; Raluca, T.A. Convergence of Romanian and European Union Agriculture-Evolution and Prospective Assessment. *Ovidius Univ. Ann. Econ. Sci. Ser.* **2018**, *18*, 239–247.
6. Reiff, M.; Surmanová, K.; Balcerzak, A.P.; Pietrzak, M.B. Multiple criteria analysis of European Union agriculture. *J. Int. Stud.* **2016**, *9*, 62–74. [CrossRef] [PubMed]
7. Nowak, A.; Kaminska, A. Agricultural competitiveness: The case of the European Union countries. *Agric. Econ.* **2016**, *62*, 507–516.
8. Dos Santos, M.L. Segmenting Farms in the European Union. *Agric. Econ.* **2013**, *59*, 49–57.
9. Carraresi, L.; Banterle, A. Agri-food Competitive Performance in EU Countries: A Fifteen-Year Retrospective. *Int. Food Agribus. Manag. Rev.* **2015**, *18*, 37–62.
10. Spicka, J. The economic disparity in European agriculture in the context of the recent EU enlargements. *J. Econ. Sustain. Dev.* **2013**, *4*, 125–133.
11. Reiff, M.; Ivanicova, Z.; Surmanova, K. Cluster analysis of selected world development indicators in the fields of agriculture and the food industry in European Union countries. *Agric. Econ.* **2018**, *64*, 197–205. [CrossRef]
12. Reiff, M.; Surmanova, K.; Balcerzak, A.P.; Pietrzak, M.B. *Quantitative Analysis of European Union Agriculture Performance*; Institute of Economic Research Working Papers No. 32/2016; 2016. Available online: http://www.badania-gospodarcze.pl/images/Working_Papers/2016_No_32.pdf (accessed on 10 January 2021).
13. Coca, O.; Ștefan, G.; Crețu, A.; Creangă, D.E. Research on the Development of Agriculture in the European Union. A Structural Analysis by Geographical Groups of Countries. *Rom. Agric. Res.* **2020**, *37*, 229–242.
14. Vanschoenwinkel, J.; Mendelsohn, R.; Van Passel, S. Do Western and Eastern Europe have the same agricultural climate response? Taking adaptive capacity into account. *Glob. Environ. Chang.* **2016**, *41*, 74–87. [CrossRef]
15. Swinnen, J.F.; Swinnen, J.; Vranken, L. *Land & EU Accession: Review of the Transitional Restrictions on New Member States on the Acquisition of Agricultural Real Estate*; CEPS: Brussels, Belgium, 2009.
16. Hartvigsen, M. Land reform and land fragmentation in Central and Eastern Europe. *Land Use Policy* **2014**, *36*, 330–341. [CrossRef]
17. Andrejovská, A.; Buleca, J.; Hudákova, M. Categorization of the EU Countries in the Context of Agricultural Production. *AGRIS On-Line Pap. Econ. Inform.* **2016**, *8*, 3–14. [CrossRef]
18. Rokicki, T.; Perkowska, A.; Ratajczak, M. Differentiation in Healthcare Financing in EU Countries. *Sustainability* **2021**, *13*, 251. [CrossRef]
19. Rokicki, T.; Perkowska, A. Changes in Energy Supplies in the Countries of the Visegrad Group. *Sustainability* **2020**, *12*, 7916. [CrossRef]
20. Ciutacu, C.; Chivu, L.; Andrei, J.V. Similarities and dissimilarities between the EU agricultural and rural development model and Romanian agriculture. Challenges and perspectives. *Land Use Policy* **2015**, *44*, 169–176. [CrossRef]
21. Popescu, G.H.; Nicoale, I.; Nica, E.; Vasile, A.J.; Andreea, I.R. The influence of land-use change paradigm on Romania's agro-food trade competitiveness—An overview. *Land Use Policy* **2017**, *61*, 293–301. [CrossRef]
22. Giannakis, E.; Bruggeman, A. Regional disparities in economic resilience in the European Union across the urban-rural divide. *Reg. Stud.* **2020**, *54*, 1200–1213. [CrossRef]
23. Szabo, L.; Grznár, M. Agriculture in the EU and Position of the Slovak Republic. *Agric. Econ.* **2015**, *61*, 493–501. [CrossRef]
24. Pietrzak, M.B.; Walczak, D. The Analysis of the Agrarian Structure in Poland with the Special Consideration of the Years 1921 and 2002. *Bulg. J. Agric. Sci.* **2014**, *20*, 1018–1039.
25. Walczak, D.; Pietrzak, M.B. Analysis of Agrarian Structure in Poland in 1921 and 2002 based on the Example of Selected Districts. In *Business Challenges in the Changing Economic Landscape*; Bilgin, M.H., Danis, H., Demir, E., Can, U., Eds.; Springer: Berlin/Heidelberg, Germany, 2016; Volume 1, pp. 461–472.
26. Nowak, A.; Różańska-Boczula, M. Differentiation in the production potential and efficiency of farms in the member states of the European Union. *Agric. Econ.* **2019**, *65*, 395–403. [CrossRef]
27. Popescu, A.; Dinu, T.A.; Stoian, E. Efficiency of the agricultural land use in the European Union. Scientific Papers Series-Management. *Econ. Eng. Agric. Rural Dev.* **2019**, *19*, 475–486.
28. Popescu, A. Considerations on utilized agricultural land and farm structure in the European Union. Scientific Papers Series Management. *Econ. Eng. Agric. Rural Dev.* **2013**, *13*, 221–226.
29. Popescu, A.; Alecu, I.N.; Dinu, T.A.; Stoian, E.; Condei, R.; Ciocan, H. Farm Structure and Land Concentration in Romania and the European Union's Agriculture. *Agric. Agric. Sci. Procedia* **2016**, *10*, 566–577. [CrossRef]
30. Bularca, E.; Toma, E. Agricultural labour productivity and its impact in farming system. *Sci. Pap. Ser. Manag. Econ. Eng. Agric. Rural Dev.* **2019**, *19*, 91–96.
31. Dorward, A. Agricultural labour productivity, food prices and sustainable development impacts and indicators. *Food Policy* **2013**, *39*, 40–50. [CrossRef]
32. Svoboda, J.; Lososová, J.; Zdeněk, R. Analysis of costs and their effectiveness in the EU agrarian sector. *Custos Agronegocio Line* **2020**, *16*, 151–173.

33. Abson, D.J. The economic drivers and consequences of agricultural specialization. In *Agroecosystem Diversity*; Academic Press: Cambridge, MA, USA, 2019; pp. 301–315. Available online: https://www.sciencedirect.com/science/article/pii/B9780128110508000194 (accessed on 10 January 2021).
34. Brunori, G.; D'Amico, S.; Rossi, A. Practices of Sustainable Intensification Farming Models: An Analysis of the Factors Conditioning Their Functioning, Expansion, and Transformative Potential. In *Agroecosystem*; Academic Press: Cambridge, MA, USA, 2019; pp. 317–333.
35. Pavlov, A.Y. Managing sustainable development of rural areas using algorithms of agricultural specialization. *Mediterr. J. Soc. Sci.* **2015**, *6*, 147. [CrossRef]
36. Pierangeli, F.; Henke, R.; Coronas, M.G. Multifunctional agriculture: An analysis of country specialization and regional differentiation. In Proceedings of the European Association of Agricultural Economists (EAAE), Ghent, Belgium, 26–29 August 2008. No. 725-2016-49366.
37. Vlasenko, T.; Vlasovets, V.; Timofeev, S.; Kravets, A. Specialization issues of production of agrarian enterprises. In *SHS Web of Conferences*; EDP Sciences: Paris, France, 2019; Volume 67, p. 06056.
38. Mushtaq, K.; Abbas, F.; Ghafoor, A. Energy use for economic growth: Cointegration and causality analysis from the agriculture sector of Pakistan. *Pak. Dev. Rev.* **2007**, *46*, 1065–1073. [CrossRef]
39. Sebri, M.; Abid, M. Energy use for economic growth: A trivariate analysis from Tunisian agriculture sector. *Energy Policy* **2012**, *48*, 711–716. [CrossRef]
40. Agheli, L. Estimating the demand for diesel in agriculture sector of Iran. *Int. J. Energy Econ. Policy* **2015**, *5*, 660–667.
41. Ceylan, Z. Assessment of agricultural energy consumption of Turkey by MLR and Bayesian optimized SVR and GPR models. *J. Forecast.* **2020**, *39*, 944–956. [CrossRef]
42. Zaman, K.; Khan, M.M.; Ahmad, M.; Rustam, R. The relationship between agricultural technology and energy demand in Pakistan. *Energy Policy* **2012**, *44*, 268–279. [CrossRef]
43. Perea, R.G.; Poyato, E.C.; Montesinos, P.; Díaz, J.R. Prediction of irrigation event occurrence at farm level using optimal decision trees. *Comput. Electron. Agric.* **2019**, *157*, 173–180. [CrossRef]
44. Qiao, F. Increasing wage, mechanization, and agriculture production in China. *China Econ. Rev.* **2017**, *46*, 249–260. [CrossRef]
45. Tanikawa, T. Mechanization of Agriculture Considering Its Business Model. In *Smart Plant Factory*; Springer: Singapore, 2018; pp. 241–244.
46. Kalita, A. Changes in labor pattern and agricultural mechanization. *J. Manag. Res. Anal.* **2019**, *6*, 9–13. [CrossRef]
47. Mihov, M.; Kr, G.; Ivanova, N. Current Problems of Agricultural Production Mechanization. *Mech. Agric. Conserv. Resour.* **2017**, *63*, 111–114.
48. Zheng, S.; Lam, C.M.; Hsu, S.C.; Ren, J. Evaluating efficiency of energy conservation measures in energy service companies in China. *Energy Policy* **2018**, *122*, 580–591. [CrossRef]
49. Raza, M.Y.; Wasim, M.; Sarwar, M.S. Development of Renewable Energy Technologies in rural areas of Pakistan. *Energy Sources Part A Recovery Util. Environ. Eff.* **2020**, *42*, 740–760. [CrossRef]
50. Karkacier, O.; Goktolga, Z.G.; Cicek, A. A regression analysis of the effect of energy use in agriculture. *Energy Policy* **2006**, *34*, 3796–3800. [CrossRef]
51. Alipour, A.; Veisi, H.; Darijani, F.; Mirbagheri, B.; Behbahani, A.G. Study and determination of energy consumption to produce conventional rice of the Guilan province. *Res. Agric. Eng.* **2012**, *58*, 99–106. [CrossRef]
52. Chandio, A.A.; Jiang, Y.; Rehman, A. Energy consumption and agricultural economic growth in Pakistan: Is there a nexus? *Int. J. Energy Sect. Manag.* **2019**, *13*, 597–609. [CrossRef]
53. Inumula, K.M.; Singh, S.; Solanki, S. Energy Consumption and Agricultural Economic Growth Nexus: Evidence from India. *Int. J. Energy Econ. Policy* **2020**, *10*, 545. [CrossRef]
54. Pamučar, D.; Behzad, M.; Božanić, D.; Behzad, M. Decision making to support sustainable energy policies corresponding to agriculture sector: Case study in Iran's Caspian Sea coastline. *J. Clean. Prod.* **2020**, 125302. [CrossRef]
55. Chang, J.; Leung, D.Y.; Wu, C.Z.; Yuan, Z.H. A review on the energy production, consumption, and prospect of renewable energy in China. *Renew. Sustain. Energy Rev.* **2003**, *7*, 453–468. [CrossRef]
56. Haberl, H.; Beringer, T.; Bhattacharya, S.C.; Erb, K.H.; Hoogwijk, M. The global technical potential of bio-energy in 2050 considering sustainability constraints. *Curr. Opin. Environ. Sustain.* **2010**, *2*, 394–403. [CrossRef]
57. Gabrielle, B.; Bamière, L.; Caldes, N.; De Cara, S.; Decocq, G.; Ferchaud, F.; Richard, G. Paving the way for sustainable bioenergy in Europe: Technological options and research avenues for large-scale biomass feedstock supply. *Renew. Sustain. Energy Rev.* **2014**, *33*, 11–25. [CrossRef]
58. Ballarin, A.; Vecchiato, D.; Tempesta, T.; Marangon, F.; Troiano, S. Biomass energy production in agriculture: A weighted goal programming analysis. *Energy Policy* **2011**, *39*, 1123–1131. [CrossRef]
59. Popp, J.; Lakner, Z.; Harangi-Rakos, M.; Fari, M. The effect of bioenergy expansion: Food, energy, and environment. *Renew. Sustain. Energy Rev.* **2014**, *32*, 559–578. [CrossRef]
60. Araújo, K.; Mahajan, D.; Kerr, R.; Silva, M.D. Global biofuels at the crossroads: An overview of technical, policy, and investment complexities in the sustainability of biofuel development. *Agriculture* **2017**, *7*, 32. [CrossRef]
61. Bilgen, S.; Keleş, S.; Sarıkaya, İ.; Kaygusuz, K. A perspective for potential and technology of bioenergy in Turkey: Present case and future view. *Renew. Sustain. Energy Rev.* **2015**, *48*, 228–239. [CrossRef]

62. Ozturk, M.; Saba, N.; Altay, V.; Iqbal, R.; Hakeem, K.R.; Jawaid, M.; Ibrahim, F.H. Biomass and bioenergy: An overview of the development potential in Turkey and Malaysia. *Renew. Sustain. Energy Rev.* **2017**, *79*, 1285–1302. [CrossRef]
63. Öztürk, M.; Gucel, S.; Armagan, B.; Kucuk, M. An overview of the biomass potential in Turkey and possibilities for the utilization of degraded areas. In Proceedings of the 1st Intern. Nuclear and Renewable Energy Conference (INREC 10), Amman, Jordan, 21 March 2010; pp. 21–24.
64. Godard, C.; Boissy, J.; Gabrielle, B. Life-cycle assessment of local feedstock supply scenarios to compare candidate biomass sources. *Gcb Bioenergy* **2013**, *5*, 16–29. [CrossRef]
65. Collotta, M.; Champagne, P.; Tomasoni, G.; Alberti, M.; Busi, L.; Mabee, W. Critical indicators of sustainability for biofuels: An analysis through a life cycle sustainabilty assessment perspective. *Renew. Sustain. Energy Rev.* **2019**, *115*, 109358. [CrossRef]
66. Dixon, P.M.; Weiner, J.; Mitchell-Olds, T.; Woodley, R. Erratum to 'Bootstrapping the Gini Coefficient of Inequality. *Ecology* **1988**, *69*, 1307. [CrossRef]
67. Dagum, C. The Generation and Distribution of Income, the Lorenz Curve and the Gini Ratio. *Econ. Appliquée* **1980**, *33*, 327–367.
68. Starzyńska, W. *Statystyka Praktyczna Wydawnictwo Naukowe*; PWN: Warszawa, Poland, 2002.
69. Abdi, H. Coefficient of variation. *Encycl. Res. Des.* **2010**, *1*, 169–171.
70. Kendall, M.G. *Rank Correlation Methods*; Griffin: London, UK, 1955; p. 19.
71. Spearman, C. The proof and measurement of association between two things. *Am. J. Psychol.* **1904**, *15*, 72–101. [CrossRef]
72. Pelletier, N.; Audsley, E.; Brodt, S.; Garnett, T.; Henriksson, P.; Kendall, A.; Kramer, K.J.; Murphy, D.; Nemecek, T.; Troell, M. Energy intensity of agriculture and food systems. *Annu. Rev. Environ. Resour.* **2011**, *36*, 223–246. [CrossRef]
73. Yang, Z.; Wang, D.; Du, T.; Zhang, A.; Zhou, Y. Total-factor energy efficiency in China's agricultural sector: Trends, disparities and potentials. *Energies* **2018**, *11*, 853. [CrossRef]
74. Fuglie, K.O.; McDonald, J.M.; Ball, V.E. *Productivity Growth in US Agriculture*; Economic Brief Number 9; Economic Research Service: Washington, DC, USA, 2007; p. 9.
75. Po-Chi, C.H.E.N.; Ming-Miin, Y.U.; Chang, C.C.; Shih-Hsun, H.S.U. Total factor productivity growth in China's agricultural sector. *China Econ. Rev.* **2008**, *19*, 580–593.
76. Gong, B. Agricultural reforms and production in China: Changes in provincial production function and productivity in 1978–2015. *J. Dev. Econ.* **2018**, *132*, 18–31. [CrossRef]
77. Yang, F.; Yang, M.; Nie, H. Productivity trends of Chinese regions: A perspective from energy saving and environmental regulations. *Appl. Energy* **2013**, *110*, 82–89. [CrossRef]
78. Ghisellini, P.; Setti, M.; Ulgiati, S. Energy and land use in worldwide agriculture: An application of life cycle energy and cluster analysis. *Environ. Dev. Sustain.* **2016**, *18*, 799–837. [CrossRef]
79. Pellegrini, P.; Fernández, R.J. Crop intensification, land use, and on-farm energy-use efficiency during the worldwide spread of the green revolution. *Proc. Natl. Acad. Sci. USA* **2018**, *115*, 2335–2340. [CrossRef] [PubMed]
80. Toma, P.; Miglietta, P.P.; Zurlini, G.; Valente, D.; Petrosillo, I. A non-parametric bootstrap-data envelopment analysis approach for environmental policy planning and management of agricultural efficiency in EU countries. *Ecol. Indic.* **2017**, *83*, 132–143. [CrossRef]
81. Rokicki, T.; Perkowska, A. Diversity and Changes in the Energy Balance in EU Countries. *Energies* **2021**, *14*, 1098. [CrossRef]
82. Miglietta, P.P.; Giove, S.; Toma, P. An optimization framework for supporting decision making in biodiesel feedstock imports: Water footprint vs. import costs. *Ecol. Indic.* **2018**, *85*, 1231–1238. [CrossRef]
83. Gomiero, T.; Paoletti, M.G.; Pimentel, D. Energy and environmental issues in organic and conventional agriculture. *Crit. Rev. Plant Sci.* **2008**, *27*, 239–254. [CrossRef]
84. Pimentel, D.; Hepperly, P.; Hanson, J.; Douds, D.; Seidel, R. Environmental, energetic, and economic comparisons of organic and conventional farming systems. *BioScience* **2005**, *55*, 573–582. [CrossRef]
85. Pergola, M.; D'Amico, M.; Celano, G.; Palese, A.; Scuderi, A.; Di Vita, G.; Pappalardo, G.; Inglese, P. Sustainability evaluation of Sicily's lemon and orange production: An energy, economic and environmental analysis. *J. Environ. Manag.* **2013**, *128*, 674–682. [CrossRef]
86. Fess, T.L.; Benedito, V.A. Organic versus conventional cropping sustainability: A comparative system analysis. *Sustainability* **2018**, *10*, 272. [CrossRef]
87. Polukhin, A.A.; Lankin, A.S.; Novoselov, E.A.; Silko, E.A. Economic assessment of energy consumption in agriculture of the Russian Federation. *Econ. Work Manag. Agric.* **2019**, *1*, 58–63.
88. Quan-le, L.I.U. The Saving Energy of Agriculture Machine Is the Only Way of Agriculture Modernization. *Chin. Agric. Mech.* **2006**, *6*.
89. Xiangxin, L.; Yali, Z.; Chao, Z.; Xianghui, Y.; Yan, Z. The Analysis of Energy Consumption of Agricultural Machinery in Our Country and the Technology of Energy Consumption. *J. Agric. Mech. Res.* **2012**, *4*.
90. Wang, X.Y.; Wang, S.; Zhang, Z.Y. Construction of energy conservation and emission reduction technology system of agricultural mechanization. *Chin. Agric. Mech.* **2010**, *4*, 6–10. [CrossRef]
91. Kondratieva, N.P.; Vladykin, I.R.; Litvinova, V.M.; Krasnolutskaya, M.G.; Bolshin, R.G. Energy-saving technologies and electric equipment applied in agriculture. *Res. Agric. Electr. Eng.* **2016**, *2*, 62–68.
92. Jawad, H.M.; Nordin, R.; Gharghan, S.K.; Jawad, A.M.; Ismail, M. Energy-efficient wireless sensor networks for precision agriculture: A review. *Sensors* **2017**, *17*, 1781. [CrossRef]

93. Zambon, I.; Cecchini, M.; Egidi, G.; Saporito, M.G.; Colantoni, A. Revolution 4.0: Industry vs. agriculture in a future development for SMEs. *Processes* **2019**, *7*, 36. [CrossRef]
94. Farajian, L.; Moghaddasi, R.; Hosseini, S. Agricultural energy demand modeling in Iran: Approaching to a more sustainable situation. *Energy Rep.* **2018**, *4*, 260–265. [CrossRef]
95. Golmohamadi, H. Agricultural Demand Response Aggregators in Electricity Markets: Structure, Challenges and Practical Solutions-a Tutorial for Energy Experts. *Technol. Econ. Smart Grids Sustain. Energy* **2020**, *5*, 1–17. [CrossRef]
96. Marks, G.; Wilcox, E.; Olsen, D.; Goli, S. *Opportunities for Demand Response in California Agricultural Irrigation: A Scoping Study (No. LBNL-6108E)*; Lawrence Berkeley National Lab. (LBNL): Berkeley, CA, USA, 2013.
97. Liu, J.; Chai, Y.; Xiang, Y.; Zhang, X.; Gou, S.; Liu, Y. Clean energy consumption of power systems towards smart agriculture: Roadmap, bottlenecks and technologies. *CSEE J. Power Energy Syst.* **2018**, *4*, 273–282. [CrossRef]
98. Mason, E.S. *Reports on the Productive Uses of Nuclear Energy: Energy Requirements and Economic Growth*; National Planning Association: Washington, DC, USA, 1955; pp. 1–51.
99. Frank, A.G. Industrial capital stocks and energy consumption. *Econ. J.* **1959**, *69*, 170–174. [CrossRef]
100. Schurr, S.H.; Netschert, B.C.; Eliasberg, V.F.; Lerner, J.; Landsberg, H.H. *Energy in the American Economy, 1850–1975: An Economic Study of Its History and Prospects*; No. HD9545 R45; Johns Hopkins Press: Baltimore, MD, USA, 1955.
101. Warren, J.C. Energy and economic advances. In *The Philippine Economic Journal Manila*; Philippine Economic Society: Quezon, Philippine, 1964.
102. de Janosi, P.E.; Grayson, L.E. Patterns of energy consumption and economic growth and structure. *J. Dev. Stud.* **1972**, *8*, 241–249. [CrossRef]
103. Solow, R.M. The economics of resources or the resources of economics. In *Classic Papers in Natural Resource Economics*; Palgrave Macmillan: London, UK, 1974; pp. 257–276.
104. Rasche, R.H.; Tatom, J.A. Energy resources and potential GNP. *Fed. Reserve Bank St. Louis Rev.* **1977**, 5910–5924. [CrossRef]
105. Kraft, J.; Kraft, A. On the relationship between energy and GNP. *J. Energy Dev.* **1978**, *3*, 401–403.
106. Humphrey, W.S.; Stanislaw, J. Economic growth and energy consumption in the UK, 1700–1975. *Energy Policy* **1979**, *7*, 29–42. [CrossRef]
107. Zilberfarb, B.Z.; Adams, F.G. The energy-gdp relationship in developing countries: Empirical evidence and stability tests. *Energy Econ.* **1981**, *3*, 244–248. [CrossRef]
108. Yu, E.S.; Choi, J.Y. The causal relationship between energy and GNP: An international comparison. *J. Energy Dev.* **1985**, *10*, 249–272.
109. Adams, F.G.; Miovic, P. On relative fuel efficiency and the output elasticity of energy consumption in western Europe. *J. Ind. Econ.* **1968**, *17*, 41–56. [CrossRef]
110. Abakah, E.M. Real incomes and the consumption of woodfuels in Ghana: An analysis of recent trends. *Energy Econ.* **1990**, *12*, 227–231. [CrossRef]
111. Shafik, N.; Bandyopadhyay, S. *Economic Growth and Environmental Quality: Time-Series and Cross-Country Evidence*; World Bank Publications: Washington, DC, USA, 1992; Volume 904.
112. Hawdon, D.; Pearson, P. Input-output simulations of energy, environment, economy interactions in the UK. *Energy Econ.* **1995**, *17*, 73–86. [CrossRef]
113. Masih, A.M.; Masih, R. Energy consumption, real income and temporal causality: Results from a multi-country study based on cointegration and error-correction modelling techniques. *Energy Econ.* **1996**, *18*, 165–183. [CrossRef]
114. Cheng, B.S.; Lai, T.W. An investigation of co-integration and causality between energy consumption and economic activity in Taiwan. *Energy Econ.* **1997**, *19*, 435–444. [CrossRef]
115. Naqvi, F. A computable general equilibrium model of energy, economy and equity interactions in Pakistan. *Energy Econ.* **1998**, *20*, 347–373. [CrossRef]
116. Apergis, N.; Payne, J.E. Energy consumption and economic growth: Evidence from the Commonwealth of Independent States. *Energy Econ.* **2009**, *31*, 641–647. [CrossRef]
117. Ozturk, I.; Aslan, A.; Kalyoncu, H. Energy consumption and economic growth relationship: Evidence from panel data for low and middle income countries. *Energy Policy* **2010**, *38*, 4422–4428. [CrossRef]
118. Ouedraogo, N.S. Energy consumption and economic growth: Evidence from the economic community of West African States (ECOWAS). *Energy Econ.* **2013**, *36*, 637–647. [CrossRef]
119. Aslan, A.; Apergis, N.; Yildirim, S. Causality between energy consumption and GDP in the US: Evidence from wavelet analysis. *Front. Energy* **2014**, *8*, 1–8. [CrossRef]
120. Huang, B.N.; Hwang, M.J.; Yang, C.W. Causal relationship between energy consumption and GDP growth revisited: A dynamic panel data approach. *Ecol. Econ.* **2008**, *67*, 41–54. [CrossRef]
121. Narayan, P.K.; Narayan, S.; Popp, S. A note on the long-run elasticities from the energy consumption–GDP relationship. *Appl. Energy* **2010**, *87*, 1054–1057. [CrossRef]
122. Kasman, A.; Duman, Y.S. CO_2 emissions, economic growth, energy consumption, trade and urbanization in new EU member and candidate countries: A panel data analysis. *Econ. Model.* **2015**, *44*, 97–103. [CrossRef]
123. Costantini, V.; Martini, C. The causality between energy consumption and economic growth: A multi-sectoral analysis using non-stationary cointegrated panel data. *Energy Econ.* **2010**, *32*, 591–603. [CrossRef]

124. Belke, A.; Dobnik, F.; Dreger, C. Energy consumption and economic growth: New insights into the cointegration relationship. *Energy Econ.* **2011**, *33*, 782–789. [CrossRef]
125. Coers, R.; Sanders, M. The energy–GDP nexus; addressing an old question with new methods. *Energy Econ.* **2013**, *36*, 708–715. [CrossRef]
126. Wolde-Rufael, Y. Energy consumption and economic growth: The experience of African countries revisited. *Energy Econ.* **2009**, *31*, 217–224. [CrossRef]
127. Kahsai, M.S.; Nondo, C.; Schaeffer, P.V.; Gebremedhin, T.G. Income level and the energy consumption–GDP nexus: Evidence from Sub-Saharan Africa. *Energy Econ.* **2012**, *34*, 739–746. [CrossRef]
128. Śmiech, S.; Papież, M. Energy consumption and economic growth in the light of meeting the targets of energy policy in the EU: The bootstrap panel Granger causality approach. *Energy Policy* **2014**, *71*, 118–129. [CrossRef]
129. Dogan, E.; Sebri, M.; Turkekul, B. Exploring the relationship between agricultural electricity consumption and output: New evidence from Turkish regional data. *Energy Policy* **2016**, *95*, 370–377. [CrossRef]
130. Raeeni, A.A.G.; Hosseini, S.; Moghaddasi, R. How energy consumption is related to agricultural growth and export: An econometric analysis on Iranian data. *Energy Rep.* **2019**, *5*, 50–53. [CrossRef]
131. Altinay, G.; Karagol, E. Electricity consumption and economic growth: Evidence from Turkey. *Energy Econ.* **2005**, *27*, 849–856. [CrossRef]
132. Lee, C.C.; Chang, C.P. Energy consumption and economic growth in Asian economies: A more comprehensive analysis using panel data. *Resour. Energy Econ.* **2008**, *30*, 50–65. [CrossRef]
133. Adebola, S.S. Electricity consumption and economic growth: Trivariate investigation in Botswana with capital formation. *Int. J. Energy Econ. Policy* **2011**, *1*, 32–46.
134. Apergis, N.; Payne, J.E. Energy consumption and growth in South America: Evidence from a panel error correction model. *Energy Econ.* **2010**, *32*, 1421–1426. [CrossRef]
135. Apergis, N.; Payne, J.E. Renewable and non-renewable energy consumption-growth nexus: Evidence from a panel error correction model. *Energy Econ.* **2012**, *34*, 733–738. [CrossRef]
136. Chen, X.; Shuai, C.; Zhang, Y.; Wu, Y. Decomposition of energy consumption and its decoupling with economic growth in the global agricultural industry. *Environ. Impact Assess. Rev.* **2020**, *81*, 106364. [CrossRef]
137. Eberhardt, M.; Vollrath, D. The effect of agricultural technology on the speed of development. *World Dev.* **2018**, *109*, 483–496. [CrossRef]
138. Marinoudi, V.; Sørensen, C.G.; Pearson, S.; Bochtis, D. Robotics and labour in agriculture. A context consideration. *Biosyst. Eng.* **2019**, *184*, 111–121. [CrossRef]
139. Rokicki, T.; Perkowska, A.; Klepacki, B.; Szczepaniuk, H.; Szczepaniuk, E.K.; Bereziński, S.; Ziółkowska, P. The Importance of Higher Education in the EU Countries in Achieving the Objectives of the Circular Economy in the Energy Sector. *Energies* **2020**, *13*, 4407. [CrossRef]
140. Fuglie, K.O. Is agricultural productivity slowing? *Glob. Food Secur.* **2018**, *17*, 73–83. [CrossRef]
141. Alene, A.D.; Manyong, V.M. The effects of education on agricultural productivity under traditional and improved technology in northern Nigeria: An endogenous switching regression analysis. *Empir. Econ.* **2007**, *32*, 141–159. [CrossRef]
142. Wang, S.L.; Schimmelpfennig, D.; Fuglie, K.O. Is agricultural productivity growth slowing in Western Europe. In *Productivity Growth in Agriculture: An International Perspective*; CABI: Oxfordshire, UK, 2012; pp. 109–125.
143. Baráth, L.; Fertő, I. Productivity and convergence in European agriculture. *J. Agric. Econ.* **2017**, *68*, 228–248. [CrossRef]
144. Kijek, A.; Kijek, T.; Nowak, A.; Skrzypek, A. Productivity and its convergence in agriculture in new and old European Union member states. *Agric. Econ.* **2019**, *65*, 1–9. [CrossRef]
145. Wu, Y. Energy intensity and its determinants in China's regional economies. *Energy Policy* **2012**, *41*, 703–711. [CrossRef]
146. Saudi, M.H.M.; Sinaga, O.; Roespinoedji, D.; Jabarullah, N.H. Industrial, Commercial, and Agricultural Energy Consumption and Economic Growth Leading to Environmental Degradation. *Ekoloji* **2019**, *28*, 299–310.
147. Zhang, L.; Pang, J.; Chen, X.; Lu, Z. Carbon emissions, energy consumption and economic growth: Evidence from the agricultural sector of China's main grain-producing areas. *Sci. Total Environ.* **2019**, *665*, 1017–1025. [CrossRef]
148. Gokmenoglu, K.K.; Taspinar, N.; Kaakeh, M. Agriculture-induced environmental Kuznets curve: The case of China. *Environ. Sci. Pollut. Res.* **2019**, *26*, 37137–37151. [CrossRef] [PubMed]
149. Nwaka, I.D.; Nwogu, M.U.; Uma, K.E.; Ike, G.N. Agricultural production and CO_2 emissions from two sources in the ECOWAS region: New insights from quantile regression and decomposition analysis. *Sci. Total Environ.* **2020**, *748*, 141329. [CrossRef] [PubMed]
150. Khan, R. Agricultural production and CO_2 emissions causes in the developing and developed countries: New insights from quantile regression and decomposition analysis. *bioRxiv* **2020**. [CrossRef]
151. Wang, C.; Zhang, L.; Zhou, P.; Chang, Y.; Zhou, D.; Pang, M.; Yin, H. Assessing the environmental externalities for biomass-and coal-fired electricity generation in China: A supply chain perspective. *J. Environ. Manag.* **2019**, *246*, 758–767. [CrossRef] [PubMed]
152. Burney, J.A.; Davis, S.J.; Lobell, D.B. Greenhouse gas mitigation by agricultural intensification. *Proc. Natl. Acad. Sci. USA* **2010**, *107*, 12052–12057. [CrossRef] [PubMed]
153. Lin, B.; Xu, B. Factors affecting CO2 emissions in China's agriculture sector: A quantile regression. *Renew. Sustain. Energy Rev.* **2018**, *94*, 15–27. [CrossRef]

energies MDPI

Article

Energy Sector Risk and Cost of Capital Assessment—Companies and Investors Perspective

Justyna Franc-Dąbrowska, Magdalena Mądra-Sawicka * and Anna Milewska

Institute of Economics and Finance, Warsaw University of Life Sciences—SGGW, 02-787 Warsaw, Poland; justyna_franc_dabrowska@sggw.edu.pl (J.F.-D.); anna_milewska1@sggw.edu.pl (A.M.)
* Correspondence: magdalena_madra@sggw.edu.pl; Tel.: +48-22-593-42-42

Abstract: This paper aims to identify the costs of capital in a group of companies from the energy sector by including an investor and market risk approach. The study also concerns the company's Weighted Average Cost of Capital (WACC) cost intra-industry analysis related to sector characteristics such as total assets, revenues, market capitalization, and companies' age. In order to assess the intergroup relationships, basic correlation relationships were compared and a nonparametric test of variance was performed. The period under study covered the years 2015–2019. The conducted research evaluates groups of companies that dedicated their activity to a particular energy intra-industry division under numerous regulations in Europe. The study contributes to assessing the level of risk among energy listed companies in European capital markets based on capital structure valuation. The study results underline the role of the cost of equity financing, which was twice as high as the cost of debt. The highest WACC was related to the Beta indicator that also expressed the political and regulatory risk over the investigated period. Across debt cost analysis, the role of effective tax rate decreased the level of WACC. The highest level of WACC was noticed among uranium and integrated oil and gas companies. The study contributes to information asymmetry theory related to the cost of capital assumptions.

Keywords: cost of capital; WACC; European energy sector; intra-industry analysis

Citation: Franc-Dąbrowska, J.; Mądra-Sawicka, M.; Milewska, A. Energy Sector Risk and Cost of Capital Assessment—Companies and Investors Perspective. *Energies* **2021**, *14*, 1613. http://doi.org/10.3390/en14061613

Academic Editor: Surender Salkuti

Received: 12 February 2021
Accepted: 11 March 2021
Published: 14 March 2021

1. Introduction

Running a business operating on an open and competitive market requires building an appropriate capital structure. All these conditions are generally the basics of corporate nature strategy challenges. Most of the research conducted in this area focuses on capital structure determinants, such as profitability, development opportunities, company size, assets, and tax shields [1]. These factors are important in assessing the effectiveness of the functioning of companies. However, the key element in assessing companies' effectiveness is the possibility of acquiring sources of financing that will not increase the cost of capital level but enhance a level of profitability. It could be assessed from two perspectives: internally by the company managers/owners and market investors, or by its competitors. From a potential investors' point of view, the decision-making process is related to selecting the most optimal investment portfolio. This process is complicated and depends on many economic factors, including the availability of financial and nonfinancial information on the market. The market's information creates positive or negative signals that impact price increase or decrease [2]. The stock market changes in a dynamic and stochastic manner, so potential investors are able to consider these risks. Therefore, the cost of capital is one of the key elements of business valuation. Another perspective of the cost of capital as a part of modern finance theory is related to investment or disinvestment decisions, economic profit prognosis, or performance efficiency [3]. The company perspective shows the cost of capital as one of the main factors influencing companies' decisions in crediting the capital structure and optimization future path of financial decisions. The cost of capital also plays an important role in valuing the invested capital. It is a key link in transforming the stream of future expected net income into present value [4]. The other side expresses the cost capital application for evaluating investment projects to define the minimum (threshold) expected and potential accepted return rate by investors [5,6]. To calculate the capital cost, information from the financial market is necessary, with the required investment return at a given risk level. Shareholders analyze companies' financial situation and their development opportunities with the cost of the capital assessment. It can reflect the potential dividend payouts or buyback of shares for redemption that can appear on the market [7].

Each group of stakeholders has specific expectations of an above-average rate of return on the capital invested. All stakeholder groups care about the company's value because they participate in its creation and consumption. The cost of capital is affected by the risk that owners and creditors bear. This risk level determines the rates of return that these groups of investors expect from their investments. The additional risk on the market should be compensated by the risk premium and risk-free interest rate [8] and impact on the cost of capital. The level of the cost of capital depends on the method of its estimation. Thus, the authors used the Weighted Average Cost of Capital (WACC) as the most popular method to increase the study comparability. WACC is one of the direct and indirect measures that is used for investment achievement evaluation. In other words, WACC impacts the return rate on capital required over a given period by owners and creditors [9]. WACC from the company side is useful as the valuation component and indicates the return rate for assessing future company projects. Therefore, WACC is an integral part of the discount rate for the Discounted Cash Flow method (DCF) and other valuation models [10–12].

Our main contribution to the literature is the firm-level approach related to the cost of capital importance in investors' and managers' decisions. A variety of studies concern the macroeconomic factors of investors' portfolio creation on stock markets related to oil price changes. Most studies refer to the energy sector to explore capital investment's cost of capital estimation, in particular technology of implementation [13–16] or cost-related issues dedicated to maintaining or operating within specific technology of production [17–19]. Analyses of the historical approach of WACC are important for the profitability of future investors' decisions. Energy companies should pay attention to risk assessments, specifically strategic sources of risk associated with developing globalization processes related to raw materials markets trends [20]. Thus, our analyses show an ex-post approach that uses a cost of annual capital value. The conducted study verified historical data for the energy sector and risk perception by capital market investors. The market WACC valuation was also investigated in a study concerning firm age and its profitability. This paper also contributes to assessing financial markets' connections in terms of quotations and the risk management process of investment portfolios to understand the European energy sector's mechanisms.

This study covers a gap between theory and empirical research study that concerns the WACC of energy companies characterizing the economy's regulated sector. The energy sector's financial performance and development strategy are closely related to government regulation and potential liberalization of prices [21]. Entities in the energy sector operate in dynamically occurring conditions, both on the regulatory and technological levels. The obtained results underline the sector companies' diversification according to the main energy source in sub-industry classification.

This paper is organized as follows. The next section presents a literature review that concerns capital cost as a factor impacting organization and investor decisions. Section 3 describes the WACC method issues. Section 4 concerns energy sector companies on capital markets performance conditions. Section 5 presents the methods and sample, and Section 6 describes the sample and the results of the study estimation. The last two sections include the discussion and conclusion of the paper.

2. Literature Review

The capital cost is an economic category that allows for combining investment decisions with the owners' income and the creditors' benefits. It can be analyzed in several dimensions, considering the interests and requirements of the capital investors, capital buyers, and potential investors looking for the best directions for allocating their available funds [22]. Since the sixties of the twentieth-century, Modigliani and Miller presented the WACC approach that reflected its usage in capital structure decision problems. The capital cost appeared as the pricing parameter in Modigliani and Miller's theory basics [23]. Modigliani and Miller (1959) showed that debt financing positively impacted companies' value on the market due to the tax shield effect [24]. The company's capital structure can be seen as a balance between the tax benefits of debt and the costs of financial distress and bankruptcy, which can be considered due to higher obligations [25]. In this view, the tax burden on equity financing limits the extent to which firms hedge against aggregate risk. Therefore, an empirically validated framework presented secondly by the Merton–Miller model was related to trade-off theory and implied the existence of an optimal capital structure. According to Meyers [26], the trade-off theory underlines the target debt level to reach the tax benefit and to account for the low cost of capital. In addition, trade-off theory considers the possibility of searching for the optimal relationship of equity and debt capital that ensures the lowest cost of capital and the firm's highest value that enhances tax benefits. This assumption should be grounded by the positive effect of financial leverage and low bankruptcy costs [11], which was included in the optimal capital structure theory by Kraus and Litzenberger [27].

Tax affects private and listed companies' debt financing decisions differently. As taxes increase in private companies, leverage increases, whereas it does not involve long-term or short-term borrowing. For listed companies, as tax increases, long-term debt financing increases while short-term decreases; however, it does not affect leverage. Thus, listed companies increase their long-term borrowing to take advantage of tax shields [28]. Tax benefits and high inflation also influence the level of a company's leverage. A tax shield is included in WACC calculation and expresses the effective tax rate, which measures the companies' tax policy's effectiveness. Properly implemented optimization solutions should contribute to lowering the effective tax rate [29–31] and consequently the cost of capital.

Inflation risk also has an impact on decisions regarding capital cost. It makes the expected cash flows from investment projects more uncertain, and hence, projects will be assessed at high discount rates [32]. As a result of such decisions, the implementation of projects will become more costly, and thus, fewer projects will be undertaken and the firm's growth will be affected [10,33].

The globalization process is progressing due to the integration of markets and faster information flow. Globalization also brings innovation in market valuation by big data support thanks to machine learning that forecasts stock return predictions with an automatic ranking list [34]. The development of financial markets allows for acceleration of the pace of assimilation of information. According to Hughes et al., information asymmetries increase the capital cost by increasing factor risk premiums [35], thus impacting mostly equity cost of capital [36]. Reducing information asymmetry helps to lower capital costs by providing less-informed investors access to information [37]. Therefore, these effects are more visible in the cost of equity capital [36]. It could also be underlined that a better-informed group of investors also appeared on the market and they could react differently from the rest [38]. Investors' information needs vary and depend on factors such as the nature of the investment knowledge and experience, and preferred method of share prices as participants of the capital market decide to buy and sell shares daily. Therefore, investors face a countless number of financial market opportunities, and using the WACC measure helps them to benchmark market alternatives [39]. Furthermore, WACC has a significant impact on the value of the firm [40]. Listed companies do not seek to optimize the capital structure by employing a leverage mechanism but seek the most available financing sources at the moment with the lowest cost.

Fernandez underlines that the capital market costs are determined by the capital market liquidity, efficiency, and risk investors [12]. However, a company's capital structure expressed by WACC level could be a cumulative result of past attempts such as issuing shares or could be affected by temporary fluctuations in equity capital cost [41]. The theoretical approach developed for the financial market efficiency level is associated with perfect and imperfect market issues. The imperfect market determinants are related to information asymmetry, transaction costs, bankruptcy costs, and administrative and legal regulations that also include tax policy. According to Harris and Raviv, the tax approach's capital structure theory could be determined by tax and non-tax theory determinants [42]. Taxation shows how companies operate, concerning which mechanisms and principles effectively manage financial activity and the tax burden [43].

3. Cost of Capital—Methods Review

The weighted average cost of capital is commonly used in models for assessing the financial efficiency of investments, business valuation, or models for estimating economic added value [44]. The WACC methodology helps to establish the level of uncertainty on financial markets (risk aversion), the cost of debt and equity capital increase, and the credit shortage issue. The optimal capital structure minimizes the value of the weighted average cost of capital and maximizes the company's value [45]. The concept of value management assumes that the company's goal is to maximize the value for owners, which can be achieved by minimizing capital cost [46]. The WACC method uses market values to express the amount of debt and equity [46]. Capital market data reflect the risk assessment for all participants and make the cost of capital calculation available.

WACC estimation is divided into two parts: debt cost and equity cost. The cost of equity capital could be investigated from multiple perspectives, given its accounting and financial research [47]. In our WACC calculation, the authors used the Capital Asset Pricing Model (CAPM) method that measures the return of an investor's portfolio. The concept of CAPM was introduced by Sharp [48], Lintner [49] and Black, and Jensen and Scholes [50] based on Markowitz portfolio theory. The main issue underlines that equity holders keep more risk than debt holders, explaining the higher equity costs. However, banks can value the increase in default risk in these countries when there is a high climate risk exposure [51]. This approach analyzes macroeconomic factors' impact base on the Arbitrage Pricing Theory (APT). The CAPM method's main difference from APT is covariance, which establishes the expected return on market portfolio statistics. The CAPM model's popularity is related to its uncomplicated structure, which allows for the relative transparency of the obtained results [52,53]. Considering this method, WACC provides a comparable capital cost valuation compared to the Gordon or APT model of assets pricing [54]. However, the Modern Portfolio Theory (MPT), CAPM, or APT models based on market efficiency assumption highlight investors' rational decisions [55]. The other perspective underlines the behavioral finance argumentation based on investors' heuristics decision on the financial market [56].

Stulz argued that the CAPM approach is the most popular model of (owner) equity valuation [57,58] and plays a central role in finance theory [59]. The cost of equity capital in the CAPM method could impact the firm differently due to industry-specific features such as revenue, profit margin, Beta, market competition, GDP industry contribution, and more [60]. Beta, based on CAPM, influences the equity cost of capital. Beta, as measured by the CAPM, is widely used for pricing stocks [61]. Beta estimation helps investors to assess the level of uncertainty and risk. Thus, the greater the risk for the investor, the higher the expected returns [32]. Thus, investors can assess risk management based on Beta and the age of the firm. The stock Beta declines with the age of the firm [62]. Young capital age companies noticed lower average returns compared with old capital age [63]. The Beta factor reflects the firm-specific systematic risk compared to the overall market risk [64]. Stock market uncertainty affects thus firms' financing costs [65]. The Beta captures stock return behavior and is time-varying among younger firms [62]. Thus, the age of companies is additionally used to assess the WACC changes. Ozcam noticed a significant relationship between the Beta coefficients of expected macroeconomic variables and asset return [66,67]. Therefore, the profitability ratios are examined in this study to assess their impact on WACC.

Market information is essential for efficient operating decisions, and the optimal capital structure became a balance between disclosure information on the market and the low cost of capital [68]. The market risk premium as an element of CAPM calculation represents the difference between the expected market return and the risk-free return rate. This measure is important for a risk-averse investor who invests in stocks and compares debt security rates [53]. A higher risk is defined in developing capital markets characterized by lower liquidity. Beta also expresses the level of liquidity cost on the market and therefore represents systematic risk. Furthermore, Beta is one of the natural measures of sector risk used by investors [69].

A Beta that expresses market uncertainty could be static [70]. Other authors confirmed that Beta level is strongly related to the return rate from particular market investments [71]. Thus, the risk of a firm's equity depends on its contribution to stock price volatility and not on the national market portfolio situation [72]. However, the market factor impacts the WACC level, and companies do not influence and manage this value fully. Researchers also use a weighted average cost of capital for risk debt and bankruptcy assessment [11,73,74]. In WACC calculation, the debt capital is related to the cost that managers could control directly. The financial risk in this area impacts the increase in the cost of capital. An increase in foreign capital causes a decrease in free cash flow (FCF) that an enterprise may have at its disposal. The shareholder's and creditors' expectations are shaped by WACC historical data [75]. The WACC interpretation is mostly related to a nonlinear relationship. The trade-off theory used the WACC leverage pattern according to which low-level debt impacts expensive equity capital cost. When, oppositely, debt increases with distress costs, then the cost of debt becomes more expensive. According to the pecking order theory, the debt capital is preferred. Prior studies, in this case, showed the linear relationship of WACC with leverage [6]. Thus, a low WACC is determined by a high debt level, and companies benefit from higher leverage. In addition, WACC rates also include credit spreads of corporate debt [76].

4. Energy Companies on Capital Markets

Spread between equity costs and debt cost represents risk allocation [25] distinguished in each industry. Access to capital markets and investment risks differs across capital markets and industries, visible at the WACC level [15]. The WACC method is more comparable when it concerns the same segment or industry. However, younger markets with shorter histories are characterized by a higher cost of capital [77]. The WACC concept is also widely used in energy cost technology identification [78].

Another approach that also includes WACC methodology implementation is vanilla WACC for regulatory price-setting purposes [79,80]. It is a weighted average of a nominal pre-tax cost of debt and a post-tax cost of equity (reflecting the corporate tax impact). Thus, it represents only the investment side of the calculation from companies' internal decisions [81]. This issue is related to the aspect of investing in the energy sector that is associated with the involvement of high expenditure in the long-term. Maintaining the stability of investing in the energy sector is important to encourage potential investors. Among European countries' regulators, there is no uniform method of determining the cost of capital; some have a nominal WACC, WACC pre-tax, or WACC vanilla level, both in gas and electricity production. However, this approach is not included in the conducted research scope.

WACC varies between countries due to the business's specific nature and the capital employed in the long-term perspective, government policies, limited access to capital, risk perception level accepted by financial institutions, and macroeconomics parameters (inflation and demand for credit). Additionally, investors' perception of risk is different in countries where the financial industry is less competitive [6,82]. However, the energy sector possesses some specific characteristics. Through EU institutions, the energy sector regulations impact the perception of these markets among investors. Market players and, mainly, consumers have been protected by European regulators. The energy system operates under changing working conditions, depending on weather factors and energy demand variability. WACC level is also important in a regulated industry such as the energy sector; thus, WACC determines the correct price. Investors need to take into account that several mechanisms impact energy security prices [83].

The stock return in the energy sector was investigated by analysis of macroeconomic variables such as inflation, money supply, exchange rate, industrial production, bond, export, import, foreign reserve, and unemployment rate by Zhu [67]. Korajczyk and Levy [84] shed new light on the fact that macroeconomic conditions are also crucial for capital structure. They argued that market conditions are significant for unconstrained firms when the issued shares decision is being made. Then, favorable macroeconomic conditions are important. For these market activities, an updated WACC level also plays a crucial role for managers.

WACC difference across regions and technologies in the electricity sector, depending on factors such as political stability or the business cycle [85]. In the energy sector, the firms' differential exposure to policy impacts the WACC level [65]. As is known, the amount of weighted average cost of capital depends not only on the cost of individual types of capital (equity and foreign capital) but also on the capital structure and the income tax rate. The WACC method is able to include taxpayer risk.

Industry-specific factors directly and indirectly affect a firm's capital structure choice and then the cost of capital. Companies tend to be more leveraged if they operate in economically significant industries [60], such as the energy sector. Comparing the capital costs of different energy sectors shows the investor's attitude to risk and technology acceptance. The realized return rate on capital in the weighted average cost method allows for assessing whether the company can create sufficient added value or not [86]. The market determines the cost of capital (interest rate). It does not depend on the preferences of a single investor but all investors in the market. If a company plans to raise capital on the financial market, it cannot independently (arbitrarily) determine the cost of that capital. The rate of return offered must be based on market information and must take into account the risk level of such equity investments. The higher the cost, the lower the present value of the company's future net cash flows and the lower its economic value. More aggressive investors who create a portfolio based on oil-sensitive stocks with higher returns may decide to buy these stocks that have higher betas (systematic risks) currently [87].

Knowledge of profitability, which corresponds to risk, enables the company to re-calculate the amount of profit. Cost of capital determines the volume of the profits [39]. According to Pouraghajan, et al., there is a significant and positive relationship between the weighted average cost of capital (WACC) and corporate performance evaluation such as Return on Assets (ROA) and Return on Equity (ROE) [88]. Thus, a change in WACC can affect the return on assets. The higher cost of capital adversely affects the profitability position of the companies [89]. The higher WACC does not necessarily relate to increased risk but is a sign of high profitability as returns on investment [90].

The main factor that impacts the energy company sector is oil price or gas production [91]. It creates a higher risk from the perspective of market efficiency due to different price anomalies. It forces managers to use proper policy implications and investment decisions on trading activities that have energy-related tendencies on the capital market [92]. Financial markets can facilitate risk diversification and can reduce financing costs due to lower asymmetric information, which affects the lower cost of capital in case of technological innovation among the energy sector [93]. The risk could be recognized by investors from the standard deviation or/and Beta coefficient. Most of the energy sector securities have a positive Beta coefficient [94]. Financial markets (equity and credit) promote biomass and non-biomass renewable energy production in the Organisation for Economic Co-operation and Development (OECD) countries, and higher innovative economies also invest in clean energy [93]. Renewable energy firms face the domestic stock market's impact on the global financial market due to international oil prices [95].

Based on the identified interdependencies, we propose testing the following three hypotheses:

Hypothesis 1 (H1). *The WACC of energy companies depends more on the size of a company, equity, total revenues, and age of settlement.*

Hypothesis 2 (H2). *The factor determining the cost of equity is the risk level resulting from the companies' general situation on the market with Beta's highest impact.*

Hypothesis 3 (H3). *There is a negative relationship between the cost of capital level and companies' profitability.*

5. Materials and Methods

5.1. Sample Description

This study concerns companies listed on European stock exchanges. The primary industry is the energy sector, distinguished according to the Global Industry Classification Standard (GICS) sector classification. On the second stage of sample formation, the Thomson Reuters Business Classification (TRBC) was implemented to include the scale of obtained revenues from basic operating activity. The researched period relates to the available time-series data in Eikon Database–Thomson Reuters (TR). The WACC methodology includes a two-step calculation of equity-based cost on the capital assets pricing model and debt cost.

A company was selected for the sample if, during the research period, data for WACC, balance, and income were reported at least for two years. We excluded a company if it was missing WACC calculations. Smaller financial markets do not collect data that could be used for WACC calculation. These observations were, therefore, omitted. The research sample constituted finally 231 companies in a 4-year study period (2016–2019) for 25 countries (according to the country of exchange). The investigated companies are listed on 41 European capital markets (more information in the Appendix A).

5.2. Methods of Data Analysis

The WACC methodology includes a two-step calculation of equity-based cost on the capital asset pricing model and debt cost. According to the WACC TR methodology, each category of capital was proportionately weighted. All capital sources, including equity stock, preferred stock, and debt, were included in the cost of capital calculation. The cost of equity was calculated by multiplying the market's equity risk premium with the Beta of the stock plus an inflation-adjusted risk-free rate. The cost of debt represents the marginal cost to the company of issuing new debt. It is calculated by adding the weighted cost of short-term debt and weighted cost of long-term debt based on the one-year and ten-year appropriate credit curve. Beta used in CAPM calculation represents how much stock moves for a given move in the market (based on the covariance of the security price movement to the market's price movement). The detailed definitions of implemented measures in the study are presented in Table 1.

WACC was calculated using the following formula:

$$\text{WACC} = (E/V) \times K_E + (D/V) \times K_D \times (1 - t_c) + (P/V) \times K_p \quad (1)$$

where E is the value of equity, D is the company's debt, P is the company's preferred stock, V = total capital ($E + D + P$), K_E is the cost of equity, K_D is the cost of debt, K_P is the cost of preferred stock, and tc is corporate tax.

Table 1. Definition of variables.

Variables	Definition
Weighted Average Cost of Capital (%)	It is calculated as an average rate that a company is expected to pay to its debt, equity, and preferred stockholders to finance its assets.
WACC Cost of Equity (%)	The cost of equity is calculated via a CAPM method.
WACC Equity Risk Premium (%)	It is the StarMine Equity Risk Premium for the company's country.
WACC Tax Rate (%)	It is the effective tax rate for the company.
WACC Cost of Debt (%)	The cost of the debt component calculates the after-tax cost of debt.
WACC Cost of Preferred (%)	The cost of preferred stock is the current preferred dividend yield on the company's preferred stock.
WACC Debt Weight (%)	It is a debt component in WACC calculation.
WACC Equity Weight (%)	It is an equity component in WACC calculation.
WACC Short-Term Debt Cost (%)	It is a short-term debt component in WACC calculation.
WACC Long-Term Debt Cost (%)	It is a long-term debt component in WACC calculation.
Beta	The Beta coefficient is calculated by considering the primary index for the country of the company's primary equity listing. The used Beta factor is calculated for a fiscal year for each company.
Age	The number of years since the company was settle
ROA	Relationship of net financial result to total assets (%)
ROE	Relationship of net financial result to share capital (%)
EBIT/Sale	EBIT (Earnings before interests and taxation) to sale ratio (%)

Source: For the Weighted Average Cost of Capital (WACC) definition, Thomson Reuters methodology; for other variables, own description.

Descriptive statistics were used to identify the WACC level in the investigated period. The distinguished groups of industries of the energy sector related to an energy company's characteristics of primary operating performance are coal, integrated oil and gas, oil and gas drilling, oil and gas exploration and production, oil and gas refining and marketing, oil and gas transportation services, oil-related services and equipment, renewable fuels, and uranium (Table 2).

Table 2. Sub-industry classification, according to TRBC (The Refinitiv Business Classification).

Sub-Industry Activity	Definition
coal (A)	Companies are primarily involved in producing and mining coal, related products, and other consumable fuels related to energy generation. Additionally, coal mining support and coal wholesale companies are in these groups.
integrated oil and gas (B)	Integrated oil companies engaged in the exploration and production of oil and gas
oil and gas drilling (C)	Drilling contractors or owners of drilling rigs that contract their services for drilling onshore or offshore

Table 2. *Cont.*

Sub-Industry Activity	Definition
oil and gas exploration and production (D)	These companies are engaged in the exploration and production of oil and gas that are not classified elsewhere.
oil and gas refining and marketing (E)	Companies engaged in the refining and marketing of oil, gas, and/or refined products: these groups also include gasoline stations and petroleum product wholesale.
oil and gas transportation services (F)	Companies engaged in the storage and/or transportation of oil, gas, and/or refined products, including diversified activities that cover pipeline transport, sea-borne tanker, and oil and gas storage
oil-related services and equipment (G)	Manufacturers of equipment and oil-related services
renewable fuels (H)	Companies that concern biodiesel production, ethanol fuels, pyrolytic and synthetic fuels, biomass and biogas fuel, and hydrogen fuels
uranium (I)	Companies for which the main activities are related to uranium mining and uranium processing

Pearson correlation was used to infer causal relationships between the WACC level of selected groups (with the highest number of observations) and basic financial market measures important for investors such as Beta, total assets, revenues, market capitalization, and company age. A similar approach was used for the identification of the most significant factor influencing the WACC level of energy companies. In empirical investigations, we also use nonparametric methods due to the nature of financial data. Nonparametric tests do not require assumptions about the type of distribution but are not without additional limitations. To examine the differences in the energy industry, nonparametric ANOVA was used (Kruskal–Wallis test). We present the ANOVA Kruskal–Wallis test and the multiple comparisons test results, with differences between energy groups of companies divided according to the TRBC classification.

6. Results

6.1. Descriptive Statistics of a Sample

Table 3 reports the descriptive statistics for the investigated companies for primary balance and income statement data as well as market capitalization value. The highest standard deviation value of a basic company's size data was noticed in the total assets value (38.79 mld EUR). The percentile analysis expresses that the market is dominated by big players in the energy sector. The rest of the companies noticed a total assets level of 0.23 mld EUR. At the same time, the mean amounted for 9.44 mld EUR. These significant differences were also repeated in the total revenues, net income, and market capitalization statistics. It reflects the dominance of big units in these sectors.

Table 3. Descriptive statistics.

Variables	N	Mean	Std Dev	25th Pctl	50th Pctl	75th Pctl
Total assets (mld EUR)	1128	9.44	39.79	0.04	0.23	2.13
Total revenues (mld EUR)	1055	6.19	27.92	0.00	0.06	0.79
Net income after taxes (mld EUR)	1131	0.32	1.79	−0.01	0.00	0.03
Market Cap (mld EUR)	1127	4.68	19.69	0.02	0.10	0.85

6.2. WACC Primary Results

The number of observations for each year was diverse due to the data availability for WACC calculations (Table 4). The average level of WACC amounts between 6.09% to 8.13%. A higher level of cost was noticed for equity capital, and its value ranged from 7.15% to 9.77%. The debt cost due to tax shield and capital structure optimization was also two-times lower than the equity capital. A tax shield's role is expressed by the WACC tax rate that presents the company's effective tax rate. Its level was in the range of 20.45% to 22.94%.

A more significant differentiation of the surveyed companies' capital cost was noted due to the classification assigned in the TR of given companies to the industry within the energy sub-industry activity (9 groups) (Table 5). The analyzed groups differ significantly in the size of the sample. The lowest WACC level was noticed in oil and gas drilling companies and amounts to 4.90%, in which the equity cost valuation was on the level of 6.75%. According to the WACC parameter, this group could be assessed by investors as being the most attractive.

Table 4. WACC, WACC equity, and WACC debt in 2015–2019.

Years	Number of Companies	WACC (%)	WACC Equity (%)	WACC Debt (%)	WACC TAX Rate (%)
2015	190	6.82	7.37	3.14	22.94
2016	198	6.09	7.15	2.56	21.62
2017	203	6.17	7.51	2.25	21.65
2018	210	8.13	9.77	3.09	20.45
2019	214	7.36	8.73	2.85	21.35

The highest level of WACC occurred in the integrated oil and gas group of companies: 9.22%. The WACC calculation is highest for effective tax rate characterized by two groups: uranium (28.38%) and integrated oil and gas companies (24.49%). The lowest tax effective rate that reflects the possibility of tax optimization was noticed in the oil and gas drilling (19.30%) and oil and gas transportation services (19.59%) groups.

Table 5. WACC, WACC equity, and WACC debt in the industry overview classification.

Industries	Number of Companies	Number of Observation	WACC (%)	WACC Equity (%)	WACC Debt (%)	WACC TAX Rate (%)
Coal (A)	10	47	8.05	8.71	4.39	20.60
Integrated Oil and Gas (B)	15	74	9.22	10.87	4.02	24.49
Oil and Gas Drilling (C)	8	38	4.90	6.75	2.83	19.30
Oil and Gas Exploration and Production (D)	99	490	7.13	7.97	2.29	20.90
Oil and Gas Refining and Marketing (E)	27	135	5.70	7.28	2.37	22.89
Oil and Gas Transportation Services (F)	20	99	6.47	7.81	3.04	19.59
Oil-Related Services and Equipment (G)	44	220	6.85	8.53	3.37	22.16
Renewable Fuels (H)	5	28	5.86	5.94	1.46	23.71
Uranium (I)	3	15	5.85	6.92	1.56	28.38

Across the presented WACC calculation elements, the WACC debt and equity weight were diverse and strongly impacted the WACC level (Figure 1). On the other hand, significant differences in the minimum and maximum values were recorded for the tax rates and the final results' WACC calculation.

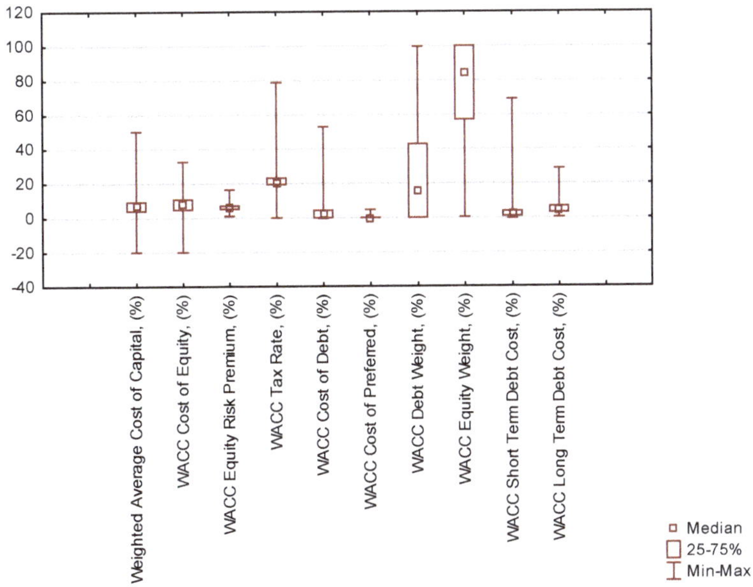

Figure 1. WACC component statistics in 2015–2019.

According to the group energy source classification, the WACC cost of capital noticed the most mixed results for group A, coal (Figure 2). The smallest volatility in the WACC level was characteristic of companies in the oil-related services and equipment group, where capital cost was also low.

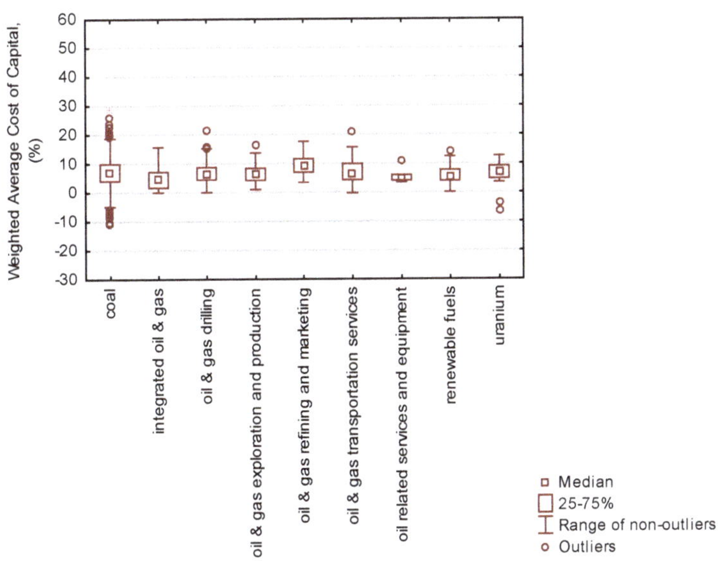

Figure 2. WACC component statistics according to intra-industry analysis.

In the next stage of the study, we excluded companies with a net loss for profitability statistics calculation (Table 6). Thus, these companies do not generate profitability. The presented results for ROE, ROA, and EBIT to sales demonstrate three levels of companies' profitability. The highest return of equity was noticed in 2015 and amounted to 36.24%, while in 2019, its value reaches the level of 11.96%. It shows high differentiation related to the return rate on equity and is associated with the high volatility of profitability in the enterprises' examined group. Lower volatility was recorded in return on assets, which, similar to ROE, also amounted to the lowest value in 2019—6.43%. However, the highest EBIT to sale relation appeared in 2016, when it amounted to 30.34%, which presents a relatively high level of operating results to reach revenues. It could be underlined that this part of the study includes only profitable companies that did not report any EBIT losses.

The average age of the company's settlement was in the investigated period between 26 and 29 years. It represents a relatively long period in which given entities operate in the energy sector, which may be the basis for assessing investors' credibility and stability. The Beta coefficient amounts between 0.88 in 2015 till 1.02. In recent years, the increased risk of doing business was noticed and underlined the investors' systematic risk measure. Regulated companies operating on the energy market reached a Beta value below 1. This means that investors assess investments in such enterprises as safer than other investments on a given capital market, which directly results in the expectation of a lower return rate on employed capital [96].

Table 6. Profitability, age, and Beta of investigated companies in 2015–2019.

Years	ROA	ROE	EBIT/Sales	Age	Beta
2015	9.20	36.24	16.54	26	0.8840
2016	6.32	12.96	30.34	27	0.9213
2017	7.96	16.69	22.72	27	0.8270
2018	8.07	21.80	20.67	28	0.9592
2019	6.43	11.96	19.71	29	1.0245

Table 7 shows the correlation dependencies of WACC in the selected groups of energy companies and parameters, indicating the market position. These variables include Beta ratio, the lower level of which is the domain of mature companies, total assets, and revenues that indicate the scale of operations and the market capitalization and age of the company calculated in years from the year of its establishment. The correlation analyses were made in a selected group of companies characterized by a higher number of observations during the investigated period. The Beta coefficient strongly impacts the WACC level in the whole group and notices a correlation relationship on the level of 0.77. This relation was even more strongly significant in the correlation analysis in selected groups in the oil and gas exploration and production group of companies (0.90) and in oil and gas refining and marketing (0.93). The total assets and revenues value do not impact so strongly on the WACC level, and this relationship was insignificant in the highlighted groups. The total assets and revenues correlation relationship was the highest in oil-related services and equipment group (0.32 and 0.33). This relation across the whole sample noticed a relatively low level of impact on WACC. It underlines that enterprises' size does not play a significant role in the valuation of listed companies' capital, resulting from the diversified scale of these companies' operations depending on a given capital commitment. A similar observation was recorded for market capitalization that noticed the highest level in the oil-related services and equipment group (0.30). In the case of the whole sample, it amounts to only 0.14. An interesting observation was noticed between company ages that across the entire sample shows a negative relationship between age and WACC, which means that the younger firm reached a higher WACC level. This indicates a more stable financial situation in the case of older companies.

Table 7. Summary statistics correlation matrix for WACC with group division and selected variables.

Variables	WACC				
	In Total	Oil and Gas Exploration and Production (D)	Oil and Gas Refining and Marketing (E)	Oil and Gas Transportation Services (F)	Oil-Related Services and Equipment (G)
Beta	0.771097 *	0.905104 *	0.934102 *	0.626490 *	0.630788 *
Total assets	0.137843 *	0.075142	0.553929 *	−0.183453	0.317630 *
Revenues	0.110416 *	0.048663	0.280272 *	−0.195659	0.331171 *
Market Cap	0.144349 *	0.076439	0.179405 *	−0.040364	0.302357 *

* values indicate significance at 5%.

Table 8 shows the correlation of individual WACC components with the horizontal WACC in the selected groups of enterprises with the highest number of observations in the analyzed period. As confirmed in the results of the descriptive statistics, the highest level of correlation was recorded in the case of WACC equity, which was on average 0.87. In entities from the oil and gas exploration and production group, it reached the level of 0.94. It shows the importance of assessing a given sector's market situation and the possibility of optimizing the WACC level. WACC cost of debt noticed a correlation relationship with WACC on the level of 0.13 for oil and gas exploration and production to 0.38 in oil-related services and equipment companies. A higher correlation relationship with WACC was noticed for short-term debt than for long-term debt costs resulting from smaller long-term engagement in the investigated sample. The highest statistical impact of the effective tax rate on WACC was noticed in oil and gas transportation services and amounted to −0.37. In the entire sample, this parameter reached the level of −0.11.

Table 8. Summary statistics correlation matrix for WACC with group division.

WACC Components	WACC				
	In Total	Oil and Gas Exploration and Production (D)	Oil and Gas Refining and Marketing (E)	Oil and Gas Transportation Services (F)	Oil-Related Services and Equipment (G)
WACC Cost of Equity	0.872811 *	0.945570 *	0.842252 *	0.810759 *	0.805891 *
WACC Cost of Debt	0.315035 *	0.135108 *	0.302357 *	0.276244 *	0.385300 *
WACC Cost of Short-Term Debt	0.314094 *	0.076126	0.289891 *	0.376047 *	0.353707 *
WACC Cost of Long-Term Debt	0.243280 *	0.079347	0.202684 *	0.255900 *	0.440429 *
WACC Tax Rate	−0.108775 *	−0.067218	−0.310043 *	−0.368650 *	−0.100062

* values indicate significance at 5%.

Table 9 presents an analysis of correlation against the specified financial data related to profitability in the selected groups of companies with the largest number of observations. ROA noticed a significant impact on WACC in the whole sample on the level of −0.1581, which expresses that higher return on assets impact lower WACC levels in energy companies. This relation was also significant in the oil and gas refining and marketing group (−0.2932). Both ROE and EBIT to sale noticed an insignificant relation to WACC level, which could be explained, but a small number of observations and more substantial impact of others not included in the study variables.

Table 9. Summary statistics correlation matrix for WACC and intra-industry activity division.

Variables	WACC				
	In Total	Oil and Gas Exploration and Production (D)	Oil and Gas Refining and Marketing (E)	Oil and Gas Transportation Services (F)	Oil-Related Services and Equipment (G)
ROA	−0.158125 *	−0.003251	−0.293192 *	−0.083323	−0.138026
ROE	−0.049386	−0.014109	−0.048931	−0.005672	0.135792
EBIT/Sale	0.044763	0.023897	0.410690 *	0.121619	−0.009494

* values indicate significance at 5%.

In order to investigate the diversity of WACC calculation, nonparametric variance was carried out among the most numerous groups of entities in the energy sector. Differences in WACC level between the investigated groups reflect changes in systematic risk [97]. Table 10 presents the summary statistics for the H Kruskal–Wallis test and post hoc test pairwise comparisons with group division and total WACC, WACC equity cost, WACC cost of debt, Beta, and tax rate. The ANOVA analysis results show a significant difference among all parameters, presenting its variation in total WACC, cost of equity, cost of debt, and tax rate. According to the intra-industry division, the level of WACC was significantly diversified in the case of companies from group D (oil and gas exploration and production) and E (oil and gas refining and marketing), which reflect the different stages of oil and gas production. These groups of companies also noticed significant mean rank differences in the case of WACC cost of equity. The level of WACC debt cost was varied significantly for mean rank in D (oil and gas exploration and production) and F (oil and gas transportation services), and D and G (oil-related services and equipment). That between D and G also noticed higher rank differences. All these energy sector activity groups underlined different characteristics of the conducted operation in this industry, impacting WACC variation to engage more debt in the capital structure.

Table 10. Summary statistics H Kruskal–Wallis test—post hoc tests pairwise comparisons with group division.

Variables	Groups	D	E	F	Chi²	H Test
WACC	E	3.322871 **			37.82999	H (8, N = 1015) = 4.92645
	F	1.209614	1.407283		df = 8	p = 0.0000
	G	0.648941	2.459835	0.666893	p − 0.0000	
WACC Cost of Equity	E	1.150129 **			31.96747	H (8, N = 991) = 39.98456
	F	0.340099	0.560647		df = 8	p = 0.0000
	G	1.529547	2.167629	1.349187	p = 0.0001	
WACC Cost of Debt	E	1.903574			49.28539	H (8, N = 1015) = 58.04512
	F	3.421469 **	1.490979		df = 8	p = 0.0000
	G	5.146366 **	2.193869	0.339302	p = 0.0000	
Beta	E	0.489634			13.43843	H (8, N = 1015) = 18.39052
	F	0.136731	0.472544		df = 8	p = 0.0185
	G	3.155637	2.815944	1.996653	p = 0.0976	
Tax rate	E	3.524048 **			69.19428	H (8, N = 1015) = 67.87653
	F	1.186811	3.571717 **		df = 8	p = 0.0000
	G	5.024795 *	0.663718	4.460214 **	p = 0.0000	

Notes: statistically significant at ** 0.05 and * at 0.10.

The Beta level was not significantly differentiated in a given group of companies. This indicates a similar relationship in terms of changes in quotations in the energy sector, impacting the WACC level. The highest difference in mean ranks (with the level of 0.05 significant) was noticed for tax rates between intra-industry company division. The rank differences were the highest between the G (oil-related services and equipment) and F groups (oil and gas transportation services) (4.4602). This reveals that tax rate implementation for WACC level tax optimization is varied and could impact company values differently, which can also be one of the key measures used by market investor valuation. The difference in rank between group D (oil and gas exploration and production) and E (oil and gas refining and marketing) and between E and F (oil and gas transportation services) were on a similar level (accordingly 3.5240 and 3.5717).

7. Discussion

The energy sector as a regulated sector expresses the importance of WACC as it can be classified as a sector under numerous regulations. Investors require a return considering current market circumstances, irrespective of past conditions [53]. The WACC gives some insights into the utilities of the most suitable financing strategy [98]. Information on capital costs is related to assessing the company's financial management [99]. Implementing the process for reducing information asymmetry can lower capital costs [35–37,100]. Other researchers' biases of WACC level between the sub-energy sectors were also observed and underlined as different regular actions and risk reduction approaches to energy production [101].

The study's stated hypothesis assumed that the WACC of energy companies depends more on the size of a company, equity, total revenues, and settlement age. The defined size from the value of assets perspective and total revenues or market capitalization were not strong determinants of WACC level across energy companies'. This approach was also underlined by Lohani that value creation is not related directly to the company's size [102]. Market capitalization is determined by multiplying the number of outstanding shares and the current market price of one share and its relation with a size measure revealed by its sales or total assets value [40]. Market capitalization was impacted significantly by WACC level; however, this relation was not significant in the case of intra-industry analysis, and its fundamental role was relatively low. For younger firms, WACC is higher than in mature firms [103,104]. It is explained by the fact that new firms' future financial performance is more uncertain to investors. This relation also appeared in a conducted study in the case of the energy sector in Europe. The highest impact of companies position on the market regarding WACC level was noticed for total assets, which is one of the main factors impacting the capital structure [105]. Hypothesis H1 was only partially confirmed.

This study's second hypothesis expects that the factor determining the cost of equity most strongly is the Beta coefficient (H2). The main factor impacting the cost of equity in the energy sector was the Beta level. This coefficient was not significantly varied across an investigated sample of companies. It could create a higher risk from market efficiency due to different price anomalies in this industry. Thus, stock market uncertainty affects firms' financing costs. A high equity cost underlines the importance of information asymmetries on the energy market. However, the role of equity in the breakdown of aggregate risks leads to the prediction that firm dividends should vary depending on macroeconomic conditions after checking the effects of relevant variables at the firm level [106].

The third hypothesis assumed a negative relationship between the cost of capital level and companies' profitability (H3). This relation was confirmed only in the case of a few groups of companies. A negative association of profitability presents the company's return to support building a company's value in investors' decisions. The negative relationship of WACC and return on assets was confirmed by Shadab and Sattar [107]. Profitable companies finance their growth from retained earnings, while less profitable companies choose debt financing [108]. More risky investment is characterized by higher WACC, i.e., higher cost of equity and debt cost. From the investor's perspective, a higher capital cost means a higher return on their investment in the form of compensation. However, firms with higher WACC should have lower values. Thus, it has a negative effect on firm performance on the market. The highest financial results are achieved by companies that are able to maintain a low WACC level [109,110].

8. Conclusions

The cost of capital assessment methods are divided into subjective and capital market-based approaches. The expected rate of return influences the risk-adjusted cost. The data from the capital market reflects the risk assessment for all transaction participants [64]. The capital cost is generally defined as the investors' expected return rate (both owners and creditors) on the invested capital at the particular risk level [111]. Other perspectives define the cost of capital as a rate that investors use to discount a firm's future cash flows. Thus, the higher the capital cost, the lower the present value of the firm's future cash flows [112]. On the other hand, companies' capital cost is inferred from market prices with current earnings and growth forecasts [113]. Costs of capital are often considered the minimum yield or the minimum expected return rate that an investor would accept [114]. The return demand by investors determines a firm's cost of equity capital [36]. The cost of debt assesses future investments and the profitability of current operations.

The high level of WACC equity capital cost shows that the market factor impacts the WACC level on the energy sector, and companies do not fully influence that. It also pictured that increased the share of debt capital will determine the firms' higher market value. It can also be summarized that companies with a high cost of equity invest less [115] and do not have as much possibility to lower their WACC using debt increase engagement. The mechanism that minimizes the WACC companies and maximizes firms' value is limited in the case of the investigated firms. The WACC of a company will be lower with an increase in debt share till the higher cost of share and debt capital forces the average up [89]. A lower WACC is supported by reduced transaction cost and risk [116]. The debt level affects the risk of default concerning bankruptcy cost. However, debt financing is considered a more aggressive strategy that can generate higher profits [44].

Companies that maximize debt share can decrease the WACC level due to the tax shield effect [117]. The tax shield affects the choice of financing sources [47], and this approach is adopted in WACC valuation methods [118]. The highest WACC tax rate was noticed in a small group of companies for which their operating activities are related to uranium. The second group was renewable fuels companies that are able to use different tax optimization tools. It thus underlines that WACC analysis compares the risks associated with the other technologies in the energy sector [32]. The WACC analysis is more relevant for financially "distressed" companies due to the significant differences between the value of the funding sources recorded in the balance sheet and their real market price [74].

This study was designed to capture the cross-country energy sector cost of capital identification, which presents the manager's attitude towards building the capital structure and, from the other side, investor perspectives of company status assessment. This study's results contribute to the information asymmetry theory related to the higher cost of equity capital due to risk premiums on the energy market that express the political, regulation, and raw trends on the global market.

The authors are aware of the WACC methodological disadvantages. The WACC's determination of industry calculation does not include all possible risks associated with a particular company or investment. A market risk premium is retained by the WACC methodology; however, no technical or techno-economic risk is directly added [119]. According to the CAPM method, the cost of equity limits the risk factor to market beta [38]; it also does not consider the international spread and equity price [120]. Nevertheless, this measure, due to popular methods, is a comparable measure for the listed companies. The study's future direction will concern the comparison of the energy sector with other industrial sectors and will include the macroeconomic factor that impacts the WACC level.

Author Contributions: Conceptualization, J.F.-D., M.M.-S. and A.M.; methodology, M.M.-S.; validation, M.M.-S.; formal analysis, M.M.-S.; investigation, M.M.-S.; writing—original draft preparation, M.M.-S., J.F.-D. and A.M.; writing—review and editing, M.M.-S.; visualization, M.M.-S. All authors have read and agreed to the published version of the manuscript.

Funding: This research received no external funding.

Conflicts of Interest: The authors declare no conflict of interest.

Appendix A

The research sample constitutes 231 companies in a 4-year study period (2016–2019) for 25 countries (country of exchange): Austria (4), Belgium (3), Bulgaria (1), Croatia (2), Cyprus (2), Denmark (3), Finland (1), France (10), Germany (7), Greece (4), Hungary (1), the Republic of Ireland (2), Italy (6), Lithuania (1), Malta (1), the Netherlands (4), Norway (39), Poland (8), Portugal (1), the Republic of Serbia (1), Romania (13), Russia (23), Spain (2), Sweden (9), Ukraine (2) and the United Kingdom (82).

The investigated companies are listed on 41 European capital markets: Ab Nasdaq Vilnius, Aim Italia—Mercato Alternativo Del Capitale, Asx—All Markets, Athens Exchange S.A. Cash Market, Belgrade Stock Exchange, Bolsa De Madrid, Budapest Stock Exchange, Bulgarian Stock Exchange, Cyprus Stock Exchange, Deutsche Boerse Ag, Euronext—Euronext Amsterdam, Euronext—Euronext Brussels, Euronext—Euronext Paris, Euronext Access Paris, Euronext Growth Paris, First North Sweden—Sme Growth Market, Hanseatische Wertpapierboerse Hamburg, Irish Stock Exchange—All Markets, London Stock Exchange, Malta Stock Exchange, Moscow Exchange—All Markets, Nasdaq Copenhagen A/S, Nasdaq Helsinki Ltd, Nasdaq Stockholm AB, Nordic Growth Market, Norwegian Over The Counter Market, Operador De Mercado Iberico de Energia—Portugal, Oslo Axess, Oslo Bors Asa, Pfts Stock Exchange, Spot Regulated Market—BVB, Spotlight Stock Market, Warsaw Stock Exchange/Equities/New Connect—Mtf, Warsaw Stock Exchange/Equities/Main Market, Wiener Boerse Ag Amtlicher Handel (Official Market), and Xetra And Zagreb Stock Exchange.

References

1. Jovanovic, B. *Determinants of Corporate Capital Structure: Evidence from Macedonian Panel Data*; Economic Analysis: Skopje, Macedonia, 2015.
2. Mądra-Sawicka, M.; Paliszkiewicz, J. Information Sharing Strategies in the Social Media Era: The Perspective of Financial Performance and CSR in the Food Industry. *Information* **2020**, *11*, 463. [CrossRef]
3. Bruner, R.F.; Eades, K.M.; Harris, R.S.; Higgins, R.C. Best practices in estimating the cost of capital: Survey and synthesis. *Financ. Pract. Educ.* **1998**, *8*, 13–28.
4. Duliniec, A. Koszt kapitału w teorii i praktyce przedsiębiorstw. *Gospod. Nar.* **2012**, *254*, 1–18. [CrossRef]
5. Damodaran, A. *Applied Corporate Finance. A User's Manual*; Wiley: New York, NY, USA, 2006.
6. Ondraczek, J.; Komendantova, N.; Patt, A. WACC the dog: The effect of financing costs on the levelized cost of solar PV power. *Renew. Energy* **2015**, *75*, 888–898. [CrossRef]
7. Franc-Dąbrowska, J.; Mądra-Sawicka, M.; Ulrichs, M. Determinants of dividend payout decisions–the case of publicly quoted food industry enterprises operating in emerging markets. *Econ. Res. Istraz.* **2020**, *33*. [CrossRef]
8. Amihud, Y.; Mendelson, H. Tthe Liquidity Route To a Lower Cost of Capital. *J. Appl. Corp. Financ.* **2000**, *12*, 8–25. [CrossRef]
9. Pagano, M.S.; Stout, D.E. Calculating a Firm's Cost of Capital. *Manag. Account. Q.* **2004**, *5*, 13–20.
10. Mian, M.A.; Vélez-Pareja, I. Applicability of the classic WACC concept in practice. *Lat. Am. Bus. Rev.* **2007**, *8*, 19–40. [CrossRef]
11. Koziol, C. A simple correction of the WACC discount rate for default risk and bankruptcy costs. *Rev. Quant. Financ. Account.* **2014**, *42*, 653–666. [CrossRef]
12. Fernandez, P. *Valuation Methods and Shareholder Value Creation*; Academic Press: London, UK, 2002.
13. Flegkas, S.; Birkelbach, F.; Winter, F.; Groenewold, H.; Werner, A. Profitability analysis and capital cost estimation of a thermochemical energy storage system utilizing fluidized bed reactors and the reaction system $MgO/Mg(OH)_2$. *Energies* **2019**, *12*, 4788. [CrossRef]
14. Osma, L.; García, L.; Pérez, R.; Barbosa, C.; Botett, J.; Sandoval, J.; Manrique, E. Benefit-cost and energy efficiency index to support the screening of hybrid cyclic steam stimulation methods. *Energies* **2019**, *12*, 4631. [CrossRef]
15. Hirth, L.; Steckel, J.C. The role of capital costs in decarbonizing the electricity sector. *Environ. Res. Lett.* **2016**, *11*. [CrossRef]
16. Tagliapietra, S.; Zachmann, G.; Fredriksson, G. Estimating the cost of capital for wind energy investments in Turkey. *Energy Policy* **2019**, *131*, 295–301. [CrossRef]
17. Choudhury, A.; Shelford, T.; Felton, G.; Gooch, C.; Lansing, S. Evaluation of hydrogen sulfide scrubbing systems for anaerobic digesters on two US dairy farms. *Energies* **2019**, *12*, 4605. [CrossRef]
18. Vogt-Schilb, A.; Meunier, G.; Hallegatte, S. When starting with the most expensive option makes sense: Optimal timing, cost and sectoral allocation of abatement investment. *J. Environ. Econ. Manag.* **2018**, *88*, 210–233. [CrossRef]
19. Stocks, K.J. Discount rate for technology assessment. An application to the energy sector. *Energy Econ.* **1984**, *6*, 177–185. [CrossRef]

20. Jonek-Kowalska, I. Efficiency of Enterprise Risk Management (ERM)systems. Comparative analysis in the fuel sector and energy sector on the basis of Central-European companies listed on the Warsaw Stock Exchange. *Resour. Policy* **2019**, *62*, 405–415. [CrossRef]
21. Paun, D. Sustainability and financial performance of companies in the energy sector in Romania. *Sustainability* **2017**, *9*, 1722. [CrossRef]
22. Andersen, T.J. Risk management, capital structure, and performance: A real options perspective. In *Global Business and Economics Anthology*; Kantarelis, D., Ed.; Business & Economics Society International: Worcester, UK, 2005; pp. 187–199.
23. Brusov, P.; Filatova, T.; Orehova, N.; Brusova, N. Weighted average cost of capital in the theory of Modigliani-Miller, modified for a finite lifetime company. *Appl. Financ. Econ.* **2011**, *21*, 815–824. [CrossRef]
24. Modigliani, F.; Miller, M.H. The cost of capital, corporation finance and the theory of investment. *Am. Econ. Rev.* **1958**, *48*, 261–297.
25. Fathi, S.; Ghandehari, F.; Shirangi, S.Y. Comparative Study of Capital Structure Determinants in Selected Stock Exchanges of Developing Countries and Tehran Stock Exchange. *Int. J. Acad. Res. Account. Financ. Manag. Sci.* **2014**, *4*, 67–75. [CrossRef]
26. Myers, S.C. Capital Structure. *J. Econ. Perspect.* **2001**, *15*, 81–102. [CrossRef]
27. Kraus, A.; Litzenberger, R.H. A State-Preference Model of Optimal Financial Leverage. *J. Finance* **1973**, *28*, 911–922. [CrossRef]
28. Bas, T.; Muradoglu, G.; Phylaktis, K. *Determinants of Capital Structure in Developing Countries*; Cass Business School: London, UK, 2009.
29. Zurita, S.; Castillo, A.; Niño, J. Inflation, tax integration and company valuation: The Latin American case. *J. Bus. Res.* **2019**, *105*, 370–380. [CrossRef]
30. Bame-Aldred, C.W.; Cullen, J.B.; Martin, K.D.; Parboteeah, K.P. National culture and firm-level tax evasion. *J. Bus. Res.* **2013**, *66*, 390–396. [CrossRef]
31. Miller, R.A. The weighted average cost of capital is not quite right. *Q. Rev. Econ. Financ.* **2009**, *49*, 128–138. [CrossRef]
32. Saługa, P.W.; Kamiński, J. The cost of equity in the energy sector. *Polityka Energy* **2018**, *21*, 81–96. [CrossRef]
33. Sinha, P.; Agnihotri, S. Macroeconomic risk and firms financing decision: An empirical panel data investigation using system GMM. *MPRA Munich Pers. RePEc Arch.* **2015**, *67088*, 1–20.
34. Mądra-Sawicka, M. Financial Management in the Big Data Era. In *Management in the Era of Big Data: Issues and Challenges*; Paliszkiewicz, J., Ed.; CRC Press, Taylor & Francis Group: Boca Raton, FL, USA, 2020; pp. 71–81.
35. Hughes, J.; Liu, J.; Liu, J.; Version, F.; Version, C. *Private Information, Diversifcation, and Asset Pricing.*; Univeristy of California: Los Angeles, CA, USA, 2005.
36. Easley, D.; O'Hara, M. Information and the cost of capital. *J. Financ.* **2004**, *59*, 1553–1583. [CrossRef]
37. Lambert, R.A.; Leuz, C.; Verrecchia, R.E. *Information Asymmetry, Information Precision, and the Cost of Capital*; NBER Working Paper Series; NBER: Cambridge, UK, 2009.
38. Botosan, C.A. Disclosure and the cost of capital: What do we know? *Account. Bus. Res.* **2006**, *36*, 31–40. [CrossRef]
39. Hussain, T.; Chakraborty, L. Is the Cost of Capital an Important Determinant of Market Performance of Private Commercial Banks in Bangladesh? *Proc. ASBBS* **2010**, *17*, 389–399.
40. Rajhans, R.K.; Kaur, K. Financial Determinants of Firms' Value Evidence From Indian Firms. *Zenith Int. J. Bus. Econ. Manag. Res.* **2013**, *3*, 70–76.
41. Albanez, T. Impact of the cost of capital on the financing decisions of brazilian companies. *Int. J. Manag. Financ.* **2015**, *11*, 285–307. [CrossRef]
42. Harris, M.; Raviv, A. The Theory of Capital Structure Milton. *J. Financ.* **1991**, *46*, 297–355. [CrossRef]
43. Vintilă, G.; Gherghina, Ş.C.; Păunescu, R.A. Study of Effective Corporate Tax Rate and Its Influential Factors: Empirical Evidence from Emerging European Markets. *Emerg. Mark. Financ. Trade* **2018**, *54*, 571–590. [CrossRef]
44. Angelopoulos, D.; Brückmann, R.; Jirouš, F.; Konstantinavičiute, I.; Noothout, P.; Psarras, J.; Tesnière, L.; Breitschopf, B. Risks and cost of capital for onshore wind energy investments in EU countries. *Energy Environ.* **2016**, *27*, 82–104. [CrossRef]
45. Fernández-Rodríguez, E.; Martínez-Arias, A. Determinants of the effective tax rate in the BRIC countries. *Emerg. Mark. Financ. Trade* **2014**, *50*, 214–228. [CrossRef]
46. Jacobs, L.; Jacobsnwuacza, E.L. Effect of debt-equity tax bias on the WACC of oil and gas companies in BRICS countries. *Int. J. Bus. Manag. Stud.* **2018**, *10*, 149–165.
47. Franc-Dąbrowska, J.; Mądra-Sawicka, M.; Bereżnicka, J. Cost of agricultural business equity capital-a theoretical and empirical study for Poland. *Economies* **2018**, *6*, 37. [CrossRef]
48. Sharpe, W.F. Capital asset prices: A theory of market equilibrium under conditions of risk. *J. Financ.* **1964**, *19*, 425–442.
49. Lintner, J. The valuation of risk assets and the selection of risky investments in stock portfolios and capital budgets. *Rev. Econ. Stat.* **1965**, *47*, 13–37. [CrossRef]
50. Black, F.; Jensen, M.; Scholes, M. The Capital Asset Pricing Model: Some Empirical Tests. In *Studies in the Theory of Capital Markets*; Praeger Publishers Inc.: New York, NY, USA, 1972.
51. Kling, G.; Volz, U.; Murinde, V.; Ayas, S. The impact of climate vulnerability on firms' cost of capital and access to finance. *World Dev.* **2021**, *137*, 105131. [CrossRef]
52. Horváthová, J.; Mokrišová, M. Determination of Cost of Equity for Selected Enterprises of the Energy Industry Applying the CAPM Model and its Comparison with the Model with Gradual Counting Risk Premium. *J. Manag. Bus. Res. Pract.* **2014**, *6*, 43–54.
53. Stubelj, I.; Dolenc, P.; Jerman, M. Estimating WACC for Regulated Industries on Developing Financial Markets and in Times of Market Uncertainty. *Manag. Glob. Transit. Int. Res. J.* **2015**, *12*, 55–77.

54. Bartholdy, J.; Peare, P. Unbiased estimation of expected return using CAPM. *Int. Rev. Financ. Anal.* **2003**, *12*, 69–81. [CrossRef]
55. Hodnett, K.; Hsieh, H.-H. Capital Market Theories: Market Efficiency Versus Investor Prospects. *Int. Bus. Econ. Res. J.* **2012**, *11*, 849. [CrossRef]
56. Bakar, S.; Yi, A.N.C. The Impact of Psychological Factors on Investors' Decision Making in Malaysian Stock Market: A Case of Klang Valley and Pahang. *Procedia Econ. Financ.* **2016**, *35*, 319–328. [CrossRef]
57. Pawlonka, T.; Franc-Dąbrowska, J. Capital Cost and Capital Structure of Businesses in the Meat Industry Listed on the Warsaw Stock Exchange–a Discussion About the Problem of Valuation. *Probl. Agric. Econ.* **2018**, *357*, 80–102. [CrossRef]
58. Kozieł, D.; Pawłowski, S.; Kustra, A. Cost of Equity Estimation in Fuel and Energy Sector Companies Based on CAPM. *E3S Web Conf.* **2018**, *35*, 1–7. [CrossRef]
59. Stulz, R.M. Pricing capital assets in an international setting: An introduction. *J. Int. Bus. Stud.* **1984**, *15*, 55–73. [CrossRef]
60. Li, L.; Islam, S.Z. Firm and industry specific determinants of capital structure: Evidence from the Australian market. *Int. Rev. Econ. Financ.* **2019**, *59*, 425–437. [CrossRef]
61. Bruner, R.F.; Li, W.; Kritzman, M.; Myrgren, S.; Page, S. Market integration in developed and emerging markets: Evidence from the CAPM. *Emerg. Mark. Rev.* **2008**, *9*, 89–103. [CrossRef]
62. Chincarini, L.B.; Kim, D.; Moneta, F. Beta and firm age. *J. Empir. Financ.* **2020**, *58*, 50–74. [CrossRef]
63. Lin, X.; Palazzo, B.; Yang, F. The risks of old capital age: Asset pricing implications of technology adoption. *J. Monet. Econ.* **2019**, *115*, 145–161. [CrossRef]
64. Pedell, B. Market-Based Assessment of the Cost of Capital for a Rate-Regulated Firm. In *Regulatory Risk and the Cost of Capital: Determinants and Implications for Rate Regulation*; Springer: Berlin, Germany, 2006; pp. 163–191.
65. Xu, Z. Economic policy uncertainty, cost of capital, and corporate innovation. *J. Bank. Financ.* **2020**, *111*, 105698. [CrossRef]
66. Özcam, M. *An Analysis of the Macroeconomic Factors that Determine Stock Return in Turkey*; Sermaye Piyasası Kurulu, İlk Baskı; Capital Market Board of Turkey: Ankara, Turkey, 1997.
67. Zhu, B. The effects of macroeconomic factors on stock return of energy sector in Shanghai stock market. *Int. J. Sci. Res. Publ.* **2012**, *2*, 326–329.
68. Vo, M.T. Capital structure and cost of capital when prices affect real investments. *J. Econ. Bus.* **2020**, 105944. [CrossRef]
69. Zarzecki, D. Koszt kapitału, płynność i ryzyko–analiza sektorowa na rynku amerykańskim. *Pr. Nauk. Uniw. Ekon. we Wrocławiu* **2014**. [CrossRef]
70. Fama, E.F.; MacBeth, J.D. Risk, return and equilibrium: Empirical tests. *J. Polit. Econ.* **1973**, *81*, 607–636. [CrossRef]
71. Chan, L.K.; Lakonishok, J. Institutional trades and intraday stock price behavior. *J. Financ. Econ.* **1993**, *33*, 173–199. [CrossRef]
72. Stulz, R.M. Globalization, Corporate Finance, and the Cost of Capital. *J. Appl. Corp. Financ.* **1999**, *12*, 8–25. [CrossRef]
73. Mari, C.; Marra, M. Valuing firm's financial flexibility under default risk and bankruptcy costs: A WACC based approach. *Int. J. Manag. Financ.* **2019**, *15*, 688–699. [CrossRef]
74. Ceron, J. Going Distress on the WACC. *SSRN Electron. J.* **2014**, 1–56. [CrossRef]
75. Pratt, S.P.; Grabowski, R.J. *Cost of Capital in Litigation: Applications and Examples*, 647th ed.; John Wiley & Sons: Hoboken, NJ, USA, 2010.
76. Mari, C.; Marra, M. Valuing Firms Under Default Risk and Bankruptcy Costs: A WACC-Based Approach. *Int. J. Bus.* **2018**, *23*, 111–130.
77. Mądra-Sawicka, M. Differences in the Cost of Capital: The Case of Food Companies From Emerging and Developed European Economies. *Acta Sci. Pol. Oeconomia* **2020**, *19*, 49–56. [CrossRef]
78. Kuckshinrichs, W.; Ketelaer, T.; Koj, J.C. Economic analysis of improved alkaline water electrolysis. *Front. Energy Res.* **2017**, *5*. [CrossRef]
79. Estache, A.; Steichen, A.S. Is Belgium overshooting in its policy support to cut the cost of capital of renewable sources of energy? *Reflets Perspect. Econ.* **2015**, *54*, 33–45. [CrossRef]
80. Foltyn-Zarychta, M.; Buła, R.; Pera, K. Discounting for Energy Transition Policies—Estimation of the Social Discount Rate for Poland. *Energies* **2021**, *14*, 741. [CrossRef]
81. Steffen, B. Estimating the cost of capital for renewable energy projects. *Energy Econ.* **2020**, *88*, 104783. [CrossRef]
82. Painuly, J.P. Barriers to renewable energy penetration: A framework for analysis. *Renew. Energy* **2001**, *24*, 73–89. [CrossRef]
83. Tucki, K.; Orynycz, O.; Wasiak, A.; Swić, A.; Dybas, W. Capacity market implementation in Poland: Analysis of a survey on consequences for the electricity market and for energy management. *Energies* **2019**, *12*, 839. [CrossRef]
84. Korajczyk, R.A.; Levy, A. Capital structure choice: Macroeconomic conditions and financial constraints. *J. Financ. Econ.* **2003**, *68*, 75–109. [CrossRef]
85. Bachner, G.; Mayer, J.; Steininger, K.W. Costs or benefits? Assessing the economy-wide effects of the electricity sector's low carbon transition–The role of capital costs, divergent risk perceptions and premiums. *Energy Strateg. Rev.* **2019**, *26*, 100373. [CrossRef]
86. Duliniec, A. *Finansowanie Przedsiębiorstwa*; PWE: Warszawa, Poland, 2011.
87. Hammoudeh, S.; Li, H. Oil sensitivity and systematic risk in oil-sensitive stock indices. *J. Econ. Bus.* **2005**, *57*, 1–21. [CrossRef]
88. Pouraghajan, A.; Tabari, N.A.Y.; Ramezani, A.; Mansourinia, E.; Emamgholipour, M.; Majd, P. Relationship between cost of capital and accounting criteria of corporate performance evaluation: Evidence from Tehran Stock Exchange. *World Appl. Sci. J.* **2012**, *20*, 666–673. [CrossRef]
89. Sharma, A. Cost of Capital and Profitability Analysis (A Case Study of Telecommunication Industry). *J. Commer. Account. Res.* **2012**, *1*, 42–50.

90. Polzin, F.; Egli, F.; Steffen, B.; Schmidt, T.S. How do policies mobilize private finance for renewable energy?—A systematic review with an investor perspective. *Appl. Energy* **2019**, *236*, 1249–1268. [CrossRef]
91. Osmundsen, P.; Asche, F.; Misund, B.; Mohn, K. Valuation of international oil companies. *Energy J.* **2006**, *27*, 49–64. [CrossRef]
92. Zhu, Z.; Ji, Q.; Sun, L.; Zhai, P. Oil price shocks, investor sentiment, and asset pricing anomalies in the oil and gas industry. *Int. Rev. Financ. Anal.* **2020**, *70*, 101516. [CrossRef]
93. Al Mamun, M.; Sohag, K.; Shahbaz, M.; Hammoudeh, S. Financial markets, innovations and cleaner energy production in OECD countries. *Energy Econ.* **2018**, *72*, 236–254. [CrossRef]
94. Parvi, R. Analysis of companies of the energy sector based on an example of the companies quoted on the Warsaw Stock Exchange in Poland and their fair value. In Proceedings of the CER-International Scientific Conference, London, UK, 27–31 October 2014; pp. 33–36.
95. Yu-Ling Hsiao, C.; Lin, W.; Wei, X.; Yan, G.; Li, S.; Sheng, N. The impact of international oil prices on the stock price fluctuations of China's renewable energy enterprises. *Energies* **2019**, *12*, 4630. [CrossRef]
96. Węgrzyn, A.A. The Long Term Regulation Model as the tool in SME Energy Enterprise value management-initial research results presentation. *Perspect. Bus. Entrep. Dev. Econ. Manag. Financ. Syst. Eng. From Acad. Pract. Views* **2015**, *15*, 161–168.
97. Zhukov, P. The impact of cash flows and weighted average cost of capital to enterprise value in the oil and gas sector. *J. Rev. Glob. Econ.* **2018**, *7*, 138–145. [CrossRef]
98. Pérez-Arriaga, I. *Regulation of the Power Sector*; Springer: London, UK, 2013.
99. Vochozka, M. Effect of the economic outturn on the cost of debt of an industrial enterprise. *SHS Web Conf.* **2017**, *39*, 01028. [CrossRef]
100. Wang, L.K. The weighted average cost of capital and sequential marginal costing: A clarification. *Eng. Econ.* **1994**, *39*, 187–192. [CrossRef]
101. Ferreira Savoia, J.R.; Securato, J.R.; Bergmann, D.R.; Lopes da Silva, F. Comparing results of the implied cost of capital and capital asset pricing models for infrastructure firms in Brazil. *Util. Policy* **2019**, *56*, 149–158. [CrossRef]
102. Lohani, V.J. Analysis of EVA for SBI and HDFC with the impact assessment of WACC and dividend over share holder value creation. *Int. J. Res. Manag. Sci. Technol.* **2016**, *4*, 4132–4141.
103. Garcia, C.S.; Saravia, J.A.; Yepes, D.A. The weighted average cost of capital over the lifecycle of the firm: Is the overinvestment problem of mature firms intensified by a higher WACC? *Corp. Board Role Duties Compos.* **2016**, *12*, 96–103. [CrossRef]
104. Molano, K.R.A.; Tanhui, K.K.; Ymasa, J.P.P.; Zambales, M.S. *The Impact of tax Avoidance, Corporate Governance, and Audit Firm Affiliation on the Weighted Average Cost of Capital (WACC) of Publicly Listed Firms in the Philippines, Malaysia, and Indonesia, Malaysia, and Indonesia*; De La Salle University: Manila, Philippines, 2015.
105. Jaworski, J.; Czerwonka, L.; Mądra-Sawicka, M. Determinants of capital structure: Evidence from Polish food manufacturing industry. *Ger. J. Agric. Econ.* **2019**, *68*, 45–56.
106. Gertler, M.; Hubbard, R.G. Corporate Financial Policy, Taxation, and Macroeconomic Risk. *RAND J. Econ.* **1993**, *24*, 286. [CrossRef]
107. Shadab, M.; Sattar, A. Cost of Capital–The Effect to the Firm Value and Profitability; Empirical Evidences in Case of Personal Goods (Textile) Sector of KSE 100 Index. *J. Poverty J.* **2015**, *17*, 24–28.
108. Ain, Q.; Jan, S.U.; Rafiq, M. Macroeconomic Factors on Capital Structure Decisions of Firm-Evidence from a Developing Country. *J. Bus. Econ. Rev.* **2011**, *3*, 64–90. [CrossRef]
109. Ivascu, E.V.; Barbuta-Misu, N. Influences of the Capital Structure and the Cost of Capital on Financial Performance. Case Study on ENGIE Group. *Risk Contemp. Econ.* **2017**, *21*, 304–320.
110. Bozec, Y.; Laurin, C.; Meier, I. The relation between excess control and cost of capital. *Int. J. Manag. Financ.* **2014**, *10*, 93–114. [CrossRef]
111. Michalak, A. The cost of capital in the effectiveness assessment of financial management in a company. *Oeconomia Copernic.* **2016**, *7*, 317. [CrossRef]
112. Sharfman, M.P.; Fernando, C.S. Environmental risk management and the cost of capital. *Strateg. Manag. J.* **2008**, *29*, 569–592. [CrossRef]
113. Jylha, P.; Ungeheuer, M. Growth Expectations out of WACC. *SSRN Electron. J.* **2020**. [CrossRef]
114. Valaskova, K.; Gregova, E. Modelling of Costs of Capital in Slovak Enterprises. *Ekon. Spektrum* **2017**, *11*, 76–89. [CrossRef]
115. Frank, M.Z.; Shen, T. Investment and the weighted average cost of capital. *J. Financ. Econ.* **2016**, *119*, 300–315. [CrossRef]
116. Habib, A. Information Risk and the Cost of Capital: Review of the Empirical Literature. *J. Account. Lit.* **2006**, *25*, 127–168. [CrossRef]
117. Budhathoki, P.B.; Rai, C.K. The Impact of the Debt Ratio, Total Assets, and Earning Growth Rate on WACC: Evidence from Nepalese Commercial Banks. *Asian J. Econ. Bus. Account.* **2020**, *15*, 16–23. [CrossRef]
118. Farber, A.; Gillet, R.L.; Szafarz, A. A General Formula for the WACC. *Int. J. Bus.* **2006**, *11*, 211–218. [CrossRef]
119. Lilford, E.; Maybee, B.; Packey, D. Cost of capital and discount rates in cash flow valuations for resources projects. *Resour. Policy* **2018**, *59*, 525–531. [CrossRef]
120. Rady, A.; Meshreki, H.; Ismail, A.; Núñez, L. Variations in Valuation Methodologies and the Cost of Capital: Evidence from MENA Countries. *Emerg. Mark. Financ. Trade* **2019**, *55*, 2106–2123. [CrossRef]

energies

MDPI

Article

The Influence of Opencast Lignite Mining Dehydration on Plant Production—A Methodological Study

Benedykt Pepliński [1,*] and Wawrzyniec Czubak [2]

1 Department of Law and Enterprise Management in Agribusiness, Faculty of Economics Sciences, Poznan University of Life Sciences, 60-637 Poznan, Poland
2 Department of Economics and Economic Policy in Agribusiness, Faculty of Economics Sciences, Poznan University of Life Sciences, 60-637 Poznan, Poland; czubak@up.poznan.pl
* Correspondence: peplinski@up.poznan.pl; Tel.: +48-61-848-71-09

Abstract: In many circles, brown coal continues to be viewed as a cheap source of energy, resulting in numerous investments in new opencast brown coal mines. Such a perception of brown coal energy is only possible if the external costs associated with mining and burning coal are not considered. In past studies, external cost analysis has focused on the external costs of coal burning and associated emissions. This paper focuses on the extraction phase and assesses the external costs to agriculture associated with the resulting depression cone. This paper discusses the difficulties researchers face in estimating agricultural losses resulting from the development of a depression cone due to opencast mineral extraction. In the case of brown coal, the impacts are of a geological, natural-climatic, agricultural-productive, temporal, and spatial nature and result from a multiplicity of interacting factors. Then, a methodology for counting external costs in crop production was proposed. The next section estimates the external costs of crop production arising from the operation of opencast mines in the Konin-Turek brown coal field, which is located in central Poland. The analyses conducted showed a large decrease in grain and potato yields and no effect of the depression cone on sugar beet levels. Including the estimated external costs in the cost of producing electricity from mined brown coal would significantly worsen the profitability of that production.

Keywords: external cost; opencast lignite; plant production; depression funnel; cereals; sugar beet; potatoes

Citation: Pepliński, B.; Czubak, W. The Influence of Opencast Lignite Mining Dehydration on Plant Production—A Methodological Study. *Energies* **2021**, *14*, 1917. https://doi.org/10.3390/en14071917

Academic Editor: Nikolaos Koukouzas

Received: 25 February 2021
Accepted: 25 March 2021
Published: 30 March 2021

1. Introduction

Lignite is widely used in the generation of electricity and is considered to be an abundant resource. World reserves of low-rank coals (LRCs) such as lignite amount to 200–280 billion Mg (Gt) and account for 20–25% of global coal reserves [1–4]. With this background, the utilization of lignite for energy production is expected to remain a common practice in the decades to come since the availability of lignite is considerable in many countries of Europe and the world [5]. It is estimated that global production of lignite could increase from the present level, which is approximately 1.05 Gt [6], and reach its peak of 2–3 Gt in the second half of the 21st century [7,8]. Lignite is accumulated in fairly shallow deposits, therefore, it is most often mined in open pits, whereas in the case of hard coal, especially in Europe, underground mines are more common. In the future, new opencast mines of lignite and hard coal will be launched, especially because open-pit coal mining is perceived as a cheaper option and the one that enables using modern technologies [9]. The development of mining will still be promoted by the fairly common belief, especially among the countries with abundant lignite resources, that lignite is cheap, if not the cheapest, source of energy [10–13]. From a global perspective, lignite is considered an energy resource with substantial security of supply; no other fossil fuel is so easily available with such certainty for the next decades [1]. In the countries with abundant fossil fuel resources, exploitation of more coal deposits is carried out or scheduled; also, new coal-fired and lignite-fired power plants are being put into operation or are planned to be launched. This applies particularly to China, India, Turkey, Vietnam, Indonesia, Bangladesh, Japan, South Africa and the Philippines [14,15].

In Poland, lignite is also perceived as a cheap energy source [16–19]. This is reflected in the draft of the latest Energy Policy of Poland [20], in which three deposits with 1.8 billion tons of lignite are assumed as a long-term energy reserve. In the case of the Złoczew deposit, with resources of 611 million tons, an environmental consent decision for the proposed open-pit mine was issued on 28 March 2018. The only requirement missing to launch the open-pit mine is the exploitation concession [21].

In the context of sustainable economic development and the concept of the European Green Deal, the perception of coal—as a cheap energy source—is only possible if the external costs of mining and combustion of coal are not taken into account. Those costs have been the subject of many analyses over the past years. One of the first attempts to value the external effects of energy consumption was made by Hohmeyer in 1988 [22]. The works by Rowe et al. [23] and Lee et al. [24] called the RFF/ORNL research, are recognized as the first comprehensive elaborations on the external costs in a fuel cycle. They focused on the entire fuel cycle of different kinds of power plants. In 1999, Krewitt et al. [25] used the bottom-up method to determine the average external costs connected with the generation of energy in the mines in Germany and Europe. The authors used the EcoSense model which originates from the ExternE methodology, as well as the CORINAR database (Core Inventory Air Emissions) [4]. They focus on the external costs associated with the combustion of coal. For example, the ExternE Model with EcoSense software package provides air quality and impact assessment models along with a database (population, use of land, agricultural production, buildings, and materials, etc.) that contains relevant input data for the whole of Europe [26]. The model determines a range of factors affecting human health, buildings, biodiversity, and crop yields using concentration-response functions, e.g., to SO_2, NO_x and their aerosols, heavy metals, and solid particles (PM2.5, PM10) [27–30].

Despite the claims that external costs of energy production have been comprehensively estimated, no analyses of external costs caused by geological damage have been performed so far [31]. There are also no studies that analyze the external costs in agriculture and forestry caused by a cone of depression, which is created as a result of draining coal deposits during their exploitation. Additionally, in the review of 20 different external cost analyses from various regions of the world, no study took account of the external costs incurred by agriculture, which may provide evidence of the above. There are only a few on the losses caused by refraining from farming as a result of the land being taken over by open-pit mines and power plants along with their supporting facilities [22,32]. Thus, there is no reflection on the full spectrum of impact, which is required to assess the sustainability of development and the European Green Deal. This study is an attempt to fill that knowledge gap and to start a discussion on the methodology of calculating agricultural external costs associated with the existing cones of depression created around open pits. In the first papers related to this subject [33–35], there were many simplifications made regarding the level of yield decline, i.e., the same level of yield decline was assumed for the entire period of the impact of open casts. By using only one period for estimating the external cost of a fall in yields, there is a risk of a significant over- or underestimation of external costs. Furthermore it was calculated by comparing the yield levels in the Wielkopolskie Voivodship before the launch of the open-cast mine, however, 30 years later, in the eastern part of the Wielkopolskie Voivodship, a multi-pit lignite open-cast mine is now operational. This makes it possible to maintain the condition of comparability of factors influencing the level of yields based on the vicinities being compared [33–35].

The problem of exploiting more deposits is also important in the context of the law of entropy discussed by Georgescu-Roegen, who stated that energy tends to degrade irreversibly to increasingly poorer qualities, i.e., from low entropy for valuable natural resources to high entropy for worthless waste and pollution. He argued that the use of exhaustible resources would result in the inevitable collapse of the world economy, leading to human extinction [36].

The study aims to assess the assumptions and methodology used to estimate the external costs in agricultural crop production located in the vicinity of open-cast lignite mines and the costs associated with the cones of depression created around the opencast pits. In the context of the above-mentioned generalizations, simplifications or the omission of agricultural external costs can achieve only results of a scientific and research novelty. In the study, an exemplary analysis is performed for the Konin-Turek lignite basin located in central Poland. The obtained research results apply to Polish conditions, however, the scientific universality of the study allows that the assumptions and methodology applied here can be adapted to the analysis of external costs incurred by agriculture in any region of the world and for any large-scale open-cast mines.

2. External Costs in Agriculture and the Difficulties Associated with Their Estimation

The variety of sources from which energy can be generated requires the knowledge of the actual production costs that make it possible to properly allocate the resources owned. Therefore, to make proper comparisons, it is necessary to take into account all types of expenses, not only investment, fuel, maintenance, and operation costs of the power plants, but most importantly, the external costs of electricity generation [37,38]. External costs are those incurred by third parties and future generations to produce energy, rather than the expenditures of direct recipients and providers of electricity [39,40]. Incorporating those costs in electricity market prices would contribute to the use of more modern and cleaner energy sources [28,41,42], while the most harmful technologies would be the first ones to be forced out of the market.

In the case of electricity generation from black coal and lignite, apart from the well-researched external costs associated with coal combustion and air pollution, in particular, the costs related to human health [43], there are also external costs associated with the exploitation operations and land reclamation after coal mining, in both underground and open-pit mines [44].

Open-cast mining is associated with the absolute and long-term necessity to drain the coal deposits to the bottom of the lowest levels of the exploited coal seams, which leads to the creation of the above-mentioned cones of depression. There are two types of cones of depression: discharge cone of depression and pressure relief cone of depression. The first one is a gravitational lowering of the groundwater table (the most critical issue in agriculture and forestry) within the area surrounding the coal deposits. In a vertical cross-section, the shape of the cone of depression resembles a funnel-shaped curve, i.e., the water table near the edge of the pit rapidly rises and it goes up more slowly as the distance grows. The Polish law requires the investor to define the estimated area around the open-pit mine where the water table will permanently be lowered by at least one meter creating an area of depressions. In the case of lignite open-cast mines, the range of the cone of depression usually varies from a few to several kilometers, starting from the edge of the open pit, and it has the shape of an ellipse. The actual impact area, i.e., the area where the water table is lowered down, is much larger and reaches up to several dozen kilometers, starting from the edge of the open pit.

In turn, the pressure relief cone of depression, which is much larger than the area of depressions, is the territory where groundwater pressure is reduced. The changes in water pressure in deeper aquifers caused by hydrogeological cracks can lower the groundwater and surface water levels because it triggers a local outflow of water to deeper ground layers, it can also reduce or generate a loss of supply of subsoil resources with water from deeper aquifers [45,46].

Various difficulties that researchers face when estimating the losses in agriculture due to the development of cones of depression as a result of open-cast mining of minerals—lignite in this case—are of geological, nature-climatic, agro-industrial, temporal-spatial nature and result mainly from the multiplicity of interactive factors. Those factors have an impact on one another and are also interdependent in many ways.

The geological factors are related to the size of the cone of depression and the changes in the water levels of subsurface water resources, which is very important in the case of agriculture. Not only the size and extent of the drainage is a crucial issue but also the pace of restoring water conditions after the drainage of the deposit is completed. The key geological factors include:

- Drainage depth—with the increase of the depth of the open pit, the area of a cone of depression becomes larger,
- Drainage period—with the increase of the drainage time, the area of a cone of depression also increases,
- Location of the opencast in the catchment area, its size and the directions of inflow of groundwater—in simplified terms, if the open-pit mine is situated in a valley, the drained area increases but the water relations are restored faster. The location of an open-cast mine on a water parting reduces the drainage area, however it is much more difficult and it takes longer for the water level to get restored after the drainage is completed because the runoff of water from the areas situated higher up is restricted,
- The geological structure of drained areas such as the shape and the direction of buried valleys, abundance in water, tectonic faults, hydrogeological cracks, the thickness of geological layers which affect the conditions of supply, circulation, and drainage of groundwater—those factors are very specific for every open cast, however, the location of thicker and impermeable layers closer to the surface reduces the risk and the area of drainage of subsurface and surface water resources,
- The amount of rainfall and surface water supply—with the increase of the abundance of rain and the level of subsurface water, the impact of the mine on the areas located further away from the open cast decreases. The increase in the share of drained agricultural land reduces the permeation of water into deeper soil layers, especially in the period from late autumn to early spring. High variability of the level of precipitation, both seasonal and during individual years, affects the changes in the level of groundwater, which makes it difficult to determine the actual impact of open-pit mines on water conditions,
- Local conditions, e.g., impermeable formations that create areas of the perched water table, hydrogeological cracks that lower the level of subsurface water below the standard level of the area,
- The initial (primary) level of groundwater, which in the case of peripheral areas of the impact land means that the mine will affect the areas with higher water levels while it will not have any influence on the surrounding areas with lower water tables, even those located further away.

In the case of open-pit mines, where there is usually a shift of the mining front and the already exploited areas of the open pit are backfilled, there is also a shift of the area of cones of depression. The dynamics of these changes mean that despite the use of more advanced econometric models, taking into account the actual conditions of the mining industry's impact on the environment, the range of a cone of depression cannot be determined [47]. Therefore, already at the very beginning of the research, there is a large obstacle that makes it difficult to estimate losses in agriculture resulting from the operation of open-pit mines.

Consequently, the amount of water that is pumped out of individual open pits varies considerably. In the case of lignite open-pit mines in Germany, an average of 6.3 m^3 of water is pumped out along with each ton of coal [48]. In Poland, over the period from 1945 to 2017, an average of 6.8 m^3 of water was pumped out per 1 ton of mined lignite, but in the second decade of the 21st century, this ratio was nearly 8.0 m^3, on average [16,18]. There were, however, large discrepancies between individual coal deposits. In the case of small deposits in the Konin-Turek Basin, the ratio exceeded 42 m^3 of water/Mg of coal, and regarding the deposits in Turów, it was 2.2 m^3 of water/Mg of coal [16].

The agro-industrial factors are also highly changeable. In the context of the studied subject, precipitation is particularly important and, as mentioned above, it is a major source of groundwater supply. Precipitation is also one of the most important factors that determine crop yield [49,50]. It is not only the overall amount of precipitation that is important but also its distribution over time, especially during the growing season. There are certain temperatures, a range of which in a given place and during a particular period is quite predictable, which is also important. In general, the observed systematic increase in global temperatures has a negative impact on agriculture. Firstly, it increases evaporation which reduces the amount of rainwater available to crops and in deeper soil layers which leads to lowering the groundwater level and consequently creates the risk of soil drought. Secondly, it causes a decrease in the agricultural efficiency of precipitation [51,52]. This efficiency is also deteriorated by a change in the nature of precipitation—from continuous to convectional rain, [53–55] and an increase in torrential rains. In the research conducted for Germany, for instance, it was estimated that every time temperature increased by 1 degree, the amount of heavy rain increased by 6.5% [56]. All of the above, in turn, lead to an increase in the dependence of crops on the level of groundwater [57,58]. However, in cooler regions such as northern Europe, climate change can have a positive impact on the yield mainly by extending the growing season [59,60].

Agro-industrial difficulties stem from, first of all, the biological nature of open space production, which is affected by many factors. Those, in turn, influence the final result, i.e., the crop the farmer wants to harvest. From the point of view of the analyzed subject, the nature-climatic factors that should be mentioned include the amount and distribution of precipitation (including snow cover), temperature distribution, groundwater levels, types and quality of soil, the topography of the land, and the length of the growing season. There are also economic factors which include, e.g., the level of agricultural development, agrarian structure, production intensity, availability of techniques and technologies, and quality of human capital [61].

Launching a new open-cast mine and lowering of groundwater level leads to a change in the conditions of production, which also affects the abundance of the yield. Sensitivity to water deficit is related to the type of crop plants [62] and the potential yield [63,64]. However, as has been indicated by numerous studies on the impact of groundwater levels on crop yield, it is difficult to estimate the extent of production losses caused by reduced water availability. Vereecken et al. [65] for instance, state that the impact of soil texture on groundwater is highly nonlinear, that is why simple texture-yield dependence is difficult to assess and understand. This is further complicated by the fact that soils with the same textural parameters can have drastically different structures and therefore water retention behaviors due to the variability in soil compaction, organic content, or aggregation [61,66]. In other studies, the key role of groundwater during dry periods is pointed out, which, if available, can account for 50–100% of total water use in the case of many crops [57,67–73]. Since the amount of used water grows in direct proportion to the increase in yield [74], in the areas of high yield where there has been a substantial reduction in the water table, crop losses will be significant, especially during the years of lower than average rainfall. The difficulties in estimating the losses are further complicated by the fact that in different parts of the same field or at different times during the growing season, depending on local conditions, differences in yield loss can vary considerably.

Additionally, the shortage of data and the difficulty in measuring and analyzing the data on groundwater and soil texture makes it necessary to use, for many studies, topographic features instead, such as the gradient of the slope or the elevation above sea level, which can cause discrepancies in the obtained results such as permanent fluctuations in yield within the same field [61,75–77]. Again, in this case, assessing the impact of changing groundwater levels on the yield is difficult because despite a fairly large number of groundwater level monitoring instruments located within the area of an open-pit it is possible to observe the trends only for a specific spot.

A significant challenge in the process of valuing the external costs of agriculture associated with open-pit mining is time. This is primarily related to the period of several decades that is required to design and launch an open-pit mine, the exploitation and the restoration of water relations.

Designing a large open-cast mining plant is a multifaceted, complicated, and time-consuming process [78]. In the case of the Tagebau Garzweiler lignite opencast mine in the Rhineland basin, it took 30 years from the decision to start a preparatory work until the proverbial "first shovel" hit the ground [35]. The process of de-watering begins with the construction of an access trench and, depending on its depth and width, it usually starts several years before the commencement of extraction of the raw material of the deposit. The main factor that determines the period of mining of deposits is the volume of the resources to be extracted and the demand for that raw material. Once the process of mining is complete, the reclamation process begins. In the majority of cases of open-pits, in the lowest parts of the excavation, water reclamation is most popular and it involves flooding the area and creating a water reservoir. The remaining area is covered with forests or it undergoes a process of recreational reclamation. Agricultural reclamation is not performed by many countries or it is implemented to a small extent because it is expensive. Farming on the reclaimed land can only be successful if the texture of topsoil is maintained based on scientific standards [79]. Moreover, the restoration of fully balanced and productive soils is a rather difficult task that lasts several decades [80]. The reclaimed areas, that represent a group of urban soils, often differ significantly from naturally formed soils in terms of many properties: lack of accumulation potential, low content of nutrients, yield instability, which makes them economically unattractive for farmers for many years [33,81,82]. In Poland, out of the three lignite basins, land reclamation is performed only in the Konin-Turek one, what is more, the share of land reclamation towards agriculture is decreasing over the subsequent years [83,84]. More and more often, industrial parks, and recently also photovoltaic power stations, are created in reclaimed post-mining areas to mitigate the social consequences of mining activities [81]. An example here is the activities of ZE PAK SA (Power Plant Group Pątnów-Adamów-Konin joint-stock company), which, in August 2021, plans to launch a 70 MWp solar farm on the reclaimed area of the Koźmin mine [85].

The process of restoring water conditions in the areas affected by a cone of depression is long and depends on many factors. Though it is one of the most important components in groundwater studies, recharge is also one of the least understood, largely because recharge rates vary widely in space and time, and rates are difficult to directly measure [86,87]. In the case of Polish lignite open-pit mines, it is estimated that the process of restoring water conditions will be equal to, and sometimes even longer than, the time it took to drain the coal bed [88]. This requires estimating the external costs in agriculture over several dozen or even 100s of years. It is also important to be aware that the full restoration of water relations in the majority of open pits will never be possible. This is related to water reclamation of the lowest-lying parts of excavation, where the water table level can be located several dozen meters below the original ground level, which is typical in the case of most open pits.

Over such a long period, huge changes take place, not only in the crop yield, which is the result of technological modifications (i.e., management practices and crop varieties) and meteorological (mainly precipitation, its structure and temperature, which is particularly important in drought-prone areas). There are also changes in the amount of agricultural land and its structure, the structure of sown areas, prices of crop raw materials and means of production, time value of money, and the volume of livestock production, which may, to a greater or lesser extent, determine the demand for fodder and the share of crop production intended for sale.

Launching an open-cast mine and the decrease in the yield caused by the impact of the mine on water resources leads to a decline in the profits of the farms operating in this area. Over time, the mine has an increasing impact on the investment capacity of those farms, which leads to large discrepancies in the level of agricultural development compared to the neighboring areas, to gradual resignation from keeping livestock and running agricultural production. In extreme cases, it can lead to a local collapse of agriculture and related industries (pre- and post-production). An example here may be the midwestern regions of the United States, [89] or the intensively irrigated North China Plain [90] and Syria [91], where agriculture has collapsed over vast areas due to overexploitation of groundwater resources.

The main reasons for spatial differences in the crop yield are technological changes (i.e., management practices and variations), meteorological factors (mainly precipitation in the areas of dry farming), types of soil and the amount of stored groundwater available to crops, and interactions among all the above-mentioned factors [92,93].

3. Materials and Methods

Calculations of external costs may be carried out at different points in the life cycle of a mining project. The optimal time is the period when a decision is being made whether to conduct (or not conduct) the mining of a given deposit as it makes it possible to estimate the actual cost of the project, not only for the entrepreneur, but also with regard to the environment, and abandon projects that do not increase the welfare of the whole society.

In order to fulfil Georgescu-Roegen's principle of absolute totality [36] for the processes associated with opencast lignite mining, it is necessary to determine the temporal and spatial extent of the impact of the opencast in three areas: coal extraction, the impact of the depression funnel created in the coal mining process on agriculture, and the impact on the rest of nature, including humans. In this study, only the second area will be analyzed.

When calculating the full external cost associated with open-pit mining of raw materials for agriculture, external costs stemming from the following must be taken into consideration:

- Use of agricultural land for an open pit, an external dump, and the necessary accompanying infrastructure, e.g., a power plant, conveyor belts, access roads, etc. (the term "open pit area" will be also used later in this paper),
- The occurrence of areas with lowered groundwater level (the term "cone of depression area" will be also used later in the paper),
- Changes in animal populations in the area impacted by the open pit (it will not be analyzed in this paper).

When calculating the external costs in crop production, one must take into account the fact that in the case of larger deposits, the exploitation of which lasts from around a dozen to several dozen years, the exclusion of land from agricultural production is gradual. For this reason, it is necessary to calculate the average agricultural area excluded from agricultural production, which can be done using the following formula:

$$Ao_{AL} = \frac{\sum_{i=1}^{n} Ac_1 \times \frac{S}{100} + Ac_2 \times \frac{S}{100} + \ldots + Ac_n \times \frac{S}{100}}{t}$$

where:

Ao_{AL}—stands for the average area of agricultural land excluded from agricultural production in the area of the open-pit mine (ha *AL*),
Ac—stands for the surface allocated for the open-pit mine, the external dump, and the necessary infrastructure, in particular years (ha),
S—stands for the share of agricultural land in the total area of the analyzed territory (%),
t—stands for the period of the impact of the open-pit mine, covering the period from the first exclusion of agricultural land until the completion of reclamation, or the entire period of the open-pit mine exerting its impact (years).

If the level of average exclusion of grounds for the open-pit mine is known, the average area of agricultural land excluded from agricultural production can be calculated from the formula:

$$Ao_{AL} = Ac_t \times E_{AL} * \frac{S}{100},$$

where

Ac_t—stands for total surface allocated for the open-pit mine, the external dump, and the necessary infrastructure (ha),
E_{AL}—is an indicator of the average exclusion of area for open-pit mining (%).

The information on the rate at which areas are being taken up by the open-pit mine is usually provided by the investor, or the indicator for similar open-cast mines can also be used. In the case of smaller open-pit mines and a shorter mining time, the indicators will be higher and will oscillate around 80–90%; in the case of larger open-pit mines with longer mining time, they may reach approx. 60% [34,94].

The external costs for the agriculture located in the area of the open-pit mine, the external dump, and the necessary infrastructure can be calculated from the formula:

$$Ec_o = \sum_{i=1}^{n} Ao_{AL} \times \frac{Sp_i}{100} \times Yo_i\text{xt} \times p_i \times P_i,$$

where

Ec_o—stands for the external cost in the area of the open-pit mine, the external dump, and the necessary infrastructure ($, €),
Sp_i—stands for the average share of the i-th crop in the structure of agricultural land (%),
Yo_i—stands for the yield of the i-th crop in the area of the open-pit mine ($\text{t} \times \text{ha}^{-1}$),
p_i—stands for the average selling price of the i-th crop ($, € $\times\ \text{t}^{-1}$),
P_i—stands for the profitability of the production of the i-th crop (%). The average profitability of the whole crop production can also be used in the calculations, but in such a case, it is necessary to use this value for all analyzed crops.

In the case of calculating external costs in the cone of depression area, the first step should also be to determine the average area of the cone of depression. At the initial stage, when drainage of the open-pit mine is commenced, the cone of depression develops; subsequently, together with the movement of the mining front and the storage of the mined part of the open-pit, the cone of depression moves as well, and after the completion of the drainage, water relationships, as well as the area of the cone of depression, are slowly restored. The average area of farmland within the cone of depression area can be calculated from the following formula:

$$Ad_{AL} = \frac{\sum_{i=1}^{n} Af_1 \times \frac{S}{100} + Af_2 \times \frac{S}{100} + \ldots + Af_n \times \frac{S}{100}}{t},$$

where

Ad_{AL}—stands for the average area of agricultural land (UR) within the area of the cone of depression (ha UR),
Af—stands for the area of the cone of depression in subsequent years.

Due to the fact that it is virtually impossible to realistically determine the area of the cone of depression in particular years, it is necessary to include the estimates of an average cone of depression area, which are uncertain.

External costs within the area of the cone of depression can be calculated from the following formula:

$$Ec_f = \sum_{i=1}^{n} Ad_{AL} \times \frac{Sf_i}{100} \times Yf_i \times t \times p_i \times P_i * \frac{cl_i}{100},$$

where

Ec_f—stands for the external cost in the area of the cone of depression,
Sf_i—stands for the share of the i-th crop in the structure of agricultural land in the area of the cone of depression (%),

Yf_i—stands for the yield of the *i*-th crop in the area of the cone of depression, in the period where the cone of depression does not exert impact (t × ha^{-1}),
Cl_i—stands for the estimated average loss of yield for *i*-th crop (%). For crops where it is not possible to estimate losses, one may use an average weighted level of losses, calculated from losses incurred in crops for which the parameter is known. The average loss in yield for the entire crop production can also be used in calculations, but in such a case, it is recommended that the value be used for all analyzed crops. The amount of lost yield when calculations are based on the level of yield that does not take into account decreased yield due to the cone of depression (mainly ex-ante analyses) can be calculated from the following formula:

$$Cl_i = 100 - \frac{Yfd_i}{Yf_i} * 100$$

When calculations are based on the level of yield taking into account lower yields due to the cone of depression (mainly ex-post analyses), the following formula can be used:

$$Cl_i = \frac{Yf_i}{Yfd_i} \times 100 - 100,$$

where
Yfd_i—stands for the yield of the *i*-th crop in the area of the cone of depression, in the period where the cone of depression exerts impact.

In the case of crops for which there are no data on yields, marketing prices, and/or prices of production profitability, one may calculate the value of external costs for crops for which full data are available, and estimate the external cost for other crops with the use of the following formula:

$$Ec_o = \sum_{i=1}^{n} Ad_{AL} \times \frac{sf_o}{100} \times V \times t \times P_i \times \frac{Cl_i}{100},$$

where
Ec_o—stands for the external cost in the area of the cone of depression,
Sf_o—stands for the average share of other crops in the structure of agricultural land (%),
V—the average value of crop production sales ($, € × ha^{-1}).

A practical simplification that allows omitting the process of estimating the area of a depression funnel is to use statistical data available for administrative units, e.g., counties or regions. In this case, the estimated yield losses will show the average yield loss for the administrative units concerned, i.e., both the areas most affected by the depression funnel and the areas with smaller or even no impact. However, care should be taken to ensure that the proportion of areas not affected by the impact of open pits is as small as possible. In this case, the Ad_{AL} area will cover farmland within these administrative units.

The total E_c external costs constitute the sum of the costs from the area of the open-pit mine and the cone of depression:

$$Ec = Ec_o + Ec_i$$

Both in the area of the open-pit mine and in the area of the cone of depression, the average cultivation area of individual crops can be calculated using the structure of sown areas in the analyzed territory in the period preceding the analysis. Alternatively, one may take into account the trends into the structure of sown areas from a longer period of time, however, it is not certain whether the trends will continue as the structure of sown area constitutes a response to changes in the profitability of particular crops.

Proper determination of the level of yield is important in the context of cost estimation. In recent decades, the development of crop cultivation (which is providing increasingly prolific types of cultivated crops), crop technology, as well as an increase in the level of mineral fertilization and the development of technological advice, have contributed to a significant increase in the level of yields in most regions of the world. However, there are very large regional differences.

In countries with a high level of yields the possibilities of their further development are already quite limited, especially in the context of the tendency to limit the use of mineral fertilizers (especially yield-forming nitrogen fertilizers) and reduce the use of plant protection products. It particularly pertains to EU countries in which, according to the latest plans of the European Commission, the use of mineral fertilizers and pesticides is to be reduced by 2030, and the use of the most dangerous pesticides is to be reduced by half [95]. It seems reasonable to use the current level of yields in calculations concerning countries of Western Europe.

However, in countries and regions where yields are poor (especially in comparison with countries characterized by similar natural and climatic conditions with the highest average yield), it is reasonable to prepare a yield forecast. To this end, one may use trends in the level of yielding of the most important crops cultivated in the area of the open-pit mine, for which the external costs are estimated. If the period of the impact of the open-mine pit is estimated for, e.g., 60 years, it is reasonable to include in the forecast the increase in yields from half of the period, that is the changes in yields from the last 30 years. A comparative analysis based on the method of spatio-temporal analogy of agricultural development can also be performed by simulating the development of agriculture in the open pit region on the example of agricultural development in a country with similar natural and climatic conditions. Considering country "X", which 30 years earlier was at the level of agricultural development close to the current level of agricultural development in the open pit area, it can be assumed that in 30 years, the assessed region will achieve production results similar to those currently present in country X. The average results from three variants can also be taken into account: the current yield level, the yield level estimated based on the trends, or the level based on spatio-temporal analysis [35].

To avoid overestimations and underestimations associated with long-term price forecasting, it seems optimal to include current prices in the calculation, using average prices from the last 3–5 years or from a period corresponding to the length of the price cycle for a given product, e.g., 4 years, which corresponds to the length of the pork cycle and the length of the cycle in the grain market [96].

The profitability of agricultural production, which is strongly correlated with purchase prices, can be estimated similarly. It is also justified to take into account the fixed costs for the profit generated on farms, as reducing the scale of production will cause a deterioration in the use of the existing fixed assets, and to adjust it to the reduced scale of production takes a long time. Taking Polish conditions into account, it is reasonable to assume the profitability level adjusted for fixed costs at 25% [35].

The last factor, but at the same time the most important one, taken into account when estimating external costs caused by the cone of depression created during opencast mining, is the decrease in the crop yield. Since the analysis of the external costs of a given open-cast mine should be performed before the commencement of the operation, it seems optimal to conduct an external cost analysis based on the studies conducted for the mines which are already closed and reclaimed or several dozen years old, where the negative effects of the cone of depression have already fully developed. In Poland, this criterion is best met by the Konin-Turek lignite basin, where large-scale mining began in 1955 and is due to be completed by 2030. As the level of the yield in individual years is characterized by a high dependence on weather conditions, primarily on rainfall, the amount and distribution of which may significantly differ between various regions and over individual years, therefore, to reduce the level of variability of the yield conditioned by weather factors, it is advisable to adopt averaging, e.g., of 5 years, which was also done for the analyses of this study.

Due to the multitude of interacting factors affecting the decrease in the level of the yield as a result of the occurrence of the cone of depression, the estimation of the losses based on the level of the yield obtained only in the area of the impact of open-cast mines is associated with high uncertainty. Therefore, a comparative analysis of the crops of grain, sugar beet, and potatoes in the area of the influence of an open pit and the yields of those crops obtained in neighboring areas will be conducted. The choice of those particular crops was dictated by their dominant role in the sown areas. Furthermore, the above-mentioned crops can be considered indicative in the context of estimating the dependency of the yield on lowering of groundwater table, as they differ in: root system, soil requirements, and demand for water during the growing season. The analysis will be conducted in two ways. First, an analysis of the yield at the voivodship level will be performed (in the analysis conducted for this work, the administrative division will be used which was in force in the years 1975–1998, when Poland was divided into 49 voivodships). The time range for this scenario will be 1956–1997. Because over the years 1956–1974, in Poland, there was a three-tier administrative division in force (voivodships—regions, poviats—districts, and gminas—local authorities), it was necessary to calculate the average yield in the area corresponding to the area of voivodships for the division in the period 1975–1998, based on the data on individual powiats. To assess the yield of a certain voivodship, poviats were taken into account, which, in 1975, were part of a given voivodeship; the weighted mean was calculated. Second, an analysis at the district level will be conducted, however, due to data availability, it will cover only the years 1960–1973. Despite the short period of the analysis, it allows determining whether, and to what extent, the decrease in the yield in the immediate vicinity of the open pit is different from the areas located further away, and it helps to assess the range of an impact of the open-cast mine.

For the above-mentioned analyses, Statistics Poland data was used, such as Statistical Yearbook of Voivodships, Statistical Yearbook of the Regions, etc. [97–107]. To determine the impact of multi-pit lignite mining on the level of the yield, the districts were divided into 5 groups:

- The first group, "up to 20 km", includes the districts of Konin and Turek, where brown lignite open-pit mines are located,
- The second group, 4 districts located at an average distance of 21–40 km away from the open pits,
- The third group, 6 districts located at an average distance of 41–60 km away from the open pits,
- The fourth group, 10 districts located at an average distance of 61–80 km away from the mines,
- The fifth group, 16 districts located at an average distance of 81–100 km away from the mines.

Łódź is not included in the analyses, and the areas which until 1974 were part of the administrative Łódź Voivodship are also excluded from the calculations, where, as a result of the expansion of the city, according to Sinclair's theory, the extensification of production could already proceed, which anticipated the urban use of agricultural land [108].

For the analyses at the voivodship level, three groups of regions were created:

- Group I, the district of Konin Voivodship, where lignite open-cast mines are located,
- Group II, the Bydgoszcz and Włocławek Voivodship, which are located closest to the open-cast mines. In this group, it was also possible to include the district of Sieradz, however, after 1980, the southern part of it was located within the range of strong influence of Bełchatów open-cast mine, which could cause discrepancies in the calculations of the yield level in this group in that period,
- Group III, the remaining 6 voivodships located at an average distance of up to 100 km away from the nearest open pits, i.e., Leszno, Kalisz, Płock, Poznań, Sieradz, and Toruń Voivodships.

The time range of the study for which external costs associated with lignite mining in the Konin-Turek Basin were estimated to cover the years 1960–2060. In order to determine the current level of external costs that should be included in electricity prices paid by current consumers, the average level of external costs for generated electricity in the Konin-Turek Basin for the years 2015–2024 was estimated.

Kruskal-Wallis one-way analysis of variance by ranks test was used to test the homogeneity of the distributions of yield change dynamics in the studied regions. This test was used to verify the hypothesis that the differences between the medians of the study variable were not significant in several populations.

The hypothesis concerns medians of consecutive populations:

H_0: $\Theta_1 = \Theta_2 = \ldots = \Theta_k$

H_1: $\exists i, j \epsilon \{1, \ldots ,k\}$ $\Theta_i \neq \Theta_j$, where

$\Theta_1, \Theta_2, \ldots \Theta_k$ is the median of the tested variable x for the *i*-th group.

Hypothesis verification was based on a statistic defined by the formula:

$$H = \frac{1}{C}\left(\frac{12}{n(n+1)}\sum_{i=1}^{k}\frac{T_i^2}{n_1} - 3(n+1)\right),$$

where

$n = n_1 + n_2 + \ldots + n_k$;

T_1 (i- 1,2, … ,k) denotes the sum of ranks in each trial;

C—correction for bind ranks $C = 1 - \frac{\sum(k^3-k)}{n^3-n}$

The p value determined on the basis of the test statistic was compared with the significance level α:

if $p \leq \alpha \Rightarrow$ we reject H_0 and accept H_1

if $p > \alpha \Rightarrow$ there are no grounds to reject H_0

In assessing yield level differences between starting years (1956–1960) and final years (1993–1997), analysis of variance (ANOVA, Analysis of variance) was also used to show statistically significant differences between the means in the three groups identified. In the analysis of variance, groups of n_i elements were compared, yielding a total of $n = \sum_{i=1}^{k} n_i$ independent observations x_{ij} for $j = 1, 2, \ldots n_i$ [109]. The presence of differences between the means indicated an association between the mean for the tested observation and the qualitative variable that was the basis for separating the groups (here: distance from the outcrop). The null hypothesis of equality of all group means μ $(1, 2, \ldots, i)$ was tested:

H_0: $\mu_1 = \mu_2 = \mu_3 = \ldots = \mu_k$, where

$\mu_{(1,2,\ldots,k)}$ denotes the mean of the dependent variable in the *k*-th group,

towards the alternative hypothesis:

H_1: at least two group means differ.

In view of this, the alternative hypothesis was that there was a significant difference between the compared groups means.

The decision to accept or reject the null hypothesis was based on the Fisher-Snedecor F test determined as:

$$F = \frac{\text{intergroup variance}}{\text{intragroup variance}}$$

If the analyzed factor of group separation is significant, then the variation within each separated group will be small (the intragroup variance will be small). The greater the difference between the groups (the intergroup variance) and the smaller the difference between the elements of each group (the intragroup variance), then the value of the F statistic is large, which argues against the null hypothesis of equality of means in the compared groups, and therefore is the basis for the rejection of H_0. The presence of statistically significant differences in yields was verified using the analysis of variance at the significance level of $\alpha = 0.05$ [109].

4. Characteristics of the Konin-Turek Lignite Basin

The Konin-Turek lignite basin is located in the Wielkopolskie Voivodship, approximately 100 km east of Poznań and 200 km west of Warsaw. In Poland, apart from the

Konin-Turek coal basin, lignite is mined on an industrial scale in three open-pit mines: Bełchatów and Szczerców located south of Łódź and Turów, situated in the south-western part of the country near the border with Germany and the Czech Republic. The characteristic feature of the Konin-Turek basin is the relatively shallow coal seam, located mainly at the depth of up to 70 m, the low abundance of coal deposits, and their geographical dispersion over the districts of Konin and Turek (Figures 1 and 2). Only the Drzewce open-pit mine is located in the neighboring district of Koło. The first open pit of the deposit in Morzysław was exploited in the period 1941–1953 and 1.04 Mg of lignite was extracted from it, which was used mainly by the local population and to supply a briquetting plant. Lignite produced by the open-pit mine in Niesłusz was managed in a similar way. The exploitation of lignite on an industrial scale started with the opening of the "Konin" power station in 1958. For the purpose of the power plant, an open-cast mine in Gosławice was launched, which operated from 1957 to 1974. The lignite was stored in shallow coal seams (up to 18.7 m), which was associated with a formation of a relatively small cone of depression. It also involved fairly low external costs incurred by agriculture. It was not until more open-pit mines were launched that lignite was mined over larger areas and from greater depths (Table 1), which justifies the need to undertake the research on the external costs incurred by agriculture in the area affected by the appearance of the outcrops. The actual depth of operation of dry wells is several meters greater. In the years 1991–2009, lignite was extracted from as many as 9 open pits.

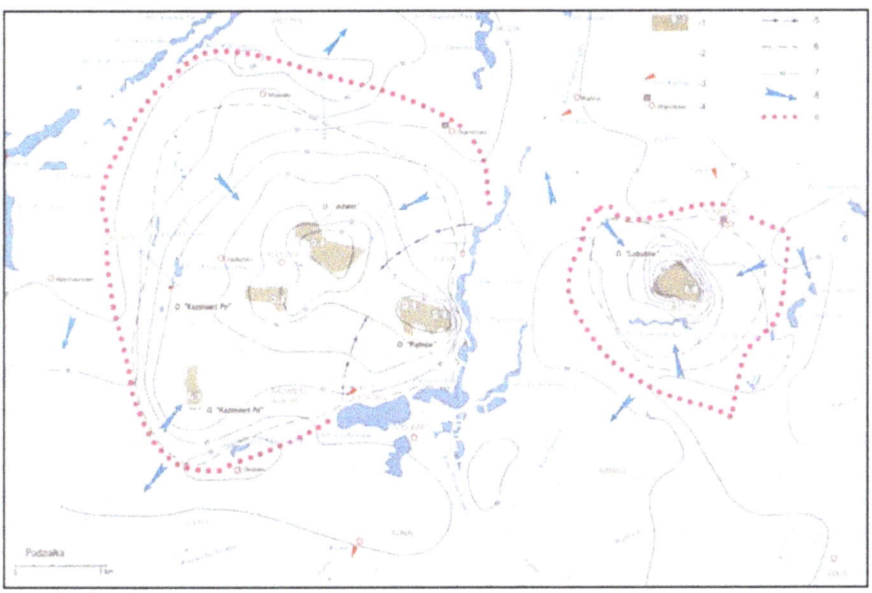

Figure 1. Location of coal deposits and cones of depression in the district of Konin as of 1996. Source: Map from [113]. Map legend: 1-lignite open-cast mines and a level of the drainage of coal bed; 2-external waste banks; 3-rivers and water level gauges of the Institute of Meteorology and Water Management; 4-rainfall and groundwater measuring stations of IMGW; 5-boundaries of the inflow zone of filtered water from lakes; 6-the range of the cone of depression in the Cretaceous-Neogene aquifer; 7-hydroisohipsa curves (meters above sea level) of the water table of the Cretaceous-Neogene aquifer as of 1996; 8-directions of groundwater runoff; 9-hydrogeological water parting of the catchment area.

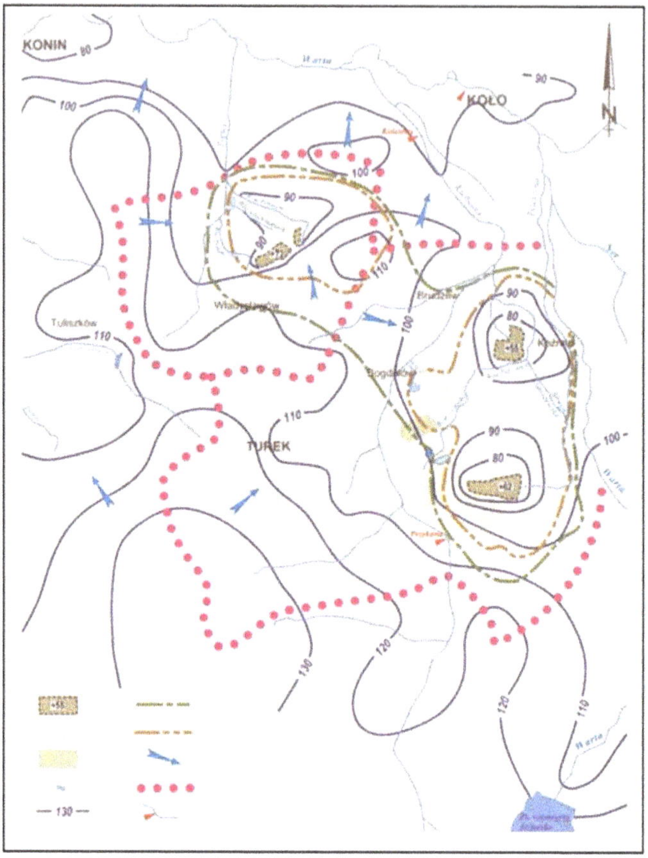

Figure 2. Location of coal deposits and cones of depression in the district of Turek as of 1996. Source: Map from [113]. Map legend as in Figure 1.

Table 1. Characteristics of lignite open-cast mines in the Konin-Turek Basin.

Coal Seam	Extraction Period (Years)	Production Volume Until 2019 (Mg)	Remaining to be Mined (Mg)	Depth of Deposit (m)	Completion of Filling in the End Reservoir (Year)
Deposits of coal in the districts of Konin and Koło					
Morzysław	1941–1953	1	-	15	-
Niesłusz	1953–1961	4.1	-	27	-
Gosławice	1958–1973	38.9	-	55	-
Pątnów	1962–2001	129.8	-	70	-
Kazimierz	1965–2011	131	-	70	2024
Jóźwin	1971–2022	146	4.9	58	2055
Lubstów	1982–2009	107	-	158	2026
Drzewce	2005–2023	31.2	4	55	2035
Tomisławice	2010–2030	15.1	26.8	67	2042
Deposits of coal in the district of Turek					
Adamów	1964–2020	109	0.8	55	2036
Bogdałów	1975–1991	38	-	50	-
Władysławów	1976–2012	38	-	55	2024
Koźmin	1991–2016	31.8		45	2023

Sources: Based on [88,110–116].

In total, from 1947 to 2019, approximately 631.1 Mg of lignite was mined from the deposits located in the districts of Konin and Koło, and approximately 216.8 Mg of lignite was extracted in the district of Turek. Only 36.5 Mg is left to be mined in the entire Konin-Turek Basin, in four oh the open-pit mines, three will have their production completed by 2023. The operation of the extraction required moving a total of over 4.8 billion m^3 of material and occupying 20.8 thousand ha of land. In the years 1945–2017, 5.34 billion m^3 of water was pumped out of the deposits located in the districts of Konin and Koło and 4.21 billion m^3 of water in the district of Turek, resulting in a waterlogging index of 8.6 m^3 water $\times$ Mg^{-1} and 19.6 m^3 water $\times$ Mg^{-1} of extracted lignite respectively, compared to an average of 6.8 m^3 $\times$ Mg^{-1} in the entire Polish lignite mining industry. The recent years of mining are characterized by an increase in the amount of pumped-out water per the amount of extracted coal. In 2017, for instance, the waterlogging index in the Konin-Turek Basin was 13.5 m^3 $\times$ Mg^{-1} and 42.3 m^3 $\times$ Mg^{-1} [16].

The amount of water pumped out and the multi-pit mining method in the Konin-Turek Basin have resulted in the formation of extensive cones of depression, some of which are combined into a regional cone of depression area. The area of cones of depression is changing due to the opening of new open-cast mines and the simultaneous completion of the exploitation of other deposits. As of 1996, in the district of Konin, the range of the cone of depression was approximately 100 km^2 in the Pliocene aquifer (discharge cone) and approximately 450 km^2, the later Tertiary period (Figure 1). In the district of Turek, discharge cones of depression within the Quaternary period (overburden) layers covered several km^2 of the area around the mines, and the cone of depression area covered approximately 90 km^2 in the Neogene formations and nearly 200 km^2 in the Cretaceous formations (Figure 2). The inflows to the drainage systems of individual open pits ranged from approximately 20 to approximately 80 m^3 $\times$ min^{-1}. During the period of maximum drainage, the total amount of water inflows into the mines in the district of Konin reached 130–150 m^3 $\times$ min^{-1}, and in the district of Turek—120–170 m^3 $\times$ min^{-1}. The range of the cones of depression created over the later Tertiary period reached 50–80 m [113]. Meanwhile, before the commencement of exploitation in the area of Pątnów, the level of the groundwater table lay at a depth of up to 7.5 m below ground level, from 2 to 4 m below ground level on average. The natural fluctuations of the level of the water table in the annual cycle ranged from 0.4 m to 3.8 m [117]. Regarding the open-cast mine in the region of Tomisławice, before the mine was launched, the groundwater table was located mostly at the depth of 0.5−1.5 m, and in the case of only 7% of the measurement stations, the level of the water table was observed on the depth of over 2.0 m [118]. The above data indicate relatively good water conditions in the period preceding the exploitation of lignite in the studied coalfield.

The expected period of restoration of water conditions is significantly varied. In the case of the analyzed open pits, the restoration of water relations is difficult because it is the region with the lowest precipitation in Poland. Filling in the end reservoirs may take from approximately 10 years in the case of the mine in Lubstów, up to 25 years for the open-pit mine in Kazimierz Północ, to approximately 40 years in the case of the Jóźwin IIB open-cast mine [88,119,120] and it also depends on the possibility of accelerating those processes through the discharge of water from the drainage of other outcrops and supplying the reservoirs with water from rivers. The estimates regarding filling in the post-mining reservoirs differ from the forecasts provided by the management of the open-cast mines and those included in Table 1. Since the water surface of mining subsidence reservoirs in post-mining areas is usually located significantly lower than the original ground level, the full restoration of water conditions within the territory of the open pit will take even longer. Therefore, it has been assumed that the reconstruction of water relations in the region will probably not be completed by 2060.

The productivity of lignite depends on its calorific value and the efficiency of the power units. The net calorific value of lignite mined in Poland in the years 2004–2019, was approximately 8437 MJ–kg^{-1} on average [121] within a range of 7400 to 10300 MJ–kg^{-1} [122]. Lignite-fired power plants are among the oldest power stations in Poland and are mostly characterized by low net/gross generation efficiency of 29.2/32.0% for Konin, 30.0/32.9% for Adamów, 31.0/33.7% for Pątnów, and 41.0/44.0% for Pątnów II [123]. Consequently, in the period 2015–2019, 1.364 Mg of lignite were consumed per net MWh × 10^6 [124–126], which, taking this efficiency into account, allows to produce approximately MWh × 10^6 of net electricity from the extracted coal, and the coal planned to be mined, in the Konin-Turek Basin.

5. Results

Agriculture in Poland, until 1990, was functioning within a centrally planned economy. Despite many attempts to reconstruct agriculture following the Soviet model, Polish agriculture was dominated by small individual farms. Until the end of the 1960s, it remained stagnant and without any prospects for development. The restrictions on access to new machinery, means of production, and fodder were particularly bothersome (the priority was given to state farms). Strong attachment to land along with the restrictions on trade associated with agricultural land limited the restructuring of Polish agriculture, e.g., the average size of a private farm increased from 5.3 ha in 1960 to 6.7 ha in 1990 [96]. Agriculture in Poland was also characterized by low productivity, including a low amount of yield. The introduction of the market economy accelerated restructuring processes in agriculture, however, the improvement of productivity of Polish farming and the pace of concentration processes are still too slow to achieve the level of the development of the Western European countries [127,128]. The above-mentioned conditions also apply to the area under examination.

Grain harvest, sugar beet, and potato yield in the analyzed region, in the years 1956–1973, was characterized by a high growth rate, however, it started from a very low level. In the districts of Konin and Turek, where coal was produced, the average level of yield in the years 1956–1960 was similar to the yield in the other districts under the study, however, grain harvest was lower by 5–10%, and in the case of sugar beet and potatoes, it was higher than in the other groups of the analyzed districts (Table 2). Over the years 1969–1973, the level of crops in the districts of Konin and Turek was less favorable compared to the other districts. This was due to the lowest growth pace, which was 4.9–8.9 percentage points lower in the case of grain and more than 10 percentage points lower for potatoes and sugar beet. The analysis of changes in the level of yield, presented in Figure 3, shows that the pace of growth of yield in the districts with coal mines was similar to the pace in other regions until the late 1960s, however, later on, there was a progressive differentiation observed regarding the growth of yield. This coincides with the period of launching new open-pit mines in the districts of Konin and Turek, i.e., Pątnów (1962), Adamów (1964), Kazimierz (1965), and Jóźwin (1971) and the development of cones of depression. It may indicate the growing negative impact of the developing cones of depression on the level of yield in the districts of Konin and Turek. No negative impact on the crop yield in the neighboring districts was noted during this period.

Table 2. The yield of selected crops and the dynamics of yield depending on the distance from the open-pit mines (according to the data from the analyzed districts).

Group	Average Yield in 1956–1960 Years [t × ha^{-1}]			Average Yield in 1969–1973 Years [t × ha^{-1}]			Dynamic [%]		
	Cereal	Potato	Sugar Beet	Cereal	Potato	Sugar Beet	Cereal	Potato	Suger Beet
up to 20 km	1.56	13.7	21.9	2.11	17.4	32.1	135.0	127.4	146.5
20–40 km	1.67	13.4	19.4	2.33	18.6	32.9	140.0	138.5	169.6
40–60 km	1.72	13.3	21.8	2.40	18.5	34.4	139.9	139.2	158.1
60–80 km	1.70	12.8	18.0	2.37	17.7	27.4	139.9	139.0	152.5
80–100 km	1.64	12.6	19.9	2.35	18.2	30.2	143.9	144.1	151.4

Sources: Calculations based on [100–106].

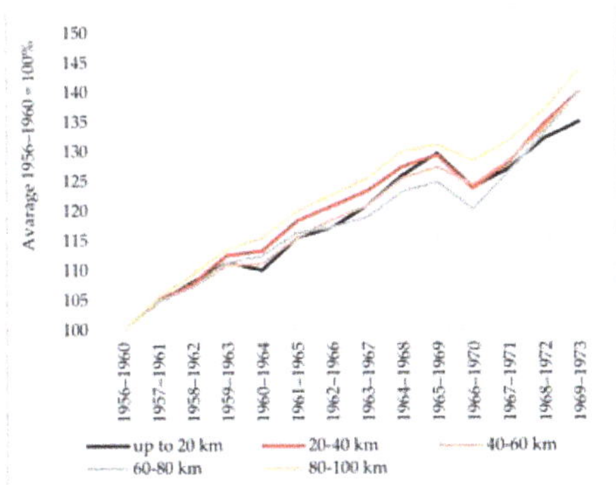

(a)

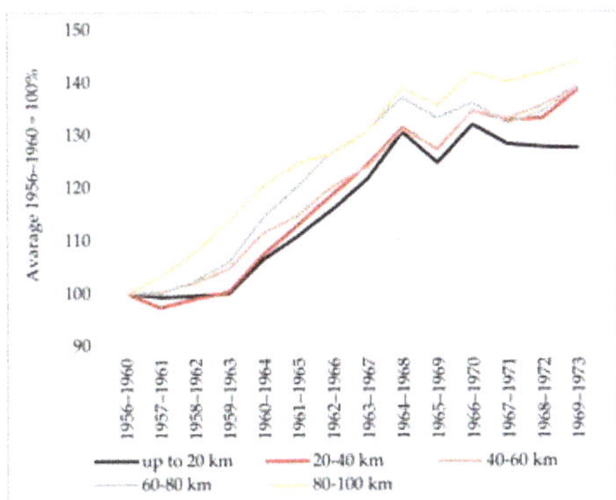

(b)

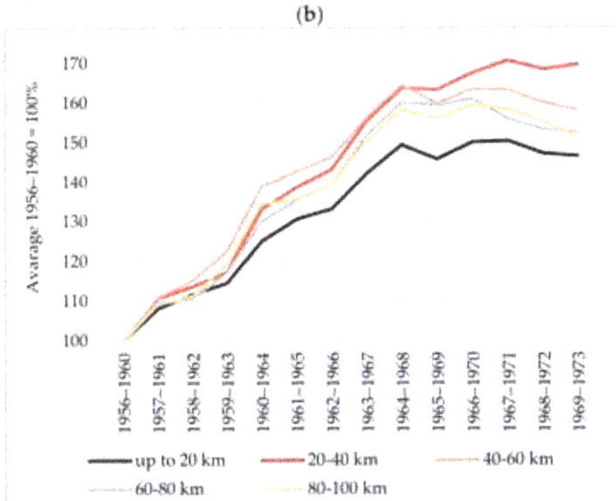

(c)

Figure 3. Yield level dynamics of selected crops depending on the distance from the open–cast mines (according to the data from the analyzed districts): (**a**) cereal; (**b**) potato; (**c**) sugar beet. Source: Based on [100–106].

The analysis at the voivodship level, due to the longer time period for which the data is available, allowed us to determine the long-term impact of the open-pit mines located in the Konin-Turek lignite basin. The average agricultural productivity of grain, potatoes, and sugar beet in 1956–1960, did not differ significantly among the three analyzed groups of voivodships. In the case of grain, the analysis of the yield, according to the data from the voivodships, also indicates an increase in the negative impact of open-pit mines on agricultural output since the late 1960s, and in the case of potatoes, since the mid-1970s. (Figure 4).

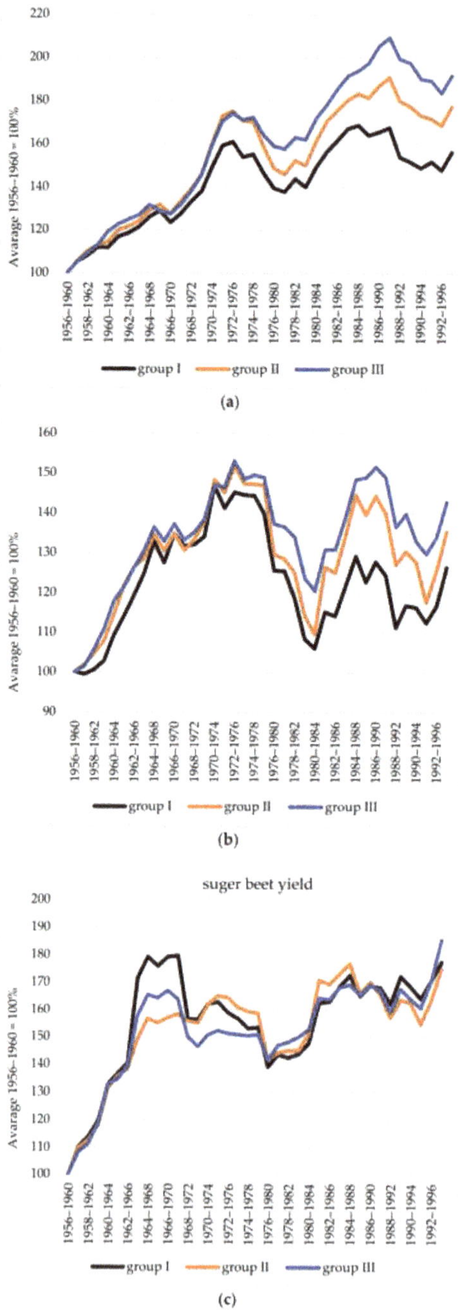

Figure 4. Yield level dynamics of selected crops depending on the distance from the open-cast mines (according to the data from the analyzed voivodships): (**a**) cereal; (**b**) potato; (**c**) sugar beet. Source: Based on [100–106].

In the case of sugar beet, this applied to a lesser extent, as in the second part of the analyzed period, for all voivodships, there was a stabilization or even a decline in the yield. The lower sensitivity of sugar beet to falling levels of groundwater may be due to the fact that the plant has a deeper root system, which allows taking water from deeper layers of soil, and the requirement for the vegetable to be cultivated on better quality soils naturally helps to reduce the sensitivity of sugar beet to changes in the levels of groundwater. This does not mean that the farmers growing this plant did not experience any losses. As can be seen from the data in Figure 3, the reduction in sugar beet yields occurred only in the vicinity of the open-pit mines, where the decrease of the level of the water table was the greatest, and it was compensated, at the voivodship level, with the highest increase of the yield in the districts located 20–40 km away from the open-cast mines. Therefore, it can be assumed that if there were no open-pit mines in the Konin Voivodship, the dynamics of the agricultural productivity of sugar beet would be higher than in the voivodships from group III.

In the case of grain and potatoes, however, the disproportions in the level of yield between particular groups of voivodships increased. Regarding grain, the agricultural output in the Konin Voivodship, in the years 1956–1961, was lower than in the voivodships from group III by 2.0%, while in the period 1993–1997 by 20.1%, which indicates a loss of 18.4% of the yield. When it comes to potatoes, instead of 4.5% higher yields, the level of agricultural productivity was lower by 7.5%, which means a decrease in the yield by 11.4%. For the voivodships from group II, the yield loss during this period was 7.4% for grain and 5.2% in the case of potatoes. Taking into account the importance of particular crops in the structure of arable land, the farmers in the Konin Voivodship lost 16.4% of their harvest, on average, and in the case of the voivodships from group II—7.0%. Considering the average of the last 10 five-year periods that are analyzed, the losses amounted to 17.2% and 7.5%, respectively.

Kruskal-Wallis one-way analysis of variance by ranks test was used to test the homogeneity of the distributions of yield change dynamics in the studied regions. This test is used to verify the hypothesis that the differences between the medians of the study variable are not significant in several populations.

As a result of the analysis, it is noted that there are no statistically significant differences in yield changes for sugar beets. The test probability level $p = 0.5084$ exceeds the critical value in this case, so there is no basis for rejecting the hypothesis of an equal rate of change in sugar beet yields. On the other hand, the test probability level $p \leq 0.05$ allows rejecting the input null hypothesis of equality of the dynamics of changes in yields of cereals ($p = 0.0047$) and potatoes ($p = 0.0005$). Thus, distance from the mine had a statistically significant effect on yield gains in the three regions studied. The multiple comparisons test indicates that the difference in the rate of increase in cereals and potato yields between Konin Voivodeship and Voivodeships from group III (the furthest from the outcrop) proved statistically significant at the 0.05 significance level. The significance of differences in yield changes between Konin Voivodeship and Voivodeships from group II was slightly weaker; these results were significant at the 0.08 significance level. The results confirm previous findings indicating that as the distance of farms from the outcrop increased, the rate of increase in yields of cereals and potatoes, shallow-rooted crops, was much faster. As explained earlier, the lack of response of sugar beets to the decrease in water table caused by the outcrop was due to the biological specificity of this crop, where the deeper root system allows the use of sub-bottom water resources located even deeper (lowered due to the loss of water due to the outcrop).

In addition to analyzing the dynamics of change in Table 3, changes in absolute values of yields of selected crops were presented.

Table 3. The yield of selected crops and the dynamics of yield depending on the distance from the open-pits mines (according to the data from the analyzed voivodships).

Group	Average Yield in 1956–1960 Years [t × ha^{-1}]			Average Yield in 1993–1997 Years [t × ha^{-1}]			Dynamic [%]		
	Cereal	Potato	Suger Beet	Cereal	Potato	Suger Beet	Cereal	Potato	Suger Beet
group I	1.63	13.5	20.13	2.56	17.0	35.7	156.4	126.5	177.5
group II	1.68	13.0	19.48	2.98	17.6	34.1	177.4	135.4	175.1
group III	1.67	12.9	20.28	3.20	18.4	37.6	191.7	142.8	185.5

Sources: Calculations based on [100–106].

The ANOVA analysis of variance indicates that at the adopted level of significance $\alpha = 0.05$, there are no grounds to reject the null hypothesis of equality of the mean values of grain, potato, and sugar beet yields in the compared groups. Nevertheless, it is worth noting changes in statistical significance when comparing the starting and final periods. The observed changes in average yield height (Table 3) are reflected in the change in the probability of significance of the differences: in the first period they were completely insignificant (cereals $p = 0.91$; potatoes $p = 0.58$; sugar beets $p = 0.62$), and in the last period the probabilities were 0.52, 0.43 and 0.49, respectively (although still significantly above the $\alpha = 0.05$ level) (Table 4). Despite the lack of significance of differences, this interpretation of the results was proposed because the observed changes are related to agricultural production. Since agricultural field production determines the livelihood of a farm, managers make every effort to offset the impact of adverse factors. Nevertheless, the presented calculations show an outlined trend of yield divergence, to the disadvantage of the regions closest to the outcrop. Even if they are small, for selected crops they can determine the profitability of production in general.

Table 4. Results of the analysis of variance test for significance of differences in yields of selected crops in three groups depending on the distance from the outcrop (according to data from voivodeships) at the beginning and end of the analysis period.

Cultivation	Time Period	Test Results			
Cereals	1956–1960	Analysis of variance	F = 0.0962		$p = 0.9096$
		Levene's test for homogeneity of variance	F = 0.5161		$p = 0.6212$
		Least significant differences test	{1}↔{2} $p = 0.6763$	{1}↔{3} $p = 0.7570$	{2}↔{3} $p = 0.8262$
	1993–1997	Analysis of variance	F = 0.7265		$p = 0.5217$
		Levene's test for homogeneity of variance	F = 2.1508		$p = 0.1976$
		Least significant differences test	{1}↔{2} $p = 0.5278$	{1}↔{3} $p = 0.2892$	{2}↔{3} $p = 0.6133$
Potatoes	1956–1960	Analysis of variance	F = 1.4104		$p = 0.3147$
		Levene's test for homogeneity of variance	F = 0.6015		$p = 0.5780$
		Least significant differences test	{1}↔{2} $p = 0.3755$	{1}↔{3} $p = 0.1549$	{2}↔{3} $p = 0.5005$
	1993–1997	Analysis of variance	F = 0.9901		$p = 0.4250$
		Levene's test for homogeneity of variance	F = 3.6032		$p = 0.0938$
		Least significant differences test	{1}↔{2} $p = 0.6946$	{1}↔{3} $p = 0.2785$	{2}↔{3} $p = 0.3745$
Sugar Beets	1956–1960	Analysis of variance	F = 0.5185		$p = 0.6199$
		Levene's test for homogeneity of variance	F = 0.5371		$p = 0.6101$
		Least significant differences test	{1}↔{2} $p = 0.6732$	{1}↔{3} $p = 0.7997$	{2}↔{3} $p = 0.3490$
	1993–1997	Analysis of variance	F = 0.8080		$p = 0.4889$
		Levene's test for homogeneity of variance	F = 3.3950		$p = 0.1032$
		Least significant differences test	{1}↔{2} $p = 0.6574$	{1}↔{3} $p = 0.6880$	{2}↔{3} $p = 0.2554$

{1}—group I; {2}—group II; {3}—group III.

Due to the long period of the impact of external costs on agriculture associated with the operation of open-pit mines in the Konin-Turek Basin, estimated until 2060, three reference periods were adopted to calculate the costs: first—for the years 1961–1987, taking into account the level of yield loss compared to the average figure from the period 1956–1960, second—for the years 1988–2033, assuming the average level of yield loss from 10 five-year periods, and third—for the years 2034–2060, considering the level of yield reduction as assumed for the first period, but in the reverse order. With such assumptions, in the whole analyzed period of 100 years, the decline of the yield in the Konin Voivodship amounted to 11.7%, and in the voivodships from group II—4.7%.

Three variants of yield level were adopted. First, the average yield from the years 2015–2019 was assumed for the whole period of the impact of the open-pit mine in Konin district, i.e., until 2060. Second, until 2019, the actual yield was assumed for each group of voivodships, and after 2019, the yield level was adjusted by the average annual increase in productivity from the period 1999 to 2019. For the third variant, the actual yield was also assumed until 2019, while for the following years it was assumed that in 2050 the level of the yield in the areas of voivodships from group III will reach the level of the output achieved by Germany in 2015.

Calculating the external costs of crop production resulting from the seizure of 20.8 thousand ha of land, the share of arable land was assumed to correspond to the share of arable land in the district of Konin, and a share of 60% of land excluded from agricultural production during the whole period of the operation of open-cast mines and after reclamation. The average share of arable land in the total area of the Konin Voivodship, in the entire analyzed period, was estimated at 68.9%.

The total of external costs for the entire period of the research, assuming that the decrease in the yield in the area affected by the operation of opencast mines is entirely the result of the cones of depression, was estimated at €5.6 billion, which, with an estimated electricity production of 648.4 TWh, equals to €8.66 per kWh (Table 5), which accounted for 16.1% of the price on the Polish market of SPOT TGE S.A. in 2019 [129].

Table 5. External costs related to the exploitation of lignite in the Konin-Turek Basin, in the period 1961–2060 (€ million).

Specification	Variant I	Variant II	Variant III	Average	€ × MWh^{-1}
Yield decline caused by the operation of open-cast lignite mines					
Open-pit mining area	19	17	17	18	0.03
Group I	3094	2752	2815	2887	4.45
Group II	2863	2603	2657	2708	4.18
Total	5976	5372	5489	5612	8.66

Source: own calculations.

6. Discussion

In Polish literature, there is no analysis of the influence of open-cast mining on agriculture. In the ongoing discussion, the most popular subject is the impact of open-pit mining on the level of soil moisture. Most of the authors [130–136] claim that there is no negative impact of cones of depression on the upper layers of the terrain and they concentrate on the negative effects of the cones on deeper layers. However, a significant decline in levels of groundwater is noticed [137]. Among the few researchers who suggest the negative impact of cones of depression on the content of moisture of farmland are Chodak et al. [138,139] and Wlodek et al. [140], however, they do not indicate the level of losses that farmers should consider.

The conducted analysis presents a huge influence of multi-pit lignite mining in the Konin-Turek Basin. The long period of exploitation and the related deficits of underground water resources led to a systematic increase in losses resulting from a slower pace of growth of agricultural productivity compared to the areas located further away from the open-pit mines. This undoubtedly contributed to a significant weakening of the economic strength of agriculture in the areas of cones of depression, which indirectly, through reduced investment and intensity of production, caused an increase in losses of the yield and insufficient use of the already limited productive capacity of soils. The loss of 16.4% of crops in the perspective of 10 years in the entire Konin Voivodship should be considered as significant, especially in the economic context, because regardless of the level of agricultural output farmers are forced to perform all cultivation and care treatments, bear the full costs of crop protection and seeds and the majority of the costs of fertilizers. Consequently, any loss of yield means a reduction in profit. It should also be assumed (in the context of the research at the district level) that in the regions located closer to open-pit mines the level of the loss of the yield will be even greater. This was confirmed by a survey conducted among farmers living in the area where open-cast mines were planned to be launched, whose farmlands were located approximately 20 km away from the spot. They expected a 40% decline in their yield [35].

The level of future yield and losses in agriculture will be determined by the level of agricultural development, production intensity, and the technologies that are used. High levels of local agricultural development, adequate know-how, capital, and the ability to adapt technology to changing conditions can minimize the losses caused by the appearance of a cone of depression.

The reason for the decline in agricultural productivity was the lowering level of the water table in large areas surrounding the open pit, which increased the dependence of the crops on rainfall during the growing season. With relatively favorable weather conditions, the effects of lowering the water table as a result of the appearance of a cone of depression are insignificant, however, in the period of deficiency in rainfall, the importance of the possibility to be able to use groundwater increases. The great decline in grain production with a limited impact of a cone of depression on sugar beet indicates that shallow-rooted crops are particularly vulnerable to drainage caused by open-pit mining. They are subject to greater damage during the years with a low amount of rainfall. Assessing the impact of drought on agricultural production is particularly difficult, despite the fact that there has been some research performed that allowed us to estimate the effect of the dry season on the soil at a regional scale [63]. A study conducted in the Czech Republic, on the effects of drought on crop yields, has shown that in the case of severe drought, the crop, depending on the species, decreases by 0.8–1.5 $t \times ha^{-1}$ (with average yield in recent years at approximately 5.8 $t \times ha^{-1}$), potato yield declined by approximately 3.0 $t \times ha^{-1}$ (with average yield in recent years at approximately 28.0 $t \times ha^{-1}$), and rapeseed yield declined by 0.55 $t \times ha^{-1}$ (with average yield over the recent years at approximately 3.3 $t \times ha^{-1}$). According to the studies, sweetcorn is characterized by relatively high tolerance to drought [141].

The optimal level of groundwater to obtain the highest yield is 0.7–1.6 m, and it is 1.5–2.5 m in the case of corn [57,142,143]. Therefore, it can be assumed that to achieve high agricultural output, the level of groundwater should be 1–3 m [61,72,144]. In turn, the level of groundwater below 4.0 m leads to a collapse of agricultural productivity. For example, according to the studies conducted on the Inland Pampas, during two growing seasons (2006/2007 and 2007/2008), the areas within these optimum bands had yields that were 3.7, 3.0, and 1.8 times larger than those where the water table was below 4 m for wheat, maize, and soybean, respectively [143]. According to research in the Hungarian Plain, the decline in groundwater level led to a stagnation in wheat and sweetcorn yield. In the case of sweetcorn, the yield loss was estimated at an average 0.65 $t \times ha^{-1}$, indicating that a 1.0 m drop in the level of groundwater, under the conditions recorded between 1986 and 2010, would result in a 2.33 $t \times ha^{-1}$ decline in corn yield in this region [57]. The level of the yield loss, according to the above-mentioned studies, is therefore similar to the losses resulting from the occurrence of cones of depression in the Konin-Turek lignite basin.

The estimated external costs of €5.6 billion, resulting in €8.66 per MWh of produced electricity must be considered as significant from an economic point of view. Current consumers of electricity generated from 2015 to 2024 from lignite from the Konin-Turek

Basin should pay even more, €14.29. Despite the numerous uncertainties regarding the impact of cones of depression on the level of yield, the conducted research proved a huge level of external costs involved, which should not be completely ignored. There is also a lot of discrepancies in the studies on external costs in the case of power generation, e.g., the costs related to health damage and internalization of emission costs [28]. If the estimated external costs were to be included in the costs of electricity production, e.g., a necessity to cover the estimated losses for farmers, the profitability of electricity production from lignite would significantly change and, consequently, its attractiveness would decrease compared to other renewable and non-renewable energy sources. The performed analysis concerned the open-pit mine terminating its operation, therefore, the research covered a period when the level of yield was much lower than it is presently. In the case of analyses of external costs for open-pit mines which are planned to be launched, and in the regions with a higher level of agricultural development, even greater losses should be expected.

The production of electricity from fossil fuels, including lignite, and its impact on the environment have also been the subject of many studies. The issue of environmental pollution caused by dust and gases from their combustion was addressed most often. The estimations of external costs differ significantly due to the variety of analyzed factors, various research methodologies, availability of data, the efficiency of power plants, combustion technology, etc. (Table 6). For example, in the case of Thailand, only the impact of the emission of PM10 and NO_x in sparsely-populated areas was analyzed, while the study conducted by Macy et al. included sulfur dioxide, nitrogen oxides, dust particles, carbon monoxide and dioxide, volatile organic compounds, polycyclic aromatic hydrocarbons, and the difference in the level of external costs was approximately 10 times higher. In the case of Bosnia and Herzegovina, lignite with different calorific values and sulfur content was analyzed.

Table 6. External costs of air pollution caused by lignite combustion € × MWh^{-1}.

Study	Georgakellos [145]	Sakulniyomporn [42]	Büke, Köne [146]	Dimitrijević [147]	Coester [39]	Máca [32]	Wang [31]	Taranto [148]
Country	Greece	Thailand	Turkey	Bosnia and Herzegovina	Germany	Czech, Hungary, Poland	China	Turkey
Year of analysis	2003–2004	2006–2008	2007	2008	1995–2003	2010	2015	2018
Health impacts	No	Yes	Yes	No	Yes	Yes	Yes	No
External costs	43.9	6.8	1.8–35.2	2.7–19.2	11.1	58.1–77.5	63.8	36.3

Source: own calculations.

One of the few works analyzing external costs at the stage of lignite extraction associated with the emission of suspended particulate matter during the mining process estimated external costs at 5.0 € × MWh^{-1} [149]. The external costs estimated in this study, on the effects of open-cast lignite mining on agricultural plant production, are similar. To determine the full amount of external costs borne by agriculture, it is still necessary to perform analyses for livestock production, which, due to its dependence on feed produced on the farmland where the animals are kept (especially the large dependence in the case of cattle and sheep), is also subject to restrictions related to cones of depression.

Conducting full research on external costs is particularly important in the context of the Energy Policy of Poland until 2040 [150], approved in January 2021, which leaves it to the discretion of potential investors to launch two more lignite open-pit mines in Złoczew and Ościsłowo, the coal seams of which, according to this document, are not to be extracted. The prices of CO_2 emission allowances, environmental conditions, and the development of new technologies are to play a key role in the management of the terrain. However, in the published strategy, as well as in other government documents, there is no mention of external costs, which should support the decision-making processes regarding the profitability of coal extraction and its combustion. To determine the total amount of external costs, it is necessary to assess the following:

- External costs associated with the emission of dust in combustion processes and the impact on human health and global warming,
- External costs associated with the emission of suspended particulates as a result of mining processes,
- External costs for agriculture (both crop and animal production), for agri-food industry and forestry, related to the drainage of open pits.

The inclusion of these external costs as essential factors in rational decision-making with regard to investments will surely contribute to abandoning the mining of other deposits in Poland and will contribute to faster improvement of air quality.

Author Contributions: Conceptualization, methodology, resources, investigation and editing, B.P.; writing, B.P. and W.C.; Kruskal-Wallis and ANOVA analysis of variance, W.C. Authorship must be limited to those who have contributed substantially to the work reported. All authors have read and agreed to the published version of the manuscript.

Funding: This research received no external funding.

Institutional Review Board Statement: Not applicable.

Data Availability Statement: The data is available from the authors upon request.

Conflicts of Interest: The authors declare no conflict of interest.

References

1. Thielemann, T.; Schmidt, S.; Gerling, J.P. Lignite and hard coal: Energy suppliers for world needs until the year 2100–An outlook. *Int. J. Coal Geol.* **2007**, *72*, 1–14. [CrossRef]
2. Federal Institute for Geosciences and Natural Resources. *Energy Study 2014: Reserves, Resources and Availability of Energy Resources*; Federal Institute for Geosciences and Natural Resources: Hanover, Germany, 2014.
3. Kashiwagi, T.; Ishino, H.; Takagi, T.; Hirose, K. Mine-mouth Power Generation System Based on Upgraded Brown Coal (UBC). *Kobelco Technol. Rev.* **2015**, *33*, 28–33.
4. Bielecki, A.; Ernst, S.; Skrodzka, W.; Wojnicki, I. The externalities of energy production in the context of development of clean energy generation. *Environ. Sci. Pollut. Res.* **2020**, *27*, 11506–11530. [CrossRef] [PubMed]
5. Violidakis, I.; Drosatos, P.; Nikolopoulos, N. Critical review of current industrial scale lignite drying technologies. In *Low-Rank Coals for Power Generation, Fuel and Chemical Production*; Luo, Z., Agraniotis, M., Eds.; Woodhead Publishing: Cambridge, UK, 2017; pp. 41–71. [CrossRef]
6. Wilczyński, M. Coal in Poland: After the Sunset, Change Partnership.org. 2016. Available online: http://www.changepartnership.org/wp-content/uploads/2016/02/Wilczynski-2016-Coal-in-Poland-after-the-sunset-translation-FINAL.pdf (accessed on 8 December 2020).
7. Mohr, S.H.; Evans, G.M. Forecasting coal production until 2100. *Fuel* **2009**, *88*, 2059–2067. [CrossRef]
8. Patzek, T.; Croft, G. A global coal production forecast with multi–Hubbert cycle analysis. *Energy* **2010**, *35*, 3109–3122. [CrossRef]
9. Mukherjee, S.; Pahari, D.P. Underground and Opencast Coal Mining Methods in India: A Comparative Assessment. *Space Cult. India* **2019**, *7*, 39–55. [CrossRef]
10. Braun, M. *Environmental External Costs from Power Generation by Renewable Energies*; Institut für Energiewirtschaft und Rationelle Energieanwendung, Universität Stuttgart: Stuttgart, Germany, 2004.
11. International Institute for Sustainable Development. *Financial Supports for Coal and Renewables in Indonesia*; International Institute for Sustainable Development: Winnipeg, MB, Canada, 2017.
12. Samadi, S. The Social Costs of Electricity Generation—Categorising Different Types of Costs and Evaluating Their Respective Relevance. *Energies* **2017**, *10*, 356. [CrossRef]
13. Karkour, S.; Ichisugi, Y.; Abeynayaka, A.; Itsubo, N. External-Cost Estimation of Electricity Generation in G20 Countries: Case Study Using a Global Life—Cycle Impact—Assessment Method. *Sustainability* **2020**, *12*, 2002. [CrossRef]
14. Adani's Coal Mine Plans, and Why Coal Is Still So Lucrative. Available online: https://www.dw.com/en/adanis-coal-mine-plans-and-why-coal-is-still-so-lucrative/a-52025626 (accessed on 4 January 2021).
15. Global Coal Exit List. Available online: https://coalexit.org/ (accessed on 4 January 2021).
16. Kasztelewicz, Z. *Report on the Condition of the Lignite Coal Industry in Poland and Germany Together with a Diagnosis of Measures for the Development of this Industry in the First Half of the 21st Century*; AGH University of Science and Technology: Krakow, Poland, 2018. (In Polish)

17. Kasztelewicz, Z.; Tajduś, A.; Cała, M.; Ptak, M.; Sikora, M. Strategic conditions for the future of brown coal mining in Poland. *Energy Policy J.* **2018**, *21*, 155–178. [CrossRef]
18. Pietraszewski, A. Polish lignite mining in 2014. *Węgiel Brunatny* **2015**, *90*, 4–14. (In Polish)
19. Widera, M.; Kasztelewicz, Z.; Ptak, M. Lignite mining and electricity generation in Poland: The current state and future prospects. *Energy Policy* **2016**, *92*, 151–157. [CrossRef]
20. Ministry of Energy. *Energy Policy of Poland Until 2040 (EPP2040)—Extract from Draft v. 2.1*; Ministry of Energy: Warsaw, Poland, 2019.
21. Pactwa, K.; Woźniak, J.; Strempski, A. Sustainable mining—Challenge of Polish mines. *Resour. Policy* **2018**. [CrossRef]
22. Faaij, A.; Meuleman, B.; Turkenburg, W.; van Wijk, A.; Bauen, A.; Rosillo-Calle, F.; Hall, D. Externalities of biomass based electricity production compared with power generation from coal in the Netherlands. *Biomass Bioenergy* **1998**, *14*, 125–147. [CrossRef]
23. Rowe, R.D.; Lang, C.M.; Chestnut, L.G.; Latimer, D.A.; Rae, D.A.; Bernow, S.M.; White, D.E. *New York State Environmental Externalities Cost Study*; Empire State Electric Energy Research Corporation, Oceana: New York, NY, USA, 1995.
24. Lee, R.; Krupnik, A.; Burtraw, D. *Estimating Externalities of Electric Fuel Cycles: Analytical Methods and Issues and Estimating Externalities of Coal Fuel Cycles*; Techical Report; McGraw—Hill/Utility Data Institute: Washington, DC, USA, 1995.
25. Krewitt, W.; Heck, T.; Trukenmüller, A.; Friedrich, R. Environmental damage costs from fossil electricity generation in Germany and Europe. *Energy Policy* **1999**, *27*, 173–183. [CrossRef]
26. EC. *ExternE—Externalities of Energy, Methodology 2005 Update*; Office for Official Publications of the European Communities: Luxembourg, 2005.
27. Rečka, L.; Ščasný, M. Brown coal and nuclear energy deployment: Effects on fuel-mix, carbon targets, and external costs in the Czech Republic up to 2050. *Fuel* **2018**, *216*, 494–502. [CrossRef]
28. Jorli, M.; Van Passel, S.; Saghdel, H.S. External costs from fossil electricity generation: A review of the applied impact pathway approach. *Energy Environ.* **2018**, *29*, 635–648. [CrossRef]
29. Ščasný, M.; Massetti, E.; Melichar, J.; Carrara, S. *Quantifying the Ancillary Benefits of the Representative Concentration Pathways on Air Quality in Europe*; Nota di Lavoro, No. 84.2015; Fondazione Eni Enrico Mattei (FEEM): Milan, Italy, 2015.
30. Preiss, P.; Klotz, V. *EcoSenceWeb V1-3*; Insitute of Energy Economics and the Rational Use of Energy: Stuttgart, Germany, 2008.
31. Wang, L.; Watanabe, T.; Xu, Z. Monetization of External Costs Using Lifecycle Analysis—A Comparative Case Study of Coal–Fired and Biomass Power Plants in Northeast China. *Energies* **2015**, *8*, 1440–1467. [CrossRef]
32. Máca, V.; Melichar, J.; Ščasný, M. Internalization of External Costs of Energy Generation in Central and Eastern European Countries. *J. Environ. Dev.* **2012**, *21*, 181–197. [CrossRef]
33. Pepliński, B. External costs of opencast brown coal mining in agriculture and agri–food industry (on the example of Wielkopolska). *Cent. Eur. Econ. J.* **2018**, *5*, 154–171. [CrossRef]
34. Pepliński, B. External costs of open-cast lignite mining for the agri–food industry (as illustrated by the example of the Oczkowice mine). *J. Agribus. Rural Dev.* **2018**, *2*, 213–223. [CrossRef]
35. Poczta, W.; Pepliński, B.; Sadowski, A.; Czubak, W. *Impact of the Planned Oczkowice Mine on the Economic, Productive and Social Potential of Agriculture and its Surroundings in the South–Western Functional Area of the Wielkopolskie Voivodeship*; Poznań University of Life Sciences: Poznan, Poland, 2017. (In Polish)
36. Georgescu-Roegen, N. *The Entropy Law and the Economic Process*; Harvard University Press: Cambridge, MA, USA, 1971.
37. Dal, E.; Koksal, M.A. Modeling the use of renewable energy sources for electricity generation based on technological, political, and environmental constraints. *Energy Procedia* **2017**, *142*, 3129–3134. [CrossRef]
38. Rafaj, P.; Kypreos, S. Internalisation of external cost in the power generation sector: Analysis with global multi-regional MARKAL model. *Energy Policy* **2007**, *35*, 828–843. [CrossRef]
39. Coester, A.; Hofkes, M.; Papyrakis, E. Economics of renewable energy expansion and security of supply: A dynamic simulation of the German electricity market. *Appl. Energy* **2018**, *231*, 1268–1284. [CrossRef]
40. Redlinger, R.Y.; Andersen, P.; Morthorst, P. *Wind Energy in the 21st Century—Economics, Policy, Technology and the Changing Electricity Industry*; Palgrave Macmillan: Basingstoke, UK, 2002. [CrossRef]
41. Becker, N.; Soloveitchik, D.; Olshansky, M. A weighted average incorporation of pollution costs into the electrical expansion planning. *Energy Environ.* **2012**, *23*, 1–15. [CrossRef]
42. Sakulniyomporn, S.; Kubaha, K.; Chullabodhi, C. External costs of fossil electricity generation: Health–based assessment in Thailand. *Renew. Sustain. Energy Rev.* **2011**, *15*, 3470–3479. [CrossRef]
43. Bozicevic Vrhovcak, M.; Tomsic, Z.; Debrecin, N. External costs of electricity production: Case study Croatia. *Energy Policy* **2005**, *33*, 1385–1395. [CrossRef]
44. Maxim, A. Sustainability assessment of electricity generation technologies using weighted multi-criteria decision analysis. *Energy Policy* **2014**, *65*, 284–297. [CrossRef]
45. Przybyłek, J. Current problems of the lignite open cast mines dewatering in the Wielkopolska region. *Górnictwo Odkrywkowe* **2018**, *2*, 5–14. (In Polish)
46. Przybyłek, J.; Górski, J. The Oczkowice lignite deposit. Voicing concern for proper hydrogeological recognition. *Przegląd Geologiczny* **2016**, *64*, 183–191. (In Polish)
47. Adamczyk, A.F.; Motyka, J. Methodical problems of determination of depression cone boundaries in the areas affected by mining drainage. *Współczesne Problemy Hydrogeologii* **1995**, *7*, 209–212. (In Polish)

48. Pusch, M.; Hoffmann, A. Conservation concept for a river ecosystem (River Spree, Germany) impacted by flow abstraction in a large post-mining area. *Landsc. Urban Plan.* **2000**, *51*, 165–176. [CrossRef]
49. Mwadzingeni, L.; Mugandani, R.; Mafongoya, P. Factors affecting the performance of Tshiombo Irrigation Scheme in Limpopo Province, South Africa. *J. Agribus. Rural Dev.* **2020**, *3*, 269–277. [CrossRef]
50. Shoko, R.R.; Chaminuka, P.; Belete, A. Maize yield sensitivity to climate variability in South Africa: Application of the ARDL-ECM approach. *J. Agribus. Rural Dev.* **2019**, *4*, 363–371. [CrossRef]
51. Rybka, K.; Nitka, Z. Modern phenotypes of cereals for growing in areas endangered with drought. *Biuletyn Instytutu Hodowli Aklimatyzacji Roślin* **2014**, *273*, 55–72. (In Polish)
52. Zawora, T.; Ziernicka–Wojtaszek, A. Extreme values of precipitation deficiencies and excesses in the light of contemporary climate change based on the example from the Podkarpackie province. *Infrastruktura Ekologia Terenów Wiejskich* **2008**, *5*, 23–29. (In Polish)
53. Sillmann, J.; Roeckner, E. Indices for extreme events in projections of anthropogenic climate change. *Clim. Chang.* **2008**, *86*, 83–104. [CrossRef]
54. Trenberth, K.E.; Dai, A.; Rasmussen, R.M.; Parsons, D.B. The Changing character of precipitation. *Bull. Am. Meteorol. Soc.* **2003**, *84*, 1205–1217. [CrossRef]
55. Easterling, D.R.; Horton, B.; Jones, P.D.; Peterson, T.C.; Karl, T.R.; Parker, D.E.; Salinger, M.J.; Razuvayev, V.; Plummer, N.; Jamason, P.; et al. Maximum and Minimum Temperature Trends for the Globe. *Science* **1997**, *277*, 364–367. [CrossRef]
56. Tomassini, L.; Jacob, D. Spatial analysis of trends in extreme precipitation events in high-resolution climate model results and observations for Germany. *J. Geophys. Res. Atmos.* **2009**, *114*, D12113. [CrossRef]
57. Pinke, Z.; Decsi, B.; Kozma, Z.; Vári, Á.; Lövei, G.L. A spatially explicit analysis of wheat and maize yield sensitivity to changing groundwater levels in Hungary, 1961–2010. *Sci. Total Environ.* **2020**, *715*, 136555. [CrossRef]
58. Jalota, S.K.; Vashisht, B.B.; Sharma, S.; Kaur, S. Climate change and groundwater. In *Understanding Climate Change: Impacts on Crop Productivity and Water Balance*; Academic Press: Cambridge, UK, 2018.
59. Ray, D.K.; Ramankutty, N.; Mueller, N.D.; West, P.C.; Foley, J.A. Recent patterns of crop yield growth and stagnation. *Nat. Commun.* **2012**, *3*, 1293. [CrossRef]
60. Olesen, J.E.; Bindi, M. Consequences of climate change for European agricultural productivity, land use and policy. *Eur. J. Agron.* **2002**, *16*, 239–262. [CrossRef]
61. Zipper, S.C.; Soylu, M.E.; Booth, E.G.; Loheide, S.P. Untangling the effects of shallow groundwater and soil texture as drivers of subfield–scale yield variability. *Water Resour. Res.* **2015**, *51*, 6338–6358. [CrossRef]
62. Chai, Q.; Gan, Y.; Zhao, C.; Xu, H.; Waskom, R.M.; Niu, Y.; Siddique, K.H.M. Regulated deficit irrigation for crop production under drought stress. A review. *Agron. Sustain. Dev.* **2016**, *36*, 3. [CrossRef]
63. Zipper, S.C.; Qiu, J.; Kucharik, C.J. Drought effects on US maize and soybean production: Spatiotemporal patterns and historical changes. *Environ. Res. Lett.* **2016**, *11*, 094021. [CrossRef]
64. Cattivelli, L.; Rizza, F.; Badeck, F.-W.; Mazzucotelli, E.; Mastrangelo, A.M.; Francia, E.; Marè, C.; Tondelli, A.; Stanca, A.M. Drought tolerance improvement in crop plants: An integrated view from breeding to genomics. *Field Crop. Res.* **2008**, *105*, 1–14. [CrossRef]
65. Vereecken, H.; Weynants, M.; Javaux, M.; Pachepsky, Y.; Schaap, M.G.; van Genuchten, M.T. Using pedotransfer functions to estimate the van Genuchten–Mualem soil hydraulic properties: A review. *Vadose Zone J.* **2010**, *9*, 795–820. [CrossRef]
66. Connolly, R.D. Modelling effects of soil structure on the water balance of soil-crop systems: A review. *Soil Tillage Res.* **1998**, *48*, 1–19. [CrossRef]
67. Talebnejad, R.; Sepaskhah, A.R. Effect of different saline groundwater depths and irrigation water salinities on yield and water use of Quinoa in Lysimeter. *Agric. Water Manag.* **2015**, *148*, 177–188. [CrossRef]
68. Wu, Y.; Liu, T.; Paredes, P.; Duan, L.; Pereira, L.S. Water use by a groundwater dependent maize in a semi–arid region of Inner Mongolia: Evapotranspiration partitioning and capillary rise. *Agric. Water Manag.* **2015**, *152*, 222–232. [CrossRef]
69. Karimov, A.K.; Šimůnek, J.; Hanjra, M.A.; Avliyakulov, M.; Forkutsa, I. Effects of the shallow water table on water use of winter wheat and ecosystem health: Implications for unlocking the potential of groundwater in the Fergana Valley (Central Asia). *Agric. Water Manag.* **2014**, *131*, 57–69. [CrossRef]
70. Ayars, J.E.; Shouse, P.; Lesch, S.M. In situ use of groundwater by alfalfa. *Agric. Water Manag.* **2009**, *96*, 1579–1586. [CrossRef]
71. Yang, J.; Wan, W.; Deng, W.; Zhang, G. Water fluxes at a fluctuating water table and groundwater contributions to wheat water use in the lower Yellow River flood plain, China. *Hydrol. Proc.* **2007**, *21*, 717–724. [CrossRef]
72. Kahlown, M.A.; Ashraf, M.; Zia-ul-Haq. Effect of shallow groundwater table on crop water requirements and crop yields. *Agric. Water Manag.* **2005**, *76*, 24–35. [CrossRef]
73. Hutmacher, R.B.; Ayars, J.E.; Vail, S.S.; Bravo, A.D.; Dettinger, D.; Schoneman, R.A. Uptake of shallow groundwater by cotton: Growth stage, groundwater salinity effects in column lysimeters. *Agric. Water Manag.* **1996**, *31*, 205–223. [CrossRef]
74. Rydałowski, M. The influence of groundwater level and year conditions on the evapotranspiration of salix viminalis l. in view of lysimetric analyses. *Woda Środowisko Obszary Wiejskie* **2016**, *16*, 73–84. (In Polish)
75. Kravchenko, A.N.; Bullock, D.G. Spatial variability of soybean quality data as a function of field topography: I. Spatial data analysis. *Crop Sci.* **2002**, *42*, 804–815. [CrossRef]
76. Kravchenko, A.N.; Bullock, D.G. Spatial variability of soybean quality data as a function of field topography: II. A proposed technique for calculating the size of the area for differential soybean harvest. *Crop Sci.* **2002**, *42*, 816–821. [CrossRef]

77. Kravchenko, A.N.; Thelen, K.D.; Bullock, D.G.; Miller, N.R. Relationship among crop grain yield, topography, and soil electrical conductivity studied with cross-correlograms. *Agron. J.* **2003**, *95*, 1132–1139. [CrossRef]
78. Kennedy, B.A. *Surface Mining*, 2nd ed.; Society for Mining, Metallurgy and Exploration Inc.: Englewood, CO, USA, 1990.
79. Shah, A.M. Reclaiming Indian Mines. *Cornerstone* **2015**, *3*, 36–39.
80. Pietrzykowski, M. Reclamation and reconstruction of terrestrial ecosystems on mine sites–ecological effectiveness assessment. In *Energy Science and Technology, Coal Energy*; Govil, J.N., Ed.; Studium Press LLC: Houston, TX, USA; New Dehli, India, 2015; Volume 2, pp. 121–151.
81. Pavloudakis, F.; Roumpos, C.; Karlopoulos, E.; Koukouzas, N. Sustainable Rehabilitation of Surface Coal Mining Areas: The Case of Greek Lignite Mines. *Energies* **2020**, *13*, 3995. [CrossRef]
82. Gruszczyński, S. Classification of soils in post mining areas reclamation. *Przegląd Górniczy* **2010**, *66*, 120–125. (In Polish)
83. Kasztelewicz, Z. Approaches to post-mining land reclamation In Polish open-cast lignite mining. *Civ. Environ. Eng. Rep.* **2014**, *12*, 55–67. [CrossRef]
84. Michalski, A. Methods of management of former opencast mining areas of "Adamów" S.A. in Turek and "Konin" S.A. in Kleczew. *Soil Sci. Annu.* **2004**, *55*, 281–290.
85. The Photovoltaic Farm in Brudzew Will be Built by ZE PAK Subsidiary. Available online: https://www.wnp.pl/energetyka/farme-fotowoltaiczna-w-brudzewie-wybuduje-spolka-corka-ze-pak,439843.html (accessed on 17 January 2021). (In Polish).
86. Szczepiński, J. The Significance of Groundwater Flow Modeling Study for Simulation of Opencast Mine Dewatering, Flooding, and the Environmental Impact. *Water* **2019**, *11*, 848. [CrossRef]
87. Healy, R. *Estimating Groundwater Recharge*, 1st ed.; Cambridge University Press: Cambridge, UK, 2010.
88. Stachowski, P. Assessment of the reconstruction of the water table in the areas transformed by brown coal mining. *Rocznik Ochrona Środowiska* **2007**, *9*, 225–238. (In Polish)
89. Basso, B.; Kendall, A.D.; Hyndman, D.W. The future of agriculture over the Ogallala Aquifer: Solutions to grow crops more efficiently with limited water. *Earths Future* **2013**, *1*, 39–41. [CrossRef]
90. Kendy, E.; Zhang, Y.; Liu, C.; Jinxia, C.; Wang, J.; Steenhuis, T. Groundwater recharge from irrigated cropland in the North China Plain: Case study of Luancheng County, Hebei Province, 1949–2000. *Hydrol. Process.* **2004**, *18*, 2289–2302. [CrossRef]
91. Aw-Hassan, A.; Rida, F.; Telleria, R.; Bruggeman, A. The impact of food and agricultural policies on groundwater use in Syria. *J. Hydrol.* **2014**, *513*, 204–215. [CrossRef]
92. Calvino, P.A.; Andrade, F.H.; Sadras, V.O. Maize yield as affected by water availability, soil depth, and crop management. *Agron. J.* **2003**, *95*, 275–281. [CrossRef]
93. Leeper, R.A.; Runge, E.C.A.; Walker, W.M. Effect of plant-available stored soil moisture on corn yields: 1. Constant climatic conditions. *Agron. J.* **1974**, *66*, 723–727. [CrossRef]
94. Kasztelewicz, Z.; Sikora, M.; Zajączkowski, M. The Poniec-Krobia deposit in extending the Konin lignite mining and power generation basin. *Energy Policy J.* **2012**, *15*, 135–146. (In Polish)
95. European Commission. *A Farm to Fork Strategy for a Fair, Healthy and Environmentally-Friendly Food System*; Communication from the Commission to the European Parliament, the Council, the European Economic and Social Committee and the Committee of the Regions; 20.5.2020 COM (2020) 381 Final; European Commission: Brussels, Belgium, 2020.
96. Pepliński, B. *Determinants of Regional Changes in the Pig Production Sector in Poland*; Poznań University of Life Sciences: Poznan, Poland, 2019. (In Polish)
97. Central Statistical Office. *Procurement and Prices of Agricultural Products in 2015–2019*; Central Statistical Office: Warsaw, Poland, 2020. (In Polish)
98. Central Statistical Office. *Statistical Yearbook 1956–1997*; Central Statistical Office: Warsaw, Poland, 1998. (In Polish)
99. Central Statistical Office. *Statistical Yearbook of Voivodships 1975–1997*; Central Statistical Office: Warsaw, Poland, 1998. (In Polish)
100. Central Statistical Office. *Statistical Yearbook of Poviats 1970–1974*; Central Statistical Office: Warsaw, Poland, 1975. (In Polish)
101. Central Statistical Office. *Poviats Statistics 1966–1970*; Central Statistical Office: Warsaw, Poland, 1971. (In Polish)
102. Central Statistical Office. *Statistical Yearbook of the Bydgoszcz Voivodship 1960–1974*; Central Statistical Office: Bydgoszcz, Poland, 1975. (In Polish)
103. Central Statistical Office. *Statistical Yearbook of the Łódź Voivodship 1960–1974*; Central Statistical Office: Lodz, Poland, 1975. (In Polish)
104. Central Statistical Office. *Statistical Yearbook of the Poznań Voivodship 1960–1974*; Central Statistical Office: Poznan, Poland, 1975. (In Polish)
105. Central Statistical Office. *Statistical Yearbook of the Warszawa Voivodship 1960–1974*; Central Statistical Office: Warsaw, Poland, 1975. (In Polish)
106. Central Statistical Office. *Economic Development of Poviats in the Years 1950–1965*; Central Statistical Office: Warsaw, Poland, 1966. (In Polish)
107. Local Date Bank. Available online: https://bdl.stat.gov.pl/BDL/start (accessed on 5 January 2021).
108. Pepliński, B. Animal production location in suburban areas of Polish metropolitan centers. *Sustainability* **2020**, *12*, 2619. [CrossRef]
109. Stanisz, A. *An Accessible Statistics Course Using STATISTICA PL with Examples from Medicine*; StatSoft: Krakow, Poland, 2007. (In Polish)

110. Plans for Reclamation of Post-Mining Areas as Part of Activities Carried Out by ZE PAK S.A. Available online: https://www.google.com/url?sa=t&rct=j&q=&esrc=s&source=web&cd=&ved=2ahUKEwiPpe6r7MruAhVNiYsKHdDzAJgQFjAEegQIBRAC&url=http%3A%2F%2Forka.sejm.gov.pl%2Fopinie9.nsf%2Fdok%3FOpenAgent%26669_20201014_2&usg=AOvVaw08GYu_jyh245tOoCSpN3-- (accessed on 2 February 2021). (In Polish)
111. Gilewska, M.; Otremba, M. Revitalisation of post-mining resions in the area of the town of Konin. *Zeszyty Naukowe Inżynieria Środowiska* **2013**, *149*, 59–67. (In Polish)
112. Kordowski, J.; Brykała, D.; Bartczak, A.; Kaczmarek, H.; Słowiński, M. *Anthropogenic and Natural Changes of Geographical Environment in the Kuyavian-Pomeranian Voivodeship–Selected Examples, Prace Geograficzne 223*; PAN: Warsaw, Poland, 2010. (In Polish)
113. Paczyński, B.; Sadurski, A. *Regional Hydrogeology of Poland: Mineral, Curative and Thermal Waters and Mine Waters*; Państwowy Instytut Geologiczny: Warsaw, Poland, 2007; Volume II. (In Polish)
114. Kasztelewicz, Z.; Czyż, J.; Dwornik, E. Mining art in the 60th anniversary of KWB "Konin". *Węgiel Brunatny* **2005**, *53*, 23–29. (In Polish)
115. Wachowiak, G.; Wachowiak, A. Reservoir at the closing excavation of the open pit "Pątnów" of the lignite opencast mine "Konin" and its water balance in 2003–2004. *Badania Fizjograficzne Polska Zachodnia Seria A Geografia Fizyczna* **2005**, *56*, 157–176. (In Polish)
116. Opencasts. Available online: http://www.kwbkonin.pl/ (accessed on 25 January 2021). (In Polish).
117. Maćkowiak, J.; Siepielska, T.; Gradecki, D. Impact of lignite drainage in the Konin region on the near-surface aquifer. *Węgiel Brunatny* **1998**, *22*, 26–28. (In Polish)
118. Mocek, A.; Owczarzak, W. *Monitoring of Soil-Soil Water Table Necessary for Delineation of Soil Degradation Zones (Mainly Grasslands) in the Area Covered by the Stage and Soil Survey in the Tomisławice Open-Pit Area*; Poznań University of Life Sciences: Poznan, Poland, March 2013. (In Polish)
119. Magdziarek, M.; Brusiło, A.; Waloch, P.; Magdziarek, M.; Szulczyk, J.; Marecki, M. *Report on Environmental Impact of the Odkrywka Ościsłowo*; Ekogeo, Przedsiębiorstwo Projektowo—Usługowe: Wargowo, Poland, 2015. (In Polish)
120. Szablewski, W.; Derkowska–Sitarz, M. Filling of Lubstów reservoir of KWB Konin S.A. *Górnictwo Odkrywkowe* **2014**, *60*, 40–46. (In Polish)
121. Gilecki, R.; Kacprowska, J.; Dembicz, E.; Nagrodkiewicz, L.; Stosio, M. *Primary Energy Balance 2004–2019*; Agencja Rozwoju Energii: Warsaw, Poland, 2020. (In Polish)
122. Kasztelewicz, Z.; Ptak, M.; Sikora, M. Lignite as the optimal energy source for Poland. *Zeszyty Naukowe IGSMiE PAN* **2018**, *106*, 61–84. (In Polish)
123. Kasztelewicz, Z.; Patyk, M. The Modern and Efficiently Coal Power Plants Strategic Challenge for Polish. *Energy Policy J.* **2015**, *18*, 46–59. (In Polish)
124. Zespół Elektrowni Pątnów-Adamów-Konin SA, 2016 Results. Available online: https://ri.zepak.com.pl/en/presentations.html?y=2016 (accessed on 31 January 2021).
125. Zespół Elektrowni Pątnów-Adamów-Konin SA, 2014 Results. Available online: https://ri.zepak.com.pl/en/presentations.html?y=2014 (accessed on 31 January 2021).
126. Zespół Elektrowni Pątnów-Adamów-Konin SA, 2012 Results. Available online: https://ri.zepak.com.pl/en/presentations.html?y=2013 (accessed on 31 January 2021).
127. Sadowski, A.; Czubak, W.; Poczta, W.; Rowiński, J. Area and economic structures of Polish agriculture and other EU countries. In *The Structure of Polish Agriculture Compared to the European Union*; Poczta, W., Rowiński, J., Eds.; CeDeWu: Warsaw, Poland, 2019; pp. 45–81. (In Polish)
128. Pepliński, B.; Poczta, W.; Rowiński, J. Structure of livestock production and its evolution. In *The Structure of Polish Agriculture Compared to the European Union*; Poczta, W., Rowiński, J., Eds.; CeDeWu: Warsaw, Poland, 2019; pp. 83–139. (In Polish)
129. Electricity Market Characteristics. 2019. Available online: https://www.ure.gov.pl/en/markets/electricity/elctricitymrket/292,2019-Electricity-Market-Characteristics.html (accessed on 3 February 2021).
130. Włodek, S.; Biskupski, A. Soil moisture dynamics in the Belchatow brown coal mine area. *Soil Sci. Annu.* **2011**, *62*, 420–427. (In Polish)
131. Owczarzak, W.; Mocek, A.; Kaczmarek, Z. Impact of the "Drzewce" open-cast mine drainage barrier on the hydrological changes in the surrounding area. *Soil Sci. Annu.* **2011**, *62*, 311–324. (In Polish)
132. Owczarzak, W.; Mocek, A.; Kaczmarek, Z.; Gajewski, P. The evaluation of natural and anthropogenic degradation of soils in the area of dehydration influence of Drzewce lignite pit. *Soil Sci. Annu.* **2008**, *59*, 177–190. (In Polish)
133. Biskupski, A.; Włodek, S.; Pabin, J. Soil moisture dynamics within the depression sink of brown coal open mine at Belchatow. *Soil Sci. Annu.* **2008**, *59*, 18–24. (In Polish)
134. Pabin, J.; Włodek, S.; Biskupski, A. The influence of the deep drainage of the brown coal strip mine at Belchatow on soil water content. *Soil Sci. Annu.* **2008**, *59*, 191–195. (In Polish)
135. Włodek, S.; Biskupski, A.; Pabin, J. Soil moisture in the region of brown coal strip mine at Belchatow. *Soil Sci. Annu.* **2008**, *59*, 230–235. (In Polish)
136. Mocek, A.; Owczarzak, W.; Kaczmarek, Z. Changes of groundwater levels in soils surrounding the "Kozmin" brown coal exploitation hollow. *Roczniki Akademii Rolniczej Poznaniu Melioracje Inżynieria Środowiska* **2002**, *23*, 331–342. (In Polish)

137. Mocek, A.; Owczarzak, W.; Kaczmarek, Z. Hydrological changes in the soils of Warta Pravalley within the influance of "Koźmin" lignite pit. *Contemp. Probl. Manag. Environ. Prot.* **2009**, *1*, 109–124.
138. Chodak, T.; Kaszubkiewicz, J.; Bogda, A.; Szerszeń, L.; Bogacz, A. Water conditions of soils located within the range of depression funnel of the existing water intake in Brzeg Głogowski and the planned one in the region of Czerna-Dobrzejowice. *Zeszyty Problemowe Postepow Nauk Rolniczych* **1995**, *418*, 169–174. (In Polish)
139. Chodak, T.; Borkowski, J.; Bogda, A.; Szerszeń, L.; Nowiński, S. Prognosis of changes in soil environment within the range of depression funnel of "Potoczek" water intake. *Zeszyty Problemowe Postepow Nauk Rolniczych* **1995**, *418*, 175–184. (In Polish)
140. Włodek, S.; Biskupski, A.; Pabin, J.; Hryńczuk, B.; Kaus, A. The dynamics of ground moisture in the region of brown coal strip mine at Bełchatów on the example of locality Łękińsko. *Roczniki Akademii Rolniczej Poznaniu* **1999**, *160*, 211–219. (In Polish)
141. Hlavinka, P.; Trnka, M.; Semerádová, D.; Dubrovský, M.; Žalud, Z.; Možný, M. Effect of drought on yield variability of key crops in Czech Republic. *Agric. For. Meteorol.* **2009**, *149*, 431–442. [CrossRef]
142. Ren, D.; Xu, X.; Engel, B.; Huang, G. Growth responses of crops and natural vegetation to irrigation and water table changes in an agro–ecosystem of Hetao, upper Yellow River basin: Scenario analysis on maize, sunflower, watermelon and tamarisk. *Agric. Water Manag.* **2018**, *199*, 93–104. [CrossRef]
143. Nosetto, M.D.; Jobbágy, E.G.; Jackson, R.B.; Sznaider, G.A. Reciprocal influence of crops and shallow ground water in sandy landscapes of the Inland Pampas. *Field Crop. Res.* **2009**, *113*, 138–148. [CrossRef]
144. Morgan, B.J.; Daughtry, C.S.; Russ, A.L.; Dulaney, W.P.; Gish, T.J.; Pachepsky, Y.A. Effect of shallow subsurface flow pathway networks on corn yield spatial variation under different weather and nutrient management. *Int. Agrophys.* **2019**, *33*, 271–276. [CrossRef]
145. Georgakellos, D.A. Impact of a possible environmental externalities internalisation on energy prices: The case of the greenhouse gases from the Greek electricity sector. *Energy Econ.* **2010**, *32*, 202–209. [CrossRef]
146. Büke, T.; Köne, A.C. Estimation of the health benefits of controlling air pollution from the Yatağan coal-fired power plant. *Environ. Sci. Policy* **2011**, *14*, 1113–1120. [CrossRef]
147. Dimitrijević, Z.; Tatić, K.; Knežević, A.; Salihbegović, I. External costs from coal-fired thermal plants and sulphur dioxide emission limit values for new plants in Bosnia and Herzegovina. *Energy Policy* **2011**, *39*, 3036–3041. [CrossRef]
148. Taranto, Y.; Köksal, M.A.; Dal, E. *The External Cost of Fossil Fuel Use in Power Generation, Heating and Road Transport in Turkey*; Shura Energy Transition Center: Istanbul, Turkey, 2020.
149. Papagiannis, A.; Roussos, D.; Menegaki, M.; Damigos, D. Externalities from lignite mining-related dust emissions. *Energy Policy* **2014**, *74*, 414–424. [CrossRef]
150. Ministry of Energy. *Energy Policy of Poland Until 2040 (EPP2040)*; Ministry of Energy: Warsaw, Poland, 2021.

energies

Article

Planning and Settlement Conditions for the Development of Renewable Energy Sources in Poland: Conclusions for Local and Regional Policy

Przemysław Śleszyński [1], Maciej Nowak [2], Agnieszka Brelik [3], Bartosz Mickiewicz [3] and Natalia Oleszczyk [3,*]

1 Institute of Geography and Spatial Organization, Polish Academy of Sciences, Twarda 51/55, 00-818 Warsaw, Poland; psleszyn@twarda.pan.pl
2 Department of Real Estate, Faculty of Economics, West Pomeranian University of Technology, Żołnierska 47, 70-310 Szczecin, Poland; Maciej.Nowak@zut.edu.pl
3 Department of Regional and European Studies, Faculty of Economics, West Pomeranian University of Technology, Żołnierska 47, 70-310 Szczecin, Poland; Agnieszka.Brelik@zut.edu.pl (A.B.); bartosz.mickiewicz@zut.edu.pl (B.M.)
* Correspondence: natalia.oleszczyk@zut.edu.pl

Abstract: The article presents an analysis of geographical-settlement and legal-planning conditions for the development of Poland's distributed generation. The choice of this country is important and interesting due to the highly dispersed settlement, which may be a factor stimulating the development of this type of energy systems. For this reason, the analysis can be a model for other countries and regions, indicating ways to analyze and evaluate settlement and planning conditions for the development of renewable and distributed energy. At the same time, Poland is struggling with a severe crisis of spatial planning. By analysing these opportunities and threats, empirical analyses try to indicate regularities in this respect in Poland's regions in a detailed approach to communes and detailed legal and planning conditions. The conclusions emphasise the usefulness of distributed generation development for peripheral and sparsely populated areas of Europe and other parts of the world and appropriate directions of changes in spatial development law.

Keywords: distributed generation; settlement systems; local development

Citation: Śleszyński, P.; Nowak, M.; Brelik, A.; Mickiewicz, B.; Oleszczyk, N. Planning and Settlement Conditions for the Development of Renewable Energy Sources in Poland: Conclusions for Local and Regional Policy. *Energies* **2021**, *14*, 1935. https://doi.org/10.3390/en14071935

Academic Editor: Maarten Wolsink

Received: 25 February 2021
Accepted: 27 March 2021
Published: 31 March 2021

1. Introduction

The currently observed strong changes in the world's energy sector make it necessary to reconstruct current local and regional development views adequately. It applies especially to Poland, where when the world is experiencing an increasingly advanced energy evolution, most of the energy produced is still obtained from non-renewable sources, including geographically concentrated large professional thermal power plants (over 70% of electricity in 2019). It is deeply embedded in this country's history, including the model of economic development chosen after World War II in the conditions of a command and distribution system, consisting in the development of particularly energy-intensive industries, and accelerated industrialisation [1,2]. Meanwhile, most of the more developed countries have been observing a reorientation towards renewable energy sources (RES) for at least two decades and a strongly related shift towards the deconcentration of energy production, especially in rural and urban-rural transition areas [3].

At the same time, apart from the development of transport systems, energy production is one of the fundamental factors determining the operation of infrastructure and the movement of people and goods, and thus economic and social development. Hence, ensuring optimal conditions for the development of energy and energy use becomes a fundamental issue that also determines appropriate spatial development.

With the problem formulated in this way, the article aims to identify barriers and stimulants for the development of renewable energy in Poland, resulting from the settlement, demographic, economic and formal-legal conditions of the spatial management system. The research objective is also an attempt to find an answer to the extent to which communes secure the areas for possible functions related to energy. It is an important issue of a demand and supply nature, related to investment opportunities on the one hand and spatial (and development) policies of local governments on the other. It is the dimension of spatial policies that significantly determines the development of renewable energy in Poland. The significant relationship of the indicated issues is confirmed by the latest Territorial Agenda of the European Union 2030 [4]. The possibilities of locating the energy sector guaranteed by the national spatial management systems are determined by the priorities of "Fair Europe" and "Green Europe" included in the Agenda (regardless of the above, renewable energy is promoted in other EU acts and documents [5]).

Considering the above goals, the following research questions were formulated:

(1) How does spatial policy in Poland determine the development of distributed renewable energy? In which ranges is it positive and stimulating, and in which negative and destimulating?
(2) In which parts of the country is the planning situation better organised, and are there any regularities related to communes' morphological and functional specificity?
(3) In which regions of the country should the settlement structure be particularly suited to developing dispersed renewable energy?
(4) What legal and planning actions should be taken to promote the development of distributed renewable energy more effectively?

To answer these questions, in the empirical part, based on data on the development of settlement in Poland, the correlation between the dispersion of settlement and the expansion and reduction of energy emissions will be determined. In this context, Poland's choice as a research field is important and interesting for several reasons. First, it is the largest country in Central Europe, undergoing the so-called political transformation after 1989 and the economic system's change from command and distribution (centrally controlled) to a free-market one. Secondly, this country struggles with a crisis in spatial management, and the scale of neglect in spatial planning is one of the largest in Europe [6,7]. It may cause significant barriers to the development of renewable energy. Finding and describing them should answer how to avoid and prevent such problems in other countries. Thirdly, Poland is characterised by a highly dispersed settlement arrangement [8], which is constantly deepening today [9,10], which translates into a different energy demand structure.

There follows a premise: the main thesis and the guiding axis of the article that Poland's settlement structure, especially in rural areas, may be a significant stimulator of distributed generation of renewable energy. In the opinion of the authors, this problem is insufficiently researched, and it can be a model for other countries and regions by indicating the ways of analysis and evaluation of settlement and planning conditions for the development of renewable and distributed energy. It is connected with better efficiency of energy systems and broadly understood spatial organization as well as with energy security. The literature highlights the mismatch between broadly defined spatial structures and the challenges of both natural processes (especially climate change) and civilization change [7]. This problem may be exacerbated by the post-COVID-19 pandemic situation, as changes in the mobility of societies and locations of residence are predicted, involving a greater desire to live in less populated areas, with lower population densities, in smaller towns and villages [11].

These issues are connected to the challenges of public policies, especially spatial policies (which is also the perspective of this article). It must be emphasized that the implementation of RES investments, as well as other demands (such as those mentioned above, included in the Territorial Agenda 2030) require adaptation and efficiency of specific spatial policy tools [12–14]. Also in this context, the Polish example is very good—because it includes a system containing numerous inefficiencies of public authorities. The article indicates the relationship between these (presented) manifestations of inefficiencies of public authorities and spatial policy tools and the real possibilities of RES investment implementation.

The findings in the first part of the article will be related to the legal framework for the location of renewable energy sources (and the planning problems that arise on this occasion). Recognition of renewable energy sources in the legal framework required modification of the scope of spatial policy tools, also covering the sphere of development of other areas. In this regard, the article identifies possible spatial conflicts between the location of renewable energy sources and the implementation of other functions of individual areas.

The article firstly analyses determinants of renewable energy development in Poland against the background of global trends. Especially the European context was taken into account. Then the methods were described. The next part includes the characteristics of planning and settlement conditions in Poland. These parts, apart from referring to the literature and legal regulations, contain the results of the study. The results are discussed. The conclusions include a summary and recommendations for spatial policies.

2. Determinants of Renewable Energy Development in Poland against the Background of Global Trends

The literature on the subject repeatedly points to the significant role of renewable energy sources in social and economic development [15]. It is also why the role of national policy and national conditions in shaping the energy market is so important [16–18]. This role must often be reduced to various public authorities activities related to the economic policy of states and regions [15–17]. In this context, it is particularly important to emphasise the energy policy's role, also related to increasing the share of renewable energy in total energy consumption [18]. Undoubtedly, the reference point on this occasion will be income and energy consumption [19,20]. Among the key objectives of RES development are both issues related to efficient management and ensuring wider environmental protection [21]. Besides, RES have a very wide potential, broadly responding to diverse forms of urban demand [22,23]. This provides a basis for further postulates related to different contexts of RES analyses and their effects [24], also in the context of linking public policies with innovations.

Depending on the specificity of individual countries, the possibilities of obtaining income and increasing the share of renewable energy directly depend on these countries. The reasons for such differentiation may be broadly understood as natural conditions and economic profitability [25–27]. Despite these differences, the very fact of the growing role of renewable energy is beyond doubt, for example, in the context of the development of rural areas, the development of the local energy value chain [28–32], as well as mitigating climate change and greater care for the environment [33]. Recent studies also highlight the important role of RES in the context of COVID-19 impacts [34].

In this context, we come to the issue of important determinants of distributed generation (DG), which include sources located independent of central planning and, at the same time, with powers in the range most often from 15 kW to 10 MW [35–37]. Renewable energy sources create specific conditions for the development of DG. They can be used by medium and low-power plants located near consumers. Currently, apart from energy production costs, there is also a technical barrier [38] to the development of distributed energy and investment risk in competitive energy markets [39].

In Poland's case, the power system's current shape, adapted to large generating units, is a serious problem (Polish coal-fired commercial power plants are among the largest in Europe) [40]. The inclusion in the DG system of small units generating energy with different technical parameters and the amount of energy production varying in time makes it difficult to control and monitor the entire system. However, their implementation may bring many technical, economic and environmental benefits, such as power improvement, reliability, system safety, reduction of high-level investment outlays, or reduction of greenhouse gas emissions [36,37,41].

The literature points to numerous problems when conducting research in the aforementioned area. These include lack of knowledge on appropriate research software, broader classifications and typologies of renewable and non-renewable sources, as well as shortages in the literature on the subject [42]. It seems all the more appropriate to approach this problem in different contexts, including those related to settlement and spatial policy. The role of public authorities can be understood very broadly, for example, by increasing the importance of the regional and local level (e.g., through appropriate legislative initiatives that give more freedom in the organisation of local energy systems and independence from the monopoly of energy suppliers). Increasing distributed energy resources at this level is equivalent to the need to reorganise centralised energy systems in many cases [43]. Also thanks to the tools of spatial policy, diversified incentives (considered crucial in the literature [44]) may be developed for the implementation—both on a national, regional and local scale—of RES investments.

The above problem is well illustrated by the changes that have taken place in Poland since accession to the European Union (2004) in connection with the increasing share of energy production from renewable sources (Figure 1). Data are available for 16 voivodships, which were grouped into 3 clusters according to urbanization level. In the less urbanized regions, the level was the highest, but it stopped at the same level since 2014. The lowest share was in the most urbanized provinces, which is due to higher energy demand, including the location of industry. Meanwhile, these provinces through conventional power plants contribute the most to environmental pollution, smog, etc. For example, the most heavily industrialized Śląskie (Silesian) voivodship produced 21.6 TWh of electricity in 2019, while the share of renewable energy was only 4.4%. This example shows how serious the problem is the structure of energy production in Poland in relation to the settlement system of the country.

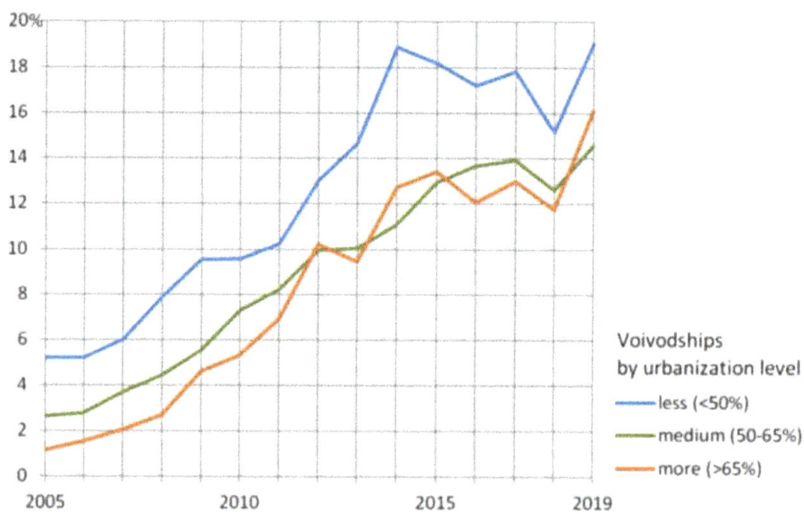

Figure 1. Share of energy production from renewable sources in Poland in 2005–2019 by level of urbanisation of provinces.Source: based on data from Local Data Bank of Central Statistical Office of Poland.

Therefore, there is no doubt that the issues mentioned above are important from both a national and local perspective. In the latter context, it is related to the issues of spatial policies [45]. As a rule, planning authorities act as energy users, participants of the competitive energy market, local energy regulators, and investors and energy producers [46–49]. The local authorities can stimulate local communities to generate renewable energy for their needs [50]. The state must support the development of the energy sector in a way that enables the achievement of specific socio-economic goals at the national, regional and local level, which is partly at odds with the principles of economic freedom and competition in the energy market [51].

The literature indicates that the location of a particular power plant itself is important from a variety of perspectives [52]. Besides the question of location, other important issues may be connected with the requirements of spatial policy. One of them may be the attempt to define comprehensive urban energy planning, especially adapting it to predictable changes in the environment [22]. However, the very aspect of integrating energy policy with environmental protection is also important [20]. The basis for an adequate connection of these spheres, apart from the demands of individual solutions [53,54], may be an adequate spatial policy. It is from the perspective of specific spatial policy actors that it will be possible to propose solutions to reconcile potential collisions of various policies and potential spatial conflicts.

Issues related to the verification of RES development for Central European countries were the subject of research. They show, among other things, that in all countries the use of renewable energy generally has a positive impact on economic growth. In some cases, the lack of some unambiguous statistical confirmation is due to the smaller scope of renewable energy in these countries compared to other EU member states [55]. Besides, internal variations also occur in the countries of Central and Eastern Europe [56]. They also result from settlement characteristics. A particularly important condition is that Poland belongs to the countries with a local settlement's dispersed structure. It applies to both typical rural areas and urban areas and, to a large extent, cities of various sizes. The dispersion conditions are different for each of these settlement types: historical [57] and contemporary [10]. The highly dispersed development is the basic reason for the high operating costs of the technical and social infrastructure intended for its operation [58]. Against the background of the Central European region, Poland has the lowest renewable energy share index (12.2). It should also be noted that this share increased between 2004 and 2019 by only 5.3 percentage points (in the other four countries, i.e., Austria, Czechia, Hungary, Slovakia—by 8.3–11.1 pp.). These results show the importance of Poland catching up with the requirements of Directive (EU) 2018/2001 of the European Parliament and of the Council of 11 December 2018. At the same time, it justifies undertaking studies that could accelerate the achievement of this objective.

3. Materials and Methods

The empirical analyses used unique data from the survey. The Central Statistical Office sends to all municipalities (gminas) in Poland (2477 units in 2019) to cooperate with the ministry responsible for spatial management. These data show the planned directions of land use in communes in two types of planning documents: studies of the conditions and directions of spatial development (commune studies) and local spatial development plans (local plans). In this, the key question about the share of land intended for industrial functions (allowing the construction of small power plants) was answered in 2019 by 2302 units (nearly 93%), and of the 2325 municipalities with at least one local plan—2302 (i.e., almost all). All larger cities were excluded from the analysis (i.e., cities with poviat rights, where the conditions for the development of energy based on RES are significantly different from those occurring in rural areas, urban-rural transition areas and smaller towns.

According to the existing regulations, there is no strict definition of production sites in the municipal study. It is treated by default as related to industrial production and facilities (warehouses, stores). On the other hand, local plans use the category of "technical and product development areas", which include "areas of production facilities and warehouses" and "mining areas".

In part concerning settlement analysis, data on the distribution of buildings and the average distances between them were used. The data was obtained from the Database of Topographical Objects (BDOT). A detailed description of this source can be found in [10].

Besides, data on the tangible effects of investments were used, i.e., the usable floor area of buildings completed, broken down into residential and non-residential buildings (data based on the Local Data Bank of the Central Statistical Office of Poland). Data exclude non-residential farm buildings.

The analyses used the communes' classification into ten types, prepared especially to monitor spatial planning [59]. This classification was developed based on the deductive-inductive method. A settlement system forms the "skeleton" with different administrative and functional hierarchy levels. The remaining types of communes are separated mainly because of their socio-economic functions and morphological features (Table 1).

4. Results—Determinants of the Location of Renewable Energy Sources in Poland

4.1. Planning Conditions

The Polish spatial planning system's numerous weaknesses have been pointed out for years [60–62]. They have recently been expressed by defining and distinguishing the specific costs of spatial chaos [10]. At the local level, three direct spatial policy tools are distinguished: studies of the conditions and directions of spatial development, local spatial development plans and decisions on building conditions and land development. The studies of the conditions and directions of spatial development are documents of a creative and conceptual nature. The regulatory action is the local spatial development plans, which bind the purpose and principles of land development. However, local plans are not obligatory: municipal authorities freely decide on their adoption. Often barriers to their enactment are concerns about the compensation consequences that municipalities would have to pay to property owners.

A problem in the Polish spatial management system is that studies of the conditions and directions of spatial development have limited impact. Their provisions are only binding for local spatial development plans. However, local plans are not obligatory (and have not been enacted in most parts of Poland). In this situation, many municipalities do not create a broader spatial policy. On the other hand, even the fact of enacting local plans is not synonymous with positive effects. Very often, local plans are constructed incorrectly (and cause doubts in interpretation [63]). Many municipalities, fearing claims of property owners, allow too wide development in their local plans.

Moreover, spatial policy tools at the local level are still not sufficiently integrated with other policy acts, such as local development strategies [64]. It is possible (and often happens) that such documents diverge completely. In the Polish spatial development law there is a principle of municipality independence. This means that the municipality may shape the local space on its own, unless the specific possibilities of interference are directly assigned to the supra-local level of government. At the same time, the Polish spatial planning system strongly (in the authors' opinion too strongly) protects the individual rights of property owners [65]. It contributes to numerous spatial conflicts and barriers to the protection of spatial order [66].

On the other hand, in a situation when a local plan does not cover a given area, an administrative decision becomes a specific counterpart of the plan (a decision on building conditions and land development, completely 'detached' from the local planning order, often independent of the content of the study of conditions and directions of spatial development [67]. Such decisions protect the spatial order even less than other spatial policy tools [68]. In order to issue such decisions, municipalities must verify whether the criteria set forth in the law are met. In practice, these criteria are formulated in very general terms and in different cases they are interpreted quite differently.

Table 1. Characteristics of the analysed communes (gminas) in Poland, for the year 2019.

Type *	Included in the Analysis	Number of Communes	Area (km^2)	Population (Thous.)
A	No	33	5004	9563
B	Yes	265	27,589	4873
C	No	55	3399	4322
D	Yes	201	21,468	2452
E	Yes	142	10,265	3783
F	Yes	137	19,964	1435
G	Yes	222	33,856	1817
H	Yes	496	62,971	3048
I	Yes	665	86,532	5179
J	Yes	261	41,657	1910
Total (Poland)		2477	312,705	38,383
Total % ("Yes")		96.4%	97.3%	63.8%

* A—functional urban areas of voivodeship capitals; B—their external zones; C—functional urban; areas of subregional centres; D—their external zones; E—multifunctional urban centres; F—gminas (communes) with developed transport functions; G—gminas with other developed non-agricultural functions (tourism and large-scale functions, including mining); H—gminas with intensively developed agricultural functions; I—gminas with moderately developed agricultural functions; J—extensively developed gminas (with forests or nature protection areas). Source: based on Śleszyński and Komornicki classification [65] and data from the Ministry of Development and Central Statistical Office of Poland.

The basic planning order shaped in this way in Poland is modified by numerous sectoral laws. The conditions for the implementation of individual investments are also influenced in a different way. This also applies to investments in renewable energy sources. Problems occur with both the use of local plans and location decisions. In the case of local plans, the investments in question are presented in an inconsistent manner. A frequent tendency is limiting the scope of RES investments, as well as including them in an unclear, undefined way. This causes specific barriers in the implementation of investments [69]. Changing investment needs (or the need to clarify them) will usually require a time-consuming amendment to the local plan.

On the other hand, potentially overly broad permissions for designated investments in local plans exacerbate spatial conflicts. The situation is even worse in the case of RES investments carried out on the basis of decisions on development conditions. Judicial decisions on the conditions of considering investments under this procedure are very diverse, which creates chaos when applying specific criteria.

The Polish legislator has separately addressed the planning basis for the implementation of wind power plants. The Act of 20 May 2016 on investments in wind power plants introduced (in Article 3) the principle that the location of wind power plants that are not micro-installations can only take place on the basis of local spatial development plans. Moreover, investments in wind turbines may be implemented at a certain distance from residential developments (residential buildings or buildings with mixed functions including residential functions). The minimum distance in this respect must be equal to ten times the height of the wind turbine. A local plan adopted for an investment specifying the location of a wind power plant must cover the entire area associated with the obligatory development restrictions.

The above solution is rigorous, especially in relation to the previous legal status. The said restriction is binding not only for the realization of an investment in the scope of a wind power plant, but also for the realization of residential buildings in a close proximity to the completed wind power plant (and on this occasion it applies not only to local plans, but also to decisions on development conditions). At the same time, it should be stressed that under the legal status before the enactment of the act in question, there were no top-down (statutorily imposed) limitations of this type (only the limitations of permissible noise levels for specific developments could be derived). As a result, power plants were built in too close proximity to buildings, especially residential buildings.

A certain problem in the context of spatial policy is also the lack of differentiation of competencies of municipalities with respect to specific types of municipalities. This means that e.g., authorities of agricultural and urban communes have exactly the same spatial policy tools at their disposal. At the same time, the role of the powiat level in this context (a higher local government unit than municipalities) is negligible.

Planning securing the areas for functions related to the energy sector depends not only on the legal framework but also on the widely understood planning practice adopted in a given country. Planning coverage with applicable local plans means conscious planning for future use and securing land for investments of various activity profiles, including RES infrastructure facilities. In Poland, at the end of 2019, there were 55.6 thousand valid local plans and another 9.0 thous. were in the design process (including 54.6% of their area related to the existing plan change). The coverage was very uneven, both between regions (Figure 2) and in the types of communes (Table 2).

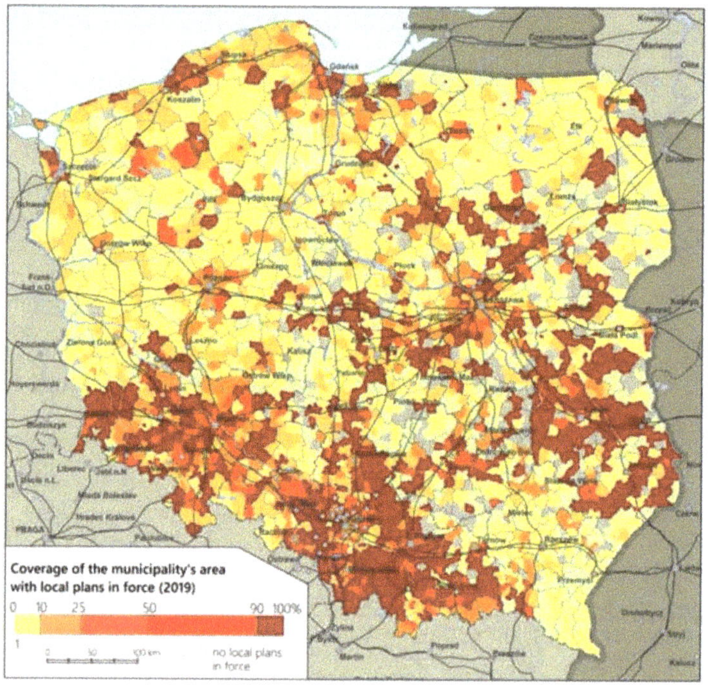

Figure 2. Planning coverage in Poland, for the year 2019. Source: based on data from the Ministry of Development and Central Statistical Office of Poland.

Table 2. Planning coverage by commune type and by population density, for the year 2019.

Commune Type	Planing Coverage (%)	Including in Communes with Population Density (Persons Per 1 sq. km)		
	Total	Below 50	50–150	Above 150
BD	42.1	13.6	39.0	63.4
EFG	29.9	21.4	36.5	42.0
HIJ	28.1	23.6	33.1	58.0
Total	31.2	22.7	35.1	54.0

Source: based on data from the Ministry of Development and Central Statistical Office of Poland.

In general, in Poland, it is visible that the planning coverage is high in the south of the country: in the Dolnośląskie, Śląskie and Małopolskie voivodeships, as well as in Lubelskie and partly Opolskie. In many communes of these regions, all communes or slightly less (90%) are covered by local law's binding document. In other voivodeships, this indicator is high only in some areas, including agglomerations (coming from the north—Szczecin, Tri-City, Poznań, Łódź, Warsaw, Kielce).

At the other extreme, there is the northern part of the country, including its part of the lake district (especially the Zachodniopomorskie, Lubuskie, Kujawsko-Pomorskie and Warmińsko-Mazurskie voivodeships). In the south of the country, it is the Podkarpackie Province. The planning coverage indices there generally do not exceed 10% of the area of communes. In 146 communes in Poland (out of 2477 existing ones), in 2019, there was not a single local plan (e.g., concentrations of such communes are in Podlasie).

If the municipalities are grouped by population density (Table 2), it turns out that there is a clear relationship with the planning coverage. It is the lowest in sparsely populated communes (less than 50 people per 1 km^2). The most disturbing fact is that the low coverage in this category of population density occurs in B and D types, i.e., suburban zones (only 13.6%). In general, coverage is also lower in typically rural areas (HIJ—28.1%) and urban areas outside agglomerations (EGF—29.9%). On the other hand, in the mentioned suburban zones (BD), the coverage is relatively high (42.1%), though certainly too small relating to the needs resulting from intensive suburbanisation processes.

However, the planning coverage itself is not the only and exhaustive indicator that would assess the protection of areas for various functions. Next, Figure 3 and Table 3 show the share of the area intended for production functions in local plans, which should illustrate the protection of the areas for energy investments. In total, there were 403,000 ha such areas. On average, it was 1.29% of communes' area for the whole country, and again the least in the least populated areas (0.68%). However, it is worth noting that such areas are anticipated in many municipalities where coverage is low (in northern Poland). But even there, these areas generally do not exceed 1% or even 0.2% of the communes' area.

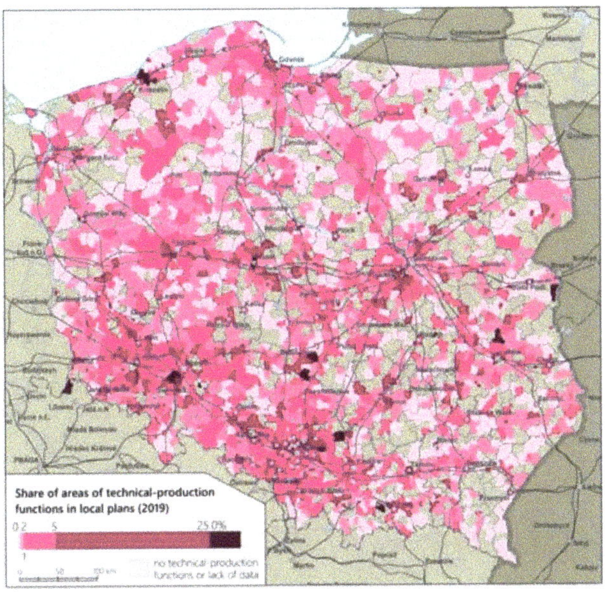

Figure 3. Share of the area of communes allocated to production functions in the spatial development local plans (local plans), for the year 2019 (relating to the area of communes). Source: based on data from the Ministry of Development and Central Statistical Office of Poland.

Table 3. Share of the area of communes with the production function planned in the local plans, for the year 2019.

Commune Type	Planing Coverage (%)	Including in Communes with Population Density (Persons Per 1 sq. km)		
	Total	Below 50	50–150	Above 150
BD	2.19	0.65	2.08	3.21
EFG	1.19	0.36	1.28	3.73
HIJ	0.93	0.77	1.04	3.34
Total	1.29	0.68	1.32	3.79

Source: based on data from the Ministry of Development and Central Statistical Office of Poland.

Besides, the documents of the study of the conditions and directions of spatial development for production functions provide for 592 thousand ha (data from 1687 communes for 68% of their number and 66% of the country's area). It is on average 2.9%, the highest in urban types A and C (7.6–8.7%), and the lowest in types D, F, G and H (less than 3%), including type G—1.2%.

All in all, it should be acknowledged that the features resulting from the planning documents are not favourable for locating the energy sector in Poland. Securing the sites is only available in a smaller part of the country. Moreover, apart from the indicated data, one should bear in mind the varied content of planning provisions. As indicated above, the mere fact of a specific destination of the area does not constitute a guarantee that a given investment will be implemented quickly and without problems (judging only from the planning perspective). Barriers related to the Polish spatial management system (also the lack of sufficient flexibility in planning) will be noticeable also in the described context. More problems may arise when using a tool alternative to local plans, i.e., the decision on development conditions. One should also bear in mind the institutional limitations in implementing investments in wind farms implemented based on local plans.

4.2. Settlement Conditions

Poland's settlement structure's uniqueness results from the fact that it was shaped as a result of special historical influences, including significant shifts of borders in the 20th century. Regions of Poland's present-day territory developed basically independently of each other, which allowed for the emergence of large cities. After the Second World War, a large city system, known as polycentric, was finally formed. However, historical processes also had an overwhelming influence when it comes to small towns and rural settlements. The foundations of the rural settlement structure were formed as early as in the Middle Ages and under the influence of feudalism, which determined agrarian relations and the distribution and size of villages [57]. The agrarian relations were quite different in individual magnate states, and additionally, this was due to the different natural conditions. The situation in the nineteenth century had a significant impact. When the European countries were undergoing the industrial revolution and intensive urbanisation, Poland was a country divided among the partitioning states that pursued various policies in this regard.

Nowadays, in Poland, we can distinguish three partition zones (Russia, Prussia, Aussie) with differently advanced urbanisation processes, as well as historically industrialised (Silesia) and peripheral lands (Lubusz, Pomerania) of the German states. Two regularities follow from this. First, the area with the highest population density has the shape of a triangle, the base of which is the south of the country, and the peak is in the Tri-City. Secondly, there are different types of rural settlement in the country. The south is dominated by large and quite densely distributed villages, and in the north—small and rare (Figure 4).

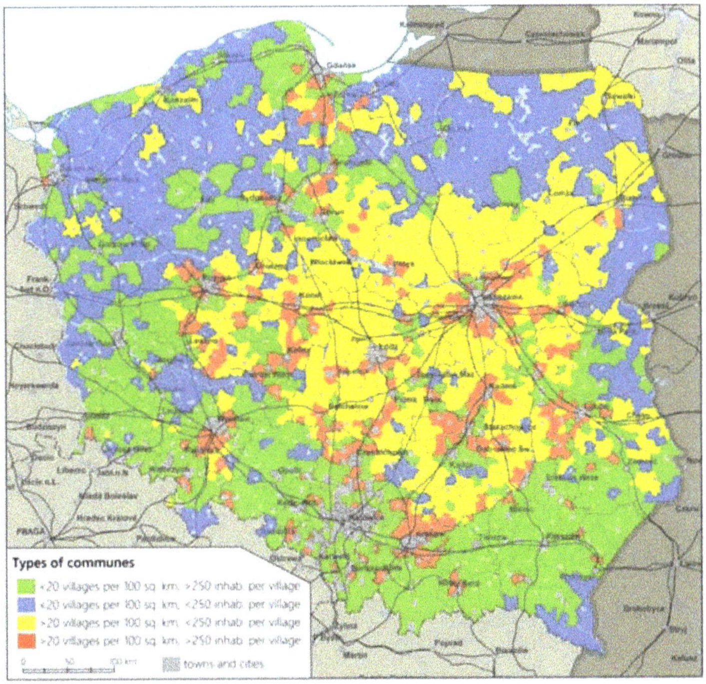

Figure 4. Typology of rural settlement in Poland in terms of village size and distribution, for the year 2019. Source: based on data of the Central Statistical Office of Poland.

In Poland, after around 1995, the process of centrifugal dispersion of buildings is progressing, and thus the deconcentration of settlement is part of the suburbanisation process [70,71]. Buildings "spread" far beyond the administrative borders of cities, as well as in many rural areas, especially attractive for tourists. However, it is not a simple "spreading" in Polish conditions, consisting of a gradual, relatively regular, centrifugal and elongated along the roads, occupying successive stretches of land, called "urban sprawl" in Western literature. In Poland, it consists of the chaotic induction of buildings in places that are often very distant from the previous settlement [72]. Between them, there are undeveloped, extensively developed spaces.

In total, the observed and growing dispersion of buildings and the deconcentration of settlement systems directly increase their functioning costs. It concerns three aspects resulting from lower population density and greater distances between places of residence, work, and services [10]:

- construction, modernisation, and maintenance costs of all line and point infrastructure,
- the costs of establishing relationships since places with different socio-economic functions are too far away,
- lower or no synergy effects and the so-called agglomeration benefits (scale).

The total costs of this are estimated at over PLN 80 billion per year.

The last data on the size structure of rural settlement in Poland comes from 2009. In this period, 41.7 thous. villages (Table 4). The vast majority of the rural population was concentrated in medium and large villages (101–3.333 inhabitants—90.2%). On the other hand, the number of villages up to 100 inhabitants was relatively large (22.1%), but they concentrated only 3.7% of the rural population. The share of villages with such a few inhabitants was characteristic, especially in the country's north-east (Figure 5).

Table 4. Size structure of villages in Poland, for the year 2009.

Size	Number of Villages	Share of Number (%)	Number of Population (thous.)	Share of Population (%)
<33	1450	3.5	22	0.1
34–100	7751	18.6	537	3.6
101–333	19,499	46.8	3822	25.4
334–1000	10,168	24.4	5575	37.0
1001–3333	2587	6.2	4178	27.7
>3333	204	0.5	939	6.2
Total	41,659	100.0	15,073	100.0

Source: based on Central Statistical Office of Poland data.

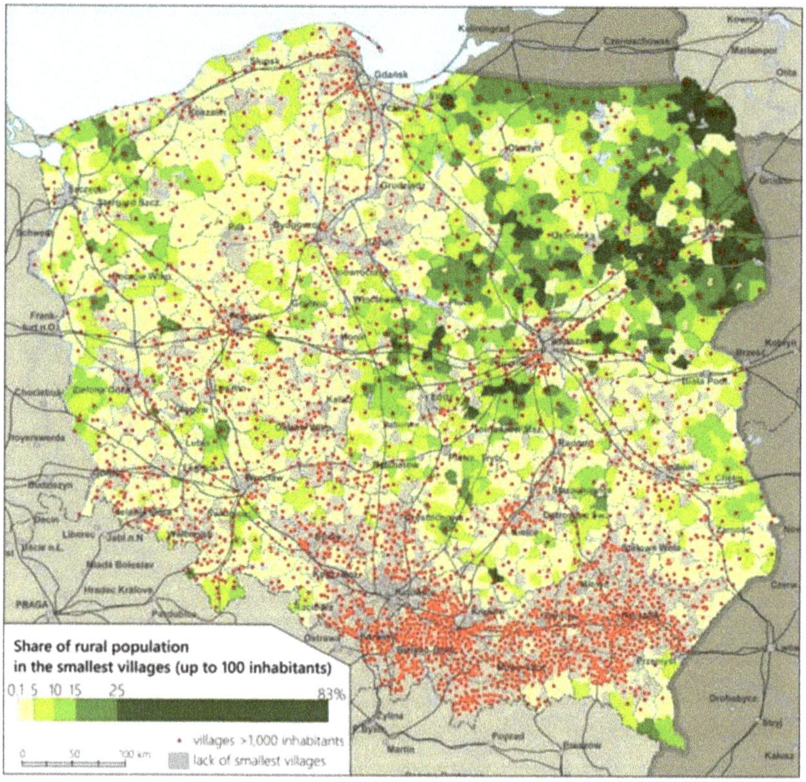

Figure 5. Share of the rural population in the smallest villages (up to 100 inhabitants). Source: based on Central Statistical Office of Poland data.

Moreover, there were very few large villages (over 1000 inhabitants) in this area. It is worth noting that this is a region with a high value for the natural environment, not accidentally known as the "Green Lungs of Poland" [73]. Hence, the use of RES-based distributed generation seems to have particularly strong foundations there.

5. Discussion

The described issues are part of settlement efficiency's fundamental problem and the optimal degree of concentration of various functions and activities. In this respect, energy production, distribution, and consumption are not yet satisfactorily identified and certainly require further research [74]. Even in very sparsely populated countries, the geographic and settlement conditions for the development of distributed energy are practically not discussed [75]. Meanwhile, the low efficiency of settlement systems in rural areas causes a reduction in "functional efficiency" and the ineffectiveness of traditional energy supply methods [76]. Examples of studies from Poland using the Minimum Spanning Tree (MST) methodology show that the indicators of the efficiency and effectiveness of the technical and settlement network in rural areas differ even ten times [77]. It creates enormous potential for more rational planning of energy systems (electric, heating). It seems that in the light of the empirical analyses in the previous section, this has been well documented.

In scattered development areas, the infrastructure network's length per person (per household) must be longer and thus more expensive. This is the case of Poland [58]. Hence, many rural areas of Poland have unfavourable conditions for the development of traditional linear infrastructure but favourable for the development of distributed generation. It is especially true of the northern part of the country, where the settlements are far apart, and the villages are small.

On the other hand, it is worth pointing to the south of the country. The problem of low emissions from traditional coal and wood-fired heating boilers is related to specific natural conditions in mountain valleys and land depression, very unfavourably exposed to smog [78]. Due to Poland's size, the importance of these issues is very great and may bring important conclusions for planning energy systems in other countries. The analysis of the contents in local spatial development plans shows that the problem may be securing land for energy infrastructure in many areas of the country.

In the context of the spatial management system's conditions, it is worth paying attention to several issues. First, the way of presenting the tools of spatial policy determines the effectiveness of the implementation of energy investments. In Poland's case, both the indeterminacy of a significant part of local spatial development plans and the unclear statutory provisions relating to development conditions' decisions determine the extension and actual blocking of this type of investment. The dependence of the clarity of planning regulations and the efficiency of energy investments implementation requires examination in other countries with different regulations and different planning practices [75]. To a limited extent, it may be a factor that determines the development of the energy sector, noticeable so far in the literature on the subject.

The issue of spatial conflicts that arise during the implementation of larger energy investments in wind farms is also important [79]. In the first place, they relate to the protection of the cultural landscape and possible collisions with the housing function. There is a discussion on how significantly different considerations regarding the protection of the cultural landscape may limit the possibilities of development (and implementation of energy investments) in the legal sphere [80]. The authors believe that this scope should be wide, and detailed analyses should be reflected as broadly as possible in legal and planning regulations (which must be considered when determining the scale and investment opportunities in the field of energy). Moreover, it can be indicated that some publications also differentiate the impact of individual RES on the environment [36,37,81]. Therefore, there is no doubt that the spatial policy perspective should balance the indicated problems and interests.

Changes in the spatial policy should contribute to solving these problems. On the one hand, it is related to the demand to specify the scope of environmental, nature and landscape protection in the Polish spatial management system. At present, the terminological diversity is too wide (also in law), which generates spatial conflicts to an even greater extent. The role of local municipal authorities seems to be important; through their own analyses, they should define threats to environment, nature and landscape much better than it is now. Better definition of threats will make it easier to correlate spatial policy with the goal of implementing RES investments. This problem is important in the context that, as shown in the Results section, the planning coverage is particularly low in areas especially predestined for renewable energy development (northern and eastern Poland).

Another sphere of significant conflicts concerns housing development. Firstly, residents' protests may lead to limitations in implementing investments, even if the formal framework theoretically does not prevent it. Hence, in this context, the postulate of deepening social participation in the development policy, including spatial planning, also seems to be right. No such blockages appear at the implementation stage [81]. It should also be remembered that these investments play an important role in improving the inhabitants' quality of life. Secondly, after 2016 large investments in wind farms are subject to significant restrictions in the development of neighbouring areas in the Polish reality. Determining the possibility of implementing such an investment in local plans is equivalent to prohibiting development around it. Especially in the case of scattered development, it may cause accusations of blocking the residential function's development. These dilemmas are interesting, especially in the context of signalling in the literature on the need to reconcile diffused energy facilities' location and operation with other purposes related to land use [82]. It makes it even more important to reconcile individual expectations when developing rural areas [30].

The above circumstance leads to the deepening of barriers in implementing investments in the field of large wind farms—even because it is often difficult to find an area that would meet all distance requirements. Thus, another dilemma is noticeable on this occasion: balancing the relationship between the scope of protection of residential areas and the effective implementation of investments in the field of wind farms. There will be much fewer problems in determining the planning foundations for implementing micro-installations (including distributed energy). Of course, the barriers mentioned above to the possible ambiguity of post-planning and ambiguities related to the interpretation of regulations related to spatial development appear on this occasion. Nevertheless, the implementation of this type of investment does not face such barriers as those resulting from the act on investments in wind farms (without causing any damage related to the exploitation of non-renewable energy sources).

Poland's case also shows how the framework related to the local spatial policy can be a derivative of specific administrative-political decisions. This issue is all the more important because local authorities' role in energy policy is strongly emphasised in the literature [83]. Based on the conducted research, it is possible to confirm the importance of local authorities in this respect [47,48] and the validity of further expectations towards the authorities [50]. In the Polish spatial management system, the change allowing for a much wider inclusion of micro-installations in local plans entered into force in 2016 (previously, micro-installations were associated only with the production purpose of the area, and a significant part of the jurisprudence indicated that they must be directly provided for in the plan). The change resulted from both the discussion on the importance of renewable energy sources for individual countries and the European Union's examples and recommendations. Thanks to this, it is possible to implement micro-installations not only in areas with assigned production purposes.

However, it should be emphasised that the interpretation dilemmas in the Polish spatial management system are related to entirely formal and legal issues. Both the problems with the interpretation of local plans and the different qualifications of micro-installations in the context of the decision on land development and development conditions do not lead to better land development (or even to a discussion in this respect), but to the problems related to the interpretation of the regulations that obscure this context. It requires correction, especially when recalling the European Parliament's recommendations regarding the transparency of administrative procedures related to a certain freedom of action by planning bodies. Wider use of soft (e.g., informal) spatial policy tools could be helpful here.

The case of Poland confirms that the quality of specific spatial policy tools and the related quality of spatial policies themselves significantly affects the implementation of other sectoral objectives. This is the case of RES investments. Problems related to the interpretation of local plans or conditions for issuing decisions on development conditions can often prolong or block the implementation of specific investments. In the long-term perspective, they may even determine the very concept regarding the scope of RES investments. This is why it seems so important to improve the current factual and legal situation in Poland. The necessity of broadening and adapting the analyses of spatial determinants to different levels of spatial policies has been already indicated [84].

Improving the quality of these analyses and adapting them to the local spatial policies is very important. It also seems important to ensure (at a later, implementation stage) some flexibility in planning. This should include at least some of the investments. The scheme would be such that the spatial policy tools based on the spatial development analyses would create certain frameworks, which would be specified in detail at the implementation stage (a similar principle is based on e.g., the British spatial planning model) [85–87].

In the context of the discussion on the integrated model of development planning, it is worth pointing to the need for a more coordinated energy policy and spatial policy integration. It can be reflected in integrated energy plans that consider spatial conditions. The recommendations contained in the 2030 Territorial Agenda can be a particular inspiration here. The more so as it is addressed to various stakeholders (not only the central level of individual countries) [88]. The specific conditions and potentials of individual countries, indicated in the literature review as the basic point of reference for energy policy issues, perfectly correspond to development policies' focus on specific places promoted in the Agenda.

The implementation of the Agenda 2030 guidelines may pose interesting challenges, also for local spatial policies [89]. Agenda 2030 defines the key directions, but its advantage is that it leaves a lot of flexibility for specific public authorities. The place-based policy proposals contained in the Agenda can therefore be applied holistically to the case of Poland (and its settlement specificity in the context of RES investment opportunities), as well as its specific areas. The concept of expanding RES investments should also be correlated with the concept of "Green Europe" included in the Agenda. While discussing public policies, it is also worth noting in passing that the COVID-19 pandemic is an opportunity to redefine many aspects of these policies [90]. It also provides a basis for developing new approaches to public policy tools. Adaptation of spatial policy tools to the indicated postulates is also fully in line with these trends.

To conclude the discussion, it is still worth noting the problem of locating renewable and distributed energy systems in the context of the social situation, including energy exclusion [91]. Other studies show that sparsely populated areas in Poland with dispersed settlement are also problem areas [92], subject to economic stagnation and depopulation. Meanwhile, the implementation of renewable energy for single households is most often associated with serious unit costs at the beginning of the investment. This requires a particularly skillful development policy at the local level, but also regional and national support programs in the form of an offer of economic instruments (e.g., subsidies). This is all the more so as the first studies from Poland indicate a widening of energy poverty due to the COVID-19 pandemic [34].

6. Conclusions

Answering the research questions posed in the introduction, based on the analyses performed, it can be indicated that dispersed development in Poland is a factor favourable for the development of a RES-based distributed generation. Its implementation, especially with the domination of non-renewable energy sources, is associated with limiting potential spatial conflicts. It is also confirmed by a relatively smaller number of spatial conflicts regarding implementing these investments (especially if we refer to wider problems of the spatial management system). Planning doubts and ambiguities do occur, of course, but on a much smaller scale. Anyway, the direction of reducing these ambiguities should be a wider adaptation of local spatial policy tools to the types of renewable energy sources (e.g., wind, solar, water). Consideration should also be given to allow for smaller power plants, particularly biomass power plants, to be located in residential areas.

The above conclusions confirm the thesis about the high homogeneousness of the Polish law of spatial development. The diagnosed barriers constitute the basis for confirming that the legal regulations do not consider the needs related to the differentiation of the applicable guidelines and their adaptation to the diverse settlement structure. Problems regarding the classification of micro-installations as public purpose investments and technical infrastructure devices confirm the thesis that in the Polish spatial management system, overly detailed, formal and legal perception of individual solutions contributes to artificial dilemmas, introducing unnecessary barriers and difficulties. Additionally, this formal and legal approach does not bring the expected results to prevent spatial conflicts. In this context, key directions for changes can be recommended. First, they should reduce to covering larger areas (constituting a whole in terms of functionality) with local plans. Such local plans should consider the guidelines related to integrated development planning, i.e., their decisions should be based on broader analyses, considering the context and needs of energy policy (the more so as it should also be different for individual areas). The direction limiting the detail of planning regulations (and the related non-substantive dilemmas) may be at least a partial introduction of flexibility in planning (based on the British spatial management system). It is also connected with broadening the scope of application of soft (e.g., informal) tools of spatial policy.

Based on the analyses carried out, the following recommendations can be made for spatial policies:

- expanding the role of analyses in spatial management—first of all on a local scale (a postulate both to local authorities and national authorities responsible for the entire spatial management system);
- wider correlation between the conditions for RES investments and the requirements of environmental, nature and landscape protection (both to the local authorities and national authorities);
- Clarification of the scope of local spatial policy tools and the introduction of greater flexibility at the implementation stage (demand to national authorities);

Include these proposals in the context of discussions on changes in public policy caused by the COVID-19 pandemic (request to local and national authorities), also in relation to the problem of energy exclusion.

It should also be emphasised that there is potential for further research. Such studies should specify which of the DG/RES types would be the most optimal from the point of view of efficiency and rationality for different areas. As has been shown, the north-eastern part of Poland is an area with a particularly large share of small villages (less than 100 inhabitants). In comparison, there are many large villages in the southern part of the country (more than 1000 inhabitants). This type of analysis should also be included in the formal and legal framework defining the spatial management system's principles, differentiated by territory. Other directions for further research should address:

- comparison of barriers when applying spatial policy tools in different countries;
- ways of solving spatial conflicts when determining the location and implementation of RES (including distributed renewable energy production) in different countries;
- public participation in RES planning and implementation process. There is no doubt that dispersed development can be a significant stimulator of distributed renewable energy production. For this to be possible, it seems of key importance to modify the current spatial policy tools, first to the extent that will guarantee a wider than currently differentiation of the conditions for the development of individual areas.

For the reasons mentioned above, it can be concluded that the securing of land for the energy-related functions by communes is insufficient. The implementation of most large wind farms in many places is indeed blocked or very difficult. Regarding distributed energy, within the framework of spatial policy tools, some grounds for further actions and facilitation can be found. However, procedural barriers (related to the interpretation of detailed provisions) are often serious obstacle.

Conclusions from the presented research for Poland are also crucial for other countries or will undergo energy transformation. The use of distributed renewable energy may prove particularly useful in peripheral and depopulating regions (the Iberian Peninsula, southern Italy, Central and Eastern Europe).

Author Contributions: P.Ś. and M.N. and A.B.; methodology, P.Ś., M.N.; software, P.Ś. and N.O.; validation, N.O. and M.N.; formal analysis, P.Ś., M.N., A.B.; investigation, P.Ś., M.N., A.B. and B.M.; resources, P.Ś. data curation, P.Ś.; writing—original draft preparation, P.Ś., M.N., A.B., B.M.; writing—review and editing, M.N., P.Ś. and A.B.; visualization, P.Ś. and N.O.; supervision, M.N.;project administration, M.N.; funding acquisition, B.M. All authors have read and agreed to the published version of the manuscript.

Funding: This research received no external funding.

Institutional Review Board Statement: Not applicable.

Informed Consent Statement: Not applicable.

Data Availability Statement: Please refer to suggested Data Availability Statements in section "MDPI Research Data Policies" at https://www.mdpi.com/ethics, accessed on 30 March 2021.

Conflicts of Interest: The authors declare no conflict of interest.

References

1. Szelenyi, I.; Beckett, K.; King, L.P. The socialist economic system. In *The Handbook of Economic Sociology*; Smelser, N.J., Swedberg, R., Eds.; Princeton University Press: New York, NY, USA, 1994; pp. 234–251.
2. Węcławowicz, G. *Contemporary Poland: Space and Society*; UCL Press: London, UK, 1996.
3. Ackermann, T.; Andersson, G.; Söder, L. Distributed generation: A definition. *Electr. Power Syst. Res.* **2001**, *57*, 195–204. [CrossRef]
4. Rezolucja Zgromadzenia Ogólnego A/RES/70/1 z dnia 25 Września 2015 r. (bez Odniesienia do Komitetu Głównego (A/70/L.1)). Agenda na Rzecz Zrównoważonego Rozwoju 2030. Available online: https://www.un.org.pl/agenda-2030-rezolucja (accessed on 29 March 2021).
5. Directive (EU) 2018/2001 of the European Parliament and of the Council of 11 December 2018 on the Promotion of the Use of Energy from Renewable Sources (Text with EEA Relevance.). Available online: https://eur-lex.europa.eu/eli/dir/2018/2001/oj (accessed on 1 March 2021).
6. Komornicki, T.; Szejgiec-Kolenda, B.; Degórska, B.; Goch, K.; Śleszyński, P.; Bednarek-Szczepańska, M.; Siłka, P. Spatial planning determinants of cohesion policy implementation in Polish regions. *Europa* **2018**, *21*, 69–87. [CrossRef]
7. Nadin, V.; Fernández Maldonado, A.; Zonneveld, W.; Stead, D.; Dąbrowski, M.; Piskorek, K.; Sarkar, A.; Schmitt, P.; Smas, L.; Cotella, G.; et al. *COMPASS—Comparative Analysis of Territorial Governance and Spatial Planning Systems in Europe Applied Research 2016–2018 Final Report*; ESPON: Luxembourg, 2018; Available online: https://www.espon.eu/planning-systems (accessed on 10 December 2020).
8. Rosner, A.; Wesołowska, M. Deagrarianisation of the Economic Structure and the Evolution of Rural Settlement Patterns in Poland. *Land* **2020**, *9*, 523. [CrossRef]
9. Stasiak, A. Changes in rural settlement in Poland up to 2000. *Geogr. Pol.* **1989**, *56*, 109–114.
10. Śleszyński, P.; Kowalewski, A.; Markowski, T.; Legutko-Kobus, P.; Nowak, M. The contemporary economic costs of spatial chaos: Evidence from Poland. *Land* **2020**, *9*, 214. [CrossRef]
11. Florida, R. The Geography of Coronavirus, City Lab, 3.04.2020. Available online: https://www.bloomberg.com/news/articles/2020-04-03/what-we-know-about-density-and-covid-19-s-spread (accessed on 13 March 2020).
12. Komornicki, T. Spatial planning in Poland the territorial perspective. In *Territorio y Estados: Elementos para la Coordinación de las Políticas de Ordenación del Territorio en el Siglo XXI*; Tirant lo Blanch: Valencia, Spain, 2018; pp. 529–562, ISBN 9788416556854.
13. Lityński, P.; Hołuj, A. Urban Sprawl Risk Delimitation: The Concept for Spatial Planning Policy in Poland. *Sustainability* **2020**, *12*, 2637. [CrossRef]
14. Kaczmarek, T.; Mikuła, Ł. The Housing Market in Poland in the Liberalized Spatial Planning System: The National Context and Metropolitan Dimension of the Poznań Agglomeration. In *Housing and Housing Politics in European Metropolises. Jahrbuch Stadtregion*; Wehrhahn, R., Pohlan, J., Hannemann, C., Othengrafen, F., Schmidt-Lauber, B., Eds.; Springer: Wiesbaden, Germany, 2019. [CrossRef]
15. Aguirre, M.; Ibikunle, G. Determinants of renewable energy growth: A global sample analysis. *Energy Policy* **2014**, *69*, 374–384. [CrossRef]
16. Burke, M.J.; Stephens, J.C. Political power and renewable energy futures: A critical review. *Energy Res. Soc. Sci.* **2018**, *35*, 78–93. [CrossRef]

17. Sareen, S. Energy distribution trajectories in two Western Indian states: Comparative politics and sectoral dynamics. *Energy Res. Soc. Sci.* **2018**, *35*, 17–27. [CrossRef]
18. Zawalińska, K.; Kinnunen, J.; Gradziuk, P.; Celińska-Janowicz, D. To Whom Should We Grant a Power Plant? Economic Effects of Investment in Nuclear Energy in Poland. *Energies* **2020**, *13*, 2687. [CrossRef]
19. Gielen, D.; Boshell, F.; Saygin, D.; Bazilian, M.D.; Wagner, N.; Gorini, R. The role of renewable energy in the global energy transformation. *Energy Strategy Rev.* **2019**, *24*, 38–50. [CrossRef]
20. Ioannou, K.; Tsantopoulos, G.; Arabatzis, G.; Andreopoulou, Z.; Zafeiriou, E.A. Spatial Decision Support System Framework for the Evaluation of Biomass Energy Production Locations: Case Study in the Regional Unit of Drama, Greece. *Sustainability* **2018**, *10*, 531. [CrossRef]
21. Lazaroiu, G.; Mihaescul, L.; Jarcu, E.A.; Stanescu, L.A.; Ciupageanu, D.A. Renewable energy employment in Romania: An environmental impact discussion. In Proceedings of the International Multidisciplinary Scientific GeoConference: SGEM, Albena, Bulgaria, 18–24 August 2020; Volume 20, pp. 177–184. Available online: https://www.sgem.org/index.php/elibrary?view=publication&task=show&id=7229 (accessed on 29 March 2021).
22. Barragán-Escandón, A.; Terrados-Cepeda, J.; Zalamea-León, E. The Role of Renewable Energy in the Promotion of Circular Urban Metabolism. *Sustainability* **2017**, *9*, 2341. [CrossRef]
23. Zhao, P.; Fang, J.; Paramati, S.R.; Jiang, K. Determinants of renewable and non-renewable energy demand in China. *Struct. Chang. Econ. Dyn.* **2020**, *54*, 202–209. [CrossRef]
24. Alvarez-Herranz, A.; Balsalobre-Lorente, D.; Shahbaz, M.; Cantos, J.M. Energy innovation and renewable energy consumption in the correction of air pollution levels. *Energy Policy* **2017**, *105*, 386–397. [CrossRef]
25. Johnstone, N.; Haščič, I.; Popp, D. Renewable Energy Policies and Technological Innovation: Evidence Based on Patent Counts. *Environ. Resour. Econ.* **2010**, *45*, 133–155. [CrossRef]
26. Dogru, T.; Bulut, U.; Kocak, E.; Isik, C.; Suess, C.; Sirakaya-Turk, E. The nexus between tourism, economic growth, renewable energy consumption, and carbon dioxide emissions: Contemporary evidence from OECD countries. *Environ. Sci. Pollut. Res. Int.* **2020**, *27*, 40930–40948. [CrossRef]
27. Gan, L.; Eskeland, G.S.; Kolshus, H.H. Green electricity market development: Lessons from Europe and the US. *Energy Policy* **2007**, *35*, 144–155. [CrossRef]
28. UNIDO. *Global Value Chains and Industrial Development: Lessons from China, South-East and South Asia*; UNIDO: Vienna, Austria, 2018. Available online: https://www.unido.org/sites/default/files/files/2018-06/EBOOK_GVC.pdf (accessed on 29 March 2021).
29. Mata, T.M.; Martins, A.A.; Sikdar, S.K.; Costa, C.A.V.; Caetano, N.S. Sustainability analysis of biofuels through the supply chain using indicators. *Sustain. Energy Technol. Assess.* **2013**, *3*, 53–60. [CrossRef]
30. Garcia-Casals, X.; Ferroukhi, R.; Parajuli, B. Measuring the socio-economic footprint of the energy transition. *Energy Transit.* **2019**, *3*, 105–118. [CrossRef]
31. Mainali, B. Sustainability of Rural Energy Access in Developing Countries. Ph.D. Thesis, Royal Institute of Technology-KTH, Stockholm, Sweden, 2014.
32. Waldenström, C.; Ferguson, R.; Sundberg, C.; Tidåker, P.; Westholm, E.; Åkersko, A. Bioenergy from agriculture: Challenges for the rural development program in Sweden. *Soc. Nat. Resour.* **2016**, *29*, 1467–1482. [CrossRef]
33. Dai, H.; Xuxuan, X.; Xie, Y.; Liu, J.; Masui, T. Green growth: The economic impacts of large-scale renewable energy development in China. *Appl. Energy* **2016**, *162*, 435–449. [CrossRef]
34. Nagaj, R.; Korpysa, J. Impact of COVID-19 on the Level of Energy Poverty in Poland. *Energies* **2020**, *13*, 4977. [CrossRef]
35. Chiradeja, P.; Ramakumar, R. An approach to quantify the technical benefis of distributed generation. *IEEE Trans. Energy Convers.* **2004**, *19*, 764–773. [CrossRef]
36. Ackerman, A.S.; Toon, O.B.; Taylor, J.P.; Johnson, D.W.; Hobbs, P.V.; Ferek, R.J. Effects of aerosols on cloud albedo: Evaluation of Twomey's parameterisation of cloud susceptibility using measurements of ship tracks. *J. Atmos. Sci.* **2000**, *57*, 2684–2695. [CrossRef]
37. Willis, H.L.; Scott, W.G. *Distributed Power Generation. Planning and Evaluation*; CRC Press: New York, NY, USA, 2018.
38. Rae, C.; Kerr, S.; Maroto-Valer, M.M. Upscaling smart local energy systems: A review of technical barriers. *Renew. Sustain. Energy Rev.* **2020**, *131*, 110020. [CrossRef]
39. Dugan, R.C.; McDermott, T.E.; Ball, G.J. Planning for distributed generation. *IEEE Ind. Appl. Mag.* **2001**, *7*, 80–88. [CrossRef]
40. Lewandowska, A.; Chodkowska-Miszczuk, J.; Rogatka, K. Development of renewable energy sources in big cities in Poland in the context of urban policy. In *Renewable Energy Sources: Engineering, Technology, Innovation*; Springer: Cham, Switzerland, 2020.
41. Herran, D.S.; Nakata, T. Design of decentralised energy systems for rural electrification in developing countries considering regional disparity. *Appl. Energy* **2012**, *91*, 130–145. [CrossRef]
42. Agarwal, U.; Jain, N. Distributed Energy Resources and Supportive Methodologies for their Optimal Planning under Modern Distribution Network. *Technol. Econ. Smart Grids Sustain. Energy* **2019**, *4*, 3. [CrossRef]
43. Koirala, B.P.; Koliou, E.; Friege, J.; Hakvoort, R.A.; Herder, P.M. Energetic communities for community energy: A review of key issues and trends shaping integrated community energy systems. *Renew. Sustain. Energy Rev.* **2016**, *56*, 722–744. [CrossRef]
44. Longo, M.; Roscia, M.; Lazaroiu, G.C.; Pagano, M. Analysis of Sustainable and Competitive Energy System. Available online: https://ieeexplore.ieee.org/abstract/document/7016516 (accessed on 29 March 2021).

45. Kabuth, A.; Dahmke, A.; Beyer, C.; Bilke, L.; Dethlefsen, F.; Dietrich, P.; Duttmann, R.; Ebert, M.; Feeser, V.; Görke, U.J.; et al. Energy storage in the geological subsurface: Dimensioning, risk analysis and spatial planning: The ANGUS+ project. *Environ. Earth Sci.* **2017**, *76*, 23. [CrossRef]
46. Lund, H.; Duić, N.; Krajac˘ić, G.; da Graça Carvalho, M. Two energy system analysis models: A comparison of methodologies and results. *Energy* **2007**, *32*, 948–954. [CrossRef]
47. Szyrski, M. *Rola Samorządu Terytorialnego w Rozwoju Odnawialnych Źródeł Energii (OZE). Analiza Administracyjnoprawna*; Wolters Kluwer: Worsaw, Poland, 2017.
48. Meschede, H.; Child, M.; Breyer, C. Assessment of sustainable energy system configuration for a small Canary island in 2030. *Energy Convers. Manag.* **2018**, *165*, 363–372. [CrossRef]
49. Dorotić, H.; Doračić, B.; Dobravec, V.; Pukšec, T.; Krajačić, G.; Duić, N. Integration of transport and energy sectors in island communities with 100% intermittent renewable energy sources. *Renew. Sustain. Energy Rev.* **2019**, *99*, 109–124. [CrossRef]
50. Gawlik, L. The Polish power industry in energy transformation process. *Miner. Econ.* **2018**, *31*, 229–237. [CrossRef]
51. Asensio, C.; Lozano, F.J.; Ortega, E.; Kikvidze, Z. Study on the effectiveness of an agricultural technique based on aeolian deposition, in a semiarid environment. *Environ. Eng. Manag. J.* **2015**, *14*, 1143–1150. [CrossRef]
52. Lazaroiu, G.; Ciupageanu, D.A.; Vatuiu, T. Highlights of renewable energy integration impact: Evolution and perspectives in Romania. In Proceedings of the 21st International Symposium on Electrical Apparatus & Technologies (SIELA), Bourgas, Bulgaria, 3–6 June 2020.
53. Lu, Y.; Khan, Z.A.; Alvarez-Alvarado, M.S.; Zhang, Y.; Huang, Z.; Imran, M. A Critical Review of Sustainable Energy Policies for the Promotion of Renewable Energy Sources. *Sustainability* **2020**, *12*, 5078. [CrossRef]
54. Ferrer, J.N.; Cătuţi, M.; Stroia, C.; Bryhn, J. *Comparative Study on the Governance Structure and Energy Policies in EU Macro-Regional Strategies*; Research Report; Thinking Ahead for Europe; CEPS: Brussels, Belgium, 2019; p. 2.
55. Alper, A.; Oguz, O. The role of renewable Energy consumption in economic growth: Evidence from assymetric causality. *Renew. Sustain. Energy Rev.* **2016**, *60*, 953–959. [CrossRef]
56. Łączak, A.; Bazan-Krzywoszańska, A.; Mrówczyńska, M.; Skiba, M. Renewable Energy Sources in the Lubusz Voivodship (Poland). The Present Conditions and Perspectives for Development. *Civ. Environ. Eng. Rep.* **2018**, *28*, 31–67. [CrossRef]
57. Szulc, H. Historical geography of rural settlement in Poland. In *Geographie Historique et Culturelle de l'Europe: Hommage au Professeur Xavier de Planhol*; Pitte, J.R., de Planhol, X., Eds.; Presses de l'Université de Paris-Sorbonne: Paris, France, 1995; pp. 381–393.
58. Heffner, K.; Gibas, P. Centra regionów a spójność regionalna w Polsce. *Res. Pap. Wrocław Univ. Econ.* **2017**, *466*, 98–110.
59. Śleszyński, P.; Komornicki, T. Functional classification of Poland's communes (gminas) for the needs of the monitoring of spatial planning [Klasyfikacja funkcjonalna gmin Polski na potrzeby monitoringu planowania przestrzennego]. *Przegląd Geogr.* **2016**, *88*, 469–488. [CrossRef]
60. Nowak, M. The Role and Effects of Case—Low in Spatial Competition. An analysis of selected cases. *Biul. Kom. Przestrz. Zagospod. Kraj. PAN* **2017**, *265*, 172–180.
61. Karwińska, A.; Böhm, A.; Kudłacz, M. The phenomenon of urban sprawl in modern Poland: Causes, effects and remedies. *Zarządzanie Publiczne/Public Gov.* **2018**, *3*, 26–43. [CrossRef]
62. Lorens, P. Trends and problems of contemporary urbanisation processes in Poland. In *Spatial Planning and Urban Development in the New EU Member States*; Altrock, U., Ed.; Routledge: Abingdon, UK, 2017; pp. 109–126. Available online: https://www.routledge.com/Spatial-Planning-and-Urban-Development-in-the-New-EU-Member-States-From/Altrock-Guntner-Peters/p/book/9780754646846 (accessed on 22 April 2017).
63. Izdebski, W.; Śleszyński, P.; Malinowski, Z.; Kursa, M. Analiza morfometryczna planów miejscowych w Polsce. *Infrastrukt. i Ekol. Teren. Wiej.* **2018**, *2*, 331–347. [CrossRef]
64. Mickiewicz, P.; Nowak, M. Environmental protection and integrated development planning in local spatial policy on the example of Poland. *Ann. Univ. Apulensis Ser. Oecon.* **2020**, *22*, 11–18. [CrossRef]
65. Śleszyński, P.; Nowak, M.; Sudra, P.; Załęczna, M.; Blaszke, M. Economic Consequences of Adopting Local Spatial Development Plans for the Spatial Management System: The Case of Poland. *Land* **2021**, *10*, 112. [CrossRef]
66. Parysek, J.J. Asking about the future of spatial management in Poland (thirteen years on from the 2003 legal regulation). *Ruch Praw. Ekon. I Socjol.* **2016**, *78*, 37–58. [CrossRef]
67. Cotella, G. Spatial planning in Poland between European influence and dominant market forces. In *Spatial Planning Systems and Practices in Europe. A Comparative Perspective on Continuity and Changes*; Reimer, M., Getimis, P., Blotevogel, H., Eds.; Routledge: New York, NY, USA, 2014.
68. Nowak, M. The Statekholders and Municipal Authorities in the Spatial Competition. *Biul. Kom. Przestrz. Zagospod. Kraj. PAN* **2017**, *265*, 22–35.
69. Nawrot, F. Budowa instalacji odnawialnych źródeł energii a planowanie przestrzenne. *Prawne Probl. Górnictwa I Ochr. Środowiska* **2017**, *1*, 71–89.
70. Biegańska, J.; Szymańska, D. The scale and the dynamics of permanent migration in rural and peri-urban areas in Poland—Some problems. *Bull. Geogr. Socio-Econ. Ser.* **2013**, *21*, 21–30. [CrossRef]
71. Gałka, J.; Warych-Juras, A. Suburbanization and migration in polish metropolitan areas during political transition. *Acta Geogr. Slov.* **2018**, *58*, 64–72. [CrossRef]
72. Gutry-Korycka, M. (Ed.) *Urban Sprawl*; Warsaw Agglomeration Case Study; Warsaw Uniwersity Press: Warsaw, Poland, 2005.

73. Stanny, M.; Czarnecki, A. Level and determinants of sustainable rural development in the Region of Green Lungs of Poland. *Rural Areas Dev.* **2010**, *7*, 197–212. [CrossRef]
74. Sharma, D.; Bartels, R. Distributed Electricity Generation in Competitive Energy Markets: A Case Study in Australia, International Association for Energy Economics. *Energy J.* **1997**, *18*. [CrossRef]
75. Miskinis, V.; Norvaisa, E.; Galinis, A.; Konstantinaviciute, I. Trends of distributed generation development in Lithuania. *Energy Policy* **2011**, *39*, 4656–4663. [CrossRef]
76. Kaundinya, D.P.; Balachandra, P.; Ravindranath, N.H.; Ashok, V. A GIS (geographical information system)-based spatial data mining approach for optimal location and capacity planning of distributed biomass power generation facilities: A case study of Tumkur district, India. *Energy* **2013**, *52*, 77–88. [CrossRef]
77. Śleszyński, P.; Sudra, P. Zastosowanie metody minimalnego drzewa rozpinającego (najkrótszego dendrytu) w ocenie efektywności i spójności sieci osadniczej województwa mazowieckiego. *Przegląd Geogr.* **2019**, *91*, 61–80. [CrossRef]
78. Wielgosiński, G.; Czerwińska, J. Smog Episodes in Poland. *Atmosphere* **2020**, *11*, 277. [CrossRef]
79. Gawlikowska, A.P. Odnawialne źródła energii w planowaniu przestrzennym na przykładzie energetyki wiatrowej. *Kwart. Archit. I Urban.* **2013**, *58*, 131–159.
80. Tomczak, A. *Otwarty Krajobraz Kulturowy z Zabytkiem w Tle. Formy Ochrony Przyrody na Przykładach z Ziemi Łódzkiej*; Politechnika Łódzka: Łódź, Poland, 2018.
81. Stremke, S.; Koh, J. Sustainable energy transition: Properties and constraints of regenerative energy systems with respect to spatial planning and design. In Proceedings of the 3rd CIB International Conference on Smart and Sustainable Built Environments, Department(s) Landscape Architecture and Spatial Planning, WIMEK, Delft, The Netherlands, 15–19 June 2009.
82. Legutko-Kobus, P. Partycypacja jako element programowania rozwoju lokalnego. *Studia Z Polityki Publicznej* **2018**, *3*, 45–58. [CrossRef]
83. Holtmeyer, M.L.; Wang, S.; Axelbaum, R.L. Considerations for decision-making on distributed power generation in rural area. *Energy Policy* **2013**, *63*, 708–715. [CrossRef]
84. Markowski, T. Main Issues in the Spatial Development of Poland in the European Context. *Eur. Spat. Res. Policy* **2007**, *14*, 7–29.
85. Berisha, E.; Cotella, G.; Janin-Rivolin, U.; Solly, A. Spatial governance and planning systems and the public control of spatial development: A European typology. *Eur. Plann. Stud.* **2020**, *29*, 181–200. [CrossRef]
86. Nadin, V.; Stead, D. Spatial planning in the United Kingdom 1990–2013. In *Spatial Planning Systems and Practices in Europe. A Comparative Perspective on Continuity and Changes*; Reimer, M., Getimis, P., Blotevogel, H., Eds.; Routledge: New York, NY, USA, 2014; pp. 198–214, ISBN 9780415727242.
87. Muñoz-Gielen, D. Urban governance, property rights, land readjustment and public value capturing. *Eur. Urban. Reg. Stud.* **2014**, *21*, 60–78. [CrossRef]
88. Zerriffi, H. *Rural Electrification Strategies for Distributed Generation*; Springer: Vancouver, BC, Canada, 2011. [CrossRef]
89. Szlachta, J. Agenda Terytorialna Unii Europejskiej 2030 jako podstawa kształtowania europejskiej polityki przestrzennej. In *Polityka Przestrzenna w Czasie Kryzysu*; Nowak, M., Ed.; Scholar: Warsaw, Poland, 2021; pp. 139–156.
90. *Geography of COVID-19 Outbreak and First Policy Answers in European Regions and Cities*; Espon Study; Policy Brief; EconomiX: Luxembourg, 2020.
91. Bouzarovski, S.; Tirado Herrero, S. Geographies of injustice: The socio-spatial determinants of energy poverty in Poland, the Czech Republic and Hungary. *Post-Communist Econ.* **2017**, *29*, 27–50. [CrossRef]
92. Śleszyński, P.; Bański, J.; Degórski, M.; Komornicki, T. Delimitation of problem areas in Poland. *Geogr. Pol.* **2017**, *90*, 131–138. [CrossRef]

energies

Article

Does Economic Structure Differentiate the Achievements towards Energy SDG in the EU?

Magdalena Cyrek * and Piotr Cyrek

Institute of Economics and Finance, University of Rzeszów, Ćwiklińskiej 2, 35-601 Rzeszów, Poland; piotrcyr@ur.edu.pl
* Correspondence: mcyrek@ur.edu.pl; Tel.: +48-17-872-16-04

Abstract: Energy sustainability constitutes an important goal for development, as declared at the global and the European levels. Some conditions decisive for energy performance, as suggested by the Environmental Kuznets Curve (EKC) hypothesis, may be specified by the sectoral structure of production, as industries vary in the intensity of energy consumption. Nevertheless, sustainability is not automatically induced along with economic development and it is important to identify its determinants. The aim of the study is to empirically verify whether the sectoral structure of an economy differentiates energy sustainability within 28 European Union member states (the EU-28). To fulfil the task, a static approach was adopted and such taxonomic methods as the Ward agglomeration method and linear ordering based on the Hellwig synthetic measure were used. The hypothesis concerning the essential role of structural features in energy achievements was verified by a one-way analysis of variance. Our results do not confirm the decisive role of economic structure in energy performance for the EU-28 states; however, they suggest some complex relationships. The interference between energy performance and sectoral structure mostly concerned primary and final energy consumptions and energy poverty, as well as the shares of agriculture, industry, traditional services and finance in total production. The findings reveal a need for further research into the potential interlinkages between different dimensions of sustainable development (SD).

Keywords: sustainable development; energy; the SDG 7; sectoral structure of production; European Union

Citation: Cyrek, M.; Cyrek, P. Does Economic Structure Differentiate the Achievements towards Energy SDG in the EU? *Energies* **2021**, *14*, 2229. https://doi.org/10.3390/en14082229

Academic Editor: Vincenzo Bianco

Received: 16 March 2021
Accepted: 12 April 2021
Published: 16 April 2021

1. Introduction

Sustainable development (SD) is a central issue in the public debate led by scientists, politicians, activists and many others, since the way in which the development process occurs has an influence on all spheres of human existence. SD is commonly defined as in the Brundtland report [1,2] (p. 20) as "development which meets the needs of the current generations without compromising the ability of future generations to meet their own needs". It assumes that human socio-economic needs can be met in harmony with environmental issues [3] and that society's productive base per capita does not decline over time [4].

The importance of achieving SD is perceived at a global level, which is reflected by numerous political declarations, of which the UN's "2030 Agenda for Sustainable Development" [2,5,6] is the most current in expressing the international pursuit of SD. At the European Union level, SD also comprises a central policy objective. It has been enshrined in its treaties since 1997, including its ascending strategies of development and policy initiatives (such as the "Europe 2020 strategy" (2010) [7], the European Commission's Communication "Next steps for a sustainable European future: European action for sustainability" (2016) [8] and "European Green Deal" (2019) [9]) [2].

SD is usually specified by its three dimensions: economic, social and environmental, with a strong interdependence [10,11] and, thus, the 2030 Agenda presents a complex and holistic attitude, one which has been adopted at the EU level as well. This is expressed by the 17 Sustainable Development Goals (SDGs), with progress towards them being regularly monitored to maintain balance between the social, economic, environmental and, additionally, institutional dimensions of sustainability. All the spheres are internally complex, which results in difficulties in both their description and assessment as well as the possible trade-offs or synergies between the SD dimensions, together making the issue even more complicated. Some efforts to identify the interlinkages between the SDGs have been already taken by researchers and institutions (e.g., European Commission's Joint Research Centre (JRC) [12], International Council for Science [13] and Interlinkages Working Group of the IAEG-SDGs [14]) [2]. Nevertheless, there is still a need to investigate the potential interlinkages between different aspects of the process of development and not just limited to those specified in the SDGs.

This study forms part of wide-ranging research into the relationships between different aspects of SD; however, it considers the relatively neglected issue of the structural foundation of the achievements in terms of the sectoral/branch-level distribution of production, which is beyond the SDGs' scope. The economic structure is understood here as proportions between contribution of each sector into total production and is expressed by sectoral shares in gross value added (GVA). This paper focuses on the importance of the structural features of economic production for the energy goals of SD as an issue appealing to the modern actions declared by the EU.

The SDG 7 pays special attention to energy use and production as an essential aspect of the environmental dimension of SD, especially considering climate change. The goal is specified as "Ensure access to affordable, reliable, sustainable and modern energy for all" and efforts towards achieving it cover improving energy efficiency and productivity, reducing energy consumption (in all economic sectors and households), increasing the share of renewable energy production, ensuring the security of energy supply and limiting energy poverty [2]. In the EU context the achievements are monitored based on the EU SDG indicator set prepared by Eurostat [15]. As reflected in the Europe 2030 climate and energy framework, the EU aimed to improve energy efficiency by 20% by 2020 [5] and by at least 32.5% by 2030 according to the revised Energy Efficiency Directive [16]. In the Europe 2020 Strategy [5], the target for the share of renewable energy sources in final energy consumption was 20% by 2020 and at least 32% by 2030, according to the revised Renewable Energy Directive [17]. The Energy Union Package [18] set the goal of the EU becoming a world leader in renewable energy sources [2]. The energy results appear to differ between EU member states and, thus, identifying the sources of such differences is an important task from the theoretical and application points of view. This study examines the issue in reference to the structural characteristics of production.

The theoretical background for setting the structural features as a determinant of energy usage and production is derived from the widely discussed environmental Kuznets curve (EKC) [19–31]. Although it is still not plausible whether a broad relationship exists between economic development and environmental quality, as the EKC suggests [24] and while many scientists perceive the EKC as a statistical artefact [26] or consider it to offer little if any empirical support for its existence [28], then the concept remains of vivid interest. Generally, it suggests that with increasing income per capita the indicators of environmental degradation first rise and then fall [19]. The main explanation for this lies in structural changes "from a clean agrarian economy to a polluting industrial economy and then to a clean service economy" [22,29]. The logic behind the EKC is that if there were no change in the structure or technology of the economy, then pure growth on the scale of the economy would result in a proportional growth in pollution and other environmental impacts [23]. This relates to the fact that each industry is specified by a different environmental effect in terms of pollution emission, energy intensity and input mix. At higher levels of development, such as those that apply to the EU member states, structural change towards information-intensive industries and services is expected to result in a gradual decline in environmental degradation [19,30]. Verification of the thesis tends to be ambiguous. Some researchers claim that although structural changes (on the input and output sides) may be important in some countries at certain times for modifying the "gross scale effect", their average contribution seems less important quantitatively than "time-related effects" [19]. For developed countries in particular, structural change is often less important than technological innovation across sectors [22]. Despite the controversies about the role of structural patterns for environmental quality in advanced economies, it is still an important task to empirically verify their validity for EU countries.

Moreover, EKC studies are based on different proxies of environmental quality. One of the most general attitudes aimed at capturing all environmental impacts is to estimate the EKCs for total energy use [19,23]. The results usually indicate a monotonic increase in energy usage along with income per capita, although this does not preclude an inverted U-shaped curve [19,20,27,31]. Some results [21] suggest long-term relationships between economic growth, energy consumption and energy pollutants and, hence, confirm the EKC hypothesis with energy consumption as a major contributor to energy pollution. Moreover, researchers [22] suggest that the high share of manufacturing in total GDP is associated with higher levels of energy consumption and, thus, a structural shift may induce the occurrence of the energy Kuznets curve.

Independently of the EKC hypothesis itself, it is worth stressing that energy goals are strictly connected with the condition of the natural environment as well as other dimensions of human existence and if achieving them is sectoral specific, then it is important to specify the existence of such relationships.

Thus, the aim of the study is to identify interferences between the sectoral structure of production and the state of energy usage and production in the context of the SDG 7 in the EU-28 states. The paper verifies an initial general hypothesis that sectoral features differentiate the energy achievements, that is the countries that differ concerning sectoral proportions in gross value added (GVA) differ also by their energy performance. It is tested adopting a static approach, basing on comparisons between national economies in 2018. The approach cannot directly prove any causality; however, it may indicate whether sectoral pattern of production is a factor interfering with differences in energy performance or not.

The aim covers:

- comparing energy usage and production across EU countries and specifying the leaders and the laggers. A detailed hypothesis verified at the research stage assumes higher energy achievements in more affluent countries, in accordance with the EKC for advanced economies.
- distinguishing groups of EU countries expressing variety of structural patterns for production. A detailed hypothesis assumes that the groups differ in relation to their levels of economic development expressed as GDP per capita and the higher the GDP per capita, the higher the share of knowledge-based services and the lower the share of agriculture, in accordance with sectoral development theory.
- identifying the differentiating potential of sectoral features for energy achievements across the dimensions of energy achievements and specific industries. More detailed hypotheses assume that relationships appear at least between the final and primary energy consumptions, energy productivity and greenhouse gas emissions as well as economic structure and that the more industrialized an economy becomes, the more severe the energy problems they encounter.

Our findings indicate no clear relationships between general energy achievements and structural features; however, some unexpected interlinkages can be identified. The results concerning the energy performance of EU countries offer some support for an increase in energy tensions along with production level and, thus, do not confirm the EKC positive assumption of declining environmental pressure in more affluent economies. Moreover, while structural patterns in production differ according to GDP per capita, revealing a typical shift from agriculture and industry towards knowledge-based services, their relations with energy achievements are not so obvious. The interference refers to primary and final energy consumptions, which may be a direct relation induced by different patterns of energy use in different industries, as well as to energy poverty, which in turn reflects indirect linkages occurring through average income level. Another finding is that the explanatory role of energy achievements is not clearly fulfilled by the "polluting" industrial sector of the economy, which unexpectedly appears to be linked positively to aggregated energy sustainability while does not reveal relations with detailed indices of energy performance. Of importance are the structural features of economies, such as the shares of agriculture, traditional services and finance in total production. What is especially important, is that the relation seems not to result from technological specificity or the energy needs of these sectors but rather expresses sectoral linkages with GDP per capita. Some trade-offs concerning sectoral influence on different energy goals seem to be a cause for the lack of a general relation between aggregated energy achievements and production structure. This ambiguity does not reject the research hypothesis; rather, it suggests a need to research the possible interlinkages between different dimensions of SD at a more disaggregated level, with more attention placed on possible trade-offs and synergies.

2. Materials and Methods

The study deals with two important aspects of development: the structural features of an economy and the characteristics of energy usage and production. It compares 28 European Union member states in year 2018, considering the possible relations between the two dimensions of development. To provide a longer horizon for the relationships, the main results were supported by comparisons for 2010.

The structural features of an economy are defined as proportions of sectoral contribution into total activity and specified in terms of the classification sections under Statistical Classification of Economic Activities in the European Community, Rev. 2 (NACE Rev. 2). Data in this field are arranged according to Eurostat grouping, with a breakdown of 10 industries:

- Agriculture, forestry and fishing (A);
- Industry (except construction) (BCDE);
- Construction (F);
- Wholesale and retail trade, transport, accommodation and food service activities (GHI);
- Information and communication (J);
- Financial and insurance activities (K);
- Real estate activities (L);
- Professional, scientific and technical activities; administrative and support service activities (MN);
- Public administration, defence, education, human health and social work activities (OPQ);
- Arts, entertainment and recreation; other service activities; activities of household and extra-territorial organizations and bodies (RSTU).

Economic activity is expressed by gross value added (GVA) in millions of euro, at current prices. The most current data were used, from the year 2018 and supplemented by data for 2010. The share of each section group in the total economic activity (totaling 100%) was calculated to enable comparability between the different scales of the economies. All data were extracted from the Eurostat database [32].

The energy usage characteristics were described in line with the European Union Sustainable Development Goals (EU SDGs). Goal 7: "Ensure access to affordable, reliable, sustainable and modern energy for all" (SDG 7) was monitored by a set of 6 indices (with one of them being presented in more detail as 2 separate indicators) and 1 multi-purpose indicator. For the purpose of the study, they were derived from the EU SDG indicator set prepared by Eurostat:

- Primary energy consumption [SDG_07_10]—tonnes of oil equivalent (TOE) per capita [33];
- Final energy consumption [SDG_07_11]—tonnes of oil equivalent (TOE) per capita [34];

- Final energy consumption in households per capita [SDG_07_20]—kilogram of oil equivalent (KGOE) [35];
- Energy productivity [SDG_07_30]—PPS per kilogram of oil equivalent (KGOE) [36];
- Share of renewable energy in gross final energy consumption by sector [SDG_07_40]—percentage [37];
- Energy import dependency by products [SDG_07_50]—percentage [38];
- Population unable to keep home adequately warm by poverty status [SDG_07_60]—percentage [39];
- Greenhouse gas emissions (source: EEA) [SDG_13_10]—tonnes per capita [40] (original indicator Greenhouse gas emissions intensity of energy consumption [SDG_13_20]—available only as index, 2000 = 100 was replaced to ensure comparability across the countries).

All data on the SDG 7 described the achievements from the year 2018 by the EU countries towards affordable and clean energy, concerning consumption, supply and accessibility, as targeted in the UN's "2030 Agenda for Sustainable Development". However, they were also supplemented by indicators for 2010 year, as the initial for the "Europe 2020 strategy".

Moreover, data on final energy consumption by sectors: industry, transport, commercial and public services and households (thousand tonnes of oil equivalent (TOE)) in 2018, extracted from Eurostat database [41], were used to give a more general view on energy usage patterns across the economies. As, in the database, agriculture is not distinguished as a separate sector, its share in total energy consumption was computed as residual. It may lead to some inaccuracy of estimation; however, allows to draw a general view of a contribution of each sector into energy consumption.

The study was based on the preliminary assumption that development is a multidimensional process and all its dimensions are mutually related. Structural features of economic activity in terms of industry breakdowns may be decisive for patterns of energy usage and production. Thus, the economic structure of production may influence the progress of a country towards its energy targets for sustainable development (SD). The aim of the paper was to identify empirically the occurrence of such relations.

We verified the hypothesis that the structural patterns of economies describe their progress towards SDG 7. To fulfil this task, we adopted a static approach and followed three steps:

1. Identify the achievements of the EU-28 states towards the SDG 7 using a synthetic measure of development.
2. Identify structural patterns in the economic activity of EU states by grouping the economies according to their branch structure of gross value added.
3. Verify differences in energy usage and production between the specified groups of economies.

For the first stage of our research taxonomic linear ordering methods, which are used to rank objects (e.g., countries) described by multidimensional characteristics, were adopted. In linear ordering empirically observed diagnostic variables are a base to calculate a synthetic indicator. The synthetic indicator may be specified adopting approach with a target model or without it. The Hellwig concept [42], which was a base for the study, assumes the former solution. In these group of ordering methods, the objects (countries) may be ranked by calculating a distance to the target. In the study, the synthetic indicator was calculated to measure development towards the SDG 7 by each of the EU-28 states. All 8 indicators of the SDG 7 were taken into account, of which 2 were specified as stimulants (SDG_07_30 and SDG_07_40) and 6 as destimulants (SDG_07_10, SDG_07_11, SDG_07_20, SDG_07_50, SDG_07_60, SDG_13_10). According to the method:

- values of the indicators were standardized (using the average value and standard deviation);

- the target model was specified with maximum values of the stimulants and minimum values of the destimulants (as well as the anti-model with minimum values of the stimulants and maximum values of the destimulants);
- distances between the objects (countries) and the target model were calculated using the Euclidean distance formula, which is a geometric mean of variables in multi-dimensional space;
- the synthetic indicator was calculated based on the formula:

$$SM_i = 1 - \frac{d_i}{d_0} \quad (1)$$

where d_i—Euclidean distance between object *i* (country) and the target model; d_0—Euclidean distance between the target model and the anti-model. It is a slight modification of the Hellwig method, however, allows to avoid negative signs of the *SM*.

The synthetic measure SM_i adopted values in the range [0;1]. The value 0 described the anti-model and the value 1 the target model and for each object the higher the value, the better results achieved.

Finally, a ranking of the EU-28 states concerning their energy performance (*SM*) was specified. To compare the results in time, similar *SM* was calculated for 2010 and Spearman's rank correlation coefficient allowed to check similarity of the rankings.

During the second stage, clustering of the EU economies according to their sectoral structures of production in 2018 was based on the Ward agglomeration method [43]. The Ward method assumes adopting a minimum variance criterion that minimizes the total within-cluster variance. It is perceived as one of the most effective classical methods of agglomeration, by leading to clustering with an equalized although not numerous quantity of objects [44,45]. For clustering we used Euclidean distance as a "natural" measure of distance between the objects. The variables were not standardized because we took into account the share of each industry in the total value added. Statistica software was used for the calculations.

During the third stage, the analysis of variance (ANOVA) was used to verify whether differences concerning energy usage and production among groups of economies were essential. The single-factor ANOVA F-test allowed a *p*-value to be specified and compared with an assumed value $\alpha = 0.05$ to verify any statistically significant differences between the groups concerning their energy performance.

The clusters were also tested for differentiation by a general level of economic development, measured by GDP per capita (GDP at market prices, current prices, euro). In this case Eurostat data for 2018 [46] were used. The relations between GDP per capita, structural features and energy usage shed some light on the character of the causality. GDP per capita for 2010 [46] were also used in the study to check stability of the results in time.

The study also tracked in detail the potential relations between sectoral features and the energy characteristics of the economies. It attempted to identify those sectors with any connection to energy achievements. To fulfil the task, the Pearson's correlation coefficient was adopted and tested at the $\alpha = 0.05$ level to examine whether the relations were statistically significant.

3. Results

The results are divided into sections to aid verification of the main hypothesis concerning the explanatory role of structural features of production in terms of energy use and production in the EU-28 states.

The preliminary assumption is supported by general data related to energy consumption in different sectors (see Figure 1).

For the EU-28 more than 30% of the energy was used by transport, ranging from 17% in Finland to 56% in Luxembourg. This sector was, thus, of most importance for achieving the energy targets. Households were responsible for nearly 27% of energy consumption, with the highest share in Croatia (34%) and the lowest in Luxembourg

(13%), indicating essential differences in lifestyle, as well as the scale of complementing—productive activities between the countries. Industry was responsible for nearly 25% of the energy usage, with the lowest amounts used in Malta (11%) and the highest in Finland (44%). The 10 percentage points lower share in energy consumption (about 14%) occurred with commercial and public services, at about 8% for Romania and 24% for Malta. This suggests that the development of a service economy typical for the most advanced stage of development should reduce energy usage and, thus, enable the fulfilling of the SDG 7. The lowest share (just above 3%) characterized agriculture, ranging from 9% in the Netherlands to merely 0.6% in Luxembourg. Agriculture also had the highest variation across the EU-28 states (coefficient of variation CV = 53%). Comparing shares of agriculture in energy consumption and GVA indicates that the sector is relatively energy intensive and may negatively influence the SDG 7; however, the feature may be very diversified.

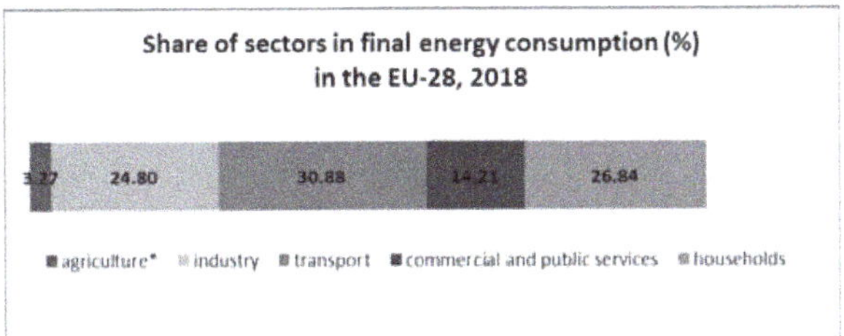

Figure 1. Share of sectors in final energy consumption (%) in the EU-28 states in 2018. Source: Own calculation based on [41]. * share of agriculture is computed as residual.

3.1. Achievements towards the SDG 7 in the European Union Member States

The first stage assumed a diagnosis of the achievements by the EU-28 states of the energy goals. A ranking of the countries in 2018 based on a synthetic measure (SM) is presented in Figure 2a. It may be compared to a ranking in 2010 (Figure 2b), as in the initial period for the "Europe 2020 strategy".

Our results revealed that the most advanced in economic development and affluent economies were not the leaders in fulfilling the energy goal. Romania, which had one of the lowest GDP rates per capita, achieved the highest score, while Luxembourg, the wealthiest state, achieved the lowest score. The position of Romania was derived from its lowest levels of primary and final energy consumption, supported by high ranks (three) in energy productivity, import dependency and greenhouse gas emissions. A relatively low production may be a reason for such results. As its economy is still catching up, it is expected that it will increase its low energy consumption and greenhouse gas emissions and, thus, its position. An important challenge for Romania is to take advantage of the newest, more environmentally friendly technologies in the course of economic development. The score of Luxembourg also resulted from primary and final energy consumption and greenhouse gas emissions, the highest across the EU-28 states. Moreover, Luxembourg also performed relatively poorly concerning energy consumption in households, import dependency and renewable energy (27th or 26th place). In addition, in this case, a high production level may be responsible for such unfavorable energy performance. Relatively poor scores were also achieved by other affluent economies, such as Finland, Belgium and the Netherlands. Moreover, Latvia and Croatia (both with relatively low GDPs per capita) were among the five best performing states in terms of the SDG 7. The correlation coefficient between GDP per capita and SM value was negative and statistically significant (−0.50) (calculation based on [33–40,46]), demonstrating that the richer countries tended not to reduce their negative influence on the environment in the energy sphere. Some countries broke this rule, such as Denmark and the UK (and to some extent also Austria and Sweden), proving that reducing energy usage was possible even with high production and that their energy solutions could be considered as potential benchmarks. The observation was that the level of production was an important although not decisive factor of energy usage, leading us to consider the structure of production as a potential explanatory factor.

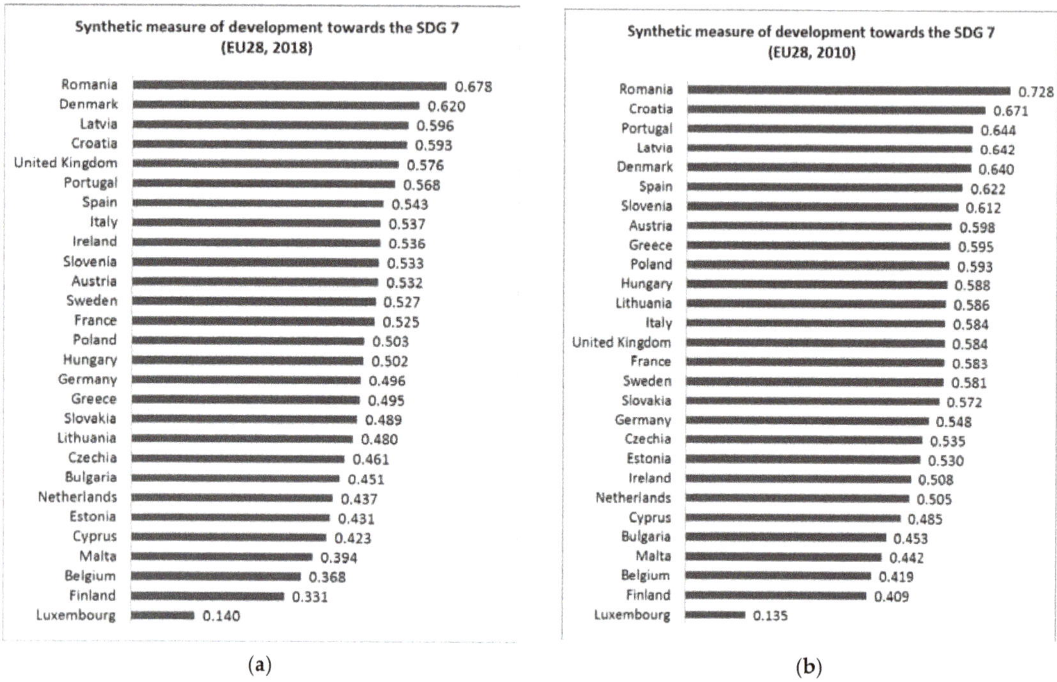

Figure 2. Synthetic measure of development towards the SDG 7 for the EU-28 states in 2018 (**a**) and in 2010 (**b**). Source: Own calculation based on [33–40].

Moreover, it appeared that the ranking of the EU-28 states is relatively stable concerning the period between 2010 and 2018. At the beginning of the initiative "Europe 2020" it revealed a similar pattern of higher energy achievements in less-developed economies and less environmentally friendly performance in the most affluent countries (with correlation coefficient −0.63). Although the values of SM in 2010 and 2018 are not directly comparable because of construction of the measure, the rank correlation was high (0.87), confirming similarity of results in energy usage and production. It also indicates that changes towards the SDG 7 are a challenging and longstanding task.

3.2. Structural Patterns of Economic Development in the European Union Member States

The sectoral structure of gross value added formed a basis to group the EU-28 states into clusters sharing similar industrial patterns in terms of economic development. The results of the cluster analysis are presented in Figure 3.

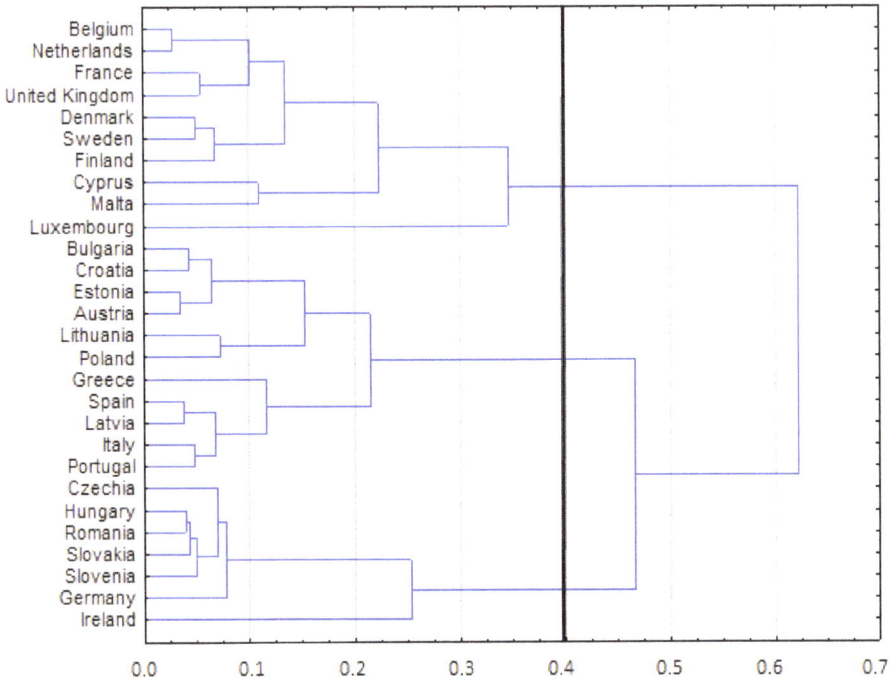

Figure 3. Clustering results—dendrogram (Ward method, Euclidean distance) for the EU-28 states concerning their sectoral structure of gross value added in 2018. Source: Own calculation based on [32].

We decided to stop the clustering at a distance 0.4 and, thus, grouped the EU countries into three clusters. Cluster 1 had 10 economies, Cluster 2 had 11 and Cluster 3 had 7. Their structural features are presented in Table 1.

Table 1. Sectoral features of the groups of EU countries.

Share of NACE Section(s) in GVA (%)	Group 1		Group 2		Group 3	
	Belgium, Denmark, France, Cyprus, Luxembourg, Malta, Netherlands, Finland, Sweden, United Kingdom		Bulgaria, Estonia, Greece, Spain, Croatia, Italy, Latvia, Lithuania, Austria, Poland, Portugal		Czechia, Germany, Ireland, Hungary, Romania, Slovenia, Slovakia	
	Avg. (%)	CV (%)	Avg. (%)	CV (%)	Avg. (%)	CV (%)
A	1.4	57	3.0	32	2.6	58
BCDE	14.1	32	19.4	17	27.6	16
F	5.8	17	5.4	33	5.4	29
GHI	19.0	16	24.6	12	17.8	18
J	6.0	19	4.5	27	6.1	49
K	8.0	83	4.6	26	3.9	27
L	9.8	26	10.7	30	8.8	17
MN	12.8	21	8.1	18	9.8	16
OPQ	19.7	11	16.6	12	15.2	16
RSTU	3.5	59	3.1	27	2.8	31
GDP pc (euro)	44,660	47	20,004	49	27,226	74

Avg.—Average/Arithmetic mean. CV—Coefficient of variation. Source: Own calculation based on [32,46].

Cluster 1 consisted of the most advanced economies, as reflected by their average GDP per capita. Their structural characteristics were also the most advanced as the average share of agriculture (A) as well as industry (BCDE) was the lowest. Their stage of development reflected the introduction of a service knowledge-based economy. This was specified, in particular, by the highest share of professional services (MN). Moreover, the economies were characterized by the strongest financial sector (K) and the essential role of welfare services (OPQ). The last feature might indicate an essential role of the state in the economies.

Cluster 2 was the most numerous one and simultaneously the least developed in terms of general economic results, expressed in GDP per capita. The structural features revealed the highest share of agriculture. Service sector development was at the initial stage as the cluster was specified by the highest share of traditional services (GHI) and the lowest was the role of ICT activities (J), as well as professional services (MN). The cluster was also characterized by the highest importance of real estate activities (L), which might reflect dynamic changes occurring within the economies.

Cluster 3 was the least numerous and its decisive feature was strong industrialization (BCDE). Moreover, industrial sector development was supported by more modern solutions, expressed by the highest share of the ICT sector (J). However, most of service activities were characterized by the lowest share among the clusters, which concerned both traditional ones (GHI) as well as financial (K) and welfare services (OPQ).

Generally, the clustering results confirmed a universal pattern in structural development along with growing GDP per capita. The least affluent countries were the most agrarian, then industrialization led to an increase in the level of production and, finally, the richest countries were characterized by knowledge-based service economies.

3.3. Achievements towards the SDG 7 and the Structural Patterns of Economic Development

Our initial hypothesis assumed that energy usage and production were influenced by the industrial structure of an economy. To verify this statement, we adopted a static approach and checked differences in each of the SDG 7 indicators between each of the three clusters of the EU-28 states. The comparisons across clusters cannot directly prove any causality; however, they may generally suggest whether sectoral pattern of production is a factor interfering with differences in energy performance or not. The general energy characteristics and the results of the test for validity of between-group differences are presented in Table 2.

Table 2. SDG 7 indicators in the 3 clusters of EU-28 states in 2018.

SDG 7 Indicator	Group 1		Group 2		Group 3		Analysis of Variance: *p*-Value
	Belgium, Denmark, France, Cyprus, Luxembourg, Malta, Netherlands, Finland, Sweden, United Kingdom		Bulgaria, Estonia, Greece, Spain, Croatia, Italy, Latvia, Lithuania, Austria, Poland, Portugal		Czechia, Germany, Ireland, Hungary, Romania, Slovenia, Slovakia		
	Avg. (%)	CV (%)	Avg. (%)	CV (%)	Avg. (%)	CV (%)	
SDG_07_10	3.97	42	2.69	29	2.94	24	0.0491 *
SDG_07_11	3.14	53	1.95	25	2.15	23	0.0455 *
SDG_07_20	639.10	37	500.55	32	541.00	22	0.2327
SDG_07_30	8.16	31	8.46	21	9.87	42	0.4437
SDG_07_40	20.65	81	24.74	34	15.99	30	0.3149
SDG_07_50	60.72	48	55.76	42	52.15	31	0.7635
SDG_07_60	5.64	107	13.74	78	4.80	51	0.0329 *
SDG_13_10	9.89	44	8.57	31	9.31	30	0.6745
SM	0.434	32	0.521	10	0.528	13	0.0834
GDP pc (euro)	44,660	47	20,004	49	27,226	74	0.0096 *

Avg.—Average/Arithmetic mean. CV—Coefficient of variation. SM—synthetic measure of development towards the SDG 7. * statistically significant at 0.05. Source: Own calculation based on [32–40,46].

Cluster 1 appeared to encounter the highest level of energy usage, concerning primary and final consumptions and consumption in households, as well as the highest greenhouse gas emissions. Moreover, it also had the lowest energy productivity, indicating that the countries were not taking advantage of the economies of scale in production and an increase in production led to growing energy usage. The cluster also faced the severe problem of energy import dependency, which may be induced by continuously growing demand for energy. Generally, it seemed that the most economically developed cluster was still far from energy neutrality, stemming from the fact that a high level of production required high energy usage.

In contrary, Cluster 2 was characterized by the lowest primary and final energy consumptions, consumption in households and greenhouse gas emission. The profile of energy consumption was also favorable as the share of renewable energy was the highest in the group. However, the cluster was coping with the severe problem of energy poverty, which was related to the more general issue of material poverty specified by low incomes.

Cluster 3 could be distinguished as having the most favorable indicators of energy dependency as well as energy poverty. Moreover, its energy productivity was the highest among the clusters. Nevertheless, its structure of energy consumption was specified by the lowest share of renewable sources and this feature might be attributed to a very traditional industrial structure of production.

Despite the above distinguishing features of each cluster, an analysis of variance confirmed essential differences between the groups only in the case of primary energy consumption, final energy consumption and energy poverty. It appeared that for most of the SDG 7 indicators the in-cluster differences were too significant to unambiguously attribute an energy specificity to the clusters. However, the clusters differed significantly concerning their economic results in terms of GDP per capita. Hence, some traditional patterns in development might be indicated, as the poorest economies were still using less energy than the wealthier ones and, thus, the former might be more environmentally friendly. This indicates that within the EU the pattern of development still seemed to reflect the initial part of the EKC and there remained much to do towards the SDG 7 concerning all European societies.

An in-depth analysis of the relations between the structural features of economic production and energy characteristics aimed to identify the specific kinds of activity that correlated with energy usage and production. The correlation matrix is presented in Table 3.

Table 3. Correlation matrix between structural features of economy and the SDG 7's indicators for the EU-28 states, 2018.

SDG 7 Indicator	Share of NACE Section(s) in GVA									
	A	BCDE	F	GHI	J	K	L	MN	OPQ	RSTU
SDG_07_10	−0.48 *	−0.17	0.31	−0.48 *	0.23	0.55 *	−0.09	0.23	0.24	−0.42 *
SDG_07_11	−0.49 *	−0.25	0.24	−0.45 *	0.21	0.71 *	−0.15	0.25	0.18	−0.37
SDG_07_20	−0.33	0.08	0.39 *	−0.43 *	0.12	0.15	0.00	0.08	0.25	−0.48 *
SDG_07_30	−0.20	0.31	−0.30	−0.30	0.39 *	0.14	−0.03	−0.06	−0.30	−0.24
SDG_07_40	0.23	0.04	0.32	0.12	−0.07	−0.33	0.23	−0.38 *	0.24	−0.10
SDG_07_50	−0.27	−0.36	−0.35	0.10	−0.08	0.42 *	−0.09	0.31	0.02	0.30
SDG_07_60	0.46 *	−0.16	−0.35	0.64 *	−0.17	−0.07	0.24	−0.47 *	−0.16	0.05
SDG_13_10	−0.45 *	−0.03	0.01	−0.28	0.31	0.65 *	−0.25	0.08	−0.16	−0.47 *
SM	0.41 *	0.38 *	−0.03	0.20	−0.17	−0.69 *	0.21	−0.32	−0.08	0.09
GDP pc (euro)	−0.74 *	−0.19	−0.06	−0.61 *	0.44 *	0.69 *	−0.17	0.46 *	0.18	−0.24

SM—synthetic measure of development towards the SDG 7. * statistically significant at 0.05. Source: Own calculation based on [32–40,46].

There were several types of economic activity that correlated significantly with some detailed indicators of energy usage and production: agriculture (A), traditional services (GHI), financial activities (K), with four indices; other services (RSTU), with three indices; professional business services (MN), with two indices; and construction (F) and ICT activities (J), with one indicator. Unexpectedly, there was no confirmed relation between any individual energy goal and the role of industry in the economy. Nevertheless, industry did correlate with the aggregated energy achievements (SM), which drew attention to the mutual interlinkages between energy goals.

Some of the identified correlations took an unexpected sign, inconsistent with the EKC assumptions of "polluting" industry and "clean" services. We examined the details to consider which were the most essential.

For agriculture, we observed that when the role of the agricultural sector was high, both primary and final energy consumptions, as well as greenhouse gas emissions were limited, indicating that agriculture was not responsible for most energy pressures. However, a high share of agriculture in an economy appeared together with energy poverty, stressing the rural character of many social problems. Deagrarization is a trend universal for economic development, as was confirmed by the correlation with GDP per capita. Simultaneously, the process of development encompasses greater energy use for production purposes and, thus, the relation with agriculture might be an indirect one. As poverty, including in the energy dimension, is induced by low income, then development and deagrarization might alleviate it. Thus, some trade-offs existed concerning the influence of agricultural production on the energy goals (limiting energy usage and greenhouse gas emissions but inducing energy poverty).

Similar considerations might be true for traditional services (trade, transport, accommodation and food service activities), where the role grew at the initial stage of service economy development, usually as supporting activities for industrial production. In more advanced economies, their share in economic production tends to decrease, probably causing the negative correlation with energy consumption (primary, final and in households), and this intermediary nature of the relation was responsible for the lack of support for the EKC thesis. Moreover, as the sector usually employs less qualified people, jobs in traditional services are typically low-paid and low-secure, inducing poverty problems.

The opposite relations may be described for financial activities (K). Their development is often perceived as a sign of economic advancement and characterizes the most affluent economies. It goes in line with high energy consumption (primary and final), greenhouse gas emissions and energy dependency. The relations are not of a causative nature but are an indirect one.

Generally, the synthetic measure of energy performance SM revealed some interlinkages with traditional kinds of activities, namely agriculture and industry, as well as with financial activity perceived as a sign of structural advancement. In the first case, the positive correlation coefficient indicated that the economies with high shares of traditional activities were also those with more environmentally friendly patterns of energy production and usage. In the second case, the coefficient was negative, indicating more severe energy tensions in economies with a high share of financial GVA. The observations suggested an indirect nature for the relationships between economic structure and energy performance, expressing the directions of structural change within the process of economic development.

In summary, the relations of the structural features to energy usage and production were mainly of an indirect nature, reflecting a connection with the general level of economic development. The connection could be observed for the SDG 7 indicators: primary energy consumption (four significant correlation coefficients), final energy consumption (three), energy consumption in households (three), energy poverty (three) and greenhouse gas emission (three). Three out of the five energy indicators distinguished the specified clusters of the EU-28 states. Unexpectedly, only one correlation coefficient detailed a relation concerning energy productivity: those having an ICT industry and which supported the EKC thesis about the positive influence of the information sector on the environment.

Moreover, the results appeared to be similar concerning 2010. As shown in Table 4, SM for 2010 was also positively correlated with a share of agriculture and industry in GVA, while it negatively correlated with financial activities. It confirms stability of potential relationships between sectoral structure of production and energy performance.

Table 4. Correlation matrix between structural features of economy and the SDG 7's indicators for the EU-28 states, 2010.

SDG 7 Indicator	Share of NACE Section(s) in GVA									
	A	BCDE	F	GHI	J	K	L	MN	OPQ	RSTU
SDG_07_10	−0.55 *	−0.28	−0.13	−0.38 *	0.33	0.59 *	−0.10	0.43 *	0.19	−0.26
SDG_07_11	−0.53 *	−0.34	−0.17	−0.31	0.30	0.70 *	−0.14	0.36	0.14	−0.27
SDG_07_20	−0.39 *	0.05	−0.34	−0.34	0.34	0.20	−0.12	0.44 *	0.25	−0.47 *
SDG_07_30	−0.21	−0.14	−0.21	−0.06	0.05	0.14	0.21	−0.07	0.23	−0.07
SDG_07_40	0.37	0.32	0.09	0.09	−0.05	−0.44 *	0.06	−0.29	0.04	−0.23
SDG_07_50	−0.29	−0.43 *	−0.11	0.21	−0.03	0.36	−0.03	0.09	−0.01	0.35
SDG_07_60	0.52 *	−0.06	0.23	0.33	−0.24	−0.04	0.17	−0.50*	−0.28	0.00
SDG_13_10	−0.50 *	−0.33	−0.18	−0.25	0.27	0.74 *	−0.15	0.25	0.04	−0.25
SM	0.47 *	0.45 *	0.18	0.19	−0.31	−0.77 *	0.17	−0.30	0.01	0.01

SM—synthetic measure of development towards the SDG 7. * statistically significant at 0.05. Source: Own calculation based on [32–40,46].

4. Discussion

The general results of the study do not confirm the hypothesis related to the explanatory role of the sectoral structure of production in terms of the energy characteristics of the EU-28 states in the context of the SDG 7. Nevertheless, the results are rather ambiguous as there are some signs of possible relations in line with economic structure to energy production and usage. First, by easing the statistical restriction and setting a significance level at $\alpha = 0.1$, the analysis of variance confirmed the differences between the clusters of EU-28 states concerning their synthetic measure of achievements towards the SDG 7. Moreover, even at significance level $\alpha = 0.05$, the differentiation is essential for some indicators of the SDG 7, namely final and primary energy consumptions and energy poverty. Finally, identification was possible for some essential correlations between the energy indicators and the sectoral characteristics, especially those concerning the share of agriculture, industry, traditional services and financial activities in GVA. All these findings suggest a need to extend the search for interlinkages between structural features and energy achievements.

Other interesting results concern the role of industry in energy performance. Detailed correlations with individual energy goals appeared to be not significant, while aggregated energy achievements correlated positively with the share of industry in the GVA. The finding stresses the importance of mutual interlinkages: synergies and trade-offs between detailed energy goals. This suggests a need for further research into specifying such interlinkages. Generally, the industrial activities did not appear to be causing the most severe environmental tensions in the EU, suggesting that new energy technologies are not sectoral specific.

The study adopts an aggregated approach to both energy goals as well as the structure of an economy. A synthetic measure of energy achievements was used to simultaneously take into account the different problems related to energy consumption, supply and accessibility. However, this may be misleading, as there are some trade-offs and unintended consequences between the individual energy goals of SD, such as between energy consumption in households and energy poverty. Nevertheless, it creates the opportunity to capture the general specificity of the economy and to compare achievements between the countries. Ranking of the EU-28 states based on their energy achievements is an important outcome of our study and indicates the generally more environmentally friendly characteristics of less affluent economies. A similar ranking was achieved by Kiselakova et al. [47]. This observation supports some theses about the monotonic growth of energy pressures on the environment within the process of economic development [23]. Once again, the relation identified in our study was not strong (correlation between the GDP per capita and the *SM* value was −0.5) and, thus, it is not able to reject the EKC hypothesis. Moreover, the causal functional relation was not intended to be tested in the study and the correlation coefficient used here merely signaled the possibility of unidentified mutual interlinkages.

Numerous studies on the existence of the EKC curve still do not achieve a common conclusion, even though sophisticated techniques have often been used with different indicators adopted to specify environmental tensions [19–31]. Some of them focused on energy consumption as a general factor of environmental pressure and tested the EKCs for total energy use, seen as a proxy indicator for all environmental impacts [23]. Shahbaz et al. [21] confirmed the existence of long-term relationship between economic growth and energy intensity; in addition, in another study [27], the relationship between economic growth, energy consumption and energy pollutants indicated energy consumption as a major contributor to energy pollution. Acaravci and Ozturk [27,48] examined the causal relationship between carbon dioxide emissions, energy consumption and economic growth for some European countries, confirming it only in a few cases. Unidirectional causality between financial development and CO_2 emissions was identified by Chang et al. [49] and related to industrial structure. Similarly, the inclusion of sectoral features identified energy consumption as an important factor for manufacturing GDP [21,50] and generally, as Dinda [22] claims, a high share of manufacturing in total GDP is associated with higher levels of energy consumption. On the contrary, Jober and Karanfil [21,51] report that there is no causality between energy consumption and economic growth at the aggregate level and at the sectoral industry level. In addition, Luzzati and Orsini [27,31] do not support an energy-EKC hypothesis after studying the relationship between absolute energy consumption and GDP per capita. A number of studies adopted a dynamic approach and tested causality between the variables. In contrast, this study takes a static approach and attempts to identify whether any interference between sectoral structure and energy performance measured at an aggregated level exists. This is checked by testing differences in the SDG 7 indices between clusters of countries specified by structural patterns of production.

In this vein, our findings suggest that there exists a relationship between the structure of production and the final and primary energy consumptions; however, the relationship is mainly induced by interlinkages with agricultural production, traditional services (with a positive environmental impact) and financial services (with a negative impact). As agricultural production is usually connected with "a clean economy", the assumption is confirmed by its positive correlation with general energy performance. It is specified not only by low energy consumption and greenhouse gas emissions, as detailed relations in our research suggest, but also, as other researchers point out [52], agriculture is expected to supply an economy with most of the energy renewable resources, thus fulfilling the climate and energy goals. The unexpected signs of the correlation coefficients in the case of traditional services and financial services indicate that the relation is not a direct one (as there is no explanation why the financial sector should itself consume a lot of energy or that traditional services such as transport should be less energy consuming), but it expresses a general level of economic development (as the advanced post-industrial knowledge-based countries with a high level of GDP per capita appear to be more energy consuming). It suggests that the EU-28 states still have to reach the level at which the environmental pressure can be essentially lowered and that there exist yet more reasons to doubt the validity of the EKC.

The study notes that important linkages can be specified between the problems of energy poverty and the structural features of production. The idea of a structural basis for economic poverty was sought, e.g., by Loayza and Raddatz [53] to indicate the alleviating role of labor-intensive sectors, such as agriculture, construction and industry. On the other hand, the service sector was specified as the most favorable for limiting the risk of poverty by Cyrek [54] and Ghani and Kharas [55], while Cyrek and Cyrek [56] found that the most socially favorable results were typical for employment in other knowledge-intensive services in contrary to market knowledge-intensive services. Thus, the results are ambiguous and do not allow definite indications as to the role played by the structural patterns of development in terms of material poverty. The findings depend on the method, data, geographical scope and period of analysis, indicating the specificity of the poverty problems faced by economies at different stages of economic development. Our study focuses on the relatively homogenous group of EU-28 states and finds that poverty in its energy dimension is induced by a low level of economic advancement expressed by a high role of agriculture and traditional services, which can be alleviated within the developmental process towards economies with a higher share of business services. Moreover, the relationship between the sectoral structure of production and energy poverty may either have a direct (connected with inter-sectoral differences in incomes and wages) or indirect character (specified by a general level of social welfare). The character of such interlinkages appears to be an interesting field for future research.

5. Conclusions

The paper empirically contributes to a relatively neglected issue of relations between sectoral features of the EU economies and their energy performance as interlinked dimensions of sustainable development. It adopts a static approach to search for such connections; however, it expresses a holistic attitude to SD. It searches for interlinkages between multi-dimensional features of these phenomena, in contrary to the more common focus on one chosen characteristic of energy performance (e.g., CO_2 emissions) or sectoral structure (e.g., a share of manufacturing in economy). It considers simultaneously shares of many sectors in production and, thus, does not limit sectoral advancement to one sector. Similarly, energy achievements are measured by a synthetic measure which aggregates detailed indices. The aggregated general approach allows to catch existence of some synergy and trade-offs between detailed dimensions and indicates that some phenomena are not simply additive. Thus, it gives different results than the research on partial relations.

The research presented in this study indicates the indirect character of the relationships between the sectoral structure of production and energy sustainability across the EU-28 states. The main hypothesis concerning the differentiating role of structural patterns for energy performance in the context of the SDG 7 was not confirmed, at least at the aggregate level. Nevertheless, it was possible to identify some detailed relationships that still suggest that certain sectoral features of an economy may influence energy achievements. These were mainly related to the primary and final energy consumptions and energy poverty, as well as the shares of agriculture, industry and finance in the GVA.

Another important conclusion indicates that the interlinkages with GDP per capita expressed by both the energy achievements and the structural characteristics. The relationships revealed, however, an unexpected result that causes some doubts as to the existence of the U-inverted EKC for the EU-28 states and, instead, suggests an increase in energy tensions in parallel with production level. Thus, the most affluent EU-28 economies need to face severe energy challenges. However, it is still important to enable spillovers concerning energy-saving or renewable resource technology usage to the catching-up countries to avoid repeating the same environment-harming phases of development and make the "time-effect" significant enough to exceed the "gross scale effect" for the benefit of all.

From practical point of view the results suggest a need to synchronize efforts in different dimensions of SD. Politicians must balance all the possible costs and benefits and their decisions should be based on as broad diagnosis as possible. Aggregated approach adopted in the study may be of help in the decision-making process. If policy is focused only on one aspect of energy performance it may lead to unexpected side-effects and, finally, miss its goal of SD. If sectoral policy supports development of a specified branch concerning possible outcomings in one sphere of energy goals, it may appear unfavorable in the other. The practical implication of the findings about indirect character of the relationships is a need to implement horizontal measures of policy aimed at stimulating SD, instead of sectoral specific ones. One must take into consideration that it is necessary to simultaneously create conditions for development of many branches.

Nevertheless, the aggregated approach reveals simultaneously strengths and weaknesses. The mutual interlinkages make it difficult to get robust unequivocal results. The ambiguity of our results suggests a need for further research into the possible interlinkages between different dimensions of SD at a more disaggregated level and to focus more attention on the possible trade-offs and synergies between the detailed goals and structural features. Moreover, a general character of the static diagnosis identifying potential relations makes it valuable to research in-depth the causality in line: economic structure–energy goals. This requires the adopting of a dynamic approach and in-time comparisons. However, it induces some problems to be resolved, such as specification of one measure of structural development that includes numerous sectors, as well as ensuring time-comparability of a synthetic measure of energy performance. Alternatively, the search for interlinkages requires development of quite new methodology of research.

Author Contributions: Conceptualization, M.C. and P.C.; methodology, M.C. and P.C.; software, M.C. and P.C.; validation, M.C. and P.C.; formal analysis, M.C. and P.C.; investigation, M.C. and P.C.; resources, M.C. and P.C.; data curation, M.C. and P.C.; writing—original draft preparation, M.C. and P.C.; writing—review and editing, M.C. and P.C.; visualization, M.C. and P.C.; supervision, M.C. and P.C.; project administration, M.C. and P.C.; funding acquisition, M.C. and P.C. All authors have read and agreed to the published version of the manuscript.

Funding: The APC was funded by the University of Rzeszów, Institute of Economics and Finance, Poland.

Institutional Review Board Statement: Not applicable.

Informed Consent Statement: Not applicable.

Data Availability Statement: All data used in the study are available in Eurostat database at https://ec.europa.eu/eurostat/data/database, accessed on 7–9 January 2021.

Conflicts of Interest: The authors declare no conflict of interest.

References

1. Our Common Future. Report of the World Commission on Environment and Development. 1987. Available online: http://www.un-documents.net/our-common-future.pdf (accessed on 17 April 2017).
2. *Sustainable Development in the European Union. Monitoring Report on Progress towards the SDGs in an EU Context*; Publications Office of the European Union: Luxembourg, 2020.
3. Bolis, I.; Morioka, S.N.; Sznelwar, I. Are we making decisions in a sustainable way? A comprehensive literature review about rationalities for sustainable development. *J. Clean. Prod.* **2017**, *45*, 310–322. [CrossRef]
4. Collins, R.D.; Selin, N.E.; De Weck, O.L.; Clark, W.C. Using inclusive wealth for policy evaluation: Application to electricity infrastructure planning in oil-exporting countries. *Ecol. Econ.* **2017**, *133*, 23–34. [CrossRef]
5. *Resolution Adopted by the General Assembly on 25 September 2015: Transforming Our World: The 2030 Agenda for Sustainable Development*; A/RES/70/1; United Nations General Assembly: New York, NY, USA, 2015.
6. *Sustainable Development in the European Union. Overview of Progress towards the SDGs in an EU Context*; Publications Office of the European Union: Luxembourg, 2020.
7. *Europe 2020—A Strategy for Smart, Sustainable and Inclusive Growth*; COM(2010) 2020 Final; European Commission: Brussels, Belgium, 2010.
8. *Next Steps for a Sustainable European Future: European Action for Sustainability*; COM(2016) 739; European Commission: Brussels, Belgium, 2016.
9. *The European Green Deal*; COM(2019) 640 Final; European Commission: Brussels, Belgium, 2019.
10. Galdeano-Gómez, E.; Pérez-Menza, J.C.; Godoy-Durán, Á. The social dimension as a driver of sustainable development: The case of family farms in southeast Spain. *Sustain. Sci.* **2016**, *11*, 349–362. [CrossRef]
11. Islam, S.M.N.; Munasinghe, M.; Clarke, M. Making long-term economic growth more sustainable: Evaluating the costs and benefits. *Ecol. Econ.* **2003**, *47*, 149–166. [CrossRef]
12. Miola, A.; Borchardt, S.; Neher, F.; Buscaglia, D. *Interlinkages and Policy Coherence for the Sustainable Development Goals Implementation: An Operational Method to Identify Trade-Offs and Co-Benefits in a Systemic Way*; Publications Office of the European Union: Luxembourg, 2019.
13. Griggs, D.J.; Nilsson, M.; Stevance, A.; McCollum, D. (Eds.) *A Guide to SDG Interactions: From Science to Implementation*; International Council for Science: Paris, France, 2017.
14. *Interlinkages of the 2030 Agenda for Sustainable Development. Background Document*; IAEG-SDGs, United Nations Statistical Commission: New York, NY, USA, 2019; Available online: https://unstats.un.org/unsd/statcom/50th-session/documents/BG-Item3a-Interlinkages-2030-Agenda-for-Sustainable-Development-E.pdf (accessed on 4 February 2021).

15. EU SDG Indicator Set 2020. Result of the Review in Preparation of the 2020 Edition of the EU SDG Monitoring Report. Final Version of 16/01/2020. Revised after Implementation of the EU SDG Indicator Set 2020 on 22 June 2020. European Commission, Eurostat, Directorate E: Sectoral and Regional Statistics, Unit E-2: Environmental Statistics and Accounts; Sustainable Development. Available online: https://ec.europa.eu/eurostat/documents/276524/12239692/SDG_indicator_set_2021.pdf/ebeb73b5-9ef5-a6d8-01ea-89c4ed17b7e4?t=1610726550972 (accessed on 4 February 2021).
16. *Directive (EU) 2018/2002 Amending Directive 2012/27/EU on Energy Efficiency*; European Parliament and Council of the European Union: Brussels, Belgium, 2018.
17. *Directive (EU) 2018/2001 on the Promotion of the Use of Energy from Renewable Sources*; European Parliament and Council of the European Union: Brussels, Belgium, 2018.
18. *A Framework Strategy for a Resilient Energy Union with a Forward-Looking Climate Change Policy*; COM(2015) 80 Final; European Commission: Brussels, Belgium, 2015.
19. Stern, D.I. The Rise and Fall of the Environmental Kuznets Curve. *World Dev.* **2004**, *32*, 1419–1439. [CrossRef]
20. Zaman, K.; Shahbaz, M.; Loganathan, N.; Raza, S.A. Tourism development, energy consumption and Environmental Kuznets Curve: Trivariate analysis in the panel of developed and developing countries. *Tour. Manag.* **2016**, *54*, 275–283. [CrossRef]
21. Shahbaz, M.; Ozturk, I.; Afza, T.; Ali, A. Revisiting the environmental Kuznets curve in a global economy. *Renew. Sustain. Energy Rev.* **2013**, *25*, 494–502. [CrossRef]
22. Dinda, S. Environmental Kuznets Curve Hypothesis: A Survey. *Ecol. Econ.* **2004**, *49*, 431–455. [CrossRef]
23. Stern, D.I. Environmental Kuznets Curve. *Encycl. Energy* **2004**, *2*, 517–525.
24. Dasgupta, S.; Laplante, B.; Wang, H.; Wheeler, D. Confronting the Environmental Kuznets Curve. *J. Econ. Perspect.* **2002**, *16*, 147–168. [CrossRef]
25. Andreoni, J.; Levinson, A. The Simple Analytics of the Environmental Kuznets Curve. *J. Public Econ.* **2001**, *80*, 269–286. [CrossRef]
26. Yandle, B.; Bhattarai, M.; Vijayaraghavan, M. Environmental Kuznets Curves: A Review of Findings, Methods, and Policy Implications. *PERC Res. Study* **2004**, *02-1*, 1–38.
27. Shahbaz, M.; Mutascu, M.; Azim, P. Environmental Kuznets Curve in Romania and the Role of Energy Consumption. MPRA Paper 2011, No. 32254. Available online: https://mpra.ub.uni-muenchen.de/32254/ (accessed on 11 February 2021).
28. Harbaugh, W.; Levinson, A.; Wilson, D. Reexamining the Empirical Evidence for an Environmental Kuznets Curve. NBER Working Paper 2000, No. 7711. Available online: http://www.nber.org/papers/w7711 (accessed on 11 February 2021).
29. Charfeddine, L.; Mrabet, Z. The impact of economic development and social-political factors on ecological footprint: A panel data analysis for 15 MENA countries. *Renew. Sustain. Energy Rev.* **2017**, *76*, 138–154. [CrossRef]
30. Panayotou, T. *Empirical Tests and Policy Analysis of Environmental Degradation at Different Stages of Economic Development*; Working Paper WP 238; Technology and Employment Programme, International Labour Office: Geneva, Switzerland, 1993.
31. Luzzati, T.; Orsini, M. Investigating the energy-environmental Kuznets curve. *Energy* **2009**, *34*, 291–300. [CrossRef]
32. Gross Value Added and Income by A*10 Industry Breakdowns [nama_10_a10], Eurostat Database. Available online: https://ec.europa.eu/eurostat/data/database (accessed on 7 January 2021).
33. Primary Energy Consumption [SDG_07_10], Eurostat Database. Available online: https://ec.europa.eu/eurostat/data/database (accessed on 9 January 2021).
34. Final Energy Consumption [SDG_07_11], Eurostat Database. Available online: https://ec.europa.eu/eurostat/data/database (accessed on 9 January 2021).
35. Final Energy Consumption in Households Per Capita [SDG_07_20], Eurostat Database. Available online: https://ec.europa.eu/eurostat/data/database (accessed on 9 January 2021).
36. Energy Productivity [SDG_07_30], Eurostat Database. Available online: https://ec.europa.eu/eurostat/data/database (accessed on 9 January 2021).
37. Share of Renewable Energy in Gross Final Energy Consumption by Sector [SDG_07_40], Eurostat Database. Available online: https://ec.europa.eu/eurostat/data/database (accessed on 9 January 2021).
38. Energy Import Dependency by Products [SDG_07_50], Eurostat Database. Available online: https://ec.europa.eu/eurostat/data/database (accessed on 9 January 2021).
39. Population Unable to Keep Home Adequately Warm by Poverty Status [SDG_07_60], Eurostat Database. Available online: https://ec.europa.eu/eurostat/data/database (accessed on 9 January 2021).
40. Greenhouse Gas Emissions (Source: EEA) [SDG_13_10], Eurostat Database. Available online: https://ec.europa.eu/eurostat/data/database (accessed on 9 January 2021).
41. Final Energy Consumption by Sector [TEN00124], Eurostat Database. Available online: https://ec.europa.eu/eurostat/data/database (accessed on 7 January 2021).
42. Hellwig, Z. Zastosowanie metody taksonomicznej do typologicznego podziału krajów ze względu na poziom ich rozwoju oraz zasoby i strukturę wykwalifikowanych kadr. *Przegląd. Stat.* **1968**, *4*, 307–327.
43. Ward, J.H., Jr. Hierarchical Grouping to Optimize an Objective Function. *J. Am. Stat. Assoc.* **1963**, *58*, 236–244. [CrossRef]
44. Stanisz, A. *Przystępny Kurs Statystyki z Zastosowaniem STATISTICA PL na Przykładach z Medycyny. Tom 3. Analizy Wielowymiarowe*; StatSoft: Kraków, Poland, 2007.
45. Electronic Statistics Textbook. Statistica. StatSoft. Available online: http://www.statistica.pl/textbook/stathome_stat.html?http%3A%2F%2Fwww.statistica.pl%2Ftextbook%2Fstcluan.html (accessed on 15 January 2021).

46. Main GDP Aggregates Per Capita [nama_10_pc], Eurostat Database. Available online: https://ec.europa.eu/eurostat/data/database (accessed on 8 January 2021).
47. Kiselakova, D.; Stec, M.; Grzebyk, M.; Sofrankova, B. A Multidimensional Evaluation of the Sustainable Development of European Union Countries—An Empirical Study. *J. Compet.* **2020**, *12*, 56–73. [CrossRef]
48. Acaravci, A.; Ozturk, I. On the relationship between energy consumption, CO_2 emissions and economic growth in Europe. *Energy* **2010**, *35*, 5412–5420. [CrossRef]
49. Chang, C.-L.; Ilomäki, J.; Laurila, H.; McAleer, M. Causality between CO_2 Emissions and Stock Markets. *Energies* **2020**, *13*, 2893. [CrossRef]
50. Soytas, U.; Sari, R. The relationship between energy and production: Evidence from Turkish manufacturing industry. *Energy Econ.* **2007**, *29*, 1151–1165. [CrossRef]
51. Jober, T.; Karanfil, F. Formation and deformation of the environmental Kuznets curve for CO_2 emissions. In *Crisis, Innovation and Sustainable Development*; Laperche, B., Levratto, N., Uzunidis, D., Eds.; Edward Elgar Publishing: Cheltenham, UK, 2012; pp. 243–258.
52. Gradziuk, P.; Gradziuk, B. Polityka klimatyczno-energetyczna Unii Europejskiej—Wyzwania i szanse dla obszarów wiejskich. *Rocz. Nauk. Stowarzyszenia Ekon. Rol. Agrobiz.* **2017**, *XIX*, 71–78. [CrossRef]
53. Loayza, N.; Raddatz, C.E. *The Composition of Growth Matters for Poverty Alleviation*; World Bank Publications: Washington, DC, USA, 2006; p. 4077.
54. Cyrek, M. Social efficiency of employment in three sectors—A comparison of Polish regions. *Equilib. Q. J. Econ. Econ. Policy* **2017**, *12*, 417–432. [CrossRef]
55. Ghani, E.; Kharas, H. The Service Revolution. In *Economic Premise*; World Bank: Washington, DC, USA, 2010; Volume 14.
56. Cyrek, M.; Cyrek, P. Economic structure and inclusive society in the integrated regions of the Central and East Europe. In *Management Rozvoje Regionů ve Středoevropské Perspective*; Dušek, J., Ed.; Vysoká Škola Evropských a Regionálních Studií: České Budějovice, Czech Republic, 2018; pp. 48–58.

energies MDPI

Article

External Costs for Agriculture from Lignite Extraction from the Złoczew Deposit

Benedykt Pepliński

Department of Law and Enterprise Management in Agribusiness, Faculty of Economics, Poznan University of Life Sciences, 60-637 Poznan, Poland; peplinski@up.poznan.pl; Tel.: +48-61-848-71-09

Abstract: In many circles, including in Poland, lignite is still viewed as a cheap source of energy, which is only possible if the external costs associated with mining and burning coal are not taken into account. In Poland, this is reflected in plans to open new Złoczew opencast lignite mines. In previous studies, the analysis of external costs has focused on the external costs of coal combustion and related pollutant emissions. This paper focuses on the extraction phase. The aim of the work here described was to estimate the external costs that agriculture may incur due to the formation of a depression funnel for the projected lignite mine in the Złoczew deposit. This paper discusses factors causing uncertainty in calculated estimates of external costs in agriculture, and characterizes the Bełchatów and Złoczew opencast mines. In the paper, a methodology for calculating external costs in livestock production is then proposed. In the next part of the study, the decrease in cereal and potato yields and in the number of cattle and pigs in the area of the cone of depression of the Bełchatów opencast mine, which has been in operation for 40 years, were estimated. The estimates obtained formed the basis for estimating external costs for the planned Złoczew lignite opencast. The analyses showed high external costs for plant production and much lower for animal production. The inclusion of the estimated external costs of 12.2 € × kWh^{-1} in the costs of electricity production will significantly worsen the profitability of launching this opencast. The paper discusses factors causing uncertainty in calculated estimates of external costs in agriculture, and characterizes the Bełchatów and Złoczew opencast mines. The discussion also shows that the level of losses incurred in crop production due to opencast coal mining is similar to the losses incurred in crop production in extremely dry years.

Keywords: external cost; opencast lignite; plant production; animal production; depression funnel

Citation: Pepliński, B. External Costs for Agriculture from Lignite Extraction from the Złoczew Deposit. *Energies* **2021**, *14*, 2660. https://doi.org/10.3390/en14092660

Academic Editor: Piotr Gradziuk

Received: 30 March 2021
Accepted: 28 April 2021
Published: 6 May 2021

1. Introduction

Poland is a country with large lignite resources. At the end of 2014, the total geological balance resources of lignite, located in 90 deposits throughout Poland, amounted to 23.5 billion Mg, of which nearly 1.5 billion Mg has already been developed [1]. Currently, the annual mining of over 60 million Mg is carried out in six opencast mines in three lignite areas. The extracted lignite generates electricity, which satisfies approximately 30–35% of Poland's demand. If new opencast mines are not put into operation, coal mining of more than 50 million Mg will continue only until 2030. By 2036, it would decrease to less than 20 million Mg, to cease around 2045 [2]. Energy companies are currently working on putting more lignite deposits into operation in Poland. To date, the most advanced mining works are performed in Złoczew and Ościsłowo deposits. In the case of the Złoczew deposit, with 611 million Mg of resources, a decision on environmental conditions for the proposed opencast mine was issued on 28 March 2018. The only requirement that is missing to launch the opencast mine is the exploitation concession [3]. For the second deposit, the implementation of the investment project was suspended due to the lack of a decision concerning environmental conditions that have to occur for the opencast mine to operate. The demands of the energy sector are partially reflected in the Energy Policy of Poland until 2040 [4], approved in February 2021, which provides for the possibility of initiating lignite mining in two new deposits, Złoczew and Ościsłowo. The strategy leaves the final decision on launching such exploitation to investors, indicating the key role of the price of CO_2 emission rights, environmental conditions, and development of new technologies. However, the said document does not mention the external costs associated with the potential exploitation of these deposits.

New mining and energy projects are being implemented in numerous developing countries with large coal reserves, such as China, India, Turkey, Vietnam, Indonesia, Bangladesh, Japan, South Africa, and the Philippines [5,6]. On one hand, their implementation is motivated by a rapidly growing demand for electricity while on the other, by very high security of supply, which no other mined energy source can match [7]. The continuously popular opinion that lignite is the cheapest or one of the cheapest sources of energy, is also crucially important [8–11]. This applies to Poland as well [12–15]. However, such a perception of lignite is only possible if external costs associated with mining and burning coal are not taken into account. They also point out that, in view of the lack of large reserves of other energy resources, further coal mining in Poland is necessary to ensure energy security [12–15].

There are many definitions of external costs. In the case of electricity production, they indicate costs incurred in connection with the production of electricity by third parties and future generations, rather than the expenditures of direct recipients and providers of electricity [16,17]. In contrast, the external cost, understood as the monetary value of the damage caused by electricity production, is today the most widely accepted common denominator for valuing environmental impacts [18–20].

Such costs are associated with coal burning and the resulting air pollution, which affects climate change (e.g., through CO_2 emissions) and human health by increasing the number of respiratory, vascular and other diseases, premature deaths, medical costs and days of medical leave, as well as reducing the productivity of the economy. Furthermore, opencast coal mining entails costs incurred by the environment, in particular related to the deterioration of the surface and underground water quality and the occurrence of a depression cone, which causes losses in agricultural and forestry production, as well as leading to surface deformation, increased dustiness, noise, etc. The environment also bears significant costs through depletion of biodiversity, elimination or obstruction of functioning of ecological corridors, as well as the loss or reduction of natural and tourist values in large areas [21,22]. Their valuation is necessary to render the costs of electricity production real, thanks to which, the undertaken decision will be as close as possible to the ecological, economic, and social optima.

Tol's review of the literature on damage caused by climate change, which includes over 588 estimates from 75 published studies, indicates that the problem of external costs associated with CO_2 emissions into the atmosphere is widely recognized [23]. Numerous studies have also been undertaken to value the costs arising in connection with deterioration of human health caused by emissions from burning coal [16,22,24–27]. The literature estimating the external costs associated with opencast coal mining is much poorer [28]. Authors of one of the few studies in this area analyzed the impact of particulate matter emissions from lignite mining performed in the South Field Mine of the Lignite Center Ptolemais-Amyntaion located in Greece [28]. A team led by Pepliński have undertaken another research project, which focused on studying the effects of a depression cone created as a result of the opencast lignite mining in Poland on agricultural production. During the initial works [29–31], the authors assumed numerous simplifications regarding the level of decrease in crops and livestock, e.g., an identical level of decrease in crops and livestock throughout the entire period of studies concerning the influence of opencast mines. It was calculated by comparing the crop level from the region of the Konin lignite field (located in the eastern part of Wielkopolskie Voivodship in Poland) with the referential crop level (i.e., from the rest of the said voivodship) before launching the opencast mines and 30 years later, which raises the risk of significant overestimation or underestimation of external costs for the intermediate periods. A subsequent paper undertook a more detailed study on the decline in the level of crops in this region [32]. However, to our knowledge there have been no studies that analyze the impact of opencast lignite mining on the level of animal production.

In connection with plans to launch the Złoczew opencast mine, it is necessary to estimate the actual costs of electricity production using coal from this deposit, including the external costs. The following study aimed to estimate the external costs associated with the exploitation of lignite in agriculture, both in plant and animal production, for the projected Złoczew opencast mine, which will complement the knowledge of the actual costs of electricity production from lignite.

2. External Costs in Agriculture and Difficulties Associated with Their Estimation

In agriculture, external costs are connected with depression cones found around the opencast lignite and other resource mines. Their formation is linked with the deposit drainage, which must be carried out to the depth at which the raw materials are mined. In the case of lignite, it is usually a range between tens and over 200 m b.g.l. There are two types of depression cones: discharge and pressure relief. The former is created due to the gravitational flow of water towards the drained deposit. This results in the creation of a depression cone, which in the vertical section is a cone-shaped curve, i.e., the water table at the edge of the opencast mine rises quickly, but as the distance increases, the water table rises at a slower pace. The Polish law obliges the investor to define the estimated area around the opencast mine where the water table will be permanently lowered by at least one meter, creating an area of depression cone. In the case of lignite opencast mines, the range of the depression cone usually varies between a few to several kilometers, starting from the edge of the opencast, and has the shape of an ellipse. However, in the case of agriculture, for which groundwater is particularly important in plant production, the impact area extends far beyond the area of the established depression cone and reaches up to several dozen kilometers from the edge of the opencast.

In turn, the pressure relief cone, which is much larger than the depression cone area, is the territory where groundwater pressure is reduced. A change in water pressure in deeper aquifers caused by hydrogeological windows can lower the groundwater and surface water levels, as it triggers a local outflow of water to deeper ground layers. It can also reduce or generate a decline in subsoil resource supply system, which uses water from deeper aquifers [32–34].

Difficulties in estimating production losses and assessing the external costs in crop production result from numerous interrelating factors. These include geological, natural and climatic, agricultural and production, as well as temporal and spatial factors.

The key geological factors include:

- Dewatering depth and time;
- The location of the opencast in the catchment area, the size of the catchment area and directions of the groundwater flow;
- Geological structure of drained areas;
- The size of supply with rainwater and surface water; and
- The initial (primary) level of groundwater, which in the case of peripheral areas of the impact zone indicates that the mine would affect areas with higher water levels and would have no influence on the surrounding areas with lower water tables, even those located further away.

In terms of natural and climatic conditions, the intensity of precipitation and its distribution during the growing season are particularly important in the context of the topic under discussion.

Agricultural and production difficulties are linked to the biological nature of production. The crop levels obtained by farmers depend, among other things, on natural factors, such as type, quality and pH level of soil, topography, length of the growing season, precipitation, temperature, and water table. There are also economic factors which include, e.g., the level of agricultural development, agrarian structure, production intensity, availability of techniques and technologies, and quality of human capital. A broader description of difficulties arising in connection with estimating production losses in crop production in the area of influence of opencast mines was presented by Pepliński and Czubak [32].

The launch of an opencast mine disrupts the existing production conditions. However, their importance varies in particular areas around the opencast due to the factors described above. Such influence would also differ in opencast mines located in different regions of a country, continent, or other areas of the world, rendering it impossible to fully apply the results of the observation from one object to another.

Many factors determine the level of impact of opencast mines on the amount of livestock, also in the case of animal production. Their direct influence is restricted to the elimination of livestock herds (or flocks) from farms located above the deposit and the area of associated infrastructure. Farms located in the vicinity of an opencast may also cease or reduce their production and subsequently lose all or part of their agricultural land, which constitutes their primary feed base. The indirect influence of opencast mines is associated with a decrease in livestock due to a decline in feed production, which in turn results from a drop in feed crops occurring in the area of a depression cone. The decrease in livestock in the depression cone area is influenced, i.e., by:

- The level of crop decline—in the case of farms located in the vicinity of an opencast, the rate of livestock decline would be higher than in regions located further away from the edge of the opencast, although it is difficult to establish the limit of crop decline that would not translate into a decline in livestock;
- Animal species—the most sensitive are ruminants, i.e., cattle and sheep that require large quantities of roughage, such as green fodder, haylage, silage and hay, which are usually produced on the farm itself or in the immediate vicinity. Due to their high economic sensitivity to transport, the possibilities of long-distance transit are limited. In the case of roughage stored in silos, its daily delivery also constitutes a restriction due to the rapid rate of feed spoilage. Pig farming is less sensitive to a decrease in feed production. In Poland, the share of own feed in pig farming reaches approx. 50% [31]. In contrast, poultry production, which relies almost entirely on industrial feed, is independent of its own feed;
- Production scale—smaller farms are more sensitive to decreases in production on their farm, as they cannot rely on discounts due to the purchase of small quantities of feed. Over time, smaller farms are more likely to cease animal production. Development of animal production is also inhibited since the increase in costs of producing own feed reduces the yield of the entire farm, thus decreasing the number of funds allocated for investment projects and development. In the case of farms operating on a large scale, emerging feed shortages are supplemented with purchased feed. In ruminant farming, it is common to purchase or lease additional land, which further inhibits the growth of these units in the long-term perspective;
- The level of production yield in the long term, which determines the economic condition of agriculture. At low profitability of animal production, the additional reduction in profit resulting from the loss of a part of own feed increases the propensity to cease a given production and the tendency to take over farms by successors;
- The level of agricultural development, including the level of plant production, the acquired expertise, wealth, and the degree of cooperation between local agriculture and the environment;
- Time of occurrence of the depression cone, since the economic conditions for agriculture in the impact area deteriorates as the period of the opencast mine's operation lengthens. This would be expressed by the reduction of investment projects and the purchase of modern technologies.

Such a large number of factors, together with the difficulty in estimating the impact of the depression cone and the associated decrease in the amount of own feed on the level of animal production, indicate that the obtained results would be characterized by a certain level of underestimation or overestimation. However, uncertainty is a well-known phenomenon associated with the estimation of external costs of electricity production, as it is better to acquire even an estimated assessment of external effects rather than completely disregard or ignore such influence. Furthermore, despite the uncertain results, they allow drawing sensible conclusions [24]. Further research may expand the knowledge and reduce the level of uncertainty, thus validating the performance of subsequent studies concerning all external costs associated with electricity production.

3. Characteristics of Bełchatów and Złoczew Opencast Mines

Since the process of coal mining from the Złoczew deposit is still in the planning phase, estimation of losses in agricultural production must be conducted based on data from an already existing opencast mine that is characterized by similar parameters in terms of the size and depth of the lignite deposit. In the case of the Złoczew deposit, these criteria are best met by the Bełchatow deposit.

The Bełchatow deposit is one of the largest opencast lignite mines in the world. The maximum depth from which coal is extracted reaches an average of 280 m b.g.l., which indicates that the original groundwater table in the mining area is lowered by an average of 280 m, but 352 m at the maximum (Table 1). Dewatering of the Bełchatów deposit began in the second half of 1975, while mining launched at the end of 1980. Originally, the plans involved exploitation of the Bełchatow field until 2020, but this term was extended until 2026 to extract small amounts of coal from the sidewalls of the excavation. In turn, it is expected that the Szczerców field will be mined until around 2038. Total resources of this deposit were estimated at 1.8 billion Mg, of which approximately 1.1 billion Mg were located in the Bełchatów field, while 720 million Mg were in the Szczerców field. By the end of 2017, 1169 Mg of coal had been mined while 618 Mg remained to be extracted [12,35]. It was necessary to remove more than 4510 million m^3 of overburden, of which 1400 million m^3 formed Kamieńsk Hill, which has an elevation of 195 m and is located next to the Bełchatów field, and a spoil tip located next to the Szczerców field, which has an elevation of 170 m and consists of 1100 million m^3 of the overburden. They occupy a total of 2640 ha and are currently being reclaimed for forestry and recreation [12,36,37].

Table 1. Parameters of the geological location of the coal seam(s) in individual deposits.

Deposit Name	Overburden Thickness (m)			Coal Thickness (m)			Depth of the Deposit Floor (m b.g.l.)		
	Min.	Average	Max.	Min.	Average	Max.	Min.	Average	Max.
Bełchatów—Bełchatów field	0.0	24.3	158.8	3.0	55.1	230.5	3.0	79.5	245.5
Bełchatów—Szczerców field	7.6	119.5	239.8	8.9	50.3	196.1	65	171.1	351.7
Złoczew	138.4	215.1	280.9	12.1	51.4	127.8	150.5	266.6	354.3

Source: Based on [35].

Since the beginning of the opencast mine's operation, 9.3 billion m^3 of water have been pumped out. As a result, the average waterlogging index amounts to 7.96 m^3 × Mg^{-1}, while the average for the entire Polish lignite mining industry is 6.8 m^3 × Mg^{-1}. In 2017 alone, 200 million m^3 of water was pumped out, which gave the waterlogging index of 4.71 m^3 × Mg^{-1} [12]. The area of the depression cone is subjected to continuous changes along with the progress of mining and dewatering depth, as well as depending on the amount of precipitation. Between 1976 and 2004, the average area of the groundwater depression cone was 438 km^2, with a maximum area of 635 km^2 in 1992. The depression cone was shaped similarly to an ellipse, measuring 40 km (W-E axis) and 20 km (S-N axis). After launching the dewatering process in the Szczerców field in 2000, the depression cone rapidly expanded towards the west, which increased the dimensions of the depression area to 45 km and 25 km, while the maximum area amounted to approx. 800 km^2 [35,38,39]. The investigation also revealed numerous areas of lowered groundwater levels outside of the main depression cone area, indicating the presence of multiple hydrological windows. They cause groundwater outflows into the deeper layers due to a decrease in water pressure (pressure relief cone) located below the discharge cone [40].

Reclamation plans provide for the creation of a lake, with a maximum depth of approx. 100 m, over an area of about 3250 ha located in the final excavations of both fields. Its filling is expected to take place over a period of 20 years and be completed around 2070. The remaining land will be reclaimed and used primarily for forestry and, to a lesser extent, recreation and industry [41]. It is estimated that in the absence of an additional supply, the restoration of water relations around the Bełchatów mine will be completed by 2110. With the use of an additional external water supply system, this time may be shortened by 15 years. It would require additional 60 years to achieve the steady-state conditions, i.e., invariability of the water flow over time [39,42].

The Złoczew deposit is located approximately 50 km (in a straight line) from the Bełchatów deposit and stretches from south-west to north-east, in the form of a narrow, 1000–1500 m wide strip for about 10 km. To accommodate an opencast mine, an external spoil tip and the necessary infrastructure, it will be necessary to occupy approx. 6100 ha of land [43]. The coal is deposited at a depth of about 300 m, while the maximum exploitation depth will amount to approx. 354 m b.g.l. It is expected that this deposit will allow 485.8 Mg of lignite to be extracted over 31 years, with a maximum annual extraction of 18 million Mg. Significant depth of the deposit and the long mining period will contribute to the creation of a depression cone over a large area. However, there are considerable discrepancies in this regard. According to the authors of the 2017 report on the environmental impact, commissioned by the investor, the maximum reach of the discharge cone would amount to 311.53 km^2 [44]. In turn, other authors estimate the maximum reach of the said discharge cone to be 14–16 km from the center of the mine at the moment of its full expansion, i.e., 615–803 km^2, which is a cone area similar to the size of the Bełchatów opencast [35,45]. In extreme cases, it may even reach 3100 km^2 [43]. Although the lignite in the Złoczew deposit, similarly to that in the Bełchatów deposit, is located in a rift valley, in Jurassic and Cretaceous formations filled with Miocene and Quaternary formations it is characterized by significantly different factors influencing the extent of the depression cone. Contrary to the Bełchatów deposit, in the case of the Złoczew deposit, attention should be focused primarily on the predominance of water-bearing formations (sand and gravel) over impermeable formations (clays and loams) in the vertical profile, which has numerous hydrogeological windows, and common, strong fractures with developed karstic features and faults conducive to the shaping of a depression cone in the Jurassic aquifer (Figure 1) [43,45,46]. Hydrogeological windows would also account for local declines in groundwater levels outside the area of the designated depression cone due to decreases in water pressure in deeper soil layers.

Within 6.5 years of accessing the deposit [47] and 31 years of exploitation, about 6300 m^3 of water will be pumped out, which, with the extraction of 485 milion Mg of coal, gives a waterlogging index of 13.1 m^3 $\times$ Mg^{-1}, i.e., more than a half higher than in the Bełchatów deposit [45]. After the coal extraction, the initial plans provided for the creation of a water reservoir on an area of 2345 ha, located in the place of the final excavation, in the eastern part of the opencast mine. The reservoir's filling would be supported by water from Warta and Oleśnica rivers extracted during the periods of increased water flow, which is expected to take place within 16 years [44].

For most opencast mines, it is estimated that after dewatering the deposit, water relations will normalize in an amount of time that is approximately equal to the dewatering period. In the case of the deposit in Bełchatów, the dewatering period is expected to last about 73 years while the restoration of water relations is planned to be completed after 72 years, provided that the final reservoir is filled naturally. A higher waterlogging index of the Złoczew deposit indicates that the said process may take longer than 38 years, particularly since this is a region of low rainfall, snowless winters, and long-lasting droughts.

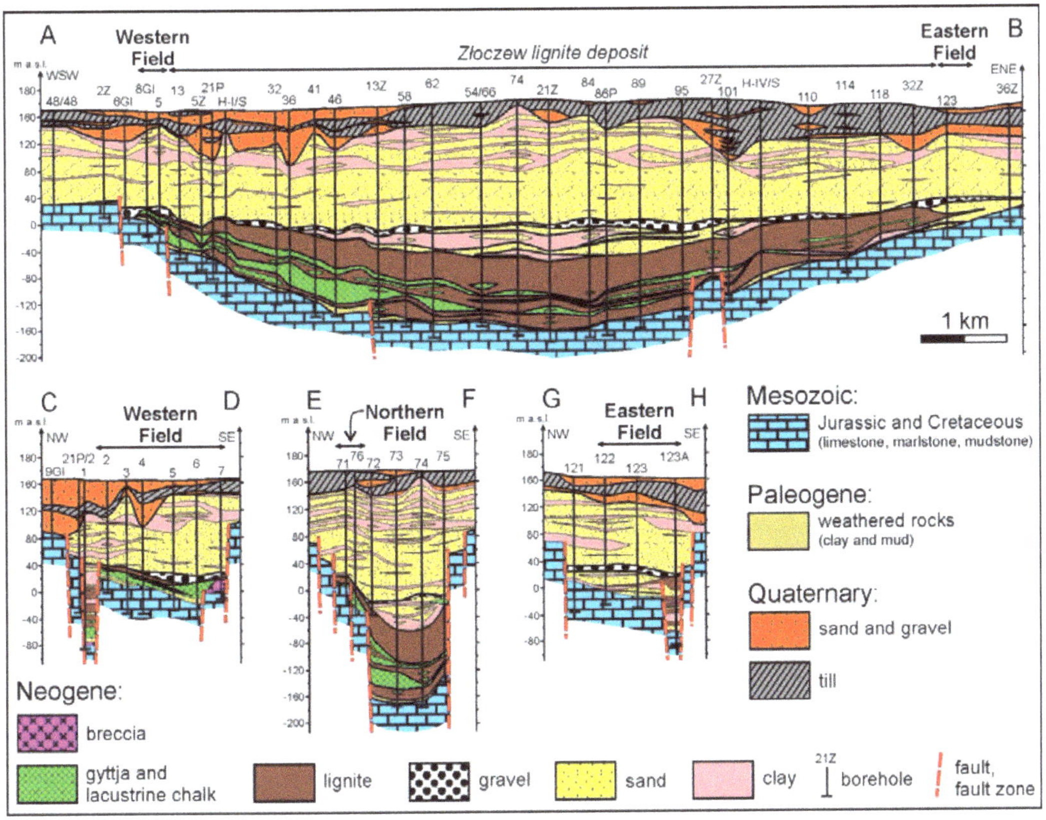

Figure 1. Geological cross sections through the Złoczew lignite deposit. Source: Based on [46].

The high complexity of the processes and the desire to obtain reliable loss estimation resulted in the adoption of fairly conservative assumptions that the impact area of the Złoczew opencast would be similar to that of the Bełchatów opencast and the period during which the Złoczew opencast will influence the surrounding area would be 76 years, including 38 years of dewatering and further 38 years of restoring water relations.

4. Materials and Methods

To estimate the external costs associated with the planned launch of the Złoczew deposit that are borne by agriculture, it is necessary to estimate the potential losses in plant and animal production, which result from:

- Occupying the area of agricultural land by the opencast, the external spoil tip and the necessary infrastructure, e.g., power plant, conveyor belts, access roads, etc., which in further parts of this paper will be referred to as the "opencast area"; and
- The presence of areas with lowered groundwater level (also those lowered by less than 1.0 m), which in further parts of this paper will be referred to as "the depression cone area" or "area of influence of the opencast".

In the case of plant production, crops of cereals and potatoes were analyzed, which was dictated by the availability of data on crops and agricultural acreage regarding the entire period under review. In turn, the sugar beet harvest was not analyzed, as in the area influenced by the Bełchatów opencast the cultivation of sugar beet has a marginal significance—e.g., in the (former) Piotrków Voivodship, it did not exceed 50 ha in several years of the analyzed period. In the case of animal production, the following types of livestock were analyzed: cattle, cows, pigs, and sows, i.e., groups of animals, whose feeding, in Polish conditions, is based primarily on the feed produced on the farm. Since the Złoczew deposit is in the planning phase, the level of production losses was estimated based on the changes in crops of the aforementioned plants and changes in selected types of livestock in the area of influence of the Bełchatów deposit. Since the period of influence of the Bełchatów opencast on the surroundings, including agriculture, is longer than the one of the Złoczew deposit, the analysis included losses in agricultural production covering a period of 38 years from the launch of dewatering of the Bełchatów opencast, i.e., from 1975 to 2013. The time necessary for the water relations to restore would cause the disappearance of the depression cone at a rate corresponding to its formation. Therefore, it was assumed that the losses in agricultural production, which occurred in particular years of the period required for water relations to normalize, would correspond to the losses from the opencast mining period in reverse order, i.e., from the last to the first year of exploitation.

Analyses estimating losses in agricultural production included data from a series of statistical yearbooks prepared by Statistics Poland, such as the Statistical Yearbook of Voivodships, the Yearbook, the Statistical Yearbook of Regions, statistical yearbooks of individual voivodships, and others [48–58].

Due to the level of detail of the available data and the administrative reforms implemented in Poland, the analysis was carried out using a multivariate approach. The basic analysis was conducted at the level of voivodships, in accordance with the administrative division effective in 1975–1998, which distinguished 49 Polish voivodships. The assessment of losses in crops and livestock occurring in the area affected by the Bełchatów opencast was based on three groups of voivodships:

- Group I, which included only the Piotrków Voivodship, where the Bełchatów lignite opencast mine is located;
- Group II, which included Sieradz and Czestochowa voivodships, located in the nearest vicinity of the opencast mines; and
- Group III, which included another six voivodships, located at an average distance of up to 100 km from the Bełchatów opencast, i.e., Kalisz, Kielce, Opole, Płock, Radom, and Skierniewice voivodships. This area will also be referred to as the reference area.

In the case of plant production, the analysis included the period until 1997 due to the fact that, from 1998 onwards, data on crops are available only according to the new administrative division, which distinguishes 16 voivodships. Such an administrative division renders it impossible to assess the impact of the Bełchatów opencast on agriculture. Therefore, an additional analysis, comparing the crop level decrease rate in areas influenced by the Bełchatów opencast with the crop level decrease rate in areas affected by opencast mines located in the Konin lignite field, was conducted and subsequently published by Pepliński and Czubak [32]. The analysis of changes in crop levels was conducted using five-year averages, thus reducing the level of crop variability resulting from weather conditions.

Regarding animal production, data on livestock are available at a district level, covering 1973, 1996, 2002, and 2010. Such data were used to calculate the number of analyzed animal species in areas corresponding to the area of voivodships distinguished according to the 1975–1998 division.

Moreover, only in the case of animal production, an analysis of changes in the head of cattle (without cows), cows, pigs (without sows), and sows around the Bełchatów opencast mine was conducted at a district level, in which 5 sectors were distinguished:

- The first sector, marked as "up to 20 km", includes the district of Bełchatów, in which the Bełchatów opencast mine is located;
- The second sector includes two districts located at an average distance of 21–40 km away from the opencast mines;
- The third sector includes five districts located at an average distance of 41–60 km away from the opencast mines;

- The fourth sector includes five districts located at an average distance of 61–80 km away from opencast mines; and
- The fifth sector includes twelve districts located at an average distance of 81–100 km from the opencast mines and constitutes the reference area for the other sectors.

Taking into consideration the assumption that the area of influence of the planned Złoczew opencast will be similar to the area of influence of the Bełchatów deposit, it is also necessary to define "Area I" and "Area II" for the Złoczew deposit. It was established that "Area I" covers the whole of the Sieradz Voivodship, where the planned Złoczew opencast is located, and 9.7% of Kalisz and Częstochowa voivodships, which in total corresponds to the area of the Piotrków Voivodship, where the Bełchatów deposit is located. The area of Kalisz, Częstochowa, Opole, and Piotrków voivodships, which corresponds to the area of voivodships from Group II for the Bełchatów opencast, was defined as "Area II". The area of five district sectors around the Złoczew opencast was determined according to similar principles.

With regard to external costs of plant production, arising in connection with the opencast lignite mining in the Bełchatów deposit, the study adopted the methodology developed by Pepliński and Czubak [32]. In the following study, external costs were calculated by distinguishing two depression cone areas (Ad_{AL}), i.e., "Area I" corresponding to the amount of agricultural land in voivodships from Group I, and "Area II" corresponding to the amount of agricultural land in voivodships from Group II.

In the case of animal production, the plan provides for the performance of an analysis comparing changes in the head of cattle, cows, pigs, and sows with the said livestock converted to a large size unit (LSU) at voivodship and district levels. The LSU is a unit of 500 kg. The conversion was based on indicators used by Eurostat [59]. The estimated losses, expressed as the LSU, will be subsequently used to calculate the external costs for each animal group. The decrease in livestock, which resulted from the launch of the Bełchatów opencast, was calculated for individual years based on the following formula:

$$SL_i = 100 - \frac{100 + LSUd_i}{100 + LSU_i} * 100 \quad (1)$$

where SL_i—the estimated loss level (%), $LSUd_i$—the change in the head of cattle (excluding cows), cows, pigs (excluding sows) and sows occurring in the examined area of impact of the opencast in the i-th year, compared to the base year, i.e., 1975, expressed as the LSU (%). In the paper, losses occurring in "Area I" and "Area II" and sectors 1–4 will be estimated separately, LSU_i—the change in the head of cattle (excluding cows), cows, pigs (excluding sows), and sows occurring in the designated reference area, i.e., "Area III" and sector 5 in the i-th year, compared to the base year, i.e., 1975, expressed as the LSU (%).

External costs in animal production in the depression cone area were calculated according to the following formula:

$$Ec_{zf} = \sum_{i=1}^{n} S_i \frac{SL_i}{100} * Pr_i * t * p_i * P_i \quad (2)$$

where Ec_{zf}—external costs in animal production, S_i—the average livestock level of the i-th group of animals in the analyzed area (number of animals), SL_i—the estimated average level of livestock losses over the entire period of impact of the opencast mine (%), Pr_i—annual productivity of a given group of animals (kg of beef/pork per animal, litres of milk per cow, piglets per sow), t—time of influence of the opencast mine (years), p_i—the average selling price of an animal product (e.g., USD, EUR $\times$ kg^{-1}). P_i—yield of production concerning the i-th group of animals (%). Regarding the calculations, it is also possible to adopt equal levels of yield for all animal production.

In the case of production losses occurring in the area occupied by the opencast mine, the external spoil tip and the necessary coexisting infrastructure, e.g., a power plant, conveyor belts, access roads, etc., together forming the external costs (Ec_{zo}), were calculated according to the following formula:

$$Ec_{zo} = \sum_{i=1}^{n} S_i * Pr_i * t * p_i * P_i \quad (3)$$

The estimation of external costs associated with the possible launch of the Złoczew opencast mine for agriculture was based on the most recent statistical data from Statistics Poland [57]. In the calculations, the study assumed:

- The average sowing structure from the 2015–2019 period for the analyzed areas;
- The acreage of the agricultural land from 2019, including subsequent changes in agricultural land acreage from the 2010–2019 period;
- Average selling prices for plant and animal products from the 2015–2019 period,
- The average yield for plant and animal production at identical, 25% level [31];
- Three variants of changes in crop levels. Variant I assumed the average level of cereal and potato crops from the 2015–2019 period. In Variant II, the crop level from the 2015–2019 period was subsequently adjusted to include trends in productivity changes occurring in the area of influence of the Złoczew opencast mine between 1981 and 2019. In turn, Variant III includes the rate of crop increase based on the rate of crop changes occurring in the two most agriculturally important German states, North Rhine-Westphalia and Lower Saxony between 1981 and 2019 [60]. Due to a large increase in crops between 1981 and 2019, it was also assumed that between the 39th and the 76th year since the launch of the Złoczew opencast, the level of crops would not change in Variant II and Variant III, remaining at the level of productivity estimated for the 39th year after the launch of the opencast;
- Three variants of changes in the population of cattle (excluding cows), cows, pigs (excluding sows) and sows and their productivity. Variant I—the animal population and productivity from 2019 was adopted, Variant II—the animal population and productivity from 2019 were adjusted according to the average pace of changes in the number of animals and productivity in the period from 2010 to 2019. Variant III—an average pace of changes in the animal population and productivity in North Rhine-Westphalia and Lower Saxony from 1981 to 2019 was adopted [60];
- The average rate of exclusion of the land for the exploitation of Złoczew opencast mine during the operation period of 60% [30,61]. Since non-agricultural reclamation of the area is planned for the region of Bełchatów [62], a similar method of reclamation of Złoczew opencast mine is also adopted. Therefore, the average shutdown rate of the opencast mine in Złoczew over 76 years is going to be 80%.

Kruskal-Wallis one–way analysis of variance by ranks test was used to test the homogeneity of the distributions of yield change dynamics in the studied regions. This test was used to verify the hypothesis that the differences between the medians of the study variable were not significant in several populations.

The hypothesis concerns medians of consecutive populations:

$$H_0\colon \Theta_1 = \Theta_2 = \ldots = \Theta_k$$

$$H_1\colon \exists i, j \in \{1, \ldots, k\}\ \Theta_i \neq \Theta_j$$

where $\Theta_1, \Theta_2, \ldots, \Theta_k$ is the median of the tested variable x for the i-th group.

Hypothesis verification was based on a statistic defined by the formula:

$$H = \frac{1}{C}\left(\frac{12}{n(n+1)}\sum_{i=1}^{k}\frac{T_i^2}{n_1} - 3(n+1)\right) \quad (4)$$

where $n = n_1 + n_2 + \ldots + n_k$; T_1 (i–1, 2, ..., k) denotes the sum of ranks in each trial; C—correction for bind ranks $C = 1 - \frac{\sum (k^3 - k)}{n^3 - n}$.

The p value determined on the basis of the test statistic was compared with the significance level α:

if $p \leq \alpha \Rightarrow$ we reject H_0 and accept H_1

if $p > \alpha \Rightarrow$ there are no grounds to reject H_0

In assessing yield level differences between starting years (1956–1960) and final years (1993–1997), analysis of variance (ANOVA) was also used to show statistically significant differences between the means in the three groups identified. In the analysis of variance, groups of n_i elements were compared, yielding a total of $n = \sum_{i=1}^{k} n_i$ independent observations x_{ij} for $j = 1, 2, \ldots, n_i$ [63]. The presence of differences between the means indicated an association between the mean for the tested observation and the qualitative variable that was the basis for separating the groups (here: distance from the outcrop). The null hypothesis of equality of all group means μ (1, 2, ... , i) was tested:

$$H_0: \mu_1 = \mu_2 = \mu_3 = \ldots = \mu_k$$

where $\mu_{(1, 2, \ldots, k)}$ denotes the mean of the dependent variable in the k-th group, towards the alternative hypothesis:

H_1: at least two group means differ.

In view of this, the alternative hypothesis was that there was a significant difference between the compared groups means.

The decision to accept or reject the null hypothesis was based on the Fisher-Snedecor F test determined as:

$$F = \frac{\text{intergroup variance}}{\text{intragroup variance}} \tag{5}$$

If the analyzed factor of group separation is significant, then the variation within each separated group will be small (the intragroup variance will be small). The greater the difference between the groups (the intergroup variance) and the smaller the difference between the elements of each group (the intragroup variance), the larger the value of the F statistic, which argues against the null hypothesis of equality of means in the compared groups, and therefore is the basis for the rejection of H_0. The presence of statistically significant differences in yields was verified using the analysis of variance at the significance level of $\alpha = 0.05$ [63].

The lignite deposits in Złoczew will be used in the Bełchatów power plant. Currently, there are 12 power units in operation, 11 out of which of a capacity of 370–390 MW and gross efficiency after modernization of approximately 38.5% with a net efficiency of approximately 36.0% were put into service in the period from 1981 to 1988. The last unit, of a capacity of 858 MW, was commissioned in 2011 and its gross efficiency was 44.4%, while net efficiency was 41.3% [64]. In 2018, in the power plant in Bełchatów, 1.355 Mg of lignite was used to produce 1 MWh of net electricity [65]. Assuming that the efficiency of the coal extracted from the deposits in Złoczew, as a result of the modernization of the units, is going to be approximately 10% more and that the least efficient units are going to be shut down, the average consumption of lignite will be approximately 1220 Mg. There will be approximately 398.2 TWh of net electricity generated.

5. Results

Agriculture in Poland is characterized by a low level of concentration, which is reflected by a small acreage of an average farm and a low average number of farm animals in farms that run livestock production. It is also characterized by low productivity, including low yield and low animal production. The pace of improvement of productivity of Polish farming and the concentration processes are still too slow to achieve the level of development of Western European countries [66–68]. Agriculture in the areas affected by the extraction of coal deposits in Bełchatów and Złoczew is characterized by smaller farms compared to other parts of Poland—25% smaller on average, and a slower than average growth rate of the size of farms. The milk yield of dairy cows and the fertility of sows are also slightly lower [58,68].

Construction of the opencast mine in Bełchatów and the formation of the depression cone had a negative effect on the level of cereal and potato yield and, out of all plants analyzed, the highest increase of the yield was observed in the voivodships of Group III, i.e., the areas not affected by the operation of lignite opencast mines (Table 2). The results obtained that show the lowest increase in the yield in the voivodships of Group II might be a consequence of poor data received from Sieradz Voivodship, where the agricultural productivity of cereal and potato in the period 1993–1997 was lower by 2.2% and by 7.7% than in 1971–1975. This is due to the impact not only of Bełchatów opencast mine, which has a negative influence on the yield in the southern part of the voivodship, but also of the opencast mines located in the Konin Basin, which negatively affect the north-western area.

Table 2. The yield of selected crops and the dynamics of the yield depending on the distance from Bełchatów opencast mines (according to the data from the analyzed voivodships).

Group	Average Yield in 1971–1975 Years [t × ha^{-1}]		Average Yield in 1993–1997 Years [t × ha^{-1}]		Dynamic [%]	
	Cereal	Potato	Cereal	Potato	Cereal	Potato
Group I	21.8	162.1	24.1	170.4	110.7	105.1
Group II	24.3	187.3	25.1	175.4	103.2	93.7
Group III	28.5	196.6	34.6	216.2	121.5	110.0

Sources: Based on [49–57].

The launching of the field drainage system in Szczerców in 2000 and the poor availability of data on the yield until 1997 make it impossible to fully estimate the impact of Bełchatów opencast mine on agriculture, especially after 2000, when the full influence of the opencast mine became obvious and the area of the depression cone reached its maximum range. To make a proper assessment of the possible range of influence of Bełchatów opencast mine after 1997, a comparative analysis was performed of the changes in the level of the yield in the area of the impact of Bełchatów opencast mine and the opencast mines located in the Konin Basin. For the Konin Basin, the base period was the average yield from 1956 to 1960, marked as year 1 in Figure 2, and for Bełchatow opencast mine it was the average yield from 1971 to 1975. The dynamics of the decline in the level of the yield in Areas I and II compared to the dynamics of the yield in Area III for individual opencast mines were different (Figure 2). In the case of the area of Bełchatów, over the period of 23 years, the yield of cereal in Area I fell by 9.9% compared to the yield in Area III, on average, while in the case of the Konin Basin, the decline in the first 23 years was 5.6%. For potatoes, the difference in agricultural productivity was 6.5% and 4.4%, respectively. The differences in the level of agricultural productivity resulted from the fact that in the case of the Bełchatów opencast mine the relative decrease in the yield occurred in the first few years, whereas in the case of the deposits in Konin the decrease in the yield occurred with a delay of several years. However, in the following 15 years, the average decline in the yield for the area of Konin was larger and amounted to 17.2% for cereal and 14.0% for potatoes, which translated into 10.2% and 8.2% decline in the yield respectively over the period of 38 years since the opencast mine in Konin was launched.

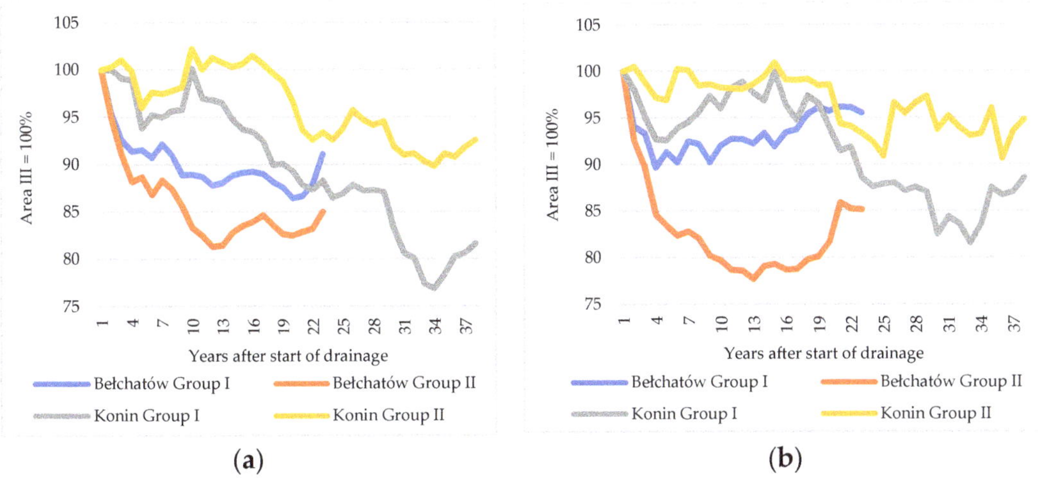

Figure 2. The dynamics of the yield of cereal and potatoes in Area I and II compared to the dynamics of the yield in Area III for Bełchatów opencast mine (base period—1971–1975 years) and the Konin Basin (base period—1956–1960 years) at the voivodship level: (**a**) cereal; (**b**) potato. Source: based on [50–58].

Therefore, it should be assumed that the decline in the yield in the area of influence of Bełchatów opencast mine was still the case after 1997. The above might be confirmed by a slower increase in the yield of cereal, by approximately 5.2%, in the period 1997–2013, in the area of present Łódzkie Voivodship (which also includes voivodships located outside the area of the influence of Bełchatów opencast mine) compared to the neighboring voivodships [58]. In the context of the trends, Figure 2, it was assumed that in the area of influence of Bełchatów opencast mine, Area I, the average decrease in the yield of cereal 24–38 years after the launch of the field drainage system, will be similar to the decrease in the yield in the area of Konin opencast mine. The average decrease in the yield of cereal for the whole period of 38 years of the drainage of the deposits in Bełchatów, adopted for calculation of external costs for Area I, was approximately 14.8%. In the case of potatoes, for 24–38 years after the launch of the drainage system there was a faster increase in the yield in the present Łódzkie Voivodship compared to the neighboring provinces; therefore, for the whole period of the impact, the level of decline in the yield was assumed to be the same as for the Konin Basin, i.e., 7.7%.

In the case of the Bełchatów deposit, the decline in the yield in Area II was larger than in Area I and it was also greater than in Area II, that is the Konin Basin. This was related, as mentioned earlier, to a decline in the level of the yield in Sieradzkie Voivodship and with a simultaneous systematic increase in the yield in the areas of the other analyzed voivodships. Due to the difficulty of determining the actual impact of Bełchatów opencast mine on the level of the yield in Area II, the level of the loss was assumed to be the same as the loss in the area of the Konin Basin. For the entire period of the impact of the opencast mine, the loss was 3.8% for cereals and 3.3% for potatoes.

The data in Figure 3 do not give a clear answer regarding the impact of Bełchatów opencast mine on cattle or pig population. The observed changeability of the population of farm animals in Area I compared to the changes in the livestock in Area III, during the period analyzed, could be assumed to be the result of ordinary changes in local agricultural conditions. In the case of Area II, a systematic decrease in the size of livestock was observed compared to Area III, which may indicate the loss in animal production.

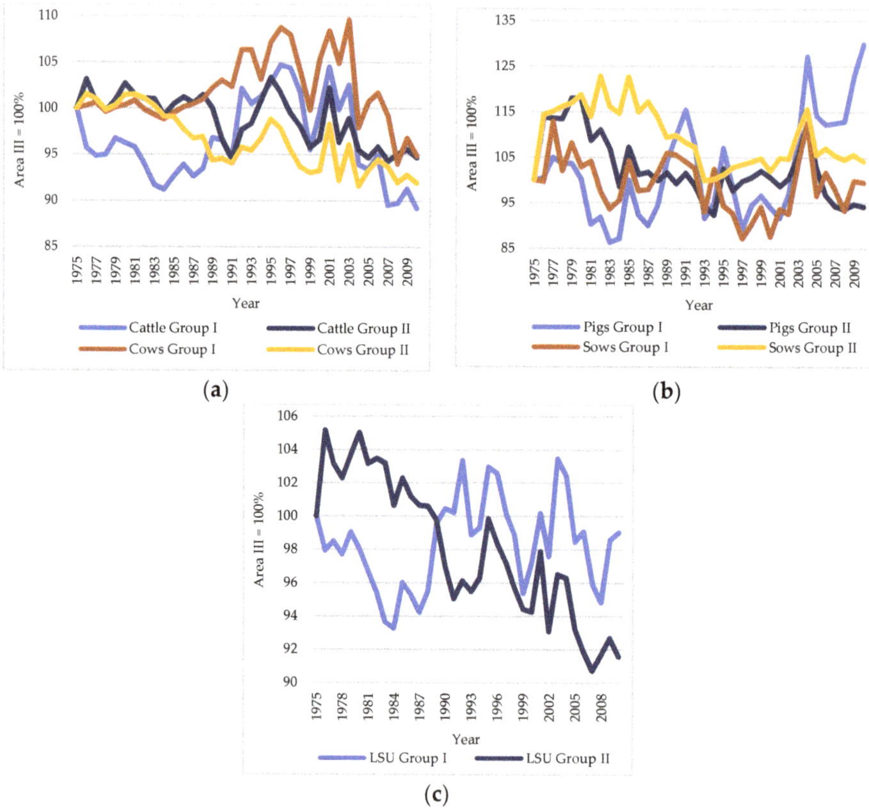

Figure 3. The dynamics of the livestock of cattle, cows, pigs, sows and the number of Large Size Units in Area I and II compared to the dynamics of the livestock in Area III for Bełchatów opencast mine (on the voivodship level): (**a**) cattle (without cows) and cows; (**b**) pigs (without sows) and sows; (**c**) large size unit (LSU). Source: based on [49–57].

A slightly different picture of the impact of Bełchatów opencast mine on animal production is presented in Figure 4. There is a visible significant impact on livestock production in the districts located up to 40 km away from the opencast mine. In the case of the districts located up to 60 km away from the opencast mine, the systematic decline in livestock compared to the areas located 80–100 km away from the open cast mine was noticed only for cattle and cows. In the case of pigs and sows, those areas were characterized by the greatest increase in the number of those animals, which was related to a large increase in the size of livestock in the area of Piotrków, mainly due to the increased importance of contract fattening. The popularity of the above contributed to the reconstruction of herds for fattening in the districts of Bełchatów and Radomsko located 20–40 km away from Bełchatów opencast mine. A detailed analysis of the changes in the livestock in individual districts located within 100 km of the Bełchatów opencast mine shows great diversity. In the case of 3 districts located up to 40 km away from the opencast mine, the decrease in the population of cattle was not compensated by an increase in the number of pigs and sows. In the case of other districts, a large decline in the population of one type of animal species was usually compensated by an increase in the number of other types of animals, which indicated the growing specialization of districts and regions in livestock production.

Figure 4. The dynamics of changes in the number of cattle, cows, pigs, sows and large size units in the districts depending on the distance from Bełchatów opencast mine compared to the dynamics of changes in the size of livestock in the districts located 80–100 km away from the mine: (**a**) cattle (without cows); (**b**) cows; (**c**) pigs (without sows); (**d**) sows; (**e**) LSU. Source: based on [50,58].

Due to the lack of proper data concerning the impact of Bełchatów opencast mine on livestock production in specific voivodships, to estimate the external costs for the deposits in Złoczew the data on the changes in animal production at the district level converted to LSU was used. For the analysis, only the areas located up to 60 km away from the opencast mine were taken into account. In the case of districts where there was an opencast mine located, the population of the animals under study decreased by 11.0% on average in the entire period under review—13.8% in the districts located 20–40 km away from the opencast mine and 6.7% in the districts located 40–60 km away from the opencast mine.

Based on the conducted analysis, it can be noticed that for the probability level $p \leq 0.05$, the initial null hypothesis of equality of the dynamics of the changes in the yield of cereal and potatoes can be rejected ($p = 0.0000$). Thus, the distance from the mine had a statistically significant effect on the yield in the three groups of voivodships under review (Table 3). The results of the multiple comparison test indicate that for cereal, the difference in the growth rate of the yield between Area I and Area II was not statistically significant. However, for both types of crops analyzed, the difference in the pace of the changes between the region located furthest from the opencast mine (Area III) and Area I and II was statistically significant and the significance level was 0.05. As was expected, the growth rate in the yield in the region located outside the area of impact of the opencast mine was significantly higher.

Table 3. The results of the Kruskal-Wallis multiple comparisons test performed to analyze the dynamics of the changes in the yield of selected crops (according to the data from specific voivodships).

Cereals		**Potatoes**	
test Kruskal-Wallis		test Kruskal-Wallis	
H (2, N = 66) = 28.3275; p = 0.0000		H (2, N = 111) = 46.25747; p = 0.0000	
post-hoc comparisons			
group I vs. II	0.19488	group I vs. II	0.00023
group I vs. III	0.00202	group I vs. III	0.01444
group II vs. III	0.00000	group II vs. III	0.00000

Based on the data received from the districts, there was no difference in the pace of changes in livestock production in the five areas, regardless of the distance from an opencast mine, for the period from 1975 to 2010, for cattle, cows, and livestock density in LSU (Table 4). The dynamics of changes was significantly different in the case of pigs and sows. Based on the analysis of the obtained data, from the point of view of the hypothesis of the impact of an opencast mine on the animal population, it can be noticed that in the case of districts located in the area near Bełchatów opencast mine, the rate of change in the population of pigs did not differ from the pace of the changes in districts located 20–40 km away from the opencast mine, however, it was significantly different in the districts located 40–60 km away and 80–100 km away from the opencast mine. The analysis of the input data indicates that the pace of changes in livestock was faster in the districts located 20–40 km away from the opencast mine than in the district of Bełchatów. Regarding the dynamics of the changes at the voivodship level, a significant difference in the rate of the changes in the livestock was recorded only for sows for the voivodships in Group II compared to the voivodships in Groups I and III (Table 5).

The results of the analysis of variance indicate that for the assumed significance level of $\alpha = 0.05$ the null hypothesis that the mean values of the number of animals in the compared groups of districts are equal should be rejected (Table 6). The increase in the significance of the differences in the period from 2006 to 2010, compared to the years 1975–1979, was mainly the case of the livestock converted to LSU and sows for the districts located close to the opencast mine compared to the districts located further away from the opencast mine. The decrease in the significance of the differences was primarily the case of the population of cattle.

Table 4. The results of the Kruskal-Wallis multiple comparisons test * performed to analyze the dynamics of the changes in livestock (according to the data from the districts).

Indicator	Cattle	Cows	Pigs	Sows	LSU
		test Kruskal-Wallis			
H (4, *N* = 175)	6.79541	5.55901	82.19775	47.57426	5.64238
p	0.1471	0.2346	0.0000 *	0.0000 *	0.2275
		post-hoc comparisons			
do 20 km vs. 20–40 km	1.0000	1.0000	1.0000	0.4393	1.0000
do 20 km vs. 40–60 km	1.0000	0.2501	0.0000 *	0.0001 *	1.0000
do 20 km vs. 60–80 km	1.0000	1.0000	0.0726 *	1.0000	1.0000
do 20 km vs. 80–100 km	1.0000	1.0000	0.0000 *	0.2707 *	1.0000
20–40 km vs. 40–60 km	1.0000	1.0000	0.0000 *	0.0000 *	1.0000
20–40 km vs. 60–80 km	1.0000	1.0000	0.0901 *	0.0037 *	0.9961
20–40 km vs. 80–100 km	0.4912	1.0000	0.0000 *	0.0002 *	0.2309
40–60 km vs. 60–80 km	1.0000	1.0000	0.0000 *	0.0314 *	1.0000
40–60 km vs. 80–100 km	0.3514	1.0000	0.3899	0.2211	1.0000
60–80 km vs. 80–100 km	1.0000	1.0000	0.1084	1.0000	1.0000

*—$p \leq 0.05$.

Table 5. The results of the Kruskal-Wallis multiple comparisons test * performed to analyze the dynamics of the changes in livestock (according to the data from the voivodships).

Indicator	Cattle	Cows	Pigs	Sows	LSU
		test Kruskal-Wallis			
H (2, *N* = 105)	0.8107	0.2670	0.8303	9.6380	0.1869
p	0.6667	0.8750	0.6603	0.0081 *	0.9108
		post-hoc comparisons			
group I vs. II	1.0000	1.0000	1.0000	0.0135 *	1.0000
group I vs. III	1.0000	1.0000	1.0000	1.0000	1.0000
group II vs. III	1.0000	1.0000	1.0000	0.0369 *	1.0000

*—$p \leq 0.05$.

Table 6. Results of the analysis of variance regarding the significance of the differences in livestock for specific districts at the beginning and the end of the analyzed period (average for five years).

Indicator	Cattle		Cows		Pigs		Sows		LSU	
	1975–1979	2006–2010	1975–1979	2006–2010	1975–1979	2006–2010	1975–1979	2006–2010	1975–1979	2006–2010
					Analysis of variance					
F	8.0281	3.7215	9.816	2.9987	4.7954	2.8436	4.6972	3.4328	8.1074	3.8956
p	0.0000	0.0068	0.0000	0.0212	0.0013	0.0271	0.0015	0.0108	0.0000	0.0052
					Least significant differences test					
{1} vs. {2}	0.0697	0.5428	0.1517	0.9491	0.0926	0.6464	0.2294	0.0683	0.5170	0.0697
{1} vs. {3}	0.6538	0.9148	0.8076	0.4921	0.2935	0.1484	0.1360	0.5483	0.5376	0.6538
{2} vs. {3}	0.0387 *	0.3032	0.0756	0.3213	0.2732	0.2206	0.8443	0.0592	0.8873	0.0387 *
{1} vs. {4}	0.0253 *	0.0601	0.0634	0.3095	0.0115 *	0.1132	0.0216 *	0.0195 *	0.0365 *	0.0253 *
{2} vs. {4}	0.7740	0.1141	0.7336	0.2164	0.3815	0.1595	0.2032	0.6843	0.0716	0.7740
{3} vs. {4}	0.0021 *	0.0007 *	0.0056 *	0.0037 *	0.0100 *	0.8078	0.1547	0.0027 *	0.0107 *	0.0021 *
{1} vs. {5}	0.8105	0.7618	0.3387	0.5899	0.5212	0.7772	0.4702	0.8654	0.7952	0.8105
{2} vs. {5}	0.0014 *	0.5732	0.0004 *	0.4031	0.0679	0.7261	0.3439	0.0019 *	0.4929	0.0014 *
{3} vs. {5}	0.1656	0.4169	0.0189 *	0.7184	0.3630	0.0162 *	0.0970	0.1188	0.4466	0.1656
{4} vs. {5}	0.0000 *	0.0012 *	0.0000 *	0.0020 *	0.0001 *	0.0073 *	0.0010 *	0.0000 *	0.0002 *	0.0000 *

Group symbol: {1}—up to 20 km; {2}—20–40 km; {3}—40–60 km; {4}—60–80 km; {5}—80–100 km. *—$p \leq 0.05$.

The levels of significance of the differences in the size of livestock in the three analyzed groups of voivodships are more noticeable. (Table 7). In both the initial and the final periods of the study, significant differences were recorded in the voivodships in Group II and III. In the case of provinces from groups I and III, at the beginning of the period, significant differences in the level of stock were obtained only for pigs, sows, and LSU. In most of the others the level of significance decreased or did not change.

Table 7. Results of the analysis of variance regarding the significance of the differences in livestock for individual voivodships at the beginning and the end of the analyzed period (average for five years).

Indicator	Cattle		Cows		Pigs		Sows		LSU	
	1975–1979	2006–2010	1975–1979	2006–2010	1975–1979	2006–2010	1975–1979	2006–2010	1975–1979	2006–2010
Analysis of variance										
F	4.4239	4.3387	3.6278	3.8527	11.5614	4.6843	15.9750	4.4968	7.0029	7.0029
p	0.0181 *	0.0194 *	0.0352 *	0.0291 *	0.0001 *	0.0146 *	0.0000 *	0.0170 *	0.0024 *	0.0066 *
Least significant differences test										
group I vs. II	0.9396	0.9582	0.2855	0.2136	0.3599	0.0815	0.1773	0.3836	0.6150	0.2739
group I vs. III	0.0658	0.0659	0.4299	0.5106	0.0214 *	0.7820	0.0144 *	0.2326	0.0472 *	0.2198
group II vs. III	0.0124 *	0.0136 *	0.0105 *	0.0083 *	0.0000 *	0.0040 *	0.0000 *	0.0056 *	0.0012 *	0.0019 *

*—$p \leq 0.05$.

The large area of impact of the opencast mine and the increasing productivity of agriculture in the regions located within the area affected by Złoczew and Bełchatów opencast mines resulted in the high level of external costs incurred due to the launch of the Złoczew opencast mine (Table 8). The average value of external costs for the three variants adopted for the analysis was approximately €4.86 billion, which, with the estimated net electricity production of 398.2 TWh, gives €12.20 × MWh^{-1}. Since in 2019, the average price of electricity in Poland was €53.74 × MWh^{-1} and to calculate the external costs incurred by agriculture, the electricity prices should be increased by 22.7%. Among the adopted variants of changes in the productivity of crops and livestock, the value of the external costs will be the lowest if the technical and economic parameters in the entire period of the impact of Złoczew opencast mine are similar to current costs, and the value of incurred costs will be the highest if agriculture in the analyzed area develops at the pace at which German agriculture has developed over the past 38 years.

Table 8. External costs of exploitation of lignite from Złoczew deposit for 76 years of the impact of the opencast mine (million €).

Specification	Variant I	Variant II	Variant III	Average	€ × MWh^{-1}
Plant production					
Open-pit mining area	61	71	97	76	0.19
Group I	2137	2529	3402	2689	6.75
Group II	1234	1445	1821	1500	3.77
Total	3432	4045	5320	4266	10.71
Animal production					
Open-pit mining area	3	3	4	3	0.01
up to 20 km	98	255	195	109	0.27
20–40 km	104	270	204	261	0.66
40–60 km	126	259	251	217	0.54
Total	331	787	654	591	1.48
All in total	3763	4833	5974	4857	12.20

6. Discussion

The estimated expected losses borne by agriculture is connected with uncertainty resulting from the multitude of influencing factors. Comparison of changes in yields and numbers of livestock in the regions affected by the impact of the opencast mines with the neighboring areas not affected by the impact of the opencast mines significantly reduces the uncertainty resulting from changes in macroeconomic, natural, and climatic conditions, which occurred similarly in the area affected by the opencast mines and regions outside of this area. However, the reaction of farmers in these areas to these changes may differ.

The risk of incorrect estimation of the impact of the opencast on agriculture also stems from the changes in the administrative division made in 1999, which does not allow for estimating the changes in the crop levels within and outside the Bełchatów opencast area. The slower growth rate in the present Łódzkie voivodship, where the Bełchatów opencast is located, as compared to the neighboring voivodships, suggests further deepening of the negative impact of the opencast after 1997, and the results obtained should be treated only as approximate. The risk of overestimation or underestimation of the obtained external costs for the Złoczew deposit also results from other uncertainties and factors described in Chapter 2. Despite these limitations and uncertainties, the cost estimates obtained should be treated as highly probable. Even a small correction of these results will not change the general conclusion that after taking into account the costs incurred in agriculture, the launching of the Złoczew open pit is not economically justified.

According to the analysis conducted, almost 88% of the external costs related to the extraction of lignite from the deposits in Złoczew are the result of the decrease in the yield of plant production. The above is the result of the specificity of production, in the case of which a decrease in the yield means not only a decline in the value of production but also a decrease of agricultural productivity by a similar value. Farmers, regardless of the level of the yield, have to perform all basic agro-technical tasks, use the standard quantity of means of production such as seeds, fertilizers, and plant protection products, and devote the same amount of time. The likelihood of this scenario is even greater due to ongoing climate changes. The area influenced by Złoczew opencast mine is located in central Poland and it is the region where steppe-formation processes are common phenomena due to very low rainfall. The region is also characterized by long periods of drought [69]. The observed systematic increase in global temperatures causes the increase in evaporation which reduces the amount of rainwater available to plants and the agricultural efficiency of precipitation [70,71]. The above-mentioned efficiency is also decreased by the latest changes in the nature of precipitation. The occurrence of convective precipitation and heavy rainfall instead of continuous precipitation has a particularly negative influence on agriculture [72–74]. According to research conducted in Germany, an increase in temperature by one-degree results in an increase in the amount of heavy rainfall by 6.5% [75]. Those processes also lead to a decline in the amount of subsurface water and a decrease in the level of the groundwater table. All of the above, in turn, lead to an increase in the dependence of crops on the level of groundwater [76,77]

Those trends are confirmed by the studies conducted in the Great Hungarian Plain, which showed a 0.21–0.60 m decline in the level of groundwater between 1986 and 2010 compared to the state from 1961 to 1985. Despite the insignificant correlation between the levels of groundwater and crop output, the potential impact of reduced level of groundwater on the production of corn was estimated to be 0.65 $t \times ha^{-1}$, i.e., 11.6 % of the average annual yield from 1986 to 2010. There was also stagnation in the level of the yield of wheat [76]. Regrettably, similar studies, despite the importance of the problem of lowering levels of groundwater in many regions of the world, are a rarely recorded element of the latest global environmental crisis, which can be a threat to food security. First of all, there is a lack of detailed data and no information collected over a long period, to analyze the progressive effects of lowering levels of groundwater on a regional scale [76]. Poland also faces the above-mentioned issues. Although, the problem of the decrease of the level of groundwater is obvious, there are no comprehensive and long-term analyses on how common and serious this phenomenon is. Thus, it is difficult to assess whether the observed decline in the yield in the area of the Great Hungarian Plain is typical for this region of the world.

The observed decrease in the level of cereal and potato yield over the period of 23 years since the launch of Bełchatów opencast mine in the area of Piotrkowskie Voivodship is similar to the decrease in the yield observed in the area affected by the opencast mine in the Konin Basin (Figure 2) and in the Great Hungarian Plain. A much larger drop in the yield, which was observed in the subsequent years of the operation of the mine in the

Konin Basin, suggests a high probability of the occurrence of similar trends also for the coal deposits in Bełchatów.

Further, in the case of the deposits in Złoczew, a similar pattern of the changes in the yield should be expected, although, in this case, there is a possibility of a greater loss. It is related to an increase in the sensitivity of crops to water deficiency along with an increase in agricultural productivity [78,79] and with the expected further increase in the yield of plants in Poland, including the region affected by the opencast mine in Złoczew. What is more, the increase in the amount of water used by arable crops, which is proportionate to the increase in the yield, means that further increase in the yield may be difficult in the case of the groundwater table lowering, even though in the neighboring areas the increase in the yield will continue [80]. The feasibility of this scenario is confirmed by a decrease in the yield of cereal and potato in the period from 1971 to 1997, in Sieradzkie Voivodship, which was affected by both Bełchatów opencast mine and the opencast mines in the Konin Basin.

Groundwater is of key importance during dry periods and for many cultivated plants water stored in the upper layers of the soil can account for 50%–100% of the total use of water [77,81–87]. The above is also important because of the increasing frequency of rain-free periods and rising average air temperature. The forecasts on climate change indicate a further increase in temperatures in Europe and a multiple increase in the frequency of centennial drought in Europe, including Poland. A centennial drought is a drought never recorded in a certain area in the 20th century [88–90]. If the forecasts prove correct, the estimated external costs caused by the construction of a new opencast mine will be even higher. According to research conducted in the Czech Republic for the years 1961–2000, during severe drought, the yield of grain decreased by 25%–35%, depending on the type of species (average yield during the period under review of approximately 3.5 t $\times$ ha^{-1}), the yield of potato decreased by approximately 20% (average yield of 16.9 t $\times$ ha^{-1}), and the yield of rapeseed by 25% (average yield of 2.2 t $\times$ ha^{-1}). The above-mentioned studies confirmed the relatively high drought tolerance of corn, and the decline in the yield was not more than 10% (average yield of 4.2 t $\times$ ha^{-1}) [91]. Since the level of the yield obtained in the period under study was low, it should be expected that for the forecasted higher level of the yield in the area of the influence of Złoczew opencast mine, the decrease in the level of agricultural productivity may be even higher during the periods of severe drought.

The drop in the yield during periods of drought depends on the type of species of the cultivated plant and its sensitivity to the lowered level of the groundwater table. In the case of wheat, for instance, the optimum growth of the plant is guaranteed by the level of the groundwater table at 0.7–1.6 m and in the case of corn at 1.0–3.0 m [76,78,86,92–94]. The level of groundwater table being lower than 4.0 m leads to a significant reduction of water stored in the upper layers of soil and the need for plants to use only rainwater. An example here, is research conducted in the Inland Pampas, during two growing seasons (2006/2007 and 2007/2008), according to which the yield of wheat, soybean and corn in the areas with the optimum level of groundwater was 3.7, 3.0, and 1.8 times larger than the yield in the regions where the water table was located below 4.0 m [93].

Studies conducted for Bełchatów opencast mine indicate a dynamic decline in the level of groundwater already in the first years of dewatering the mine, which explains, to some extent, the significant decrease in the yield that was noticed since the launch of the opencast mine. In the years 1982–1985, i.e., 6–9 years after the launch of the drainage system of the opencast mine near Brudzice, which is a town located approximately 6 km away from the southern edge of the opencast mine, the level of groundwater was lowered by approximately 5.22 cm per month, with the initial level of groundwater at 356 cm, and in Woźniki—a town located approximately 10 km away from the edge of the mine—the level of groundwater dropped at the rate of 1.1 cm per month; the initial level was at 244 cm. The sub clay water-bearing level was dewatered at the same time. In Ligota Wielka, a town located approximately 9 km away from the edge of the mine, the decrease in the level of groundwater table was at the pace of 9.14 cm per month [95]. According to another study, in places located at the same distance from the opencast mine, the level of groundwater table, even during the period of high precipitation, was lower than 200 cm, although there were places where the level of groundwater table was at 200 cm [96–99], which might have been caused by the irregularity of the area of the depression cone and the impermeable layers which hold water. The above-mentioned areas are used to conduct comparative analyses of moisture conditions of the soil. For the majority of the analyses, there is one conclusion: "There is no negative impact of Bełchatów opencast mine on the level of moisture of arable land, regardless of the distance of the place from the opencast mine" [96–101] or "the impact is insignificant" [102]. According to the review of the results of the research, in the case of soil layers located up to 200 cm below the surface, where there was groundwater, the level of soil moisture at a depth of 25 cm and 45 cm was approximately 25% and it was higher by approximately 10 percentage points than in the layers where there was no groundwater [97–99]. To investigate the above, there was an experiment performed, the results of which are very interesting. It was a laboratory assessment on the effect of the capillary action in soil on the moisture of the top layers. It was proved that groundwater is important for the productivity of soil and plants can access water via their root system thanks to the capillary action, additionally, within the range of a depression cone, where there is no groundwater, the possibilities of using rainwater by plants are also limited [103].

A similar level of humidity of soil may be the evidence that dewatering of a mine does not, or may not, pose a threat to water management and agricultural losses such as reduction of the yield of cultivated plants, and the decrease in the yield depends on many other factors, including the direction of the water flow and the amount of rainfall [99,104–111]. Moreover, lower levels of groundwater might be a consequence of deforestation, intensification of agriculture, river regulation land reclamation, climate change and higher temperatures, or the increase of the level of precipitation in winter at the expense of the amount of precipitation in summer [99,112,113]. The above are not dominant factors that affect the level of groundwater table in the areas surrounding opencast mines.

There are also limited options for large-scale irrigation as it involves drawing water from deeper layers of soil and high costs associated with the construction and operation of irrigation systems. Launching irrigation systems on a larger scale delays the process of restoration of water surface and, in extreme cases, prevents it, which permanently limits the possibility of intensification of agricultural production. The above was confirmed by various research conducted in different regions of the world where overexploitation of underground resources led to a decline in the level of groundwater. The examples here are the Midwestern regions of the United States [114], the intensively irrigated North China Plain [115], and Syria [116], where there was a decrease in the level of groundwater by several dozen meters and a collapse of agricultural production in vast areas. Moreover, agriculture in those regions depends on costly irrigation and if it is not possible to organize it the yield varies according to the amount of rainwater.

The research conducted on external costs show that opencast lignite mining has a negative impact on the yield in large areas located around the mine, which is confirmed by analyses of the deposits in Bełchatów and in the Konin Basin. Nevertheless, it is necessary to conduct more research for other opencast mines.

According to the analysis conducted, there are relatively low costs incurred by livestock producers. This is mainly due to the decline in the number of cows and pigs over the last 10 years and the stabilization of cattle population in the areas affected by the opencast mines, as well as a much smaller area (almost 50% smaller), where livestock production is affected by Bełchatów opencast mine. It might be assumed that in the case of a forecasted slight decrease in the yield, the tendency of farmers to reduce or eliminate livestock production is the same as in the areas located further away from the opencast mines. The loss of profits from plant production is so small that it does not affect the investment decisions regarding livestock production. The tendency of some farmers to increase the number of animals may be a response to the decline in profits from plant production, which is compensated for by the development of livestock production.

Interesting results are provided by a survey conducted in Poland among 190 farmers in a region of high intensity of agriculture, where an opencast lignite mine was planned to be launched. The farmers did not plan to reduce their livestock even if the level of the yield was going to fall to 10%. Every fifth farmer declared a reduction of livestock production by at least 20% if the yield of fodder fell by 10–20%, and another one third of the farmers would reduce the production if the yield fell by 20–30%. Complete abandonment of livestock production with a yield drop to 30% was confirmed by every fourth farmer, and another 44% of farmers would stop the production if the yields fell by 30–50% [29]. The survey indicates a relatively high sensitivity of animal production to a decrease in the yield by at least 10%, which is usually caused by the appearance of a depression cone. The above may be partly related to the dominance of small and very small farms, with a standard production of less than €25,000, in the case of which it is important to produce own fodder.

The location of the Złoczew opencast mine is unfavorable from the point of view of the distance to large urban centers. It is located 72 km away from the center of Łódź, i.e., the area that, according to Sinclair's theory, should be designated for intensive agricultural production, including animal production [117]. According to the studies conducted for Poland, the areas located 50–100 km away from major urban centers are characterized by the highest cow and pig density [68,117,118]. The launch of an opencast mine in such areas disrupts the system, forcing an increased concentration of production in the neighboring regions.

The analysis conducted gives only the approximate external costs related to the planned launching of the opencast mine in Złoczew. The demonstrated loss of 22.7% of the estimated net value of electricity production and the expected further increase in the costs related to the emission of CO_2 indicate that the plans to launch an opencast mine in Złoczew should be abandoned. In order to fully estimate the costs, it is still necessary to estimate other external costs, primarily those related to coal combustion. Studies by various authors show that depending on the range of factors analyzed, study methodology, data availability, power plant efficiency, combustion technology, land population, and others, the spread of external costs varies considerably (Table 9). For example, in the case of Thailand, where only the impacts of PM10 and NOx emissions in sparsely populated areas were analyzed, while the Macy et al. study included sulfur dioxide, nitrogen oxides, dust particles, carbon monovide and dioxide, volatile organic compounds, polycyclic aromatic hydrocarbons, and heavy metals, resulting in a difference in external costs of about 10 times. In the case of Bosnia and Herzegovina, brown coals with different calorific and sulfur contents were analyzed [32]. There was less disproportion in the case of the study of external costs associated with opencast coal mining and these were lower than in the case of the projected external costs for the Złoczew opencast.

Table 9. External costs of air pollution caused by lignite combustion and opencast coal mining € × MWh^{-1}.

Study	Georgakellos [119]	Sakulniyomporn [120]	Büke, Köne [121]	Dimitrijević [122]	Coester [16]	Máca [26]	Wang [25]	Taranto [123]	Papagiannis [28]	Pepliński, Czubak [32]
Country	Greece	Thailand	Turkey	Bosnia and Herzegovina	Germany	Czech, Hungary, Poland	China	Turkey	Greece	Poland
Year of analysis	2003–2004	2006–2008	2007	2008	1995–2003	2010	2015	2018	2014	2021
Health impacts	No	Yes	Yes	No	Yes	Yes	Yes	No	Yes	No
External costs	43.9	6.8	1.8–35.2	2.7–19.2	11.1	58.1–77.5	63.8	36.3	5.0	8.7

Source: Based on [32].

This analysis is the first comprehensive estimate of external costs in crop and livestock production associated with the depression funnel created by opencast coal mining. It is necessary to carry out further studies of external costs for currently operating and planned opencast coal and other natural resources in other countries and on other continents, which will extend the knowledge of factors affecting crop production losses and the magnitude of external costs in different regions of the world and with different agricultural structures. It will also reduce uncertainty about the value of estimated external costs.

Funding: This research received no external funding.

Institutional Review Board Statement: Not applicable.

Data Availability Statement: The data is available from the authors upon request.

Conflicts of Interest: The authors declare no conflict of interest.

References

1. Wilczyński, M. Coal in Poland: After the Sunset. Change Partnership.Org. 2016. Available online: http://www.changepartnership.org/wp-content/uploads/2016/02/Wilczynski-2016-Coal-in-Poland-after-the-sunset-translation-FINAL.pdf (accessed on 8 December 2020).
2. Kasztelewicz, Z.; Ptak, M.; Sikora, M. Lignite as the optimal energy source for Poland. *Zesz. Nauk. IGSMiE PAN* **2018**, *106*, 61–84. (In Polish)
3. Pactwa, K.; Woźniak, J.; Strempski, A. Sustainable mining—Challenge of Polish mines. *Resour. Policy* **2018**, 101269. [CrossRef]
4. Ministry of Energy. *Energy Policy of Poland Until 2040 (EPP2040)*; Ministry of Energy: Warsaw, Poland, 2021.
5. Adani's Coal Mine Plans, and why Coal is still so Lucrative. Available online: https://www.dw.com/en/adanis-coal-mine-plans-and-why-coal-is-still-so-lucrative/a-52025626 (accessed on 4 January 2021).
6. Global Coal Exit List. Available online: https://coalexit.org/ (accessed on 4 January 2021).
7. Thielemann, T.; Schmidt, S.; Gerling, J.P. Lignite and hard coal: Energy suppliers for world needs until the year 2100–An outlook. *Int. J. Coal Geol.* **2007**, *72*, 1–14. [CrossRef]
8. Braun, M. *Environmental External Costs from Power Generation by Renewable Energies*; Institut für Energiewirtschaft und Rationelle Energieanwendung, Universität Stuttgart: Stuttgarrt, Germany, 2004.
9. International Institute for Sustainable Development. *Financial Supports for Coal and Renewables in Indonesia*; International Institute for Sustainable Development: Winnipeg, MB, Canada, 2017.
10. Samadi, S. The Social Costs of Electricity Generation—Categorising Different Types of Costs and Evaluating Their Respective Relevance. *Energies* **2017**, *10*, 356. [CrossRef]
11. Karkour, S.; Ichisugi, Y.; Abeynayaka, A.; Itsubo, N. External-Cost Estimation of Electricity Generation in G20 Countries: Case Study Using a Global Life—Cycle Impact—Assessment Method. *Sustainability* **2020**, *12*, 2002. [CrossRef]
12. Kasztelewicz, Z. *Report on the Condition of the Lignite Coal Industry in Poland and Germany Together with a Diagnosis of Measures for the Development of this Industry in the First Half of the 21st Century*; AGH University of Science and Technology: Kraków, Poland, 2018. (In Polish)
13. Kasztelewicz, Z.; Tajduś, A.; Cała, M.; Ptak, M.; Sikora, M. Strategic conditions for the future of brown coal mining in Poland. *Energy Policy J.* **2018**, *21*, 155–178. [CrossRef]
14. Pietraszewski, A. Polish lignite mining in 2014. *Węgiel Brunatny* **2015**, *90*, 4–14. (In Polish)
15. Widera, M.; Kasztelewicz, Z.; Ptak, M. Lignite mining and electricity generation in Poland: The current state and future prospects. *Energy Policy* **2016**, *92*, 151–157. [CrossRef]
16. Coester, A.; Hofkes, M.; Papyrakis, E. Economics of renewable energy expansion and security of supply: A dynamic simulation of the German electricity market. *Appl. Energy* **2018**, *231*, 1268–1284. [CrossRef]
17. Redlinger, R.Y.; Andersen, P.; Morthorst, P. *Wind Energy in the 21st Century—Economics, Policy, Technology and the Changing Electricity Industry*; Palgrave Macmillan: Basingstoke, UK, 2002. [CrossRef]
18. Bickel, P.; Reiner, F. *EC (2005) ExternE—Externalities of Energy*; Office for Official Publications of the European Communities: Luxembourg, 1995; Volume 6.
19. Oak Ridge National Laboratory and Resources for the Future. *Estimating Externalities of Fuel Cycles*; McGraw-Hill/Utility Data Institute: Washington, DC, USA, 1992–1998; Volume 8.
20. Rowe, R.D.; Lang, C.M.; Chestnut, L.G.; Latimer, D.A.; Rae, D.A.; Bernow, S.M.; White, D.E. *New York State Environmental Externalities Cost Study*; Empire State Electric Energy Research Corporation, Oceana: New York, NY, USA, 1995.
21. Wilczyński, M. *Coal. After the Sunset*; European Climate Foundation: Warsaw, Poland, 2015. (In Polish)
22. *What ZE PAK?* Foundation "Development YES—Open-pit Mines NO": Warsaw, Poland, 2017.
23. Tol, R.S.J. Targets for global climate policy: An overview. *J. Econ. Dyn. Control* **2013**, *37*, 911–928. [CrossRef]
24. Jorli, M.; Van Passel, S.; Saghdel, H.S. External costs from fossil electricity generation: A review of the applied impact pathway approach. *Energy Environ.* **2018**, *29*, 635–648. [CrossRef]
25. Wang, L.; Watanabe, T.; Xu, Z. Monetization of External Costs Using Lifecycle Analysis–A Comparative Case Study of Coal–Fired and Biomass Power Plants in Northeast China. *Energies* **2015**, *8*, 1440–1467. [CrossRef]
26. Máca, V.; Melichar, J.; Ščasný, M. Internalization of External Costs of Energy Generation in Central and Eastern European Countries. *J. Environ. Dev.* **2012**, *21*, 181–197. [CrossRef]

27. WWF European Policy Office, Sandbag, CAN Europe and HEAL. *Europe's Dark Cloud*; WWF European Policy Office, Sandbag, CAN Europe and HEAL: Brussels, Belgium, 2016.
28. Papagiannis, A.; Roussos, D.; Menegaki, M.; Damigos, D. Externalities from lignite mining-related dust emissions. *Energy Policy* **2014**, *74*, 414–424. [CrossRef]
29. Poczta, W.; Pepliński, B.; Sadowski, A.; Czubak, W. *Impact of the Planned Oczkowice Mine on the Economic, Productive and Social Potential of Agriculture and its Surroundings in the South–Western Functional Area of the Wielkopolskie Voivodeship*; Poznań University of Life Sciences: Poznań, Poland, 2017. (In Polish)
30. Pepliński, B. External costs of opencast brown coal mining in agriculture and agri–food industry (on the example of Wielkopolska). *CEEJ* **2018**, *5*, 154–171. [CrossRef]
31. Pepliński, B. External costs of open-cast lignite mining for the agri–food industry (as illustrated by the example of the Oczkowice mine). *J. Agribus. Rural Dev.* **2018**, *2*, 213–223. [CrossRef]
32. Pepliński, B.; Czubak, W. The Influence of Opencast Lignite Mining Dehydration on Plant Production—A Methodological Study. *Energies* **2021**, *14*, 1917. [CrossRef]
33. Przybyłek, J. Current problems of the lignite open cast mines dewatering in the Wielkopolska region. *Górnictwo Odkryw.* **2018**, *2*, 5–14. (In Polish)
34. Przybyłek, J.; Górski, J. The Oczkowice lignite deposit. Voicing concern for proper hydrogeological recognition. *Przegląd Geol.* **2016**, *64*, 183–191. (In Polish)
35. Wilczyński, M. *What Coal for Bełchatów Power Plant? Resources and Quality of Lignite in the "Bełchatów" Deposit (Bełchatów and Szczerców Fields) and in the "Złoczew" Deposit*; Fundacja Client Earth Prawnicy dla Ziemi: Warszawa, Poland, 2019. (In Polish)
36. Kasztelewicz, Z.; Sypniowski, S. Reclamation in Polish opencast lignite mining on selected examples. *Górnictwo i Geoinżynieria* **2011**, *34*, 289–304. (In Polish)
37. Malina, G.; Niezgoda, G. Redevelopment concept of open pits following lignite exploitation in the area of Belchatow. *Ochrona Środowiska* **2017**, *39*, 19–30. (In Polish)
38. Paczyński, B.; Sadurski, A. *Regional Hydrogeology of Poland, Mineral, Curative and Thermal Waters and Mine Waters*; Państwowy Instytut Geologiczny: Warszawa, Poland, 2007; Volume 2. (In Polish)
39. Szczepiński, J. *Forcasting of Groundwater Table Recovery after Closure of the Bełchatów Lignite Open Cast Mine*; Poltegor: Wrocław, Poland, 2001.
40. Gasiński, J.; Kaczmarek, T. Groundwater observation system in the lignite opencast Bełchatów—Advancement of the system. *Wiert. Naft. Gaz* **2008**, *25*, 277–283. (In Polish)
41. Reclamation. Available online: https://kwbbelchatow.pgegiek.pl/Ochrona-srodowiska/Rekultywacja (accessed on 3 February 2021).
42. Szczepiński, J. The Significance of Groundwater Flow Modeling Study for Simulation of Opencast Mine Dewatering, Flooding, and the Environmental Impact. *Water* **2019**, *11*, 848. [CrossRef]
43. Kraśnicki, S. Scientific opinion on the impact of dewatering of the projected lignite mine Złoczew on the quality and quantity of groundwater and surface water. Ludów Polski, Poland, May 2016.
44. Poltegor-Projekt Sp. z o.o. *Report on Environmental Impact for the Project "Mining of Brown Coal From Złoczew Deposit"*; Poltegor-Projekt Sp. z o.o.: Wrocław, Poland, 2017.
45. Sawicki, J. Hydrogeological conditions and mining operation brown coal deposits "Złoczew". *Pr. Nauk. Inst. Górnictwa Politech. Wrocławskiej* **2010**, *123*, 127–148. (In Polish)
46. Urbański, P.; Widera, M. Is the Złoczew lignite deposit geologically suitable for the first underground gasification installation in Poland? *Geologos* **2020**, *26*, 113–125. [CrossRef]
47. Kuliński, M.; Pomorski, A. *Development Project for the Złoczew Lignite Deposit*; Proxima, Poltegor: Wrocław, Poland, 2015. (In Polish)
48. Central Statistical Office. *Procurement and Prices of Agricultural Products in 2015–2019*; Central Statistical Office: Warszawa, Poland, 2016–2020. (In Polish)
49. Central Statistical Office. *Statistical Yearbook 1970–1997*; Central Statistical Office: Warszawa, Poland, 1971–1998. (In Polish)
50. Central Statistical Office. *Statistical Yearbook of Voivodships 1975–1997*; Central Statistical Office: Warszawa, Poland, 1976–1998. (In Polish)
51. Central Statistical Office. *Statistical Yearbook of Poviats 1970–1974*; Central Statistical Office: Warszawa, Poland, 1971–1975. (In Polish)
52. Central Statistical Office. *Poviats Statistics in 1970*; Central Statistical Office: Warszawa, Poland, 1971. (In Polish)
53. Central Statistical Office. *Statistical Yearbook of the Katowice Voivodship 1970–1974*; Central Statistical Office: Katowice, Poland, 1971–1975. (In Polish)
54. Central Statistical Office. *Statistical Yearbook of the Łódź Voivodship 1970–1974*; Central Statistical Office: Łódź, Poland, 1971–1975. (In Polish)
55. Central Statistical Office. *Statistical Yearbook of the Opole Voivodship 1970–1974*; Central Statistical Office: Opole, Poland, 1971–1975. (In Polish)
56. Central Statistical Office. *Statistical Yearbook of the Poznań Voivodship 1970–1974*; Central Statistical Office: Poznań, Poland, 1971–1975. (In Polish)

57. Central Statistical Office. *Statistical Yearbook of the Warszawa Voivodship 1970–1974*; Central Statistical Office: Warszawa, Poland, 1971–1975. (In Polish)
58. Local Date Bank. Available online: https://bdl.stat.gov.pl/BDL/start (accessed on 5 February 2021).
59. Glossary: Livestock_unit_(LSU). Available online: https://ec.europa.eu/eurostat/statistics-explained/index.php/Glossary:Livestock_unit_(LSU) (accessed on 15 February 2021).
60. Wirtschaftsbereiche. Available online: https://www-genesis.destatis.de/genesis/online (accessed on 15 January 2021). (In German).
61. Kasztelewicz, Z.; Sikora, M.; Zajączkowski, M. The Poniec-Krobia deposit in extending the Konin lignite mining and power generation basin. *Energy Policy J.* **2012**, *15*, 135–146. (In Polish)
62. Kasztelewicz, Z. Approaches to post-mining land reclamation in Polish open-cast lignite mining. *Civ. Environ. Eng. Rep.* **2014**, *12*, 55–67. [CrossRef]
63. Stanisz, A. *An Accessible Statistics COURSE Using STATISTICA PL with Examples from medicine*; StatSoft: Kraków, Poland, 2007. (In Polish)
64. Kasztelewicz, Z.; Patyk, M. The Modern and Efficiently Coal Power Plants Strategic Challenge for Polish. *Energy Policy J.* **2015**, *18*, 46–59. (In Polish)
65. From Page-History. Available online: https://elbelchatow.pgegiek.pl/O-oddziale/Z-kart-historii (accessed on 2 March 2021). (In Polish).
66. Sadowski, A.; Czubak, W.; Poczta, W.; Rowiński, J. Area and economic structures of Polish agriculture and other EU countries. In *The Structure of Polish Agriculture Compared to the European Union*; Poczta, W., Rowiński, J., Eds.; CeDeWu: Warszawa, Poland, 2019; pp. 45–81. (In Polish)
67. Pepliński, B.; Poczta, W.; Rowiński, J. Structure of livestock production and its evolution. In *The Structure of Polish Agriculture Compared to the European Union*; Poczta, W., Rowiński, J., Eds.; CeDeWu: Warszawa, Poland, 2019; pp. 83–139. (In Polish)
68. Pepliński, B. *Determinants of Regional Changes in the Pig Production Sector in Poland*; Poznań University of Life Sciences: Poznań, Poland, 2019. (In Polish)
69. Igliński, B. Hydro energy in Poland: The history, current state, potential, SWOT analysis, environmental aspects. *Int. J. Energy Water Resour.* **2019**, *3*, 61–72. [CrossRef]
70. Rybka, K.; Nitka, Z. Modern phenotypes of cereals for growing in areas endangered with drought. *Biuletyn Instytutu Hodowli i Aklimatyzacji Roślin* **2014**, *273*, 55–72. (In Polish)
71. Zawora, T.; Ziernicka–Wojtaszek, A. Extreme values of precipitation deficiencies and excesses in the light of contemporary climate change based on the example from the Podkarpackie province. *Infrastruktura i Ekologia Terenów Wiejskich* **2008**, *5*, 23–29. (In Polish)
72. Sillmann, J.; Roeckner, E. Indices for extreme events in projections of anthropogenic climate change. *Clim. Chang.* **2008**, *86*, 83–104. [CrossRef]
73. Trenberth, K.E.; Dai, A.; Rasmussen, R.M.; Parsons, D.B. The Changing character of precipitation. *Bull. Am. Meteorol. Soc.* **2003**, *84*, 1205–1217. [CrossRef]
74. Easterling, D.R.; Horton, B.; Jones, P.D.; Peterson, T.C.; Karl, T.R.; Parker, D.E.; Salinger, M.J.; Razuvayev, V.; Plummer, N.; Jamason, P.; et al. Maximum and Minimum Temperature Trends for the Globe. *Science* **1997**, *277*, 364–367. [CrossRef]
75. Tomassini, L.; Jacob, D. Spatial analysis of trends in extreme precipitation events in high-resolution climate model results and observations for Germany. *J. Geophys. Res. Atmos.* **2009**, *114*, D12113. [CrossRef]
76. Pinke, Z.; Decsi, B.; Kozma, Z.; Vári, Á.; Lövei, G.L. A spatially explicit analysis of wheat and maize yield sensitivity to changing groundwater levels in Hungary, 1961–2010. *Sci. Total Environ.* **2020**, *715*, 136555. [CrossRef]
77. Jalota, S.K.; Vashisht, B.B.; Sharma, S.; Kaur, S. Climate change and groundwater. In *Understanding Climate Change. Impacts on Crop Productivity and Water Balance*; Academic Press: Cambridge, UK, 2018.
78. Zipper, S.C.; Soylu, M.E.; Booth, E.G.; Loheide, S.P. Untangling the effects of shallow groundwater and soil texture as drivers of subfield–scale yield variability. *Water Resour. Res.* **2015**, *51*, 6338–6358. [CrossRef]
79. Connolly, R.D. Modelling effects of soil structure on the water balance of soil-crop systems: A review. *Soil Tillage Res.* **1998**, *48*, 1–19. [CrossRef]
80. Rydałowski, M. The influence of groundwater level and year conditions on the evapotranspiration of salix viminalis l. in view of lysimetric analyses. *Woda Środowisko Obsz. Wiej.* **2016**, *16*, 73–84. (In Polish)
81. Talebnejad, R.; Sepaskhah, A.R. Effect of different saline groundwater depths and irrigation water salinities on yield and water use of Quinoa in Lysimeter. *Agric. Water Manag.* **2015**, *148*, 177–188. [CrossRef]
82. Wu, Y.; Liu, T.; Paredes, P.; Duan, L.; Pereira, L.S. Water use by a groundwater dependent maize in a semi–arid region of Inner Mongolia: Evapotranspiration partitioning and capillary rise. *Agric. Water Manag.* **2015**, *152*, 222–232. [CrossRef]
83. Karimov, A.K.; Šimůnek, J.; Hanjra, M.A.; Avliyakulov, M.; Forkutsa, I. Effects of the shallow water table on water use of winter wheat and ecosystem health: Implications for unlocking the potential of groundwater in the Fergana Valley (Central Asia). *Agric. Water Manag.* **2014**, *131*, 57–69. [CrossRef]
84. Ayars, J.E.; Shouse, P.; Lesch, S.M. In situ use of groundwater by alfalfa. *Agric. Water Manag.* **2009**, *96*, 1579–1586. [CrossRef]
85. Yang, J.; Wan, W.; Deng, W.; Zhang, G. Water fluxes at a fluctuating water table and groundwater contributions to wheat water use in the lower Yellow River flood plain, China. *Hydrol. Proc.* **2007**, *21*, 717–724. [CrossRef]

86. Kahlown, M.A.; Ashraf, M.; Zia-Ul-Haq. Effect of shallow groundwater table on crop water requirements and crop yields. *Agric. Water Manag.* **2005**, *76*, 24–35. [CrossRef]
87. Hutmacher, R.B.; Ayars, J.E.; Vail, S.S.; Bravo, A.D.; Dettinger, D.; Schoneman, R.A. Uptake of shallow groundwater by cotton: Growth stage, groundwater salinity effects in column lysimeters. *Agric. Water Manag.* **1996**, *31*, 205–223. [CrossRef]
88. Parry, M.L.; Canziani, O.F.; Palutikof, J.P.; van der Linden, P.J.; Hanson, C.E. *Climate Change 2007: Impacts, Adaptation and Vulnerability. Contribution of Working Group II to the Fourth Assessment Report of the Intergovernmental Panel on Climate Change*; Cambridge University Press: Cambridge, UK, 2007.
89. Field, C.B.; Barros, V.; Stocker, T.F.; Dahe, Q.; Dokken, D.J.; Ebi, K.L.; Mastrandrea, M.D.; Mach, K.J.; Plattner, G.-K.; Allen, S.K.; et al. *Managing the Risks of Extreme Events and Disasters to Advance Climate Change Adaptation. Special Report of the Intergovernmental Panel on Climate Change*; IPCC, Cambridge University Press: Cambridge, UK, 2012.
90. Wawer, R. Agricultural water management in a changing climate. The prospect of transition to irrigated agriculture versus equitable and sustainable water use in the light of Spanish solutions and advances in information technology. *Pol. J. Agron.* **2020**, *41*, 38–48. (In Polish) [CrossRef]
91. Hlavinka, P.; Trnka, M.; Semerádová, D.; Dubrovský, M.; Žalud, Z.; Možný, M. Effect of drought on yield variability of key crops in Czech Republic. *Agric. For. Meteorol.* **2009**, *149*, 431–442. [CrossRef]
92. Ren, D.; Xu, X.; Engel, B.; Huang, G. Growth responses of crops and natural vegetation to irrigation and water table changes in an agro–ecosystem of Hetao, upper Yellow River basin: Scenario analysis on maize, sunflower, watermelon and tamarisk. *Agric. Water Manag.* **2018**, *199*, 93–104. [CrossRef]
93. Nosetto, M.D.; Jobbágy, E.G.; Jackson, R.B.; Sznaider, G.A. Reciprocal influence of crops and shallow ground water in sandy landscapes of the Inland Pampas. *Field Crop. Res.* **2009**, *113*, 138–148. [CrossRef]
94. Morgan, B.J.; Daughtry, C.S.; Russ, A.L.; Dulaney, W.P.; Gish, T.J.; Pachepsky, Y.A. Effect of shallow subsurface flow pathway networks on corn yield spatial variation under different weather and nutrient management. *Int. Agrophys.* **2019**, *33*, 271–276. [CrossRef]
95. Jokiel, P.; Maksymiuk, Z. The impact of the "Bełchatów" open pit on run-off changes. In Antropogenic Determinants of Changes in River Run-Off and Regimen in Different Regions of Poland. *Dok. Geogr.* **1988**, *4*, 79–95. (In Polish)
96. Włodek, S.; Biskupski, A. Soil moisture dynamics in the Belchatow brown coal mine area. *Soil Sci. Annu.* **2011**, *62*, 420–427. (In Polish)
97. Biskupski, A.; Włodek, S.; Pabin, J. Soil moisture dynamics within the depression sink of brown coal open mine at Belchatow. *Soil Sci. Annu.* **2008**, *59*, 18–24. (In Polish)
98. Pabin, J.; Włodek, S.; Biskupski, A. The influence of the deep drainage of the brown coal strip mine at Belchatow on soil water content. *Soil Sci. Annu.* **2008**, *59*, 191–195. (In Polish)
99. Biskupski, A.; Włodek, S. Influence of the Bełchatów brown coal open-cast mine on soil moisture. *Soil Sci. Annu.* **2011**, *62*, 32–39. (In Polish)
100. Ślusarczyk, E. The influence of lignite mines on water relations of soils in the Bełchatów Industrial District. Part I: The impact of hydrological depression funnel on groundwater and soil moisture. *Pamiętniki Puławskie* **1985**, *85*, 31–44. (In Polish)
101. Sarnacka, S.; Sokołowski, W.; Lesiak, J. *Wpływ Głębokości Odwodnienia Spowodowanego Przez Kopalnię Bełchatów na Stosunki Wodne Gleb. Synteza badań Przeprowadzonych w Latach 1979–1985, Seria S, 55*; Wydawnictwo IUNG: Puławy, Poland, 1987. (In Polish)
102. Włodek, S.; Biskupski, A.; Pabin, J. Soil moisture in the region of brown coal strip mine at Belchatow. *Soil Sci. Annu.* **2008**, *59*, 230–235. (In Polish)
103. Tołoczko, W. Assessment of sandy soils with a rainfall-retention type of water supply based on an experimental study on capillary rise. *Folia Geogr. Phys.* **2019**, *18*, 67–75. [CrossRef]
104. Chodak, T.; Kaszubkiewicz, J.; Bogda, A.; Szerszeń, L.; Bogacz, A. Water conditions of soils located within the range of depression funnel of the existing water intake in Brzeg Głogowski and the planned one in the region of Czerna-Dobrzejowice. *Zesz. Probl. Post. Nauk Roln.* **1995**, *418*, 169–174. (In Polish)
105. Chodak, T.; Borkowski, J.; Bogda, A.; Szerszeń, L.; Nowiński, S. Prognosis of changes in soil environment within the range of depression funnel of "Potoczek" water intake. *Zesz. Probl. Post. Nauk Roln.* **1995**, *418*, 175–184. (In Polish)
106. Mierosławska, A. Regression in the depression funnel. *Nowocz. Rol.* **1995**, *8*, 40–95. (In Polish)
107. Stachowski, P. Shaping of the ground water table on the internal heap of the "Pątnow" opencast mine. *Soil Sci. Annu.* **2004**, *55*, 385–395. (In Polish)
108. Stolarska, M. Influence of the Belchatow mine on water relations. *Aura* **2004**, *4*, 8–10. (In Polish)
109. Włodek, S.; Biskupski, A.; Pabin, J.; Hryńczuk, B.; Kaus, A. The dynamics of ground moisture in the region of brown coal strip mine at Bełchatow on the example of locality Łękińsko. *Melior. Inż. Środ.* **1999**, *20*, 211–219. (In Polish)
110. Kozaczyk, P.; Przybyła, C. Differentiation of soil water management in Pojezierze Poznańskie under various meteorological conditions. *Soil Sci. Annu.* **2004**, *55*, 251–258. (In Polish)
111. Szafrański, C.; Stachowski, P. Changes of moisture postmining grounds waste heap of the "Kazimierz Północ" open cast. *Soil Sci. Annu.* **2004**, *55*, 427–433. (In Polish)
112. Wojcik, R. The relationship between the Koppen climatic index and water levels of the main rivers of the northern part of the Kalisz Voivodeship. *Melior. Inż. Środ.* **1997**, *19*, 311–317. (In Polish)

113. Kędziora, A. Water balance of Konin strip mine landscape in changing climatic conditions. *Soil Sci. Annu.* **2008**, *59*, 104–118. (In Polish)
114. Basso, B.; Kendall, A.D.; Hyndman, D.W. The future of agriculture over the Ogallala Aquifer: Solutions to grow crops more efficiently with limited water. *Earths Future* **2013**, *1*, 39–41. [CrossRef]
115. Kendy, E.; Zhang, Y.; Liu, C.; Jinxia, C.; Wang, J.; Steenhuis, T. Groundwater recharge from irrigated cropland in the North China Plain: Case study of Luancheng County, Hebei Province, 1949-2000. *Hydrol. Process.* **2004**, *18*, 2289–2302. [CrossRef]
116. Aw-Hassan, A.; Rida, F.; Telleria, R.; Bruggeman, A. The impact of food and agricultural policies on groundwater use in Syria. *J. Hydrol.* **2014**, *513*, 204–215. [CrossRef]
117. Pepliński, B. Animal production location in suburban areas of Polish metropolitan centers. *Sustainability* **2020**, *12*, 2619. [CrossRef]
118. Pepliński, B. Changes in the location of cow herds in Poland in the light of Sinclair's location theories. *Ann. PAAAE* **2020**, *12*, 1–10. [CrossRef]
119. Georgakellos, D.A. Impact of a possible environmental externalities internalisation on energy prices: The case of the greenhouse gases from the Greek electricity sector. *Energy Econ.* **2010**, *32*, 202–209. [CrossRef]
120. Sakulniyomporn, S.; Kubaha, K.; Chullabodhi, C. External costs of fossil electricity generation: Health-based assessment in Thailand. *Renew. Sustain. Energy Rev.* **2011**, *15*, 3470–3479. [CrossRef]
121. Büke, T.; Köne, A.C. Estimation of the health benefits of controlling air pollution from the Yatağan coal-fired power plant. *Env. Sci. Policy* **2011**, *14*, 1113–1120. [CrossRef]
122. Dimitrijević, Z.; Tatić, K.; Knežević, A.; Salihbegović, I. External costs from coal-fired thermal plants and sulphur dioxide emission limit values for new plants in Bosnia and Herzegovina. *Energy Policy* **2011**, *39*, 3036–3041. [CrossRef]
123. Taranto, Y.; Köksal, M.A.; Dal, E. *The External Cost of Fossil Fuel Use in Power Generation, Heating and Road Transport in Turkey*; Shura Energy Transition Center: Istanbul, Turkey, 2020.

energies

MDPI

Article

Energy as a Factor of Investment Attractiveness of Regions for Agricultural Enterprises

Hanna Godlewska-Majkowska [1] and Agnieszka Komor [2,*]

1 Institute of Enterprise, Collegium of Business Administration, Warsaw School of Economics, 02-554 Warszawa, Poland; hgodle@sgh.waw.pl
2 Department of Management and Marketing, Faculty of Agrobioengineering, University of Life Sciences in Lublin, 20-950 Lublin, Poland
* Correspondence: agnieszka.komor@up.lublin.pl; Tel.: +48-81-461-0061

Abstract: The aim of the article is to identify and assess the relationship between the investment attractiveness of regions for agricultural enterprises and the energy factor. Classical theories of the location of agriculture emphasise the importance of the market factor. The energy factor has so far been ignored, despite the global trend related to the increasing importance of production scales and rising energy consumption in agriculture. There are also no methodological proposals that allow a comprehensive assessment of the investment attractiveness of regions for agricultural enterprises, taking into account the leading location factors. The article presents the author's methodological model based on the weight-correlation method of valorisation of investment attractiveness of regions for economic entities that invest in agricultural production. It contains a sub-aggregate describing the energy factor. This proposal is a contribution to the theory of the location of agriculture in the field of location factor analysis. The developed methodological model is used to explain location decisions of agricultural enterprises at the regional level. Access to energy as well as energy management increase locational advantages and reduce the economic risk of carrying out agricultural activities in economic units, which contributes to an increase in the sustainability of agricultural production. This is especially true in areas dominated in the past by state-owned and cooperative enterprises, which are the dominant group of enterprises in this area after privatization. The proposed methodology was positively verified on the example of Polish regions, as a significant influence of the energy factor on investment attractiveness at the local level was demonstrated.

Keywords: investment attractiveness of regions; energy; agriculture; economic entities; sustainability

Citation: Godlewska-Majkowska, H.; Komor, A. Energy as a Factor of Investment Attractiveness of Regions for Agricultural Enterprises. *Energies* **2021**, *14*, 2731. https://doi.org/10.3390/en14092731

Academic Editors: Piotr Gradziuk, Bogdan Klepacki, Mariusz J. Stolarski and Josef Maroušek

Received: 10 March 2021
Accepted: 7 May 2021
Published: 10 May 2021

1. Introduction

Investment attractiveness of regions has been a subject of interest for researchers dealing with the analysis of locational advantages of potential industrial or service investment locations. Papers that take into account the location needs of enterprises, whose main economic activity is agricultural production, are rarely published.

The need for studies on this subject arises from changes in the ownership structure within the agricultural sector, as individual farms are falling out of land use. This is linked to an ageing population in rural areas. Migration of the rural population to cities causes succession problems in regard to individual farms. This gives rise to a farmland trade.

A second reason for the legitimacy of undertaking studies on this subject is the privatisation of former state farms and agricultural cooperatives. Privatised farms of this type manage their estates according to changes in the market for agricultural and food products. For this reason, there is an emerging offer of real property in rural areas that can be the subject of location decisions by agricultural enterprises.

Agricultural enterprises can achieve competitive advantages in the form of low business costs and can also compete with high product quality. One of the most important elements influencing the possibility of obtaining competitive advantages by these units is the lack of interference in the course of operational processes. The course is the result of numerous elements of a technical and organisational nature. The energy factor plays a fundamental role among them. This is shown by the 12.9% increase in energy use in agriculture and forestry in the EU 28 countries over the period 2009–2018, based on EUROSTAT data. It was particularly high in Germany, United Kingdom, Albania, Romania, and Hungary. In Poland, this increase reached 9.7% of the 2009 value. Agricultural enterprises achieve economies of scale and diversity benefits through the use of modern technologies. Therefore, the process of selecting the location of agricultural enterprises should be considered in an analogous way to the location analysis of industrial and service enterprises.

It is worth noting that, nowadays, investment activity is a prerequisite for increasing the competitiveness of agricultural production, increasing its innovativeness, increasing its export potential, low-carbon emissions, and climate neutrality, thus contributing to the development of the bioeconomy in the area. Investments by agri-enterprises can thus, in the long term, support the socioeconomic development of areas traditionally relying on the agricultural sector in order to ensure sustainable regional and local development.

The Polish regions can serve as an example of a location environment for agricultural enterprises representative of other countries and regions in Central and Eastern Europe. Following the privatisation of state or cooperative farms, opportunities are being sought to improve the functioning of agriculture by, among other things, introducing the principle of spatial order, according to which the right activity should be located in the right place. Regions with a strong agricultural function can therefore be included in the location analysis, not only for the domestic real estate market, but also for foreign investors. Therefore, the study of locational determinants of agricultural enterprises is a research gap that the authors of this study wish to fill.

The goal of this paper was to identify and assess the relationship between the investment attractiveness of regions for agricultural enterprises and the of energy factor.

The following research hypothesis was verified in the process of execution of the above mentioned goal: there is a positive relationship between the energy factor and the investment attractiveness of regions for enterprises from Section A (agriculture, forestry, and fishing) (H1).

The specific objectives are as follows:

1. Identification of the essence of investment attractiveness of regions in the light of literature, with particular emphasis on the investment attractiveness of regions for agricultural enterprises (O1).
2. Development of a methodological model for evaluating investment attractiveness of regions for Section A enterprises (agriculture, forestry, hunting, and fishing), taking into account the energy factor (O2).

The paper poses the following research questions related to the specific objectives:

I. How is the specificity of investment attractiveness of regions for agricultural enterprises (Section A) expressed?
II. How can the investment attractiveness of regions be evaluated for investments by economic entities from Section A, taking into account the energy factor?

2. Literature Review

2.1. Theories of Agricultural Location and Current Trends in Practice

The theoretical basis for analysing the distribution of agricultural production is contained in the pioneering work of Johann Wolfgang von Thünen [1]. According to the agricultural location theory formulated by the author, commodity agricultural production was concentrated in the form of concentric rings around a market formed by a large city. The rings were characterised by production specialisation determined by the amount of the land rent, which was the resultant of transport costs and the difference between the unit price of a given agricultural product and its unit production cost [2]. Due to the limited shelf life of fruit and milk, which was reflected in high transport rates, this type of production was located close to the market. While extensive livestock production occupied areas further afield, representing the outermost ring of agricultural production. It is worth noting that, according to the theory, the forest ring was closest to the town due to the fact that in the 18th century firewood was the main energy resource. The disadvantage of this model is that it assumes that the space is homogeneous and isotropic, whereas in reality it is heterogeneous. Therefore, the author applied a modification of his model, extending the range of zones in that part of the suburban zone where areas with above-average soil values occurred, as assumed. However, one can agree with U. Mäki that J. H. von Thünen was a realist who deliberately used unrealistic assumptions to pursue a truthful account of an important aspect of determining agricultural land use patterns [3].

This model, despite being developed at the end of the 18th century, is an interesting starting point for a contemporary explanation of the relationship between the distribution of agricultural enterprises and the locational pattern of large markets. Although a long time has passed since this theory was formulated, it can be noted that to this day the presence of large markets influenced the distribution of agricultural commodity production [4].

The theory of rings has been the basis for the development of numerous theoretical modifications aimed at explaining the regularities of agricultural development and distribution, as well as other economic theories [5]. An example of such approaches may be, for example, the Sinclair's inverted food rings models [6], according to which production results and productivity per unit area increase with distance from the city. At the same time, the value of agricultural land near cities is inversely proportional to its market price [7]. It is also important that industrialisation contributes to the contamination of food produced in suburban areas. It was therefore suggested that a forest ring should be maintained around the town, followed by a strip of meadow and pastureland for grazing animals. The suburban forest belt can also act as a local forest landscape with cultural value for the community living in large cities [8].

The theory of rings, although it originated from observations made in southern Germany, has gained universal value because the actual spatial structures of agricultural production were formed according to this model. This is evidenced, for example, by the very frequent direction of special agricultural production in the form of greenhouse cultivation of vegetables, fruits, and flowers in the suburban zone. To this day, provision zones can still be seen around large cities [9], resulting from the desire to compete on the quality of agricultural products for direct consumption by the inhabitants of large cities. The significant influence of large and spatially concentrated markets on the location of the agricultural commodity sector has therefore been maintained. However, it should be noted that in agriculture the outlets created by the processing industry as well as by wholesale trade (for example, commodity exchanges) are also important [10]. Consequently, the distribution of agricultural enterprises is influenced by the spatial arrangement of outlets created in all three of their types mentioned above.

A prerequisite for obtaining the benefits of proximity to markets is communication accessibility for agricultural enterprises, as well as a sufficiently high level of production concentration to obtain economies of scale in production and rationalisation of energy management. The proximity of markets does not exclude multiple transports of agricultural raw materials to urban food processing plants, nor does it eliminate the need for multiple transports of food delivered to residents or other customers forming the first tier of supply chains [11].

The organisation of transport also brings with it the issue of energy efficiency. The location of agricultural enterprises is linked to the energy factor not only because of the need to secure the energy needs of the enterprise, but also has an impact on rationalising the use of fuels necessary for organising the transport of both raw materials and products. Furthermore, from the point of view of the public good, the energy aspects of losses and damage to raw materials and agricultural products caused by the irrational organisation of shipping or transport arrangements must be taken into account. The resulting losses and damages during transport are a waste of the energy that was used to produce them [12]. Therefore, when considering the energy determinants of the location of agricultural enterprises, account must be taken not only of ensuring a sufficient supply of energy, but also of reducing the various sources of energy wastage. In developing countries, it is also important to streamline energy management and exploit all opportunities to obtain additional energy sources. For example, the use of digestate from biogas plants helps to reduce energy consumption in the production of certain industrial products in developing countries, such as cement, which also has a positive impact on energy management and helps to reduce the impact on the natural environment [13].

The locational advantages relevant for agricultural enterprises have another context related to energy management. In addition to the already mentioned great importance of access to energy supply and energy savings in the logistical sphere, it is worth mentioning the energy savings resulting from closed-loop energy management. The opportunities in this regard are related to the use of production waste or agricultural by-products for energy purposes. Products hitherto regarded as sources of environmental pollution can be used as sources of energy, especially heat. These include, for example, manure, slurry, straw, and other by-products generated in agriculture. Another possibility is to produce energy from purpose-grown phytomass, which has a rapid growth rate [14]. They can be a source of energy for the distributed energy system.

With this solution, it is possible to reduce the consumption of energy in agricultural production, as well as to reduce the nuisance to the natural environment resulting from the activities of agricultural enterprises. The closed loop circuit applied to electricity management is an element, which will allow for greater self-sufficiency on the part of agricultural enterprises, provided that they can be integrated into the national or regional energy system. The electricity from biogas combustion creates an additional source of income. This is especially important in developing countries [15].

It is worth noting that initiatives to prevent waste production or to maximise its treatment and use in order to reduce its quantity and harmfulness are part of the bioeconomy model promoted by the European Union. In addition, with careful energy management, it is possible to use energy from other renewable sources that are inextricably linked to the advantages of rural locations (e.g., wind or hydroelectric power stations, as well as biogas and biomass). An example that illustrates this point well is the use of photovoltaic cells in rural areas with a high number of sunny days. The installed photovoltaic panels provide a source of energy that is independent of other energy suppliers, thus allowing economic activity to be carried out in an environmentally friendly manner. The use of renewable energy sources by agricultural production entities contributes to increasing the sustainability of agricultural production, to increasing the self-sufficiency of entities in the sphere of energy supply, which reduces the economic risk of doing business.

2.2. New Technologies as a Basis for New Location Trends in Agriculture

Taking into account the fact that agricultural enterprises are in terms of technological advancement similar to industrial and service enterprises, the application of modern technologies in their activity is also of significance. This type of technology requires an adequate supply of energy to be secured, as well as a high level of energy security.

Despite the fact that agriculture is perceived as a traditional branch of the economy, modern technologies are being used in it with great success [16]. Key enabling technologies have been presented by the European Commission [17]. Among these, there are solutions based on nanotechnology [18], micro-technology [19], and technologies derived from life sciences. Agricultural enterprises can develop genetically modified food. This issue is admittedly very controversial [20] because the impact of genetic modification on the entire food chain is not fully known yet. However, without the work of geneticists in agriculture, it would not be possible to achieve improvements in the functional traits of many crop and livestock species, and consequently to reduce world hunger. Thus, the literature presents both positive and negative effects of GM foods [21]. Agricultural production units dealing with the genetics and breeding of crops and livestock may have similar location requirements to those of research and development units accompanying industrial or service activities.

Another new technology being used in agriculture is artificial intelligence in the context of agriculture automation [22] and getting value in agriculture [23]. Modern technologies related to connectivity are also important, for example related to the use of mobile phones [24], especially SMS information [25]. Another new technology is the Internet of Things [26], also in the context of sustainable rural development [27]. They contribute to a significant acceleration of information flows in modern managed farms through artificial intelligence based solutions. One example is computer control of physical and chemical parameters of livestock habitat. Other new technologies related to agriculture that can be mentioned here include precision agriculture [28] (considering factors influencing the adoption of precision agriculture technologies [29]) and Industry 4.0 [30].

Modern technologies are therefore finding more and more widespread use especially in agriculture in highly economically developed countries, where human labour has long been replaced by machines. An investment in an agricultural enterprise by a foreign investor may therefore need to take into account not only traditional factors of agricultural development and distribution [31]. This is particularly important for entrepreneurs who are diversifying their business activities. Consequently, they can invest not only in agricultural real estate, but also in industry or services that are linked to agricultural activity through the value chain. This is a feature that significantly distinguishes agricultural enterprises from individual farms. These farms are bound to households and not to industrial or service enterprises. All of this makes it necessary to learn about the method of assessing location advantages for agricultural economic entities, not only from the theoretical point of view, but also from the practical one.

To conclude, the specificity of the location requirements of agricultural enterprises compared to individual farms lies in the following elements:

- The large scale of agricultural production within individual agricultural enterprises.
- Greater opportunities and the associated benefits of using modern technologies.
- Cooperation with research facilities in the event of specialisation in plant and/or animal breeding.
- More frequent ownership and organisational links with industrial and/or commercial entities.

Agricultural enterprises depend on the energy factor because of:

- Greater dependence on energy supplies, due to the large scale of production and greater mechanisation of work than on traditional individual farms.
- Greater opportunities to use distributed renewable energy sources in a closed loop system.

2.3. The Essence of Investment Attractiveness of Regions

It is worth noting that in the literature the concept of investment attractiveness functions both in relation to enterprises [32], including agricultural enterprises [33], as well as spatial units. The subject of the research in this study was investment attractiveness of spatial units. There are numerous scientific elaborations in the literature on investment attractiveness of spatial units, but there is no unambiguous definition of this concept. A number of studies examine investment attractiveness of spatial units of various taxonomic levels (e.g., countries, regions, communes) without providing a concrete definition of this concept. In such a situation, investment attractiveness is often interpreted as a set of factors, e.g., geographic, economic, natural, political, social, the knowledge of which enables an investor to imagine the possibility of investing in a given country, region, or industry [34]. The multitude of factors influencing the level of investment attractiveness testifies to the multifaceted nature of this concept.

In a general sense it can be stated that investment attractiveness is the ability to induce investors to choose a region as a location for an investment [35]. According to E. Czerwieniec, investment attractiveness can be understood as a set of advantages of a given place (country, region), as certain areas show relatively better conditions for investment activity than others [36]. A similar interpretation of the concept of investment attractiveness is presented by T. Kalinowski, who states that it is a combination of location-specific advantages which can be achieved in the course of business activity and result from the specific features of the area in which the activity takes place [37]. Other researchers understand investment attractiveness as the purposefulness of making capital investments in the expansion and technical reconstruction of operating enterprises, defined as potential investment objects [38]. Investment attractiveness is also considered as a combination of signs, factors and conditions affecting the intensity of investment inflows [39], and as a system of existing opportunities, and measures affecting the demand for investment in a particular location or industry [40]. Investment attractiveness for agro-industrial production requires the creation of appropriate conditions (e.g., economic–financial, organisational–regulatory, and social) aimed at minimising the investment risk, which consequently contributes to increasing the investment value [41].

To sum up, investment attractiveness can be understood as the possibilities of satisfying the expectations of investors investing in a given region. Due to the fact that investor's satisfaction can be considered from the point of view of the locational advantages of a given place as well as the expected effects of an investment, two types of investment attractiveness are distinguished, i.e., potential investment attractiveness and actual investment attractiveness. Whereas actual investment attractiveness concerns the expected effects of an investment—it is the ability of a region to absorb financial and physical capital in the form of an investment [42]; and potential investment attractiveness of regions is understood as a set of regional location-specific advantages, which influence the attainment of the investor's objectives [43]. On the basis of literature studies, it has been demonstrated that there is no unambiguous and universally accepted definition of investment attractiveness in the literature. To sum up the above considerations, it should be stated that investment attractiveness is a multifaceted notion.

The subject of research in this study was potential investment attractiveness. The starting point is an analysis of the locational advantages of a place, understood as the elements of the geographical environment accumulated in a given part of space (natural, human, cultural resources, etc.), which become locational factors if they constitute the grounds for the investor's locational decision. Hence, for a particular enterprise, only those location advantages are relevant which at the same time constitute important location factors for that enterprise.

The concept of location factors was introduced to the literature by Alfred Weber. According to this author's concept, the location factor is a precisely defined benefit in the form of a reduction in production costs associated with conducting industrial activity in a precisely specified location (the transport factor was fundamental, and the labour and agglomeration factors were modifying) [44]. It should be noted that, in the modern economy, the scope of location factors has expanded and their role has been re-evaluated [45].

Thus, currently in the literature there are many divisions of location factors, among others a division into factors forcing investment (strategic) and encouraging investment (institutional) [46], factors stimulating to undertake activity, factors de-stimulating (discouraging) to undertake activity and factors that are indifferent [47], or a division into environmental, spatial, economic, sociocultural, political, legal–administrative, and technical–technological factors [48]. In addition, a distinction is made between hard factors (objective—they directly affect the enterprise's operations and are easily measurable) and soft factors (subjective—they are difficult to measure or estimate, but play an important role in the operation of the enterprise as they shape the investment environment) [49]. When making location decisions, investors should consider the specifics of the location and incorporate various elements, such as earnings management [50].

It is worth noting that the issues related to investment attractiveness are often considered in the literature in the context of an inflow of foreign direct investment, which is related to the progressing processes of globalisation, the development of new forms of integration of economic entities and the formation of corporate structures. In the literature, investment attractiveness of spatial units is relatively often analysed in various taxonomic scales, i.e., national [51], regional [52], and local [53].

Investment attractiveness of cities [54] and, to a lesser extent, of rural areas [55], is relatively often examined. The researchers are probably less interested in the issue of investment attractiveness of rural areas due to the fact that they are characterised by a significantly lower economic and social potential in comparison with urban areas, where the accumulation of economic activity takes place. However, due to the fact that in the modern economy the location factors considered important are evolving, it should be stressed that rural areas can offer potential investors important advantages, e.g., in the form of the quality of the natural environment, and a greater supply of free land for investment than in cities [56]. Studies on investment attractiveness of rural areas most often concern location factors characteristic for various types of service and industrial activities, which is connected with the idea of multifunctional development of rural areas and creation of non-agricultural jobs. In literature it is relatively rare to find studies focusing on investment attractiveness of rural areas for agricultural activity.

Therefore, taking into account the above scientific discussion on the essence of investment attractiveness, the authors propose to apply a definition of potential investment attractiveness understood as a set of regional location advantages that influence achieving investors' goals (such as costs of running a business, sales revenues, net return on investment, and investment's competitiveness) [57]. These are the elements of the geographical environment accumulated in a given part of space (natural, human, cultural resources, etc.), which become locational factors if they constitute the grounds for the investor's locational decision. Hence, for a particular enterprise, only those location advantages are relevant, which, at the same time, constitute important location factors for that enterprise. Location advantages important for agricultural enterprises include not only socioeconomic characteristics but also natural features of the location environment.

3. Materials and Methods

Empirical analyses were based on indices of potential investment attractiveness of local government units for Section A businesses based on the methodology developed by the team led by H. Godlewska-Majkowska within the framework of statutory research conducted at the Collegium of Business Administration of the Warsaw School of Economics [58]. The analysis of potential investment attractiveness for Sections A businesses was conducted on two taxonomic levels, i.e., on the regional level—for all Polish voivodeships (16 units) and on the local level—for all rural counties (314 units). Urban counties were excluded from the analyses due to their urban nature [59]. (Since 1 January, 1999, Poland has had a three-tier administrative (territorial) division, under which the Polish territory has been divided into voivodeships, then into counties (including municipalities with the status of cities with county rights), and the smallest units—communes. As of 1 January, 2021, the administrative division of Poland included: 16 voivodeships, 314 counties (comprising several or more neighbouring communes—so-called rural counties), and 66 cities with county rights (a county can also be a spatial unit consisting of one city with county rights)).

The data for calculating the potential investment attractiveness index for businesses classified under Section A was obtained from the Central Statistical Office (GUS) and the Institute of Soil Science and Plant Cultivation in Puławy (IUNG) (data concerning the assessment of agricultural production space). All data are for 2019. The research methods used for this article are the weight-correlation method and cartographic analysis.

The weight-correlation method [60] used in this article makes it possible to standardise the individual input variables through the use of standardisation of the variables which avoids the problem of comparability of the sub-measures. It consists in transforming the univariate variables into pseudo univariate variables and further the set of pseudo univariate indices into a synthetic measure. The pseudo univariate indices make up microclimates and the microclimates make up the synthetic index of potential investment attractiveness.

This article presents a proposal for calculating investment attractiveness indicators for agricultural enterprises in two variants—one at the regional and one at the local level. This is dictated by the different level of availability of statistical data—more statistical data are available at the regional level than at the local level. The research procedure is the same in both cases.

The research procedure consisted of the following steps:

- Selection of input variables for investment attractiveness assessment.
- Determination of the type of variables (stimulant or destimulant).
- Standardization of input variables.
- Division of variables into microclimates.
- Selection of microclimate weights to be included in the final index.
- The iterative part of calculation of the final index of investment attractiveness.
- Definition of final indicators of investment attractiveness.
- Division of statistical units into classes of investment attractiveness.

In the first step, statistical data for investment attractiveness assessment for regional and local levels were selected and transformed into input indices, taking into account the size of the studied statistical units (per capita, unit area, etc.).

In the second step, for the purpose of evaluating investment attractiveness, the variables, which unambiguously constitute the stimulants or destimulants of investment attractiveness, were used, i.e.:

- Destimulant (D): if the level of the variable arises, the potential attractiveness of a region will rise too.
- Stimulant (S): if the level of the variable arises, the potential attractiveness of a region will shrink.

In this step, some of the data that did not allow a clear assignment to stimulant or destimulant was also rejected.

In the third step, the input variables were standardised based on the following formulas.

- For stimulants:

$$x'_{ij} = \frac{x_{ij} - x_{\min.j}}{x_{\max.j} - x_{\min.j}} \times 100 \quad (1)$$

- For destimulants:

$$x'_{ij} = \frac{x_{\max.j} - x_{ij}}{x_{\max.j} - x_{\min.j}} \times 100 \quad (2)$$

where:

j—number of the next attribute sequential number of the spatial unit.
x'_{ij}—normalised j-variable in i-spatial unit.
x_{ij}—value of j-variable in i-spatial unit.
$x_{\min.j}$—minimum of j-variable.
$x_{\max.j}$—maximum of j-variable.

In the fourth step, the input variables were divided into microclimates corresponding to the main factors of the location of agricultural enterprises. The number of microclimates and the list of input variables constituting them were determined according to access to statistical data. For the study of investment attractiveness for Section A enterprises at the regional level, 61 variables were used, while a total of 36 variables were used for the analyses at the local level. For analysis at the regional level, these variables were grouped into eight microclimates: i.e., energy, labour resources, technical infrastructure, social infrastructure, market, administration, agricultural production intensity, and microclimate—quality and determinants of agricultural production. At the local level, this included 7 microclimates, as, due to the lower availability of statistical data at this taxonomic level, the intensity of agricultural production as well as quality and conditions of agricultural production microclimates were abandoned, and the agricultural values microclimate consisting of one index (i.e., quality of agricultural productive area) was introduced.

The table below presents input variables for evaluating potential investment attractiveness of agricultural enterprises at the regional level (voivodeship) in Poland (PAI A REGION), which form particular microclimates, together with their character (stimulant/destimulant) (Table 1).

Table 1. Input variables for estimating potential investment attractiveness of Section A businesses at regional level (voivodeships) in Poland in 2019 (PAI A REGION).

No.	Microclimate/Input Variables	Character
	Microclimate—Energy	
1.	Heat energy produced per year from the treatment of landfill gas in degassing installations per 1000 inhabitants	S
2.	Electrical energy produced per year from the treatment of landfill gas in degassing installations per 1000 inhabitants	S
3.	Electricity consumption in rural areas per 1 inhabitant	S
4.	Piped gas consumption in rural households per 1 inhabitant	S
5.	Fluctuation of gas consumption in rural households per 1 inhabitant	D
6.	Gas users in rural areas as % of total population	S
7.	Distribution network in km per 100 km^2—gas distribution infrastructure in rural areas	S
8.	Density of heat transmission and distribution network in rural areas in km/km^2	S
9.	Electricity production from renewable sources per 1 inhabitant	S
10.	Electricity consumption in agriculture per 100 ha of arable land	S
11.	Rural electricity consumption including consumption for agricultural production per 1 inhabitant	S
12.	Expenditure on fixed assets for environmental protection—energy saving per 1 inhabitant	S
13.	Municipal waste management—weight of collected municipal waste for thermal treatment with energy recovery per 1000 inhabitants (tons/person)	S
	Microclimate—Labour Resources	
14.	Percentage of the population in non-productive age per 100 people in productive age	D
15.	Labour force participation rate	S
16.	Population in post-productive age per 100 people in pre-productive age	D
17.	Share of population in productive age	S
	Microclimate—Technical Infrastructure	
18.	% share of the population covered by the water supply system	S
19.	% share of dwellings connected to the gas pipeline	S
20.	% share of the population covered by the sewage system	S
21.	Density of the water supply network in km per 100 km^2	S
22.	Density of the gas pipeline network in km per 100 km^2	S
23.	Density of the sewage network in km per 100 km^2	S
24.	Sludge previously stored (accumulated) on the premises of the treatment plant—as of 31.12 (tons of dry matter) per 1000 inhabitants	D
25.	Waste generated during the year—disposed of/waste generated during the year	S
26.	Share of treated wastewater in wastewater requiring treatment	S

Table 1. *Cont.*

No.	Microclimate/Input Variables	Character
	Microclimate—Social Infrastructure	
27.	Medical practices in the countryside and in the city per 100,000 inhabitants	S
28.	Number of health care facilities per 100 thousand inhabitants	S
29.	Number of pharmacies per 100 thousand inhabitants	S
30.	Usable floor area of apartments per capita	S
31.	The number of viewers in stationary cinemas per 100 inhabitants	S
32.	Number of visitors to museums with branches per 1000 inhabitants	S
33.	Length of bicycle paths per 1000 inhabitants	S
	Microclimate—Market	
34.	Population density per km^2	S
35.	Value added per person employed in Section A	S
36.	Share of value added in Section A in relation to share of those employed in Section A	S
	Microclimate—Administration	
37.	Funds for financing own tasks obtained from other sources per capita	S
38.	Share of own revenues in total revenues	S
39.	Expenditure on municipal management and environmental protection and on safety and fire protection per capita	S
	Microclimate—Agricultural Production Intensity	
40.	Value of final agricultural production per 1 ha of arable land (PLN/ha)	S
41.	Share of agricultural commodity production in the final agricultural production (%)	S
42.	Livestock (cattle, sheep, goats, horses, pigs, poultry) in livestock units (LSU) per 1 ha of arable land	S
43.	Livestock production per ha of arable land—milk (litres)	S
44.	Livestock production per 1 ha of arable land—slaughter livestock converted into meat (kg)	S
45.	Crop production—yield per ha—total cereals (dt/ha)	S
46.	Agricultural crops—yield per hectare—field vegetables + tree fruit + berry fruit (dt/ha)	S
47.	Other agricultural crops—yield per ha	S
48.	Purchase of products per 1 ha of arable land—basic cereals	S
49.	Purchase of products per 1 ha of arable land—potatoes	S
50.	Purchase of products per 1 ha of arable land—sugar beets	S
51.	Purchase of products per 1 ha of arable land—cow's milk	S
52.	Mineral fertiliser consumption (nitrogen, phosphorus, potassium) per 1 ha of arable land in good cultivation (kg/ha)	S
53.	Calcium fertiliser consumption per 1 ha of arable land in good cultivation (kg/ha)	S
	Microclimate—Quality and Determinants of Agricultural Production	
54.	Samples disqualified as % of total tested—fruit	D
55.	Share of arable land area in certified organic farms in the total arable land (%)	S
56.	Expenditure on fixed assets for environmental protection—protection and restoration of use value of soil, protection of underground and surface water per 1 ha of total area	S

Table 1. *Cont.*

No.	Microclimate/Input Variables	Character
57.	Share of devastated and degraded land which has been rehabilitated or developed for agricultural purposes in the total area of arable land (%)	S
58.	Share of area of fires on meadows, stubble and wasteland in the total area of arable land (%)	D
59.	Quality index for agricultural production area	S
60.	Share of agricultural tax revenue in tax revenues	S
61.	Arable land area per inhabitant	S

Note: the quality index for agricultural production area consists of the following components: soil quality and agricultural suitability, agroclimate, land topography, and soil water relations [61]. Source: own study based on data from the Central Statistical Office (GUS) and the Institute of Soil Science and Plant Cultivation in Puławy (IUNG).

Due to the lower availability of statistical data at the county level in Poland, a slightly smaller number of input variables were used to assess the potential investment attractiveness for agricultural enterprises on the local level (counties) (PAI A LOCAL) (Table 2).

Table 2. Input variables for assessing potential investment attractiveness of Section A businesses on the local level (counties) in Poland in 2019 (PAI A LOCAL).

No.	Microclimate/Input Variables	Character
	Microclimate—Energy	
1.	Electrical energy produced per year from the treatment of landfill gas in degassing installations per 1000 inhabitants	S
2.	Electricity consumption in rural areas per 1 inhabitant	S
3.	Piped gas consumption in rural households per 1 inhabitant	S
4.	Fluctuation of gas consumption in rural households per 1 inhabitant	D
5.	Gas users in rural areas as % of total population	S
6.	Distribution network in km per 100 km^2—gas distribution infrastructure in rural areas	S
7.	Density of heat transmission and distribution network in rural areas in km/km^2	S
	Microclimate—Labour Resources	
8.	Percentage of the population in non-productive age per 100 people in productive age	D
9.	Labour force participation rate	S
10.	Internal permanent migration rate per 1000 inhabitants	S
11.	Population in post-productive age per 100 people in pre-productive age	D
12.	Share of population in productive age	S
No.	**Microclimate/Input Variables**	**Character**
	Microclimate—Technical Infrastructure	
13.	% share of the population covered by the water supply system	S
14.	% share of dwellings connected to the gas pipeline	S
15.	% share of the population covered by the sewage system	S
16.	Density of the water supply network in km per 100 km^2	S
17.	Density of the gas pipeline network in km per 100 km^2	S
18.	Density of the sewage network in km per 100 km^2	S
19.	Sludge previously stored (accumulated) on the premises of the treatment plant—as of 31.12 (tons of dry matter) per 1000 inhabitants	D
20.	Waste generated during the year—disposed of/waste generated during the year	S
21.	Share of treated wastewater in wastewater requiring treatment	S

Table 2. *Cont.*

	Microclimate—Social Infrastructure	
22.	Medical practices in the countryside and in the city per 100,000 inhabitants	S
23.	Number of health care facilities per 100 thousand inhabitants	S
24.	Number of pharmacies per 100 thousand inhabitants	S
25.	Usable floor area of apartments per capita	S
26.	The number of viewers in stationary cinemas per 100 inhabitants	S
27.	Number of visitors to museums with branches per 1000 inhabitants	S
28.	Length of bicycle paths per 1000 inhabitants	S
	Microclimate—Market	
29.	Population density per km^2	S
30.	Revenue of commune budgets from personal income tax (PIT) per 1000 inhabitants	S
31.	Share of social welfare expenditure in commune budget expenditure	D
32.	Agricultural tax revenue per inhabitant	S
	Microclimate—Administration	
33.	Funds for financing own tasks obtained from other sources per 1 inhabitant	S
34.	Share of own revenues in total revenues	S
35.	Expenditure on municipal management and environmental protection and on safety and fire protection per 1 inhabitant	S
	Microclimate—Agricultural Values	
36.	Quality index for agricultural production area	S

Note: the quality index for agricultural production area consists of the following components: soil quality and agricultural suitability, agroclimate, land topography, and soil water relations [61]. Source: own study based on data from the Central Statistical Office (GUS) and the Institute of Soil Science and Plant Cultivation in Puławy (IUNG).

In the case of the regional level, the agricultural factors were included in two microclimates according to the availability of data allowing for additional identification of the microclimates: agricultural production intensity and quality and determinants of agricultural production. However, in the case of the investment attractiveness index at the local level, the availability of data allowed only the use of the Quality index for agricultural production area developed by IUNG (hence, the original name of this index as a microclimate was retained). Therefore, when this method is used at the local level by researchers from other countries, we recommend the use of analogous indices describing the valorisation of agricultural production space, developed by relevant scientific institutions. In both tables, the energy microclimate was introduced as an original proposal that allows to demonstrate the influence of the energy factor on the investment attractiveness of the regions, which is directly related to the research hypothesis H1. Due to the different availability of statistical data, the structure of the energy microclimate at the regional level consists of a greater number of input variables than compared to the local level.

In the fifth step, the microclimate weights included in the final index were selected. For each microclimate, an aggregated vector of standardised sums was determined according to the formula:

$$q_{i,n} = \frac{1}{m_n} \times \sum_{j=1}^{m_n} x'_{ij} \quad (3)$$

where:

$q_{i,n}$—evaluation of n-microclimate in i-spatial unit.
m_n—number of explanatory variables comprising the microclimate in question.
n—number of microclimate.

The sixth step is the iterative part of the calculation of the final index of investment attractiveness. It consisted in the calculation of the output correlation vector r_n (using Pearson's correlation coefficient) between the value of each microclimate and the vector of microclimates sums. The correlation vector r'_n between the microclimate and the vector of microclimate sums was then iteratively determined until the changes of correlation coefficients between following iterations became irrelevant.

The seventh step was to identify the final index of investment attractiveness. The final correlation coefficients, r'_n, thus determined, constitute weights for the individual microclimates, reflecting their strength of influence on the synthetic index, according to the formula:

$$PAI_i = \frac{1}{n}\sum_{k=1}^{n} r'_n \times q_{i,n} \quad (4)$$

where:

PAI_i—summary evaluation.

$q_{i,n}$—evaluation of n-microclimate in i-spatial unit.

n—number of microclimate.

r'_n—final correlation between each microclimate and the sum of all microclimates.

By an iterative route, multiple recalculations yielded stabilised correlation coefficients between individual microclimates and the aggregated potential investment attractiveness index (PAI A). According to the methodology used in the article, the presented correlation indices are the weights with which the individual microclimates are accounted for in the aggregated investment attractiveness index (according to formula (4)). These indices are presented in the figures below.

The results presented in Figures 1 and 2 allow to assess the strength of the relationship between the energy factor and the investment attractiveness of territorial units for agricultural enterprises, which is the basis for testing the research hypothesis H1. The results also enable the identification of other important localization factors. It is important that the preliminary research results allow the identification of factors with significant importance (the correlation index should be at least 0.6). In addition, none of the correlation indices should be negative. However, if any of the correlation indices has a negative value, the input variables should be checked and the whole testing procedure should be repeated.

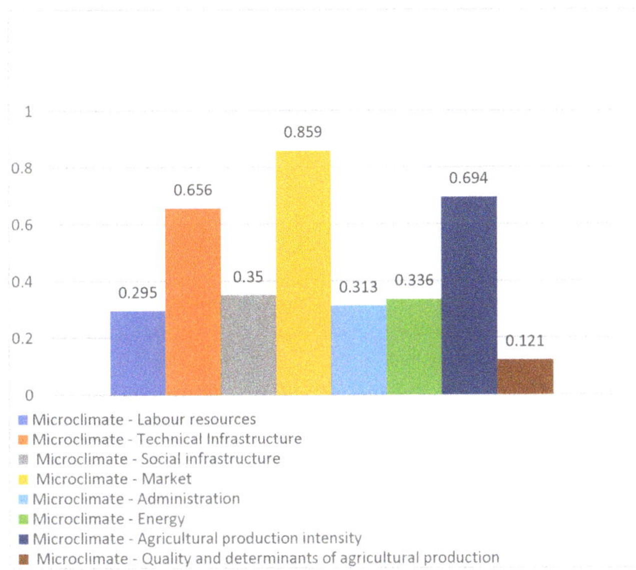

Figure 1. Correlation coefficient between individual microclimates and the aggregated potential investment attractiveness index at the regional level (PAI A REGION) in Poland. Source: own research.

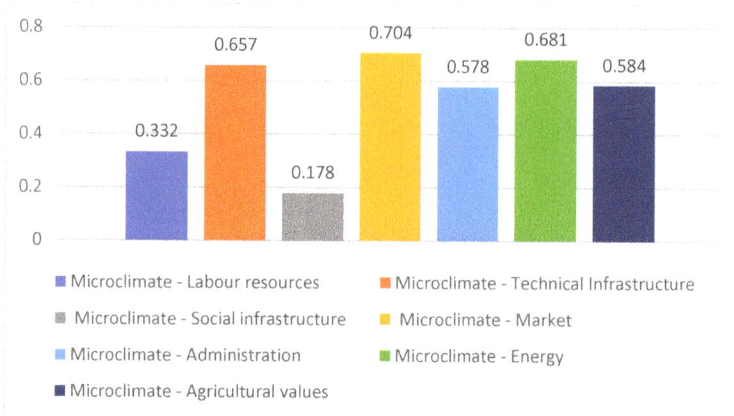

Figure 2. Correlation coefficient between individual microclimates and the aggregated potential investment attractiveness index at the local level (PAI A LOCAL) in Poland. Source: own research.

In the eighth step, the statistical units were classified into investment attractiveness classes. The PAI index determined on the basis of the presented procedure constitutes a basis for the division of spatial units into classes A–F, where class A stands for the highest investment attractiveness, while class F stands for the lowest investment attractiveness. The class range was determined in accordance with formula (5). In other words, class A contains spatial units with the PAI level higher than sum of average and standard deviation derived from the PAI measure (inclusively). Respectively, class B ranges from sum of average and half of standard deviation (inclusively) to sum of average and standard deviation (exclusively).

$$W_{PAI} = \begin{cases} A: \ PAI_i \geq \overline{x}_{PAI} + S_{PAI} \\ B: \ PAI_i \geq \overline{x}_{PAI} + 0.5 \times S_{PAI} \wedge PAI_i < \overline{x}_{PAI} + S_{PAI} \\ C: \ PAI_i \geq \overline{x}_{PAI} < \overline{x}_{PAI} + 0.5 \times S_{PAI} \\ D: \ PAI_i \geq \overline{x}_{PAI} - 0.5 \times S_{PAI} \wedge PAI_i < \overline{x}_{PAI} \\ E: \ PAI_i \geq \overline{x}_{PAI} - S_{PAI} \wedge PAI_i < \overline{x}_{PAI} - 0.5 \times S_{PAI} \\ F: \ PAI_i < \overline{x}_{PAI} - S_{PAI} \end{cases} \quad (5)$$

where:

A–F—investment attractiveness classes, where *A* stands for the highest level of investment attractiveness, and *F* for the lowest level of investment attractiveness.

PAI—potential investment attractiveness index.

$\overline{x}_{PAI}$—average of *PAI* for each level type of regions.

S_{PAI}—standard deviation of *PAI* for each level type of regions.

4. Results

The spatial analysis is based on the cartograms presented in Figures 3 and 4. The first refers to the spatial differentiation of potential investment attractiveness of NUT2 level regions for investment in agriculture (Section A). As shown in Figure 3, voivodeships characterised by the highest investment attractiveness (Class A) are those that are provisioning regions for big cities and/or characterised by long-term food surpluses, obtained as a result of high level of agricultural culture, with a tradition of food export and above-average level of production intensity. That is why the list of highest ranked voivodeships opens with the Greater Poland voivodeship (investment attractiveness class A), which fulfils all of the above mentioned conditions. The Mazowieckie and Pomorskie voivodeships were also highly rated (Class A). The Zachodniopomorskie and Kujawsko-Pomorskie voivodeships also achieved high class B. Voivodeships: Dolnoslaskie, Opolskie, Warminsko-Mazurskie, and Slaskie are also ranked above average (class C).

Figure 3. Spatial diversification of the potential investment attractiveness index for Section A enterprises at the regional level (voivodeships) in Poland in 2019 (PAI A REGION). More about the regions of Poland: https://www.paih.gov.pl/publications/regions. (accessed on 30 April 2021). Source: own research (using the program Map Viewer).

The specificity of investment attractiveness of regions for enterprises in agriculture is already expressed in higher assessment of investment attractiveness of regions in which in the past agriculture was dominated by state farms or production cooperatives. Agricultural enterprises are already located in such regions as Wielkopolskie (10,641 economic entities out of 70,347 in Poland in 2019), Mazowieckie (8301), Pomorskie (4117), and Slaskie voivodeships (4833). The majority of the regions, due to their location in western Poland, used to provide provisions not only for the home regions, but also as a provisioning area for Germany, especially in the period before the Second World War. The legacy of this period is an extensive road network, a higher level of agricultural culture and better education of the rural population compared to eastern Poland.

Due to the fact that regions show great internal differentiation, the analysis of investment attractiveness at the regional level provides a source of directional guidance in the choice of location options. Therefore, a valuable supplement to the regional analysis is the presentation of indices based on the analysis of investment attractiveness on a local scale (counties), which is presented in Figure 4.

As can be seen from Figure 4, the regions with the highest attractiveness classes are often located in the provisioning zones of large cities, which is particularly evident in the provisioning zones of Warsaw (districts surrounding Warsaw), also the suburban areas of the agglomerations of Kraków, Wrocław, and other cities serving as regional capitals. The role of natural factors is also evident, as well as that of wholesale or industrial markets in regions located in the Lubelskie, Podkarpackie, and Zachodniopomorskie voivodeships. In addition, attention is drawn to the southern provisioning zone of the Slaskie voivodeship (Silesia) that is more distant from the immediate vicinity.

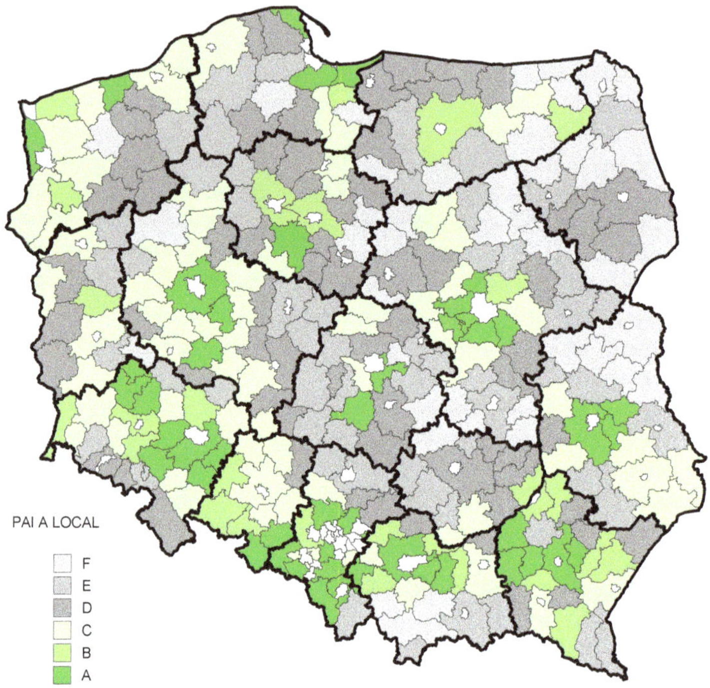

Figure 4. Spatial diversification of the potential investment attractiveness index for Section A enterprises at the local level (counties) in Poland in 2019 (PAI A LOCAL). White colour means cities with county rights, which are not covered by the study due to the lack of agricultural functions. Source: own research (using the program Map Viewer).

Analyses carried out using econometric and cartographic methods indicate that there is a directly proportional (positive) relationship between the investment attractiveness of regions and energy management, which is reflected in the high correlation coefficient between the energy microclimate and the aggregated investment attractiveness index at a local level (0.681), and not so high at regional level (0.336)—see Figures 1 and 2. This confirms the research hypothesis H1: there is a positive relationship between the energy factor and the investment attractiveness of regions for enterprises from Section A. This relationship is stronger at the local level, which is due to the reduced level of differentiation of locational assets at the local scale compared to the regional scale.

Taking into account the fact that the assessment of the energy factor comprised components describing access to energy infrastructure as well as variables describing the rationality of the use of energy sources, it can also be concluded that good energy supply conditions and the rationalisation of energy management favour the creation of well-organised supply chains for agriculture. This applies both to the supply of inputs for agricultural production and to the distribution of agricultural products. The existence of such locational advantages is indicated by the high rating of the energy microclimate in voivodeships such as: Śląskie (which results from the specialisation in coal and steel production associated with the traditions of hard coal mining and steel production, as well as the largest urban complex in Poland). In addition, the Małopolska voivodeship received a high rating, which is related to the presence of the second largest city in Poland after Warsaw, i.e., Krakow. The city of Krakow is an important centre with good electricity supply conditions as a consequence of its industrial specialisation. The lowest rating was given to the Warminsko-Mazurskie voivodeship, which is located in north-eastern Poland, far from large Polish cities.

At the county level, counties that are part of the suburban zones of Polish urban agglomerations, as well as individual industrial centres, were rated highest in terms of energy values. The following counties belong to the most prominent ones in terms of energy values favourable for agriculture: Pruszkowski, Grodziski, Warszawski Zachodni, Legionowski, Piaseczyński, which surround the capital city of Poland and at the same time constitute its direct provisioning base. In addition, counties surrounding other large cities in Poland, notably Krakow, Wroclaw, Poznan, and Szczecin, received high scores based on this criterion. Interestingly, these cities form a network of regional level centres which occur in Poland with a regularity corresponding to the assumptions of Thünen's theory or the creator of the theory of central places, W. Christaller. This can therefore be linked to the type of hierarchical settlement network specific to Central Europe, which was the starting point for the formulation of assumptions about isotropic space in location theories.

Both the cartographic analysis and the level of location indices demonstrate the high validity of Thünen's rings theory. This is evidenced by the high weight of the market microclimate in the regional and local model of potential investment attractiveness. (Figures 1 and 2). In addition, the distribution of the most attractive spatial units shows a strong correlation with the location of markets both created by large cities and their provisioning zones.

To summarise: based on the analysis of the spatial diversification of the proposed index, it can be concluded that:

- The most attractive regions for investment in the agricultural sector are those with well-developed market outlets and well-equipped with technical infrastructure, characterised by a high level of agricultural culture,
- The historical factor and the new location arrangement of agricultural enterprises formed as a result of privatisation of state and cooperative farms have a great influence on the formation of location advantages for agricultural enterprises,
- The spatial diversification of investment attractiveness for agricultural enterprises is explained by the spatial diversification of settlement network distribution, in particular, the proximity of big cities, and also shows a relation to the spatial diversification of natural values (the quality of agricultural production space.

The next step was to analyse the results of the evaluation of the investment attractiveness and the actual distribution of agricultural enterprises in Poland. The analysis of correlation indices between the number of Section A enterprises operating in individual voivodeships and the potential attractiveness index (PAI REGION) showed a high correlation (0.75). The deeper analysis of the data also showed a significant importance of specialization in livestock production as a factor influencing the actual distribution of agricultural enterprises at the regional level. At the local level, this correlation could not be found due to the lack of data on the production specialization of each region.

It is worth emphasising that the Polish regions can serve as an example of a location environment for agricultural enterprises representative of other countries and regions in Central and Eastern Europe. This is the result of similar historical determinants of agricultural development in these countries, as well as contemporary trends. These include the determinants associated with the privatisation of agricultural state enterprises, as well as similar demographic problems in rural areas. Furthermore, the example of Poland allows for generalisations due to the fact that the long-term effects of the loss of sovereignty (the partitions of Poland) persist to this day in the Polish economic space, including in particular the division of space between three countries, i.e., Germany, Russia, and Austria.

Consequently, the settlement network and agricultural development were shaped by the settlement and agricultural policies of the three countries. This has led to significant disparities in Poland's regional development, which remain to this day.

5. Discussion

The issue of investment attractiveness of regions for Section A enterprises (agriculture, forestry, hunting and fishing) is relatively rarely studied in the literature. It is tackled, for example, in the Russian literature, but the conditions of the agricultural sector in Russia (operating under international economic sanctions and import substitution) are different from those typical of free-competitive economies. In addition, sometimes the analyses lack a proposal for an aggregate measure of investment attractiveness of regions for agricultural enterprises, they are descriptive in nature presenting the volume of production of agricultural crops, indicators of land use and livestock and some indicators of the socioeconomic development [41].

M. Vasilchenko, E. Derunova, in turn, analyse investment attractiveness in terms of agricultural organization, rather than spatial units. They define investment attractiveness as "ability to realise its innovative potential by attracting additional sources of investment and introducing new innovative forms and methods of investment policy" [62] (p. 513). The authors propose to use for assessment of investment attractiveness of agriculture an integrated indicator based on evaluation of innovative, scientific, intellectual, production and technological potential, natural resources, as well as taking into account the risks of financial and economic activity [62]. Their research approach refers to the entrepreneurial and sectoral context of institutional theory, not to the location theory.

To assess the investment attractiveness of the agro-industrial complex of the Lipetsk region (using a balanced scorecard), it was proposed factors characterizing the level of investment attractiveness of regional agribusiness are grouped into ten groups: the current state of investment activity, the potential of the agro-industrial complex, the group of indicators of staffing, production factors, the innovative development opportunities of the agro-industrial complex, the group of infrastructural factors of the investment process, the financial conditions for investment, the institutional conditions for making investments, consumer restrictions and social restrictions. One of the factors belonging to the group of infrastructural factors of the investment process was the level of energy supply, but it was not explained how this level should be measured at regional or local level [63].

Ukrainian researchers suggest using SWOT analysis to assess the investment attractiveness of the region for the agricultural sector, which is a useful proposition for analysing the investment attractiveness of a single selected region (it was presented on the basis of Cherkasy region) [64].

The literature also suggests internal and external factors determining investment attractiveness of agricultural enterprises at the national level without taking into account regional and local differences [65]. With regard to the quantitative measurement of investment attractiveness, the analyses look at the country's place in world attractiveness rankings and the level of the Total Support Estimate (TSE), which presents the amount of gross transfers received from taxpayers and consumers so that the government can implement policies to support agriculture [65]. In the research on investment potential of agriculture in Kazakhstan, the correlation–regression model was used, in which indicators that influence investment attractiveness of agriculture include: employed population in agriculture, agriculture expenses form state budget, GDP, gross output of agriculture, and number of agricultural enterprises [66]. The analysis of the Investment Attractiveness of the Agricultural Sector in Republic of Moldova was based mainly on financial data, i.e., financial statements of commercial banks and other financial institutions determining the involvement of these entities in investments in agriculture [67]. This approach also does not provide a basis for analysing the spatial differentiation of investment attractiveness, including the energy factor.

On the basis of the conducted literature research, it can be concluded that the issue of investment attractiveness of regions for agricultural enterprises has not yet been the subject of in-depth scientific analyses, which indicates the existence of a research gap in this area.

The presented research streams on the investment attractiveness of regions or the agricultural sector do not provide guidelines for measuring investment attractiveness as a spatially diversified phenomenon, nor do they provide insights into the current location factors of agricultural enterprises.

This gap is both theoretical (connected with a small number of scientific publications on investment attractiveness of regions for agricultural enterprises, which implies poor recognition of the problem) and methodological. Moreover, the lack of recognition in the literature of the influence of the energy factor on the investment attractiveness of regions for agricultural enterprises, as well as the lack of emphasis on the significance of this factor in shaping the level of investment attractiveness of regions was also demonstrated. This study contributes to reducing the identified research gap.

Therefore, the conducted analysis allows for generalizations regarding the directions of changes in spatial structure not only in the Polish economic area, but also in other European regions that underwent the system transformation in the 1990s. The long-term effects of the urbanization processes, denaturalization of food consumption, and increasing energy consumption are visible. The current importance of location factors for agricultural enterprises is related to historically-shaped sales markets created by large agglomerations, but natural values, access to technical infrastructure, and energy are also important at the local level.

It should be noted that the research results obtained have certain limitations. They are mainly related to the lower availability of data at the local level than at the regional level. Therefore, in this study, the construction of the investment attractiveness index at the local level is based on a smaller number of input variables than at the regional level. However, due to the smaller area (i.e., greater internal similarity), the investment attractiveness indices for local units are subject to a smaller cognitive error than the regional ones. Therefore, at the local level, the dependence of investment attractiveness on factors related to the proximity of markets, the quality of natural factors, access to technical infrastructure and the energy factor is more visible.

On the other hand, the investment attractiveness index at the regional level is based on a larger number of input variables than at the local level. While a limitation of the regional index is the greater internal differentiation of regions due to their larger area. Therefore, the regional indices take into account the internal differentiation of the location values for agriculture to a lesser extent.

Another limitation is the frequency of collecting detailed statistical data on agriculture based on the Agricultural Census, which is conducted every 10 years in all EU Member States. Currently, data from the last census, conducted in 2020, is not yet available. Moreover, it should be noted that some of the indices used to characterise agriculture are a generalisations of data collected on the basis of a selected sample, while maintaining its representativeness.

It should be stressed that the research method used has both advantages and disadvantages. The advantages of the adopted research method include the elimination of mutually correlated features, as well as taking into account the impact strength of individual diagnostic features on the final result without the need for the authors of the analysis to assign ranks subjectively. Omitting the ranks of sub-variables could risk distorting the assessment of investment attractiveness of spatial units.

The limitation of the weight correlation method is the lack of comparability of the computational results for the time series, since the vector of weights is different for each period of analysis. In addition, like most methods based on multicriteria analysis, this method also has drawbacks in the form of: discretion in the selection of input variables, lack of full access to spatial data and incomparability of data for units with changed administrative boundaries.

Another methodological limitation is the availability of regional and local databases, which varies across countries. The proposed model uses the quality index for agricultural production area developed by the IUNG (the index is available in Poland for all territorial units at both regional and local levels). Due to the fact that its development requires the assessment of soil quality, topography, and agroclimate, analogous indices may not be available for other European regions. Therefore, the possibility of valorisation of investment attractiveness for agricultural enterprises may be hampered by the limited availability of data.

Despite these limitations, the weight-correlation method is also gaining interest among researchers who create multi-criteria indices for example to assess lake trophic status [68] or monitoring and managing aquatic environment quality in regional eutrophic lakes around the world [69]. However, this is hardly the case for spatial analyses, especially those carried out at different taxonomic levels taking into account sectoral specificities.

Further research directions should be associated with the expansion of the scope of diagnostic variables and the verification of the statistical model with the application of other methods, which will allow the study of this phenomenon for European regions (including their verification based on the actual course of capital inflow into the sector of agricultural enterprises, taking into account the attractiveness class and specialization of production regions and development paths). It may be particularly interesting to look for development paths of regions with agricultural specialization that combine agricultural specialization with sustainable development and bioeconomy as a basis for complex regional development.

6. Conclusions

Current location trends in the activities of agricultural enterprises are still closely linked to the distribution of markets. This is especially true of enterprises similar in character to industrial enterprises, taking into account the use of equipment, which makes it possible to increase the scale of operations. This means that the J.H. Thünen agricultural location theory is still applicable. In this theory, the city was the main element of the organization of the spatial structure of agricultural production. Taking into account the growing importance of electricity in agricultural production—the availability, prices, and cost of obtaining energy should be included in the land rent.

In the location theory, attempts have been made to create a model that would allow optimization of decisions on the location of entrepreneurial activity. However, there is still no model approach that would allow for rationalisation of location decisions under the conditions of the global economy and the creation of extensive spatial structures by agricultural enterprises. There is, therefore, a need to develop a model that would allow the investment attractiveness of regions to be assessed at different taxonomic levels.

In light of the research results presented in this study, it can be concluded that the application of the valorisation model of investment attractiveness of regions developed by the authors allows the simultaneous comparison of any number of statistical units at the regional level at the scale of a country or a group of countries, subject to a harmonised statistical reporting system. On the basis of the model used for the location analysis of investment attractiveness of Polish regions, it can be stated that the location advantages and location factors can be explained on the basis of the path dependence theory. The region, which in the past was a food base for large markets, still has high investment attractiveness for agricultural activity due to the historically high level of agricultural culture and specialization in livestock breeding.

The research model based on the proposed sub-aggregates (microclimates) separately for the regional and local level, subject to the availability of data, is an original solution that so far has no counterparts in the literature. The correlation-weight method is also relatively rarely used. The results of the presented analysis indicate the validity of its wider application in spatial studies. An important argument in favour of its application is the achievement of non-modal solutions, which are nowadays the subject of exploration in decision-making analyses based on multi-criteria models.

When looking for model approaches, it is worth verifying the results of in-depth research on the example of a single country, provided that it is representative of spatial structures that go beyond the political borders of a single country. It is worth emphasising that the Polish regions can serve as an example of a location environment for agricultural enterprises representative of other countries and regions in Central and Eastern Europe. This is the result of the similarity between the historical and contemporary conditions of agricultural development in these countries.

Further work on finding links between the location decisions made and the investment attractiveness of regions for agricultural enterprises requires the standardisation of basic concepts. Due to the fact that there is no unambiguous, universally accepted definition of investment attractiveness in the literature. Investment attractiveness is a multifaceted notion, which translates into a lack of unanimity in its definition. Moreover, it has been proven that the issue of investment attractiveness of regions for agricultural enterprises has not been the subject of in-depth scientific analyses so far, which indicates the existence of a research gap in this field of theoretical and methodological nature. This is the achievement of the first specific objective (O1).

The authors therefore propose to define investment attractiveness as: understood as a set of regional location advantages that influence achieving investors' goals (such as costs of running a business, sales revenues, net return on investment and investment's competitiveness), understood as the elements of the geographical environment accumulated in a given part of space (natural, human, cultural resources, etc.), which become locational factors if they constitute the grounds for the investor's locational decision. Hence, for a particular enterprise, only those location advantages are relevant, which, at the same time, constitute important location factors for that enterprise. In the case of locational advantages, not only the socioeconomic characteristics, but also the natural features of the locational environment will be important for agricultural enterprises.

The author's potential investment attractiveness index for agricultural enterprises (PAI A) proposed in the article was enhanced in comparison with other proposals of the H. Godlewska-Majkowska's team by adding the following microclimates: energy, agriculture values (index for local units), as well as quality and determinants of agricultural production microclimates, and microclimate of agricultural production intensity (regional level). This is the achievement of the second specific objective (O2).

The applied components of the assessment of energy values, while showing a directly proportional relationship between investment attractiveness and the energy factor, indicate that the use of renewable energy sources by agricultural production entities contributes to increasing the sustainability of agricultural production, to increasing the self-sufficiency of entities in the sphere of energy supply, which reduces the economic risk of doing business. Due to the high internal differentiation of locational values at the regional level, this relationship is weaker in the case of NUTS2 regions, while it is stronger for local units. This confirms the research hypothesis (H1), which indicates that the main research objective has been achieved.

The results of the research can be used by local government units to develop and implement regional and local agricultural development programmes, as well as to shape investment attractiveness for agricultural enterprises. They can provide valuable practical guidance on what type of actions should be implemented to achieve synergies. This is particularly important in view of the fact that investments by agricultural enterprises can support the socioeconomic development of areas traditionally relying on the agricultural sector, in order to ensure sustainable regional and local development based on the bioeconomy.

Further research directions may be based on an attempt to generalise the location theory approach, which is based on the use of elements of the path dependence theory and complex region theory.

Author Contributions: Conceptualization, H.G.-M. and A.K.; Methodology, H.G.-M. and A.K.; Software, H.G.-M.; Validation, H.G.-M. and A.K.; Formal, Analysis, H.G.-M. and A.K.; Investigation, H.G.-M. and A.K.; Resources, H.G.-M. and A.K.; Data Curation, H.G.-M. and A.K.; Writing—Original Draft Preparation, H.G.-M. and A.K.; Writing—Review and Editing, A.K.; Visualization, H.G.-M.; Supervision, H.G.-M.; Project Administration, A.K.; Funding Acquisition, H.G.-M. and A.K. All authors have read and agreed to the published version of the manuscript.

Funding: The APC was funded by SGH Warsaw School of Economics (budget line KNoP/S19/14/19) and the University of Life Sciences, Faculty of Agrobioengineering (grant number RKZ/S/29/2021).

Institutional Review Board Statement: Not applicable.

Informed Consent Statement: Not applicable.

Data Availability Statement: Not applicable.

Conflicts of Interest: The authors declare no conflict of interest.

References

1. Thünen, J.H. *Der Isolierte Staat in Beziehung auf Landwirtschaft und Nationalökonomie*; Perthes: Hamburg, Germany, 1826.
2. Angelsen, A. *Forest Cover Change in Space and Time: Combining the Von Thünen and Forest Transition Theories*; World Bank Policy Research Paper 4117; World Bank: Washington, DC, USA, 2007.
3. Mäki, U. Realism and the Nature of Theory: A Lesson from J H von Thünen for Economists and Geographers. *Environ. Plan. A Econ. Space* **2004**, *36*, 1719–1736. [CrossRef]
4. O'Kelly, M.; Bryan, D. Argicultural location theory: Von Thunen's contributions to economic geography. *Prog. Hum. Geogr.* **1996**, *20*, 457–475. [CrossRef]
5. Fujita, M. Thünen and the New Economic Geography. *Reg. Sci. Urban. Econ.* **2012**, *42*, 907–912. [CrossRef]
6. Sinclair, R. Von Thünen and urban sprawl. *Ann. Am. Geogr. Assoc.* **1967**, *57*, 72–87. [CrossRef]
7. Wigier, M. Czas i przestrzeń w długookresowym rozwoju rolnictwa i obszarów wiejskich—ujęcie retrospektywne. *J. Agribus. Rural Dev.* **2012**, *2*, 263–272.
8. Roos, A.; Eggers, J.; Mark-Herbert, C.; Lindhagen, A. Using von Thünen rings and service-dominant logic in balancing forest ecosystem services. *Land Use Policy* **2018**, *79*, 622–632. [CrossRef]
9. Wagner, K. *Funktionen der Landwirtschaft in Stadtnahen Grünstrukturen. EU-COST Aktion C11 Greenstructure and Urban Planning*; AWI Bundesanstalt für Agarwirtschaft: Vienna, Austria, 2005.
10. Sroka, W. Resources and use of agricultural land in Polish cities according to chosen theories of location of agricultural production. *Acta Sci. Pol. Oeconomia* **2015**, *14*, 135–147.
11. Kuciński, K. *Geografia Ekonomiczna. Zarys Teoretyczny*; SGH: Warszawa, Poland, 2000; pp. 76–77.
12. Pepliński, B. Location of Cows and Pigs in Suburban Areas of Polish Metropolitan Centers. *Sustainability* **2020**, *12*, 2619. [CrossRef]
13. Maroušek, J.; Maroušková, A.; Kůs, T. Shower cooler reduces pollutants release in production of competitive cement substitute at low cost. *Energy Sources Part A* **2020**, 1–10. [CrossRef]
14. Maroušek, J.; Bartoš, P.; Filip, M.; Kolář, L.; Konvalina, P.; Maroušková, A.; Moudrý, J.; Peterka, J.; Šál, J.; Šoch, M.; et al. Advances in the agrochemical utilization of fermentation residues reduce the cost of purpose-grown phytomass for biogas production. *Energy Sources Part A Recover. Util. Environ. Eff.* **2020**. [CrossRef]
15. Maroušek, J.; Rowland, Z.; Valášková, K.; Král, P. Techno-economic assessment of potato waste management in developing economies. *Clean Technol. Environ.* **2020**, *22*, 937–944. [CrossRef]
16. Ugochukwu, A.I.; Phillips, P.W.B. Technology Adoption by Agricultural Producers: A Review of the Literature. In *From Agriscience to Agribusiness. Innovation, Technology, and Knowledge Management*; Kalaitzandonakes, N., Carayannis, E., Grigoroudis, E., Rozakis, S., Eds.; Springer: Cham, Switzerland, 2018.
17. European Commission. Key Enabling Technologies. Available online: https://ec.europa.eu/info/research-and-innovation/research-area/industrial-research-and-innovation/key-enabling-technologies_en (accessed on 26 February 2021).
18. Chen, H.; Yada, R. Nanotechnologies in agriculture: New tools for sustainable development. *Trends Food Sci. Technol.* **2011**, *22*, 585–594. [CrossRef]
19. Nazzaro, F.; Fratianni, F.; Coppola, R. Microtechnology and nanotechnology in food science. In *Green Technologies in Food Production and Processing*; Boye, J., Arcand, Y., Eds.; Springer: Boston, MA, USA, 2012.
20. Uzogara, S.G. The impact of genetic modification of human foods in the 21st century: A review. *Biotechnol. Adv.* **2000**, *18*, 179–206. [CrossRef]
21. Islam, R.; Parvin, A.; Billah, M.M.; Islam, M.; Imran, M.A.S.; Sarker, R.K.; Uddin, M.E.; Alam, M.S.; Abedin, M.Z. Assessment of the Effects of Genetically Modified (GM) Foods: A Brief Study on Health and Environmental Concerns. *J. Mater. Environ. Sci.* **2020**, *11*, 1676–1688.
22. Jha, K.; Doshi, A.; Patel, P.; Shah, M. A comprehensive review on automation in agriculture using artificial intelligence. *Artif. Intell. Agric.* **2019**, *2*, 1–12. [CrossRef]

23. Smith Matthew, J. Getting value from artificial intelligence in agriculture. *Anim. Prod. Sci.* **2018**, *60*, 46–54. [CrossRef]
24. Gupta, A.; Ponticelli, J.; Tesei, A. *Information, Technology Adoption and Productivity: The Role of Mobile Phones in Agriculture. National Bureau of Economic Research*; NBER Working Paper Series, Working Paper 27192; National Bureau of Economic Research: Cambridge, MA, USA, 2020.
25. Fafchamps, M.; Minten, B. Impact of SMS-based Agricultural Information on Indian Farmers. *World Bank Econ. Rev.* **2012**, *26*, 383–414. [CrossRef]
26. Tzounis, A.; Katsoulas, N.; Bartzanas, T.; Kittas, C. Internet of Things in agriculture, recent advances and future challenges. *Biosyst. Eng.* **2017**, *164*, 31–48.
27. Dlodlo, N.; Kalezhi, J. The internet of things in agriculture for sustainable rural development. In Proceedings of the International Conference on Emerging Trends in Networks and Computer Communications (ETNCC), Windhoek, Namibia, 17–20 May 2015; pp. 13–18.
28. Zhang, N.; Wang, M.; Wang, N. Precision agriculture—A worldwide overview. *Comput. Electron. Agric.* **2002**, *36*, 113–132. [CrossRef]
29. Tey, Y.S.; Brindal, M. Factors influencing the adoption of precision agricultural technologies: A review for policy implications. *Precis. Agric.* **2012**, *13*, 713–730.
30. Trivelli, L.; Apicella, A.; Chiarello, F.; Rana, R.; Fantoni, G.; Tarabella, A. From precision agriculture to Industry 4.0: Unveiling technological connections in the agrifood sector. *Br. Food J.* **2019**, *121*, 1730–1743. [CrossRef]
31. Pechenevsky, V.; Zakshevsky, G. Prospects for the spatial development of agricultural production in the region. In Proceedings of the International Scientific and Practical Conference on Agrarian Economy in the Era of Globalization and Integration, Moscow, Russia, 25–28 October 2018.
32. Akhmetshin, E.M.; Artemova, E.I.; Vermennikova, L.V.; Shichiyakh, R.A.; Prodanova, N.A.; Kuchukova, N.M. Management of Investment Attractiveness of Enterprises: Principles, Methods, Organization. *Int. J. Appl. Bus. Econ. Res.* **2017**, *15*, 71–82.
33. Davydenko, N.; Skrypnyk, H. Evaluation Methods of Investment Attractiveness of Ukrainian Agricultural Enterprises. *Balt. J. Econ. Stud.* **2017**, *3*, 103–107. [CrossRef]
34. Vologdin, E.V. Methodical and Practical Aspects of Assessing the Investment Attractiveness of the Region, Novosibirsk. Available online: www.searc.rsl.ru (accessed on 8 January 2021).
35. Gawlikowska-Hueckel, K.; Umiński, S. *Ocena Konkurencyjności Województw*; Instytut Badań nad Gospodarką Rynkową: Gdańsk, Poland, 2000; p. 7.
36. Czerwieniec, E. Bezpośrednie inwestycje zagraniczne w środowisku lokalnym. *Zesz. Nauk. Pr. Katedr. Mikroekonomii Akad. Ekon. W Pozn.* **2007**, *94*, 7–23.
37. Kalinowski, T. *Atrakcyjność Inwestycyjna Województw i Podregionów Polski 2006*; Instytut Badań nad Gospodarką Rynkową: Gdańsk, Poland, 2006; p. 13.
38. Shilov, E.; Kukhlenko, O.; Goyko, A. Методы Оценки Эффективности иПривлекательности ИнвестиционныхПроектов; Kitusa: Kiev, Ukraine, 1996; p. 14.
39. Jumaev, N.X.; Rakhmonov, D.A.; Sulaymanov, S.A. Ways to Attract Investment with Assessment of Investment Potential of the Regions. *Int. J. Econ. Commer. Manag.* **2019**, *7*, 664–673.
40. Ponin, A.S. *Management of Investment Attractiveness in Region*; RAGS: Moscow, Russia, 2000; p. 175.
41. Semin, A.; Kibirov, A.; Rassukhanov, U. Problems and Main Mechanisms to Increase Investment Attractiveness of Agricultural Production. *Eur. Res. Stud. J.* **2018**, *21*, 378–400. [CrossRef]
42. Godlewska–Majkowska, H. *Atrakcyjność Inwestycyjna Polskich Regionów. W Poszukiwaniu Nowych Miar*; Studia i Analizy Instytutu Przedsiębiorstwa, Oficyna Wydawnicza SGH: Warszawa, Poland, 2008.
43. Godlewska-Majkowska, H. *Atrakcyjność Inwestycyjna Regionów Polski a Kształtowanie Lokalnych i Regionalnych Specjalizacji Gospodarczych*; Studia i Analizy Instytutu Przedsiębiorstwa, Oficyna Wydawnicza SGH: Warszawa, Poland, 2009.
44. Weber, A. *Über den Standort der Industrien. I Teil, Reine Theorie des Standorts*; Verlag von J. C. B. Mohr (Paul Siebeck): Tübingen, Germany, 1909.
45. Kramin, M.V.; Safiullin, L.N.; Kramin, T.V.; Timiryasova, A.V. Drivers of economic growth and investment attractiveness of Russian regions. *Life Sci. J.* **2014**, *11*, 526–530.
46. Zachorowska, A. *Ryzyko Działalności Inwestycyjnej Przedsiębiorstw*; PWE: Warszawa, Poland, 2006; pp. 115–116.
47. Dzikowska-Zawirska, J.; Włodarczyk-Guzek, W. Napływ zagranicznych inwestycji bezpośrednich ze szczególnym uwzględnieniem Łódzkiej Specjalnej Strefy Ekonomicznej. *Acta Univ. Lodz. Folia Oeconomica* **2002**, *158*, 53–83.
48. Budner, W. *Lokalizacja Przedsiębiorstw. Czynniki Ekonomiczno-Przestrzenne i Środowiskowe*; Wydawnictwo Akademii Ekonomicznej w Poznaniu: Poznań, Poland, 2003; pp. 28–30.
49. Dziemianowicz, W. *Bezpośrednie Inwestycje Zagraniczne—Między Teorią a Praktyką*; Regionalne Centrum Informacji Integracji Europejskiej przy Akademii Ekonomicznej w Krakowie: Kraków, Poland, 2002.
50. Kliestik, T.; Valaskova, K.; Nica, E.; Kovacova, M.; Lazaroiu, G. Advanced methods of earnings management: Monotonic trends and change-points under spotlight in the Visegrad countries. *Oeconomia Copernic.* **2020**, *11*, 371–400. [CrossRef]
51. Dorożyński, T.; Kuna-Marszałek, A. Investment Attractiveness. The Case of the Visegrad Group Countries. *Comp. Econ. Res.* **2016**, *19*, 119–140. [CrossRef]

52. Valinurova, L.S.; Kazakova, O.B.; Sulimova, E.A. Evaluation of Investment Attractiveness and Prediction of Investment Volumes to the Region. *Mediterr. J. Soc. Sci.* **2015**, *6*, 371–379. [CrossRef]
53. Godlewska-Majkowska, H.; Perło, D. Zastosowanie metody wagowo-korelacyjnej i modelowania miękkiego do analizy atrakcyjności inwestycyjnej powiatów w Polsce. *Zarządzanie I Finans. J. Manag. Financ.* **2017**, *15*, 187–214.
54. Snieska, V.; Zykiene, I. City attractiveness for investment: Characteristics and underlying factors. *Procedia Soc. Behav. Sci.* **2015**, *213*, 48–54. [CrossRef]
55. Zarębski, P. Atrakcyjność inwestycyjna obszarów wiejskich w Polsce. *Zesz. Nauk. Sggw Warszawie. Probl. Rol. Światowego* **2012**, *12*, 169–178.
56. Komor, A. Atrakcyjność inwestycyjna obszarów wiejskich województwa lubelskiego. *Zesz. Probl. Postępów Nauk Rol.* **2009**, *542*, 1039–1046.
57. Godlewska-Majkowska, H.; Komor, A.; Turek, D.; Zarębski, P.; Czernecki, M.; Typa, M. *Regional Investment Attractiveness 2016, Report Prepared for the Polish Information and Foreign Investment Agency at the Institute of Enterprise*; Warsaw School of Economics: Warsaw, Poland, 2016. Available online: https://www.paih.gov.pl/publications/regions (accessed on 2 April 2021).
58. Godlewska-Majkowska, H. *Atrakcyjność Inwestycyjna Regionów Polski Jako Źródło Przedsiębiorczych Przewag Konkurencyjnych*; Studia i Analizy Instytutu Przedsiębiorstwa, Oficyna Wydawnicza SGH: Warszawa, Poland, 2012.
59. Central Statistical Office (GUS). Available online: www.stat.gov.pl (accessed on 5 February 2021).
60. Brzozowski, T.; Pogorzelski, W. Naturalne wartości wskaźników jakości, metoda wagowo-korelacyjna. *Wiadomości. Statystyczne GUS* **1992**, *4*, 42–44.
61. Quality Index for Agricultural Production Area (Wskaźnik Waloryzacji Rolniczej Przestrzeni Produkcyjnej—WWRPP), Institute of Soil Science and Plant Cultivation in Puławy (IUNG). Available online: http://onw.iung.pulawy.pl/specyficzne/wwrpp (accessed on 2 April 2021).
62. Vasilchenko, M.; Derunova, E. Factors of Investment Attractiveness of Russian Agriculture in the Context of Innovative Structural Adjustment. *Sci. Pap. Ser. Manag. Econ. Eng. Agric. Rural Dev.* **2020**, *20*, 511–522.
63. Esina, Y.L.; Stepanenkova, N.M. Methodology for Assessing Investment Attractiveness of AIC of the Lipetsk Region. Available online: https://www.elibrary.ru/item.asp?id=42869373 (accessed on 2 April 2021).
64. Davydenko, N.; Buriak, A.; Demianenko, I. Investment Attractiveness of Agricultural Sector of Ukrainian Region in Modern Conditions of Social and Economic Development. *Balt. J. Econ. Stud.* **2018**, *4*, 106–110. [CrossRef]
65. Kulynych, Y.M.; Shirinyan, L.V. The Investment Attractiveness of the Agricultural Sector of Ukraine. Економіка **2018**, *12*, 211–216.
66. Tankiyeva, A.K. Myth and Reality of Investment Attractiveness of Agriculture in Kazakhstan. *Econ. Agric.* **2016**, *6*, 149–155.
67. Timofti, E.; Popa, D.; Petrascu, S.; Osborne, P. The Investment Attractiveness of the Agricultural Sector in Republic of Moldova in Terms of European Integration. *Probl. World Agric.* **2018**, *18*, 315–322.
68. Maofei, N.; Qiushi, G.; Siyue, L.; Zhikang, W.; Yunjie, W. Trophic state index linked to partial pressure of aquatic carbon dioxide in atypical karst plateau lake. *Ecol. Indic.* **2021**, *120*, 106912.
69. Liu, D.; Du, Y.; Yu, S.; Luo, J.; Duan, H. Human activities determine quantity and composition of dissolved organic matter in lakes along the Yangtze River. *Water Res.* **2020**, *168*, 115132. [CrossRef] [PubMed]

energies

MDPI

Article

The Patterns of Energy Innovation Convergence across European Countries

Tomasz Kijek [1,*], Arkadiusz Kijek [2], Piotr Bolibok [3] and Anna Matras-Bolibok [1]

[1] Department of Microeconomics and Applied Economics, Maria Curie-Sklodowska University, Plac Marii Curie-Sklodowskiej 5, 20-031 Lublin, Poland; anna.matras@umcs.pl
[2] Department of Statistics and Econometrics, Maria Curie-Sklodowska University, Plac Marii Curie-Sklodowskiej 5, 20-031 Lublin, Poland; akijek@poczta.umcs.lublin.pl
[3] Department of Economic Policy and Banking, The John Paul II Catholic University of Lublin, Al. Raclawickie 14, 20-950 Lublin, Poland; piotr.bolibok@kul.pl
* Correspondence: tomasz.kijek@umcs.pl

Abstract: Energy innovation is critical for addressing climate change and the ecological transitions of both developed and emerging economies. The present paper aims at the identification and assessment of patterns in energy innovation convergence across a sample of 27 European countries over the period 2000–2018. The research is based on data covering a broad category of patents related to climate change mitigation technologies in the energy sector, including combustion inventions with mitigation potential (e.g., using biomass), extracted from the Organisation for Economic Co-operation and Development (OECD) Statistical Database. Using a nonlinear time-varying factor model, the paper demonstrates that energy innovation efforts in the examined sample follow a pattern of club convergence. The findings allow the identification of three convergence clubs characterised by distinct disparities in energy patent intensity, as measured by the number of patent applications per 10 million inhabitants. Moreover, the results of an ordered logit model demonstrate that the emergence of the identified convergence clubs might be attributable to initial differences in per capita environmental research and development (R&D) expenditure, human resources in science and technology (HRST), and environmental policy stringency. The findings have important policy implications as they suggest the need for more tailored policies based on smart development and specialization frameworks designed to boost the energy innovation performance of the laggard countries, more fully exploiting the potential of their less technologically advanced sectors, such as agriculture.

Keywords: energy innovation; energy patents; convergence; club convergence; R&D expenditure; HRST; environmental policy stringency

Citation: Kijek, T.; Kijek, A.; Bolibok, P.; Matras-Bolibok, A. The Patterns of Energy Innovation Convergence across European Countries. *Energies* **2021**, *14*, 2755. https://doi.org/10.3390/en14102755

Academic Editor: Dalia Štreimikienė

Received: 30 March 2021
Accepted: 7 May 2021
Published: 11 May 2021

Publisher's Note: MDPI stays neutral with regard to jurisdictional claims in published maps and institutional affiliations.

1. Introduction

Facing the constantly growing consumption of energy in the world on the one hand, and the scarcity of natural resources and the looming perspective of climate change on the other, the search for new sources of energy, increasing the use of renewable ones, and improving their efficiency inevitably become the central issues of sustainable development and ecological transition of both developed and emerging economies. Reduction of the harmful environmental impacts of energy production and consumption is of crucial relevance from the standpoint of policies aiming at mitigation of adverse consequences of climate change. Not surprisingly, therefore, the issues related to those processes constitute the most important dimensions of contemporary environmental protection frameworks [1].

The intensity and efficiency of innovative activities in the field of energy become critical for addressing key challenges related to environmental protection and ensuring a more sustainable consumption of natural resources, such as energy security, combating pollution or limiting global warming. Other vital challenges in the area of energy include improving access to modern energy carriers, in particular electricity, and the security and resiliency of energy supply and distribution systems [2]. The development of more efficient and less polluting technologies related to energy use, supply and conversion is, therefore, undoubtedly one of the most important and socially desirable directions of international technological progress. Given the complexity and turbulent nature of the contemporary socio-economic environment, successful energy innovations often result from collective learning processes combining knowledge, skills, R&D and the deployment efforts of suppliers and users of particular technologies. It is worth pointing out, however, that such processes are usually possible only in specific contexts and within particular incentive structures [3]. Mutual relationships and feedback between the economic, environmental, and political dimensions of energy efficiency and sustainability render the ecological transition of the energy sector a particularly difficult issue. Every energy strategy must accommodate a multitude of often conflicting goals related to security, reliability, ecological performance, and costs of possible energy sources [4].

Energy innovation processes are often impeded by intrinsic structural weaknesses which tend to hamper both demand for the new technologies and the short-term business prospects of their potential providers. Firstly, the large scale of necessary investment outlays as well as significant technological and regulatory inertia of existing energy systems render the lead times needed to provide new technologies to mass market use particularly long. Secondly, new energy technologies are usually more expensive and not necessarily more effective than the existing substitutes, which likely slows down the pace of market penetration. Moreover, in the particular context of eco-innovations in the field of energy, the direct benefits accrue primarily to society as a whole, rather than the final users. Finally, energy innovations typically have to confront a multitude of barriers to entry, including, in particular, incompatibility of existing network infrastructure, extensive market power of key competitors, price controls or unstable regulatory frameworks. In the light of the above difficulties, successful market implementation of energy innovations largely depends on public policy support [5] (p. 3).

The so-called Porter hypothesis claims that "the properly designed environmental standards can trigger innovation that may partially or more than fully offset the costs of complying with them" [6] (p. 98). What is worth pointing out, however, is that innovation is likely driven not only by the quantity of regulations but primarily by their stringency. Additionally, as Fabrizi et al. [7] demonstrate, the effectiveness of environmental regulation policies can be increased by combining them with appropriate innovation policies. The actual impact of environmental regulation on innovation performance has been explored by an increasing bulk of studies, e.g., [8,9]. Furthermore, as a type of an environmental innovation, energy innovation has a "double externality" nature. As Rennings [10] (pp. 325–326) stresses, environmental innovation reduces negative environmental externalities and it is subject to externalities arising from knowledge spill-overs involving both environmental and standard innovation processes. Both these externalities, however, result in sub-optimal investment in environmental innovation, thus indicating the importance of the regulatory framework.

The general directions for ecological transition of energy sectors worldwide result from the Paris Agreement on climate change adopted in 2015 by nearly 200 countries [11]. The challenges related to the mitigation of adverse consequences of climate change increase the importance of innovation in all the major areas of contemporary energy policies, i.e., energy conversion, distribution and use. Effectiveness of energy innovation impacts a broad spectrum of energy development policy goals, including energy security, access, cost, international competitiveness, modernization of energy systems and reduction of adverse environmental impact [12]. At the national level the development of energy innovation policy is not only constrained by existing institutional, economic and social factors, but also involves multiple stakeholders, often with conflicting interests. In turn, policy guidelines shape each country's energy innovation development and deployment models. In the context of the EU, a policy framework for energy research and innovation activities is outlined in several strategic documents: the Strategic Energy Technology Plan (originally issued in 2007 [5] and revised in 2015 [13]), and the 'Accelerating Clean Energy Innovation' communication from the European Commission, adopted as an integral part of the 'Clean energy for all Europeans' package [14], following the Paris Agreement [11]. Given the fact that contemporarily energy is responsible for more than 75% of the EU's greenhouse gas (GHG) emissions [15], energy innovations become critically important for successful transition towards climate neutrality. To tackle the key environmental and climate-related challenges by decoupling economic growth from resource use and achieving climate neutrality (no net emissions of GHG) by 2050, a new EU strategy, the *European Green Deal*, was designed [16]. The strategy strongly emphasizes the role of cross-border and regional cooperation in achieving the benefits of clean energy at affordable prices, as well as the need for efficient regulatory framework and financing schemes to foster the deployment of innovative energy technologies and infrastructure. The research and innovation efforts in the field of energy are to be supported by the full range of instruments available under the *Horizon Europe* programme [17]. Given the specificity and the aforementioned structural weaknesses of energy innovation processes, the programme aims at fostering initiatives designed to combine societal pull and technology push effects.

Although energy innovation leading to transformational changes in energy sector is vital for limiting the adverse consequences of global climate change, no single country seems capable of addressing all the related energy and environmental challenges alone [18]. As demonstrated by Costantini et al. [19], the speed of innovation in the renewable energy sector is higher if more countries are engaged in R&D and invention activities. Innovative capacity, however, is not uniformly distributed across countries, which in turn results in significant disparities both in the actual effectiveness of R&D efforts, as well as in general approach to the creation of new knowledge. This problem is particularly important in the context of the European Union, which has set convergence across the Member States as one of its key priorities, and recognized innovation policy as a fundamental instrument in reaching this goal [20]. Moreover, as argued by Archibugi and Coco [21], reduction of cross-country disparities in innovative capacity is also a vital condition for boosting the global competitiveness of the EU's economy.

As economic growth is driven primarily by technological progress and innovation [22], long-run economic convergence is largely dependent on technological convergence. According to Jungmittag [23], given varying production technologies across countries, the convergence of national innovation capabilities (i.e., adoption and accumulation of technologies) is a *sine qua non* condition of the convergence in terms of labour productivities and per capita incomes. The convergence of labour productivities is largely driven by the diffusion of technologies, which in turn becomes a crucial determinant of economic growth for the catching-up countries. At the same time, for the advanced economies, transferable technological knowledge is the level of Ricardian technological specialization. In turn, larger differences in the level of technological specialisation are likely to impede the process of convergence.

Although economic integration fosters dissemination of innovative infrastructure, it may also exert the exactly opposite effect on the very creation of new knowledge and innovations, which tend to agglomerate in the most developed regions [20]. The return on investment in technological research usually increases in the areas where other research activities take place [24], in particular due to "agglomeration effects" and other kinds of positive spillovers and externalities resultant from geographical proximity [25]. Inventive firms and researchers are, therefore, often attracted to locations of intense innovative activities in a given field, where the returns on new knowledge tend to be much larger than in a less competitive environment of laggard regions [26].

Following Sharp [27], convergence in terms of innovation performance becomes an important driver of successful integration, as innovations foster not only economic performance, but also general socio-political cohesion. The latter notion is particularly important from the standpoint of overcoming aforementioned structural weaknesses of energy innovation.

Given the above, patterns of energy innovation convergence might shape the progress in reaching the policy goals regarding mitigation of the adverse consequences of climate change in Europe. Investigation of these patterns in the long run not only becomes an interesting research problem, but also might have important policy implications.

An assessment of the outcomes of innovation activities in the field of energy oriented towards mitigation of the adverse consequences of climate change is, however, not an easy task. One of the approaches to the above issue most commonly adopted in the relevant literature is based on the analyses of patent intensity, see e.g., [26,28,29], as measured by the number of energy patent applications per a given number of inhabitants [30].

The convergence in terms of patenting activity implies that countries exhibiting lower initial levels of patent intensity over time increase their innovative capacity, achieving higher rates of growth in per capita patent applications than their counterparts in the examined sample. This in turn allows them to gradually reduce their distance from the leaders.

The fact that knowledge is considered to be largely a public good might however render the issue of convergence in patent activity less important, since many countries may simultaneously benefit from their creation in one of them. Notwithstanding the above notion, several arguments of political and economic nature supporting the view that such a process is desirable might be brought up [17].

From an economic perspective, convergence in terms of patenting activity might indicate the improvement of innovation absorption capacity across the examined sample of countries, i.e., their ability to successfully adopt, adapt and implement knowledge created elsewhere. This capacity is, in turn, crucial not only from the standpoint of individual economies, as it enables them to guide their innovation efforts with respect to the conditions of the local markets for factors of production and improve their innovative productivity, but it also determines the directions and scale of international technology flows, see e.g., [31,32]. Following Cohen and Levinthal [33], it is worth pointing out, however, that the potential gains from technological spillovers are largely determined by the given country's past experience in relevant R&D. The improvement of innovation absorption capacity is also of crucial importance for the less technologically advanced economies, as it allows them to strengthen and expand their innovative potential and improves their resilience to external shocks.

The political importance of convergence in energy patent intensity, and in particular in the area of climate change mitigation technologies, results from its potential negative relationship with the scale of free-riding on innovation between countries. As demonstrated by Bosetti et al. [34], international knowledge spillovers typically encourage free-riding on already developed technologies, which likely crowds out domestic R&D investments. Higher convergence in energy patent intensity in the area of climate change mitigation technologies may therefore contribute to the limitation of innovation free-riding across countries. It may also reflect both the increasing engagement in the ecological transition of their energy sectors and public acceptance for the necessary costs of this process. In contrast, lower convergence within a largely homogenous regulatory environment suggests that some countries tend to free-ride on environmental-friendly solutions developed elsewhere. This in turn increases the risk that the innovation leaders might become discouraged from bearing disproportionately high costs of ecological transformation, which would make the achievement of the established energy policy targets even more difficult [35].

A vast majority of studies addressing convergence in the area of innovative capacity investigate the general dimension of these processes, abstracting from their course in particular technology fields, see e.g., [36–39]. Even though the relevant literature on energy innovation seems quite extensive (among others: [40–44]), to date only a couple of studies have directly addressed the problem of convergence in this area.

Using the data for 13 EU member countries over the period of 1990–2012, Grafström [26] found the evidence of conditional β- and σ-divergence in renewable energy innovation capabilities (patent applications per capita).This means that both the gap in patent intensity between innovation leaders and laggard countries and its dispersion increased in the examined period. It also implies that some EU countries tend to free-ride on the development efforts of other Member States. More recently, Bai et al. [45] examined trends in the renewable energy technology innovation (RETI) levels, as measured by the number of patents granted, adjusted for technology depreciation and diffusion, across the provinces of China over the period of 1997–2015 and found the evidence of club convergence. Their results demonstrate that over the examined period thirty provinces converged to three clubs characterized with significant disparities both in the level and the annual growth rate of RETI.

Given the above considerations and largely limited prior empirical evidence, the present study aims at the identification and assessment of patterns in energy innovation convergence in the area of climate change mitigation technologies across European countries.

The paper contributes to the relevant literature in three ways.

First, bearing in mind the complexity and multidimensionality of energy innovation, the study investigates a broader and more comprehensive category of patent applications related to climate change mitigation technologies in the energy sector that have sought protection in at least two jurisdictions. Such an approach allows reflection upon the relevant outcomes of R&D in the field of clean and energy saving technologies, irrespective of the industry in which they are introduced, which makes it a useful, direct and comprehensive proxy of the inventive activities oriented towards energy, e.g., [1]. Moreover, the paper examines a larger set of countries and a longer time span than prior research in the European context.

Second, the significant disparities in the innovative capacity between European countries, and the specificity of their individual development paths, render absolute convergence in terms of energy patent intensity in the field of climate change mitigation technologies highly unlikely. Therefore, given the historical, political, and socio-economic factors shaping the directions of technological progress in Europe, it can be hypothesized that patent intensity in the above area is characterised by the presence of convergence clubs. Given the above, the study makes an original attempt to delineate the related convergence clubs using the regression t test proposed by Phillips and Sul [46].

Third, the paper identifies and assesses the key determinants of the hypothesized club convergence. Given the evidence in the prior studies, it is likely that the energy innovation convergence paths are driven primarily by initial levels of the following factors: R&D, human capital and environmental policy-related measures. Therefore, the paper attempts to explain the emergence of the convergence clubs using the logit model by McKelvey and Zavoina [47].

The obtained results allowed the identification of three convergence clubs characterised by distinct disparities in energy patent intensity. The paper also demonstrates that the emergence of the identified convergence clubs might be attributable to the initial differences in per capita environmental R&D expenditure, HRST, and environmental policy stringency.

The remainder of the paper is organised as follows. Section 2 presents the methodological framework of the study and the details of the data selection procedures. Sections 3 and 4 present and discuss the results of the empirical analyses. The paper ends with conclusions recapitulating its main findings along with policy recommendations and suggestions for future research.

2. Materials and Methods

The examined sample covers 27 European countries, including 24 EU Member States, GB, Norway, and Switzerland, over the period 2000–2018, as determined by the availability of data on energy patent applications in the OECD Patent Database. [48]. Although other patent databases (see e.g., the World Intellectual Property Organization Patent Database) offer more recent data, the patent statistics presented in the OECD Patent Database are constructed using algorithms, which allows for the precise identification of climate change mitigation technologies related to energy generation, transmission or distribution. These technologies pertain to renewable energy generation, energy generation from fuels of non-fossil origin, nuclear energy, combustion inventions with mitigation potential (e.g., using biomass), inventions for efficient electrical power generation, transmission or distribution, and inventions with potential or indirect contribution to GHG emission mitigation. Therefore, relying on a single data source allows for the avoidance of potential issues related to data comparability. The number of inventions related to energy generation, transmission or distribution was identified by:

- Inventor country—fractional counts by country of residence of the inventor(s).
- Family size—"2 and greater", which counts only the higher-value inventions that have sought patent protection in at least two jurisdictions.
- Priority date—the first filing date worldwide.

Regarding factors potentially affecting the process of convergence club formation, the Eurostat and OECD datasets were used. The former includes R&D related to environmental protection per capita and human resources in science and technology (i.e., persons with tertiary education as percentage of active population). The latter relates to the Environmental Policy Stringency Index (EPS). It measures the degree to which environmental policies set a real or shadow price on environmentally undesirable activities primarily related to climate and air pollution. The index is scaled from zero to six, where six indicates the highest degree of stringency. The data on initial conditions refers to 2000.

To find convergence patterns in energy patent intensity across European countries, a regression t test proposed by Phillips and Sul [46] was applied. The test is based on the time varying factor representation of the convergence variable:

$$X_{it} = \delta_{it}\mu_t, \tag{1}$$

where μ_t is the common factor and δ_{it} is the time varying idiosyncratic distance from the common factor. In this study, X_{it} refers to energy patent intensity, as measured by the number of patent applications per 10 million inhabitants. The time varying element δ_{it} is modelled in semi-parametric form as:

$$\delta_{it} = \delta_i + \sigma_i\xi_{it}L(t)^{-1}t^{-\alpha}, \tag{2}$$

where δ_i is the time-invariant part of δ_{it}, σ_i is the idiosyncratic scale parameter, ξ_{it} is iid(0, 1) across i and weakly dependent over t, and $L(t)$ is a slowly varying function for which $L(t) \to \infty$ as $t \to \infty$.

Relative loading coefficient:

$$h_{it} = \frac{X_{it}}{N^{-1}\sum_{i=1}^{N} X_{it}} = \frac{\delta_{it}}{N^{-1}\sum_{i=1}^{N} \delta_{it}}, \tag{3}$$

measures the relation of the loading coefficient δ_{it} to the panel average at time t. As the cross sectional mean of h_{it} is unity, its variance is given by:

$$H_t = \frac{1}{N}\sum_{i=1}^{N}(h_{it} - 1)^2. \tag{4}$$

The convergence is present if $H_t \to \infty$ as $t \to \infty$.

Considering the approach of Philips and Sul [46], the null hypothesis of the convergence test is formulated as follows:

$$H_0 : \delta_i = \delta \text{ and } \alpha \geq 0 \text{ against } H_1 : \delta_i \neq \delta \text{ for all } i \text{ or } \alpha < 0. \quad (5)$$

The testing procedure consists of the following steps:

1. Calculation of cross-sectional variance ratios H_1/H_t ($t = 1, 2, \ldots, T$).
2. Estimation of the following regression:

$$\log\left(\frac{H_1}{H_t}\right) - 2\log L(t) = a + b\log t + u_t, \text{ for } t = (rT), (rT)+1, \ldots, T, \quad (6)$$

   where $r \in (0, 1)$. Following the results of their simulations, Philips and Sul [46] recommend the use of $r \in (0.2, 0.3)$. When T is small, $r = 0.2$ is preferred, and if T is large, $r = 0.3$ is better choice.
3. Application of autocorrelation and a heteroskedasticity robust one-sided t test to verify the null hypothesis $\alpha \geq 0$ using $\hat{b} = 2\hat{\alpha}$ and a HAC standard error. At a standard significance level (0.05), the null hypothesis is rejected if $t_{\hat{b}} < -1.65$.

Rejection of the null hypothesis means that there is no convergence in the group of all panel units. It does not imply, however, that there is no evidence of convergence in subgroups of units (i.e., club convergence). Philips and Sul [46] propose a specific procedure for testing club convergence. The algorithm includes four steps. First, the units are arranged in descending order with respect to the last period. Next, a core group is formed by adding countries one after another to a group of the two highest-patent countries at the start and performing the log t test up until the $t_{\hat{b}}$ for this group is larger than -1.65. Then, the log t test is performed again for this group and all the other units (one after another) forming the sample to determine if they converge. If they do not converge, the first three steps are performed for the all the other units. In the case that no clubs are identified, it means that those units diverge.

In order to explain the process of club formation within the sample of European countries, an ordered logit model pioneered by McKelvey and Zavoina [47] was used. This model designates every country to a particular club and allows for explaining variation in an ordered categorical dependent variable (i.e., belonging to alternative clubs ranked in line with the steady-state energy patent intensity of every club) as a function of independent variables.

3. Results

The log t test used for the whole sample indicates that the hypothesis of overall convergence can be rejected at the 5% significance level (-6.2339). As a consequence, the procedure for testing club convergence was applied. Table 1 shows summary results for the clustering and merging test procedures (i.e., the number of clubs and countries belonging to the particular club, the estimated parameters, and the standard errors).

Table 1. Summary results for the log t test.

Club	No. of Countries	$\hat{b}$	SE	t
1	8	0.2321	0.6459	0.3594
2	11	−0.2362	0.2122	−1.1127
3	6	−0.2888	0.2546	−1.1347

The results of the analysis allowed the identification of 3 clubs and two non-converging countries (Denmark and Romania). Club 1, with the lowest energy patent intensity, includes: Bulgaria, Czech Republic, Greece, Hungary, Latvia, Lithuania, Luxembourg, and Poland. Club 2 is composed of medium energy patent active countries such as: Belgium,

Estonia, Ireland, Italy, Netherlands, Norway, Portugal, Slovak Republic, Slovenia, Spain, and United Kingdom. The last club, with the highest energy patent intensity, is comprised of Austria, Finland, France, Germany, Sweden, and Switzerland. Figure 1 provides a visualization of club membership. Interestingly, club 1 is dominated by Central and Eastern European countries, whereas club 2 is more dispersed geographically and covers most parts of Europe. The smallest club (club 3) is formed of Western and Northwestern European countries. Geographic effects seem to be evident for club 1 and club 3.

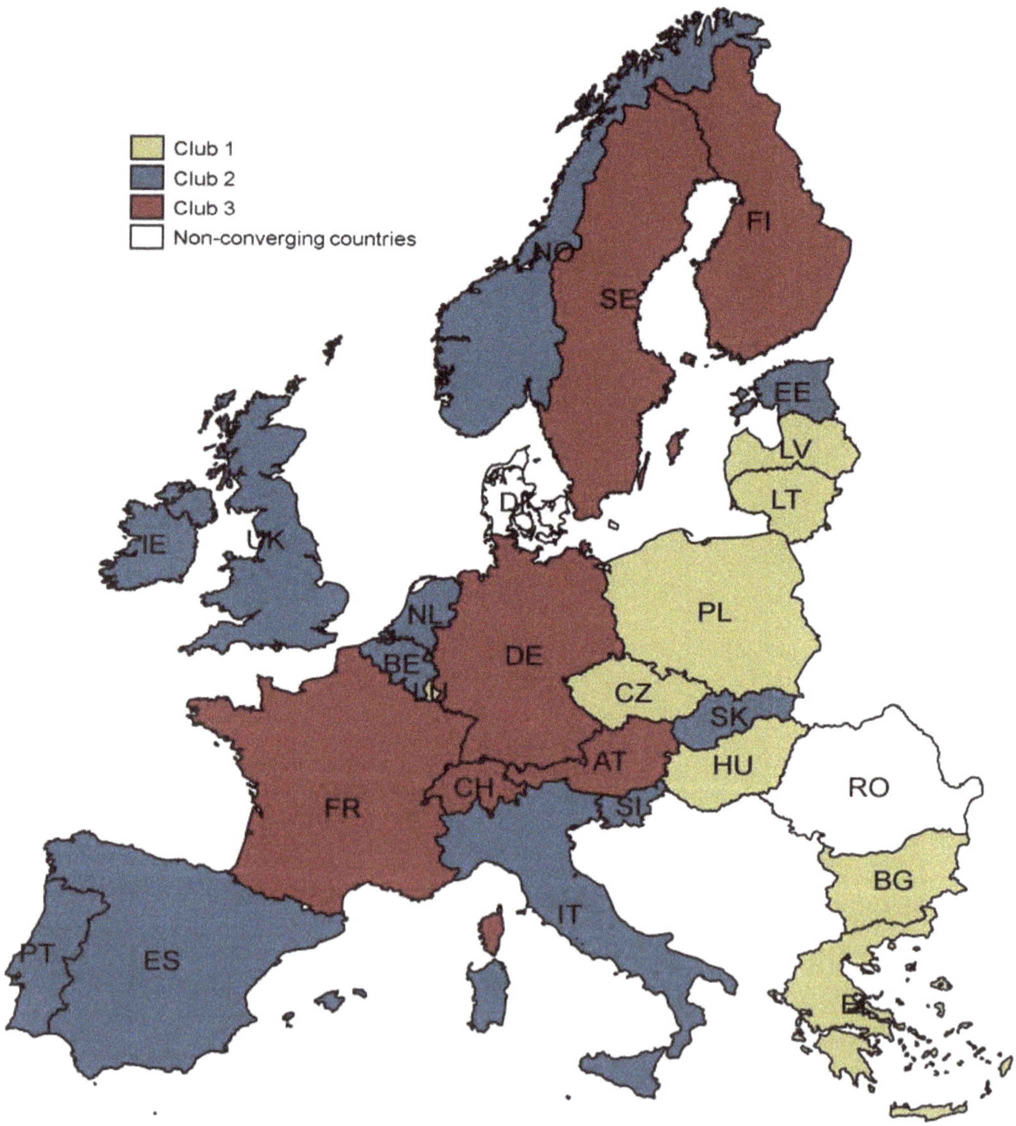

Figure 1. Spatial distribution of club members.

Figure 2 illustrates the change of energy patent intensities of the countries (in logs) belonging to particular clubs over the research period. As can be seen, there exists a catch-up effect, which is especially visible within club 2 and club 3, where countries with low energy patent intensities in 2000 are characterised by higher growth rates (i.e., the distances between points and the 45 degree line) than countries with medium and high energy patent intensities. Interestingly, the points representing countries of each club are distributed horizontally. Such a pattern of energy patent intensity distribution indicates indirectly the convergence processes to different steady states in each individual club. It is worth noting that in the case of club 1 the observed tendency is distorted by Luxembourg that significantly reduced patent intensity in the research period. This situation may result from the fact that Luxembourg's energy system is characterised by high import dependence.

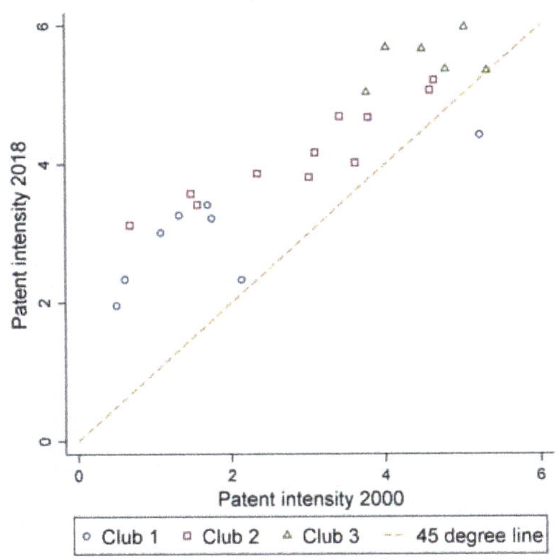

Figure 2. Shifts in energy patent intensity in clubs.

To find the factors influencing membership of a certain club, the ordered logit model was used with a three-level club membership—CM-outcome variable (coded 1, 2, 3) and three predictors: R&D related to environmental protection per capita—RD, human resources in science and technology—HRST, and the Environmental Policy Stringency Index—EPS. For the reasons of data availability, 5 countries were excluded from the analyses (i.e., Bulgaria, Estonia, Latvia, Lithuania, and Luxembourg). Due to non-intuitive interpretation of estimated coefficients of the ordered logit model, Table 2 presents the marginal effects, which show how the probabilities of each outcome (club membership) change with respect to changes in RD, HRST, and EPS. The marginal effects were computed as an average of the marginal effects at each value of covariates.

In the next step, the variations in marginal effects in response to the changes in the level of club membership determinants were examined (Figure 3). It should be noted that for higher levels of HRST, marginal effects increase for club 1 and club 3, but in the former case they remain negative. A similar trend is visible for the EPS variable and to some extent to the RD variable. In the case of club 2 the sign of marginal effects of the HRST variable and the EPS variable changes when we move from low values to high values of covariates, which results in the insignificance of marginal effect averages (see Table 2).

Table 2. Marginal effects on probabilities.

Variable		dy/dx	SE	z	P > z
RD					
	Club 1	−0.0213	0.010	−2.06	0.039
	Club 2	−0.008	0.005	−1.41	0.159
	Club 3	0.029	0.012	2.41	0.016
HRST					
	Club 1	−0.015	0.0047	−3.19	0.001
	Club 2	−0.005	0.005	−1.03	0.304
	Club 3	0.020	0.009	2.30	0.021
EPS					
	Club 1	−0.172	0.096	−1.80	0.071
	Club 2	−0.062	0.051	−1.21	0.228
	Club 3	0.234	0.123	1.91	0.056
Pseudo R2 = 0.2997, LR chi2(3) = 12.34					

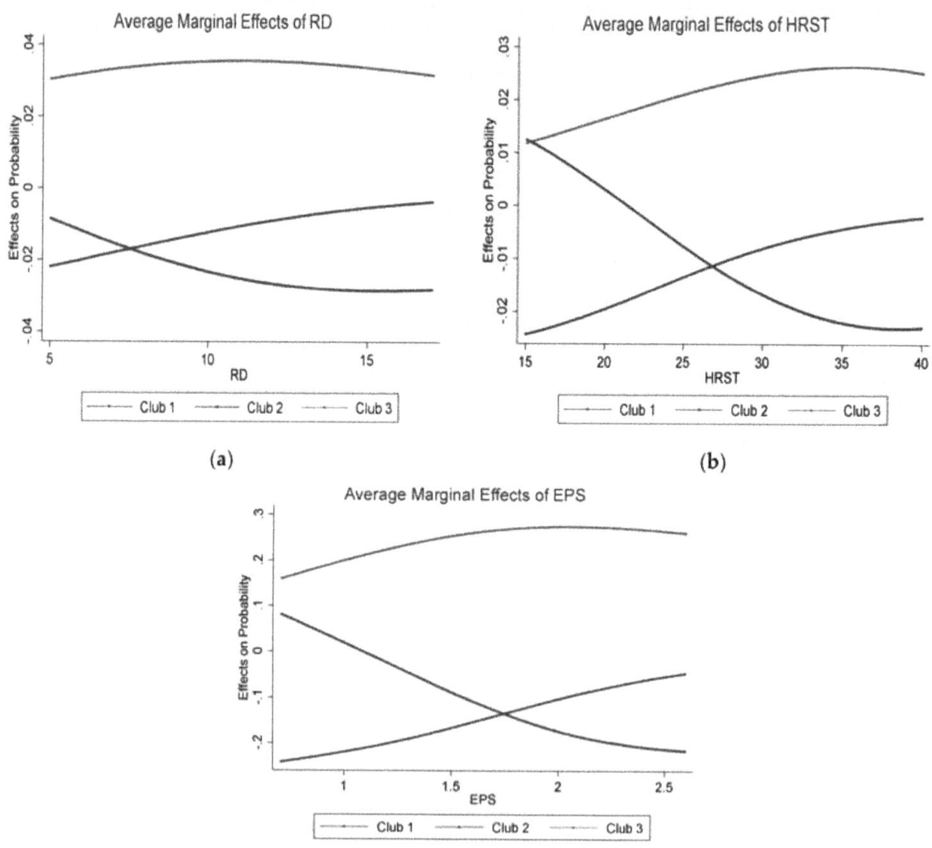

Figure 3. Changes of marginal effects: (**a**) Marginal effects of RD; (**b**) Marginal effects of HRST; (**c**) Marginal effects of EPS.

The sign of the marginal effects of the RD variable indicates that a one-unit increase in R&D related to environmental protection increases the probability of belonging to the high energy patent intensity club. The opposite holds true for club 1. These findings are consistent with results of many general studies on the drivers of eco-innovation, where renewable energy patenting is regarded as a function of public R&D expenditures and the remaining factors [49–51]. On the other hand, R&D investment is a necessary but not sufficient condition to generate high quality inventions, since the effect of R&D expenditure is inherently uncertain and depends on the cumulative R&D capacity (learning-by-searching). For example, Nesta et al. [52] report the statistically insignificant effect of R&D on innovation activities in renewable energy and explain it by the omission of a patent quality dimension.

Concerning the marginal effects of HRST, the probability of club membership is explained well for club 1 and club 3. As with the marginal effects of RD, a one-unit increase in persons with tertiary education increases the probability of belonging to the high energy patent intensity club. The opposite effect can be observed for club 1. This means that specialised human capital is an important driver for countries patenting activities in energy. As suggested by Beise and Rennings [53] and Keller [54], a country's potential to become a leader in a particular field of technology results from its inventive and absorptive capacity formed by skilled human capital. In particular, tertiary education is often considered as one of the most valuable inputs into the inventive process in the field of eco-innovation. As reported by the OECD [55], several European countries (e.g., Germany and Sweden) have tailored their curricula or vocational training to environmental issues and eco-innovation.

Consistent with prior evidence in the relevant literature [1,45], the results of the present study indicate that the stringency of environmental policies plays an important role in shaping the trajectories of energy patent intensity across the European countries. In particular, higher stringency of environmental instruments increases the probability of being a member of the high energy patent intensive club. This finding supports the so-called Porter hypothesis. However, the interpretation of the results should take into account the fact that that the analysis was based on an aggregate measure of the stringency of environmental policy instruments. Therefore, the impact of its particular components on energy inventions trajectories remains unexplored and may vary according to the instrument type (i.e., market-based or non-market-based instrument) [7].

4. Discussion

The results of the research indicate the presence of club convergence in energy innovation across European countries over the years 2000–2018. The empirical evidence indicates that, over the analysed period, 25 out of the 27 examined countries have converged to three clubs characterised with significant disparities in energy patent intensity, as measured by the number of energy patent applications per 10 million inhabitants. These findings are generally in line with Bai et al. [45] who found evidence of club convergence in renewable energy technology innovation (RETI) across Chinese provinces and also identified three distinct clubs. Regarding the European context, delineation of the convergence clubs allowed the identification of a set of countries that are potentially most prone to free-riding on energy innovation efforts developed abroad. These results add value to the evidence provided by Grafström [26], who found conditional β- and σ-divergence in renewable energy invention capabilities across the 13 EU countries, suggesting that some of them tend to free-ride on the development efforts of other Member States.

Bearing in mind the complexity and multidimensionality of climate change mitigation challenges in the energy sector, unlike the prior studies on convergence in energy innovation that focused primarily on patents related to renewable energy technologies, the present research explores the patterns of energy innovation convergence using a broad category of patent applications in the field of climate change mitigation technologies.

Regarding patent applications as a proxy for innovation, it is important to keep in mind some drawbacks of using such a measure, arising, in particular, from the large disproportions in actual economic and technological performance of individual patents. In fact, many patented inventions have no or marginal economic value and quite short market life [56], as they turn out to be unattractive for the intended users, for instance due to technological underperformance or incompatibility with the existing infrastructure and complementary technologies. In contrast, a relatively small fraction of patents are often able to capture even over 90% of total monetary returns available in a given market (see e.g., [57] or [58]). Additionally, many patent applications are unsuccessful or do not ever become genuine innovations, which makes the linkages between patenting and the actual technological progress even harder to capture [59].

Moreover, given the complexity and difficulties inherent in patent application procedures, many smaller firms actually employ the effects of their research activities in production, attempting to veil them from competitors as trade secrets [60], without even trying to obtain a formal patent protection [61]. In addition, as pointed out by Schetino and Sterlacchini [62], the propensity to apply for patent protection is largely dependent on the individual firm's size, strategy, or ability to enforce patent rights, and thus varies significantly both across and within particular industries.

Due to the specificity of individual climate change-related technologies, both the effectiveness of patent protection rights and the propensity to patent differ significantly across diverse technological fields [59]. Moreover, different countries develop and apply different green technologies, basing on their suitability for a given geographical location, compatibility with a county's industrial structure and stage of development.

Notwithstanding the above limitations, a broad and comprehensive measure of patent applications employed in the present study allowed the capture of general patterns in energy innovation convergence in the European context.

The study has also identified three factors contributing to the emergence of the convergence clubs: i.e., per capita environmental R&D expenditures, HRST and environmental policy stringency. Given the above, the results indicate that the convergence paths in energy innovation intensity across the examined countries are determined by the initial levels of each of the above factors. These findings seem to be largely consistent with the results presented by Bai et al. [45], according to whom the convergence paths of individual Chinese provinces are shaped, in particular, by historical intensity of both R&D investment and environmental regulation. The results of the present study suggest that, due to the large gaps in the initial levels of the identified determinants between the weakest and the strongest countries, the former ones were largely unable to reduce the distance dividing them from technological leaders in the field of energy innovation.

Given the large distance still dividing many European economies from the established climate change mitigation goals [63,64] the success of the envisioned ecological transition depends critically on joint innovative effort and stronger inclusion of the laggard countries in the processes of technological convergence in the field of energy.

The emergence of the energy innovation convergence clubs might also be linked to the technological and industrial composition of particular economies. In the light of the so-called Porter hypothesis [6], the observed disparities in the relative energy innovation performance, as measured by patent intensity, might result from cross-country differences in the effective reach of environmental regulations. As demonstrated by [65], unregulated enterprises tend to exhibit a relatively low propensity to innovate in comparison to regulated ones. Additionally, in light of prior studies [64], willingness to engage in the development of climate change mitigation technologies appears to be driven by the actual costs of polluting. If such costs are relatively low, enterprises typically lack incentives to invest in environmental-friendly solutions.

Moreover, since the private sector appears to be generally more reluctant to innovate in the field of energy, trying to postpone costly ecological transition and reinforce the existing fossil-based paradigm, boosting the energy innovativeness of the laggard countries seems to be crucially dependent on public support for the related research, development, and deployment of innovative technologies [66].

The presence of convergence clubs in terms of energy innovation has several important implications of economic, environmental, and political nature.

Given the fact that energy is an essential input in almost every productive activity and that technological progress and innovations play a crucial role in economic growth, the patterns of technological convergence in the energy sector likely affect the course of overall economic convergence in Europe. The revealed disparities in energy innovation performance within each of the identified convergence clubs might, therefore, shape the paths of economic growth of the corresponding countries [22]. The process of ecological transition generates a substantial demand for innovative environmental-friendly technologies and complementary investments. It also leads to the emergence of new market arenas, products, and services, as well as creation of new job opportunities and broader structural shifts in the labour markets [17]. As the global market for eco-innovation is currently estimated at about one trillion euro per annum and expected to triple its size by 2030, the area of eco-innovation is naturally offering the EU economy a unique opportunity to improve competitiveness and job creation [67]. This opportunity seems particularly important in the wake of the COVID-19 pandemic, as intensification of research, development, and technology deployment activities related to energy innovation might also become an important driver of economic recovery.

The existence of convergence clubs suggests a persistently uneven contribution of their 'members' to the collective effort of combating climate change. Such disproportions might, in turn, increase the overall costs of achieving the energy-related goals of environmental policy adopted by the European countries [35].

From a political perspective, a persistently uneven burden of energy innovation efforts poses a more general threat to the fulfillment of the adopted policy goals. While combating climate change depends critically on collective international effort, the countries belonging to the least innovative club appear to be more prone to free-riding on innovations developed abroad [26]. Given the above, the innovation leaders may gradually become discouraged by an unsatisfactory engagement of other countries in the development of climate change mitigation technologies related to energy [35].

Given the above, club convergence in the field of energy innovation suggests the need for more tailored policies, based on smart development and specialization strategies, rather than 'one-size-fits-all' frameworks. Such policies should take into account both the specificity of individual economies, as well as the existence of apparent path dependence in their long-run energy innovation performance. Therefore, the results seem to be in line with Tödtling and Trippl [68] who argue that there is no 'ideal model' for innovation policy as innovation activities differ strongly between central, peripheral, and old industrial areas.

As the results of the present study attribute the emergence of the identified convergence clubs to the initial differences in environmental R&D expenditure, HRST and environmental policy stringency, it seems that the suggested revision of the relevant policies may be focused precisely on these areas. Additionally, the identified positive impact of the above variables on energy innovation performance seems to corroborate the findings of Fabrizi et al. [7] who demonstrate that the effectiveness of environmental regulation policies might be improved by an appropriate innovation policy.

In particular, boosting the relative innovation performance of the countries belonging to the least-innovative club might require the development and implementation of special economic incentives and financing schemes allowing them to more fully exploit the innovative potential of their less technologically advanced sectors, such as agriculture, and to increase R&D efforts and HRST engaged in the search for innovative solutions in the field of energy. Properly designed policies and incentives may therefore allow them to reduce the distance from the innovation leaders faster.

Theoretically, the same goal can be achieved by increasing the stringency of the relevant environmental policies or their reach. However, given the fact that weaker innovative performance is usually associated with a lower level of overall economic and technological development, such a solution would imply that the less advanced countries would have to comply with more stringent policies. This, in turn, could likely raise doubts about the fairness of such an approach and cause an increasing reluctance towards its adoption. Moreover, given the prior empirical evidence, suggesting the existence of optimal limits to the regulation stringency, the latter solution bears the risk of overregulation, which would likely impede the innovative performance of the laggard countries. Given the above, the identified club convergence and the related problem of free-riding on energy innovation should be addressed primarily by properly designed incentives oriented towards increasing the R&D expenditure and HRST in that field.

5. Conclusions

The present study aimed at the identification and assessment of patterns in energy innovation convergence across a sample of 27 European countries, including 24 current EU Member States, GB, Norway and Switzerland, over the period 2000–2018. The results of the conducted analyses indicate that energy innovation efforts in the area of climate change mitigation technologies, as measured by the number of patent applications per 10 million inhabitants, follow the pattern of club convergence.

The novelty of the paper arises from the following aspects. First, unlike previous energy innovation convergence studies that focused on renewable energy, it investigates a broad and comprehensive category of patent applications in the area of climate change mitigation technologies related to energy production, transmission or distribution. Moreover, the examined sample covers a larger set of countries and a longer time span than prior research in the European context. Second, to the Authors' knowledge, the present study is the first to find and delineate energy innovation convergence clubs in Europe. Third, the conducted analyses allowed the identification and assessment of the key factors that had contributed to the emergence of the identified clubs.

Consistent with prior research, the findings suggest a lack of overall convergence in energy innovation performance in the European context. The present study, however, enhances the existing literature on convergence patterns in the field of energy-related innovation by the identification of three distinct convergence clubs. The strongest energy innovation performance is observed in the club composed of the advanced economies of Western and North-Western Europe, while the 'laggard' one is dominated by Central and Eastern European countries. The observed disparities in energy patent intensity suggest a risk of free-riding on energy innovation.

As the mitigation of adverse consequences of climate change requires collective engagement of the European countries, the observed disparities may be addressed by proper policy actions. Since the obtained results attribute the emergence of energy innovation convergence clubs to the initial gaps in per capita environmental R&D expenditure, HRST, and environmental policy stringency between the countries exhibiting the lowest patent intensity and the innovation leaders the revision of policy should focus particularly on these areas. Therefore, the above findings suggest the need for more tailored policies based on smart development and specialization strategies designed to boost the energy innovation performance of the laggard countries, more fully exploiting the potential of their less technologically advanced sectors, such as agriculture. Given the risk of overregulation resultant from implementation of more stringent policies, fostering R&D and HRST by economic incentives and financing schemes oriented towards laggard countries seems to be the preferred direction of policy revision.

The main limitation of the study results from the incompleteness of the long-run statistical data that has rendered the exploration of the patterns of energy innovation convergence in the field of climate change mitigation technologies and their determinants across a larger set of European countries not possible.

Given the importance of the formulated research problem and its policy implications, the conducted analyses could be further extended by assessing the impact of a broader set of determinants shaping the course of convergence in the area of energy innovation.

Author Contributions: Conceptualization, A.K., T.K., P.B., A.M.-B.; theoretical framework; A.M.-B., P.B.; methodology, A.K., T.K.; software, A.K.; results analysis, A.K., T.K.; discussion and conclusions, P.B., A.M.-B. All authors have read and agreed to the published version of the manuscript.

Funding: The APC was funded by the John Paul II Catholic University of Lublin, Faculty of Social Sciences, internal grant no. 05-0501-2-0592, Maria Curie-Sklodowska University, Faculty of Economics, and the vouchers.

Data Availability Statement: Data supporting reported results can be found at: https://ec.europa.eu/eurostat/data/database; https://stats.oecd.org/ (accessed on 25 February 2021).

Conflicts of Interest: The authors declare no conflict of interest.

References

1. Sterlacchini, A. Trends and Determinants of Energy Innovations: Patents, Environmental Policies and Oil Prices. *J. Econ. Policy Reform* **2020**, *23*, 49–66. [CrossRef]
2. Wilson, C.; Grubler, A. The Energy Technology Innovation System. In *Energy Technology Innovation: Learning from Historical Successes and Failures*; Grubler, A., Wilson, C., Eds.; Cambridge University Press: Cambridge, UK, 2013; pp. 11–29. ISBN 978-1-107-02322-2.
3. Gallagher, K.S.; Grübler, A.; Kuhl, L.; Nemet, G.; Wilson, C. The Energy Technology Innovation System. *Annu. Rev. Environ. Resour.* **2012**, *37*, 137–162. [CrossRef]
4. Holdren, J.P. The Energy Innovation Imperative: Addressing Oil Dependence, Climate Change, and Other 21st Century Energy Challenges. *Innov. Technol. Gov. Glob.* **2006**, *1*, 3–23. [CrossRef]
5. European Commission. *Communication from the Commission to the Council, the European Parliament, the European Economic and Social Committee and the Committee of the Regions—A European Strategic Energy Technology Plan (SET-Plan)—"Towards a Low Carbon Future"*; European Commission: Brussels, Belgium, 2007; Volume 723.
6. Porter, M.E.; van der Linde, C. Toward a New Conception of the Environment-Competitiveness Relationship. *J. Econ. Perspect.* **1995**, *9*, 97–118. [CrossRef]
7. Meliciani, V. Green Patents, Regulatory Policies and Research Network Policies. *Res. Policy* **2018**, *47*, 1018–1031. [CrossRef]
8. Ambec, S.; Cohen, M.; Elgie, S.; Lanoie, P. The Porter Hypothesis at 20: Can Environmental Regulation Enhance Innovation and Competitiveness? *Rev. Environ. Econ. Policy* **2013**, *7*, 2–22. [CrossRef]
9. Dechezleprêtre, A.; Sato, M. The Impacts of Environmental Regulations on Competitiveness. *Rev. Environ. Econ. Policy* **2017**, *11*, 183–206. [CrossRef]
10. Rennings, K. Redefining Innovation—Eco-Innovation Research and the Contribution from Ecological Economics. *Ecol. Econ.* **2000**, *32*, 319–332. [CrossRef]
11. United Nations. *Paris Agreement*; United Nations: San Francisco, CA, USA, 2015.
12. International Renewable Energy Agency (IRENA). *Renewable Energy Innovation Policy: Success Criteria and Strategies*; International Renewable Energy Agency (IRENA): Abu Dhabi, United Arab Emirates, 2013; p. 48.
13. European Commission. *Communication from the Commision: Towards an Integrated Strategic Energy Technology (SET) Plan: Accelerating the European Energy System Transformation*; European Commission: Brussels, Belgium, 2015; Volume 6317.
14. European Commission. *Communication from the Commission to the European Parliament, the Council, the European Economic and Social Committee, the Committee of the Regions, and the European Investment Bank: Accelerating Clean Energy Innovation*; European Commission: Brussels, Belgium, 2016; Volume 763.
15. European Commission Energy Research and Innovation. Why the EU Supports Energy Research and Innovation. Available online: https://ec.europa.eu/info/research-and-innovation/research-area/energy-research-and-innovation_en (accessed on 29 March 2021).
16. European Commission. *Communication from the Commission to the European Parliament, the European Council, the Council, the European Economic and Social Committee and the Committee of the Regions: 'The European Green Deal'*; European Commission: Brussels, Belgium, 2019; Volume COM/2019/640.
17. *Horizon Europe: Strategic Plan 2021–2024*; Publications Office of the European Union: Luxembourg, 2021.
18. World Energy Outlook 2019—Analysis. Available online: https://www.iea.org/reports/world-energy-outlook-2019 (accessed on 29 March 2021).

19. Costantini, V.; Crespi, F.; Palma, A. Characterizing the Policy Mix and Its Impact on Eco-Innovation: A Patent Analysis of Energy-Efficient Technologies. *Res. Policy* **2017**, *46*, 799–819. [CrossRef]
20. Archibugi, D.; Filippetti, A. Is the Economic Crisis Impairing Convergence in Innovation Performance across Europe? *J. Common Mark. Stud.* **2011**, *49*, 1153–1182. [CrossRef]
21. Archibugi, D.; Coco, A. Is Europe Becoming the Most Dynamic Knowledge Economy in the World? *J. Common Mark. Stud.* **2005**, *43*, 433–459. [CrossRef]
22. Aghion, P.; Howitt, P. *Endogenous Growth Theory*; MIT Press: Cambridge, MA, USA, 1998.
23. Jungmittag, A. Innovations, Technological Specialisation and Economic Growth in the EU. *Int. Econ. Econ. Policy* **2004**, *1*, 247–273. [CrossRef]
24. Porter, M.E. Location, Competition, and Economic Development: Local Clusters in a Global Economy. *Econ. Dev. Q.* **2000**, *14*, 15–34. [CrossRef]
25. Rosenthal, S.S.; Strange, W.C. The Determinants of Agglomeration. *J. Urban Econ.* **2001**, *50*, 191–229. [CrossRef]
26. Grafström, J. Divergence of Renewable Energy Invention Efforts in Europe: An Econometric Analysis Based on Patent Counts. *Environ. Econ. Policy Stud.* **2018**, *20*, 829–859. [CrossRef]
27. Sharp, M. Competitiveness and Cohesion—Are the Two Compatible? *Res. Policy* **1998**, *27*, 569–588. [CrossRef]
28. Haščič, I.; Migotto, M. *Measuring Environmental Innovation Using Patent Data*; OECD Environment Working Papers; OECD Publishing: Paris, France, 2015. [CrossRef]
29. Dechezleprêtre, A.; Glachant, M.; Haščič, I.; Johnstone, N.; Ménière, Y. Invention and Transfer of Climate Change—Mitigation Technologies: A Global Analysis. *Rev. Environ. Econ. Policy* **2011**, *5*, 109–130. [CrossRef]
30. Organisation for Economic Co-operation and Development (Ed.) *OECD Patent Statistics Manual*; OECD: Paris, France, 2009; ISBN 978-92-64-05412-7.
31. Antonelli, C.; Quatraro, F. The Effects of Biased Technological Change on Total Factor Productivity: Empirical Evidence from a Sample of OECD Countries. *J. Technol. Transf.* **2010**, *35*, 361–383. [CrossRef]
32. Mancusi, M. International Spillovers and Absorptive Capacity: A Cross-Country Cross-Sector Analysis Based on Patents and Citations. *J. Int. Econ.* **2008**, *76*, 155–165. [CrossRef]
33. Cohen, W.M.; Levinthal, D.A. Innovation and Learning: The Two Faces of R&D. *Econ. J.* **1989**, *99*, 569–596.
34. Bosetti, V.; Carraro, C.; Massetti, E.; Tavoni, M. International Energy R&D Spillovers and the Economics of Greenhouse Gas Atmospheric Stabilization. *Energy Econ.* **2008**, *30*, 2912–2929.
35. Corradini, M.; Costantini, V.; Mancinelli, S.; Mazzanti, M. Interacting Innovation Investments and Environmental Performances: A Dynamic Impure Public Good Model. *Environ. Econ. Policy Stud.* **2015**, *17*, 109–129. [CrossRef]
36. Archibugi, D.; Pianta, M. Aggregate Convergence and Sectoral Specialization in Innovation. *J. Evol. Econ.* **1994**, *4*, 17–33. [CrossRef]
37. Martín, C.; Mulas-Granados, C.; Sanz, I. Spatial Distribution of R&D Expenditure and Patent Applications across EU Regions and Its Impact on Economic Cohesion. *Investig. Reg. J. Reg. Res.* **2005**, *6*, 41–61.
38. Jungmittag, A. Innovation Dynamics in the EU: Convergence or Divergence? A Cross-Country Panel Data Analysis. *Empir. Econ.* **2006**, *31*, 313–331. [CrossRef]
39. Žižmond, E.; Novak, M. Controversies of Technology Convergence within the European Union. *Ind. Manag. Data Syst.* **2007**, *107*, 618–635. [CrossRef]
40. Bayer, P.; Dolan, L.; Urpelainen, J. Global Patterns of Renewable Energy Innovation, 1990–2009. *Energy Sustain. Dev.* **2013**, *17*, 288–295. [CrossRef]
41. Johnstone, N.; Haščič, I.; Popp, D. Renewable Energy Policies and Technological Innovation: Evidence Based on Patent Counts. *Environ. Resour. Econ.* **2010**, *45*, 133–155. [CrossRef]
42. Dalmazzone, S.; Corsatea, T. *A Regional Analysis of Renewable Energy Patenting in Italy*; Social Science Research Network: Rochester, NY, USA, 2012.
43. Balsalobre, D.; Álvarez, A.; Cantos, J.M. Public Budgets for Energy RD&D and the Effects on Energy Intensity and Pollution Levels. *Environ. Sci. Pollut. Res. Int.* **2015**, *22*, 4881–4892. [CrossRef] [PubMed]
44. Dauda, L.; Long, X.; Mensah, C.N.; Salman, M. The Effects of Economic Growth and Innovation on CO2 Emissions in Different Regions. *Environ. Sci. Pollut. Res. Int.* **2019**, *26*, 15028–15038. [CrossRef]
45. Bai, C.; Feng, C.; Du, K.; Wang, Y.; Gong, Y. Understanding Spatial-Temporal Evolution of Renewable Energy Technology Innovation in China: Evidence from Convergence Analysis. *Energy Policy* **2020**, *143*, 111570. [CrossRef]
46. Phillips, P.C.B.; Sul, D. Economic Transition and Growth. *J. Appl. Econom.* **2009**, *24*, 1153–1185. [CrossRef]
47. McKelvey, R.D.; Zavoina, W. A Statistical Model for the Analysis of Ordinal Level Dependent Variables. *J. Math. Sociol.* **1975**, *4*, 103–120. [CrossRef]
48. Patents by Technology: Patents in Environment-Related Technologies. Available online: https://stats.oecd.org/index.aspx?queryid=29068 (accessed on 30 March 2021).
49. Verdolini, E.; Galeotti, M. At Home and Abroad: An Empirical Analysis of Innovation and Diffusion in Energy Technologies. *J. Environ. Econ. Manag.* **2011**, *61*, 119–134. [CrossRef]
50. Peters, M.; Schneider, M.; Griesshaber, T.; Hoffmann, V.H. The Impact of Technology-Push and Demand-Pull Policies on Technical Change—Does the Locus of Policies Matter? *Res. Policy* **2012**, *41*, 1296–1308. [CrossRef]

51. Dechezleprêtre, A.; Glachant, M. Does Foreign Environmental Policy Influence Domestic Innovation? Evidence from the Wind Industry. *Environ. Resour. Econ.* **2014**, *58*, 391–413. [CrossRef]
52. Nesta, L.; Vona, F.; Nicolli, F. Environmental Policies, Competition and Innovation in Renewable Energy. *J. Environ. Econ. Manag.* **2014**, *67*, 396–411. [CrossRef]
53. Beise, M.; Rennings, K. Lead Markets and Regulation: A Framework for Analyzing the International Diffusion of Environmental Innovations. *Ecol. Econ.* **2005**, *52*, 5–17. [CrossRef]
54. Keller, W. International Technology Diffusion. *J. Econ. Lit.* **2004**, *42*, 752–782. [CrossRef]
55. Eco-Innovation in Industry: Enabling Green Growth—OECD. Available online: https://www.oecd.org/sti/ind/eco-innovationinindustryenablinggreengrowth.htm (accessed on 30 March 2021).
56. Pakes, A. On Patents, R&D, and the Stock Market Rate of Return. *J. Political Econ.* **1985**, *93*, 390–409.
57. Scherer, F.M.; Harhoff, D. Technology Policy for a World of Skew-Distributed Outcomes. *Res. Policy* **2000**, *29*, 559–566. [CrossRef]
58. Harhoff, D.; Scherer, F.M.; Vopel, K. Citations, Family Size, Opposition and the Value of Patent Rights. *Res. Policy* **2003**, *32*, 1343–1363. [CrossRef]
59. Hall, B.; Helmers, C. *The Role of Patent Protection in (Clean/Green) Technology Transfer*; Centre for the Study of African Economies, University of Oxford: Oxford, UK, 2010.
60. Cohen, W.M.; Nelson, R.R.; Walsh, J.P. *Protecting Their Intellectual Assets: Appropriability Conditions and Why U.S. Manufacturing Firms Patent (or Not)*; NBER Working Papers; National Bureau of Economic Research, Inc.: Cambridge, MA, USA, 2000.
61. Adams, S. *Information Sources in Patents*, Completely Rev. ed.; De Gruyter Saur: Boston, MA, USA, 2020; ISBN 978-3-11-055006-1.
62. Schettino, F.; Sterlacchini, A. Reaping the Benefits of Patenting Activities: Does the Size of Patentees Matter? *Ind. Innov.* **2009**, *16*, 613–633. [CrossRef]
63. Renewable Energy Statistics—Statistics Explained. Available online: https://ec.europa.eu/eurostat/statistics-explained/index.php/Renewable_energy_statistics (accessed on 30 March 2021).
64. Kasztelan, A. How Circular Are the European Economies? A Taxonomic Analysis Based on the INEC (Index of National Economies' Circularity). *Sustainability* **2020**, *12*, 7613. [CrossRef]
65. Calel, R.; Dechezleprêtre, A. Environmental Policy and Directed Technological Change: Evidence from the European Carbon Market. *Rev. Econ. Stat.* **2016**, *98*, 173–191. [CrossRef]
66. Rhodes, A.; Skea, J.; Hannon, M. The Global Surge in Energy Innovation. *Energies* **2014**, *7*, 5601–5623. [CrossRef]
67. Climate Action, Environment, Resource Efficiency and Raw Materials. Available online: https://ec.europa.eu/programmes/horizon2020//en/h2020-section/climate-action-environment-resource-efficiency-and-raw-materials (accessed on 30 March 2021).
68. Tödtling, F.; Trippl, M. One Size Fits All?: Towards a Differentiated Regional Innovation Policy Approach. *Res. Policy* **2005**, *34*, 1203–1219. [CrossRef]

energies

Article

Renewable Energy Attitudes and Behaviour of Local Governments in Poland

Joanna Rakowska * and Irena Ozimek

Institute of Economics and Finance, Warsaw University of Life Sciences—SGGW, 02-787 Warsaw, Poland; irena_ozimek@sggw.edu.pl
* Correspondence: joanna_rakowska@sggw.edu.pl

Abstract: The deployment of renewable energy at the local level can contribute significantly to mitigating climate change, improving energy security and increasing social, economic and environmental benefits. In many countries local authorities play an important role in the local development, but renewable energy deployment is not an obligatory task for them. Hence there are two research questions: (1) Do local governments think investments in renewable energy (RE) are urgent and affordable within the local budgets? (2) How do they react to the public aid co-financing investments in renewable energy? To provide the answer we performed qualitative analysis and non-parametric tests of data from a survey of 252 local authorities, analysis of 292 strategies of local development and datasets of 1170 renewable energy projects co-financed by EU funds under operational programs 2007–2013 and 2014–2020 in Poland. Findings showed that local authorities' attitudes were rather careful, caused by financial constraints of local budgets and the scope of obligatory tasks, which made renewable energy investments not the most urgent. Public aid was a factor significantly affecting local authorities' behavior. It triggered local authorities' renewable energy initiatives, increasing the number and scope of renewable energy investments as well cooperation with other municipalities and local communities. Despite this general trend, there were also considerable regional differences in local authorities' renewable energy behavior.

Keywords: renewable energy; local authorities; European Union funds

Citation: Rakowska, J.; Ozimek, I. Renewable Energy Attitudes and Behaviour of Local Governments in Poland. *Energies* **2021**, *14*, 2765. https://doi.org/10.3390/en14102765

Academic Editor: Piotr Gradziuk

Received: 30 March 2021
Accepted: 7 May 2021
Published: 12 May 2021

1. Introduction

Do local governments think investments in renewable energy are urgent and affordable within the local budgets? How do they react to the public aid co-financing investments in renewable energy? The crucial and still growing role of renewable energy sources in meeting the urgent need of mitigating climate change, improving energy security and increasing social, economic and environmental benefits has been well recognised and acknowledged [1–6]. Despite all the drawbacks of fossil fuel energy use and benefits of renewable energy (RE), the deployment of renewable energy happens neither on its own nor fast enough [7–9]. It needs to be policy-driven [10–14], supported by adequate legal regulations, international and national strategies, and public aid and incentives [15–19] addressed to relevant beneficiaries. Local governments are considered very important entities in the process of increasing renewable energy production and use at the local level, in many economic, social and political contexts. However, the literature lacks cohesive conclusions on the renewable energy attitudes and behavior of local governments in Poland. This study aims to fill this gap.

2. Literature Review

The European Union began to build its policy framework for renewable energy in 1997 with the White Paper for a Community Strategy and Action Plan Energy for the future: Renewable sources of energy [20], setting basis for the policy on renewable energy. The policy has significantly evolved afterwards [21–26]. The recent legal framework for the promotion of renewable energy sources until 2020 was laid down in directive adopted in 2009 [27], while renewable energy production and consumption goals until 2030 are set in the Regulation on the Governance of the Energy Union and Climate Action (EU) 2018/1999 [28], for the EU as a whole and for its individual member states [29].

To promote renewable energy and thus achieve climate goals, the EU applies Cohesion policy instruments, including structural and cohesion funds. They are said to be the most important funding sources for promoting renewable energy among the EU spending programmes, of continusly increasing value—from only 600 million euro during the 2000–2006 programming period, through approximately 4.7 billion euro in the 2007–2013, up to 27 billion euro in 2014–2020 [30]. The EU policy is based on the already well verified assumption that decarbonisation requires solutions at all levels of governance and collaboration—global, regional, national and especially local [31–37].

The focus on the deployment of renewable energy at the local level results from its positive impact on local communities, economies and environment. Renewable energy production and use provide new job opportunities [38–40], cause income generation, diversification of economic activities, use of endogenous resources [41], contribute to satisfying local energy demands [42,43], are drivers of economic recovery in peripheral or remote areas [44–46] and drivers of relevant business opportunities in large metropolitan areas [47]. Renewable energy enables transfer to community-owned energy sources [48,49], decentralisation of energy production and supplies [50,51], even local and regional energy autarky [52–55]. Renewable energy can cause such externalities as positive health effects [56]. However, achieving these benefits requires social capital [57] and the involvement of local actors. Local authorities are of particular significance in this process.

Many studies, providing insight into different political, economic and administrative backgrounds argue that local authorities can play a key role in promoting renewable energy production and use [58–62]. Local authorities can initiate, invest, produce and be the end users of renewable energy [63–66]. Due to the importance of renewable energy the EU addresses its structural funds also to potential beneficiaries who can invest in renewable energy. Local authorities implement some of the state's tasks on a local scale because they have the best knowledge of local factors and development conditions as well as of the needs of local communities. This enables a more accurate adjustment of the supply of public goods and services to local demand. Such approach is in line with the principle of subsidiarity enforced in the cohesion policy of the EU [67]. It orders the decentralization of activities and the delegation of decision-making and executive powers to the lowest possible level of administration, capable of implementing them, in order to ensure the most effective use of public funds. The principle of subsidiarity is anchored in Art. 4, Section 3 of the European Charter of Local Self-Government [68]. Despite this general rule, local authorities in individual member states of the EU have different obligatory and facultative tasks to perform, as well as different measures to obtain their goals. Thus they also play different roles in supporting renewable energy deployment at the local level.

In the Polish legal system, there are currently three main statutory acts that directly regulate the production and use of renewable energy. These are: the Act of 10 April 1997 Energy Law [69], the Act of 20 February 2015 on renewable energy sources [70], and the Act of 20 May 2016 on investments in wind farms [71]. The Act of 20 February 2015 on renewable energy sources defines renewable energy sources as renewable, non-fossil energy sources, including wind energy, energy solar radiation, aerothermal energy, geothermal energy, hydrothermal energy, hydropower, wave, current and tidal energy, energy obtained from biomass, biogas, agricultural biogas and bioliquids. The Energy Law Act specifies, among other, the development of the use of renewable energy installations and provides for certain tasks and competences of administrators, also in relation to local authorities (Articles 17–19 of the EPA). According to Polish law local authorities in Poland are to create and implement the development policy and monitor its effects, using the instruments available and effective in the given conditions [72]. They can support the deployment of renewable energy within their own public tasks, in order to meet the needs of the local community, as pursuant to the Act of 8 March 1990 on the commune self-government [73], art. 7 sec. 1, they are responsible for 'meeting the collective needs of the community'. Electricity, heat and gas supplies are mentioned expressis verbis among the tasks of the local government.

In Poland, local authorities of municipalities (called in Polish 'gmina' and classified by the European Union as lower LAUs [74]) which are the basic local government unit, can play a special role on the energy market and in the field of renewable energy sources. On one hand, they can independently create energy through their own activities, and on the other hand, they can support the development of energy in their area, due to the applicable regulations [75]. According to some authors [76], a commune and its local authority may act as 'an energy user and a participant in the competitive energy market, as a local energy regulator, as an investor and producer of energy, and as an entity responsible for planning and financing the lighting of public places and roads within its area'. It should be stressed however, that Polish law does not give local authorities too many tools to effectively stimulate the development of renewable energy. The law does not introduce a transparent system of tasks of competences of the administrative entities, in particular in case of the local government [77]. Summing up, Polish local authorities are allowed to, but do not have to support renewable energy.

The European Union structural and cohesion funds allocated in eligible member states in consequent budget perspectives under operational programmes of regional policy were to suport member states also in deploying renewable energy and in achieving overall national targets in this field [78]. In 2007–2013 EU funds co-financing renewable energy investments were available in Poland from the Operational Program Infrastructure and Environment 2007–2013, which received the highest EU funding for any operational program in the history of the EU regional and cohesion policy and from 16 regional operational programs [79]. In financial perspective of 2014–2020 renewable energy projects were supported by EU funds under 16 regional operational programmes. In both these financial perspectives EU funds for renewable energy projects have been addressed to many groups of potential beneficiaries, including local governments.

The review of literature provides theoretical assumptions on the role of local governments in deployment of renewable energy, while the European Union and national legal frameworks together with the EU financial support give local governments opportunities to actively participate in renewable energy deployment. Despite all the above, the literature lacks a broader insight into renewable energy attitudes of local governments in Poland. It also lacks comparative analysis of local governments behavior related to the deployment of renewable energy, supported by EU funds in 2007–2013 and 2014–2020 financial perspectives. The aim of this study is to fill in this gap.

3. Materials and Methods

The main aim of the study was to answer the following research questions on the attitudes and behavior of local governments: (1) Do local governments think investments in renewable energy are urgent and affordable within the local budgets? (2) How do they react to the public aid co-financing investments in renewable energy? To answer these questions, the study was based on qualitative and quantitative data analyses.

The attitudes of local governments to renewable energy investments were analysed based on the answers to the first research questions, i.e., whether local governments think that investments in renewable energy are urgent and affordable within the local budgets. The data used to answer these questions came from two sources: (i) a survey of local governments of municipalities of Mazovian Voivodship and (ii) a qualitative analysis of data obtained from the strategies of development of municipalities of Mazovian Voivodship. Both analyses were carried out for the municipalities of Mazovian, which is currently NUTS 1 by Eurostat [80]. It was selected for the survey, as it is the largest voivodship in Poland, both in terms of population and the area. At the same time it is a voivodship with the biggest number of municipalities and consequently the largest number of local governments, which are a subject of this research. As the local governments have been functioning within the borders of the same region, they have been addressees of the same regional operational programmes for 2007–2013 and 2014–2020. On the other hand, municipalities of Mazovian Voivodship represent very different types concerning population, population density, total annual budget revenues, rural-urban categories, etc. All these give rationale for the studying the attitudes of local governments to renewable energy investments based on municipalities of Mazovian Voivodship.

The survey questionnaire was sent by mail to governments of all 314 municipalities of Mazovian Voivodship in 2017. The response rate reached 62%. The questionnaire included questions that were compiled to avoid suggesting any answers. The respondents were asked: (i) what investments, that could be carried out by the self-government, were most urgent? (ii) what investments were necessary to improve the quality of natural environment in the municipality? (iii) how they assessed the quality of natural environment in the municipality? (iv) what were the biggest obstacles in supporting the development of the municipalities by local governments?

Strategies of local development of 292 municipalities of Mazovian Voivodship were another source of qualitative data used for the investigation into the attitudes of local self-governments to renewable energy investments. The strategies were retrieved from the websites of municipalities between 3 January 2021 and 20 January 2021. Based on these, we studied 93% of all 314 municipalities of this voivodship. Strategies of local development for the remaining 22 municipalities were not available. In Poland, pursuant to the Act of 8 March 1990 on the municipality government [73], the municipality may develop a development strategy. Although strategies are not obligatory, they are a great help in effective acquisition of funds from European funds and other external sources, including funds for the development of energy from renewable sources.

The retrieved strategies were analysed to find answer to the following, set a priori, research questions: (i) does the strategy include any plans for renewable energy investments? (ii) If so, is the type of renewable energy investment defined? (iii) What type is it (solar, wind, biomass etc.), if defined? (iv) Is the source of financing for renewable energy investments indicated (domestic, EU etc.)? (v) Are the conditions for renewable energy investments favorable in the municipality? (vi) If so, what kind of renewable energy investments are the conditions good for?

The behaviour of local self-governments in the field of renewable energy was analyzed based on qualitative and quantitative data on projects (investments) in renewable energy made by all municipalities in Poland under the 2007–2013 and 2014–2020 operational programs co-financed by the European Union funds (Table 1).

The TERYT codes, names and types of administrative units, as well as addresses and codes of municipality offices were used to verify, complete and merge the datasets, as well as to carry out data curation. To analyze qualitative data we used standard qualitative analysis tools [81–83].

To analyse quantitative data we first applied selected methods of descriptive statistics and tested the distribution of quantitative variables using Kolmogorov-Smirnov test. As the results showed that the distribution of the majority of variables is non-normal, the further analysis was performed using non-parametric tests, including and Kruskal-Wallis H test, Mann-Whitney U test together with Wilcoxon rank-sum test and Wilcoxon signed-rank test (Table 2). We applied these non-parametric tests as they are appropriate to obtain answers to our research questions based on variables of non-normal distribution.

Table 1. Data categories and sources.

Data Categories	Description and Sources
For 2007–2013: Number of renewable energy projects Total value of RE projects (mln Polish zloties) EU funds for RE projects (mln Polish zloties) Types of projects by renewable energy sources Location of renewable energy investments Leading beneficiaries and partners	Qualitative and quantitative data on projects carried out in Poland under Operational Programmes 2007–2013, obtained from the National Information System SIMIK 07–13 [84] run by the Ministry of Regional Development (name of the Ministry as of 31 January 2016, when the data was obtained), the data shows the state of arts as of 31 December 2015, according to the n+2 EU regional policy rule [85], that allows the managing institutions in eligible member states to spend the EU funds allocations till the end of the second year after the year of receiving the funds, i.e., in case of 2007–2013 financial perspective till the end of 2015; from database containing 150,000 entries, we extracted entries describing all 264 investments in renewable energy, carried out by municipalities.
For 2014–2020: Number of renewable energy projects Total value of renewable energy projects (mln Polish zloties) EU funds for renewable energy projects (mln Polish zloties) Types of projects by renewable energy sources Location of renewable energy investments Leading beneficiaries and partners	Qualitative and quantitative data on projects carried out in Poland under Operational Programmes 2014–2020, as of September 2020, obtained from the Central Teleinformation System SL 2014 run by the Ministry of Funds and Regional Policy, and retrieved on 1 December 2020 from [86], from the database containing 177,458 entries, we extracted 1161 entries describing 909 investments carried out by 638 municipalities, defined as leading beneficiaries.
Types of municipalities by degree of urbanisation (DEGURBA)	The degree of urbanization (DEGURBA) classification categorizes municipalities into the three categories: Code 1—cities, or: densely populated areas Code 2—towns and suburbs, or: intermediate density areas Code 3—rural areas, or: thinly populated areas retrieved on 30 June2020 from [87]
Tax ID of municipalities	National Court Register [88] accessed on 15 January 2021.
TERYT codes, names and types of administrative units	Database retrieved from the National Official Register of the Territorial Division of the Country (TERYT), retrieved from [89].
addresses and codes of municipality offices	Data retrieved from the Public Information Bulletin [90].

Table 2. Statistical methods and their application.

Method	Applied to Test:
Kolmogorov-Smirnov test	to what degree the distribution of the analyzed data categories was skewed vs. normally distributed
Kruskal-Wallis H test	whether the total value of renewable energy projects and the share of EU funds in the total value of renewable energy projects in 2007–2013 and 2014–2020 were significantly affected by the types of municipalities
Mann-Whitney U test Wilcoxon rank-sum test	whether there were significant differences in total value of renewable energy projects, value of EU funds cofounding renewable energy projects and the share of EU funds in total value of renewable energy projects between different types of municipalities, i.e., those classified as cities, towns and suburbs, and rural.
Wilcoxon signed-rank test	whether the total value of renewable energy projects and the value of EU funds cofounding renewable energy projects differed between 2007–2013 and 2014–2020 in case of the municipalities that carried out renewable energy projects in both financial perspectives.

We looked into relations between various qualitative and quantitative features of objects assigned to one category (intra-case analysis) and attempted to define relations between different categories selected from the research sample (cross-case analysis) [91].

4. Results

4.1. *The Attitudes of Local Governments to Renewable Energy Investments*

The first part of the study on attitudes of local governments to renewable energy investments was based on the analysis of 292 local development strategies, retrieved from the websites of the municipalities in 2021. Therefore, the qualitative data set contains qualitative information on 93% of all municipalities of Mazovian Voivodship.

The findings show that 85% of local governments declared the deployment of renewable energy. The types of renewable energy sources were indicated in 53.5% of the analyzed strategies. Other strategies contained only general information, emphasizing that renewable, also called alternative or ecological, energy sources should be deployed.

Among the 248 municipalities which accentuated the need to develop particular types of renewable energy, many referred to several different renewable energy sources at the same time. As many as 94.7% of them pointed out deployment of solar renewable energy, which is the most widespread renewable energy source in Poland, used for electricity production and heating purposes. Since solar energy is processed with the use of solar collectors and photovoltaic panels, 72.3% of the analysed strategies declared installation of solar collectors, and 22.3% photovoltaic panels.

Biomass and its particular types were indicated in 68.1% of strategies, which articulated the need to deploy renewable energy. Biomass consists of products, waste or residues from forest and agricultural production, which are biodegradable. It can also be biogas and some fractions of municipal and industrial waste. So, in 50% of strategies local authorities stated that biomass is in general important as renewable energy sources, sometimes also biogas (6.4%), energy plants (7.4%) or a specific type of an energy plant, i.g. willow (4.3%).

Referring to the source of renewable energy, 40.4% of analyzed strategies declared investments in wind energy, 20.2% use of heat pumps, and 17% geothermal energy. In addition, biofuels were mentioned in 16.0% of them. No strategy involved the analysis or description of conditions for renewable energy investments. No strategy defined what funds will be used to cover the costs of investments in renewable energy, although SWOT (strengths, weaknesses, opportunities, and threats) analyses of nearly all of them stated that the possibility of obtaining EU funding is a vital development opportunity.

In the next stage, the study on attitudes of local self-governments to renewable energy investments was based on qualitative datasets obtained from a survey of local governments carried out in 2017. The response rate at the level of 62% provided answers of 195 local authorities. When asked about the most urgent investments in general, only 18% of respondents indicated renewable energy projects. However, they listed investments in renewable energy together with other urgent projects, such as construction or modernisation of roads, sewage systems, water system, support for home sewage system plants, and less of them together with water treatment plants and landfills. Other respondents (82%) listed only road, water supply and treatment, sewage and landfill investments as urgent.

Next, the respondents assessed the quality of the natural environment: 2% as very bad, 13% as bad, 56% as good and 29% as very good. Answering the question on investments necessary to improve the quality of natural environment in the municipality, 18% of respondents indicated renewable energy projects. However, most of them did not list the renewable energy projects as the most urgent for their municipality answering the earlier question. Cross-tabulation of so-far findings showed that all self-governments that listed investments in renewable energy as very urgent assessed the quality of natural environment in their municipalities either as good (61%) or very good (39%). Interestingly enough, none of those who listed investments in renewable energy as very urgent assessed the quality of natural environment as bad or very bad.

All respondents pointed at the lack of funding as the biggest obstacle in supporting the development of municipalities, both those who considered investments in renewable energy project as urgent, and those who did not.

4.2. Local Governments' Renewable Energy Investments under Operational Programs 2007–2013 and 2014–2020

According to the assumptions of this study, the behaviour of local authorities was assessed based on their investments in renewable energy under operational programs 2007–2013 and 2014–2020, shown in SIMIK database for 2007–2013 (2015) financial perspective and in SL 2014 database for 2014–2020, as a still on-going financial perspective due to $n + 2$ rule.

Over the analysed time local governments became more and more active in investing in renewable energy co-financed by the EU funds under operational programmes. In 2014–2020 compared to 2007–2013 there was a nearly threefold increase in the number of municipalities in Poland that carried out investments in renewable energy. In 2007–2013 it was 9.2% out of all 2479 municipalities in Poland, while in 2014–2020 it was 636 municipalities making 25.6% of all municipalities in the country (Table 3). Investigating rural-urban differences in the behaviour of local authorities we applied classification by the degree of urbanization (DEGURBA). The findings show that in 2007–2013 the shares of cities, towns and suburbs as well as rural municipalities, investing in renewable energy, were quite similar, respectively 11%, 10% and 9%. In 2014–2020 these shares doubled for cities and towns and suburbs, as 20% of each of these groups invested in renewable energy. But rural municipalities became even more active. The share of rural municipalities who invested in renewable energy almost tripled in 2014–2020 compared to 2007–2013.

The increased activity of local governments in deploying renewable energy was also proved by the fact that more and more of them carried out more than only one project. Although in 2007–2013 only 11% of municipalities carried out two projects and 1.7% municipalities three projects, in 2014–2020 the group of municipalities that carried out two and three projects increased significantly, up to 23.4% and 6.9% respectively. Moreover some of municipalities carried out four, six or even nine projects (1.2%; 0.2%; 0.2%, respectively). The increased activity of 111 local governments also shows in carrying out renewable energy investments co-financed by EU funds under operational programmes of both financial perspectives.

The increasing activity of local governments in deploying renewable energy was reflected in much higher mean total value of renewable energy projects carried out in 2014–2020 (Table 4), proving their bigger extent. This effect occurred most strongly in case of cities, which in 2014–2020 invested on average 10 mln Polish zloties more in each renewable energy project than in 2007–2013. Rural municipalities invested on average 2.2 mln more, which was the least increase, however they tripled the number of renewable energy projects as mentioned before.

The projects used mainly solar power and their share increased from 80% to 95%. Other sources of renewable energy were less popular and their share even decreased: in case of projects using hydro-, geothermal and other renewable energy sources from 14% to 3% and in case of biomass from 5% to 1%. Wind renewable energy projects made less than 1% of all renewable energy investments by municipalities. The investments resulted in modernization of heating systems for public buildings, installation of water heating systems for swimming pools, installation of economic lighting of municipal roads and squares, including hybrid lighting, biomass boiler networks together with the installation of solar collectors, installation of photovoltaics, Uniejów thermal baths.

The behavior of local authorities changed positively considering the collaboration in renewable energy projects. During the first financial perspective a prevailing majority of projects (96%) was carried out by individual municipalities, only 4% of projects were carried out in collaboration at that time. This has changed during the on-going financial perspective as 28.7% of projects were carried out in collaboration between from two to 10 municipalities, and in one case even by 41 municipalities. Analysis of qualitative data shows that investing in renewable energy municipalities collaborated with other municipalities listed as partners, and with local residents mentioned in the descriptions of projects.

Table 3. Number of municipalities-investors and renewable energy projects under operational programs of 2007–2013 and 2014–2020.

Types of Municipalities	Financial Perspective 2007–2013				Financial Perspective 2014–2020			
	Municipalities Investing in Renewable Energy		Renewable Energy Projects		Municipalities Investing in Renewable Energy		Renewable Energy Projects	
	No	% of All Municipalities of This Type	No	% of All Renewable Energy Projects by Municipalities	No	% of All Municipalities of This Type	No	% of All Renewable Energy Projects by Municipalities
Cities (1)	8	11	13	5	20	26	48	5
Towns and suburbs (2)	54	10	61	23	103	20	165	18
Rural (3)	167	9	188	72	513	27	695	77
All	229	9.2	262	100	636	25.6	908	100

Table 4. Descriptive statistics for RE investments carried out by municipalities in Poland in 2007–2013 and 2014–2020 EU financial perspectives, by types of municipalities.

	Type of Municipalities *	N	Mean	Std. Error	Median	Std. Dev.	Min	Max	Range	Skewness	Kurthosis
Total value of renewable energy projects (Polish zloty) 2007–2013	1	8	3.7	1.6	1.9	4.4	0.25	13.4	13.1	1.86	3.35
	2	54	3.2	0.5	1.8	3.6	0.07	14.9	14.8	1.85	3.02
	3	167	3.9	0.2	3.0	3.1	0.05	14.4	14.4	1.12	0.99
	all	229	3.7	0.2	2.9	3.3	0.05	14.9	14.9	1.33	1.48
Value of EU funds for projects (Polish zloty) 2007–2013	1	8	2.2	0.7	1.3	2.1	0.15	5.9	5.7	1.10	−0.22
	2	54	2.4	0.4	1.3	2.6	0.04	11.7	11.7	1.81	2.88
	3	167	2.9	0.2	2.3	2.4	0.03	11.9	11.9	1.28	1.77
	all	229	2.8	0.2	2.1	2.5	0.03	11.9	11.9	1.38	1.87
The share of EU funds in total value of renewable energy projects (%) 2007–2013	1	8	68%	4%	71%	12%	44%	85%	41%	−1.11	2.31
	2	54	72%	2%	76%	13%	25%	85%	60%	−1.63	3.44
	3	167	73%	1%	77%	10%	26%	85%	59%	−1.91	4.91
	all	229	73%	0.7%	76%	11%	25%	85%	60%	−1.79	4.19
Total value of renewable energy projects (Polish zloty) 2014–2020	1	20	13.6	4.1	7.1	18.3	0.9	68.5	67.6	2.24	4.69
	2	103	11.3	1.8	5.1	18.2	0.2	87.6	87.4	2.77	7.22
	3	503	6.1	0.5	3.2	10.1	0.1	77.7	77.5	4.58	24.56
	all	626	7.2	0.5	3.4	12.3	0.1	87.6	87.5	4.01	17.63
Value of EU funds for projects (Polish zloty) 2014–2020	1	20	10.7	3.7	5.0	16.7	0.7	61.3	60.6	2.46	5.46
	2	103	8.0	1.3	3.2	13.0	0.1	71.6	71.5	2.91	8.53
	3	503	4.3	0.3	2.1	7.0	0.1	67.9	67.8	4.79	28.67
	all	626	5.1	0.4	2.4	8.9	0.1	71.6	71.5	4.26	20.74
The share of EU funds in total value of renewable energy projects (%) 2014–2020	1	20	69%	4%	70%	17%	41%	95%	54%	−0.29	−0.80
	2	103	70%	1%	73%	13%	31%	95%	64%	−0.47	0.16
	3	503	68%	1%	70%	11%	32%	88%	56%	−0.80	0.03
	all	626	69%	0.5%	70%	12%	31%	95%	64%	−0.66	0.07

* type of municipalities by degree of urbanization (DEGURBA): code 1—cities (densely populated areas); code 2—towns and suburbs (intermediate density areas), code 3—rural areas (thinly populated areas).

In case of projects carried out in collaboration, only one municipality was registered in adequate databases as the beneficiary, which significantly influences the interpretation of data on the number of beneficiaries and the number of renewable energy projects locations, shown in Figure 1 for the financial perspective 2007–2013 and in Figure 2 for 2014–2020.

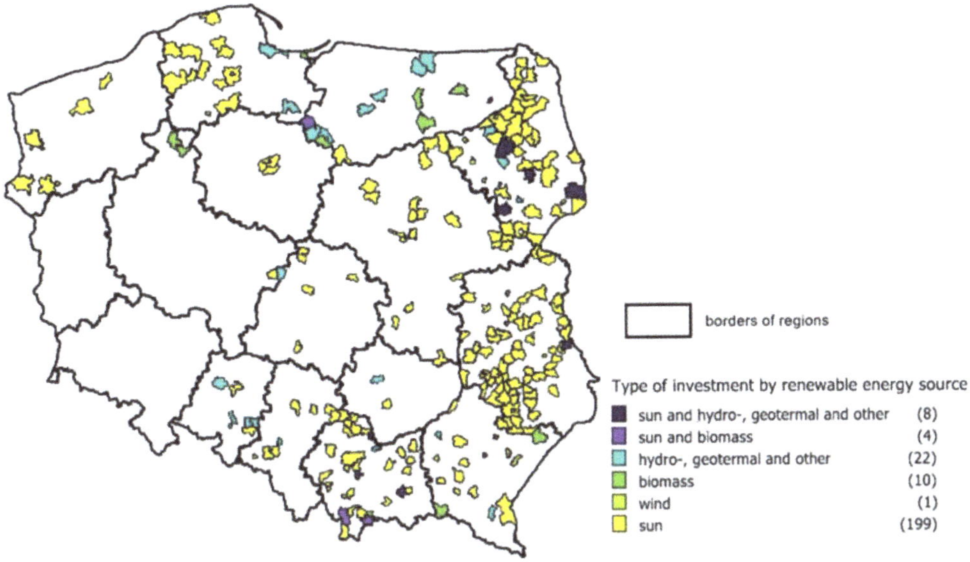

Figure 1. Location of renewable energy projects carried out by local authorities under operational programs 2007–2013, by the source of renewable energy.

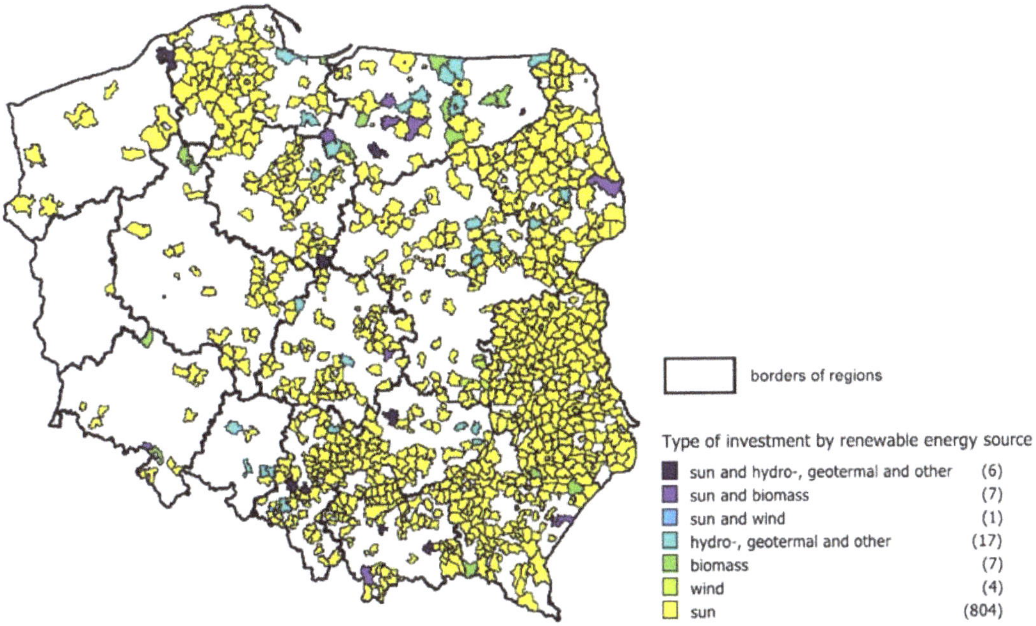

Figure 2. Location of renewable energy projects carried out by local authorities under operational programs 2014–2020 by the source of renewable energy.

The increased collaboration means that there are more municipalities, actual beneficiaries of EU funding for renewable energy projects, than those shown in databases as 'leading beneficiaries'.

There are significant regional differences in the behavior of local authorities in the field of renewable energy investments under operational programs. More renewable energy projects were carried out in Podlaskie and Lubelskie Voivodships of Eastern Poland starting from 2007–2013, and the largest increase was also noted in those two regions. However, the number of renewable energy projects carried out by local governments also increased in regions of southern and northern parts of the country. Local governments of central regions were moderately active, and those in eastern Poland the least active in carrying out renewable energy projects co-financed by EU funds.

In both financial perspectives the renewable energy projects were co-financed by EU funds from the European Regional Development Fund. In 2007–2013 local authorities obtained these funds from the Operational Programme Infrastructure and Environment 2007–2013 (OPIE), which received the highest EU funding for any operational program in the history of EU regional and cohesion policy and from 16 regional operational programmes. In 2014–2020 the EU funds were obtained from 16 regional operational programmes. We tested whether the type of municipalities affected the total value of renewable energy projects and the share of EU funds in the total value of renewable energy projects using Kruskal-Wallis test (Table 5). The results show that the total value of renewable energy projects was not significantly affected by the types of municipalities in 2007–2013, but it was significantly affected in 2014–2020. The shares of EU funds in the total value of renewable energy projects were significantly affected by the types of municipalities neither in 2007–2013 nor in 2014–2020.

Table 5. Kruskal-Wallis tests for the type of municipalities vs. total value of renewable energy projects and the share of EU funds in the total value of renewable energy projects in 2007–2013 and 2014–2020.

Kruskal-Wallis Tests Outcomes on	$H(2)$	p
The types of municipalities affecting the total value of renewable energy projects		
in 2007–2013	4.918	>0.86
in 2014–2020	12.629	>0.002
The types of municipalities affecting the share of EU funds in the total value of renewable energy projects		
in 2007–2013	1.991	>0.37
in 2014–2020	2.157	>0.34

To investigate whether the total value, the value of European Union funds and their share in the total value of renewable energy projects differed significantly between cities, towns and suburbs, and rural municipalities in the consecutive financial perspectives, we ran Mann-Whitney tests, and Wilcoxon rank-sum test (Table 6), illustrated with descriptive statistics for these variables in (Tables 3 and 4).

Based on the results shown in Table 4 we conclude that these variables for cities (1) did not differ significantly from those for towns and suburbs (2), both in 2007–2013 and in 2014–2020. Most of the analyzed variables describing renewable energy projects by cities (1) did not differ significantly from those describing this kind of projects by rural municipalities (3). There are only two exceptions. The first is that the value of EU funds for the analyzed projects by cities was significantly different from that obtained for projects by rural municipalities in 2014–2020. The second is that the share of EU funds in the total value of renewable energy projects carried out by the cities was significantly different from the share of EU funds in the total value of renewable energy projects carried out by rural municipalities in 2007–2013.

Table 6. Mann-Whitney U test for analyzed variables.

Tests for Groups of Municipalities ***	Total value of Renewable Energy Projects in		European Union Funds Cofounding Renewable Energy Projects in		% of European Union Funds in Total Value of Renewable Energy Projects in	
	A *	B **	A *	B **	A *	B **
(1) and (2)						
Mann-Whitney (*U*)	203.000	890.000	204.000	925.000	175.000	970.500
Wilcoxon rank-sum (*W*)	1688.000	6246.000	1689.000	6281.000	211.000	6326.500
Z	−0.273	−0.960	−0.252	−0.720	−0.861	−0.408
Effect size (*r*)	−0.020	−0.071	−0.019	−0.053	−0.063	−0.030
Asympt. Sig. (*2-tailed*)	0.785	0.337	0.801	0.472	0.389	0.683
Mt. Carlo Sig. (*1-tailed*) b	0.794	0.335	0.815	0.468	0.394	0.687
(1) and (3)						
Mann-Whitney (*U*)	576.000	549.000	449.500	3426.500	3505.500	4882.000
Wilcoxon rank-sum (*W*)	612.000	585.000	485.500	130,182.500	130,261.500	5092.000
Z	−0.607	−0.802	−1.522	−2.419	−2.300	−0.223
Effect size (*r*)	−0.023	−0.030	−0.057	−0.091	−0.086	−0.008
Asympt. Sig. (*2-tailed*)	0.544	0.422	0.128	0.016	0.021	0.823
Mt. Carlo Sig. (*1-tailed*) b	0.540	0.419	0.132	0.014	0.020	0.826
(2) and (3)						
Mann-Whitney (*U*)	3571.000	3520.000	4350.000	21,382.000	21,146.000	23,530.000
Wilcoxon rank-sum (*W*)	5056.000	5005.000	5835.000	148,138.000	147,902.000	28,886.000
Z	−2.187	−2.313	−0.260	−2.794	−2.939	−1.467
Effect size (*r*)	−0.076	−0.080	−0.009	−0.097	−0.102	−0.051
Asympt. Sig. (*2-tailed*)	0.029	0.021	0.795	0.005	0.003	0.142
Mt. Carlo Sig. (*1-tailed*) b	0.029	0.021	0.801	0.005	0.004	0.147

A *—2007–2013 financial perspective; B *—2014–2020 financial perspective; b—based on a sample of 10,000 tables at the starting number 92,208,573 of the generator of random numbers, *** types of municipalities coded as: (1) cities, (2) towns and suburbs and (3) rural municipalities. Grey cells show significant differences.

Some significant differences appeared between rural municipalities and those classified as towns and suburbs in the two financial perspectives. The total value of renewable energy projects carried by towns and suburbs was significantly different from that obtained by rural municipalities in 2007–2013 and similarly in 2014–2020. Alike in case of cities and rural municipalities, in 2014–2020 the value of EU funds for renewable energy projects by towns and suburbs was significantly different from that obtained by rural municipalities.

We tested whether the total value, the value of EU funding and the share of EU funding in the total value varied in the analyzed two financial perspectives using a Wilcoxon signed-rank test (Table 7). Its results show, that both the total value of projects and the value of EU funds were higher in 2014–2020 than in 2007–2013, while the share of EU funds in the total value of renewable energy projects decreased in 2014–2020, which is explained by the fact that the increase in total value was higher than the increase in the value of EU funds.

Table 7. Wilcoxon signed-rank test for analyzed variables.

	Median 2007–2013 (mln Polish Zloties)	Median 2014–2020 (mln Polish Zloties)	T	*p*-Value	Change r
The total value of renewable energy projects	2.9	3.5	2021	<0.05	−0.29
European Union funds value	2.1	2.4	2217	<0.05	−0.22
The share of EU funds in the total value of renewable energy projects	76%	70%	1632	<0.05	−0.49

5. Discussion

Local authorities of 85% of the analyzed municipalities declared deployment of renewable energy in their municipalities as one of the aims of development strategies. However, the strategies themselves contained rather general information on the type of renewable energy that could be used. Additionally, none of the strategies involved analysis of the conditions for renewable energy development or the sources of funds for such investments. It shows that, although the strategies can be an important instrument of encouraging renewable energy development at the local and regional level in other countries [92–95], they are not used this way in Poland yet.

Although such a large share of local authorities declares deployment of renewable energy, many less of them find it the most urgent need. Knowing the local conditions and being obliged by the law to perform obligatory tasks, local authorities indicate construction or modernisation of roads, sewage systems, water system, support for home sewage system plants, and less of them together with water treatment plants and landfills as the most urgent. This hierarchy is also caused by local budgets constraints that do not allow local authorities to extend their investments beyond the obligatory tasks. These obstacles in developing renewable energy are also observed in other countries [96–98]. Referring to this problem, local authorities pointed at EU funds as a crucial factor moderating budget constraints. This is in line with many studies which prove that the success in introducing renewable energy largely depends on public aid [13].

Findings of our study prove that there is a substantial progress in the number and scope of investments in renewable energy carried out by local authorities in Poland under operational programmes 2007–2013 and 2014–2020. Taking into consideration rural-urban classification of municipalities, the cities, towns and suburbs doubled their renewable energy projects co-financed by the EU funds, while rural municipalities even tripled them. This is an important outcome as renewable energy is said to be an under-utilised resource both in urban and rural areas [99–101]. The increase in the intensity of renewable energy projects is reflected also by the fact that during 2014–2020 more communities carried out more than 1 renewable energy project.

In both analysed financial perspectives there were more renewable energy investments in the north and east of Poland. The numerous renewable energy projects in voivodships of Eastern Poland confirm that smaller and more remote communities may be more willing to deploy renewable energy [102], as regions of Eastern Poland are among the poorest in the EU.

During 2014–2020 there was a more widespread collaboration between local authorities and between local authorities and residents in carrying out renewable energy projects. The collaboration with the local population may increase their acceptance and support for developing renewable energy sources. Local acceptance is recognised as one of the main determinants of deploying renewable energy at the local level and transition towards decentralized energy systems and achieving regional energy self-sufficiency [32,45,103,104]. Mutual projects carried out by local authorities and municipality residents can have an added value. They may enable to 'experience benefits' and start a 'local participatory process' [105], by encouraging others to benefit from renewable energy and increase their support and commitment to renewable energy development. This is important as resistance or unwillingness of local population to new energy infrastructure may cause conflicts [106] and hinder achieving the EU goals [107]. None of the projects showed collaboration between local authorities and university, industry, or government, which could be favorable to increase renewable energy at local level [108].

Renewable energy projects carried out by communities used mainly sun renewable energy, which is the most easily accessible and relatively least controversial. The effects of projects improved energy efficiency of public buildings and public utilities, as well as family homes in case of projects carried out in collaboration with residents. Such effects are also observed in other countries [109]. The effects improved the quality of life which is are considered success factors in case of renewable energy projects carried out by local governments [104].

The use of EU funding (international aid) by local authorities proves that multilevel governance is effective and necessary to achieve the renewable energy and consequently sustainable development goals [110].

6. Limitations

As the study is based on a survey carried out in one of Polish regions and on the data for the EU supported projects, we think there may be two limitations to it. We sent the survey to all municipalities in Mazovian Voivodship and we achieved response rate of 62% for this largest region in Poland. However, local governments in other regions were not surveyed and their opinions may be different.

The other limitation may result from the fact that findings on the analysed effects of projects under operational programs tell an important, but possibly not the whole story on the behaviour of local governments towards renewable energy investments. Other sources of financing for such investments can be available to some local governments and used by them. Thus other studies should look into this matter.

7. Conclusions

According to the law, deployment of renewable energy is not an obligatory task for Polish local authorities. However, they are allowed to support its development. Based on the findings we claim that a majority of local authorities declare deployment of renewable energy as one of goals in the strategies of local development, but they do not perceive it the most urgent matter. In local governments' experience the catalogue of obligatory tasks together with budget constraints make renewable energy deployment less feasible and less urgent. Thus we conclude that the careful declarations on renewable energy deployment reflecting local authorities' attitudes are caused by the awareness of financial (dis)abilities of local budgets. However, possibilities of obtaining public aid, here EU funds, are a stimulus affecting local authorities' behaviour positively, resulting in the increasing number and scope of investments in renewable energy between consecutive financial perspectives.

There were significant differences in the number of local governments' renewable energy investments carried out in the regions of Western and Eastern Poland, but there were no regional differences in the sources—a prevailing majority of the analyzed projects used sun renewable energy. As many projects were carried out jointly by either different local governments or by local governments and residents, they improved energy efficiency of not only public buildings and public utilities, but also family homes. So it is recommended to promote such practices and achieved benefits of local collaboration to start or strengthen participatory processes in more local communities.

Drawing up on the findings and discussion, it is recommended that more public aid should be addressed to co-finance renewable energy projects carried out by local authorities. The principles of co-financing should be evidence-based and should promote practices that are best in a given social, economic and environmental context. Thus, further research into the attitudes and behaviour of local authorities and other actors should be carried out to provide a basis for multilevel decisions on the deployment of renewable energy.

Author Contributions: Conceptualization, J.R.; methodology, J.R. and I.O.; validation, J.R.; formal analysis, J.R. and I.O.; investigation, J.R. and I.O.; data curation, J.R.; writing—original draft preparation, J.R. and I.O.; writing—review and editing J.R.; visualization, J.R.; supervision, J.R. and I.O.; project administration, J.R.; funding acquisition, J.R. and I.O. Both authors have read and agreed to the published version of the manuscript.

Funding: This research received no external funding.

Institutional Review Board Statement: Not applicable.

Informed Consent Statement: Not applicable.

Data Availability Statement: Not applicable.

Conflicts of Interest: The authors declare no conflict of interest.

References

1. UNFCCC. United Nations Framework Convention on Climate Change. 1994. Available online: https://unfccc.int/process-and-meetings/the-convention/what-is-the-united-nations-framework-convention-on-climate-change (accessed on 15 February 2021).
2. Kyoto Protocol. Reference Manual on Accounting of Emissions and Assigned Amount, UNFCCC. 1997. Available online: https://unfccc.int/sites/default/files/08_unfccc_kp_ref_manual.pdf (accessed on 15 December 2020).
3. Paris Agreement. United Nations 2015. Available online: https://unfccc.int/sites/default/files/english_paris_agreement.pdf (accessed on 30 September 2020).
4. European Council. *2030 Climate and Energy Policy Framework*; European Council: Brussels, Belgium, 2014.
5. Lucas, J.N.V.; Francés, G.E.; González, E.S.M. Energy security and renewable energy deployment in the EU: Liaisons Dangereuses or Virtuous Circle? *Renew. Sustain. Energy Rev.* **2016**, *62*, 1032–1046. [CrossRef]
6. Benedek, J.; Tihamér-Tibor, S.; Bartók, B. Evaluation of renewable energy sources in peripheral areas and renewable energy-based rural development. *Renew. Sustain. Energy Rev.* **2018**, *90*, 516–535. [CrossRef]
7. *COM(2010) 639 Final, COM(2010) 639 Final of 10 November 2010 'Energy 2020: A Strategy for Competitive, Sustainable and Secure Energy'*; SEC(2010) 1346; European Commission: Brussels, Belgium, 10 November 2010.
8. Negro, S.O.; Alkemade, F.; Hekkert, M.P. Why does renewable energy diffuse so slowly? A review of innovation system problems. *Renew. Sustain. Energy Rev.* **2012**, *16*, 3836–3846. [CrossRef]
9. Alonso, P.M.; Hewitt, R.; Díaz-Pacheco, J.; Bermejo, L.R.; Hernández-Jiménez, V.; Guillén, J.V.; Bressers, H.; de Boer, C. Losing the roadmap: Renewable energy paralysis in Spain and its implications for the EU low carbon economy. *Renew. Energy* **2016**, *89*, 680–694. [CrossRef]
10. OECD. *Linking Renewable Energy to Rural Development, OECD Green Growth Studies*; OECD Publishing: Paris, France, 2016. [CrossRef]
11. Leikučs, J.; Strīķis, V. An overview of legislative and institutional barriers of Latvia's renewable energy production. *Manag. Theory Stud. Rural Bus. Infrastruct. Dev.* **2011**, *2*, 146–155.
12. Rennkamp, B.; Haunss, S.; Wongsa, K.; Ortega, A.; Casamadrid, E. Competing coalitions: The politics of renewable energy and fossil fuels in Mexico, South Africa and Thailand. *Energy Res. Soc. Sci.* **2017**, *34*, 214–223. [CrossRef]
13. Stokes, L.C.; Breetz, H.L. Politics in the U.S. energy transition: Case studies of solar, wind, biofuels and electric vehicles policy. *Energy Policy* **2018**, *113*, 76–86. [CrossRef]
14. Stokes, L.C. The politics of renewable energy policies: The case of feed-in-tariffs in Ontario, Canada. *Energy Policy* **2013**, *56*, 490–500. [CrossRef]
15. Prados, M.-J. Renewable energy policy and landscape management in Andalusia, Spain: The facts. *Energy Policy* **2010**, *38*, 6900–6909. [CrossRef]
16. Cansino, J.M.; Pablo-Romero, M.P.; Román, R.; Yñiguez, R. Promoting renewable energy sources for heating and cooling in EU-27 countries. *Energy Policy* **2011**, *39*, 3803–3812. [CrossRef]
17. Peidong, Z.; Yanli, Y.; Jin, S.; Yonghong, Z.; Lisheng, W.; Xinrong, L. Opportunities and challenges for renewable energy policy in China. *Renew. Sustain. Energy Rev.* **2009**, *13*, 439–449. [CrossRef]
18. Boon, F.P.; Dieperink, C. Local civil society based renewable energy organisations in the Netherlands: Exploring the factors that stimulate their emergence and development. *Energy Policy* **2014**, *69*, 297–307. [CrossRef]
19. White, W.; Lunnan, A.; Nybakk, E.; Kulisic, B. The role of governments in renewable energy: The importance of policy consistency. *Biomass Bioenergy* **2013**, *57*, 97–105. [CrossRef]
20. EC. *Energy for the Future: Renewable Sources of Energy*; White Paper for a Community Strategy and Action Plan, COM(97) 599 Final; EC: Brussels, Belgium, 1997.
21. Blok, K. Renewable energy policies in the European Union. *Energy Policy* **2006**, *34*, 251–255. [CrossRef]
22. Klessmann, C.; Held, A.; Rathmann, M.; Ragwitz, M. Status and perspectives of renewable energy policy and deployment in the European Union—What is needed to reach the 2020 targets? *Energy Policy* **2011**, *39*, 7637–7657. [CrossRef]
23. Helm, D. The European framework for energy and climate policies. *Energy Policy* **2014**, *64*, 29–35. [CrossRef]
24. D'Adamo, R.P. Current state of renewable energies performances in the European Union: A new reference framework. *Energy Convers. Manag.* **2016**, *121*, 84–92. [CrossRef]
25. Veum, K.; Bauknecht, D. How to reach the EU renewables target by 2030? An analysis of the governance framework. *Energy Policy* **2019**, *127*, 299–307. [CrossRef]
26. Tagliapietra, S.; Zachmann, G.; Edenhofer, O.; Glachant, J.-M.; Linares, P.; Loeschel, A. The European Union energy transition: Key priorities for the next five years. *Energy Policy* **2019**, *132*, 950–954. [CrossRef]
27. Directive 2009/28/EC of the European Parliament and of the Council of 23 April 2009 on the Promotion of the Use of Energy from Renewable Sources and Amending and Subsequently Repealing Directives 2001/77/EC and 2003/30/EC ('RES Directive') (OJ L 140, 5.6.2009). Available online: https://eur-lex.europa.eu/legal-content/EN/ALL/?uri=celex%3A32009L0028 (accessed on 30 September 2020).

28. Regulation (EU) 2018/1999 of the European Parliament and of the Council of 11 December 2018 on the Governance of the Energy Union and Climate Action, amending Regulations (EC) No 663/2009 and (EC) No 715/2009 of the European Parliament and of the Council, Directives 94/22/EC, 98/70/EC, 2009/31/EC, 2009/73/EC, 2010/31/EU, 2012/27/EU and 2013/30/EU of the European Parliament and of the Council, Council Directives 2009/119/EC and (EU) 2015/652 and repealing Regulation (EU) No 525/2013 of the European Parliament and of the Council. Available online: https://eur-lex.europa.eu/legal-content/EN/TXT/?uri=uriserv%3AOJ.L_.2018.328.01.0001.01.ENG (accessed on 30 September 2020).
29. European Commission 2014. *A Policy Framework for Climate and Energy in the Period from 2020 to 2030*; COM(2014) 15 Final; European Commission: Brussels, Belgium, 2014; Available online: https://eur-lex.europa.eu/legal-content/EN/ALL/?uri=celex%3A52014DC0015 (accessed on 30 September 2020).
30. European Court of Auditors. *Cohesion Policy Funds Support to Renewable Energy Generation—Has It Achieved Good Results?* Publications Office of the European Union, European Union: Luxembourg, 2014. [CrossRef]
31. IRENA. *Global Energy Transformation: A Roadmap to 2050*; International Renewable Energy Agency: Abu Dhabi, United Arab Emirates, 2019.
32. *IEA—International Energy Agency: Cities, Towns & Renewable Energy*; Yes in My Front Yard; IEA: Paris, France, 2009.
33. Bulkeley, H.; Kern, K. Local Government and the Governing of Climate Change in Germany and the UK. *Urban Stud.* **2006**, *43*, 2237–2259. [CrossRef]
34. Müller, M.O.; Stämpfli, A.; Dold, U.; Hammer, T. Energy autarky: A conceptual framework for sustainable regional development. *Energy Policy* **2011**, *39*, 5800–5810. [CrossRef]
35. Burgess, P.J.; Rivas Casado, M.; Gavu, J.; Mead, A.; Cockerill, T.; Lord, R.; van der Horst, D.; Howard, D.C. A framework for reviewing the trade-offs between, renewable energy, food, feed and wood production at a local level. *Renew Sustain. Energy Rev.* **2012**, *16*, 129–142. [CrossRef]
36. Emelianoff, C. Local Energy Transition and Multilevel Climate Governance: The Contrasted Experiences of Two Pioneer Cities (Hannover, Germany, and Växjö, Sweden). *Urban Stud.* **2014**, *51*, 1378–1393. [CrossRef]
37. Schnidt, B.; Meister, T.; Klagge, B.; Seidl, I. Energy Cooperatives and Municipalities in Local Energy Governance Arrangements in Switzerland and Germany. *J. Environ. Dev.* **2020**, *29*, 123–146.
38. Río, P.; Burguillo, M. An empirical analysis of the impact of renewable energy employment on local sustainability. *Renew. Sustain. Energy Rev.* **2009**, *13*, 1314–1325.
39. Poggi, F.; Firmino, A.; Amado, M. Planning renewable energy in rural areas: Impacts on occupation and land use. *Energy* **2018**, *155*, 630–640. [CrossRef]
40. Dvořák, P.; Martinát, S.; Van der Horst, D.; Frantál, B.; Turečková, K. Renewable energy investment and job creation; a cross-sectoral assessment for the Czech Republic with reference to EU benchmarks. *Renew. Sustain. Energy Rev.* **2017**, *69*, 360–368. [CrossRef]
41. Shirley, R.; Kammen, D. Renewable energy sector development in the Caribbean: Current trends and lessons from history. *Energy Policy* **2013**, *57*, 244–252. [CrossRef]
42. Schneider, D.R.; Duić, N.; Bogdan, Ž. Mapping the potential for decentralized energy generation based on renewable energy sources in the Republic of Croatia. *Energy* **2007**, *32*, 1731–1744. [CrossRef]
43. Nepal, R. Roles and potentials of renewable energy in less-developed economies: The case of Nepal. *Renew. Sustain. Energy Rev.* **2012**, *16*, 2200–2206. [CrossRef]
44. Gonda, G. Utilization of renewable energies and changing energy structure at local government, as drivers of economic recovery: Hungarian case study. *Int. J. Soc. Sci. Humanit. Stud.* **2011**, *3*, 267–276.
45. Karanasios, K.; Parker, P. Explaining the Diffusion of Renewable Electricity Technologies in Canadian Remote Indigenous Communities through the Technological Innovation System Approach. *Sustainability* **2018**, *10*, 3871. [CrossRef]
46. Creutzig, F.; Goldschmidt, J.C.; Lehmann, P.; Schmid, E.; von Blücher, F.; Breyer, C.; Fernandez, B.; Jakob, M.; Knopf, B.; Lohrey, S.; et al. Catching two European birds with one renewable stone: Mitigating climate change and Eurozone crisis by an energy transition. *Renew. Sustain. Energy Rev.* **2014**, *38*, 1015–1028. [CrossRef]
47. García-Gusano, D.; Iribarren, D.; Dufour, J. Towards Energy Self-sufficiency in Large Metropolitan Areas: Business Opportunities on Renewable Electricity in Madrid. In *Renewable Energies*; García Márquez, F., Karyotakis, A., Papaelias, M., Eds.; Springer International Publishing: Cham, Switzerland, 2018. [CrossRef]
48. Li, L.W.; Birmelea, J.; Schaicha, H.; Werner, K. Transitioning to community-owned renewable energy: Lessons from Germany. *Procedia Environ. Sci.* **2013**, *17*, 719–728. [CrossRef]
49. Lowitzsch, J.; Hoicka, C.E.; van Tulder, F.J. Renewable energy communities under the 2019 European Clean Energy Package—Governance model for the energy clusters of the future? *Renew. Sustain. Energy Rev.* **2020**, *122*. [CrossRef]
50. Lilliestam, J.; Hanger, S. Shades of green: Centralisation, decentralisation and controversy among European renewable electricity visions. *Energy Res. Soc. Sci.* **2016**, *17*, 20–29. [CrossRef]
51. McKenna, R.; Bertsch, V.; Mainzer, K.; Fichtner, W. Combining local preferences with multi-criteria decision analysis and linear optimization to develop feasible energy concepts in small communities. *Eur. J. Oper. Res.* **2018**, *268*, 1092–1110. [CrossRef]
52. Tröndle, T.; Pfenninger, S.; Lilliestam, J. Home-made or imported: On the possibility for renewable electricity autarky on all scales in Europe. *Energy Strategy Rev.* **2019**, *26*, 100388. [CrossRef]

53. Kostevsek, A.; Cizelj, L.; Petek, J.; Cucek, L.; Varbanov, P.S.; Klemes, J.J.; Pivec, A. Use of renewables in rural municipalities' integrated energy systems. *Chem. Eng. Trans.* **2013**, *35*, 895–900.
54. Petersen, J.P. Energy concepts for self-supplying communities based on local and renewable energy sources: A case study from northern Germany. *Sustain. Cities Soc.* **2016**, *26*, 1–8. [CrossRef]
55. Engelken, M.; Römer, B.; Drescher, M.; Welpe, I. Transforming the energy system: Why municipalities strive for energy self-sufficiency. *Energy Policy* **2016**, *98*, 365–377. [CrossRef]
56. Mathiesen, B.V.; Lund, H.; Karlsson, K. 100% Renewable energy systems, climate mitigation and economic growth. *Appl. Energy* **2011**, *88*, 488–501. [CrossRef]
57. Fraser, T. Does social capital boost or block renewable energy sitting? South African solar politics in comparison. *Energy Res. Soc. Sci.* **2021**, *71*, 101845. [CrossRef]
58. Comodi, G.; Cioccolanti, L.; Polonara, F.; Brandoni, C. Local authorities in the context of energy and climate policy. *Energy Policy* **2012**, *51*, 737–748. [CrossRef]
59. Cheung, G.; Davies, P.J.; Trück, S. Transforming urban energy systems: The role of local governments' regional energy master plan. *J. Clean. Prod.* **2019**, *220*, 655–667. [CrossRef]
60. Fouché, E.; Brent, A. Journey towards Renewable Energy for Sustainable Development at the Local Government Level: The Case of Hessequa Municipality in South Africa. *Sustainability* **2019**, *11*, 755. [CrossRef]
61. Mey, F.; Diesendorf, M.; MacGill, I. Can local government play a greater role for community renewable energy? A case study from Australia. *Energy Res. Soc. Sci.* **2016**, *21*, 33–43. [CrossRef]
62. *United Nations Sustainable Development Goals Report 2016*; U.N. Publications: New York, NY, USA, 2016.
63. Sperling, K.; Hvelplund, F.; Mathiesen, B.V. Centralisation and decentralisation in strategic municipal energy planning in Denmark. *Energy Policy* **2011**, *39*, 1338–1351. [CrossRef]
64. Gronkowska, J. Model energy cluster—Special energy zone delivering integrated territorial energy. *Geomat. Landmanag. Landsc.* **2017**, *3*, 47–57. [CrossRef]
65. Węglarz, A.; Winkowska, E.; Wójcik, W. *Gospodarka Niskoemisyjna Zaczyna się w Gminie. Podręcznik dla Polskich Samorządów*; Adelphi: Berlin, Germany, 2015.
66. Kosiński, E.; Trupkiewicz, M. Gmina jako podmiot systemu wspierania wytwarzania energii elektrycznej z odnawialnych źródeł energii (A municipality as a part of the suport system for generetion of electricity from renewable energy sources). *Ruch Praw. Ekon. Socjol.* **2016**, *78*, 93–107.
67. *Kompendium Wiedzy o Unii Europejskiej*; Małuszyńska, E.; Gruchman, B. (Eds.) Wyd. Naukowe PWN: Warsaw, Poland, 2005.
68. European Charter of Local Self-Government. 1985. Available online: https://www.coe.int/en/web/conventions/full-list/-/conventions/rms/090000168007a088 (accessed on 30 September 2019).
69. The Act of 10 April 1997, Energy Law, Ustawa z dnia 10 kwietnia 1997 r. Prawo energetyczne. Tekst jedn.: DzU z 2020 r. poz. 833 z późn. zm. Available online: http://isap.sejm.gov.pl/isap.nsf/DocDetails.xsp?id=wdu19970540348 (accessed on 30 September 2020).
70. The Act of 20 February 2015 on Renewable Energy Sources, Ustawa z dnia 20 lutego 2015 r. o odnawialnych źródłach energii. Tekst jedn.: DzU z 2020 r. poz. 261 z późn. zm. Available online: http://isap.sejm.gov.pl/isap.nsf/DocDetails.xsp?id=wdu20150000478 (accessed on 30 September 2019).
71. The Act of May 20, 2016 on Investments in Wind Farms. Ustawa z dnia 20 maja 2016 r. o inwestycjach w zakresie elektrowni wiatrowych. Tekst jedn.: DzU z 2020 r. poz. 981 z późn. zm. Available online: https://isap.sejm.gov.pl/isap.nsf/DocDetails.xsp?id=WDU20160000961 (accessed on 30 September 2019).
72. Rakowska, J. *Fundusze unijne jako czynnik rozwoju obszarów wiejskich w Polsce w świetle teorii rozwoju lokalnego*; Wydawnictwo SGGW: Warsaw, Poland, 2019.
73. The Act of March 8, 1990 on the Municipality Self-Government, JoL of 2020, Item 713. Available online: https://isap.sejm.gov.pl/isap.nsf/DocDetails.xsp?id=WDU19900160095 (accessed on 30 September 2019).
74. Local Administrative Units (LAU). Available online: https://ec.europa.eu/eurostat/web/nuts/local-administrative-units (accessed on 30 September 2020).
75. Szyrski, M. *Rola samorządu terytorialnego w rozwoju odnawialnych źródeł energii (OZE). Analiza administracyjnoprawna*; Wolters Kluwer: Warsaw, Poland, 2017.
76. Swora, M. Organy właściwe w sprawach energetyki (charakter, zadania i kompetencje). In *System Prawa Administracyjnego, t. 8B, Publiczne prawo gospodarcze*; Hauser, H., Niewiadomski, Z., Wróbel, A., Eds.; C.H. Beck: Warsaw, Poland, 2013.
77. Szyrski, M. Ocena realizacji konstytucyjnej zasady pomocniczości w prawie odnawialnych źródeł energii. *Samorząd Terytorialny* **2018**, *5*, 21–31.
78. *Directive (EU) 2018/2001 of the European Parliament and of the Council of 11 December 2018 on the Promotion of the Use of Energy from Renewable Sources (Text with EEA Relevance) OJ L 328*; European Commission: Brussels, Belgium, 21 December 2018; pp. 82–209.
79. Rakowska, J. European Union Regional Policy Support for Investments in Renewable Energy in Rural Areas of the Mazovian Voivodship. *Roczniki Naukowe Stowarzyszenia Ekonomistów Rolnictwa i Agrobiznesu* **2020**, *22*, 279–288. [CrossRef]
80. NUTS 1 Classification. Available online: https://ec.europa.eu/eurostat/documents/345175/7451602/NUTS1-changes-2016-to-2021.pdf (accessed on 30 September 2020).
81. Silverman, D. *Doing Qualitative Research*, 2nd ed.; Sage Publications: London, UK, 2005.
82. Charmaz, K. *Teoria Ugruntowana. Praktyczny Przewodnik po Analizie Jakościowej*; Wydawnictwo PWN: Warsaw, Poland, 2009.

83. Glaser, B.G.; Strauss, A.L. *Odkrywanie Teorii Ugruntowanej. Strategie Badania Jakościowego*; Nomos: Cracov, Poland, 2009.
84. National Information System SIMIK. Available online: https://archiwum.miir.gov.pl/strony/zadania/fundusze-europejskie/wykorzystanie-projekty-i-efekty-funduszy-europejskich/fundusze-europejskie-na-lata-2007-2013/projekty-zrealizowane-z-funduszy/ (accessed on 15 January 2016).
85. The n + 2 Explanation. Available online: https://ec.europa.eu/regional_policy/en/policy/what/glossary/n/nplus2 (accessed on 21 January 2021).
86. *Database of Implemented EU Projects*; Ministry of Development Funds and Regional Policy: Warsaw, Poland. Available online: https://www.funduszeeuropejskie.gov.pl/strony/o-funduszach/projekty/lista-projektow/lista-projektow-realizowanychz-funduszy-europejskich-w-polsce-w-latach-2014-2020/ (accessed on 22 September 2020).
87. DEGURBA Classification. Available online: https://ec.europa.eu/eurostat/ramon/miscellaneous/index.cfm?TargetUrl=DSP_DEGURBA (accessed on 21 January 2021).
88. National Court Register. Available online: https://www.gov.pl/web/sprawiedliwosc/krajowy-rejestr-sadowy (accessed on 21 January 2021).
89. National Official Register of the Territorial Division of the Country (TERYT). Available online: https://eteryt.stat.gov.pl/eTeryt/rejestr_teryt/udostepnianie_danych/baza_teryt/uzytkownicy_indywidualni/wyszukiwanie/wyszukiwanie.aspx?contrast=default (accessed on 10 January 2021).
90. Biuletyn Informacji Publicznej (BIP). Available online: https://www.bip.gov.pl (accessed on 21 January 2021).
91. Berkowitz, S. Using Qualitative and Mixed Method Approaches. In *Needs Assessment: A Creative and Practical Guide for Social Scientists*; Reviere, R., Berkowitz, S., Carter, C.C., Graves-Ferguson, C., Eds.; Taylor & Francis Publishing House: Washington, DC, USA, 1996.
92. Jones, E.; Leach, M. Devolving Residential Energy Efficiency Responsibility to Local Government: The case of HECA. *Local Environ.* **2000**, *5*, 69–81. [CrossRef]
93. Terrados, J.; Almonacid, G.; Hontoria, L. Regional energy planning through SWOT analysis and strategic planning tools: Impact on renewables development. *Renew. Sustain. Energy Rev.* **2007**, *11*, 1275–1287. [CrossRef]
94. Petersen, J.P. The application of municipal renewable energy policies at community level in Denmark: A taxonomy of implementation challenges. *Sustain. Cities Soc.* **2018**, *38*, 205–218. [CrossRef]
95. Thellufsen, J.K.; Lund, H. Roles of local and national energy systems in the integration of renewable energy. *Appl. Energy* **2016**, *183*, 419–429. [CrossRef]
96. Collier, U. Local authorities and climate protection in the European Union: Putting subsidiarity into practice? *Local Environ.* **1997**, *2*, 39–57. [CrossRef]
97. Morell, I.A. Can Renewable Energy Contributo to Poverty Reduction? A Case Study on Romafa, a Hungarian LEADER. In *Evaluating the European Approach to Rural Development: Grass-Roots Experiences of the LEADER Programme*; Granberg, L., Adersson, K., Kovách, I., Eds.; Routledge Taylor & Francis Group: London, UK; New York, NY, USA, 2015.
98. Busch, H.; McCormick, K. Local power: Exploring the motivations of mayors and key success factors for local municipalities to go 100% renewable energy. *Energy Sustain. Soc.* **2014**, *4*, 1–15. [CrossRef]
99. Villamor, E.; Akizu-Gardoki, O.; Azurza, O.; Urkidi, L.; Campos-Celador, A.; Basurko, I.; Barcena Hinojal, I. European Cities in the Energy Transition: A Preliminary Analysis of 27 Cities. *Energies* **2020**, *13*, 1315. [CrossRef]
100. Hoesen, V.J.; Letendre, S. Evaluating potential renewable energy resources in Poultney, Vermont: A GIS-based approach to supporting rural community energy planning. *Renew. Energy* **2010**, *35*, 2114–2122. [CrossRef]
101. Sait, M.A.; Chigbu, U.E.; Hamiduddin, I.; De Vries, W.T. Renewable Energy as an Underutilised Resource in Cities: Germany's 'Energiewende' and Lessons for Post-Brexit Cities in the United Kingdom. *Resources* **2019**, *8*, 7. [CrossRef]
102. St. Denis, G.; Parker, P. Community energy planning in Canada: The role of renewable energy. *Renew. Sustain. Energy Rev.* **2009**, *13*, 2088–2095. [CrossRef]
103. Michalena, E.; Angeon, V. Local challenges in the promotion of renewable energy sources: The case of Crete. *Energy Policy* **2009**, *37*, 2018–2026. [CrossRef]
104. Allman, L.; Fleming, P.; Wallace, A. The progress of English and Welsh local authorities in addressing climate change. *Local Environ.* **2004**, *9*, 271–283. [CrossRef]
105. Süsser, D.; Kannen, A. 'Renewables? Yes, please!': Perceptions and assessment of community transition induced by renewable-energy projects in North Frisia. *Sustain. Sci.* **2017**, *12*, 563–578. [CrossRef]
106. Walker, G. Renewable energy and the public. *Land Use Policy* **1995**, *12*, 49–59. [CrossRef]
107. Azarova, V.; Jed, C.; Friedl, C.; Reichl, J. Designing local renewable energy communities to increase social acceptance: Evidence from a choice experiment in Austria, Germany, Italy, and Switzerland. *Energy Policy* **2019**, *132*, 1176–1183. [CrossRef]
108. Lerman, L.V.; Gerstlberger, W.; Ferreira Lima, M.; Frank, A.G. How governments, universities, and companies contribute to renewable energy development? A municipal innovation policy perspective of the triple helix. *Energy Res. Soc. Sci.* **2021**, *71*, 101854. [CrossRef]
109. Fiaschi, D.; Bandinelli, R.; Conti, S. A case study for energy issues of public buildings and utilities in a small municipality: Investigation of possible improvements and integration with renewables. *Appl. Energy* **2012**, *97*, 101–114. [CrossRef]
110. Bulkeley, H.; Betsil, M.E. Rethinking Sustainable Cities: Multilevel Governance and the 'Urban' Politics of Climate Change. *Environ. Politics* **2005**, *14*, 42–63. [CrossRef]

energies

MDPI

Article

Energy Policy of European Union Member States in the Context of Renewable Energy Sources Development

Wiesław Musiał [1], Monika Zioło [2,*], Lidia Luty [2] and Kamila Musiał [3]

1 Department of Economics and Food Economy, University of Agriculture in Krakow, al. Mickiewicza 21, 31-120 Kraków, Poland; wieslaw.musial@urk.edu.pl
2 Department of Statistics and Social Policy, University of Agriculture in Krakow, al. Mickiewicza 21, 31-120 Kraków, Poland; lidia.luty@urk.edu.pl
3 Department of Production Systems and Environment, National Research Institute of Animal Production, ul. Krakowska 1, 32-083 Balice, Poland; kamila.musial@iz.edu.pl
* Correspondence: monika.ziolo@gmail.com

Abstract: As a consequence of increasing air pollution, the European Commission has decided to introduce special directives laying down the measures to achieve climate and energy neutrality. Renewable energy (RE) sources play an important role in the pursuit of these goals, which has been taken into account in the 2030 Agenda for Sustainable Development. The aim of this article is to describe patterns and trends in the achievements of the energy policy of European Union (EU) countries in the field of renewable energy in sustainable development. The identification of leaders in this field gives the possibility to analyse actions taken by the governments of these countries and the possible implementation of the introduced solutions on the ground of individual Member States at the regional and national levels. At the beginning Main goal of energy policy on the field of renewable energy sources (RES) is to increase production from environmentally friendly sources that is why trends were determined in order to assess the rate of achievement of the national target for changes the share of energy from renewable sources in total gross energy consumption. Groups of similar countries were then identified on the basis of three indicators corresponding to the targets set in the climate and energy package. In the group of analysed countries, 14 have achieved the 2020 targets and 4 have exceeded the 2030 targets. The main renewable energy sources (RES) are biofuels, wind, and hydropower. In the assessment of the achievement of energy policy targets, the best situation was observed in the case of Denmark, Ireland, and the United Kingdom. These countries have significantly increased the share of renewable energy in total energy consumption. Compared to other EU countries, they have reduced the economy's energy consumption and greenhouse gas emissions the most.

Keywords: renewable energy sources; sustainable development; sustainable energy; energy policy

Citation: Musiał, W.; Zioło, M.; Luty, L.; Musiał, K. Energy Policy of European Union Member States in the Context of Renewable Energy Sources Development. *Energies* **2021**, *14*, 2864. https://doi.org/10.3390/en14102864

Academic Editors: Piotr Gradziuk, Vincenzo Bianco, Bogdan Klepacki and Mariusz J. Stolarski

Received: 29 March 2021
Accepted: 10 May 2021
Published: 15 May 2021

1. Introduction

Sustainable energy management is undoubtedly related to increased use of renewable energy sources, which ensure energy security, diversify energy supplies, and maintain and improve the environmental and life quality of local communities. Renewable energy sources play an important role in the concept of sustainable development and sustainable energy.

Most studies in the field of renewable energy sources (RES) deal with the research on determinants of the development of this type of energy in the EU countries ([1–3]. The subject of numerous articles is the search for a connection between Gross Domestic Product (GDP) and renewable energy (RE) [4–6]. Among those, who have carried out research on the grouping of the EU countries in terms of different variables characterising the development of RES, are the studies by Neizel [7], Śmiech and Papież [8], Kasman and Duman [9].

The aim of this study was to show the differences between EU countries in terms of the implementation of policies for the development of renewable energy sources (RES), as well as to identify the countries that most effectively implement these policies. The determination of leaders in this field gives a possibility to analyse actions taken by the governments of these countries and possible implementation of the introduced solutions on the ground of individual Member States on regional and national levels. The paper asks the following research questions: What are the most common types of RES used in the analysed countries? Will the EU countries achieve their national RES targets? Which countries are most effective in introducing energy policies to take environmentally friendly measures?

At the preliminary stage, groups of countries similar in terms of the structure of renewable energy production in 2018 were defined using Czekanowski's method [10]. The similarity of countries in terms of the structure of renewable energy production (by source of origin) was presented. The main goal of the energy policy in the field of RES is to increase the production of energy from environmentally friendly sources. Thus, at the second stage, the trends assessing the rate of achieving of the national targets by each of the analysed Member States were determined and the countries which at the current rate of development will meet the targets provided in the climate and energy packages for the coming years were identified. Next, a typological division of countries in terms of the level of achieving of energy policy targets using the model method of linear ordering of objects (the Technique for Order of Preference by Similarity to Ideal Solution-TOPSIS) was carried out.

The selection of variables for the assessment was based on the strategic targets of EU climate and energy policy referred to as "20-20-20", which should have been achieved by 2020. The analysis identified homogeneous groups of countries with respect to the three energy policy-specific variables included in the 2020 climate and energy package. This allowed us to compare the Member States in terms of the targets listed in the package, taking into account the effects of changes in the structure of energy production assessed in terms of three variables: the rate of change in greenhouse gas (GHG) emissions reduction, changes in the share of renewable energy sources (RES) in gross final energy consumption, and the modernisation of economies (reduction of energy intensity—EI). Information about the most successful countries in terms of achieving national targets may be helpful when formulating energy policy for the next few years in Member States. Research carried out on new technologies will make it possible in the future to obtain energy from ecological sources not only more effectively, but also more economically. Observing the solutions proposed by leaders in this field will make it possible in the future to diversify sources of environmentally friendly energy.

1.1. The Concept of Sustainable Development

Sustainable development to be regarded as a type of socio-economic development rejecting egocentric approach to development, but also departing from extreme anthropocentrism, in particular in the short-term perspective. Currently, when analysing this concept, researchers tend to focus on a new, also supra-environmental, approach emphasising intergenerational equity, the sustainability of human living environment and the quality of human life. The Brundtland Report [11], which might be regarded as an important contribution to organising the terminology relating to sustainable development, draws attention to three important implications of the proposed definition of this type of development, i.e., environmental obligations towards future generations, intra- and inter-species equity, and viewing sustainability not as a state but as a process. However, Haughton [12], when analysing the aspect of intergenerational equity, mainly focused on natural-environmental-capital. Hence, he focused in particular on the economic management of natural resources, the recirculation of resources, maintaining an appropriate relationship between consumption and investment, and ensuring demographic sustainability which is often neglected in such considerations.

Kates, Parris, and Leiserowitz [13] propose four alternative methods of defining sustainable development through goals, indicators, values and economic practice (Table 1). However, this approach does not take into account the changes in social needs, spatial components and cultural differences. It focuses on economic aspects, which undoubtedly perform a decisive role in terms of contributing to human welfare and foster or even determine the quality of life.

Table 1. Overview of approaches to the definition of sustainable development.

	Specification
Goals	Sustainable Development Goals (SDG) Sustainable development goals for people and planet—6 goals 2030 Agenda for Sustainable Development—17 goals
Indicators	Global SDG Indicators—220 indicators EU Sustainable Development Indicators (SDI)-Eurostat 2017—130 indicators Sustainable Development Indicator—56 indicators
Values	The Earth Charter
Practice	A program to improve the social determinants of health in Latin America Sustainable supply chains in production Eco-innovation

Source: Own study based on: [14–23].

Sustainable development can be defined as economic development that is stimulated by the societal demands and that is carried out with appropriate economic calculation taking environmental aspects into account. Among the main goals of sustainable development, Janka [24] lists:

- Ensuring equal opportunities in terms of access to natural assets (taking future generations into account);
- Maintaining the sustainability of all natural processes and ecosystems;
- Conservation of non-renewable resources and the possibility for renewable resources to regenerate;
- Increasing the share of environmentally-friendly projects;
- Using renewable energy sources in global economies while improving the quality of the environment and human lives. Konstańczak [25] emphasises that the idea of sustainable development is to improve both the condition of our planet and the comfort of human life through consistent action in specific areas.

In 2015, the United Nations adopted the 2030 Development Strategy. All UN member states unanimously adopted the resolution "Transforming our world: the 2030 Agenda for Sustainable Development" containing 17 Sustainable Development Goals (SDGs) to be achieved by 2030.

Focusing on the sustainable development strategy, it emphasises the need to promote a modern way of life taking into account appropriate environmental policies and philosophies that will counteract past practices of non-prospective exploitation of the earth's resources [26].

Public authorities play an important role in the implementation of the Sustainable Development Goals, as they set the targets for the protection of the environment and its resources [27]. It should be stressed that in market economy conditions public authorities do not own most of the factors necessary for the implementation of this strategy, which is why they have to use the appropriately selected instruments to trigger the activity of specific entities, and thus contribute to the implementation of environmental policy targets.

The concept of sustainable development includes many references to the management of resources, including energy resources. In many countries, in particular those with coal, but also oil and even natural gas, such resources are a factor triggering various environmental and socio-economic imbalances. Therefore, the concept of sustainable development was also transferred to the energy sector, which gave rise to the term "sustainable energy development", the most important rule of which is effective use of energy, human, economic, and natural resources [28].

Sustainable development in relation to the energy sector (sustainable energy) should be defined as the conversion of primary energy into secondary energy, i.e., electricity and heat, and its delivery to the end consumer in such a way as to meet the needs of present and future generations taking into account the economic, social, and environmental aspects of human development [29].

Renewable energy sources play an increasingly important role in the concept of sustainable development and sustainable energy. They offer hope for a green transformation with regard to energy, as well as for satisfying the demand for energy in countries without their own energy resources.

Promoting and supporting the development of the use of energy from renewable sources contributes to the compliance with sustainable development principles, such as:

- The principle of integration of environmental policy with sectoral policy-through the development of RES, environmental goals are taken into account to the same extent as economic and social objectives;
- The principle of equal access to the natural environment-the development of RES offers equal opportunities in terms of the use of natural resources and human needs;
- The socialisation principle-through the development of RES, environmental education is carried out to stimulate ecological sensitivity and to build new environmental ethics;
- The prevention principle-the development of RES imposes on the investor the obligation to assess the environmental impact of the planned project and to monitor it after project completion;
- The principle of applying best available techniques-solutions for generating energy from RES allow implementation of the best reasonable and available technologies, e.g., in the form of wind farms.

Sustainable energy development in the context of RES occurs when it concerns activities integrated on various levels, including global, national, and local, as well as in individual areas of such development: economic, social, psychological, environmental, technological, informational, political, and legal [30].

Ensuring the implementation of the above objectives and Sustainable Development Goals is a necessity resulting from the obligations of each Member State towards the EU.

1.2. Renewable Energy Sources

In the 2030 Agenda for Sustainable Development, one of the goals is to ensure universal access to "affordable, reliable, sustainable and modern energy". It is to be achieved by "increasing substantially the share of renewable energy in the global energy mix" and "promoting investment in energy infrastructure and clean energy technology" [31].

The measures taken in this respect are intended to increase the effectiveness of the fight against the progressive degradation of the environment and the increasing greenhouse gas emission, which pose a serious threat to humanity in the form of environmental pollution and adverse climate change [32]. In order to halt these processes, global developed countries are implementing the United Nations Framework Convention on Climate Change (UNFCCC). It describes the basic framework for global cooperation in this complex area. This document was supplemented by the provisions of the 2030 Agenda, the Kyoto Protocol (1997), and the Copenhagen Accord (2009). These arrangements identify measures to address the deteriorating quality of the environment [33].

The European Union also developed a strategy aimed at fulfilling international commitments in the fight against climate change as well as introducing the concept of sustainable energy [34]. The strategy is being implemented progressively in three major stages through the achievement of the programme targets set out in each of them [35,36] (Table 2).

Table 2. Energy policy targets in the European Union.

Specification	Target
20-20-20 package	→ 20% cut in greenhouse gas emissions (compared to 1990 levels) in 2020 → 20% improvement in energy efficiency 20% cut in greenhouse gas emissions (from 1990 levels) in 2020 → 20% share of EU energy from renewables in total energy consumption by 2230
Green Paper	→ 40% cut in greenhouse gas emissions (compared to 1990 levels) in 2030 → 32% share of EU energy from renewables in total energy consumption by 2030
Low Carbon Economy 2050	→ achieving climate neutrality in the EU by 2050

Source: Own study based on [27,28,37].

A key event for the development of RES in EU Member States was the agreement [38] reached during the 21st session of the Conference of the Parties held in Paris.

Its aims include combating climate change and supporting economic development to achieve more sustainable development and lower greenhouse gas emissions. The main objective of the agreement is to keep the temperature at the level between 1.5 and 2 °C higher than in the pre-industrial period. The agreement, having entered into force in 2016, was ratified by 187 countries. Signatories are required to prepare their NDC (National Determined Contribution), in which they outline methods to reduce GHG emissions, and methods to monitor the progress of its implementation.

Consequently, the EU adopted a plan under the 2030 Framework for Climate and Energy [39] to create a sustainable energy system. The plan consists of the following components:

- Improving energy efficiency;
- Providing access to affordable energy for all consumers;
- Increasing energy independence, which is important in light of the information that 55% of the energy consumed in the Member States was produced from resources originating outside the EU;
- Introducing a fully integrated common energy market (energy union);
- EU Member States becoming world leaders in producing energy from renewable sources.

It was therefore planned that renewable energy sources will play an important role in the future EU energy system.

The European Union's actions are important for achieving global energy equilibrium, as it is composed of countries with a high level of consumption, well-developed economies, and strong urbanisation.

An important role in the future EU energy system.

In 2018, energy production in the EU was at 634.751 TOE, of which 35.2% came from fossil fuels, 30.8% from nuclear power plants, and 34% from RES. The value of energy produced from RES in EU Member States was at 217.388 TOE, of which 40% was from solid biofuels, 14% from wind and hydropower, and 9% from liquid biofuels. The share of energy derived from solar and geothermal sources and from biogas was less than 5%.

In the course of strategy implementation, between 2020 and 2050 the energy mix is to be substantially modified so that by the end of the reference period (2050) 20% of energy will come from fossil fuels, 25% from nuclear power plants, and 55% from RES.

Renewable energy sources are of key importance due to rising CO_2 emissions. Therefore, economists and analysts are focusing on the particular importance of the use of renewable energy sources instead of conventional resources [40]. In their opinion, the path to a sustainable environment should lead to a reduced use of traditional energy sources, which should be replaced by RES, characterised by lower GHG emissions and environmentally-friendly technological processes [41,42]. In addition, RES support the implementation of the majority of the energy policy targets adopted by many European countries, i.e., increased diversification of supplies makes it possible to reduce the demand for imported energy. The implementation of the energy and climate package through the reduction of greenhouse gas and dust emissions will support the development of competitive markets as well as the growth of innovation and entrepreneurship of the human capital. In addition, the more widespread the sources of distributed generation and the use of regional resources for its production, the greater the likelihood of ensuring local energy security and reducing transmission losses. The main energy policy objectives in the area of RES include:

- Minimum use of forests for biomass production and maximising the use of agricultural areas for diversified RES generation;
- Significant increase in the share of biofuels in total transport fuels;
- Creating an appropriate framework for the sustainable development of distributed generation sources.

According to sustainable development standards, one of the priorities for politicians at every level should be to reduce the negative impact of the energy sector on the environment. Hence the importance of measures to mitigate the adverse effects of climate and biotic changes.

In the context of energy sources, the problem of sustainable development can be limited to the use of sources which are not significantly depleted by continued use, and whose use does not result in large-scale emissions of pollutants or other substances hazardous to the environment. It is also important that their use does not perpetuate significant risks to human life and health, but also social injustice [43].

Prandecki points out that there is no doubt that renewable sources are an example of the most sustainable energy generation technologies. However, their use must take into account many factors and conditions at the national and regional level, e.g., generating capacity, access to energy resources, as well as the demand for energy in a given area and the size of its distribution [44]. Suska-Szczerbicka emphasises that renewable energy sources are characterised by a special property, as their use in a given location does not limit the generally available energy resources; instead, the level of RES remains constant, and they are not depleted [45]. Non-renewable resources can thus be treated as an energy reserve of sorts, which can be used in "better times", i.e., when production and energy technologies are radically improved.

Therefore, the inexhaustibility, universality, and general availability of renewable energy resources, combined with an effectively pursued energy policy, encourage the increasing use of RES in energy production worldwide [46]. In particular, this is due to the fact that the potential offered by renewable energy sources is considerable and includes wind energy, solar energy, aerothermal energy, geothermal energy, hydrothermal energy, hydropower, wave, current and tidal energy, energy from biomass, biogas, agricultural biogas, and bioliquids. Society is slowly becoming aware of the existence of so many alternative energy sources in relation to conventional energy sources; however, the necessity to meet international obligations resulting from the sustainable development strategy, as well as the United Nations Framework Convention on Climate Change and the 1997 Kyoto Protocol on the reduction of GHG emissions into the atmosphere, results in intensified measures to develop the RES sector.

The importance of clean and sustainable energy to which all citizens have universal access is underscored by the fact that in its Agenda for Sustainable Development, the UN decided to highlight an area that focuses on energy. The key targets of this goal include:

- Substantially increasing the share of renewable energy in the global energy mix.
- Ensuring universal access to affordable, reliable and modern energy services, including RES.

Among the technologies available on the market, there are solutions that are financially available only to the largest entities, such as governments or enterprises, but there are also solutions dedicated to households, corresponding to their financial capacity. These are state-of-the-art technologies under continuous development, which constantly reduces their price and improves their reliability.

2. Materials and Methods

The statistical material used in the study was obtained from the Eurostat [47], World Bank [48], and International Renewable Energy Agency databases [49]. The time frame of the study covered the period from 2010 to 2019. Three countries, Cyprus, Malta, and Luxembourg, were omitted from part of the analyses due to their negligible renewable energy generation (less than 10PJ in 2018). The direction and rate of change of the share of renewable energy in gross final energy consumption was determined by identifying trend models on the basis of time series $(y_t)_{t=1,\ldots,10}$. The fitting of the trend line resulted from the analysis of the increments of the studied characteristic in the studied time. Statistical verification of trend models was also performed.

Czekanowski's method, [50–54] was used to indicate the similarities of countries in terms of the structure of renewable energy production (by source). The following components of the structure of renewable generation were selected for the analysis: hydropower, solar (thermal, photovoltaic-PV), wind, biofuels (solid, liquid), biogas, geothermal, ren-waste, and other renewable. The Manhattan metric was used as a measure of distance. Ordering of the diagram (graphical representation of the distance matrix of objects) was performed with the use of MaCzek software, version 3.3.49 [55]. Jan Czekanowski invented the original method at the beginning of the 20th century, twenty years ahead of subsequent works on classification problems. The advantage of a Czekanowski's diagram is that it presents the relationships and similarities between the studied objects and at the same time emphasises all the connections between variables. A Czekanowski's diagram, being a kind of a similarity map of objects, unlike dendrograms, also preserves information about the relationships between all pairs of objects in a series.

In the process of creating a synthetic measure for assessing the level of implementation of energy policy goals resulting from the National Renewable Energy Action Plans, the synthetic variable proposed by Hwang was applied [56]. The choice of the country ordering algorithm was preceded by the Kukuła, Luty [57] procedure supporting the selection of the linear ordering method. Under the Technique for Order of Preference by Similarity to Ideal Solution method (TOPSIS), a three-stage algorithm was adopted. Firstly, the characteristics describing the analysed phenomenon were selected using Sustainable Development Goals indicators (Table 3). The basis for the selection of variables for the assessment was the strategic EU climate and energy policy objectives referred to as "20-20-20", which should have been achieved by 2020.

Table 3. Variable description.

Variable		Full Name
X_1	RES	The change share of renewable energy in gross final energy consumption in 2019 compared to 2010 (%)
X_2	EI	The change of energy intensity level of primary energy (MJ/$2011 PPP GDP) [1] in 2015 compared to 2010 (%)
X_3	GHG	The change of greenhouse gas emissions intensity of energy consumption in 2018 compared to 2010 (%)

[1] Energy intensity level of primary energy is the ratio of energy supply to gross domestic product measured by purchasing power parity.
Source: Own study based on [47–49].

The selected variables indicate the percentage by which the figure increased (share of renewable energy in gross final energy consumption) or decreased (energy intensity level of primary energy, greenhouse gas emissions intensity of energy consumption) in the given year compared to 2010.

In the second step, the features were normalised using the feature standardisation method according to the following formula:

$$z_{ij} = \frac{x_{ij} - \overline{x}_j}{S_j} i = 1, \ldots, n, j = 1, 2, \ldots, N \quad (1)$$

where x_{ij},z_{ij}– the actual and normalised value of the feature X_j for the object i, respectively; $\overline{x}_j$,S_j– the arithmetic mean and standard deviation of the feature X_j, respectively.

Then, for each variable, vector coordinates of the pattern (z_j^+) and the anti-pattern (z_j^-) of development were determined, defined as follows:

$$z_j^+ := \max_i \{ z_{ij} \} \text{ or } z_j^- := \min_i \{ z_{ij} \} \quad (2)$$

In the third step, the values of the synthetic variable Q_i for each object were determined according to the following formula:

$$Q_i = \frac{d_i^-}{d_i^- + d_i^+} \quad (3)$$

where d_i^-, d_i^+ are the Euclidean distances-between the objects and the pattern and the anti-pattern of development defined as:

$$d_i^- = \sqrt{\sum_{j=1}^{m} (z_{ij} - z_j^-)^2} \text{ or } d_i^+ = \sqrt{\sum_{j=1}^{m} (z_{ij} - z_j^+)^2} \quad (4)$$

The highest value Q_i indicates the best object.

The value of the synthetic variable made it possible to divide the analysed objects into groups according to the following principle:

Group 1: $Q_i \in \left(\overline{Q} + S_Q, \max_i Q_i\right]$
Group 2: $Q_i \in (\overline{Q}, \overline{Q} + S_Q]$
Group 3: $Q_i \in (\overline{Q} - S_Q, \overline{Q}]$
Group 4: $Q_i \in \left[\min_i Q_i, \overline{Q} - S_Q\right]$

where $\overline{Q}$, S_Q are respectively the arithmetic mean and the standard deviation of the synthetic variable Q_i defined according to Formula (3).

The TOPSIS method has found great recognition in many fields, among others in economics or management. This method is a counterpart of Hellwig's taxonomic method of ordering objects, which takes into account both the best and the worst alternatives for measuring the adopted diagnostic variables.

3. Results

3.1. Differentiation of the European Union Member States in Terms of Renewable Energy Production

Achieving the long-term targets set in the National Renewable Energy Action Plans requires a shift from conventional resources towards low-carbon energy sources. This implies the need to invest in an energy infrastructure that generates energy from environmentally-friendly sources. The disparities in RES energy production between EU Member States were, and still are, very large (Figure 1; Table 4). In particular, these differences become visible when the Member States are divided into countries that joined the EU before 2004 (EU-14) and others (EU-11). The clear leader in the EU is Germany, where the amount of energy generated in 2018 was 1800 PJ. Among the EU-11, Poland produced the most energy from RES in 2018 (372PJ). In 2018, in the case of eight countries (Croatia, Estonia, Czechia, Hungary, Latvia, Lithuania, Slovakia, Slovenia) RES production did not exceed 100 PJ.

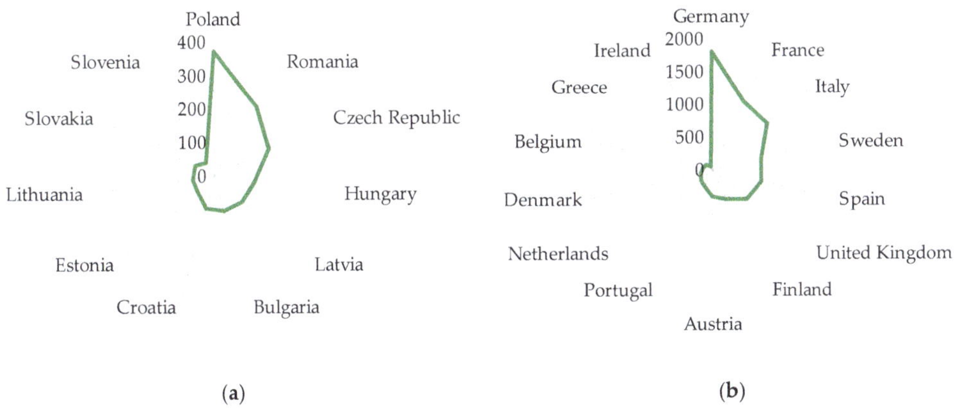

Figure 1. Renewable energy production in EU countries 2018 (PJ): (**a**) Countries that joined the EU in 2004 or later, (**b**) Countries that joined the EU before 2004. Source: Own study based on [49].

Table 4. Numerical characteristics of renewable energy production by source in EU country groups 2019 (PJ).

Source of Renewable Energy	Specification									
	(a)					(b)				
	Min	Max	Median	Mean	CV *	Min	Max	Median	Mean	CV *
Hydropower	0.1	63.6	8.8	14.9	116.9	0.1	236.8	46.3	78.4	104.6
Wind	0.0	46.1	2.3	8.2	164.9	21.0	395.8	47.7	90.5	112.4
Solar	0.0	9.4	2.4	3.0	97.0	0.4	196.8	14.8	42.0	131.8
Biofuels	23.0	295.3	69.1	102.4	69.9	11.5	632.7	211.3	248.4	75.4
Biogas	0.6	25.3	3.1	5.5	127.5	2.1	319.5	10.4	45.4	181.8
Geothermal	0.0	5.9	0.4	1.2	141.6	0.0	226.9	0.6	19.5	296.3
RenWaste	0.0	4.1	0.8	1.3	107.6	0.0	130.5	17.0	29.7	109.5
Others [1]	0.0	7.2	0.4	1.4	153.1	1.9	108.7	25.5	36.2	92.8

[1] other sources of renewable energy (not already itemised), * CV-Coefficient of Variation, (a) Countries that joined the EU in 2004 or later, (b) Countries that joined the EU before 2004. Own study based on [49].

Basic characteristics of the volume of renewable energy generation by source also indicate notable differences between the EU-14 and the EU-11 (Table 4). Biofuels are definitely an important source of renewable energy in both analysed groups, with the average production volume in the EU-14 amounting to more than double the production in the EU-11.

The second major renewable energy source in the EU-11 was hydropower, with the average national production in 2018 amounting to 14.9 PJ. This figure was 78.4 PJ in the EU-14, which was lower than the average national production from wind power (90.5 PJ).

The similarity of Member States in terms of the structure of renewable energy production (by source) using the Czekanowski method and the Manhattan metric, taking into account 2018 data, is presented in Figure 2. This figure is a graphic visualisation of the distance matrix (similar). For pairs of countries with identical structures, the distance is zero. The increasing structural differences of the compared spatial objects are accompanied by an increase in the value of the measure of similarity.

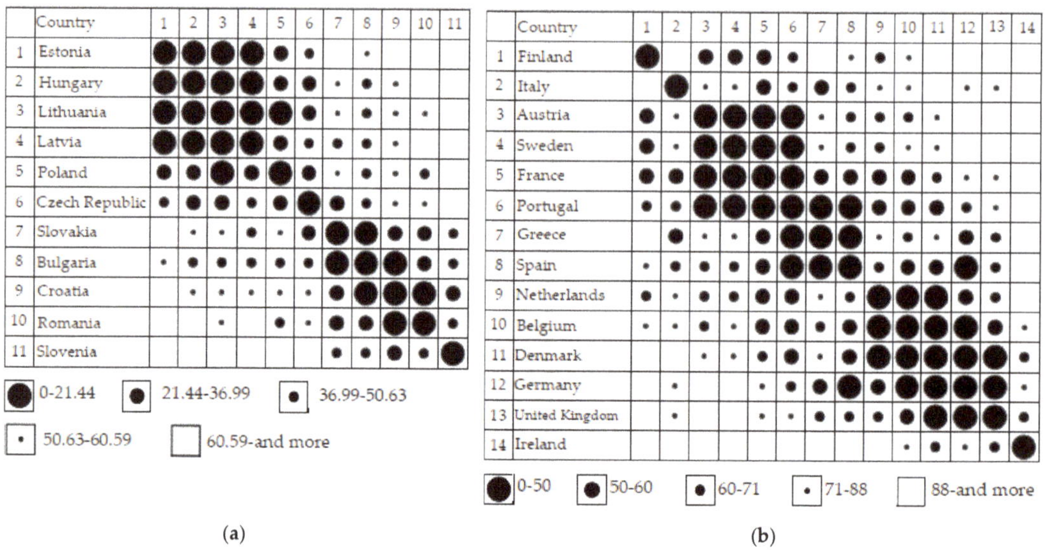

Figure 2. Similarity of EU Member States in terms of renewable energy production structure (by source) as of 2018: (**a**) Countries that joined the EU in 2004 or later, (**b**) Countries that joined the EU before 2004. Own study based on [49].

In the EU-11, countries such as Estonia, Hungary, Lithuania, and Latvia form a homogeneous group. In this group, biofuels are the main source of energy. The structure of renewable energy generation in Poland also shows significant similarity to the Member States from this group. Thus, in Estonia as much as 95% of RES is generated from biofuels. A slightly smaller percentage of energy from this source is generated in Latvia (89%), Hungary (86%) Lithuania (86%), and Poland (79%). The second largest renewable energy source in these countries in terms of share is respectively: wind (Poland 12%, Lithuania 6%, Estonia 3%), water (Latvia 7%), and geothermal (Hungary 5%).

Seventy-two percent of RES production in Czechia comes from biofuels. The second largest source in this country in terms of share is biogas (13%).

Slovakia, Bulgaria, Croatia, and Romania are another group showing significant similarity of the analysed structure. In Bulgaria and Slovakia, renewable energy is generated from biofuels (64% and 67%, respectively) and hydropower (17% and 19%, respectively). In such countries as Croatia and Romania, almost 62% of RES comes from biofuels, and 28% and 26% from hydropower, respectively.

Slovenia stands out among the EU-11 for having the highest share of renewable energy production from hydropower (38%). The country generates 51% of its renewable energy from biofuels.

In the EU-14, countries such as Finland, Italy, and Ireland are distinguished by the structure of RES generation by source. In Finland, this energy is mainly derived from biofuels (77%) and hydropower (10%). Italy stands out from all EU Member States due to significant percentage of renewable energy from geothermal sources (20%). Furthermore, 29% of renewable energy in Italy comes from biofuels and 16% from hydropower. Ireland derives 56% of its energy from wind and 21% from biofuels.

Countries such as Austria, Sweden, France, and Portugal generate about 50% of their renewable energy from biofuels. The second largest source of this energy is hydropower, whose share in total production is 33%, 28%, 20%, and 17%, respectively. For Portugal, energy generated from wind also accounts for 17% of total renewable energy.

Greece and Spain have a renewable energy generation structure similar to Portugal. In these countries, the main sources of renewable energy are biofuels, wind, solar, and hydropower.

Another group in the EU-14 are the Netherlands, Belgium, Denmark, Germany, and the United Kingdom. In these countries, the main RES are biofuels and wind. In Germany and the UK, 18% and 17% of this energy is derived from biogas, respectively.

3.2. Achieving EU Energy Policy Targets Relating to Renewable Energy Sources

In 2019, the share of RES in gross energy consumption in the EU was 18.9%, which means that the 2020 target is 1.1 pp. short of being met. Back in 2010, the share in the EU energy mix was only 13.2%. Between 2010 and 2019, the share of RES in national energy mixes increased in all EU Member States (Figure 3). The clear EU leader in terms of RES share in gross energy consumption is Sweden (56.4%), followed by Finland (43.1%), Latvia (41.0%), and Denmark (37.2%). On the other hand, in 2019 the Member States with the lowest share of RES were the Netherlands, with a share of only 8.8%, followed by Belgium (9.9%), Malta (8.5%), and Luxembourg (7.0%).

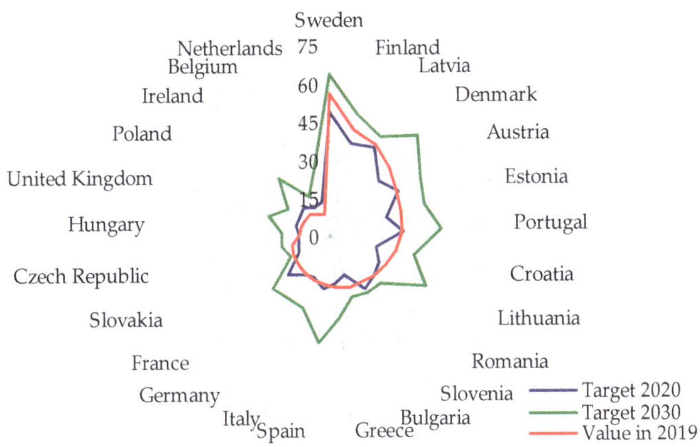

Figure 3. Share of renewable energy in gross final energy consumption in year 2019 and target 2020 and 2030 (%). Source: Own study based on [47].

In 2019, fourteen EU Member States reached the targets they had committed to achieve by 2020: Bulgaria, Czechia, Denmark, Estonia, Greece, Croatia, Italy, Latvia, Lithuania, Finland, Romania, Slovakia, Sweden, and Cyprus. Four countries, Hungary, Austria, Portugal, and Germany, were close to meeting their 2020 commitment, falling short by less than 0.7 pp. The widest gap to meet the national 2020 targets was observed for France (5.8 pp.), the Netherlands (5.2 pp.), Ireland (4.0 pp.), Belgium (3.1 pp.), and Slovenia (3.0 pp.). Other Member States fall short by 1.6 pp. (Spain) to 2.8 pp. (Poland).

According to the target set by the European Union by 2030 as part of the so-called "Winter Package", the share of RES in the EU energy mix should increase to 32%. It is worth noting that three countries have set national targets for the share of renewable energy in gross final energy consumption for at least 50%: Sweden (64%), Denmark (55%), and Finland (50%). The nine countries which have set a target of increasing their share of energy from renewable sources to no more than 25% include Poland (21%), Bulgaria (25%), Luxembourg (23%), Czechia (21%), Hungary (20%), Cyprus (19%), Slovakia (18%), Belgium (18%), and Malta (11%).

For the majority of EU Member States (excluding Hungary and Slovenia), between 2010 and 2019 there were continuous and regular changes in the share of renewable energy in gross final energy consumption (%), which allowed for the determination of trend models (Table 5). By extrapolating the fitted linear trends, forecasts for 2020 and 2030 were developed. The quality of these forecasts will depend on the stability of the economic regularity of the phenomenon over time, as they are built on the assumption that the trend observed so far will not change. In the next step, they were compared with the targets set in the national plans of respective Member States.

Table 5. Development trend models share of renewable energy in gross final energy consumption (%) and EU country projections estimated from time series 2010–2019 $(y_t)_{t=1,\ldots,10}$.

Country	$\hat{y}_t = a + bt$		Forecast (F)		Target (T)		Difference (F–T)	
	a	b	2020	2030	2020	2030	2020	2030
Denmark	20.27	1.72 ***	39.22	56.45	30	55	9.22	1.45
Finland	31.27	1.19 ***	44.39	56.32	38	50	6.39	6.32
Sweden	46.45	0.99 ***	57.32	67.21	49	64	8.32	3.21
United Kingdom	2.15	0.99 ***	12.99	22.85	15	27	−2.01	−4.15
Latvia	31.76	0.95 ***	42.23	51.74	40	45	2.23	6.74
Greece	10.12	0.93 ***	20.30	29.55	18	32	2.30	−2.45
Portugal	23.34	0.87 ***	32.94	41.67	31	47	1.94	−5.33
Estonia	23.22	0.78 ***	31.82	39.63	25	42	6.82	−2.37
Bulgaria	13.61	0.78 ***	22.14	29.89	16	25	6.14	4.89
Lithuania	19.50	0.72 ***	27.47	34.72	23	45	4.47	−10.28
Ireland	5.09	0.66 ***	12.32	18.90	16	31	−3.68	−12.10
France	11.23	0.60 ***	17.78	23.75	23	32	−5.22	−8.25
Italy	13.19	0.59 **	19.67	25.55	17	30	2.67	−4.45
Germany	11.37	0.57 ***	17.65	23.37	18	30	−0.35	−6.63
Czech Republic	10.82	0.57 ***	17.07	22.76	13	21	4.07	1.76
Slovakia	8.59	0.56 **	14.79	20.42	14	18	0.79	2.42
Spain	12.90	0.56 ***	19.09	24.71	20	42	−0.91	−17.29
Netherlands	3.19	0.46 ***	8.26	12.86	14	27	−5.74	−14.14
Belgium	5.66	0.43 ***	10.41	14.72	13	18	−2.59	−3.28
Croatia	25.64	0.32 *	29.19	32.42	20	36	9.19	−3.58
Romania	22.35	0.26 *	25.25	27.88	24	28	1.25	−0.12
Austria	31.53	0.25 **	34.30	36.83	34	45	0.30	−8.17
Poland	10.03	0.21 *	12.32	14.41	15	21	−2.68	−6.59
UE-28	12.65	0.63 ***	19.53	25.78	20	32	−0.47	−6.22

*, **, *** test significant at $p < 0.05, 0.01, 0.001$, respectively. Source: Own study based on [47].

The rate of changes to date indicates that not all Member States will achieve the target without adjusting their policies. Bulgaria, Czechia, Denmark, Finland, Latvia, Slovakia, and Sweden are likely to meet the 2030 targets.

The leaders in the implementation of energy policies in the context of RES are Denmark and Finland, where the average annual increase in 2010–2019 was 1.72 pp. and 1.19 pp., respectively. The least intensive measures among the EU Member States were taken in Poland, Austria, and Romania, where year-on-year indicators increased by 0.21 pp., 0.25 pp., and 0.26 pp., respectively.

In Hungary and Slovenia, the trends in renewable energy in gross final energy consumption over the years under study were not stable. The coefficients of the average rate of change of the phenomenon, which were estimated respectively at the level of 0.999 and 1.005, indicate that the changes of the examined characteristic over the ten years only slightly. Thus, in the case of Hungary there was a decrease of 0.13 pp and for Slovenia an increase of 0.89 pp.

3.3. Achievement of EU Energy Policy Targets

The aim of this part of the analysis was to identify similarities between Member States in terms of taking action to meet the EU energy policy targets set out in the 2020 climate and energy package. Groups were identified by comparing the dynamics of the three variables: reduction of the energy intensity of the economy (EI), reduction of greenhouse gas emissions (GHG) and share of energy from renewable sources (RES). Groups of similar countries were created using a synthetic variable proposed under the TOPSIS method.

The final breakdown into four country groups is presented in Table 6. Descriptive statistics for the variables in the groups are listed in Table 7.

Table 6. Groups of EU Member States similar in terms of achievement of energy policy targets as of 2019.

Specification	Country
Group 1	Denmark, Ireland, United Kingdom
Group 2	Estonia, Finland, Latvia, Lithuania, Netherlands, Slovakia, Sweden
Group 3	Belgium, Bulgaria, Croatia, Czech Republic, France, Germany, Greece, Hungary, Italy, Poland, Romania, Slovenia
Group 4	Austria, Portugal, Spain

Source: Own study based on [47,48].

Table 7. Numerical characteristics of indicators of the level of achievement of energy policy targets in groups of Member States.

Specification		Group 1	Group 2	Group 3	Group 4
RES [1]	mean	132.23	51.16	37.32	22.41
	min	69.97	21.02	−1.00	7.75
	max	219.42	123.84	95.27	32.72
EI [1]	mean	25.70	17.84	10.87	4.87
	min	19.33	12.22	−4.83	1.55
	max	34.03	21.76	19.72	8.60
GHG [1]	mean	15.32	15.11	8.55	3.36
	min	8.81	3.39	3.54	−0.71
	max	21.58	26.33	15.95	5.76

[1] designations according to Table 1 Source: Own study based on [47–49].

The first group included Denmark, Ireland, and the United Kingdom. The share of energy derived from renewable sources in 2019 compared to 2010 increased by 219% in the UK, by 107% in Ireland and by 69% in Denmark (Table 7). This group also saw the largest decrease in the energy intensity of the economy, averaging 25.70%. There was also a significant, although not the largest, reduction in greenhouse gas emissions compared to 2010.

Countries from the second group increased their share of low-carbon energy sources on average by 21%, with the largest change in the Netherlands (123%) and Slovakia (86%). Significant progress in this group could be observed in relation to GHG emission reductions, as the average result was similar to that of the first group at 15.11%. The leaders in this respect were Finland (26.33%), Sweden (23.96%), and Lithuania (18.09%).

In the third group, which included the largest number of countries, the share of energy from renewable sources increased on average by 37% in the analysed period. The most intensive measures in this respects were taken in Greece (95.27%), Belgium (65%), Bulgaria (54%), and Czechia, where the share of renewable energy increased by 54%. Energy intensity reduction measures in certain countries did not bring the desired effects, in particular in Greece, where the energy intensity of the economy increased in 2015 compared to 2010. The rate of changes in GHG emissions reduction was almost one half lower compared to the first and the second group.

Austria, Portugal, and Spain formed the fourth group. These were countries where measures to increase energy generation from environmentally-friendly sources were the least effective. The energy intensity of the economy was reduced only slightly, on average by 4.87%. Changes in the reduction of greenhouse gas emissions were also very low in this group, with the average result of 3.36% in 2018 compared to 2010. It is worth noting that these Member States did not manage to reduce greenhouse gas emissions by 20%; in fact, these values increased compared to 1990.

The relationship between energy generation from RES and economic growth assessed in terms of GDP growth is of great interest to the economists. Authors of numerous studies [58–61] analyse different countries and use different modelling methods to verify hypotheses on the existence of causal connections between energy consumption, including energy obtained from renewable sources, and GDP growth. Figure 4 presents the descriptive statistics of real GDP in thousands of euro per capita in 2019, and the trend coefficients of the share of renewable energy in gross final energy consumption (%) presented in Table 5 in the groups of similar countries in terms of the level of achievement of energy targets in 2019 adopted variables to assess the level of achievement of energy policy targets resulting from the National Renewable Energy Action Plans.

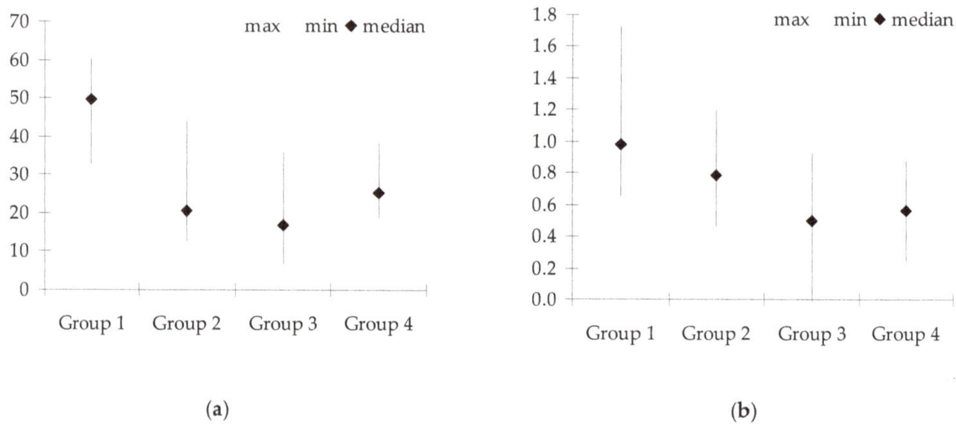

Figure 4. Figure characteristics of real GDP in thousands of euro per capita in 2019 (**a**) and trend coefficients of the share of renewable energy in gross final energy consumption (%) (**b**) in groups of similar countries in terms of the level of achievement of energy targets in 2019, respectively. Own study based on Table 5, [47].

All Group 1 countries had real GDP per capita in 2019 significantly higher than the EU-wide figure of GBP 27.970. Ireland and the UK have set national targets for the share of renewable energy in gross final energy consumption for both 2020 and 2030 that are lower than those indicated in the European Commission Directive. At the current rate of change, these countries will not meet the policy (Table 5). Denmark, on the other hand, has set targets significantly higher than those set for the EU, which it consistently fulfils. In all Group 1 countries the main sources of environmentally friendly energy are biofuels and wind.

Among the countries classified in the second group, only the Netherlands is predicted to miss its national target of share of renewable energy in gross final energy consumption for 2020, although it is one of the lowest in the EU. This indicator for Group 2 countries has on average increased from 0.46 pp to 1.19 pp year on year. The GDP per capita variation was 48.4% for the countries in this group. The three countries of this group that joined the EU in 2004 had GDP per capita significantly lower than the others. The main source of renewable energy for all countries in this group is biofuels.

The most populous group of similar countries in terms of measures taken to achieve the EU energy policy objectives set out in the climate and energy package for 2020 is characterised by the lowest average rate of change of the share of renewable energy in gross final energy consumption. Countries in this group have generally (except Slovenia, Romania, France, and Croatia) set this indicator at the level of no more than 18%. Half of them have a GDP per capita of less than EUR 17000. Among the countries in this group, that joined the EU in 2004 or later, biofuels are the main source of renewable energy. Among the EU-14 Group 2 countries, other resources are also used to diversify energy sources.

The three countries classified in Group 4 set their national targets for share of renewable energy in gross final energy consumption for 2020 and 2030 at a higher than average for EU level. Portugal, with the lowest GDP per capita in this group, had the highest (also in this group) average annual increase of renewable energy in gross final energy consumption. The opposite was true for Austria. Important sources of renewable energy in these countries are biofuels, hydropower, and wind.

Increasing requirements, in particular the environmental ones, for the construction of various renewable energy production plants have and will have a direct impact on RES development in the coming years. This includes new environmental requirements for the construction of hydropower plants, as well as various national restrictions on the location of biogas plants and wind farms. The latter are also accused and sometimes stigmatised for being harmful to the environment, including bioflora, in an almost hidden yet significant way. Agriculture, a potentially important producer of resources for renewable energy generation [62] also seems to be in decline in this respect. Although the issue of supporting the production of energy resources in agriculture is one of the components of the EU Common Agricultural Policy, between 2014 and 2020 the "enthusiasm" in this area has clearly faded [63]. Most of the instruments and mechanisms of financial support for the production of energy resources have been withdrawn, both on arable land and grassland, as well as with regard to new energy plantings. Production of these resources is now carried out on a strictly commercial basis with the EU direct payments applied on a general basis. Hence, maize processed in biogas plants has become the most important commodity in many Member States. However, there are still great opportunities for RES production in rural areas, in particular on farms. This is reflected in the Green New Deal, which will be implemented between 2021 and 2027 in EU Member States [64]. The Green New Deal is a new appeal to the Member States to respect the environment in the conditions of sustaining economic development and raising the standard and quality of life. It thus meets both the general conditions for sustainable development and for promoting the level of economic development. It formulates ambitious EU climate targets for 2030 and beyond to 2050. It also indicates that, on the basis of research and by stimulating innovation, it is possible to provide clean energy at affordable prices on a much larger scale. It also draws attention to the need to accelerate the achievement of zero pollution in the operation of energy equipment (and chemical plants). Rural areas, including agricultural areas, are to play a greater role in photovoltaic installations, heat pumps [65], biogas plants, and wind farms. RES production should also have a more local nature, i.e., be based on relatively small plants generating energy for the needs of local businesses and households of the local population. The development of prosumer energy, i.e., photovoltaic micro-plants, is also a great opportunity to increase renewable energy production. It can be assumed at some risk that the future of RES development lies in the popularisation, even at the massive scale, of small photovoltaic plants which, by producing energy from sunlight instead from heat, can be installed in almost all EU Member States. It is an open question how the economies of EU Member States will react to the upcoming economic changes caused by the COVID-19 pandemic. Its various effects, also those relating to the economy, including energy, will probably become apparent in its wake. The level of economic development measured by GDP (per capita) may be reviewed. Changes in the economy caused by the pandemic, their rate, but also the direction of necessary and possible transformation will be important, also from the cognitive point of view. In view of the economic slowdown and the threat of recession, and thus a drop in the demand for energy, will the priorities in the area of RES support be maintained? It can be expected, however, that the demand for electricity will increase, which will undoubtedly be strongly influenced by the accelerating conversion of the European car fleet from liquid fuels to hybrid or purely electric vehicles. Moreover, countries whose energy production is mainly or significantly based on coal, such as Poland and Germany, will be encouraged by pro-environmental EU regulations and the system of financial transfers to restructure their energy portfolios, which will undoubtedly involve the development of RES.

4. Conclusions

Renewable energy sources (RES) are playing an increasingly important role in the energy supply structure, and certain RES technologies have reached the level of competitiveness similar to technologies based on fossil fuels. The process of gradual transformation from a coal-based economy to an economy using green, low-carbon technologies that meet social needs and ensure energy security not only locally, but also regionally and in the long term, is being initiated by the growing number of EU Member States.

Over the past decade or so, the importance of energy from renewable sources in Europe has significantly increased. EU Member States are highly diversified in terms of generating energy from these sources. In the group of countries that joined the EU after 2004, the problem of generating energy from renewable sources is related, on the one hand, to the necessity of fulfilling national obligations in the ratified energy and climate package and, on the other hand, to the specific nature of the economies of those countries, which are mainly based on conventional energy.

The low share of renewable energy sources is also due to the fact that investments in environmentally-friendly energy solutions require corresponding large financial outlays.

Research has confirmed that the share of renewable energy in in gross final energy consumption in the analysed EU Member States is highly diverse. This is mainly due to the resources for energy production available in these countries and the existing and still efficient systems for their acquisition and generation of non-renewable energy. It also depends on the level of energy demand in the economy, including households. In these analyses, it is also important to refer the level and structure of the produced energy to such country characteristics as its area and population. The environmental conditions in a given country are also important here, in particular climate conditions and those affecting the productivity of ecosystems, which can produce resources for RES generation. The use of RES technologies depends largely on natural circumstances. The ongoing research on modern technologies gives the possibility to use various sources of energy. Some of them, with the use of modern solutions, are or will be able to be introduced in many countries. Hence, the presentation of a group of similar countries in terms of the main sources of energy. This information will give the opportunity to compare a given country with others of similar environmental and geographical conditions and, if such opportunities arise, to diversify the sources of energy.

Among the leaders in generating energy from renewable sources are Germany, France, and Italy. The structure of energy production from different sources is highly dependent on natural predestination, including in particular hydropower, wind energy, solar energy, and biomass. The highest share of RES from biofuels is reported in Estonia, Latvia, Lithuania, and Hungary. With regard to hydropower, Austria, Sweden, Croatia, and Romania are the leaders among the analysed countries. On the other hand, Ireland, United Kingdom, Denmark, and Germany have a high share of wind energy production.

In the case of all analysed EU Member States, there is an upward trend in the share of renewable energy in gross final energy consumption, which is mainly due to the need to achieve the national targets set out in the national plans which are a direct result of EU directives. Most countries (14) have met the 2020 target and 4 have exceeded the 2030 target. The most dynamic changes in 2019 compared to 2010 occurred in Denmark, Finland, and Sweden, which allowed these countries to significantly exceed their 2020 targets by 9.22%, 6.39%, and 8.32%, respectively.

Some Member States made efforts to increase production of (energy from) RES, and some probably decided to purchase the volumes of energy they require to meet their national RES target within the so-called Statistical Transfer. Virtual green energy may be supplied by the countries which have already met their targets.

The leaders in implementing environmentally friendly energy policies are Denmark, the UK, and Ireland. These countries show the highest rate of change in terms of achieving the targets set in their National Renewable Energy Action Plans. The decrease in energy intensity of these economies and efforts to reduce greenhouse gas emissions are also positive. In addition to environmental conditions, which Ireland undoubtedly has at its disposal, solutions for the implementation of new technologies may be implemented in the countries where the dynamics of changes in the aforementioned areas is lower than the EU average. This refers to organisational as well as technological solutions.

In the struggle to improve the quality of the climate, it is important to reduce the energy intensity of the economy and greenhouse gas emissions. The least effective measures in this respect were introduced by Austria, Portugal, and Spain, where the energy intensity of the economy was reduced only slightly, by 4.87% on average, and the reduction of greenhouse gas emissions was very low, at 3.36% on average in 2018 compared to 2010.

Institutional aspects have an impact on the development of the RES market, including the country's level of development, its innovativeness, or its openness to change. It should also be considered that each EU Member State has its own established sources of non-renewable energy, previously built power plants, as well as financial and human capital involved in the energy sector. Moreover, important is the political situation in respective countries where restructuring the energy mix requires large capital outlays, the decline of part of the old classic power stations and hence a reduction in employment or even regression in sub-regional development. However, the increase in environmental awareness of societies and various pro-environmental activities of the state and non-governmental organisations are strong determinants for the continuation of sustainable development, in which RES are a very important link. The EU policy has a significant impact on the activities undertaken in the Member States in the field of introducing technologies that generate energy from RES. National authorities supervised by the European Commission in many countries guarantee the implementation of policies favouring positive climate change. It is also connected with the implementation of the recommendations contained in the EU Directives regarding the support of RES development. Removal of government subsidies for the extraction of fossil fuels affects the relative increase in the competitiveness of RES, which is an example of an effective policy encouraging the use of this technology. The possibility of implementing RES technologies also depends on natural conditions; however, the overwhelming majority of countries are able to effectively implement RES. Identifying the leading countries makes it possible to compare the actions taken by the member states, both on the legislative and organisational level. Determining the pace of changes in the implementation of the national goals for 2030 will allow to determine which countries will be able to achieve the goals regarding the share of renewable energy sources in the total energy mix. Identifying the similarities will allow countries to implement policies that have been proven effective, in relation to the introduced EU criteria, in countries with similar environmental and institutional conditions.

Author Contributions: Conceptualization, W.M., M.Z., L.L. and K.M.; methodology, L.L.; formal analysis, L.L. and M.Z.; investigation, M.Z., W.M. and L.L.; resources, M.Z, W.M., K.M. and L.L.; data curation, L.L. and M.Z.; writing—original draft preparation, W.M., M.Z., L.L. and K.M.; writing—review and editing, W.M.; visualisation, L.L.; supervision, W.M.; funding acquisition, L.L., W.M. and M.Z. All authors have read and agreed to the published version of the manuscript.

Funding: The research was cofinanced from resources of Faculty of Agriculture and Economics University of Agriculture in Krakow.

Institutional Review Board Statement: Not applicable.

Informed Consent Statement: Not applicable.

Data Availability Statement: Not applicable.

Conflicts of Interest: The authors declare no conflict of interest.

References

1. Cadoret, I.; Padovano, F. The political drivers of renewable energies policies. *Energy Econ.* **2016**, *56*, 261–269. [CrossRef]
2. Lucas, J.N.V.; Francés, G.E.; González, E.S.M. Energy security and renewable energy deployment in the EU: Liaisons dangereuses or virtuous circle? *Renew. Sustain. Energy Rev.* **2016**, *62*, 1032–1046. [CrossRef]
3. Papież, M.; Śmiech, S.; Frodyma, K. Determinants of the renewable energy development in the EU countries. A 20-year perspective. *Renew. Sustain. Energy Rev.* **2018**, *91*, 918–934. [CrossRef]
4. Menegaki, A.N. Growth and renewable energy in Europe: A random effect model with evidence for neutrality hypothesis. *Energy Econ.* **2011**, *33*, 257–263. [CrossRef]
5. Alper, A.; Oguz, O. The role of renewable energy consumption in economic growth: Evidence from asymmetric causality. *Renew. Sustain. Energy Rev.* **2016**, *60*, 953–959. [CrossRef]
6. Kula, F. The long-run relationship between renewable electricity consumption and GDP: Evidence from panel data. *Energy Sources Part B Econ. Plan. Policy* **2014**, *9*, 156–160. [CrossRef]
7. Neitzel, D. Examining Renewable Energy and Economic Growth: Evidence from 22 OECD Countries. Honors Program Theses, 2017. Available online: https://scholarship.rollins.edu/honors/46 (accessed on 25 February 2021).
8. Śmiech, S.; Papież, M. Energy consumption and economic growth in the light of meeting the targets of energy policy in the EU: The bootstrap panel Granger causality approach. *Energy Policy* **2014**, *71*, 118–129. [CrossRef]
9. Kasman, A.; Duman, Y.S. CO_2 emissions, economic growth, energy consumption, trade and urbanization in new EU member and candidate countries: A panel data analysis. *Econ. Model.* **2015**, *44*, 97–103. [CrossRef]
10. Czekanowski, J. *Zarys Metod Statystycznych w Zastosowaniu do Antropologii*; Prace Towarzystwa Naukowego Warszawskiego: Warsaw, Poland, 1913.
11. Brundtland, G.H. Report of the World Commission on Environment and Development: Our Common Future towards Sustainable Development 2; Part II. Common Challenges Population and Human Resources 4. 1987. Available online: https://sustainabledevelopment.un.org/content/documents/5987our-common-future.pdf. (accessed on 24 February 2021).
12. Haughton, G. Environmental Justice and the Sustainable City. *J. Plan. Educ. Res.* **1999**, *18*, 233–243. [CrossRef]
13. Kates, R.; Parris, T.M.; Leiserowita, A.A. What is sustainable development? Goals, indicators, values, and practice. *Environ. Sci. Policy Sustain. Dev.* **2005**, *47*, 8–21.
14. Sachs, J.D. From Millennium Development Goals to Sustainable Development Goals. *Lancet* **2012**, *37*, 2206–2211. [CrossRef]
15. Griggs, D.; Stafford-Smith, M.; Gaffney, O.; Rockström, J.; Öhman, M.C.; Shyamsundar, P.; Steffen, W.; Glaser, G.; Kanie, N.; Noble, I. Policy: Sustainable Development Goals for People and Planet. *Nature* **2013**, *495*, 305–307. [CrossRef]
16. Przekształcamy Nasz świat: Agenda na Rzecz Zrównoważonego Rozwoju 2030 (We are Transforming Our World: The 2030 Agenda for Sustainable Development) Rezolucja Zgromadzenia Ogólnego ONZ A/RES/70/1, Paryż, 25 Września 2015. Available online: http://www.unic.un.org.pl/files/164/Agenda%202030_pl_2016_ostateczna.pdf. (accessed on 24 February 2021).
17. Cele Zrównoważonego Rozwoju (Sustainable Development Goals). Available online: www.un.org.pl (accessed on 11 January 2021).
18. Eurostat, Indicators—Sustainable Development Indicators. Available online: https://ec.europa.eu/eurostat/data/database (accessed on 19 March 2021).
19. Główny Urząd Statystyczny. *Wskaźniki Zrównoważonego Rozwoju Polski (Sustainable Development Indicators of Poland) Katowice*; Główny Urząd Statystyczny: Warsaw, Porland, 2021.
20. 27 Zasad Zrównoważonego Rozwoju—Deklaracja z Rio (27 Principles of Sustainable Development—Rio Declaration). Available online: https://www.ekologia.pl/artykul/inne/27-zasad-zrownowazonego-rozwoju,7035.html (accessed on 19 February 2021).
21. De Andrade, L.O.M.; Filho, A.P.; Solar, O.; Rígoli, F.; De Salazar, L.M.; Serrate, P.C.F.; Ribeiro, K.G.; Koller, T.S.; Cruz, F.N.B.; Atun, R. Social Determinants of Health, Universal Health Coverage, and Sustainable Development: Case Studies from Latin American Countries. *Lancet* **2015**, *385*, 1343–1351. [CrossRef]
22. Hassini, E.; Surti, C.; Searcy, C. A Literature Review and a Case Study of Sustainable Supply Chains with a Focus on Metrics. *Int. J. Prod. Econ.* **2012**, *140*, 69–82. [CrossRef]
23. Del Río, P.; Carrillo-Hermosilla, J.; Könnölä, T. Policy Strategies to Promote Eco-Innovation An Integrated Framework. *J. Ind. Ecol.* **2010**, *14*, 541–557. [CrossRef]
24. Janka, R.M. Emisja Zanieczyszczeń. In *Podstawy Programowania Wielkości Emisji Oraz Opłat za Wprowadzanie Zanieczyszczeń do Atmosfery*; (Emission of Pollution. Basics of Programming the Amount of Emissions and Charges for Discharging Pollutants into the Atmosphere); Wydawnictwo KURIER: Opole, Poland, 2006.
25. Konstańczak, S. Racjonalność strategii zrównoważonego rozwoju (Rationality of the sustainable development strategy). In *Etyka Biznesu i Zrównoważony Rozwój. Interdyscyplinarne Studia Teoretyczno-Empiryczne*; Śląskie Centrum Etyki Biznesu i Zrównoważonego Rozwoju: Katowice, Poland, 2016; Volume 1, pp. 58–70.
26. Lewandowski, W.M. *Proekologiczne Odnawialne Źródła Energii (Pro-Ecological Renewable energy Sources)*; Wydawnictwo Naukowo-Techniczne: Warsaw, Porland, 2007.
27. Gradziuk, P.; Gradziuk, B. Próba oceny efektów absorpcji Środków z Funduszy Europejskich na rozwój wykorzystania odnawialnych źródeł energii w województwie lubelskim (An attempt to assess the effects of the absorption of funds from European Funds on the development of the use of renewable energy sources in the Lublin Province). *Rocz. Nauk. Ekon. Rol. i Rozw. Obsz. Wiej.* **2017**, *104*, 95–105. [CrossRef]

28. Ciepielewska, M. Rozwój Odnawialnych Źródeł Energii w Polsce w Świetle Unijnego Pakietu Klimatyczno-Energetycznego Oraz Ustawy o Odnawialnych Źródłach Energii (Development of renewable energy sources in Poland in the light of the EU Climate and Energy Package and the Act on Renewable Energy Sources). *Gospod. Prakt. Teor.* **2016**, *43*, 7–18. [CrossRef]
29. Prandecki, K. Theoretical Aspects of Sustainable Energy. *Energy Environ. Eng.* **2014**, *2*, 83–90. [CrossRef]
30. Pultowicz, A. Mechanizm wsparcia rozwoju energetyki wiatrowej w Polsce na tle rozwiązań europejskich (Mechanism to support the development of wind energy in Poland compared to European solutions). In *Problemy Regionalne i Globalne We Współczesnej Gospodarce Światowej*; Rymarczyk, J., Michalczyk, W., Eds.; Tom 2, Katedra Międzynarodowych Stosunków Gospodarczych, Akademia Ekonomiczna we Wrocławiu: Wrocław, Poland, 2005; pp. 128–135.
31. Wąs, A.; Sulewski, P.; Krupin, V.; Popadynets, N.; Malak-Rawlikowska, A.; Szymańska, M.; Skorokhod, I.; Wysokiński, M. The Potential of Agricultural Biogas Production in Ukraine—Impact on GHG Emissions and Energy Production. *Energies* **2020**, *13*, 5755. [CrossRef]
32. Mahjabeen, N.; Shah, S.Z.; Chughtai, S.; Simonetti, B. Renewable energy, institutional stability, environment and economic growth nexus of D-8 countries. *Energy Strategy Rev.* **2020**, *29*, 100484. [CrossRef]
33. Li, X. Diversification and localization of energy systems for sustainable development and energy security. *Energy Policy* **2005**, *33*, 2237–2243. [CrossRef]
34. European Commission, Energy Union Package—Communication from the Commission to the European Parliament, the Council, the European Economic and Social Committee, the Committee of the Regions and the European Investment Bank: A Framework Strategy for a Resilient Energy Union with a Forward-Looking Climate Change Policy, COM(2015) 80 Final. 2015. Available online: https://eur-lex.europa.eu/resource.html?uri=cellar:8263aedc-5ab2-11e9-9151-01aa75ed71a1.0009.02/DOC_1&format=PDF (accessed on 12 February 2021).
35. Gokgoz, F.; Guvercin, M.T. Energy security and renewable energy efficiency in EU. *Renew. Sustain. Energy Rev.* **2018**, *96*, 226–239. [CrossRef]
36. Chalvatzis, K.J.; Ioannidis, A. Energy supply security in the EU: Benchmarking diversity and dependence of primary energy. *Appl. Energy* **2017**, *207*, 465–476. [CrossRef]
37. European Commission, Commission Communication (COM(2018)773) A Clean Planet for All: A European Strategic Long-Term Vision for a Prosperous, Modern, Competitive and Climate Neutral Economy, Brussels. 2018. Available online: https://eur-lex.europa.eu/legal-content/EN/TXT/PDF/?uri=CELEX:52018DC0773&qid=1616667870433&from=EN (accessed on 12 February 2021).
38. Nations Union Paris Agreement. In Proceedings of the Conference of the Parties to the United Nations Framework Convention on Climate Change, Paris, France, 30 November–12 December 2015.
39. European Commission. Communication from the Commission to the European Parliament, the Council, the European Economic and Social Committee and the Committee of the Regions. A Policy Framework for Climate and Energy in the Period from 2020 to 2030. Available online: https://eur-lex.europa.eu/legal-content/EN/ALL/?uri=CELEX:52014DC0015 (accessed on 22 February 2021).
40. Shahbaz, M.; Raghutla, C.; Chittedi, K.R.; Jiao, Z.; Vo, X.V. The Effect of Renewable Energy consumption on Economic Growth: Evidence from the Renewable Energy Country Attractive Index. *Energy* **2020**, *207*, 118162. [CrossRef]
41. Karakosta, C.; Pappas, C.; Marinakis, V.; Psarras, J. Renewable energy and nuclear power towards sustainable development: Characteristics and prospects. *Renew. Sustain. Energy Rev.* **2013**, *22*, 187–197. [CrossRef]
42. Pultowicz, A. Przesłanki rozwoju rynku odnawialnych źródeł energii w Polsce w Świetle idei zrównoważonego rozwoju (Premises for the development of the renewable energy market in Poland in the light of the idea of sustainable development). *Probl. Ekorozw. Probl. Sustain. Dev.* **2009**, *4*, 109–115.
43. Boyle, G.; Everett, B.J.; Ramage, J. *Energy Systems and Sustainability. Power for a Sustainable Future*; Oxford University Press: Oxford, UK, 2004.
44. Prandecki, K. Teoretyczne podstawy zrównoważonej energetyki (Theoretical foundations of sustainable energy). In *Polityka Gospodarcza w Okresie Transformacji i Kryzysu*; Studia Ekonomiczne Uniwersytet Ekonomiczny w, Katowicach; Barteczek, A., Rączaszek, A., Eds.; Uniwersytet Ekonomiczny w Katowicach: Katowice, Poland, 2014; p. 166.
45. Suska-Szczerbicka, M. Wykorzystywanie Odnawialnych Źródeł Energii w Strategii Zrównoważonego Rozwoju (The Use of Renewable Energy Sources in the Strategy of Sustainable Development). VII Konferencja Samorządowa Dylematy Rozwoju Zrównoważonego, Uniwersytet Szczeciński, Wydział Nauk Ekonomicznych i Zarządzania, Kalisz. 2011. Available online: http://global.am-wind.pl/czytaj/za%C5%82acznik_nr_2_kalisz.pdf (accessed on 12 December 2020).
46. Chodkowska-Miszczuk, J.; Środa-Murawska, S.; Biegańska, J. Znaczenie edukacji w zakresie odnawialnych źródeł energii w upowszechnianiu zasad rozwoju zrównoważonego The importance of education in the field of renewable energy sources in the dissemination of the principles of sustainable development). *Ekonomia i Środowisko* **2016**, *1*, 34–49.
47. Eurostat. Available online: https://ec.europa.eu/eurostat (accessed on 25 January 2021).
48. World Development Indicators. Available online: http://www.worldbank.org/data/onlinedatabases/onlinedatabases.html (accessed on 20 January 2021).
49. Intentational Reanwable Energy Agency. Available online: https://www.irena.org/statistics (accessed on 20 February 2021).

50. Sołtysiak, A.; Jaskulski, P. Czekanowski's diagram a method of multidimensional clustering. In *New Techniques for Old Times. CAA 98. Computer Applications and Quantitative Methods in Archaeology, Proceedings of the 26th Conference, Barcelona, March 1998*; Barceló, J.A., Briz, I., Vila, A., Eds.; BAR International Series 757; BAR Publishing: Oxford, UK, 1999; pp. 175–184.
51. Homa, M.; MoŚcibrodzka, M. Application of diagram methods and hierarchical agglomerative procedures to assess the risk of investment funds on The Warsaw Stock Exchange. *Nauki o Finans.* **2016**, *4*, 21–34. [CrossRef]
52. Krakowiak-Bal, A. Multidimensional comparative analysis of other gainful activities of agricultural holdings in EU countries. *Infrastruktura i Ekologia Terenów Wiejskich* **2009**, *7*, 129–137.
53. Västerlund, A. Czekanowski's Diagram: Implementing and Exploring Czekanowski's Diagram with Differeneriation Methods. Ph.D. Thesis, Division for Statistics and Machine Learning, Department of Computer and Information Science, Linköping University, Linköping, Sweden, 2019.
54. Warzecha, K. The use of quantitative methods in research on selected behavioral addictions of young people. *Stud. Ekon.* **2015**, *247*, 121–139.
55. MaCzek-program Do Grupowania Obiektów. Available online: http://www.antropologia.uw.edu.pl/MaCzek/maczek.html#instal (accessed on 23 February 2021).
56. Hwang, C.L.; Yoon, K. *Multiple Attribute Decision Making: Methods and Applications*; Springer: Berlin/Heidelberg, Germany, 1981.
57. Kukuła, K.; Luty, L. Propozycja procedury wspomagającej wybór metody porządkowania liniowego (Proposition of procedure supporting choice of linear ordering metod). *Przegląd Statystyczny (Stat. Rev.)* **2015**, *62*, 219–231.
58. Apergis, N.; Payne, J.E. Renewable energy consumption and economic growth: Evidence from a panel of OECD countries. *Energy Policy* **2010**, *38*, 656–660. [CrossRef]
59. Belke, A.; Dobnik, F.; Dreger, C. Energy consumption and economic growth: New insights into the cointegration relationship. *Energy Econ.* **2011**, *33*, 782–789. [CrossRef]
60. Coers, R.; Sanders, M. The energy–GDP nexus; addressing an old question with new methods. *Energy Econ.* **2013**, *36*, 708–715. [CrossRef]
61. Rafindadi, A.A.; Ozturk, I. Impacts of renewable energy consumption on the German economic growth: Evidence from combined cointegration test. *Renew. Sustain. Energy Rev.* **2017**, *75*, 1130–1141. [CrossRef]
62. Rokicki, T.; Perkowska, A.; Klepacki, B.; Bórawski, P.; Bełdycka-Bórawska, A.; Michalski, K. Changes in Energy Consumption in Agriculture in the EU Countries. *Energies* **2021**, *14*, 1570. [CrossRef]
63. Plan Strategiczny dla Wspólnej Polityki Rolnej (Strategic Plan for the CAP), MRiRW, Warszawa, 2020. pp. 1–148. Available online: https://www.gov.pl/attachment/e2a25e58-8e6d-435e-bb3e-b29c888128b9 (accessed on 21 February 2021).
64. Communication from the Commission to the European Parliament, the European Council, the Council, the European Economic and Social Committee and the Committee of the Regions the European Green Deal, COM (2019) 640; Brussels. 2019. Available online: https://eur-lex.europa.eu/resource.html?uri=cellar:b828d165-1c22-11ea-8c1f-01aa75ed71a1.0002.02/DOC_1&format=PDF (accessed on 12 February 2021).
65. Gradziuk, P.; Gradziuk, B. Perspektywy rozwoju rynku pomp ciepła w Polsce w kontekŚcie strategii Unii Europejskiej „Czysta Planeta Dla Wszystkich (Prospects for the development of the heat pumps market in Poland in the context of the new EU strategy "A Clean Planet for All"). *Polish Assoc. Agric. Econ. Agribus. Stow. Ekon. Rol. Agrobiznesu* **2018**, *20*, 77–82. [CrossRef]

energies

Article

The Role of Renewable Energy Sources in Alleviating Energy Poverty in Households in Poland

Agnieszka Biernat-Jarka, Paulina Trębska * and Sławomir Jarka

Institute of Economics and Finance, Warsaw University of Life Sciences—SGGW, Nowoursynowska 166, 02-787 Warsaw, Poland; agnieszka_biernat_jarka@sggw.edu.pl (A.B.-J.); slawomir_jarka@sggw.edu.pl (S.J.)
* Correspondence: paulina_trebska@sggw.edu.pl

Abstract: Energy poverty is a problem that affects all member states of the European Union to a varying degree, including Poland, where about 9% of the population is at risk of energy poverty. The article aims to show the changes in energy poverty in Poland in 2010–2018. The specific goal, however, is to evaluate government measures aimed at reducing energy poverty through investments based on renewable energy sources. To present changes in the level of energy poverty in 2010–2018, the authors proposed a new synthetic measure that unifies several different measures used by researchers and allows for a comprehensive assessment of this phenomenon. The conducted research showed that in 2010–2018 there was a slow but visible decrease in the level of energy poverty in Poland. In addition, the article indicates investments in renewable energy sources that may have a positive impact on reducing the scale of energy poverty in Poland. The programs implemented with national and EU public funds, which finance investments in renewable energy sources in Poland, are also presented.

Keywords: energy poverty; renewable energy sources; Polish households

Citation: Biernat-Jarka, A.; Trębska, P.; Jarka, S. The Role of Renewable Energy Sources in Alleviating Energy Poverty in Households in Poland. *Energies* **2021**, *14*, 2957. https://doi.org/10.3390/en14102957

Academic Editor: Dimitrios Katsaprakakis

Received: 19 April 2021
Accepted: 17 May 2021
Published: 20 May 2021

1. Introduction

Research on energy poverty conducted in the region of Central and Eastern Europe is an extremely important issue. The article is an important voice in the analysis of this phenomenon, because Poland, like other countries in this region, is characterized by frosty winters, which means that energy security related to heating the apartment at an appropriate level is of particular importance. In addition, in Poland, as in other countries of Central and Eastern Europe, the consequences of a centrally planned economy can also be observed, such as dependence on polluting energy sources, the dominance of coal in energy production, ineffective housing resources and heating systems, and a small share of RES in energy production [1]. In addition, the main goal of the EU climate policy is to achieve at least 32% share of energy from renewable sources in total energy consumption by 2030. Undertaking intensive activities in this area is the last moment that allows achieving or at least getting closer to this goal. The above premises make the importance of renewable energy sources in reducing energy poverty in Poland a particularly important issue.

Energy poverty is a problem that, to varying degrees, affects all member states of the European Union [2–6], including Poland, where, according to the "high costs–low income" indicator, popular in international practice, energy poverty affects about 9% of citizens. Most of the people affected by energy poverty live in rural areas and small towns, usually in single-family houses, and use a solid fuel boiler or stove as their primary heat source. The social group most exposed to energy poverty is Polish farmers. One in three of them is energy-poor and they make up almost a fifth of all those affected by the problem. Rural residents, including farmers, usually live in single-family houses, characterized by low energy performance (non-insulated walls and roofs, improperly sealed windows, old stoves) and large floor area, which contributes to energy poverty. Energy poverty also significantly affects Polish pensioners and disability pensioners (together they represent 25% of all people affected by energy poverty), namely the elderly [7].

The aim of the paper is to show the changes in the phenomenon of energy poverty in Poland over the period of 2010–2018. However, the specific objective is to assess the government's efforts to reduce the phenomenon of energy poverty through investments based on renewable energy sources. The issues discussed in this paper are important for several reasons. Primarily, the problem of energy poverty is important in terms of economy, energy security and social security, as well as in terms of the health of those affected.

The added value of the article is the development of a synthetic indicator that unifies several different measures used by researchers and allows for a comprehensive assessment of the phenomenon of energy poverty. Based on the literature review and their research, the authors formulated the concept of energy poverty in terms of micro and macroeconomics. It can be used for research conducted in other countries. Another important issue raised in the article is air pollution and smog. The burning of low-quality fuels and municipal waste is the main factor responsible for air pollution in Poland. The Polish energy sector, based primarily on domestic hard coal and lignite deposits, is one of the most expensive ones and deviating from European trends. This can be attributed to the insufficient use of renewable energy sources and low social awareness of energy poverty prevention. Following the above, the direction of changes that should be taken by the authorities in order to reduce energy poverty in Poland was proposed.

1.1. Definition of the Energy Poverty Phenomenon and Methods of Its Measurement

The existence of worldwide energy poverty poses a major challenge to the global energy system [8]. Energy poverty is a problem related to income situation, inadequate house quality and energy prices. To date, no single definition or indicator to measure energy poverty has been developed. EU Energy Poverty Observatory [9] defines energy poverty as lacking the basic energy needs of households. Bouzarovski also (2013) stated that energy poverty should be understood as the inability of a household to provide the required level of energy services in the home [10]. Fuel poverty is the difficulty that households face in maintaining adequate temperature in the home, as well as in using other basic energy services in homes [11]. When defining the concept of energy poverty, it can be stated that it means the lack of sufficient choice in access to proper, cheap, reliable, high-quality, secure and environmentally friendly energy services [12,13]. The lack of access to energy may mean deprivation not only of basic household services, such as cooking and heating in the home but also other possibilities such as access to education, health or information. A vital element is the reliability of the technologies used, guaranteeing a continuous supply of electricity. Technologies should also be environmentally friendly and should not contribute to the excessive extraction of existing resources at the same time having a negative impact on the environment. In the context of increasing climate change and too extensive use of resources, the utilisation of renewable energy sources can prove to be extremely important. The causes of energy poverty vary depending on the level of development of a country. Namely, in developed countries, there is a situation characterised by high energy costs [14–16] while in developing countries, there is no access to modern energy [17,18]. Energy poverty, which is reflected, among others, in the inadequate heating of spaces and the consequent development of harmful microorganisms, etc., results in a greater likelihood of respiratory diseases, allergies (in the case of excessively humid and mouldy homes), hormonal disorders, cardiovascular disorders and deterioration of mental well-being [19–22]. Resolving the problem of energy poverty, or at least reducing it, would largely contribute to reducing expenditure on medical treatment for citizens, a better quality of life or increased economic activity. It is estimated that over 50 million households in the European Union are affected by energy poverty [14].

Interesting research is conducted on energy poverty and gender inequality. Women are more often concerned with the problem of energy poverty than men, but women are more often interested in modern energy [23–25]. Energy poverty is closely related to income poverty. The people most at risk of energy poverty are the poor and this problem is seen all over Europe. According to studies by Bouzarowski, 64.5% of Bulgarian households in 2010 were unable to keep their home "sufficiently warm", while 32.1% said they had arrears in paying for utilities [1]. Energy poverty, and thus difficult housing conditions, insufficient heating of the apartment or the lack of access to electronic devices, may be one of the many reasons for the stigmatization of certain national and social groups, such as Romani people living in many countries, e.g., Hungary or Romania [26,27].

Energy poverty also affects life satisfaction and the environment [28]. Energy consumption in homes contributes to the quality of life of households through lighting, cooking, heating and cooling [29]. However, increasing energy consumption levels can be contrary to the need to reduce carbon dioxide levels, thus using renewable energy sources can be an ideal solution to this problem.

There are three types of causes of energy poverty that are being indicated [30]:

- Technical causes—they occur when a residence has a low level of energy efficiency, making it more expensive to maintain an optimal standard of heat. Another cause of a technical nature is the malfunctioning of heating systems, not allowing the home to be properly heated. Higher energy consumption for heating entails higher expenditure and thus reduced possibilities of financing other expenses, often of a living nature.
- Economic causes—they occur when financial resources are too low, which can lead to arrears on energy bills, cutting off energy supply or saving on heating to reduce the cost of energy bills.
- Causes related to attitudes towards efficient and proper use of energy—they occur when improper use of appliances leads to significant energy losses and consequently increased energy expenses, higher than what the household can afford.

Based on the literature review and their research, the authors formulated the concept of energy poverty in terms of micro and macroeconomics. In microeconomic terms, energy poverty can be defined as the failure to meet the energy needs of the household, which negatively affects the quality of family life. In macroeconomic terms, energy poverty contributes to environmental pollution, which causes negative social, economic, and health problems for citizens.

In addition to defining energy poverty, it is necessary to refer to the determinants and indicators measuring its level [31,32]. Studies showing levels of energy poverty regarded both individual countries [33] and the EU as a whole [34]. EU member states are required to assess the scale of energy poverty in their country. In 1991, B. Boardman [11] formulated the following definition: "energy poverty occurs when a household does not receive adequate energy services for 10% of its income". The definition of B. Boardman is based on the definition adopted for national use by the UK, which is a precursor in research on energy poverty. The 10% measure was used in energy poverty research by Heindl and Schüssler [35], Phimister et al. [36], Okushima [37] and Pachauri et al. [38]. In Poland, the first nationwide research on energy poverty based on the British absolute definition of "10% of income" was carried out by Kurowski [39].

In 2012, a landmark report by J. Hills, commissioned by the UK Department of Energy and Climate Change (DECC), was published and has contributed to changing the definition and measures of energy poverty in the United Kingdom. The definition of energy poverty in the Hills report has been formulated as follows: "households are considered energy poor" if:

- their energy costs are above the median cost for all households and
- if they bore them at that level the rest of their disposable income would be below the official poverty line [40].

One measure of energy poverty is subjective indicators based on respondents' personal opinions, interpretations, viewpoints and judgements. They are generally constructed on the basis of answers given by household members to questions included in a survey conducted by social researchers (e.g., is the home warm enough in winter?). Subjective variables were used, among others, by Gordon et al. [41], Healy and Clinch [16], Healy [42], Petrova et al. [43] or Thomson and Snell [34].

According to the European Energy Poverty Observatory, the main indicators of energy poverty are: low absolute energy consumption, arrears on utility bills, spending a high proportion of income on energy and the inability to maintain an adequate temperature in the home [44].

In recent publications on energy poverty, authors and researchers also present more complex measures of energy poverty, designed as a compromise between the simplicity of unidimensional indicators and the need to take account of the multidimensional nature of the energy poverty problem. They are an attempt to overcome the shortcomings of unidimensional indicators and at the same time, they produce a result that condenses the information into single values that are easy to interpret [45].

Energy poverty is also measured using Chakravarty and D'Ambrosio's poverty measures [46]. It is this methodology that was the foundation for many studies measuring energy poverty according to a multidimensional framework. Nussbaumer et al. developed the Multidimensional Energy Poverty Index (MEPI) that takes into account both the occurrence and intensity of energy poverty and provides a new tool to support policy making [47]. Bouzarovski and Tirado Herrero [48] developed an energy poverty index that took into account the EU-SILC population percentages of people who have reported (i) being unable to keep their homes adequately warm (Inability); (ii) having arrears in utility bills (Arrears); and (iii) living in a home with a leaking roof, or the presence of damp and rot (Housing faults): Energy poverty index = (0.5 × % Inability + 0.25 × % Arrears + 0.25 × % Housing faults) × 100.

The energy poverty index proposed by Alkire and Foster [49] takes into consideration five dimensions of energy deprivation: two objective indicators "low income, high costs" and "high share of energy expenditure in income", as well as three subjective indicators: "inability to keep the home adequately warm", "presence of leaks, damp, or rot" and "difficulties paying utility bills". Households that experience at least two forms of deprivation are considered energy-poor. Using the methodology proposed by Alkire and Foster, based on the Household Budget Survey, a multidimensional energy poverty index for Poland was developed [50]. Szamrej-Baran [51] also used a multidimensional approach in conducting her research on energy poverty in Poland.

1.2. Investment in Renewable Energy Sources in the Context of Caring for the Environment and Counteracting Energy Poverty

Studies on the relationship between energy poverty and renewable energy sources carried out by researchers from many countries indicate the great importance of renewable energy sources in combating energy poverty [52]. Even if it may seem expensive to some economic experts, access to energy sources of all kinds brings social benefits, improving the quality of life—which in this case is a real victory in the fight against energy poverty. Poles are already bearing enough costs by paying with their deteriorating health and feeling the effects of climate change. Renewable energy is one of the crucial elements of sustainable development, and in protecting and improving air quality [53–55].

Generation of energy from renewable sources is an important part of the efforts to lower carbon intensity as well as to diversify energy and meet the growing demand therefor. It is an expression of care for the natural environment and a response to the need to promote sustainable development and enhance the strength of regions and local communities in the European Union [56]. Sustainable energy is the golden thread that will wave environmental sustainability [57]. Investments in renewable energy sources can lower energy costs and thus reduce the scale of energy poverty [58]. Energy generated from renewable sources is called "fuel of the future". Research shows that the production of renewable energy is associated with a positive and statistically significant impact on economic growth in both developed and developing countries [59]. As part of its climate policy until 2030, the European Union plans to achieve the following targets [60]:

- reducing greenhouse gas emissions by at least 40% with respect to 1990 (in the sectors covered by the EU ETS Directive (including energy and district heating sectors) the reduction is to be 43% compared to 2005);
- improving energy efficiency by 32.5% compared to the 2007 forecast;
- increasing the share of renewable energy in gross final energy consumption in the European Union to 32%.

This trend shows that Europe will seek to reduce the use of natural resources (including coal and oil) in favour of developing alternative sources. Failure to comply with EU requirements may lead to higher electricity and heat prices. Financial penalties imposed by the European Commission can be expected if EU member states fail to meet the targets of the climate and energy package. Within the framework of achieving the EU-wide target for 2030, Poland declares achieving 21–23% share of RES in gross final energy consumption (total consumption in electricity, district heating and cooling sectors, and for transport purposes) by 2030.

Eurostat data shows that in 2018 in Poland, the share of renewable energy in the consumption of electricity, heat and in transport was 11.3% (Figure 1). This was an increase of 2 p.p. compared to the index in 2010 (9.3%).

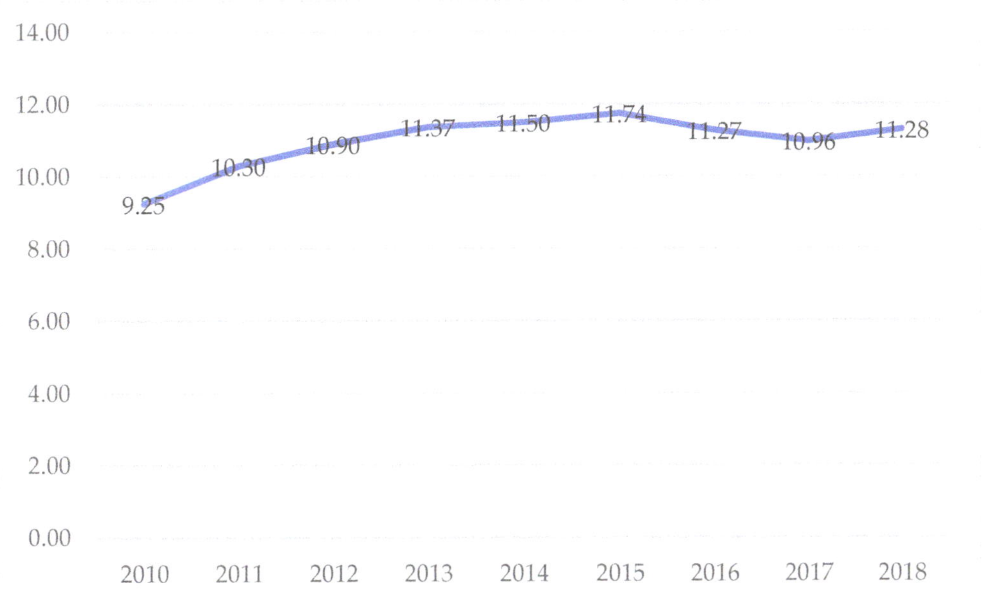

Figure 1. Share of energy from renewable sources [%]. Source: Eurostat.

In the structure of energy consumption by households in Poland, solid fuels, mainly hard coal (which is an exception in the European Union) and firewood, are of the greatest importance. They were most often used for space heating (by 45.4% of households). Firewood was used by 29.9% of households—it was the only renewable energy carrier massively used in households. It was usually burnt in the same boilers and stoves as hard coal, either simultaneously with coal or alternatively. In addition to firewood, households also consumed other types of biomass, but the prevalence of their use was much lower than firewood [61].

In the case of renewable energy, households had a 52.8% share of domestic wood consumption, and geothermal energy, including heat from heat pumps—71.4%. In 2018, a small group of households was equipped with solar panels (1.77%), and the quantity of solar energy obtained in this way compared to the total national solar energy consumption accounted for 60.0%. Solar collectors were used by one household in 52 and heat pumps by only one in 200 [61].

The increase in solar and geothermal energy consumption shows that solar equipment and heat pumps are used more increasingly for space heating (Table 1).

Table 1. Consumption of individual renewable energy carriers in households in Poland, in 2010–2018.

	Years								
	2010	2011	2012	2013	2014	2015	2016	2017	2018
Energy from water and wind [TJ]	250	300	380	460	1255	1655	1943	2033	2129
Geothermal energy [TJ]	440	430	510	561	608	674	705	2406	2531
Peat and wood [thous. m^3]	11,868	12,105	12,300	12,300	11,100	11,100	11,730	11,550	11,370

Source: Eurostat.

Only a small number of households, just 4.8%, install devices for their own energy production (solar collectors, heat pumps). Only 7.2% of households considered the issue of self-generation of energy for their own needs, where the following financial support opportunities were taken into account when deciding on such a solution: subsidies for the purchase or construction of generating plant, subsidised loan or attractive price at which generators will be able to sell electricity [61].

Renewable energy is becoming an increasingly competitive way to meet new power generation needs. Renewable power generation costs have fallen sharply over the past decade, driven by steadily improving technologies, economies of scale, competitive supply chains and growing developer experience. The International Renewable Energy Agency notes in its latest report titled "Renewable Power Generation Costs in 2019" another decrease in wind and solar energy generation costs. The costs of all commercially available renewable energy generation technologies have fallen in 2019. Solar and wind power costs have continued to fall, complementing the more mature bioenergy, geothermal and hydropower technologies. Solar photovoltaics (PV) shows the sharpest cost decline over 2010–2019 at 82%, followed by concentrating solar power (CSP) at 47%, onshore wind at 40% and offshore wind at 29%. Continuing cost declines confirm that competitive renewables are a low-cost climate and decarbonisation solution that aligns short-term economic needs with medium- and long-term sustainable development goals [62]. According to the International Renewable Energy Agency, the average energy production costs of more than half of the wind and solar plants built in 2019 are lower than the costs of energy generation of even the least expensive new coal-fired plants. As shown by the calculations of Mrowiec [63], in Poland, taking into account the LCOE (Levelized Cost of Electricity), the lowest energy production costs are achieved by onshore wind power plants. On the other hand, when analysing the years 2010–2018, there is a tendency to decrease the average unit cost of electricity obtained from renewable energy sources. This is particularly evident in photovoltaic investments, which are becoming increasingly cost-competitive.

2. Materials and Methods

The research material used as the basis for the calculation of the indicators was statistical data from GUS, from household budget surveys in 2010–2018. Statistical data analyses were carried out using Excel and Python.

In order to achieve the research objective and accomplish the tasks presented in the paper, the researchers had to apply a number of measures and statistical methods to calculate the synthetic measure of energy poverty, i.e., the synthetic indicator method, such as the Principal Components Analysis (PCA), which involves the reduction of a large number of variables to a few uncorrelated factors that retain as much as possible of the information about the phenomenon under study contained in the primary variables [64]. Poland was chosen as the research object, where studies on energy poverty in the EU measured using indirect measures confirm that Poland belongs to the group of countries affected by this problem to a greater extent than other countries [10,48]. This can be attributed not only to the socio-cultural environment but also to the spread of income poverty or the significant burden on household budgets with energy. Research shows that energy poverty in the Central and Eastern Europe region is widespread and depends on economic, socio-political and environmental issues [65,66].

Collecting information on energy poverty is part of the reporting obligations of European Union Member States resulting from the Regulation of the European Parliament and of the Council on the Governance of the Energy Union [67]. Energy poverty occurs when a household is unable to afford adequate heating, cooling, lighting and energy to run appliances as a result of a combination of low income, high energy expenditure and poor energy performance of the building.

The most commonly used indicators to measure the level of energy poverty include [51,61]:

1. Low Income, High Cost (LIHC)—high required energy costs (i.e., above the national median level) and low income (i.e., disposable income below the officially defined poverty line).
2. Double median energy expenditure (2M)—the share of actual energy expenditure in income is higher than the double median of this value in the population.
3. Ability to pay bills on time (Bills)—Problems with arrears on energy bills or inability to pay them.
4. A building with a leaking roof, damp walls, floors, foundations, rotting windows or floors (Leaks)—problems with the condition of the building.
5. Inadequate thermal comfort (Thermal)—a declared inability to sufficiently heat the house/flat.

During the period analysed by the authors, all these five measures of energy poverty had a decreasing trend [61]. The highest value was taken by the double median energy expenditure, according to which in 2018, 17.2% of households were energy poor, which means a decrease of 2.3 percentage points from 2012. The second of the objective indicators—Low Income, High Costs—which includes both the technical condition of buildings by determining the required energy costs and the material status by including income, decreased from 11.1% in 2012 to 9.4% in 2018. The main reason for this state of affairs can be found in the improving material situation of society as a result of economic development and social policy. Subjective indicators that also relate to the quality (severity) of energy poverty include the ability to pay bills on time—only a small percentage (1.7% in 2018) of households were unable to pay bills on time. The two indicators relating to the technical and functional qualities of buildings (buildings with a leaking roof and insufficient thermal comfort), took a similar value, which in 2018 was 8.2% and 10.7%, respectively [61].

The choice of a set of indicators describing the level of energy poverty is a challenge similar to that of defining energy poverty: on the one hand, the indicators should take into account the multidimensional aspect of energy poverty and, on the other hand, they should allow for efficient and relatively simple planning and application of energy policies.

The authors of the paper decided to build a synthetic measure of energy poverty in order to show a comprehensive description of the energy poverty phenomenon. The differences between the levels of individual indicators are really big and there are even several percentage points of discrepancy between them. Composite measures are easier to assess the extent of energy poverty. Synthetic measures unify several different measures and allow for a full evaluation of the phenomenon. This is particularly important in the case of energy poverty, as many indicators are used that measure the scale of the phenomenon in different ways, some may show an increase and others a decrease, depending on the methodology used. The choice of indicators for constructing a synthetic measure of energy poverty was based on a review of scientific literature, a review of relevant methodologies and an assessment of data that is available in Polish and European datasets. On this basis, the authors chose the energy poverty indicator that took into account the EU-SILC population percentages of people who have reported:

- being unable to keep their homes adequately warm (Inability);
- having arrears in utility bills (Arrears);
- living in a home with a leaking roof, or the presence of damp and rot (Housing faults).

Figure 2 shows the changes in these three indicators in 2010–2018 for Poland. For the period under analysis, a decreasing trend can be seen for the Arrears indicator, while the other two have varied significantly over the period.

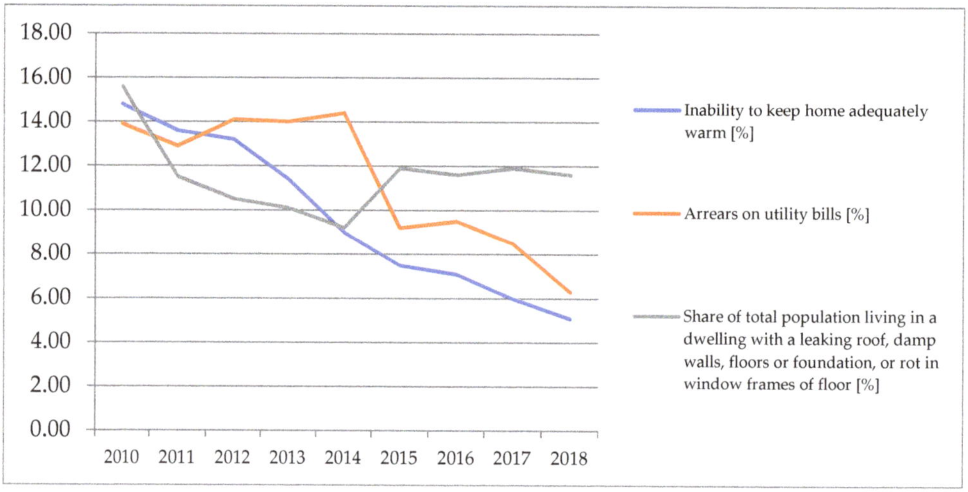

Figure 2. Selected energy poverty indicators for Poland in 2010–2018. Source: Eurostat.

Poland pursues an active climate and energy policy and takes measures in all dimensions of the energy union. The Polish power system is one of the largest within the European Union (it ranks in the top ten in terms of the main macro-energy indicators.

The basis for the production of electricity in Poland is hard coal and lignite, which ensure an adequate level of energy security and stability of generation, but is one of the more expensive sources of energy and deviates from European trends. At present, about 77% of electricity in Poland is produced from hard coal and lignite. Due to the decommissioning of worn-out generating units, the need to meet restrictive environmental protection requirements and the deteriorating market situation (including, above all, an increase in CO_2 emission allowance prices), the share of coal in the electricity generation structure will systematically decrease. The target assumes the reduction of the share of coal in electricity generation to 56–60% by 2030 [68].

Changes in the domestic power sector result in an increased share of renewable energy sources (RES) in the structure of capacities installed in the National Power System (NPS) and in electricity generation. In 2018, the share of RES in electricity generation was 12.7%. The installed capacity of renewable source-based generation in 2018 increased to about 8.5 GW of the total installed capacity in the NPS at a level of about 44.3 GW in 2018. Over the period 2010–2018, the installed RES capacity increased fourfold and the electricity generation from these sources doubled. In 2018, Poland surpassed 16 EU countries in the volume of installed capacity in RES. The main objective of the state energy policy is energy security, while ensuring competitiveness of the economy, energy efficiency and reducing the environmental impact of the energy sector, with the optimum use of own energy resources.

3. Results

3.1. *The Level of Energy Poverty in Poland—A Synthetic Measure of Energy Poverty*

Before proceeding to the calculation of the synthetic measure of energy poverty, a preliminary analysis of the descriptive statistics of the studied variables was carried out (Table 2). During the study period (2010–2018), the average percentage of people reporting an inability to adequately heat their home was 9.74%, the percentage of people in arrears on utility bills was 11.42%, and the percentage of the population living in homes with a leaking roof, damp walls, floors or foundations, or rot in window frames was 15.4%. Based on standard deviations, coefficients of variation (V) were calculated, ranging from 36.9% for the first indicator to 15.4% for the last indicator. This shows that the indicators are characterized by moderate or fairly low variability.

Table 2. Descriptive statistics of the variables used to measure energy poverty levels.

	MEAN	STD	V
Inability	9.74	3.59	36.9%
Arrears	11.42	3.05	26.7%
Housing faults	11.54	1.78	15.4%

Source: own calculations.

Then, the linear correlation coefficients between the indicators were calculated. Based on the graph shown above, one would expect a fairly strong dependency between the first and second indicators, with a lower dependency with the third one. This is confirmed by the linear correlation coefficients (Table 3). For the first two indicators, the correlation is 0.85, which means that the indicators are highly correlated with each other and show similar directions of change. This is different for the third indicator, which is fairly low correlated with the other two. Therefore, it can be considered that it describes slightly different aspects of energy poverty (the first two are more strongly related to income and wealth levels, while the second indicator measures the quality of housing infrastructure, which is not subject to the same dynamics as income).

Table 3. Correlation matrix between input indicators of energy poverty.

	Inability	Arrears	Housing Faults
Inability	1.00	0.85	0.26
Arrears	0.85	1.00	−0.13
Housing faults	0.26	−0.13	1.00

Source: own calculations.

When constructing the synthetic indicator, the indicator proposed by Bouzarovski and Tirado Herrero [48] was adopted as a baseline version, which is a weighted average of three indices. As a result, this indicator (S1) is calculated as:

$$S1 = (I1 + I2 + I3)/3$$

where I1, I2, I3 are the individual indicators from Table 4.

This method can be modified by giving different weights to the different input indicators. There are various methods for calculating these weights. One is to relate the weights to the inverse of the multiple correlation coefficient. In this method, the weight of an individual indicator is inversely proportional to the indetermination coefficient in the models explaining that individual indicator with the other indicators. An approach similar to the VIF (Variance Inflation Factor) is used here. As a result, more weight is given here to the indicator that is less related to the others.

$$S2 = a1I1 + a2I2 + a3I3$$

where

$$ai = (1 - R2i)/(3 - R21\text{-}R22\text{-}R23)$$

where R2 is the determination coefficient

The following results were obtained in the case analysed here:

Table 4. Determination and indetermination coefficients used to construct synthetic indicator S2.

	I1	I2	I3
R2	0.863	0.855	0.522
1−R2	0.137	0.145	0.478
waga	0.180	0.190	0.630

Source: own calculations.

Another variant of the weighted average method is where the weight of a given indicator is related to the share of its variance in the total variance of the set. It is worth noting here that the variables should be expressed on the same scale (have the same range of variation, and thus similar variances). In the case analysed here, this requirement is met (the variables are percentages of the population), if it were not met, some type of scaling would be required. Thus, the indicator takes the form of:

$$S3 = a1I1 + a2I2 + a3I3$$

where

$$ai = var(Ii)/(var(I1) + var\ (I2) + var(I3))$$

The following results were obtained in the analysed case (Table 5):

Table 5. Variances and weights used to construct synthetic indicator S3.

	I1	I2	I3
Var	12.910	9.302	3.173
waga	0.509	0.366	0.125

Source: own calculations.

The third, and most advanced method, is to extract common information from all factors using the principal components approach (PCA). The method consists of linear transformation of the input variable matrix in such a way as to create synthetic indicators (components) of maximum variance of the input set and maximally uncorrelated with each other. First, the number of components to extract must be determined. It is determined by analysing the stock of extracted common variability. The results are shown in Table 6.

Table 6. Extracted variance values for each component.

	PC 1	PC 2	PC 3
Extracted variance values	81.2%	16.5%	2.3%

Source: own calculations.

The first component (PC 1) explains 81.2% of the input set, which is a high indication, so it can be considered that one component is sufficient to capture most of the information of the set.

The calculation of the individual indicator weights in the formation of the component is shown in Table 7. There is a strong association with the first two indicators and a weaker association with the third one.

Table 7. Charges for the PCA method.

	I1	I2	I3
Charges	−0.7733	−0.6321	−0.0491

Source: own calculations.

In the next step, the comparison of all obtained energy poverty indicators can be carried out. In order to better compare them, they were all corrected by the mean (i.e., the mean values were subtracted), resulting in centering around zero. Table 8 presents the numerical values of the indicators, while their evolution is shown in Figure 3.

Table 8. Synthetic poverty indicators S1–S4 and individual input indicators.

	S1	S2	S3	S4	Inability	Arrears	Housing
2010	3.86	3.94	3.99	5.67	14.80	13.90	15.60
2011	1.76	0.95	2.50	3.91	13.60	12.90	11.50
2012	1.70	0.47	2.61	4.31	13.20	14.10	10.50
2013	0.93	-0.12	1.61	2.84	11.40	14.00	10.10
2014	−0.04	−1.04	0.42	1.19	9.00	14.40	9.20
2015	−1.37	−0.60	−1.91	−3.12	7.50	9.20	11.90
2016	−1.50	−0.81	−2.04	−3.26	7.10	9.50	11.60
2017	−2.10	−1.01	−2.93	−4.73	6.00	8.50	11.90
2018	−3.24	−1.78	−4.23	−6.83	5.10	6.30	11.60

Source: own calculations.

It can be observed that indicators S1 and S3 show a very similar pattern. Indicator S4 shows a more dynamic decline in energy poverty (it is more strongly related to indicators I2 and I1). Indicator S2, on the other hand, shows a weaker decline in poverty (through a stronger link with indicator I3). Thus, it seems that indicators S2 and S3 are the most balanced for the studied set.

Synthetic indicators S2 and S3 of energy poverty show a slow but visible decline in the level of energy poverty over the study period.

3.2. Methods for Reducing Energy Poverty in Poland

As part of the "Clean Energy for All Europeans" package, the European Commission has proposed a series of measures to combat energy poverty through energy efficiency, securing against disconnection and better definition and monitoring of the problem at a member state level through integrated national energy and climate plans [69].

Research has shown that in recent years there has been a decline in energy poverty indicators, which means a reduction in the number of households at risk of energy poverty in Poland. However, with the current situation resulting from the COVID-19 pandemic, the reduction or even loss of income, the loss of many jobs, it may turn out that the problems of energy poverty will continue to affect many households. Therefore, it should be emphasised that counteracting energy poverty in Poland requires actions on the part of the government administration, in particular by providing an institutional framework and financing allowing communes to assess the technical condition of flats and buildings, providing energy advice and planning and implementing thermal modernisation measures.

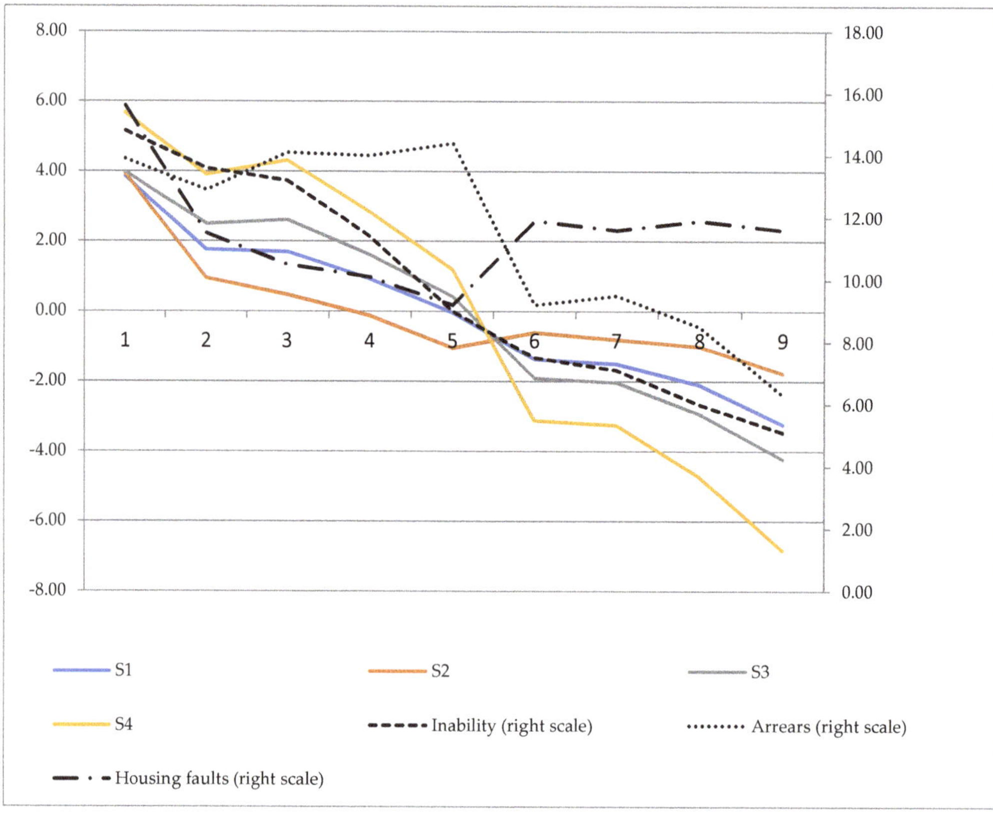

Figure 3. Crude and synthetic indicators of energy poverty. Source: own calculations.

Measures to counteract energy poverty in Poland include the government's "Czyste powietrze" Programme, implemented in 2017 and aimed directly at the group of energy-poor households (Table 9). Measures under the Programme include: stricter regulations concerning the quality of solid fuels, stricter regulations concerning the standards of boilers for solid fuels, a program for thermal modernisation of single-family houses, income tax relief for thermal modernisation investments. By 11 December 2020, a total of over 155 thousand agreements on investment co-financing were signed under the Programme, for the amount of PLN 2.7 billion [70]. Further support programmes are still under discussion in central public administration. They are aimed at reducing air pollution (popularly known as smog), which has been publicised by numerous non-governmental organizations, above all, including numerous local smog alerts. As Owczarek's research has shown, smog is closely related to the phenomenon of energy poverty in Poland [71]. The link is that buildings with poor energy performance need more fuel to heat—mainly coal supplemented by firewood. Often, however, the problem lies in the fact that the energy-poor cannot afford to buy enough fuel, so they supplement (or even completely replace) the full-value fuel with rubbish or, e.g., wood waste collected in the immediate vicinity of the house.

Table 9. Instruments for combating energy poverty used in Poland.

Selected Measures	Type of Measure	Organisation	Result
Clean Air program	Building insulation, Heating system	National government	The Clean Air program provides financing to improve heating systems in households.
Energy lump sum	Energy bill support	National government	This measure provides financial assistance for energy bills to people that were involved in military operations or wars.
Energy allowance/Housing allowance	Energy bill support	Local government	This measure provides financial assistance to households to pay their electricity bills.
National support system for energy efficiency and RES	Information and awareness	National government	This project aims to support different stakeholders in Poland to improve energy efficiency by providing guidance and information. Advisors are available that can give households information on how to improve energy efficiency.
Procurement subsidies for small RES installations	Renewable energy	National government, Local government	This measure includes subsidies for small-scale RES generation.
Special purpose allowance	Social support	National government	This measure can be given in certain cases to meet basic needs, including fuel and energy expenses.

Source: https://www.energypoverty.eu/ (accessed on 19 February 2021).

The objective of government action in the context of smog and energy poverty is to eliminate the burning of rubbish and reduce the burning of wood waste, as well as to improve the energy performance of buildings to keep fuel consumption as low as possible (which would generate fewer costs and thus reduce the scale of energy poverty) or to lead to a situation where the burning of fuel for heating is as little harmful to the environment as possible (e.g., by connecting to a district heating network powered by combined heat and power plants using advanced technologies allowing for a significant reduction of harmful substances when burning coal). Therefore, to a large extent, the fight against air pollution and energy poverty should be pursued through integrated public policy instruments [19].

The burning of low-quality fuels and municipal waste (even rubbish) and the use of low-quality boilers are the main factors responsible for air pollution in Poland [56,72]. The situation is aggravated by the low energy performance of buildings and efficiency of heating systems, which impose much higher fuel consumption (in Poland, it is mainly coal). This state of affairs can be attributed to the insufficient use of renewable energy sources and low social awareness of energy poverty prevention. Particulate matter, which is a measure of air pollution, is produced, among others, by the burning of solid fuels in households [73]. The complex interactions between residential energy consumption, climate change and thermal performance of buildings in Eastern Europe have been identified in studies by Urge-Vorsatz and Tirado-Herrero [74].

To significantly reduce the scale of energy poverty in Poland, and thus meet the limits resulting from air pollution, public policy instruments should be directed towards the diversification of energy sources and investments in renewable energy sources, especially generating electricity for own needs in its place. consumption, i.e., in households, which improves energy security.

Many factors contribute to the favorable conditions for the development of renewable energy sources. Some of them are various financial instruments such as loans, grants, and others. Until now, the most important source of financing for small-scale RES investments in Poland has been the EU structural funds managed at the level of individual regions. ROPs, or Regional Operational Programs, are co-financed from two funds: ERDF—European Regional Development Fund and ESF—European Social Fund and from national funds. ROPs are aimed at the development of, inter alia, energy that uses renewable energy sources (RES) in many dimensions, including energy production, effective distribution, support for enterprises operating in the field of servicing the renewable energy sector, and others. Photovoltaic prosumer installations are the most popular technology supported in Regional Operating Programs.

Taking into account the climatic conditions in Poland, the following can be used in single-family houses: solar collectors to heat utility water, photovoltaic cells or home wind farms generating electricity, biomass boilers or heat pumps used in the heating system, wastewater heat recovery system, ground heat exchangers and the so-called hybrid solutions, combining various RES [75]. Solar collectors recover heat energy by heating the medium (e.g., glycol) with sunlight. This energy is sufficient to heat domestic water. When designing an installation based on solar collectors, it is necessary to take into account the number of inhabitants and their need for hot water, as well as the level of sunlight and the possibility of properly positioning the collectors concerning the directions of the world, directing them to the south [75]. An interesting, although expensive solution is home wind farms. However, it should be remembered that the collectors, cells, or wind farms themselves will not fully cover the demand for heat and electricity in a single-family house. However, they are an important supplement, thanks to which it is possible to reduce electricity bills and thus reduce energy poverty [76,77].

Effective home heating is an activity that significantly influences the thermal comfort of residents. One possible solution is biomass heating boilers. The fuel can be used i.a. biomass in the form of briquettes, pellets, or wood chips. An argument that additionally supports the use of biomass boilers is the fact that the CO_2 generated during its combustion is absorbed by vegetation and does not accumulate in the atmosphere [78]. Another solution for the heating system can be a heat pump [79,80]. It is a device that allows you to recover thermal energy from soil, air, or water. As with the use of other renewable energy sources, the pump selection should also be based on the actual needs of the household members and the specificity of the building. It is possible to combine more than one renewable energy source. The most popular solution is hybrid collectors using photovoltaic cells and solar collectors. At the same time, they heat water and generate electricity. Often, a combination of renewable and conventional sources is also used.

In existing multi-family buildings, it is impossible to meet all energy consumption needs by generating energy from renewable sources in installations located within the building or property. The obstacles are the current legal status, but also the technical possibilities. Renewable energy installations can be built on designated areas and the energy they generate can be purchased for buildings. For the production of energy in multi-family buildings, can be used [81]: wind farms, photovoltaic farms, biomass heating plants, biogas plants, etc. In Poland, solar collectors, photovoltaic panels, and air-to-water heat pumps can also be installed in already functioning multi-family buildings. In some cases, it is possible to use biomass boilers located in a building or a district heating plant (CHP). Heat and electricity can also be stored within the building. Unfortunately, storage technologies are still expensive, which often makes this process unprofitable [82].

The increase in the production of energy from renewable sources should be large enough to meet the growing demand and additionally allow to reduce production from conventional sources [83]. Thanks to the improvement of the income situation in Poland, households are replacing coal stoves with gas, biomass and other more expensive fuels, which is confirmed by the research results [84].

In Poland, at the end of 2020, the installed capacity of all renewable energy sources in the power system was almost 10 GW, of which over 183 MW in small RES installations [85].

Small installations include installations with a total installed electrical power greater than 50 kW and less than 500 kW, connected to a power grid with a rated voltage lower than

110 kV, or with a combined heat output greater than 150 kW and not greater than 900 kW with the total installed electrical capacity greater than 50 kW and less than 500 kW [85].

In 2020, energy was produced in a total of 898 small installations. Their total installed capacity was over 183 MW. The most numerous were small installations using water energy (343), with a total installed capacity of 51.96 MW. Next, in terms of the number of installations (328), but at the same time, the largest in terms of the total installed capacity (66.86 MW), were photovoltaic sources. The least frequent were small installations producing energy from biomass. At the end of 2020, there were only two such installations in Poland (Table 10).

Table 10. RES installation entered in the register of energy producers in a small installation, by source type (as at the end of 2020).

Type of RES Installation	Number of Installation	Installed Capacity [MW]
Using hydropower (WO)	343	51.96
Using the energy of solar radiation (PV)	328	66.86
Using biogas (BG)	117	32.10
Using wind energy (WI)	108	31.71
Using biomass (BM)	2	0.47
Total	898	183.10

Source: Energy Regulatory Office in Poland.

In 2020, the most energy—over 146 GWh—was produced by small hydropower plants. The second-largest source in 2020 was biogas plants using non-agricultural biogas. They produced over 107 GWh of energy (Figure 4).

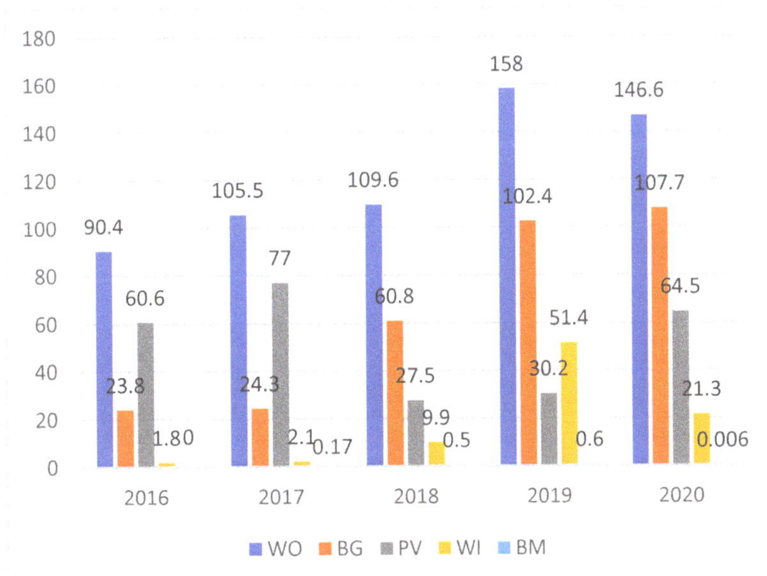

Figure 4. Energy production in small RES installations in 2016–2020 by type of source (in GWh). Type of renewable energy installations: WO—hydropower, WI—wind energy, PV—solar radiation energy, BG—non-agricultural biogas, BM—biomass. Source: Energy Regulatory Office in Poland.

The photovoltaic (PV) market is the fastest growing of all renewable energy sectors in Poland [86]. Currently, the largest increase in new capacity is observed in the micro-installation segment, which means high activity of individual and business prosumers. In

2020, over 64 GWh of energy was generated in small photovoltaic installations—as much as 114% more than in the previous year. This source accounts for 19% of the total production.

4. Discussion

So far, many studies have been conducted on energy poverty in different countries. They indicate how important problems in the modern world are the access to energy and the appropriate level of its prices. Government policies position energy consumption as a basic need or right to be provided for citizens, and the lack of it as a form of deprivation to be dealt with [13]. This is an important problem that requires political action contributing to its reduction. Research on energy poverty has been undertaken in EU and non-EU countries e.g., Japan [87], Australia [88], India [18] and Brazil [89]. Many studies have been devoted to the impact of energy poverty on increased human mortality. For example, studies in the UK and Ireland have shown increased levels of death during the winter months and a link has been established between insufficiently heated homes and increased rates of morbidity and higher incidence of various cardiovascular and respiratory diseases [13,16]. Whereas, in developing countries (Ghana), studies indicated the high dependence of households on traditional biomass for cooking. Traditional biomass, which is usually burned in houses, exposes householders to indoor air pollution. The use of candles and kerosene lamps as light sources also pollutes indoor air, which contains high levels of particulate matter (PM) and toxins that are dangerous to the respiratory system. Inhalation of PM causes asthma, lung cancer, cardiovascular and respiratory diseases [17]. The importance of modern renewable sources (water, wind, sun) should also be stressed here, as they can contribute to limiting environmental pollution and reducing the number of serious diseases, thus reducing the costs of treating the sick or of their increased mortality. In the case of Poland, this is a huge problem, as our country has the most polluted air of all the member states of the European Union. It is therefore important to emphasise that renewable sources can help to reduce energy poverty and reduce many serious diseases resulting from air pollution. The World Health Organization (WHO) pays particular attention to the problems of environmental pollution. A WHO report showed that in 2010 air pollution contributed to the premature deaths of 48.5 thousand Poles. The burning of conventional fuels in households between 2010 and 2016 caused 19 to 22 thousand premature deaths annually [90]. Deaths related to air pollution in Poland generate costs of USD 101,826 million [91] or as much as 12.9% of gross domestic product (GDP). This amounts to over PLN 800 per month per Polish citizen. This data is also confirmed by Greenpeace, which estimates the costs of air pollution caused by fossil fuels at PLN 113 billion per year [92]. Many authors point to the importance of renewable energy sources in reducing energy poverty. Neacsa et al. showed the use of renewable energy in Romanian households [52]. Romania has adopted a solution consisting of government subsidies for the purchase of solar panel systems, geothermal energy as well as wind energy for households. This example shows support for investment from public funds, which, on the one hand, is seen as measures reducing energy poverty and, on the other hand, reducing pollution.

An interesting example of the use of renewable energy in district heating is the Organic Rankine Cycle (ORC) system powered by biomass used in Spain [93]. The CCHP configuration used in Spain is worth attention in regions with mild winters and medium-hot summers, as well as the availability of biomass, which could not be applicable in all regions in Poland.

The use of renewable energy as a solution to the problem of energy poverty is presented by researchers from South Korea [94]. They propose the use of solar PV systems for low-income families in social housing. This solution has several advantages: systems balance energy costs, reduce environmental impact and CO_2 emissions, and contribute to energy independence.

In turn, Luderer et al. [95] have developed scenarios to assess the role of renewable energy in climate change mitigation and also in heat production. Almost all scenarios show strong possible growth in renewable energy production, with a significant increase in the use of wind and solar power. Wind power is competitive even without climate policy, while the prospects for photovoltaics (PV) are highly dependent on climate policy assumptions.

Many experts believe that the use of renewable energy is a solution that could help reduce the problem of energy poverty, despite the challenges and limitations associated with the use of this energy source. Even if the generation of such energy in some cases entails more costs than conventional energy generation, experts underline the short- and long-term benefits of its use [96]. The implementation of renewable energy projects generates not only disputes related to the costs and financing of investments, but also to their acceptance by local communities (due to the negative external effects generated in the form of aesthetics, noise, biodiversity degradation, etc.) [52,97]. However, investment in renewable energy production should become an "engine" for rural area development as, on the one hand, it provides required energy and contributes to reducing energy poverty and, on the other, it attracts local labour and creates new jobs [52,98].

5. Conclusions

Energy poverty is a dynamic and complex phenomenon, generated by a combination of factors such as low household incomes, high energy prices, and difficult access to the energy system.

To show the level of energy poverty in Poland, the authors developed a synthetic measure that unifies several different measures used by researchers and allows for a comprehensive assessment of the phenomenon. On its basis, it can be concluded that in 2010–2018 there was a slow but visible decrease in the level of energy poverty in Poland. Investments in renewable energy that can contribute to the greatest extent to reducing the energy poverty of households include solar collectors to heat utility water, photovoltaic cells or home wind farms generating electricity, biomass boilers or heat pumps used in the heating system, wastewater heat recovery system, ground heat exchangers and the so-called hybrid solutions, combining various RES. It should be remembered that the collectors, cells, or wind farms themselves will not fully cover the heat and electricity demand of a single-family house. However, they are an important supplement, thanks to which it is possible to reduce electricity bills and thus reduce energy poverty. It is possible to combine more than one renewable energy source. The most popular solution is hybrid collectors using photovoltaic cells and solar collectors. At the same time, they heat water and generate electricity. Often, a combination of renewable and conventional sources is also used. However, it should be remembered that all investments in renewable energy require appropriate financing. The problem of energy poverty often concerns poor people who are not able to generate funds for these purposes. Therefore, various types of publicly funded programs supporting such investments and alleviating the problem of energy poverty are an important activity. It must be remembered that renewable energy is a long-term solution that could solve the problem of energy poverty, despite the challenges associated with the difficulties of implementing these unconventional energy sources. Investments in renewable energy sources contribute to the fight against smog by reducing the emission of dust and other pollutants released into the atmosphere. Investing in renewable energy sources not only helps to protect the environment and combat smog but also allows you to increase your home budget. It is possible thanks to financial support for the purchase and installation of heat pumps and photovoltaics, as well as constant savings on energy costs. As the analyses show, newly installed renewable energy costs less and becomes more competitive compared to traditional methods of energy production. Therefore, investments in renewable energy sources are a legitimate measure contributing to reducing energy poverty in Poland. Public institutions should support investments in renewable energy sources for environmental reasons, limiting the processes of climate warming, health, as well as the slow depletion of fossil sources, which unfortunately will increase the cost of energy production in the future. The assumptions of the EU climate policy indicate that striving to increase the share of renewable energy sources in total consumption, reducing greenhouse gas emissions, and improving energy efficiency are goals that all member states must pursue.

The authors, however, encountered some limitations in carrying out the analyzes that were the basis for the preparation of the article. The limitation of this study was the range of available GUS statistical data as well as the selection of measures in the statistical analysis. A challenge for other researchers in the field of conducted research may be the application of the developed indicators in other countries at risk of energy poverty. In the current situation, further research should be undertaken to show the impact of the coronavirus pandemic on the level of energy poverty in various countries.

Author Contributions: Conceptualization, A.B.-J., P.T., S.J.; data curation, A.B.-J., P.T., S.J.; formal analysis, A.B.-J., P.T., S.J.; methodology, A.B.-J., P.T., S.J.; resources, A.B.-J., P.T., S.J.; visualization, A.B.-J., P.T., S.J.; writing—original draft, A.B.-J., P.T., S.J.; writing—review and editing, A.B.-J., P.T., S.J.; supervision, A.B.-J., P.T., S.J.; funding acquisition, A.B.-J., P.T., S.J. All authors have read and agreed to the published version of the manuscript.

Funding: This research received no external funding.

Institutional Review Board Statement: Not applicable.

Informed Consent Statement: Not applicable.

Data Availability Statement: The data presented in this study are available on request from the corresponding author.

Conflicts of Interest: The authors declare no conflict of interest.

References

1. Bouzarovski, S.; Petrova, S.; Sarlamanov, R. Energy poverty policies in the EU: A critical perspective. *Energy Policy* **2012**, *49*, 76–82. [CrossRef]
2. EEN. *EnR Position Paper on Energy Poverty in the European Union*; European Energy Network: Roma, Italy, 2019.
3. Thomson, H.; Bouzarovski, S.; Snell, C. Rethinking the measurement of energy poverty in Europe: A critical analysis of indicators and data. *Indoor Built Environ.* **2017**, *26*, 879–901. [CrossRef] [PubMed]
4. Thomson, H.; Snell, C.; Bouzarovski, S. Health, Well-Being and Energy Poverty in Europe: A Comparative Study of 32 European Countries. *Int. J. Environ. Res. Public Health* **2017**, *14*, 584. [CrossRef] [PubMed]
5. Maxim, A.; Mihai, C.; Apostoaie, C.M.; Maxim, A. Energy poverty in Southern and Eastern Europe: Peculiar regional issues. *Eur. J. Sustain. Dev.* **2017**, *6*, 247. [CrossRef]
6. Maxim, A.; Mihai, C.; Apostoaie, C.-M.; Popescu, C.; Istrate, C.; Bostan, I. Implications and Measurement of Energy Poverty across the European Union. *Sustainability* **2016**, *8*, 483. [CrossRef]
7. Sałach, K.; Lewandowski, P. Pomiar ubóstwa energetycznego na podstawie danych BBGD—Metodologia i zastosowanie (Measurement of energy poverty on the basis of BBGD data—Methodology and application). *IBS Res. Rep.* **2018**, *1*, 1–21. (In Polish)
8. Che, X.; Zhu, B.; Wang, P. Assessing global energy poverty: An integrated approach. *Energy Policy* **2021**, *149*, 112099. [CrossRef]
9. European Commission. Indicators and Data. EU Energy Poverty Observatory. Available online: https://www.energypoverty.eu/indicators-data (accessed on 20 January 2021).
10. Bouzarovski, S. Energy poverty in the European Union: Landscapes of vulnerability. *WIREs Energy Environ.* **2014**, *3*, 276–289. [CrossRef]
11. Boardman, B. *Fuel Poverty: From Cold Homes to Affordable Warmth*; Belhaven Press: London, UK, 1991.
12. Reddy, A. Energy and social issues. In *World Energy Council and UNEP*; Energy and the challenge of sustainability: New York, NY, USA, 2000.
13. Day, R.; Walker, G.; Simcock, N. Conceptualising energy use and energy poverty using a capabilities framework. *Energy Policy* **2016**, *93*, 255–264. [CrossRef]
14. Charlier, D.; Kahouli, S. From Residential Energy Demand to Fuel Poverty: Income-induced Non-linearities in the Reactions of Households to Energy Price Fluctuations. *Energy J.* **2019**, *40*. [CrossRef]
15. Curwen, M. Excess winter mortality: A British phenomenon? *Health Trends* **1991**, *22*, 169–175.
16. Healy, J.D.; Clinch, J. Quantifying the severity of fuel poverty, its relationship with poor housing and reasons for non-investment in energy-saving measures in Ireland. *Energy Policy* **2004**, *32*, 207–220. [CrossRef]
17. Adusah-Poku, F.; Takeuchi, K. Energy poverty in Ghana: Any progress so far? *Renew. Sustain. Energy Rev.* **2019**, *112*, 853–864. [CrossRef]

18. Bhide, A.; Monroy, C.R. Energy poverty: A special focus on energy poverty in India and renewable energy technologies. *Renew. Sustain. Energy Rev.* **2011**, *15*, 1057–1066. [CrossRef]
19. Boguszewski, R.; Herudziński, T. *Ubóstwo Energetyczne w Polsce (Energy Poverty in Poland)*; Pracownia Badań Społecznych SGGW: Warsaw, Poland, 2018. (In Polish)
20. Howden-Chapman, P.; Viggers, H.; Chapman, R.; O'Sullivan, K.; Barnard, L.T.; Lloyd, B. Tackling cold housing and fuel poverty in New Zealand: A review of policies, research, and health impacts. *Energy Policy* **2012**, *49*, 134–142. [CrossRef]
21. Baudu, R.; Charlier, D.; Legendre, B. Fuel poverty and health: A panel data analysis. *FAERE Work. Pap.* **2020**, *4*, 2274–5556.
22. Sagar, A.D. Alleviating energy poverty for the world's poor. *Energy Policy* **2005**, *33*, 1367–1372. [CrossRef]
23. Pachauri, S.; Rao, N.D. Gender impacts and determinants of energy poverty: Are we asking the right questions? *Curr. Opin. Environ. Sustain.* **2013**, *5*, 205–215. [CrossRef]
24. Gonzalez Pijuan, I. *Gender Inequality and Energy Poverty. A Forgotten Risk Factor*; Engineering Without Borders: Geneva, Switzerland, 2018.
25. Clancy, J.S.; Skutsch, M.; Batchelor, S. *The Gender-Energy-Poverty Nexus. Finding the Energy to Address Gender Concerns in Development. DFID Project CNTR998521*; UK Department for International Development (DFID): London, UK, 2002.
26. Creţan, R.; Málovics, G.; Méreiné-Berki, B. On the perpetuation and contestation of racial stigma: Urban Roma in a disadvantaged neighbourhood of Szeged. *Geogr. Pannonica* **2020**, *24*, 294–310. [CrossRef]
27. Creţan, R.; Turnock, D. Romania's Roma Population: From Marginality to Social Integration. *Scott. Geogr. J.* **2008**, *124*, 274–299. [CrossRef]
28. Special Report. *Air Pollution: Our Health Still Insufficiently Protected*; European Court of Auditors: Luxembourg, Luxembourg, 2018.
29. Welsch, H.; Biermann, P. Energy Affordability and Subjective Well-Being: Evidence for European Countries. *Energy J.* **2017**, *38*, 159–176. [CrossRef]
30. Węglarz, A.; Kubalski, G.; Owczarek, D. *Propozycje Mechanizmów Wsparcia Procesu Przeciwdziałania Zjawisku Ubóstwa Energetycznego w Polsce (Proposed Mechanisms to Support the Process of Counteracting the Phenomenon of Energy Poverty in Poland)*; Instytut na Rzecz Ekorozwoju: Warszawa, Poland, 2014. (In Polish)
31. Recalde, M.; Peralta, A.; Oliveras, L.; Herrero, S.T.; Borrell, C.; Palència, L.; Gotsens, M.; Artazcoz, L.; Marí-Dell'Olmo, M. Structural energy poverty vulnerability and excess winter mortality in the European Union: Exploring the association between structural determinants and health. *Energy Policy* **2019**, *133*, 110869. [CrossRef]
32. Sareen, S.; Thomson, H.; Herrero, S.T.; Gouveia, J.P.; Lippert, I.; Lis, A. European energy poverty metrics: Scales, prospects and limits. *Glob. Transit.* **2020**, *2*, 26–36. [CrossRef]
33. Papada, L.; Kaliampakos, D. Being forced to skimp on energy needs: A new look at energy poverty in Greece. *Energy Res. Soc. Sci.* **2020**, *64*, 101450. [CrossRef]
34. Thomson, H.; Snell, C. Quantifying the prevalence of fuel poverty across the European Union. *Energy Policy* **2013**, *52*, 563–572. [CrossRef]
35. Heindl, P.; Schüssler, R. Dynamic properties of energy affordability measures. *Energy Policy* **2015**, *86*, 123–132. [CrossRef]
36. Phimister, E.; Vera-Toscano, E.; Roberts, D.J. The Dynamics of Energy Poverty: Evidence from Spain. *Econ. Energy Environ. Policy* **2015**, *4*, 153–166. [CrossRef]
37. Okushima, S. Measuring energy poverty in Japan, 2004–2013. *Energy Policy* **2016**, *98*, 557–564. [CrossRef]
38. Pachauri, S.; Mueller, A.; Kemmler, A.; Spreng, D. On Measuring Energy Poverty in Indian Households. *World Dev.* **2004**, *32*, 2083–2104. [CrossRef]
39. Kurowski, P. Zagrożenie ubóstwem energetycznym. Próba ustalenia zjawiska (The threat of energy poverty. An attempt to determine the phenomenon). *Bull. Energy Regul. Off.* **2012**, *79*, 3–8. (In Polish)
40. Hills, J. Getting the measure of fuel poverty. In *Final Report of the Energy Poverty Review, Case Report No. 72, Commissioned by the UK Department of Energy and Climate Change*; UK (DECC): London, UK, 2012.
41. Gordon, D.; Adelman, L.; Ashworth, K.; Bradshaw, J.; Levitas, R.; Middleton, S.; Pantazis, C.; Patsios, D.; Payne, S.; Townsend, P.; et al. *Poverty and Social Exclusion in Britain*; Joseph Rowntree Foundation: York, UK, 2000.
42. Healy, J. *Housing, Fuel Poverty and Health: A Pan-European Analysis*; Ashgate: Aldershot, UK, 2004.
43. Petrova, S.; Gentile, M.; Mäkinen, I.H.; Bouzarovski, S. Perceptions of Thermal Comfort and Housing Quality: Exploring the Microgeographies of Energy Poverty in Stakhanov, Ukraine. *Environ. Plan. A Econ. Space* **2013**, *45*, 1240–1257. [CrossRef]
44. Thema, J.; Vondung, F. *EPOV Indicator Dashboard: Methodology Guidebook*; Wuppertal Institut für Klima, Umwelt, Energie GmbH: Wuppertal, Germany, 2020.
45. Piekut, M. Patterns of Energy Consumption in Polish One-Person Households. *Energies* **2020**, *13*, 5699. [CrossRef]
46. Bossert, W.; Chakravarty, S.R.; D'Ambrosio, C. Measuring Multidimensional Poverty: The Generalized Counting Approach. In *Analyzing Multidimensional Well-Being: A Quantitative Approach*; John Wiley: New Jersey, NJ, USA, 2020.
47. Nussbaumer, P.; Bazilian, M.; Modi, V. Measuring energy poverty: Focusing on what matters. *Renew. Sustain. Energy Rev.* **2012**, *16*, 231–243. [CrossRef]
48. Bouzarovski, S.; Herrero, S.T. The energy divide: Integrating energy transitions, regional inequalities and poverty trends in the European Union. *Eur. Urban Reg. Stud.* **2017**, *24*, 69–86. [CrossRef]
49. Alkire, S.; Foster, J. Counting and multidimensional poverty measurement. *J. Public Econ.* **2011**, *95*, 476–487. [CrossRef]

50. Bouzarovski, S.; Kiełczewska, A.; Lewandowski, P.; Sokołowski, J. Measuring energy poverty in Poland with the Multidimensional Energy Poverty Index. *Energy Sources Part B Econ. Plan. Policy* **2020**, *15*, 2451–4373.
51. Szamrej-Baran, I. Identyfikacja przyczyn ubóstwa energetycznego w Polsce przy wykorzystaniu modelowania miękkiego. (Identification of the causes of energy poverty in Poland using soft modeling). *Sci. Pap. Wrocław Univ. Econ.* **2014**, *23*, 343–352. (In Polish)
52. Neacsa, A.; Panait, M.; Muresan, J.D.; Voica, M.C. Energy Poverty in European Union: Assessment Difficulties, Effects on the Quality of Life, Mitigation Measures. Some Evidences from Romania. *Sustainability* **2020**, *12*, 4036. [CrossRef]
53. Kaya, O.; Klepacka, A.M.; Florkowski, W.J. Achieving renewable energy, climate, and air quality policy goals: Rural residential investment in solar panel. *J. Environ. Manag.* **2019**, *248*, 109309. [CrossRef]
54. Edenhofer, O.; Seyboth, K.; Creutzig, F.; Schlömer, S. On the Sustainability of Renewable Energy Sources. *Annu. Rev. Environ. Resour.* **2013**, *38*, 169–200. [CrossRef]
55. Piwowar, A.; Dzikuć, M. Development of Renewable Energy Sources in the Context of Threats Resulting from Low-Altitude Emissions in Rural Areas in Poland: A Review. *Energies* **2019**, *12*, 3558. [CrossRef]
56. Gielen, D.; Boshell, F.; Saygin, D.; Bazilian, M.D.; Wagner, N.; Gorini, R. The role of renewable energy in the global energy transformation. *Energy Strategy Rev.* **2019**, *24*, 38–50. [CrossRef]
57. Abu-Hamed, T.; Peric, K. The Role of Renewable Energy Resources in Alleviating Energy Poverty in Palestine. *Renew. Energy Focus* **2020**, *35*, 97–107. [CrossRef]
58. Mshelia, H.I. Energy Access for All. The role of clean energy in alleviating energy poverty. *Energy Access All* **2012**, *33*, 43.
59. Singh, N.; Nyuur, R.; Richmond, B. Renewable Energy Development as a Driver of Economic Growth: Evidence from Multivariate Panel Data Analysis. *Sustainability* **2019**, *11*, 2418. [CrossRef]
60. Climate & Energy Framework. Available online: https://ec.europa.eu/clima/policies/strategies/2030_en (accessed on 20 October 2020).
61. Walkowska, K. *Energy Consumption in Households in 2018*; Statistics Poland: Warsaw, Poland, 2019.
62. IRENA. *Renewable Power Generation Costs in 2019*; International Renewable Energy Agency: Abu Dhabi, United Arab Emirates, 2020.
63. Mrowiec, D. Jednostkowy koszt wytwarzania energii elektrycznej (LCOE) jako wskaźnik porównawczy kosztów produkcji różnych źródeł (Unit Cost of Electricity Generation (LCOE) as a benchmark for production costs of different sources). *PSE Innowacje* **2019**, *2*, 71–76. (In Polish)
64. Jolliffe, I.T. Principal Component Analysis Series. In *Springer Series in Statistics*, 2nd ed.; Springer: New York, NY, USA, 2002.
65. Bouzarovski, S. *Energy Poverty (Dis)Assembling Europe's Infrastructural Divide*; Springer International Publishing AG: Cham, Switzerland, 2018.
66. Dubois, U.; Meier, H. Energy affordability and energy inequality in Europe: Implications for policymaking. *Energy Res. Soc. Sci.* **2016**, *18*, 21–35. [CrossRef]
67. Regulation (EU) 2018/1999 of the European Parliament and of the Council on Governance of the European Energy Union and Climate Action. Available online: https://ec.europa.eu/transparency/regdoc/rep/3/2020/PL/C-2020-9600-F1-PL-MAIN-PART-1.PDF (accessed on 15 February 2021).
68. Ministry of State Assets. *National Energy and Climate Plan 2021–2030. Assumptions and Objectives and Policies and Measures*; Ministry of State Assets: Warsaw, Poland, 2019.
69. European Committee of the Regions. Time to Eradicate Energy Poverty in Europe. Press Release 27/06/2019 | Energy Efficiency, Market and Technology. Available online: https://cor.europa.eu/pl/news/Pages/time-to-eradicate-energy-poverty-in-europe.aspx (accessed on 8 April 2021).
70. Information from: Clean Air Department of the National Fund for Environmental Protection and Water Management, ul. Konstruktorska 3A, 02-673 Warsaw, Poland. Available online: www.nfosigw.gov.pl (accessed on 17 December 2020).
71. Owczarek, D. *Czyste Powietrze Dzięki Walce z Ubóstwem Energetycznym (Clean Air Thanks to Fighting Energy Poverty)*; Institute of Public Affairs: Warsaw, Poland, 2017.
72. Dzikuć, M.; Kułyk, P.; Dzikuć, M.; Urban, S.; Piwowar, A. Outline of Ecological and Economic Problems Associated with Low Emission Reductions in Poland's Lubuskie Voivodeship. *Pol. J. Environ. Stud.* **2018**, *28*, 65–72. [CrossRef]
73. Karpinska, L.; Śmiech, S. Invisible energy poverty? Analysing housing costs in Central and Eastern Europe. *Energy Res. Soc. Sci.* **2020**, *70*, 101670. [CrossRef]
74. Ürge-Vorsatz, D.; Herrero, S.T. Building synergies between climate change mitigation and energy poverty alleviation. *Energy Policy* **2012**, *49*, 83–90. [CrossRef]
75. Graczyk, A.M. Wykorzystanie odnawialnych źródeł energii w gospodarstwach domowych na Dolnym Śląsku/The use of renewable energy sources in households in the Lower Silesia region. *Pr. Nauk. Uniw. Ekon. Wrocławiu* **2016**, *453*, 190–198. [CrossRef]
76. TPA Horwath. *Wind Energy in Poland*; TPA Horwath: Warsaw, Poland, 2014.
77. Żabicki, D. Przydomowe elektrownie wiatrowe (Home wind farms). *Czysta Energia* **2013**, *11*, 1–4. (In Polish)
78. Ślusarz, G.; Gołębiewska, B.; Cierpiał-Wolan, M.; Gołębiewski, J.; Twaróg, D.; Wójcik, S. Regional Diversification of Potential, Production and Efficiency of Use of Biogas and Biomass in Poland. *Energies* **2021**, *14*, 742. [CrossRef]

79. Gromada, A.; Trębska, P. Acquisition and Consumption of Renewable Energy Used by Heat Pumps in 2010–2017 in Poland. *Ann. Pol. Assoc. Agric. Agribus. Econ.* **2019**, *XXI*, 97–106. [CrossRef]
80. Vanden Borre, A. Definition of heat pumps and their use of renewable energy sources. *REHVA J.* **2011**, *4*, 38–39.
81. Wolańksa, K. Efektywność Energetyczna i Odnawialne źródła Energii (Energy Efficiency and Renewable Energy Sources). Fundacja Instytut NA Rzecz Ekorozwoju, "Eko-Lokator" 2017. Available online: https://www.pine.org.pl/wp-content/uploads/2019/11/3.-efektywnosc-energetyczna-i-odnawialne-zrodla-energii.pdf (accessed on 20 February 2021). (In Polish).
82. Institute for Renewable Energy. *RES and Heat Storage in the Polish Heating Sector*; EC BREC: Warsaw, Poland, 2019.
83. Brodny, J.; Tutak, M.; Saki, S.A. Forecasting the Structure of Energy Production from Renewable Energy Sources and Biofuels in Poland. *Energies* **2020**, *13*, 2539. [CrossRef]
84. Lis, M.; Ramsza, M.; Miazga, A. *Dynamiczne Własności Miar Ubóstwa Energetycznego. (Dynamic Properties of Energy Poverty Measures)*; IBS Research Report 01/2016; Institute for Structural Research: Warsaw, Poland, 2016. (In Polish)
85. Energy Regulatory Office. *Wytwarzanie Energii Elektrycznej w Polsce w MałYch Instalacjach Oze. Raport Prezesa Ure Za 2020 Rok*; Energy Regulatory Office: Warsaw, Poland, 2021. (In Polish)
86. Institute for Renewable Energy. *Photovoltaics in Poland*; EC BREC: Warsaw, Poland, 2020.
87. Okushima, S. Gauging energy poverty: A multidimensional approach. *Energy* **2017**, *137*, 1159–1166. [CrossRef]
88. Chester, L.; Morris, A. A new form of energy poverty is the hallmark of liberalised electricity sectors. *Aust. J. Soc. Issues* **2011**, *46*, 435–459. [CrossRef]
89. Pereira, M.G.; Freitas, M.A.V.; da Silva, N.F. Rural electrification and energy poverty: Empirical evidences from Brazil. *Renew. Sustain. Energy Rev.* **2010**, *14*, 1229–1240. [CrossRef]
90. Adamkiewicz, Ł. Zewnętrzne Koszty Zdrowotne Emisji Zanieczyszczeń Powietrza z Sektora Bytowo-Komunalnego. In *Szacunki na Postawie Dostępnych Danych (External Health Costs of Air Pollutant Emissions from the Household and Municipal Sector. Estimates Based on Available Data)*; Ministerstwo Przedsiębiorczości i Technologii: Warszawa, Polska, 2018; pp. 1–13. (In Polish)
91. WHO Regional Office for Europe, OECD. *Economic Cost of the Health Impact of Air Pollution in Europe: Clean Air, Health and Wealth*; WHO Regional Office for Europe: Copenhagen, Denmark, 2015.
92. Wąsowski, M. Zanieczyszczenia Powietrza Kosztują Świat Miliardy Dolarów Dziennie. Polska Wśród Krajów Ponoszących Największe Koszty. Available online: https://businessinsider.com.pl/finanse/makroekonomia/koszty-zanieczyszczenia-powietrza-zwiazane-z-paliwami-kopalnymi-polska-i-swiat-raport/pwcs9bv (accessed on 20 February 2021).
93. Uris, M.; Linares, J.I.; Arenas, E. Feasibility assessment of an Organic Rankine Cycle (ORC) cogeneration plant (CHP/CCHP) fueled by biomass for a district network in mainland Spain. *Energy* **2017**, *133*, 969–985. [CrossRef]
94. Lee, J.; Shepley, M.M. Benefits of solar photovoltaic systems for low-income families in social housing of Korea: Renewable energy applications as solutions to energy poverty. *J. Build. Eng.* **2020**, *28*, 101016. [CrossRef]
95. Luderer, G.; Krey, V.; Calvin, K.; Merrick, J.; Mima, S.; Pietzcker, R.; Van Vliet, J.; Wada, K. The role of renewable energy in climate stabilization: Results from the EMF27 scenarios. *Clim. Chang.* **2014**, *123*, 427–441. [CrossRef]
96. Dusmanescu, D. Aspects Regarding Implementation of Renewable Energy Sources in Romania up to 2050. *Int. J. Sustain. Econ. Manag.* **2013**, *2*, 1–21. [CrossRef]
97. Llera-Sastresa, E.; Scarpellini, S.; Rivera-Torres, P.; Aranda, J.; Zabalza-Bribián, I.; Aranda-Usón, A. Energy Vulnerability Composite Index in Social Housing, from a Household Energy Poverty Perspective. *Sustainability* **2017**, *9*, 691. [CrossRef]
98. Cebotari, S.; Benedek, J. Renewable Energy Project as a Source of Innovation in Rural Communities: Lessons from the Periphery. *Sustainability* **2017**, *9*, 509. [CrossRef]

energies

MDPI

Article

Economic and Social Aspects of Using Energy from PV and Solar Installations in Farmers' Households in the Podkarpackie Region

Ryszard Kata, Kazimierz Cyran, Sławomir Dybka, Małgorzata Lechwar and Rafał Pitera *

Institute of Economics and Finance, University of Rzeszow, ul. M. Ćwiklinskiej 2, 35-601 Rzeszów, Poland; rdkata@ur.edu.pl (R.K.); kcyran@ur.edu.pl (K.C.); slawekd@ur.edu.pl (S.D.); lechwarm@ur.edu.pl (M.L.)

* Correspondence: rpitera@ur.edu.pl

Abstract: The growing energy needs of agriculture, the need to reduce the burning of fossil fuels, and, on the other hand, the increasing technical efficiency are contributing to the wider use of solar energy technology in agriculture. The aim of the article is to identify factors determining farmers' investments in solar photovoltaic and solar thermal installations for electricity and heat production, to establish the proportion between the consumption of such energy for the needs of the farmer's family and for the needs of the farm, and to identify the drivers of solar energy use in agricultural production. Empirical materials were collected through surveys of farmers conducted at the end of 2020 in south-eastern Poland, in the Podkarpackie region. It is a region characterized by significant land fragmentation. Producing energy from renewable sources can be an opportunity for farmers not only to reduce household expenses, but also to increase agricultural income. As a result, it can be a driver of sustainable agricultural development in the region. The article presents the most important economic and social determinants that stimulate the adoption of solar photovoltaic and solar thermal technologies by farmers for the needs of their households as well as for agricultural production.

Keywords: photovoltaic installations; solar installations; renewable energy; farmers' households; agricultural holding; agricultural production

Citation: Kata, R.; Cyran, K.; Dybka, S.; Lechwar, M.; Pitera, R. Economic and Social Aspects of Using Energy from PV and Solar Installations in Farmers' Households in the Podkarpackie Region. *Energies* **2021**, *14*, 3158. https://doi.org/10.3390/en14113158

Academic Editor: Vincenzo Bianco

Received: 12 April 2021
Accepted: 24 May 2021
Published: 28 May 2021

1. Introduction

In rural areas, investments in renewable energy sources (RESs) can be of great economic, social, and environmental importance (reduction in low-stack gas and dust emissions, reduction in CO_2 emissions) [1]. However, this depends mainly on the scale of such investments. RESs have the potential to improve the quality of life of the population and the quality of the environment, and they may constitute an additional source of income or a source of savings for economic entities, local government units, households, and agricultural holdings [2].

Agriculture is one of the sectors that can and should make intensive use of RESs [3–6]. The development of renewable energy sources in agriculture is necessary for several reasons. From the perspective of the sector, this necessity results from the need to reduce the dependence of agriculture on fossil fuels and to achieve specific objectives related to the reduction in CO_2 emissions and the so-called low-stack emissions. It is also an opportunity to improve the multifunctionality of agriculture and to use biomass from agricultural production, waste, or roof surfaces.

In Poland, the use of innovative, small-scale renewable energy technologies classified as distributed generation that have given rise to the development of community energy is a relatively new phenomenon but is becoming increasingly important for the development of the renewable energy sector [7]. An example of this may be the significant increase in the use of solar energy initially for water heating and since 2015 also for electricity generation as part of a prosumer system. The growth of this sector is fostered by the increasing energy efficiency of installations in relation to their price, as well as by favorable financing conditions in addition to the introduction of favorable legal regulations concerning the prosumer status [8,9]. Thanks to such changes (legal and financing method), the number of investments generating green energy will be increased several times [10]. The report by R. Fu et al. [11] also confirms the growing number of the discussed investments. Research conducted by P. Gradziuk is also worth mentioning. In his research, the author lists the decrease in the unit costs of photovoltaic installations, both pro-consumer and commercial, among the factors having a significant impact on the development of PV and solar installations [12]. However, despite such a significant development, the photovoltaic market in Poland, according to many opinions, still has great potential for development [13]. The confirmation is, among others, the results of the research conducted by A. Ciechomska, which indicate still insufficient support for the RES sector [14]. On the other hand, the most frequently implemented investments include photovoltaic and solar panels. It is for their assembly that funds from co-financing are allocated to the greatest extent [15,16].

Solar photovoltaic (PV) and solar thermal systems, i.e., the technology used to convert solar energy into electricity or heat, are gaining increased attention among farmers. For farmers, it is not only an opportunity to reduce household expenses, but above all a chance to reduce the cost of agricultural production. It is also an opportunity to draw agricultural production to a larger extent from ecological solutions [17–19].

Investments in renewable energy sources by rural households and local governments receive support from European Union funds and national budgets [20]. In order to benefit from the financial instruments available to support RES investments, the potential beneficiaries must have certain knowledge and financial competence [21]. Awareness of environmental considerations and appreciation of their importance is also a major factor behind the investment. Therefore, the most important determinants of RES sector development in rural areas in Poland are connected to human resources, as well as to the attitudes of the local population, including farmers, towards RESs. Activities related to the use of renewable energy sources are relatively novel and innovative, and their implementation is associated with business risks.

There is a gap in research on farmers' investment in solar PV and solar thermal installations. In the case of family farms, it is not possible to clearly separate the household from the farm perceived as the workplace of the farmer's family. The household needs are combined with the needs of the agricultural production, which impacts the energy consumption structure [22–25].

The purpose of the study is to identify economic and social factors determining farmers' investments in solar PV and solar thermal installations. The paper also aims to demonstrate what part of the energy produced from RESs is used for the needs of farmers' families and what part is used for agricultural production and then to identify the factors determining this proportion. Achieving these objectives can contribute to a better alignment of policies and tools to support investment in green energy in agriculture.

2. Factors Determining the Adoption of PV and Solar Installations in the Agriculture

The adoption of solar PV and solar thermal systems by rural households has been addressed in the literature. Although the studies cover different countries and regions, they most often refer to the following two aspects:

(1) The first aspect that studies often refer to is the reasons why households make the decision to invest in generating energy through PV and solar panel technologies [3]. Labay et al. draw attention to demographic factors in their research [26]. Sidiras and Koukios point to a number of economic, socio-cultural, and political factors [27]. Faiers and Neame show the low importance of environmental features in relation to, inter alia, economic factors [28]. Bollinger et al. indicate, on the other hand, that the appearance of PV and solar installations is increasing among households that already have such installations [29]. On the other hand, Zhai and Williams point out that despite the importance of reducing costs in the case of decisions related to the

installation of PV panels, with the passage of time, users appreciated the aspect of environmental protection more and more [30]. As the main factor determining the choice of renewable energy investments, Fleiß et al. indicate the economic factor. It is the main determinant of the choice of investments in renewable energy among many others. Factors such as energy autonomy, the belief in environmental protection, or the prestige of having renewable energy investments are still less important than the profitability of investments. [31]. Therefore, the visible trends that determine the choice of this type of installation are still economic factors. Hence, proposals call for a more active policy of disseminating this type of investment and raising awareness of the importance of other factors as equally important [32,33]. The dominance of PV panels among this type of investment is also visible, despite the growing number of other types. PV and solar installations still dominate [34]. The importance of various institutional factors has also been explored in this context. It has been confirmed that RES support policies are important in the rapid diffusion of solar PV and solar thermal technologies. Such policies include in particular provisions for guaranteed tariffs for energy produced in household installations and financial support through grants, subsidies, low-interest loans, and credits. Wustenhagen emphasizes the importance of public policy in promoting renewable energy [35]. Guidolin and Mortarino also emphasize the importance of supporting energy policy [36]. Research by Kwan [37] and Cherrington confirms that without proper support from regulations and subsidies, it will not be possible to efficiently and increasingly popularly invest in RES-type solutions [38]. Jenner et al. emphasize that this policy should be effective and adjusted to the specific needs of potential recipients [39]. In turn, Bauner emphasizes that despite many incentives from the state policy, there is still much room for improvement. The conclusions that he formulates are therefore convergent with the previously cited results speaking of a better and more effective regulatory system [40].

(2) The second aspect that studies often refer to is the features of the households and decision-makers that made solar PV and solar thermal investments [37]. Chodkowska-Miszczuk emphasizes socio-demographic features such as the age of the farm manager, which is one of the most important in this respect. [41]. The socio-demographic characteristics of farms were also the main subject of research by the team led by Brudermann [42]. Similarly, Ba-kundukize et al. have conducted research for Rwanda [43]. In turn, for Indian households, similar studies were carried out by Irfan et al. [44]. In most of the studies cited, the age of the farm manager was found to be of great importance; the approaches taken by local authorities and the qualifications and level of education of farmers were also found to be important factors. In this respect, the attitudes of rural residents towards RES technologies and the possibilities of changing these attitudes towards greater acceptance and implementation have also been studied [45–47].

Relatively few studies have attempted to identify the reasons behind the adoption of solar PV and solar thermal micro-installations by farmers.

The price of conventional electricity is a key stimulus for investment in solar PV and solar thermal micro-installations [48]. This factor is important for all households investing in green energy, but it is critical for farmers because in their case a large amount of energy is consumed by agricultural production in addition to household use. This is especially true for large farms (in terms of farmland area and operating surplus) generating significant energy costs [42].

Previous works show that the adoption of renewable energy in agricultural holdings depends on factors related to the farmer as the farm manager, factors related to the farm, and socio-economic factors [49,50]. As far as the factors related to the farmer are concerned, the propensity to invest in RESs is mainly influenced by the farmer's age and education. Most studies show an inverse relationship between the farmer's age and his/her involvement in RES adoption [49,51–54]. Education, in turn, is a major driver behind investment in renewable technologies [51–54]. The psycho-social characteristics of the person managing the farm, such as risk aversion or openness to innovations, are of great importance as well [49,50].

The determinants related to farm characteristics include the size and legal status of the agricultural holding. The probability of undertaking investments in solar PV and solar thermal systems is higher for large area farms [55]. This is due to the increased need for energy and the fact that large agricultural holdings earn relatively more compared to small farms. The legal status of the holding concerns, for example, the ownership or tenancy of the agricultural property. Agricultural property ownership and long-term lease encourage investment in renewable energy production [49].

Farmers' economic interests are an important motivation to invest in RESs [42,56]. Environmental reasons seem to have less influence on farmers' decisions [31,56,57]. However, this may change as a result of environmental education programs, among other things, as pointed out by Shi et al. [58]. It is also worth noting that research most often focuses on farmers' investments in bioenergy production installations exclusively for agricultural production. Investment in solar energy installations is rarely addressed. An additional problem that is often overlooked is that energy generated in such installations is divided into energy for household and farm needs. In such cases, the distribution of farmers' motivation to invest may be different than in the case of investments in biomass or biogas facilities which use media harvested from agricultural production and are used entirely to cover the energy needs of the production process.

Other determinants of farmers' investment in RESs include the farm's agricultural type as defined by its agricultural production structure and the type of economic activity of the farmer [42,52,54,59,60]. Using the example of the United States, Borchers et al. [53] showed that organic farms are about five times more likely to adopt renewable energy generation technologies than conventional farms. The type and scale of RES investments in agriculture are also influenced by the biophysical characteristics of the farm such as the amount of sunshine in the area, average wind strength, soil erosion, slope, or precipitation [49,61].

Many studies indicate that household income is the dominant predictor of green energy investments [37,49,53,54,62,63]. In the case of agricultural holdings, investments in RESs are more likely if the income from the family farm is high or if there are additional non-farm earnings [51].

Farmers who are well informed about available energy technologies can adopt them faster [42,49,53,61]. Brudermann et al. [42] confirm the importance of social and behavioral factors for the adoption of solar PV and solar thermal technologies in agriculture. Such actions are motivated by the desire to strengthen the farmer's position in the local community and by emulation. According to the literature on innovation diffusion, about 3% of initial adoption is driven by innovators, and later on, adoption is spurred by the imitation effect [64]. However, the scale of investment in solar micro-installations must exceed a certain threshold of prevalence in households for the imitation effect to become clearly visible [47]. Kim and Lee [65] indicate that there is an imitation effect that is part of the local process of learning and applying solar PV and solar thermal technologies.

Research suggests a growing importance of institutional factors closely linked to fiscal and energy policy measures established by individual countries and regions that stimulate the adoption of renewable energy generation systems by agricultural holdings [60]. In the European Union, support for the energy transformation of agriculture towards a sustainable model takes the form of various RES subsidy programs and different support instruments. Poland, for example, has the AgroEnergy (Polish: AgroEnergia) program intended for farmers. It was launched in 2019 and it gives farmers the opportunity to obtain grants from public funds to finance investments in RES micro-installations (with a capacity between 10 and 50 kW). The grant may amount to up to 20% of the project costs, but the subsidized installation should serve to satisfy the beneficiary's own energy needs. Farmer households, just like other households, can also benefit from subsidies for PV installations under the My Electricity (Polish: Mój Prąd—more in the explanations) program or the Clean Air (Polish: Czyste Powietrze—more in the explanations) program and, from 2019, they can write off the installation costs from their personal income tax under the so-called thermal modernization relief (after deduction of any subsidies). My Electricity is a program of co-financing photovoltaic micro-installations with an installed capacity of 2 to 10 kW. With the number of applications submitted (actually over 220,000), the total capacity of these installations reaches 1.2 GW, with an annual production of approx. 1200 GWh/year [66]. My Electricity program is very popular, especially in the southern regions of Poland. The Podkarpackie region, analyzed in this article, has special, favorable conditions related to insolation. Detailed data also indicate that the Podkarpackie region has a fairly average level of introduced PV power installation [67] while simultaneously having highest average subsidies to power (expressed in kWp) [68]. Meanwhile, the Clean Air program strives to improve air quality through the removal of obsolete stoves and boilers and insulation of buildings; so far, it has reached 247,275 entities with support for a total amount of PLN 4.4 billion. Actually, the framework of the next edition of the program has been modified, which should make its implementation more dynamic [69].

In addition, farmers could receive a grant for an RES investment under the EU 2014–2020 farm modernization support funds. The support will continue under the new EU 2021–2027 financial perspective. Many countries have similar programs to facilitate RES investments in agriculture, but there is still little research investigating their effectiveness in this regard.

The Polish economy needs improvement in terms of energy efficiency and developing a distributed energy system based on renewable sources. Social expectations and Poland's energy obligations resulting from the objectives set by the European Union in its 2030 climate and energy framework [8] are the key drivers for undertakings in this area. They include the following:

- Reducing a minimum of 40% of greenhouse gas emissions (compared to 1990 levels);
- Increasing the share of energy from renewable sources in the total energy consumption to a minimum of 32%;
- Increasing energy efficiency by a minimum of 32.5% [20,70].

These targets, although aggregated for the entire EU, impose certain obligations on each member state, which in the case of Poland have been included in the National Energy and Climate Plan for the years 2021–2030. It sets the goal of achieving a 21–23% share of RESs in gross final energy consumption by 2030 and reducing the share of coal in electricity production to 56–60% [71].

Poland is still lagging behind the EU average in terms of meeting the targets set in the earlier Directive 2009/28/EC of the European Parliament and of the Council [72] regarding the share of energy from renewable sources in overall energy consumption by 2020 despite the progress in recent years. To be precise, the share of renewable energy in total primary energy generation increased from 12.12% to 14.31% (EU average: from 26.1% to 29.9%) between 2014 and 2018. At the same time, the share of renewable energy in gross final energy consumption in 2018 was 11.16%, with a target of 15% by the end of 2020 [73].

Poland's energy transition and the achievement of the climate and energy targets for 2030 and beyond will not succeed without a widespread development of energy based on renewable energy sources at the local level, including in rural areas and in agriculture.

In Poland, energy consumption differs between rural and urban households. Between 2009 and 2018, the average annual household electricity consumption in rural areas was more than 30% higher compared to urban areas (Statistics Poland, 2019). This was a direct result of the larger average dwelling size and the larger average number of persons in a household in the rural areas. Additionally, agricultural households consumed on average 24.3% more electricity for household needs than other households in rural areas [5]. The energy demand of all of the buildings, machinery, and equipment located on the farm usually far exceeds that of regular houses. With regard to solar PV or solar thermal installations, however, this is equivalent to an increase in investment costs due to the required plant capacity. Thus, the economic benefits of an RES installation can be much higher for a farmer than for a non-agricultural household, but the investment costs are higher as well. The results of a study by Klepacka et al. [3] indicate that farmers in Poland attach more importance to energy costs as a rationale for RES investments compared to other rural residents.

In Poland, 88.4% of rural households still use solid fuels, mainly coal and wood, for heating rooms and domestic hot water [73]. The use of fossil fuels causes not only CO_2 emissions but also the so-called low-stack emission of many gases and forms of dust that are harmful to health, which is especially true of hard coal combustion. It is estimated that 44,000 people in Poland die prematurely each year as a result of smog caused by low-stack emissions of gases and dust emitted during coal burning in rural and urban household furnaces [74]. Improving air quality is, therefore, a very important part of public health policy.

Increased use of solar energy in rural households reduces their dependence on fossil fuels, lowers the amount of ash discharged to landfills, improves air quality for the local community, and saves money on the monthly household energy bill [5]. There is a need to determine how many other aspects besides economic ones, in particular environmental and health factors, influence farmers' willingness to make investments in green energy. This is particularly important because hard coal and wood are still cheaper than other energy sources.

The issue of reducing the consumption of low-quality fuels and introducing new technologies in agriculture, which serves to reduce air pollution problems and improve the living conditions of farmers, is topical in many countries [75–77]. One solution to this problem is the development of solar PV and solar thermal systems in rural areas and agriculture. In Poland, the rationale behind investments in such systems includes the following:

- Growing demand for electricity on agricultural holdings;
- Reduction in energy consumption from conventional sources, for example as a result of an increase in the price of such energy;
- Reduction in agricultural production costs and in the farmer's family costs associated with the consumption of electricity and heat;
- Increased energy self-reliance and reduced dependence on energy prices;
- Obtaining financial benefits, i.e., grants for the implementation of investments, agricultural tax, and personal income tax reliefs for investment.

3. Materials and Methods

The research involved a literature and documentary analysis, as well as analysis of public statistics data. Empirical analyses were based on the results of a diagnostic survey conducted among farmers—owners of family farms in the Podkarpackie region. The research tool was a questionnaire that included questions about RES installations and their use in the respondents' households and farms, factors determining farmers' investment in renewable energy, assessment of the benefits of RESs, and limitations to the development of installations on farms. Targeted selection was used to choose the group of research subjects comprising farmers who had a renewable energy installation or had declared that they would start investing in solar PV or solar thermal installations in 2021. The intention to undertake such an investment was evidenced by the farmer's application for financial support for the investment in a publicly funded program supporting solar PV or solar thermal installations. Respondents were selected from the ODR Boguchwała database (Agricultural Advisory Center in Boguchwała), which keeps a register of the farms in the Podkarpackie region and monitors their functioning and investments. The research was carried out on farms from all 21 poviats of the Podkarpackie region, assumed representativeness for agriculture in the Podkarpackie region, and assumed proportionality in the context of the number of farms in poviats. The minimum sample was set at 226 units and was realized. In case of refusal to participate in the research, another respondent was selected. The study was conducted in Q4 2020. The following research hypotheses were adopted in this study:

Hypotheses (H1). *Factors stimulating the adoption of solar PV and solar thermal installations on farms include the reduction in energy expenses and access to grants for such investments. The young age of the farmer and a larger farm area are also contributing factors.*

Hypotheses (H2). *On agricultural holdings, a high share of agricultural production in renewable energy consumption reflects farmers' preferences for the economic benefits of RESs.*

Hypotheses (H3). *Farmer's focus on livestock production and the commercial nature of the farm is a factor that promotes a higher share of agricultural production in the consumption of energy obtained from solar PV and solar thermal installations.*

Hypotheses (H4). *Younger farmers managing large-scale commercial and specialized farms attach more importance to the reduction in agricultural production costs resulting from the adoption of solar PV and solar thermal installations.*

The research hypotheses were verified using logistic regression and multiple regression methods. The chi-squared test (χ^2) and Cramér's V were also applied.

A logistic regression model was used to verify hypothesis H1. In this model, the dependent variable is dichotomous, i.e., it assumes the value 1 when the desired event occurs or the value 0 when such an event does not occur. For a given case *i*, the probability of the variable y taking the value 1 or 0 is

$$P(y_i = 1) = p_i,\ P(y_i = 0) = 1 - p \tag{1}$$

The probability is a function of the vector of explanatory variables x_i and the parameter vector β, and therefore,

$$p_i = P(y_i = 1) = F\left(x_i^T \beta\right);\ \text{for } i = 1, 2, \ldots, n \tag{2}$$

The logit model assumes that the probability p_i corresponds to the distribution of the logistic distribution [78]:

$$P_i = F(x_i^T) = \frac{1}{1 + \exp(-x_i^T \beta)} = \frac{\exp(x_i^T \beta)}{1 + \exp(x_i^T \beta)} \tag{3}$$

The parameters (coefficients) of the logistic regression model are estimated using the maximum likelihood method [79]. Parameters $\beta_0, \beta_1, \ldots, \beta_k$ for known values $y_i, x_{1i}, \ldots, x_{ki}$ must be estimated in such a way that they provide the maximum value of the logarithm of the reliability function.

The logistic regression model can be defined in more detail:

$$P(Y = 1 | x_i, \ldots, x_k) = \frac{exp^{\beta_0 + \beta_1 x_1 + \ldots + \beta_k x_k}}{1 + exp^{\beta_0 + \beta_1 x_1 + \ldots + \beta_k x_k}} \tag{4}$$

where $\beta_1, \ldots, \beta_k$ are logistic regression coefficients.

In the analysis under consideration, the fact that the household owns/does not own a solar PV and/or solar thermal installation was interpreted as an explanatory variable (denoted by Y_1). The variable Y_1 was defined as follows:

$$Y_{1i} = \begin{cases} 1 \text{ if the } i_{th} \text{ household has a solar PV and / or solar thermal energy system} \\ 0 \text{ if the } i_{th} \text{ household does not have such an installation} \end{cases} \tag{5}$$

where $i = 1, 2, \ldots, n$ and is the number of surveyed households.

The dependent variable (Y_1) was determined for all 226 studied units. The input set of independent variables for the estimation of the logistic regression model consisted of variables marked with symbols from X_1 to X_{10} (Table 1). The selection of explanatory variables was based on a correlation matrix, and the selected variables were significantly associated with the dependent variable Y_1.

Table 1. Explanatory variables used for the estimation of logistic and multiple regression models.

Variable	Symbol	Regression	
		Logit Y_{1i}	Multiple Y_{2i}
Having other RES installations (0/1)	X_1	+	+
Age of the farm manager (years)	X_2	+	+
Farm area (ha of agricultural land)	X_3	+	+
Using repayable funds (loans, leasing) to finance investments in renewable energy (0/1)	X_4	+	+
Use of renewable energy subsidies (0/1)	X_5	+	+
Specialized or targeted holding (0/1)	X_6	+	+
Saving energy costs is of great importance (points 1–3)	X_7	+	+
Environmental responsibility is of great importance (0/1)	X_8	+	+
It is of great importance to increase the quality of life (0/1)	X_9	+	+
Neighborly prestige is of great importance (0/1)	X_{10}	+	+
High importance of tax benefits (pkt 1–2)	X_{11}	–	+
Use of undeveloped space (0/1)	X_{12}	–	+
The farm is focused on animal production (0/1)	X_{13}	–	+
The farm functions as a special department (0/1)	X_{14}	–	+
The farm carries out organic production (0/1)	X_{15}	–	+
The farm runs agritourism production (0/1)	X_{16}	–	+
The farm sells agricultural products (0/1)	X_{17}	–	+
A farm associated in a production group (0/1)	X_{18}	–	+
VAT on general terms (0/1)	X_{19}	–	+

"+" use of a variable for modeling, "–" variable omission. Source: own survey.

Analysis involving the multiple regression method was used to verify hypotheses H2 and H3. Y_{2i} is the share of energy obtained from RESs used for agricultural production (in %) and was adopted as the dependent variable. The multiple linear regression model looks as follows [80]:

$$Y_{2i} = \beta_0 + \beta_{1\times 1} + \beta_2 X_2 + \ldots + \beta_k X_k + \varepsilon t \tag{6}$$

where Y_{2i} is the dependent variable explained by the model; X_1, X_2, ..., X_k are independent (explanatory) variables; B_0, β_1, β_2, ..., β_k are parameters; and ε_t is a random (residual) component.

The coefficients for the model are estimated using the classic ordinary least squares (OLS) method. Based on these values, the direction and strength of the influence of the independent variables on the dependent variable can be inferred. Student's *t*-test was used to determine the statistical significance of individual regression coefficients for independent variables. The quality of the multiple regression model was assessed using an F test, the variance of the random component (εt), a normality test of the residual component, and the coefficient of determination R^2.

The estimation of the multiple regression model was preceded by a collinearity analysis of the variables, which removed relatively highly correlated preselected explanatory variables from the analysis (Pearson's correlation coefficient $r_{xy} > 0.7$). Backward stepwise regression was used in the estimation of the regression model.

The dependent variable (Y_2) was determined for 150 units, i.e., those that had solar PV and solar thermal installations. Variables marked with symbols from X_1 to X_{19} were initially qualified for the analysis (Table 1). The estimation of the regression model was preceded by an analysis of the interdependence of the independent variables, as a result of which the variables X_4 and X_{19} were eliminated.

The chi-squared test of independence (χ^2) and Cramér's V coefficient were used to verify hypothesis H4. The independence test allowed us to verify the hypothesis of the independence of the two variables X_i and Y_i measured on nominal scales [81]: H0, features X_i and Y_i are independent; H1, features X_i and Y_i are dependent, with an assumed significance level of $\alpha = 0.05$. In this analysis, the benefits from an RES installation were shown as X_i features (8 variables were selected), while 5 features describing the farmer and the farm were shown as Yi features.

To verify the hypotheses, a χ^2 statistic, expressed as the following formula, was used:

$$\chi^2 = \sum_{i=1}^{r}\sum_{j=1}^{k} \frac{\left(n_{ij} - \hat{n}_{ij}\right)^2}{\hat{n}_{ij}}$$

In order to determine the strength of the relationship between the studied features, Cramér's V coefficient was used, which was calculated as follows [82]:

$$V_C = \sqrt{\frac{\chi^2}{N \cdot (\min(k, r) - 1)}}$$

where V_c is Cramér's V coefficient, χ^2 is chi-squared, k is the number of rows, and r is the number of columns in the correlation table.

Cramér's V coefficient assumes values [0,1], where V = 0 indicates independence of features and V = 1 indicates a strong relationship.

4. Results of Empirical Studies

4.1. PV and Solar Installations at Farmers' Households in the Podkarpackie Region

The research carried out in the Podkarpackie region on a purposefully selected sample of 226 farms indicates that 66.4% of the units had an installation for generating energy from renewable sources and 73% intended to invest in this type of installation in the future (Figure 1). As the criterion for the inclusion of households in the sample was the ownership of a solar PV/solar thermal installation or the intention to implement such an installation within the following year, these results may not come as a surprise. Two-thirds of the surveyed farmers had already made such an investment, while the remaining group (33.6%) intended to do so. At the same time, as many as 39.4% of the farmers who already had an RES installation (e.g., solar thermal) intended to invest in another installation (e.g., a solar PV installation).

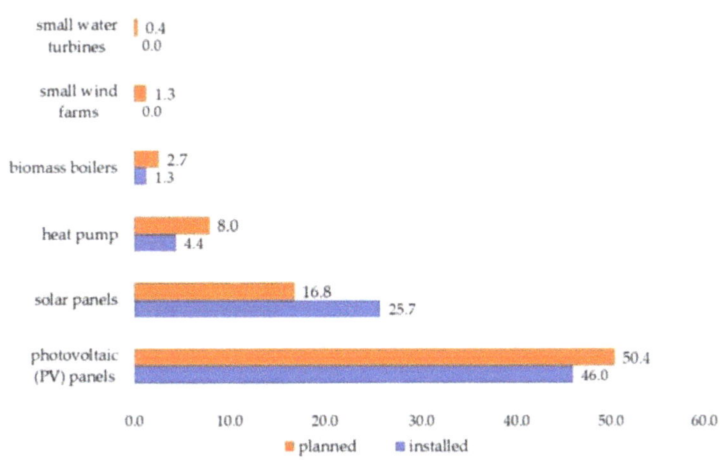

Figure 1. Percentage of researched farms owning and planning RES installations (%). Source: own survey.

The research was not carried out on a random sample; hence, it was not possible to determine what percentage of farms in the region have an RES installation, but the high dynamics of change that took place between 2015 and 2020 in this area are clearly visible. Research conducted in the Podkarpackie region in 2014 [83] showed that the vast majority of residents of municipalities with a dominant agricultural profile did not use renewable energy sources. The respondents showed potential interest in solar collectors and photovoltaic systems; however, most of them stated that they could not afford such investments due to the investment costs in relation to the respondents' income and the long period of return on investment. The dynamic technical progress that took place between 2015 and 2020, especially in the segment of solar PV installations, the emergence of financial support programs for this type of investment, the personal income tax reliefs from 2019, and finally the improvement of farmers' income [84] have made PV technology much more accessible to farmers, and the number of such installations has increased significantly. In Poland, particularly dynamic growth took place in the prosumer PV micro-installation segment between 2018 and 2020. By 2020, electricity was generated in 458,600 PV micro-installations, and their total capacity was over 3000 MW. In comparison, at the end of 2019, there were 155,100 such installations; at the end of 2018, there were only 51,000, and the capacity of the micro-installations was 344 MW [85]. There are no detailed data on how many of these micro-installations are located in rural areas and how many of them are found in agricultural holdings. It can be estimated that about three-quarters of prosumer PV installations are located in rural areas, and of these at least 30% can be found in farmers' households. This means that between 2018 and 2020, the number of PV installations in farmers' households increased from about 11,500 to 103,000 nationwide. A similar growth dynamic was also observed in the Podkarpackie region.

The surveyed farms had mainly photovoltaic installations (46.0% of farms in total), solar collectors (25.7%), heat pumps (4.4%), and biomass boilers (1.3%). PV installations also prevail in the case of RES investments planned to be implemented. Over 50% of farm-ers declared making such an investment within a year. Solar collectors are planned to be installed by almost 26% of respondents, heat pumps by 8%, and biomass boilers by 2.7%. Very few farmers planned to adapt a small wind power plant or a small water turbine, while no respondent declared investment in a micro gas plant.

The structure of farms in terms of the use of energy generated from eco-energy installations is presented in Table 2.

Table 2. The direction of using energy from renewable sources in the researched farms (%).

No.	Direction of Using Energy from RES	Percentage of Farms	
		With a Photovoltaic Installation	With Solar Installation
1	Only in the household	41.4	84.5
2	Only for agricultural production	6.3	8.6
3	Both in agricultural production and in the household	52.3	6.9

Source: own survey.

Among the agricultural holdings equipped with PV systems, almost 42% use the electricity generated by the systems only for their household needs (Table 2). Photovoltaic panels were usually placed on residential buildings on these farms, and the farmers were prosumers. On the other hand, 6.3% of respondents used energy only for broadly defined agricultural production (including drying, cooling, and storage of agricultural products, as well as packaging and preparing products for sale). In this case, PV panels were usually placed on buildings and outbuildings (e.g., warehouses, storage facilities). The largest group of households with PV installations (52.3%) used the obtained energy both for household purposes and for agricultural production. In the case of units with a solar thermal installation, the vast majority (84.5%) used the acquired thermal energy only in the household. Only 8.6% of the farmers used the solar thermal installation for agricultural production, while 6.9% divided the acquired energy into consumption related to agricultural production and consumption related to the family's household needs.

The highest percentage of units with solar PV installations was found among mixed-production farms and farms focused on livestock production (over 60%). These groups of holdings, as well as holdings operating as so-called special branches of agricultural production (e.g., crops grown in greenhouses and foil tunnels, poultry farms) also included the largest number of units equipped with solar thermal installations (Table 3). The smallest number of solar PV and solar thermal installations was declared by holdings that perform agricultural activities related only to the maintenance of agricultural land in good agricultural condition (Table 3).

Table 3. The use of energy from renewable sources in the researched farms, taking into account the type of farm (%).

No.	Farm Profile	Percentage of Farms in the Group with an Installation			Share of Farms in the Group by Area of Energy Use for the Purposes of: Only Agricultural Production			Agricultural and Living Production of the Family		
		PV	Solar	Other RES	PV	Solar	Other RES	PV	Solar	Other RES
1.	Plant production oriented [1]	48.6	33.8	8.3	2.9	-	-	62.9	22.6	33.3
2.	Animal production oriented [2]	60.7	43.1	14.3	5.9	28.6	-	70.6	28.6	25.0
3.	Special department [3]	50.0	50.0	-	-	-	-	100.0	100.0	-
4.	Multidirectional farms [4]	62.2	47.6	4.0	6.5	4.0	-	47.8	40.0	-
5.	Only those keeping the land in good agricultural condition [5]	47.6	25.0	4.8	10.0	10.0	-	20.0	10.0	-
6.	Ecological [6]	50.0	25.0	25.0	50.0	100.0	-	-	-	-
7.	Conducting agritourism activities [7]	50.0	-	50.0	-	-	-	-	-	-

Number of respondents: [1] 72, [2] 28, [3] 4, [4] 74, [5] 21, [6] 4, [7] 2. Source: own survey.

In terms of agricultural production purposes, energy from solar PV installations is used for lighting farm premises, powering machinery and equipment on the farm, and heating water, as well as cooling or drying products, while energy from solar thermal installations is mainly used for heating water (Table 4).

Table 4. Ways of using energy from renewable sources in the researched farms.

No.	Objectives of the Use of Renewable Energy	Type of RES Installation: Photovoltaic	Solar	Other
		Household		
1	Lighting of living quarters	3.8	-	0.5
2	Domestic water heating	2.5	3.0	1.5
3	Heating of living quarters	1.8	0.7	1.7
4	Air conditioning, ventilation	0.7	-	0.2
5	Other	0.1	-	-
		Agricultural Production		
1	Room lighting	2.5	-	0.2
2	Heating water for agricultural production	1.3	2.0	0.4
3	Product cooling	1.0	-	0.1
4	Drying of agricultural produce	0.8	-	0.1
5	Space heating or cooling	0.6	0.5	0.5
6	Drive of agricultural vehicles and machines	0.5	-	-
7	Irrigation or drainage of land	0.3	-	0.1
8	Other	0.2	-	-

Use on a scale of 0–5, where 0—none, 1—small scale, 2—medium scale, 3—large scale, 5—very large use. Source: own survey.

Respondents' statements indicate that the most important determinant of investment in various types of RESs was the desire to reduce household maintenance costs (mean score of 4.6 on a 5-point scale). The availability of grants and other low-cost sources of funding and the reduction in the environmental burden of agricultural production were identified as factors of high importance. On the other hand, respondents attributed moderate importance to such benefits as reduction in agricultural production costs, popularity of RESs, and increase in farm income (Figure 2). The wish to utilize residues (waste, by-products) from agricultural production was not an important reason for farmers to invest in RESs. This last observation points to a major challenge in the context of a wider inclusion of agriculture in Poland in the implementation of the concept of a circular economy [86]. It should also be noted that the primacy of household benefits over farm benefits as a determinant of farmers' RES investments may be due to two reasons:

(1) Low importance of farm income in the disposable income structure of many farming families, which particularly applies to small farms that dominate in the studied region (the average area of farmland in 2020 here was only 4.9 ha, compared to the national average of 11 ha) [87];
(2) Underestimation by some farmers of the potential benefits of green energy for agricultural production.

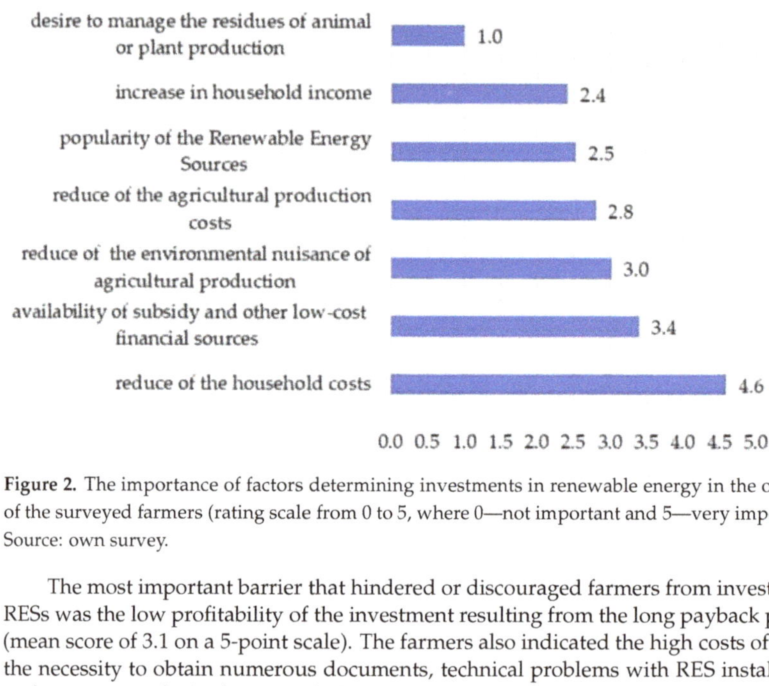

Figure 2. The importance of factors determining investments in renewable energy in the opinion of the surveyed farmers (rating scale from 0 to 5, where 0—not important and 5—very important). Source: own survey.

The most important barrier that hindered or discouraged farmers from investing in RESs was the low profitability of the investment resulting from the long payback period (mean score of 3.1 on a 5-point scale). The farmers also indicated the high costs of loans, the necessity to obtain numerous documents, technical problems with RES installation and operation, a lack of time to deal with new tasks, and a lack of knowledge on RESs. The farmers attributed moderate importance to these barriers, while other constraints identified by them were rated as insignificant (Figure 3). Looking at the average ratings of determinants and barriers to RES investment, it can be seen that barriers were assigned lower weights. According to the farmers, it was worthwhile to undertake such projects despite the difficulties and constraints.

The farmers financed investments in RESs mainly from their own funds and grants (Table 5). In their financing structure, own funds accounted for 59% on average, but in the case of 73.4% of the surveyed holdings, own funds covered at least three-fourths of the amount of investment expenditure. On the other hand, grants and tax reliefs accounted for 34.4% on average in the structure of investment expenditures. The share of bank credits and loans was very low (5.9%), and leasing was completely marginal. The financing structure of the planned investments looked slightly different (Table 5). The farmers expected that about half of the investment costs would be covered from grants and tax relief, which were planned to be used by as many as 88.1% of respondents. The farmers were also slightly more willing to finance such ventures through loans and leasing, which were chosen by 21.7% of prospective investors (Table 5). In general, however, the farmers conditioned the implementation of investments in photovoltaic and solar installations on access to financial support from public funds.

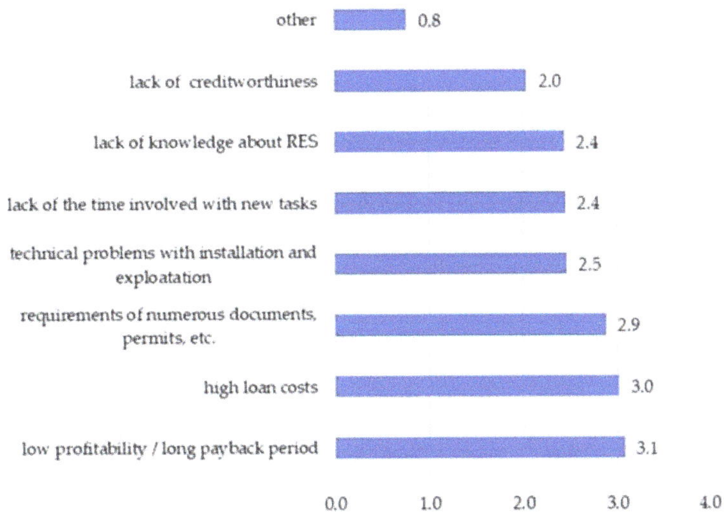

Figure 3. Barriers limiting farmers' interest in investing in PV and solar installations (rating scale from 0 to 5, where 0—not important and 5—very important). Source: own survey.

Table 5. Sources of financing for RES installations in the researched farms.

No.	Type of Funding Sources	Structure of Financing Investments in Renewable Energy (%)		Percentage of Farmers Engaging Specific Sources of Financing (%)	
		Implemented	Planned	Completed Investments	Planned Investments
1	Own funds	59.0	42.8	87.3	88.8
2	Bank credit, loans	5.9	7.5	12.0	18.2
3	Subsidies, tax breaks	34.4	47.8	69.3	88.1
4	Leasing	0.7	1.9	0.7	3.5

Source: own survey research.

The assessment of the benefits of solar PV or solar thermal systems is largely a reflection of the factors determining investment in such technologies. Respondents ranked saving expenses on electricity consumption as the most important benefit, as well as hedging against conventional energy price increases and the associated increase in energy independence of the household (Figure 4). Respondents also gave high priority to environmental aspects. This indicates high environmental awareness of farmers who had invested or intended to invest in RES installations. Increased quality of life through the use of green energy technologies was also of great importance for the respondents, which can also be associated with environmental benefits.

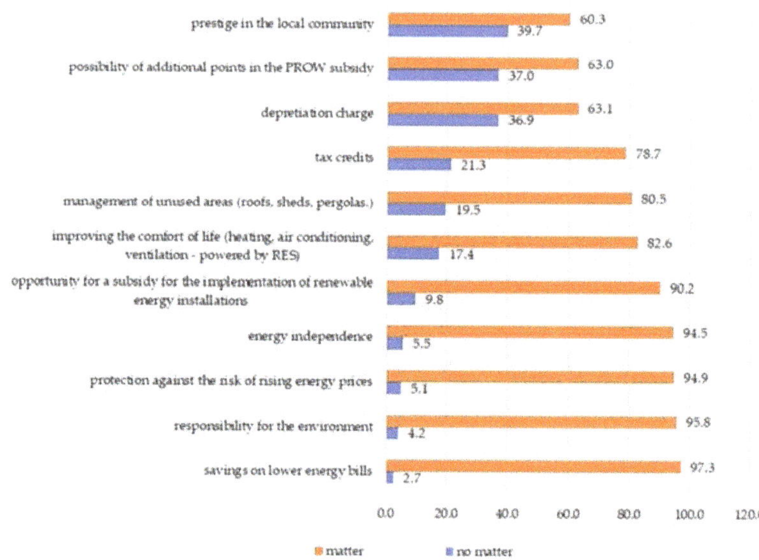

Figure 4. Assessment of benefits resulting from PV and solar installations in farmers' households (%). Source: own survey.

In addition to the above-mentioned economic benefits, the farmers also pointed to the possibility of obtaining grants for solar PV and solar thermal installations, thus reducing their own investment outlays. Slightly fewer farmers recognized the importance of tax benefits (personal income and agricultural tax reliefs, depreciation write-offs) or indicated the prestige in the local community associated with the ownership of RES installations.

4.2. Statistical Analysis

The results of the estimation of the logistic regression model establishing the determinants of farmers' investment in solar PV and solar thermal installations are presented in Table 6. The significance of the statistical parameters of the model was verified based on Student's *t*-test, and the adopted level of significance was $\alpha = 0.05$. The model parameters for the variables X_1, X_2, X_3, and X_5 were found to be statistically significant, so the factors mentioned above have a significant effect on the probability of owning PV panels or solar thermal installations. A positive parameter estimate for the variable X_1 means that the household's ownership of an RES installation positively influences the probability of getting another solar PV or solar thermal installation. Another statistically significant variable is the age of the farmer acting as the farm manager. In this case, the older the farmer, the less likely he/she is to have a first or subsequent RES installation on the farm. The area of a farm and farmer's eligibility for RES grants were also indicated as important determinants of investment in solar PV or solar thermal installations.

The estimated logistic regression model enables, to a large extent, the adoption of hypothesis H1, which assumes that the young age of the farmer, a larger farm area, and the possibility to benefit from RES grants will stimulate farmers' investments in solar PV and solar thermal systems. At this stage of the research, only the high importance of energy expenditure savings as a stimulant for such investments was not confirmed.

Table 6. Parameters of the logit regression model for the Y_1 variable determining the probability of having a PV or solar installation.

Variable	Factor	Standard Error	*t*-Statistic	*p*-Value
Const	3.82243	1.5108	2.5300	0.0391
X_1—Having other RES installations (0/1)	2.91298	0.7041	4.1370	0.0434
X_2—Farmer's age (years)	−2.29318	1.1629	−1.9720	0.0172
X_3—Farm area (ha of agricultural land)	0.95583	0.4779	2.0001	0.0161
X_5—Use of renewable energy subsidies (0/1)	1.96388	0.8834	2.2230	0.0331
Number of observations 226, *p*-value = 0.05 Number of cases of correct prediction 91.6% Chi-square 93.27, Corrected R^2 0.37, McFadden R^2 0.55				

Source: own survey.

The multiple regression method was used to analyze the determinants of the share of agricultural production in the RES energy consumption structure. The estimated regression model included five independent variables that showed a statistically significant effect on the dependent variable (at $p < 0.05$). The regression equation explains the studied phenomenon relatively well, as evidenced by the coefficient R^2 of 77.4% (Table 7). White's test indicates that heteroscedasticity does not occur, which proves the statistical significance of the regression model. Similarly, the distribution of the residuals of the model has the characteristics of a normal distribution (Table 7).

Table 7. Parameters of the multiple regression model for the Y_2 variable describing the share of energy obtained from PV and solar installations used for agricultural production.

Variable	Factor	Standard Error	*t*-Statistic	*p*-Value
Const	1.7389	2.1971	0.7915	0.0041
X_2—Farmer's age (years)	−2.1109	0.2378	−8.8768	0.0178
X_3—Farm area (ha of agricultural land)	1.2119	0.1781	6.8046	0.0206
X_7—Saving energy costs is of great importance	1.7927	2.4493	0.7319	0.0083
X_{13}—The farm is focused on animal production	0.3762	6.7281	0.0560	0.0439
X_{17}—The farm sells agricultural products	0.9781	0.8291	1.1797	0.0349
Number of observations 150, *p*-value = 0.05, F 0.0076 R 0.879, R^2 0.774, Corrected R^2 0.683 AIC 143.479, White's test 0.454, Test for the normality of the distribution of residuals 0.0799				

Source: own survey.

The regression model shows that the age of the farm manager has a significant impact on the share of energy obtained from RESs used for agricultural production. The relationship is inverse, which means that as the farmer's age increases, the share of green energy used for agricultural production in the energy consumption structure decreases. A positive effect on the explained variable was shown for farm area (X_3). The high importance of savings in energy expenditure as perceived by the farmer is associated with greater use of RES energy for farm purposes, which allows us to accept hypothesis H2. A similar relationship concerns farm features such as commercial nature and focus on livestock production, which in turn allows us to accept hypothesis H3.

The χ^2 test of independence was used to determine the relationship between the farmer and farm features and the farmers' declared benefits of implementing solar PV and solar thermal installations. When the null hypothesis H0 indicating independence of features was rejected in favor of the alternative hypothesis H1, the strength of the relationship between the features was assessed using Cramér's V coefficient. The results of the conducted testing are shown in Table 8.

Table 8. The results of the χ^2 and V-Cramer (VC) tests describing the relationships between the characteristics of farmers and their farms and the benefits of PV and solar installations.

Benefits of RES Installations	Holding Area	Farmer's Age	Farm Commodity	Direction of Agricultural Production	Form of Taxation
Environmental benefits	0.32260	0.40162	0.03642 VC = 0.244	0.00898 VC = 0.250	0.29074
Reducing household expenses	0.07938	0.87685	0.74990	0.80779	0.39668
Reduction in agricultural production costs	0.02028 VC = 0.258	0.12588	0.00500 VC = 0.287	0.04700 VC = 0.230	0.25008
Tax benefits	0.15375	0.73025	0.013474 VC = 0.210	0.46465	0.047089 VC = 0.157
The possibility of selling surplus energy	0.67912	0.24905	0.03024 VC = 0.257	0.37347	0.66543
Availability of grants and other low-cost sources of funding	0.04533 VC = 0.229	0.12665	0.00410 VC = 0.297	0.20131	0.029346 VC = 0.185
Prestige, recognition in the local community	0.31922	0.65037	0.04219 VC = 0.237	0.24749	0.28013
Increasing the quality of life	0.12832	0.56699	0.1245	0.2232	0.02402 VC = 0.168

p-value of less than 0.05 indicates rejection of the independence hypothesis. Source: own survey.

The results indicate that all of the farmers recognized the high importance of economic benefits, followed by environmental and other benefits, regardless of age. As far as the commercial character of the farm is concerned, the surveyed units were divided into two categories: commercial farms, i.e., those that directed their production to the market, and non-commercial farms. The concordance of answers in both groups concerned only the benefit from RESs, which are the increase in living comfort and reduction in household expenses. Farmers managing commercial farms, as opposed to their counterparts in the alternative group, placed significantly more importance on the economic benefits associated with using green energy for agricultural production, potential opportunities to sell surplus energy (not applicable to prosumer installations), and tax benefits. They also appreciated the importance of the environmental benefits and the opportunity to build their own position and recognition in the local community.

In relation to the agricultural area of farms, differences in farmers' answers occurred in the case of benefits such as the reduction in agricultural production costs. Such a benefit was definitely more often declared by farmers managing relatively larger farms. Farmers in this group also acknowledged the benefit of access to grants or other low-cost sources of financing for RES installations.

In terms of the direction of agricultural production, the differences concerned the assessment of the importance of two RES benefits, namely environmental benefits and the reduction in agricultural production costs. Farmers managing farms focused on plant or animal production, as well as organic farms and the so-called special branches, attributed greater importance to both types of benefits.

The relationships between the form of VAT taxation of a farm and the importance of tax benefits, access to grants for financing investments in RESs, and increasing the comfort of a farmer's family were noticed. In this case, the first two of these benefits were more often declared by farmers taxed on a general basis (i.e., as entrepreneurs), and benefits related to the quality of life were more often indicated by farmers taxed as so-called flat-rate farmers. This latter tax status is usually adopted by farmers with small holdings that either do not produce goods for sale at all or produce goods predominantly for the self-supply of the farmer's family.

The observed relationships measured with Cramér's V coefficient are not particularly strong, which is due to the multidimensionality of the benefits of RES installations. The strongest relationship was observed between the commercial nature of the farm and the economic benefits and access to RES grants. The results of χ^2 testing give rise to a partial acceptance of hypothesis H4. It was confirmed that farmers managing large-scale commercial and more specialized farms attach more importance to the reduction in agricultural production costs resulting from the adoption of solar PV and solar thermal installations.

5. Summary and Conclusions Remarks

Over the past few years, the scale of interest in renewable energy technologies in Poland has significantly increased, and the type of green energy projects implemented has changed, mainly towards photovoltaic systems. The increase in investments in solar PV micro-installations results from the wide spectrum of use of the energy obtained from this source, the increase in micro-installation efficiency, and the growing affordability of this technology. Programs supporting RES investments from public funds are also an important driver, as from 2019 funding started to be more widely available to beneficiaries. As a result, the number of solar PV installations in Poland grew nearly 9 times between 2018 and 2020, and the installation capacity increased 8 times. A significant proportion of solar PV and solar thermal installations are installed on residential or farm buildings in rural areas, including farms. However, it can be estimated that at the end of 2020, only 7.6% of all farms in Poland had such installations. The potential for the popularization of solar PV and solar thermal installations in agriculture is, therefore, very high; hence, it is worthwhile to conduct research on the factors determining farmers' investments in such systems. Considering the coexistence of household needs and agricultural production needs on a farm, it is also worth examining the factors favoring the increase in energy consumption for farm needs in the structure of total energy consumption.

Numerous authors have studied the drivers of and barriers to farm investment in renewable energy production technologies, often focusing on some specific technology or type of energy. This paper focuses on the characteristics of a farm and farmers, including their attitudes towards the benefits of RESs. Such studies were also conducted by other authors [50,52,54] but were focused on the determinants of renewable energy installations for the purposes of agricultural production. This article analyzes the factors that influence the adoption of RES installations by farmers, both for the needs of agricultural production and for the needs of the household. The research covered a region with fragmented agriculture, where most farming families combine farm income with income from non-agricultural sources. As a result, the RES installation is equally important to meet the living needs of the farmer's family and to reduce the costs of energy consumption in agricultural production. Research in this context is rare; hence, the article fills a gap in the literature.

The research conducted in the Podkarpackie region shows that farmers' decisions to invest in RESs are conditioned by many factors. Among these, economic benefits, including savings on energy costs and access to favorable sources of investment financing (mainly efficiency of investments are also important [2–4]. Environmental efficiency is a particularly new concept [5,6]. Environmental efficiency was already defined slightly earlier, but it referred to the agricultural sector [7–9] or industry [10–12]. Today, only some renewable energy technologies have achieved a competitive level similar to fossil-based technologies. The most important feature of renewable energy sources (RES) is providing energy with zero or almost zero emissions of both air pollutants and greenhouse gases [13–15].

Investments in renewable energy and the resulting CO_2 emissions are usually the most important investment evaluation criteria. Efficiency is the most frequently used technical criterion for assessing energy systems. Additionally, attention is paid to the power of energy devices, investment cost, operation and maintenance costs, land use, job creation and social acceptance [16–20]. The achievement of the assumed goals concerning the share of renewable energy in the total energy consumption requires financial support. Investments in renewable energy and the coordination of these activities require a good use of the Structural Funds and framework programs allocated for this purpose. Funding should come from multiple sources, such as the European Investment Bank and other public financial institutions, grants, loan support schemes, etc. Better coordination of the community and national funding and other forms of support is also needed. Actions to support initiatives to invest in renewable energy should also be coordinated at the national level [21–24]. Different EU countries use different combinations of instruments supporting the development of the use of energy from renewable sources. The primary support instruments include feed-in laws, TGCs certificates, and tendering. In turn, the secondary support instruments include: investment subsidies, fiscal incentives, soft loans [25–27]. From the investor's perspective, investment outlays in renewable energy include investment costs, i.e., costs of technology, land, construction and project development (permits, grid connection agreements, consultancy, etc.). The second group is the cost of financing, i.e., the cost of capital determined by the debt interest rate, the required return on equity. The cost of producing renewable energy also includes operating expenses, i.e., fuel and maintenance costs as well as the costs of service contracts, guarantees and insurance after the start-up of the power installation [28–31].

There are differences in the definition of rural areas from country to country. In Australia, rural areas include small towns with a population ranging from 200 to 999 people. There are three types of rural areas in China. They are a major village (1000–3000 inhabitants), medium village (300–1000 inhabitants), small village (up to 300 inhabitants). In the USA, rural areas are all territory outside of defined urbanized areas and urban clusters, that is, open country and settlements with fewer than 2500 residents, with population densities as high as 386 people km^2 [32–34]. The definition of rural areas proposed by the Organization for Economic Cooperation and Development (OECD) has been adopted in the few national and regional RDP programs (Rural Development Program). According to OECD, a rural area should be understood as an area where over 50% of the population lives in rural municipalities. Rural communes are those where the population density does not exceed 150 inhabitants per km^2 [35]. In Poland, the Central Statistical Office determines rural areas on the basis of the administrative division of the state, and rural areas are areas located outside the city limits, i.e., rural communes and rural areas of urban-rural communes. This approach is related to the administrative division of Poland and three types of communes: rural (which consist only of villages), urban (in which territory is occupied by the city) and urban-rural (those that have at least one city within their territory) [36]. It is the administrative division that will be the basis for the research in the presented study.

Renewable energy is an energy source that is constantly regenerated. Such sources include solar energy, wind energy, water, including river currents, sea and ocean waves, energy from biomass, biogas, or bioliquids. Renewable energy is also the heat obtained from the ground (heat pumps, geothermal energy), air (aerothermal), and water (hydrothermal) [37–42]. The share of renewable energy is systematically growing. There is no one universal source of renewable energy. Different countries and regions use various renewable energy sources. For example, in Poland in 2018, the primary carrier of renewable energy used in heating was biomass (90.5%), and the share of heat pumps was only 0.4% [43]. This is strange because using a ground source heat pump to heat the building was more economically effective than biomass and a system powered by fuel oil [44]. Investments in photovoltaic installations grew particularly rapidly, as their payback period was short.

Additionally, such investments were subsidized from various types of European funds [45]. Local governments play a unique role in such undertakings, and, apart from promoting renewable energy, they should also invest in installations, especially in public utility facilities [46]. Due to the dispersed nature and use of local resources, RSE may be an element enabling, to some extent, increasing energy security (especially in the regional perspective) and reducing energy costs. The RSE share in the energy balance of individual communes and even voivodships is significant [47]. Due to their quantitative and qualitative potential, rural areas are predestined to produce energy raw materials or energy; hence, the Rural Development Program (RDP) provides funds for the development of renewable energy [48]. The undertaken research topic is vital due to the enormous possibilities of renewable energy production in rural areas. For this purpose, investments are necessary, especially by local government units such as municipalities.

The main aim of the research was to identify the factors related to investments in renewable energy sources in basic local government units in rural areas. Additional objectives were to define the conditions for the development of renewable energy sources in Poland, to determine the directions of changes on the use of renewable energy, as well as the importance of renewable energy in Poland, to present the relationship between the level of expenditure on renewable energy and budget components in rural and rural-urban communes. A research hypothesis was formulated in the work: the amount of investment expenditure in renewable energy sources was corelated with budget expenditure and property expenditure of the commune.

2. Materials and Methods

The first stage of the research focused on Poland. Historical conditions for the consumption of individual energy sources are presented, as well as assumptions concerning the development of renewable energy. Changes in the production of energy from various sources were shown. Legal conditions in the field of renewable energy and strategic documents for the development of this area of energy production were also examined. Particular attention was paid to Regional Operational Programs offering the possibility of supporting investments in renewable energy. The differences between individual voivodeships in terms of supporting investments in renewable energy are also presented.

In the second stage, the Świętokrzyskie voivodeship was selected for research using the purposeful selection method. Its characteristics in terms of socio-economic parameters are presented. Opportunities and barriers in the development of renewable energy sources are shown. This voivodeship is a relatively less developed region, with a large share of non-urbanized areas. The large share of rural areas was the main reason for selecting this voivodeship for research. However, the voivodeship was not a leader in the development of renewable energy in Poland. However, it did have the potential to develop renewable energy in rural areas. It is an area naturally predestined for the construction of renewable energy installations, due to its upland and mountainous nature (good wind conditions to drive windmills and sunny slopes for the use of photovoltaic installations). The research was carried out in 27 basic units, i.e., rural and urban-rural communes of the Świętokrzyskie voivodeship, which in 2016–2019 made investments in renewable energy sources. All rural and urban-rural communes that did not invest in renewable energy were omitted, as well as urban communes.

In the last stage, non-parametric tests were used to establish the correlation between the variables. The main barrier to the development of renewable energy is high investment expenditure [48]. Therefore, the parameters related to investment expenditure on RES, total expenditure of the commune and the amount of investment support from public

energies

MDPI

Article

Solid Biomass Energy Potential as a Development Opportunity for Rural Communities

Mariusz Jerzy Stolarski *, Paweł Dudziec, Michał Krzyżaniak and Ewelina Olba-Zięty

Department of Genetics, Plant Breeding and Bioresource Engineering, Faculty of Agriculture and Forestry, Centre for Bioeconomy and Renewable Energies, University of Warmia and Mazury in Olsztyn, 10-724 Olsztyn, Poland; pawel.dudziec@student.uwm.edu.pl (P.D.); michal.krzyzaniak@uwm.edu.pl (M.K.); e.olba-ziety@uwm.edu.pl (E.O.-Z.)

* Correspondence: mariusz.stolarski@uwm.edu.pl; Tel.: +48-89-5234838

Abstract: Conventional energy sources often do not fully satisfy the needs of a modern economy, especially given the climate changes associated with them. These issues should be addressed by diversification of energy generation, including the development of renewable energy sources (RES). Solid biomass will play a major part in the process in Poland. The function of rural areas, along with a well-developed agricultural and forest economy sector, will be a key aspect in this as these areas are suitable for solid biomass acquisition in various ways. This study aimed to determine the solid biomass energy potential in the commune of Goworowo to illustrate the potential in the smallest administrative units of Poland. This research determined the environmental and natural conditions in the commune, which helped to identify the crucial usable solid biomass resources. The total energy potential of solid biomass resources in the commune of Goworowo amounted to 97,672 GJ y^{-1}. The highest potential was accumulated in straw surplus (37,288 GJ y^{-1}) and the lowest was in wood from roadside maintenance (113 GJ y^{-1}). This study showed that rural areas could soon play a significant role in obtaining solid biomass, and individual communes could become spaces for the diversification of energy feedstock.

Keywords: solid biomass; bioenergy potential; rural communities; forest residues; agricultural residues; straw; energy crops

Citation: Stolarski, M.J.; Dudziec, P.; Krzyżaniak, M.; Olba-Zięty, E. Solid Biomass Energy Potential as a Development Opportunity for Rural Communities. *Energies* **2021**, *14*, 3398. https://doi.org/10.3390/en14123398

Academic Editor: Robert H. Beach

Received: 5 May 2021
Accepted: 3 June 2021
Published: 9 June 2021

Publisher's Note: MDPI stays neutral with regard to jurisdictional claims in published maps and institutional affiliations.

1. Introduction

A region's economic development depends on access to energy and conventional resources such as coal, natural gas and oil are no longer sufficient to satisfy the increasing demand from the economy. Moreover, non-renewable energy sources contribute to climate changes, making it necessary to seek alternative options, including renewable solutions (RES) [1–3]. Energy generation in Poland is based on hard coal and brown coal, which boosts the greenhouse effect and, since fossil fuel resources are highly likely to be exhausted, measures should be taken to promote RES, whose supply is unlimited. Since coal accounted for 77% of all energy carriers used in Poland in 2019, it is claimed that the proportion of RES in the energy mix in Poland is insufficient [1,4]. Pressure from both society and the international community necessitates changes in the energy source structure, and the removal of coal from power generation in Poland is still too slow [3,4]. However, pressure is growing, as the solutions proposed by the European Commission at the UN COP25 climate summit in Madrid assume that the European economy will have achieved climatic neutrality by 2050 (The European Green Deal). The realization of this idea would require the implementation of multiple measures covering all aspects of EU citizens' lives, including bioeconomy—a topic which is not covered by any Polish strategic document [5,6].

Energy from RES includes energy from biomass, solar energy, energy from water, wind, geothermal sources and energy from the environment obtained using heat pumps.

The proportion of energy from RES in the structure of energy generation in Poland has been increasing in recent years, but it is still small. RES accounted for 16.0% of the total primary energy in 2019; more than in 2014–2018 (by 3.9, 2.9, 2.4, 1.7 and 1.5 percentage points, respectively). At the same time, the average proportion of energy from renewable sources in the total primary energy in the EU-28 increased much faster—from 26.1% in 2014 to 32.8% in 2019 [7–10]. This situation necessitates a greater use of renewable energy to increase the proportion of RES in energy generation.

Solid biomass is the dominant RES in Poland and it accounted for 65.6% of the total in 2019, with 55% of the amount being consumed by end-users without being converted to another energy carrier. Solid biomass is the leader in heat generation from RES in Poland (90.1% in 2019) and it also accounted for 25.1% of electricity generation in 2018, with only wind energy having a larger share [11]. The proportion of biomass can increase since individual solid fuel-fired boilers do not increase total operating costs compared to fossil fuel-fired boilers. Nearly 60% of the EU-28 population live in houses that require heating boilers, many of which have to be replaced [12].

Diversified renewable energy is now the most promising sustainable energy system instead of non-renewable and centralized systems. Local generation and distribution of energy increases the system reliability and reduces the distribution related loss, which is the case with the centralization of energy sources [13]. Therefore, a single region should be regarded as a not-fully-used energy system considering the RES resources present in it. Such activities align with the concept of sustainable development and circular economy [1] because seeking local and renewable energy sources guarantees their development, which is important given the crisis resulting from the exhaustion of conventional energy sources [4]. Therefore, sustainable development can be achieved through the development of society, which is guided by a comprehensive approach to products and/or services concerning materials and energy and ensures raw material effectiveness and economic growth throughout the product life cycle [14].

The territory of Poland is divided into regions, with the administrative division being the most common manifestation of regionalization. Poland is divided into 16 voivodships, 314 districts and 2477 communes as the smallest administrative units. There are 302 urban communes, 642 urban–rural communes and 1533 rural communes [15]. Considerable parts of rural areas contain potential sources of solid biomass which could play a special role in the local energy system, especially since the average commune size is 12,500 ha [16]. It should be noted that rural communes have good conditions for the diverse use of green energy. This applies particularly to agriculture, which can facilitate the transition from a fossil fuel-based economy to an economy based on renewable energy [4,6]. Significant benefits from solid biomass use as energy feedstock include regional energy independence, the prevention of low emissions, the creation of new jobs, the use of marginal land, agricultural and forest residue management, local communities becoming motivated to act for the benefit of the environment and, primarily, the opportunity to obtain clean energy [1]. Agriculture in EU countries is consuming increasing amounts of energy, which has resulted in the diversification of its resources and the growing importance of RES. Introducing green energy presents an opportunity for agriculture modernization without intensifying its adverse impact on the environment [17].

To date, the energy potential of agricultural residues in 294 countries of the world has been determined [18], as well as the solid biomass energy potential in Switzerland [19] and Turkey [20]. Stolarski et al. [21] determined the bioenergy potential of the countries bordering the Baltic Sea, including Poland. Smaller regions, i.e., districts, were dealt with by Kowalczyk-Juśko et al. [22]. However, there were no detailed papers in the literature that provided a comprehensive methodology of the conducted studies or analyses of the biomass energy potential in individual communes considering their specific local conditions. Although such studies are of significance to the diversification of local energy sources which ultimately affects the country's energy balance, the number of in-depth scientific papers on the topic is low. Therefore, taking up this subject was justified since this forecast could affect the investment decisions concerning energy generation and consequently increase the RES proportion in the local generation structure. This study aimed to determine the solid biomass resources and the energy potential in the Commune of Goworowo. This analysis should be used as the basis for changes in the local strategies of regional development, raising social awareness of the potential and local use of RES in rural areas.

2. Materials and Methods

The research was conducted in the commune of Goworowo (the commune is the basic unit in the administrative division of Poland). Goworowo is a typical rural commune in the north Mazovian Lowland, whose conditions are representative of those of central-eastern Poland. The methodological work was started by determining the environmental and natural conditions in the commune with respect to the solid biomass acquisition potential. The materials that were originally accumulated were used to identify the most significant and usable sources within the administrative unit under study. Data were mainly gathered using the official, up-to-date administrative databases. It is noteworthy that there are often no data solely concerning the administrative area of a commune in Poland as there may be several managing bodies whose authority extends over a commune, as is the case with forest lands.

The next stage involved the determination of the solid biomass amount from various sources based on the processed data and the methodology presented further in the paper. As a consequence, the total solid biomass energy potential could be referred to hard coal—the most common energy source in Poland, including rural areas [9]. The specific local conditions in the commune were considered when the solid biomass resources in it were determined.

2.1. Characteristics of the Commune

2.1.1. Administrative Affiliation and General Information

The commune of Goworowo covered an area of 21,909.44 ha in 2020 [23]. Administratively, it is part of the Mazowieckie Voivodship and is situated in the north-east of the voivodship, in the southern part of the Ostrołęcki District, 52°54′ N 21°33′ E [24]. The relative position of the commune in the administrative units is shown in Figure 1.

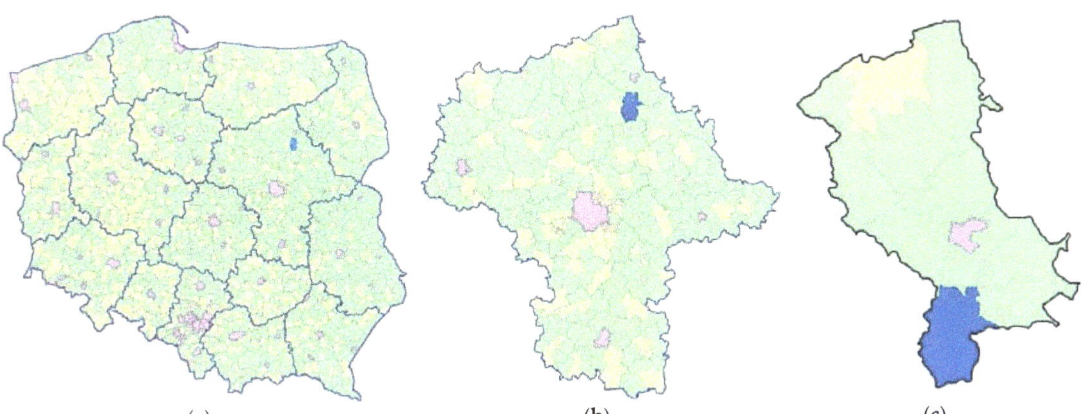

(a) (b) (c)

Figure 1. Position of the commune of Goworowo (blue): (**a**) in Poland; (**b**) in the Mazowieckie Voivodship; (**c**) in the Ostrołęcki District.

It had a population of 8455 people at the end of December 2019, which accounted for 9.5% of the population of the Ostrołęcki District. The population density in the commune was 39 people per km^{-2}. The unit under analysis was a commune with a medium-sized population and area when compared to the other communes in the Ostrołęcki District [23,25]. When compared to all 2478 communes in Poland, Goworowo had a nearly twice smaller population (an average of 15,500 people) and its area was nearly twice larger than the average (12,500 ha) [16,23].

It is an agricultural commune, with dominating individual farming oriented towards dairy cattle breeding. Swine are also bred. In crop production, grain crops, corn for silage and root crops dominate. Non-agricultural business activities include services—mainly food industry related, forestry related and in the repair and construction and transport area. Industry includes only production and service facilities. Currently, the agriculture, construction and processing industries enjoy the greatest development opportunity in the commune, along with rural tourism, including agrotourism and organic farms. Therefore, using RES would be in line with this development trend. There were no industrial facilities within the commune that posed an increased or high risk of industrial failure [23,26].

2.1.2. Land Use

The commune land was mainly covered by forest or used for agriculture, with farmland accounting for 61.87% of the total area and forest for 32.36% (Table 1). The other lands accounted for 5.77% of the commune area [23].

Table 1. Land use structure in the commune of Goworowo.

Item	Proportion of the Area (%)
Farmland	61.87
Farmland with trees and bushes	0.80
Forest land	32.36
Wasteland	0.97
Built-up and urbanized area	2.86
Land under waters	1.05
Other land	0.09

Residential buildings occupied the greatest part of the built-up area (41%) followed by farmstead buildings (39%). Approximately 8% of the area was occupied by recreational buildings and 2% by service and industrial facilities [23,26].

The soil classification system in Poland includes soils from class I (best) to class VI (poorest). Arable land class V and VI in the commune of Goworowo (excluding orchards) altogether accounted for 56.1%. Permanent grassland was dominated by poor and very poor quality soils (62.8%), with larger complexes of such land occurring in the north and in the south of the commune, where disadvantageous moisture content in the soil (requiring soil melioration) dominated [27]. Class IV soils dominated in orchards.

Better soils were formed on loams and glacial dust and were present in vast and compact areas in the commune center. These included medium and good soils, mainly acidic brown soils with occasional podsolic or pseudo-podsolic soils. However, poor and very poor quality soils, brown acidic soils and podsolic soils formed on glacial sand dominated in the south and the north of the commune [28]. The commune was among the areas of the Mazowieckie Voivodship with medium-advantageous conditions for agriculture development.

2.1.3. Structure of Agricultural Land

Arable land (9571.03 ha) dominated the farmland. Grassland—pastures (1784.00 ha) and meadows (1543.16 ha)—occupied a much smaller area. Orchards covered the smallest area (21.96 ha) [23].

The major crop area was determined based on the average crop production area structure in the Mazowieckie Voivodship [29]. Therefore, the area where crops were produced accounted for 77% of the entire arable land. The remaining 23% was occupied by agriculture or horticulture supporting facilities and structures, fallow land and areas occupied by ornamental tree plantations and ornamental tree or bush nurseries [29–31].

The cultivated area of species constituting the most important, primary source of straw was 5556.15 ha (Table 2).

Table 2. Pattern of crops—exclusively main sources of straw.

Crop	Area (ha)
Triticale	1180.31
Wheat	1039.30
Cereal mixtures	968.70
Rye	913.76
Oat	526.98
Grain maize	367.29
Barley	297.95
Rape and turnip-like rape	261.86

2.2. *Determination of Biomass Resources and Their Energy Potential*

2.2.1. Straw

The amount of straw depends on the production area of crops which produce straw as a by-product, grain yield, crop species, fertilization, agricultural practices and climate and soil conditions [21]. The grain–straw coefficient index (Table 3) enables the theoretical determination of the straw amount per 1 ha of crop cultivation area [32–34].

Table 3. Yield and grain/straw coefficient for crops.

Crop	Mean Grain Yield (Mg ha^{-1})	Grain/Straw Coefficient
Wheat (mean for winter and spring yield)	4.47	0.93
Triticale (mean for winter and spring yield)	3.69	1.16
Rye (as winter crop)	3.24	1.45
Barley (mean for winter and spring yield)	4.22	0.78
Cereal mixtures (mean for winter and spring yield)	3.42	1.10
Oat	3.07	1.05
Rape and turnip-like rape	3.15	1.00
Grain maize	11.95	1.40

However, not all straw can be collected during harvest due to the field conditions or the height at which the harvester cuts down the crop. Moreover, some straw is lost while being collected, baled and transported. Since the analyses conducted in straw-fired boilers also showed that the available straw amount calculations based on the grain/straw coefficient produce excessive results, the technical and practical potential for straw acquisition in cereal and oily crop production was calculated to be 60% (coefficient 0.6). This means that the average yield of straw collected from the field in bales corresponded to 0.6 of the grain yield [21]. Therefore, the total straw yield was determined based on the yield of cereals, rape and turnip-like rape from the Equation (1):

$$Y_S = 0.6 \cdot Y_G \quad (1)$$

where:

Y_S—straw yield (Mg y^{-1}),

Y_G—grain yield (Mg y^{-1}).

Moreover, large amounts of straw are used in animal production as fodder and bedding. Straw should also be ploughed under and returned to the soil to maintain the organic matter balance. The straw surplus in Poland, usable as an energy feedstock, accounts for 33% of the total straw produced [35]. However, the surplus was adjusted in the current study due to the local conditions. The straw consumption coefficient was increased because cattle and swine breeding in the bedding system dominated in the commune of Goworowo. Therefore, it was assumed for these analyses that the demand for straw as bedding, fodder and for ploughing-under in the commune was: 26%; 16% and 39% of the total straw yield, respectively, which gave a total of 81%. Therefore, the straw surplus unused in agriculture in the commune of Goworowo was only 19%. This figure was considered in calculations of the straw surplus potential, which can be used as energy feedstock (2), assuming that the mean lower heating value (LHV) was 14 GJ Mg^{-1} [36] (3):

$$Y_{Se} = 0.19 \cdot Y_S \quad (2)$$

where:

Y_{Se}—straw yield for energy purposes (Mg y^{-1}),
Y_S—straw yield (Mg y^{-1}).
Subsequently, the straw energy value was determined:

$$Q_S = Y_{Se} \cdot LHV \quad (3)$$

where:

Q_S—energy potential of straw (GJ y^{-1}),
Y_{Se}—straw yield for energy purposes (Mg y^{-1}),
LHV—lower heating value (GJ Mg^{-1}).

2.2.2. Orchard Wood Residue

Wood residue in orchards is produced mainly by annual pruning of trees [37], done to shape the tree crown and obtain the optimum fruit yield. Trees are usually pruned in winter and spring, depending on the local climate and the tree species [38]. The amount of wood collected after pruning ranges from 1.9 to 5 Mg ha^{-1} in six-year-old orchards. The apple tree, which dominates in Polish orchards, produces 3.5 Mg of wood biomass per 1 ha from annual pruning. Part of the biomass is mulched and scattered as litter in orchards (in situ mulching). The Mazowieckie Voivodship has the greatest energy potential accumulated in wood residue from pruning apple orchards [38,39].

The majority of orchards in the commune of Goworowo were occupied by fruit tree nurseries in which pruning or soil mulching was not done. They were assumed to account for 60% of the total area. The remaining part was occupied by apple tree orchards, usually not large-scale orchards. They were not mulched, and the wood obtained in them was used as fuel or for recreational purposes. Therefore, since orchards were often situated on medium quality soils, it was assumed that 3.0 Mg of fresh wood biomass was produced by pruning 1 ha of an orchard.

Moreover, many small orchards (up to 2 ha) could not be cultivated properly or not pruned at all. Consequently, they were excluded from the potential calculations and assumed to account for 10% of the total. Trees in the commune were pruned manually, so no such limitations were observed as with machine pruning when losses in wood yield are caused by machine use or uneven ground [38]. It was calculated in this study that the potential loss associated with collecting the wood and its transport amounted to 5%. The weight of wood biomass from tree pruning was calculated from the Equation (4):

$$M_O = 0.4 \cdot A_O \cdot M \cdot 0.86 \quad (4)$$

where:

M_O—mass of wood from the pruning of orchards (Mg y^{-1}),
0.4—orchards, excluding fruit nurseries,
A_O—orchards area,
M—mass of wood from pruning (Mg^{-1} ha^{-1} y^{-1}),

0.86—factor taking into account orchards not cultivated properly or excluded from regular pruning (0.90) and taking into account losses associated with harvest and transport (0.95).

Orchard biomass does not differ significantly from forest biomass in terms of its energy properties, which makes it usable as the substitute for the latter [39]. Fresh wood LHV is assumed to be 8.0 GJ Mg^{-1} [40] (5):

$$Q_O = M_O \cdot LHV \tag{5}$$

where:

Q_O—energy potential of wood from orchards (GJ y^{-1}),
M_O—mass of wood from the pruning of orchards (Mg y^{-1}),
LHV—lower heating value (GJ Mg^{-1}).

2.2.3. Energy Feedstock from Forests

Forest land owned by the state in Poland is managed by the State Forests, divided into forest districts—basic forest economic units. The commune of Goworowo is located within three such units: Ostrołęka, Wyszków and Pułtusk with the part managed by each of them occupying 82.37%, 1.38% and 16.25% of its area, respectively. The annual increment of wood resources in Poland in 2019 amounted to 9.42 m^3 ha^{-1} $year^{-1}$ [41]. It was lower in the commune of Goworowo (as a weighted average from these forest districts)—5.9 m^3 ha^{-1} $year^{-1}$ [42–44].

Logging residue is one of the most commonly used energy feedstock from forests. Wood obtained from forests in Poland, including logging residue, is classified into quality groups, with smallwood (M2) being one of them. M2 wood accounts for 8% to 15% of the whole above-ground wood biomass obtained from a logging area unit (in different administrative units) [45]. Apart from M2, middle-sized wood and logging residue, such as treetops and branches, can be used as fuel. Altogether, this energy feedstock accounts for 17% of the total wood yield in Poland [46]. It has been stressed that the proportion could be much higher if this category also included M1—wood of a different quality class, now used in industry [1,47].

The total forest biomass potential was determined under the assumption that 74.8% of the annual wood increment is obtained every year, as it is in Poland [48], with the energy feedstock accounting for 15% of the total wood yield. The wood loss during logging and transport to the final destination was not taken into account in the yield. Forest land in Poland also includes non-afforested land and land associated with forest management. This fact was considered in the calculations, with the average proportion of the afforested land in forest land calculated to be 0.97 [31,42–44].

The energy potential of logging residues was calculated using the following Equation (6):

$$V_F = A_F \cdot I \cdot H \cdot E \cdot F \tag{6}$$

where:

V_F—volume of energy resources from forests (m^3 y^{-1}),
A_F—forest area (ha),
I—wood resources increase (m^3 ha^{-1} y^{-1}),
H—timber harvest in relation to timber increment (0.748),
E—share of energy resources (0.15),
F—share of afforested land in the forest land area (0.97).
Fresh wood LHV was assumed to be 7.5 GJ m^{-3} [49] (7):

$$Q_F = V_F \cdot LHV \tag{7}$$

where:

Q_F—energy potential of energy resources from forests (GJ y^{-1}),

V_F—volume of energy resources from forests (m^3 y^{-1}),
LHV—lower heating value (GJ m^{-3}).

2.2.4. Solid Biomass from Roadside Maintenance

Roadsides are in intensive use and are heavily polluted by transport. They are easily accessible and have to be maintained regularly, for example, to keep the traffic safe. All this makes them a promising source of biomass [50]. The road infrastructure in the commune of Goworowo comprised [26]:

- supralocal roads: 13.5 km of the trunk road and 101.3 km of district roads;
- local roads: 123.7 km of communal roads.

The roadside trees and bushes in the commune were not pruned regularly, although there is 0.4 ha of roadside maintenance area per 1 km of road, which shows their energy potential [51]. The roadside maintenance area was assumed as 0.2 ha km^{-1} for local roads. Approximately 4 Mg ha^{-1} of fresh woody biomass can be obtained from roadsides annually, but only 25% of this amount can be acquired in Poland [51], which is why 1 Mg ha^{-1} $year^{-1}$ was taken for the calculations. Under the Polish Nature Conservation Act [52], only intervention pruning, such as removing dead boughs to improve safety, is allowed, but regular maintenance pruning is not.

It was assumed in the current research that 20% of the roadside with the maintenance area as given above was overgrown with trees and/or bushes and the yield resulted from all work done, including collection and transport. The following Equation was applied (8):

$$M_R = 0.2\ (L_{SL} \cdot A_{SL} + L_L \cdot A_L) \cdot H_R \quad (8)$$

where:
M_R—mass of wood from roadsides (Mg y^{-1}),
0.2—factor taking into account roadsides covered with shrubs and/or trees,
L_{SL}, L_L—length of supralocal and local roads (km),
A_{SL}, A_L—area of supralocal and local roadsides (ha km^{-1}),
H_R—timber harvest from roadsides (Mg ha^{-1} y^{-1}).
Fresh wood LHV is assumed to be 8.0 GJ Mg^{-1} [40] (9):

$$Q_R = M_R \cdot LHV \quad (9)$$

where:
Q_R—energy potential of wood from roadsides (GJ y^{-1}),
M_R—mass of wood from roadsides (Mg y^{-1}),
LHV—lower heating value (GJ Mg^{-1}).

2.2.5. Biomass from Perennial Energy Crops

Perennial energy crops should be grown mainly on soils of poorer quality, unusable for growing edible crops. Marginal soils, including sandy soils and/or those highly susceptible to erosion, are recommended for growing perennial grasses and short-rotation woody crops. However, poor quality soils are not often used for growing energy crops, although they should be used primarily for this purpose [53]. Growing energy crops can reduce water and wind erosion and enable carbon sequestration in soil [54]. It is therefore prudent to use land of lower agricultural productivity for energy crop plantations. Popular woody energy crops include basket willow (*Salix viminalis* L.) and grasses such as giant miscanthus (*Miscanthus* × *giganteus* J.M. Greef and M. Deuter) [55–57]. Miscanthus is a C4 plant which can be grown successfully in various climate conditions. It is cultivated on marginal soils and does not need irrigation or intensive fertilization. Owing to its deep root system, it uses water effectively and prevents soil erosion. These properties make its cultivation recommended in areas threatened with erosion and with poor water availability [58].

Aerenchyma cells present in stems and roots also improve the gas exchange effectiveness, enabling it to grow on wetlands [54]. *Miscanthus* × *giganteus* is regarded as one of the best choices for low-cost bioenergy production in Europe [59]. It requires the minimum amount of nutrients and its cultivation is perceived as an advantageous way to use soils of low usability for food crop growing [60]. Willow can be grown on many types of agricultural land as it is highly tolerant of environmental conditions, with wetlands being preferred for this purpose [61–63]. Szczukowski et al. [64] demonstrated the potential for willow biomass production on excessively damp soil with a high groundwater table level. *Salix viminalis* can grow even when the soil profile is filled with water [65].

The present study showed that the commune of Goworowo has the right environmental conditions for growing perennial energy crops. There are 147 ha of usable marginal soils, with half of them being excessively humid and suitable for basket willow cultivation. The other half are sandy soils with a low groundwater table intended for giant miscanthus plantations. Based on the result of long-term experiments conducted by the University of Warmia and Mazury in Olsztyn [57,66], annual harvest rotations were used for the calculations. Based on the cited studies, it was assumed that willow and miscanthus would yield 15 and 12 Mg ha^{-1} y^{-1} of fresh biomass, respectively. The corresponding LHV was 8 and 12 GJ Mg^{-1}, respectively. It should be noted that the yield depends on the soil and weather conditions, planting density and agricultural procedures. The energy potential of biomass from energy crop plantations for annual harvest rotations is (10):

$$Q_E = A_E \cdot Y_E \cdot LHV \tag{10}$$

where:

Q_E—energy potential of biomass from *Salix viminalis* and *Miscanthus* × *giganteus* energy crops (GJ year^{-1}),

A_E—energy crops area (ha),

Y_E—assumed average annual yield of fresh biomass (Mg ha^{-1} y^{-1}),

LHV—lower heating value (GJ Mg^{-1}).

2.2.6. Hay from Meadows and Pastures

The high biodiversity of semi-natural grasslands can be maintained only by continuous management. The hay yield from meadows in Poland amounts to 4.9 Mg ha^{-1} y^{-1} and from pastures to 3.6 Mg ha^{-1} y^{-1} [67]. The demand for pasture fodder, hay and silage for ruminants is decreasing, which is why a surplus can be used as energy feedstock [68–70]. The demand for fodder from grasslands in Poland is dropping due mainly to the changing system of farm animal (mainly cattle) feeding and the decreasing profitability of their breeding with the following reduction of their stock. As a result, many meadows and pastures remain unused and the limited extent of their use causes damage to nature (e.g., soil degradation) and economic loss (unused production potential). Hay from grassland is a considerable biomass resource which can be used as energy feedstock [71,72]. It is noteworthy that biomass from unused grassland for energy generation may prevent its natural succession [73]. It is recommended that cattle grazing in pastures be replaced with mowing to maintain and increase biodiversity, with no interference in traditional management methods (e.g., no fertilizers). Hay produced in this way can also be used as energy feedstock [74].

Semi-natural mesophilic mesotrophic grassland dominates in the commune of Goworowo. The meadows are used extensively, with two cuts annually, and virtually all hay is used in animal breeding. However, some farms were found to harvest three cuts while other farms mowed pastures in the face of insufficient amounts of hay or silage. Therefore, it was assumed that three instead of two cuts could be obtained from 5% of the meadows, and the surplus produced amounting to 2 Mg ha^{-1} year^{-1} could be used as energy feedstock [75,76]. The potential of hay from meadows was calculated with the lower heating value taken as 13.5 GJ Mg^{-1} [77] (11). The assumed yield was regarded as the amount of hay obtained from mowing and collection, including possible baling.

$$Q_M = 0.05 \cdot A_M \cdot S_H \cdot LHV \tag{11}$$

where:

Q_M—energy potential of hay from meadows (GJ y^{-1}),

0.05—share of area, which can be used for energy purposes,

A_M—area of meadows (ha),

S_H—surplus hay (Mg ha^{-1} y^{-1}),

LHV—lower heating value (GJ Mg^{-1}).

It was also assumed that 5% of pastures are not fully used (excluding fallows) and one cut of hay can be obtained annually as energy feedstock. The annual average energy potential of hay from pastures was taken as 50 GJ ha^{-1} $year^{-1}$ [68,74] and it was included in the Equation (12). Late cut biomass from pastures can be used, which is beneficial as late swath is more flexible than early because of the weather conditions. Moreover, the hay quality had not deteriorated much by that time [70].

$$Q_P = 0.05 \cdot A_P \cdot E_P \tag{12}$$

where:

Q_P—total energy potential of hay from pastures (GJ y^{-1}),

0.05—share of area, which can be used for energy purposes,

A_P—area of pastures (ha),

E_p—energy potential of hay from pastures (GJ ha^{-1} y^{-1}).

2.2.7. Landfilled Sludge and Municipal Waste

The possibility of using landfilled sludge and municipal biodegradable waste produced in the commune was analyzed. Sludge surplus will make it necessary to seek new solutions for its proper management and use [78]. Currently, fluidized bed technologies enable thermal conversion of mechanically dehydrated or partially dried sludge [79]; co-combustion of dehydrated (not dried) sludge is possible, e.g., in CHPs [80]. The importance of sludge from rural wastewater also increases as it does not contain excessive amounts of heavy metals [81], which is the case with sludge from heavily industrialized areas [82]. Mechanical–biological wastewater treatment of communal wastewater in 2019 produced 51 Mg (expressed as dry weight) of hydrated non-stabilized sludge (excessive sludge) [83], and 41 Mg of the sludge was transported to Ostrołęka where it was subjected to anaerobic stabilization followed by mechanical dehydration (the sludge was not dried—the construction of a drier is planned [84]). Ten megagrams of sludge was stored in the commune [83]. Collected municipal waste—apart from mixed waste—included biodegradable waste, comprising mainly tree branches and shrub branches, sawdust and bark, mowed grass, leaves, flowers, fruit and vegetable waste. A total of 1,454.98 Mg of mixed municipal waste and 10.76 Mg of biodegradable waste was collected from households in the commune in 2019 [85]. The biodegradable fraction accounts for 48% of the mixed municipal waste in rural areas [86]. The biodegradable waste amount was determined, assuming that it comprised the biodegradable fraction mentioned above (13).

$$M_T = M_{Bio} + 0.48\ M_M \tag{13}$$

where:

M_T—total mass of biodegradable waste (Mg y^{-1}),

M_{Bio}—mass of biodegradable fraction (Mg y^{-1}),

0.48—factor taking into account share of biodegradable fraction in mixed waste,

M_M—mass of mixed waste (Mg y^{-1}).

3. Results and Discussion

3.1. *Straw Potential*

Given the local conditions in the commune of Goworowo, the mass of straw as energy feedstock amounted to 2663 Mg y^{-1} (Table 4). It was a small amount, as it accounted for only 0.05% of the straw energy resources in Poland (5.1 million Mg y^{-1}) [21]). However, it accounted for the largest part (38.18%) of the local solid biomass energy potential, 37,288 GJ y^{-1} (Table 4, Figure 2). Moreover, maintaining food supply and soil quality while acquiring the straw necessary to replace fossil fuels is of key importance (straw use for energy generation is enabled by pellet or briquet technology) [87,88]. For this reason, the current study (quantitatively determining energy potential) took into account the straw residue depending on its use in agriculture—as straw for fodder, for bedding or to be ploughed under.

Table 4. The amount and energy potential of each solid biomass type.

Solid Biomass	Amount (Mg y^{-1})	Theoretical Energy Potential (GJ y^{-1})
Straw surplus	2663	37,288
Residue from orchards	23	180
Hay	475	6543
Plantations of energy crops	1985	19,404
Logging residue	4553 [a]	34,144
Roadside wood	14	113
Biodegradable waste	709 [b]	-
Landfilled sludge	10 [c,b]	-
Total	-	97,672

[a] m^3 y^{-1}, [b] is not taken into account in the determination of the total potential (GJ y^{-1}), [c] expressed per dry weight.

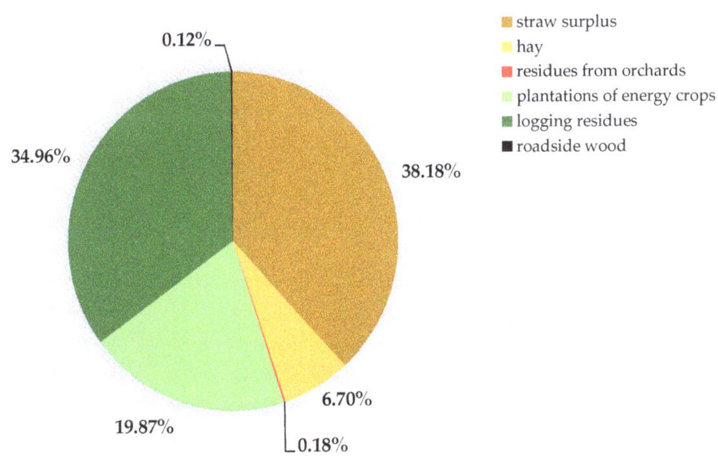

Figure 2. The proportion of individual solid biomass resources in the total energy potential of the commune of Goworowo (%).

Crop production diversity in the commune of Goworowo was not high because of the soil and climate conditions, with the local straw resources most affected by triticale, wheat, cereal mixes and rye, in accordance with their cultivation area. Managing crop residue is particularly important for biomass utilization and energy generation for solving environmental problems and using alternative energy sources. More importantly, the cereal production continuity ensures the permanent availability of straw resources, whose overproduction is caused by a decreasing number of farm animals with an increasing proportion of cereals in the structure of crops [20,22].

Cereal straw resources are regarded as the greatest unused energy potential in agriculture. Unlike the area used for the cultivation of rape and turnip-like rape in the Mazovian Voivodship, the area that was used for cereal cultivation has been growing in recent years. Since the yield per ha of cereals as well as rape and turnip-like rape is continuously increasing [89], the straw energy potential in the commune of Goworowo is expected to increase, or at least to remain unchanged.

3.2. *Orchard Potential*

The estimated energy wood resources from orchards in Poland amount to 88,700 Mg y^{-1} [90]. The small orchards area in the agricultural land in the commune of Goworowo—merely 21.96 ha—resulted in a small orchard wood biomass amount, i.e., 23 Mg y^{-1} (Table 4). The orchard waste energy potential amounted to 180 GJ y^{-1}, which accounted for merely 0.18% of the total (Figure 2). The biomass obtained by pruning fruit trees was often used in households, but this applies mainly to biomass from backyard gardens. The energy potential of the considerable amounts of waste from typical orchards is usually lost due to storing or burning it in the field [91], while it could be used as fuel [22]. Proper management of these resources may increase the effectiveness of biomass use in the future, as orchard biomass waste used for energy generation does not compete with food production [92], which may be the case with cereal straw or hay.

Moreover, pollution emissions from burning pruned biomass in appropriate boilers are low [92]. However, the orchard area in the commune decreased by over half over the 2017–2020 period [23,93] and the orchard area also decreased in the Mazowieckie Voivodship (by over 1100 ha over the 2017–2020 period) [30,94–96]. Therefore, the local resources of this biomass source are likely to decrease in future, especially given the drop in the proportion of fruit trees in orchards in the voivodship, which are being replaced by bushes and berry plantations [89,97].

It should be noted that pruning waste cannot always be used for energy production, although its impact on the greenhouse effect is smaller than in situ mulching. This happens when the absence of plant cover in the inter-row space exceeds 80% and the soil structure has a tendency for compaction, for becoming muddy or for surface runoff, and when the orchard is susceptible to erosion. The ground biomass should be mulched in the first and second case and ploughed under in the third. The local potential cannot then be fully utilized [98].

3.3. *Forest Potential*

Local resources of forest residues biomass in the commune of Goworowo were determined as 4553 m^3 y^{-1}, which was equivalent to the energy potential of 34,144 GJ y^{-1} (34.96% of the whole) (Table 4, Figure 2). The annual increment ratio was assumed at the level given for Poland—74.8% [48], although it was 116% in the Pułtusk Forest District and 114% in the Wyszków Forest District [43,44]. The wood harvest in a calendar year depends on the area ratio of the final cutting sites to intermediate cutting sites. The energy potential of logging residue largely depends on the development phase of the forests in the area. When final cutting sites (with much thicker and slower growing trees than in intermediate cutting sites) dominate the forest structure, the harvest-to-increment ratio can exceed 100% and the logging residue amount then increases considerably. However, these indices are given for a whole forest district area and they do not provide data for individual communes. There are also data for 10 year periods, which is why a lower value was adopted to avoid overestimating the local resources.

The optimum wood amounts to be left in local forests should be determined to maintain biodiversity and conserve forest ecosystems. Perceiving the forest as a source of energy feedstock must be accompanied by ecological thinking [1]. Taking this factor into account in research is called "potential with increased biodiversity conservation" [99]. It is important because the forest area in Europe with satisfactory amounts of deadwood has been very small during the past few decades [100].

The forest area in the commune of Goworowo was larger in 2020 compared to 2017 by over 1000 ha, i.e., by 16% [23,93]. This increase reflects the increasing forestation level in Poland [101] and ensures the stability of forest residues as a source of solid biomass. The issues related to forest management are often left out of development strategies, due to which forests are not regarded as a factor of the region's socio-economic development [102]. Mentioning the bioenergy aspects of forests in strategic documents may prevent its marginalization in the political spheres and increase society's interest in it.

3.4. Roadside Potential

Roadside maintenance produces biomass for bioenergy generation, but only a small part of it is used due to dispersion [103]. This study has shown that roadsides in the commune of Goworowo constitute a local space concerning the solid biomass energy potential. About 14 Mg of wood biomass, with an energy potential of 113 GJ y^{-1}, can be obtained from the roadsides in the commune (Table 4). The presence of trees on roadsides ensures that biomass can be acquired by pruning, which is usually wasted [104]. The potential in this regard also increases as the process may contribute to roadside naturalization when the biomass of an invasive species is obtained, which can be replaced by native flora, thereby improving biodiversity [105].

A better choice of tree species and developing more diverse resources improves the ecosystem advantages provided by roadsides [106]. Managing wood from roadsides provides more benefits, e.g., ensuring the patency of roadside draining ditches [105]. Roads must be accessible to society, be convenient and safe, and integrate with their surroundings. Their development is accompanied by an increase in the roadside area, which connects forests, farms and traffic networks. The main purpose of roadside management is to ensure traffic safety (improving visibility and providing a space for emergency stops). It can produce biomass from tree and bush pruning, which can be used as fuel to maintain a positive energy balance. The promotion of ecosystem services from roadsides is a future line of sustainable management in such areas [105–107]. Growing urbanization will increase the roadside area, which makes proper management crucial. The roadside wood potential can help transition from conventional road management to a form based on a circular economy [105]. Even leaves from roadside trees can be used in the production of high-quality solid biofuels [108].

3.5. Potential of Perennial Energy Crops

The land in the commune of Goworowo allocated for perennial energy crop plantations could yield 1103 Mg of fresh willow biomass and 882 Mg of fresh miscanthus biomass annually (Table 4). The total energy potential of the plantations was 19,404 GJ y^{-1}, accounting for 19.87% (Figure 2). Only 1.5% of arable land was proposed for the plantations in the current study, and all of it belonged to the poorest quality class (class V and VI). Currently, winter rye and serradella are grown on the poorest quality soil in the commune and the cultivation effectiveness is low. Growing perennial energy crops on such soils, where the nutrient abundance is low, where biophysical restrictions exist and where the crop yield is low, is a more sustainable alternative than traditional food crops [109]. Since competition in soil used for food crop production requires spatial land segregation [110], choosing the poorest quality soils with the lowest effectiveness in food crop cultivation was justified. Growing perennial crops on this land could improve the soil properties by increasing its total organic carbon content, and decreasing its susceptibility to erosion [109,110].

Sustainable development stimulates an interest in bioenergy generation from renewable sources, and biofuel production from energy crops in Europe is expected to grow to meet the policy goals (The European Green Deal) [5,111]. Energy crops in Poland are grown in an area of approximately 17,900 ha [21]. No such plantations exist in the commune of Goworowo, even though it has appropriate potential for them. The use of 143 ha of marginal land for this purpose could considerably increase lignocellulosic biomass production, especially since Poland is one of the major producers of SRC willow in Europe [61]. The crop species chosen for the commune of Goworowo—miscanthus and willow—are among the main candidates for lignocellulosic crop plantations in Europe. Their impact on biodiversity in the field is often regarded as beneficial compared to conventional food crops [112]. A similar impact is also expected in the commune as agricultural land was selected for such plantations, while permanent grassland and wasteland with high biodiversity (part of which is protected under the NATURE 2000 program) were excluded.

3.6. Potential of Permanent Grassland

The commune had a large area of permanent grassland—over 3327 ha—which can be used for energy generation despite some limitations connected with animal breeding [23]. The possibilities of the annual hay harvest amounted to 475 Mg of hay (321 Mg from pastures and 154 Mg from meadows) annually with an energy potential of 6543 GJ (4460 GJ from pastures and 2083 GJ from meadows) (Table 4). The difference in biomass yield from pastures and meadows was a consequence of local conditions. The meadows were mainly used extensively, not only at Nature 2000 sites but all over the commune. Therefore, increasing the number of cuts from two to three gave a surplus which could be used as fuel. Many pastures were not used to the full extent, reflecting the situation around the country, where most pastures are not used because of the drop in the farm animal stock [67].

Apart from the considerable benefits of using the hay surplus, certain difficulties are connected with obtaining it. These resources are not often easily accessible (wetlands) and their quality varies, depending on the moisture and mineral content [22]. Since the commune of Goworowo is situated in two NATURE 2000 Special Bird Protection sites—Dolina Dolnej Narwi PLB140014 and Puszcza Biała PLB140007—the grassland in them has to be mowed to maintain the bird populations covered by the Birds Directive. Abandoning the mowing results in natural succession and biocenotic evolution and poses a threat to the objects of protection in the commune. Mowing can prevent it, especially since protective measures include removing the produced biomass [113,114]. The permanent grassland area has been decreasing in recent years, both in the Mazowieckie Voivodship and in Poland [89,115,116]. The hay resources in the commune under study are expected to stay at the same level since they have remained unchanged over the years [23,93].

3.7. Potential of the Landfilled Sludge and Municipal Waste

The analysis has shown that the sludge landfilled in the commune of Goworowo (10 Mg of dry weight) is stored only to recirculate to bioreactors when biogenic elements for microorganisms are absent (Table 4). Therefore, its storage helps to maintain the continuity of the wastewater treatment process. The sludge was therefore excluded as a potential source of solid biomass. There is 165,000 Mg y^{-1} of municipal sludge in Poland intended for energy recovery by combustion [21]. Typical stabilized and dehydrated sludge has a calorific value of approximately 0.5 GJ Mg^{-1} [117], which increases to 11 GJ Mg^{-1} after drying [118]. Any surplus of recirculated sludge from the commune of Goworowo which was not utilized in agriculture can be used for thermal conversion.

The annual production of biodegradable waste in the commune amounted to 709 Mg, but all of it was collected by external companies and transported to municipal waste processing installations (Table 4). In future, determination of the biodegradable waste potential will require an analysis of what part of mixed and separately collected municipal waste is stored despite being intended for processing and the acquisition of data on its amount. Annually, 2,724,000 Mg of waste is used as an energy source in Poland [21]. Potential biodegradable/green waste in the commune could be used for thermal conversion with energy recovery or biogas plant feedstock, increasing the RES use indices.

3.8. Commune Potential

The total energy potential of solid biomass resources in the commune of Goworowo amount to 97,672 GJ y^{-1} (Table 4). The amounts of individual types of solid biomass and their potential varied significantly, which resulted from the local conditions. The solid biomass resources in the commune were equivalent to 4192 Mg of hard coal (the most commonly used heat source in the commune), assuming its calorific value of 23.3 GJ Mg^{-1} (for eco-pea coal) [36] (Figure 3). This shows that the solid biomass potential in the commune of Goworowo is high. Surplus straw accounted for the greatest part and wood from roadside maintenance accounted for the smallest part of the energy potential in the commune (Figure 3).

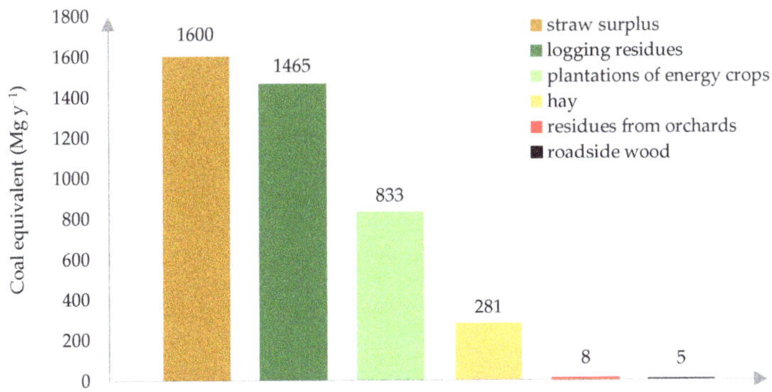

Figure 3. The potential of energy accumulated in solid biomass resources expressed as the coal equivalent (Mg y^{-1}).

Kowalczyk-Juśko et al. [22] determined the energy potential of the Bialski District in the southeast of Poland, where the energy potential of communes ranged from 7690 to 707,933 GJ y^{-1}. However, that study also included solid biomass residues from wood industry facilities. It was not mentioned in the current research as there is no such facility in the commune. Moreover, the Bialski District is one of the three largest districts in Poland [119]. Moreover, 10% of marginal arable soils and permanent grassland were allocated for energy crop plantations in the majority of the communes in the cited research [22], whereas in the current study, it was only 1.5% arable land, all belonging to the poorest quality. However, the current findings for the commune of Goworowo confirm that the smallest territorial units (communes) have large solid biomass resources, which can (and should) be of great importance in satisfying energy needs, especially as fuel in local boilers and for individual recipients.

A high value of solid biomass must also be emphasized as referring to the price of hard coal (eco-pea coal). The retail price of this fuel, depending on its origin and quality, ranged from EUR 146 Mg^{-1} to even more than EUR 225 Mg^{-1} (average PLN/EUR exchange rate in 2020—4.4448/1.0). Assuming the average eco-pea coal price of EUR 180 Mg^{-1}, the value of energy accumulated in solid biomass, referred to as the eco-pea coal price, amounted to EUR 754,487 y^{-1}. Therefore, this is a high value when referred to the scale of the commune of Goworowo, which can potentially be used continuously for many years. It is particularly important from the point of view of the local supply of energy feedstock. There are no fossil fuel resources in the commune of Goworowo, so all the money spent to purchase it flows out of the commune and often out of the country. When biomass is used as an energy feedstock, the money for the biofuel remains in the commune, in local circulation. Owing to this system, the entities producing fuel or energy from biomass have funds for investment and development and can employ new personnel in the RES industry, which contributes to local development. Therefore, biomass use as energy feedstock can result in continuous stimulation for the infrastructure development in rural areas and can help to implement modern technologies of biomass conversion to energy and, in future, to various high-value bioproducts. This issue is extremely important as biomass—as feedstock, and unlike other RES—requires the involvement of many entities to organize the whole logistics chain, from production and acquisition of biomass, through to its storage, warehousing, transport, preparation for technological processes, conversion and final use. Therefore, one can say that—unlike other RES—biomass is a "demanding" fuel. However, paradoxically, this demanding energy feedstock can provide a positive development stimulus for the commune and the region since it engages a considerable workforce.

4. Conclusions

Much research into local biomass potential is needed to connect its management effectively with the development of renewable energy sources. This is of particular importance for further development of research in this domain and the subsequent application of the results in science and practice. Firstly, the current study shows that the smallest administrative unit in Poland—a rural commune—has solid biomass resources with considerable energy and financial potential for use. Secondly, the study also emphasizes the challenges, such as sustainable management of local solid biomass resources, taking into account the protection of biodiversity and the environment, e.g., the role of the space where solid biomass is present as a natural habitat. Thirdly, it was demonstrated that the commune could be a place where energy sources are diversified and where links are created between the economy and ecology by obtaining biomass. Fourthly, research should be continued to determine the technical potential of biomass resources and the link between local development and local community welfare.

Rural areas, such as the commune of Goworowo that has an agriculture and forestry sector in its area, could soon play a major role in producing and using solid biomass. Local communities and fragmented farms, which do not develop on an industrial scale, could be provided with many opportunities. Further research is also necessary in this field. Local communities must be given information on the resources in their area and about their value. They must be informed about effective technologies and their use and cost-effective logistics and management systems.

Moreover, policymakers and local governments should undertake appropriate measures to allow for the most effective use of these local resources. However, this study also has some limitations since it refers to a single commune existing under certain geographic and demographic conditions. Furthermore, it is necessary to have access to reference data to identify specific local features, which may be diverse even for neighboring communes.

Author Contributions: Conceptualization, M.J.S. and P.D.; methodology, M.J.S. and P.D.; validation, M.J.S., E.O.-Z. and M.K.; formal analysis, M.J.S., P.D., M.K. and E.O.-Z.; investigation, M.J.S., P.D., M.K. and E.O.-Z.; data curation, M.J.S. and P.D.; writing—original draft preparation, M.J.S. and P.D.; writing—review and editing, M.J.S., M.K. and E.O.-Z.; visualization, P.D.; supervision, M.J.S.; project administration, M.J.S. All authors have read and agreed to the published version of the manuscript.

Funding: The results presented in this paper were obtained as part of a comprehensive study financed by the University of Warmia and Mazury in Olsztyn, Faculty of Agriculture and Forestry, Department of Genetics, Plant Breeding and Bioresource Engineering (grant No. 30.610.007-110) and has been co-financed by the Interreg Baltic Sea Region Programme 2014–2020 co-funded by the European Regional Development Fund under the project "Unlocking the Potential of Bio-based Value Chains in the Baltic Sea Region" (BalticBiomass4Value-BB4V), No. #R095 and co-financed from the funds of the Ministry of Science and Higher Education under the program "PMW" in the years 2019–2021; No. 5047/INTERREG BSR/2019/2.

Acknowledgments: We would like to thank the staff of the Goworowo Municipal Office for providing the necessary information for the preparation of this manuscript.

Conflicts of Interest: The authors declare no conflict of interest.

References

1. Kluczkowski, A. The use of wood biomass in the regional system of renewable energy sources as a chance for the region. *J. Vasyl Stefanyk Precarpathian Natl. Univ.* **2017**, *4*, 67–73. [CrossRef]
2. Rosales-Calderon, O.; Arantes, V. A review on commercial-scale high-value products that can be produced alongside cellulosic ethanol. *Biotechnol. Biofuels* **2019**, *12*, 240. [CrossRef]
3. Brodny, J.; Tutak, M.; Saki, S.A. Forecasting the structure of energy production from renewable energy sources and biofuels in Poland. *Energies* **2020**, *13*, 2539. [CrossRef]
4. Marks-Bielska, R.; Bielski, S.; Pik, K.; Kurowska, K. The importance of renewable energy sources in Poland's energy mix. *Energies* **2020**, *13*, 4624. [CrossRef]
5. Brodny, J.; Tutak, M. Analyzing similarities between the european union countries in terms of the structure and volume of energy production from renewable energy sources. *Energies* **2020**, *13*, 913. [CrossRef]
6. Sefeedpari, P.; Pudełko, R.; Jędrejek, A.; Kozak, M.; Borzęcka, M. To what extent is manure produced, distributed, and potentially available for bioenergy? A step toward stimulating circular bio-economy in Poland. *Energies* **2020**, *13*, 6266. [CrossRef]
7. Statistics Poland. Energy from Renewable Sources in 2018. 2019. Available online: https://stat.gov.pl/obszary-tematyczne/srodowisko-energia/energia/energia-ze-zrodel-odnawialnych-w-2018-roku,10,2.html (accessed on 12 February 2021).
8. Eurostat. Energy Data—2020 Edition. 2020. Available online: https://ec.europa.eu/eurostat/web/products-statistical-books/-/KS-HB-20-001 (accessed on 13 February 2021).
9. Statistics Poland. Energy Statistics in 2018 and 2019. 2020. Available online: https://stat.gov.pl/obszary-tematyczne/srodowisko-energia/energia/gospodarka-paliwowo-energetyczna-w-latach-2018-i-2019,4,15.html (accessed on 12 February 2021).
10. Eurostat. 2019. Available online: https://ec.europa.eu/eurostat/data/statistics-a-z/abc (accessed on 2 April 2021).
11. Statistics Poland. Energy from Renewable Sources in 2019. 2020. Available online: https://stat.gov.pl/obszary-tematyczne/srodowisko-energia/energia/energia-ze-zrodel-odnawialnych-w-2019-roku,3,14.html (accessed on 2 April 2021).
12. Stolarski, M.J.; Warmiński, K.; Krzyżaniak, M.; Olba-Zięty, E.; Stachowicz, P. Energy consumption and heating costs for a detached house over a 12-year period—Renewable fuels versus fossil fuels. *Energy* **2020**, *204*, 117952. [CrossRef]
13. Vezzoli, C.; Ceschin, F.; Osanjo, L.; M'Rithaa, M.K.; Moalosi, R.; Nakazibwe, V.; Diehl, J.C. Distributed/decentralised renewable energy systems. In *Designing Sustainable Energy for All*, 1st ed.; Vezzoli, C., Ceschin, F., Osanjo, L., M'Rithaa, M.K., Moalosi, R., Nakazibwe, V., Diehl, J.C., Eds.; Springer: Cham, Switzerland, 2018; pp. 23–39.
14. Kluczkowski, A.; Wyrostkiewicz, M. Circular economy as an important subject of environmental education in the era of energy demand. *J. Vasyl Stefanyk Precarpathian Natl. Univ.* **2018**, *5*, 88–94. [CrossRef]
15. Statistics Poland. Area and Population in the Territorial Profile in 2020. 2020. Available online: https://stat.gov.pl/obszary-tematyczne/ludnosc/ludnosc/powierzchnia-i-ludnosc-w-przekroju-terytorialnym-w-2020-roku,7,17.html (accessed on 14 March 2021).
16. Kaczmarek, T. Communal administrative division in the light of 25 years of operation of local government in Poland. *Political Sci. Rev.* **2016**, *1*, 63–80.
17. Rokicki, T.; Perkowska, A.; Klepacki, B.; Bórawski, P.; Bełdycka-Bórawska, A.; Michalski, K. Changes in energy consumption in agriculture in the EU countries. *Energies* **2021**, *14*, 1570. [CrossRef]
18. Rocha-Meneses, L.; Bergamo, T.F.; Kikas, K. Potential of cereal-based agricultural residues available for bioenergy production. *DataBrief* **2019**, *23*, 103829. [CrossRef] [PubMed]
19. Steubing, B.; Zah, R.; Waeger, P.; Ludwig, C. Bioenergy in Switzerland: Assessing the domestic sustainable biomass potential. *Renew. Sustain. Energy Rev.* **2010**, *14*, 2256–2265. [CrossRef]
20. Avcıoğlu, A.O.; Dayıoğlu, M.A.; Türker, U. Assessment of the energy potential of agricultural biomass residues in Turkey. *Renew. Energy* **2019**, *138*, 610–619. [CrossRef]
21. Stolarski, M.J.; Warmiński, K.; Krzyżaniak, M.; Olba-Zięty, E.; Akincza, M. Bioenergy technologies and biomass potential vary in Northern European countries. *Renew. Sustain. Energy Rev.* **2020**, *133*, 110238. [CrossRef]
22. Kowalczyk-Juśko, A.; Listosz, A.; Mazur, K.; Macig, M.; Pochwatka, P.; Mazur, A. The state and the perspectives of the eco-energy infrastructure development in Biała Podlaska County (Poland). Part II Estimation of solid biomass resources for energy purposes. *E3S Web Conf.* **2020**, *171*, 01005. [CrossRef]
23. Mayor of the Goworowo Commune. Report on the State of the Goworowo Commune. 2020. Available online: http://goworowo.pl/wp-content/uploads/2020/05/Goworowo_raport_o_stanie_gminy_2019.pdf (accessed on 21 January 2021).
24. Poland's Open Data Portal. Open Spatial Data Geoportal. 2021. Available online: https://polska.e-mapa.net/ (accessed on 2 April 2021).
25. Statistics Poland. Statistical Vademecum of a Local Government. 2020. Available online: https://warszawa.stat.gov.pl/statystyczne-vademecum-samorzadowca/ (accessed on 12 March 2021).
26. Mayor of the Goworowo Commune. Environmental Protection Program for the Goworowo Commune. 2019. Available online: http://goworowo.pl/wp-content/uploads/2019/09/Program-Ochrony-%C5%9Arodowiska.pdf (accessed on 20 January 2021).
27. Mayor of the Goworowo Commune. *The Basic Ecophysiographic Study Prepared for the Needs of Study of the Conditions and Directions of Spatial Development in the Commune of Goworowo (Local Document Provided by the Goworowo Commune Office on Request)*; Goworowo Commune Office: Goworowo, Poland, 2013.

28. Marshal's Office of the Mazowieckie Voivodeship. Spatial Information System for the Mazowieckie Voivodeship. 2021. Available online: https://msip.wrotamazowsza.pl/msip/Full.aspx (accessed on 13 January 2021).
29. Statistics Poland. Land Use and Sown Area in 2019. 2020. Available online: https://stat.gov.pl/obszary-tematyczne/rolnictwo-lesnictwo/rolnictwo/uzytkowanie-gruntow-i-powierzchnia-zasiewow-w-2019-roku,8,15.html (accessed on 7 December 2020).
30. Marshal's Office of the Mazowieckie Voivodeship. A Summary of the Data from the Land and Building Register of the Mazowieckie Voivodship. 2019. Available online: https://gis.wrotamazowsza.pl/bezplatne-dane/ (accessed on 7 December 2020).
31. Minister of Investment and Development. The Ordinance on Land and Building Register. 2019. Available online: https://isap.sejm.gov.pl/isap.nsf/DocDetails.xsp?id=WDU20190000393 (accessed on 7 December 2020).
32. Statistics Poland. The Resulting Estimate of the Main Agricultural and Horticultural Crops. 2020. Available online: https://stat.gov.pl/obszary-tematyczne/rolnictwo-lesnictwo/uprawy-rolne-i-ogrodnicze/wynikowy-szacunek-glownych-ziemioplodow-rolnych-i-ogrodniczych-w-2020-roku,5,19.html (accessed on 11 January 2021).
33. Warmia and Mazury Agricultural Advisory Center. Review of Polish Maize Varieties. 2018. Available online: https://wmodr.pl/files/tnUefXPeFS3uEU5QOPGyrGyiNp0JhC3jzRTxCAJz.pdf (accessed on 11 January 2021).
34. Hryniewicz, M.; Grzybek, A. Available straw surplus for use for energy purposes in 2016. *Probl. Agric. Eng.* **2017**, *3*, 15–31.
35. Bieranowski, J.; Olkowski, J. The concept of an energy self-sufficient farm system compatible with sustainable development in a selected region in Poland. *Pol. J. Environ. Stud.* **2016**, *25*, 529–544. [CrossRef]
36. Stolarski, M.J.; Rybczyńska, B.; Krzyżaniak, M.; Lajszner, W.; Graban, Ł.; Peni, D.; Bordiean, A. Thermophysical properties and elemental composition of agricultural and forest solid biofuels versus fossil fuels. *J. Elem.* **2019**, *24*, 1215–1228. [CrossRef]
37. Roman, M.; Roman, K. Energy use of wood waste from fruit trees on the example of an agritourism farm. *Ann. Pol. Assoc. Agric. Agribus. Econ.* **2018**, *XX*, 158–161.
38. Dyjakon, A.; den Boer, J.; Bukowski, P.; Adamczyk, F.; Frąckowiak, P. Wooden biomass potential from apple orchards in Poland. *Wood Res. Pap. Rep. Announc.* **2016**, *59*, 73–86.
39. Cichy, W.; Witczak, M.; Walkowiak, M. Fuel properties of woody biomass from pruning operations in fruit orchards. *BioResources* **2017**, *12*, 6458–6470. [CrossRef]
40. Swithenbank, J.; Chen, Q.; Zhang, X.; Sharifi, V.; Pourkashanian, M. Wood would burn. *Biomass Bioenergy* **2011**, *35*, 999–1007. [CrossRef]
41. Statistics Poland. Statistical Yearbook of Forestry. 2020. Available online: https://stat.gov.pl/obszary-tematyczne/roczniki-statystyczne/roczniki-statystyczne/rocznik-statystyczny-lesnictwa-2020,13,3.html (accessed on 5 March 2021).
42. Forest District Ostrołęka. Forest Management Plan. 2012. Available online: https://bip.lasy.gov.pl/pl/bip/dg/rdlp_olsztyn/nadl_ostroleka/plan_urzadzania_lasu/czesc_opisowa (accessed on 5 March 2021).
43. Forest District Pułtusk. Forest Management Plan. 2014. Available online: https://bip.lasy.gov.pl/pl/bip/dg/rdlp_warszawa/nadl_pultusk/plan_urzadzania_lasu/czesc_opisowa (accessed on 5 March 2021).
44. Forest District Wyszków. Forest Management Plan. 2017. Available online: https://bip.lasy.gov.pl/pl/bip/dg/rdlp_warszawa/nadl_wyszkow/plan_urzadzania_lasu/czesc_opisowa (accessed on 5 March 2021).
45. Gendek, A.; Węzyk, P.; Moskalik, T. Share and accuracy of estimation of logging residues in the total volume of harvested timber. *Sylwan* **2018**, *162*, 679–687.
46. Różański, H.; Jabłoński, K. Prospects for forest biomass harvesting for energy purposes in Poland. *J. Civ. Eng. Environ. Archit.* **2015**, *XXXII*, 351–358. [CrossRef]
47. Order of the General Director of State Forests. Technical Conditions—Division, Terminology and Symbols Used in the Turnover of Wood Raw Material. 2019. Available online: http://drewno.zilp.lasy.gov.pl/drewno/Normy/ (accessed on 6 March 2021).
48. National Forest Holding the State Forests. Report on the Condition of Forests in Poland. 2020. Available online: https://www.lasy.gov.pl/pl/informacje/publikacje/informacje-statystyczne-i-raporty/raport-o-stanie-lasow/raport-o-stanie-lasow.pdf/view (accessed on 6 March 2021).
49. Abolins, J.; Gravitis, J. Sustainable supply of energy from biomass. *Latv. J. Phys. Tech. Sci.* **2010**, *47*, 57–63. [CrossRef]
50. Voinov, A.; Arodudu, O.; Duren, I.; Morales, J.; Qin, L. Estimating the potential of roadside vegetation for bioenergy production. *J. Clean. Prod.* **2015**, *102*, 213–225. [CrossRef]
51. Oniszk-Popławska, A.; Krasuska, E. *Energy from Organic Waste*, 1st ed.; Research and Innovation Centre Pro-Akademia Branch of Polish Academy of Sciences: Łódź, Poland, 2014; pp. 1–48.
52. Act on Nature Conservation. Act of 16 April 2004, Journal of Laws 2020 Item 55. Available online: https://isap.sejm.gov.pl/isap.nsf/DocDetails.xsp?id=WDU20040920880 (accessed on 22 January 2021).
53. Jezierska-Thöle, A.; Rudnicki, R.; Kluba, M. Development of energy crops cultivation for biomass production in Poland. *Renew. Sustain. Energy Rev.* **2016**, *62*, 534–545. [CrossRef]
54. Blanco-Canqui, H. Growing dedicated energy crops on marginal lands and ecosystem services. *Soil Sci. Soc. Am. J.* **2016**, *80*, 845–858. [CrossRef]
55. Dubis, B.; Jankowski, K.J.; Załuski, D.; Sokólski, M. The effect of sewage sludge fertilization on the biomass yield of giant miscanthus and the energy balance of the production process. *Energy* **2020**, *206*, 118189. [CrossRef]
56. Dubis, B.; Jankowski, K.J.; Załuski, D.; Bórawski, P.; Szempliński, W. Biomass production and energy balance of Miscanthus over a period of 11 years: A case study in a large-scale farm in Poland. *GCB Bioenergy* **2019**, *11*, 1187–1201. [CrossRef]

57. Stolarski, M.J.; Śnieg, M.; Krzyżaniak, M.; Tworkowski, J.; Szczukowski, S. Short rotation coppices, grasses and other herbaceous crops: Productivity and yield energy value versus 26 genotypes. *Biomass Bioenergy* **2018**, *119*, 109–120. [CrossRef]
58. Mehmood, M.A.; Ibrahim, M.; Rashid, U.; Nawaz, M.; Ali, S.; Hussain, A.; Gull, M. Biomass production for bioenergy using marginal lands. *Sustain. Prod. Consum.* **2017**, *9*, 3–21. [CrossRef]
59. Matyka, M.; Kuś, J. Influence of Soil Quality for Yielding and Biometric Features of Miscanthus x Giganteus. *Pol. J. Environ. Stud.* **2016**, *25*, 213–219. [CrossRef]
60. Shepherd, A.; Clifton-Brown, J.; Kam, J.; Buckby, S.; Hastings, A. Commercial experience with miscanthus crops: Establishment, yields and environmental observations. *GCB Bioenergy* **2020**, *12*, 510–523. [CrossRef]
61. Mantziaris, S.; Iliopoulos, C.; Theodorakopoulou, I.; Petropoulou, E. Perennial energy crops vs. durum wheat in low input lands: Economic analysis of a Greek case study. *Renew. Sustain. Energy Rev.* **2017**, *80*, 789–800. [CrossRef]
62. Pray, T.J.; Guidi Nissim, W.; St-Arnaud, M.; Labrecque, M. Investigating the effect of a mixed mycorrhizal inoculum on the productivity of biomass plantation willows grown on marginal farm land. *Forests* **2018**, *9*, 185. [CrossRef]
63. Matyka, M.; Radzikowski, P. Productivity and biometric characteristics of 11 varieties of willow cultivated on marginal soil. *Agriculture* **2020**, *10*, 616. [CrossRef]
64. Szczukowski, S.; Stolarski, M.; Tworkowski, J. Yield of willow biomass produced by the Eco-Salix system. *Fragm. Agron.* **2011**, *28*, 104–115.
65. Jurczuk, S.; Chrzanowski, S.; Jaszczyński, J. Yielding of the energy willow under differentiated soil-water conditions. *Probl. Agric. Eng.* **2010**, *18*, 113–121.
66. Stolarski, M.J.; Krzyżaniak, M.; Warmiński, K.; Olba-Zięty, E.; Penni, D.; Bordiean, A. Energy efficiency indices for lignocellulosic biomass production: Short rotation coppices versus grasses and other herbaceous crops. *Ind. Crop. Prod.* **2019**, *135*, 10–20. [CrossRef]
67. Igliński, B.; Buczkowski, R.; Cichosz, M. Biogas production in Poland—Current state, potential and perspectives. *Renew. Sustain. Energy Rev.* **2015**, *50*, 686–695. [CrossRef]
68. Melts, I.; Heinsoo, K.; Nurk, L.; Pärn, L. Comparison of two different bioenergy production options from late harvested biomass of Estonian semi-natural grasslands. *Energy* **2013**, *61*, 6–12. [CrossRef]
69. Ketzer, D.; Rösch, C.; Haase, M. Assessment of sustainable Grassland biomass potentials for energy supply in Northwest Europe. *Biomass Bioenergy* **2017**, *100*, 39–51. [CrossRef]
70. Boob, M.; Elsaesser, M.; Thumm, U.; Hartung, J.; Lewandowski, I. Harvest time determines quality and usability of biomass from lowland hay meadows. *Agriculture* **2019**, *9*, 198. [CrossRef]
71. Jarosz, Z. Energy potential of agricultural crops biomass and their use for energy purposes. *Probl. World Agric.* **2017**, *17*, 81–92.
72. Igliński, B.; Cichosz, M.; Skrzatek, M.; Buczkowski, R. Technical potential of waste biomass for energy in Poland. *Eng. Prot. Environ.* **2018**, *22*, 109–118. [CrossRef]
73. Lipińska, H.; Kurzepa, K.; Kościk, S. Economically unexploited permanent grassland in Chełm County as a potential source of biomass production for energy generation. *Grassl. Sci. Pol.* **2017**, *20*, 107–119.
74. Heinsoo, K.; Melts, I.; Holm, B. The potential of Estonian semi-natural grasslands for bioenergy production. *Agric. Ecosyst. Environ.* **2010**, *137*, 86–92. [CrossRef]
75. Agricultural Advisory Center in Brwinów. Rational Use of Grassland on an Ecological Farm. 2010. Available online: https://www.cdr.gov.pl/images/wydawnictwa/2010/2010-racjonalne-wykorzystanie-uzytkow-zielonych-w-gospodarstwie-ekologicznym.pdf (accessed on 16 January 2021).
76. von Cossel, M.; Bauerle, A.; Boob, M.; Thumm, U.; Elsaesser, M.; Lewandowski, I. The performance of mesotrophic arrhenatheretum grassland under different cutting frequency regimes for biomass production in southwest Germany. *Agriculture* **2019**, *9*, 199. [CrossRef]
77. Djomo, S.N.; Knudsen, M.T.; Martinsen, L.; Andersen, M.S.; Ambye-Jensen, M.; Møller, H.B.; Hermansen, J.E. Green proteins: An energy-efficient solution for increased self-sufficiency in protein in Europe. *Biofuels Bioprod. Biorefining* **2020**, *14*, 605–619. [CrossRef]
78. Chalamoński, M.; Szymczak, M. The dried sludge as alternative fuel. *J. Civ. Eng. Environ. Archit.* **2017**, *XXXIV*, 35–42.
79. Szydełko, A.; Pawlak, M. Combustion and co-incineration of sewage sludge with mineral additives in terms of chlorine binding and chloride corrosion. *Energetic Noteb.* **2014**, *1*, 165–174.
80. Werle, S. Thermal methods of sewage sludge management. Energy from wastewater. *Therm. Prof. Energy* **2010**, *9*, 28–32.
81. Czyżyk, F.; Kozdraś, M. Chemical properties and composting of sludge from a rural wastewater treatment plant. *Water Environ. Rural Areas* **2004**, *4*, 559–569.
82. Ministry of the Environment. The Strategy for Dealing with Municipal Sewage Sludge for the Period 2019–2022. 2018. Available online: https://www.gov.pl/web/srodowisko/planowanie-gospodarki-odpadami (accessed on 8 March 2021).
83. Statistics Poland. Local Data Bank. 2021. Available online: https://bdl.stat.gov.pl/BDL/dane/teryt/jednostka (accessed on 8 March 2021).
84. Mayor of the City of Ostrołęka. Environmental Protection Program of the City of Ostrołęka for 2017–2020 with a Perspective until 2024. 2017. Available online: https://bip.um.ostroleka.pl/artykul/117/4275/program-ochrony-srdoowiska-na-lata-2017-2020-z-perspektywa-do-2024-roku (accessed on 8 March 2021).

85. Mayor of the Goworowo Commune. Analysis of the State of Municipal Waste Management in the Commune. 2020. Available online: http://goworowo.pl/gmina/gospodarka-odpadami (accessed on 9 March 2021).
86. The minister of Environment. The Ordinance on Landfill Limitation Levels for Biodegradable Municipal Waste Mass. 2017. Available online: http://isap.sejm.gov.pl/isap.nsf/DocDetails.xsp?id=WDU20170002412 (accessed on 9 March 2021).
87. Cuong, T.T.; Le, H.A.; Khai, N.M.; Hung, P.A.; Linh, L.T.; Thanh, N.V.; Tri, N.D.; Huan, N.X. Renewable energy from biomass surplus resource: Potential of power generation from rice straw in Vietnam. *Sci. Rep.* **2021**, *11*, 792. [CrossRef]
88. Hansen, J.H.; Hamelin, L.; Taghizadeh-Toosi, A.; Olesen, J.E.; Wenzel, E. Agricultural residues bioenergy potential that sustain soil carbon depends on energy conversion pathways. *GCB Bioenergy* **2020**, *12*, 1002–1013. [CrossRef]
89. Statistics Poland. Statistical Yearbook of Agriculture. 2020. Available online: https://stat.gov.pl/obszary-tematyczne/roczniki-statystyczne/roczniki-statystyczne/rocznik-statystyczny-rolnictwa-2019,6,13.html (accessed on 24 March 2021).
90. Ministry of Climate and Environment. National Energy and Climate Plan for 2021–2030. 2019. Available online: https://www.gov.pl/web/klimat/krajowy-plan-na-rzecz-energii-i-klimatu (accessed on 4 December 2020).
91. Dyjakon, A. The influence of apple orchard management on energy performance and pruned biomass harvesting for energetic applications. *Energies* **2019**, *12*, 632. [CrossRef]
92. Dyjakon, A.; den Boer, J.; Gebresenbet, G.; Bosona, T.; Adamczyk, F. Economic analysis of the collection and transportation of pruned branches from orchards for energy productions. *Wood Res. Pap. Rep. Announc.* **2020**, *63*, 125–140.
93. Mayor of the Goworowo Commune. Study of the Conditions and Directions of Spatial Development. 2017. Available online: http://bip.goworowo.pl/?k=247 (accessed on 22 January 2021).
94. Marshal's Office of the Mazowieckie Voivodeship. A Summary of the Data from the Land and Building Register of the Mazowieckie Voivodship. 2017. Available online: https://gis.wrotamazowsza.pl/bezplatne-dane/ (accessed on 19 March 2021).
95. Marshal's Office of the Mazowieckie Voivodeship. A Summary of the Data from the Land and Building Register of the Mazowieckie Voivodship. 2018. Available online: https://gis.wrotamazowsza.pl/bezplatne-dane/ (accessed on 19 March 2021).
96. Marshal's Office of the Mazowieckie Voivodeship. A summary of the Data from the Land and Building Register of the Mazowieckie Voivodship. 2020. Available online: https://gis.wrotamazowsza.pl/bezplatne-dane/ (accessed on 19 March 2021).
97. Statistics Poland. Statistical Yearbook of Agriculture. 2019. Available online: https://stat.gov.pl/obszary-tematyczne/roczniki-statystyczne/roczniki-statystyczne/rocznik-statystyczny-rolnictwa-2018,6,12.html (accessed on 24 March 2021).
98. Dyjakon, A.; den Boer, J.; Szumny, A.; den Boer, E. Local energy use of biomass from apple orchards—An LCA study. *Sustainability* **2019**, *11*, 1604. [CrossRef]
99. Verkerk, P.J.; Fitzgerald, J.B.; Datta, P.; Dees, M.; Hengeveld, G.H.; Lindner, M.; Zudin, S. Spatial distribution of the potential forest biomass availability in Europe. *For. Ecosyst.* **2019**, *6*, 5. [CrossRef]
100. Holeksa, J.; Kapusta, P.; Budziakowska-Kubik, E.; Izworska, K.; Kurek, P.; Piechnik, Ł.; Szarek-Łukaszewska, G.; Wojterska, M.; Zielonka, T.; Żywiec, M. Stock of deadwood in the Niepołomice Forest as a result of long-term forest use and short-term protection within the Natura 2000 network. *Fragm. Florist. Et Geobot. Pol.* **2020**, *27*, 119–139.
101. Sierota, Z.; Grodzki, W.; Szczepkowski, A. Abiotic and biotic disturbances affecting forest health in Poland over the past 30 years: Impacts of climate and forest management. *Forests* **2019**, *10*, 75. [CrossRef]
102. Kaliszewski, A. Forest policy goals in Poland in light of the current forestry aims in Europe Part 3. European priorities for the forest policy in Polish programmes and strategies. *For. Res. Pap.* **2018**, *79*, 211–227. [CrossRef]
103. Van Meerbeek, V.; Ottoy, S.; De Meyer, A.; Van Schaeybroeck, T.; Van Orshoven, J.; Muys, B.; Hermy, M. The bioenergy potential of conservation areas and roadsides for biogas in an urbanized region. *Appl. Energy* **2015**, *154*, 742–751. [CrossRef]
104. Rocchi, L.; Paolotti, L.; Fagioli, F.F.; Boggia, A. Production of insulating panel from pruning remains: An economic and environmental analysis. *Energy Procedia* **2018**, *147*, 145–153. [CrossRef]
105. Bautista, S.; Camargo, M.; Bachmann, C. Sustainable roadside management from an innovative approach to ecosystem services and bioenergy generation. In Proceedings of the 2020 IEEE International Conference on Engineering, Technology and Innovation (ICE/ITMC), Cardiff, UK, 15–17 June 2020.
106. O'Sullivan, O.S.; Holt, A.R.; Warren, P.H.; Evans, K.L. Optimising UK urban road verge contributions to biodiversity and ecosystem services with cost-effective management. *J. Environ. Manag.* **2017**, *191*, 162–171. [CrossRef] [PubMed]
107. Meyer, A.K.P.; Ehimen, E.A.; Holm-Nielsen, J.B. Bioenergy production from roadside grass: A case study of the feasibility of using roadside grass for biogas production in Denmark. *Resour. Conserv. Recycl.* **2014**, *93*, 124–133. [CrossRef]
108. Mudryk, K.; Jewiarz, M.; Wróbel, M.; Niemiec, M.; Dyjakon, A. Evaluation of urban tree leaf biomass-potential, physico-mechanical and chemical parameters of raw material and solid biofuel. *Energies* **2021**, *14*, 818. [CrossRef]
109. Ciria, C.S.; Sanz, M.; Carrasco, J.; Ciria, P. Identification of arable marginal lands under rainfed conditions for bioenergy purposes in Spain. *Sustainability* **2019**, *11*, 1833. [CrossRef]
110. Fernando, A.L.; Costa, J.; Barbosa, B.; Monti, A.; Rettenmaier, N. Environmental impact assessment of perennial crops cultivation on marginal soils in the Mediterranean Region. *Biomass Bioenergy* **2018**, *11*, 174–186. [CrossRef]
111. Pandey, V.C.; Bajpai, O.; Singh, N. Energy crops in sustainable phytoremediation. *Renew. Sustain. Energy Rev.* **2016**, *54*, 58–73. [CrossRef]
112. Dauber, J.; Miyake, S. To integrate or to segregate food crop and energy crop cultivation at the landscape scale? Perspectives on biodiversity conservation in agriculture in Europe. *Energy Sustain. Soc.* **2016**, *6*, 25. [CrossRef]

113. Regional Director of Environmental Protection in Warsaw and Regional Director of Environmental Protection in Białystok. The Plan of Protection Tasks for the Natura 2000 PLB140014. 2016. Available online: http://crfop.gdos.gov.pl/CRFOP/widok/viewnatura2000.jsf?fop=PL.ZIPOP.1393.N2K.PLB140014.B (accessed on 19 March 2021).
114. Regional Director of Environmental Protection in Warsaw. The Plan of Protection Tasks for the Natura 2000 PLB140007. 2016. Available online: http://crfop.gdos.gov.pl/CRFOP/widok/viewnatura2000.jsf?fop=PL.ZIPOP.1393.N2K.PLB140007.B (accessed on 19 March 2021).
115. Golimowski, W.; Butlewski, K.; Gracz, W.; Marcinkowski, D.; Konieczny, R. Energy and environmental potential of grasslands in Poland. In *Renewable Energy Sources: Engineering, Technology, Innovation*, 1st ed.; Mundryk, K., Werle, K., Eds.; Springer: Cham, Switzerland, 2018; pp. 107–117.
116. Statistics Poland. Statistical Yearbook of Agriculture. 2021. Available online: https://stat.gov.pl/obszary-tematyczne/roczniki-statystyczne/roczniki-statystyczne/rocznik-statystyczny-rolnictwa-2020,6,14.html (accessed on 28 March 2021).
117. Duda, J.; Wasilewski, M. Innovative technology for disposal of sewage sludge. In *Innovation in Management and Production Engineering*, 1st ed.; Knosala, R., Ed.; Publishing House of the Polish Production Management Society: Opole, Poland, 2014; pp. 68–77.
118. Czop, M.; Jarząbkowska, N. Testing fuel properties of municipal sewage sludge. *Arch. Waste Manag. Environ. Prot.* **2016**, *18*, 51–62.
119. Listosz, A.; Kowalczyk-Juśko, A.; Mazur, A.; Jóźwiakowski, K.; Gizińska-Górna, M.; Pytka, A.; Marzec, M. The state and the perspectives of the ecoenergy infrastructure development in Biała Podlaska county p. i. state analysis. *Water-Environ. Rural Areas* **2017**, *17*, 81–93.

energies

MDPI

Article

Sources of Greenhouse Gas Emissions in Agriculture, with Particular Emphasis on Emissions from Energy Used

Piotr Gołasa [1], Marcin Wysokiński [1], Wioletta Bieńkowska-Gołasa [1], Piotr Gradziuk [2], Magdalena Golonko [1], Barbara Gradziuk [3], Agnieszka Siedlecka [4] and Arkadiusz Gromada [1,*]

1 Institute of Economics and Finance, Warsaw University of Life Sciences, 02-787 Warsaw, Poland; piotr_golasa@sggw.edu.pl (P.G.); marcin_wysokinski@sggw.edu.pl (M.W.); wioletta_bienkowska@sggw.edu.pl (W.B.-G.); magdalena_golonko@sggw.edu.pl (M.G.)
2 Poland Economic Modelling Department, Institute of Rural and Agricultural Development, Polish Academy of Sciences, 00-330 Warsaw, Poland; pgradziuk@irwirpan.waw.pl
3 Poland Department of Management and Marketing, Faculty of Agrobioengineering, University of Life Sciences in Lublin, 22-033 Lublin, Poland; barbara.gradziuk@up.lublin.pl
4 Institute of Economics, Pope John Paul II State School of Higher Education in Biała Podlaska, 21-500 Biała Podlaska, Poland; a.siedlecka@dydaktyka.pswbp.pl
* Correspondence: arkadiusz_gromada@sggw.edu.pl

Citation: Gołasa, P.; Wysokiński, M.; Bieńkowska-Gołasa, W.; Gradziuk, P.; Golonko, M.; Gradziuk, B.; Siedlecka, A.; Gromada, A. Sources of Greenhouse Gas Emissions in Agriculture, with Particular Emphasis on Emissions from Energy Used. *Energies* **2021**, *14*, 3784. https://doi.org/10.3390/en14133784

Academic Editors: Mejdi Jeguirim and Rajender Gupta

Received: 11 May 2021
Accepted: 22 June 2021
Published: 23 June 2021

Abstract: The relationship between agriculture and climate change is two-sided. Agriculture is the branch of the economy most affected by the ongoing processes. It is also a large emitter of greenhouse gases and there are more and more voices about the need to reduce emissions. The purpose of the study was, based on FADN (Farm Accountancy Data Network) data, to determine the structure of greenhouse gas emissions in farms and to identify types of farms where it is possible to reduce GHG (greenhouse gas) emissions through better energy use. The emission volume was determined on the basis of the IPCC (Intergovernmental Panel on Climate Change) methodology modified for the FADN data. The emissions related to the production of energy were found to be of minor importance compared to other emission sources. Only in the horticultural crop type is the emission from the Energy section the dominant stream of GHG emission. The greatest emissions come from livestock production. Therefore, the emphasis on reducing emissions should not be placed on the Energy sector because, except for the type of horticultural farm, there is not much potential for reduction. The introduction of taxes for GHG emissions at the level of 27.31 EUR/t would reduce farm income from 21% for the type of field crops to 40% for the type of herbivorous animals. The exception is low-emission permanent crops, where the decrease in income would be only 3.85%.

Keywords: GHG; agriculture; energy consumption; farms; FADN

1. Introduction

Over the past several hundred years, human activities have had a huge, mostly negative, impact on the environment. As a result, the area of forests was reduced, biodiversity was reduced, species died out, and many harmful substances were introduced into the environment. However, in the opinion of experts, the main threat to the environment is the climate change caused by anthropogenic heating of the atmosphere, as a result of the increasing concentration of greenhouse gases, mainly CO_2.

It is worth emphasizing that the concept of the greenhouse effect and climate change caused by GHG emissions is not new [1–3]. Pioneering scientific works appeared as early as the end of the 19th century [4]. After the Second World War, there was a breakthrough in climate research [5]. There is now an almost full scientific consensus that we are dealing with rapid climate change and that people are responsible for it [6,7]. In recognized scientific journals, one can find publications that indicate that many positive feedback loops were activated in the world, which resulted in the violation of the so-called tipping points. This could mean that climate change will be rapid, over decades, not linearly as previously thought, but abruptly [8–13]. The environmental, social, and economic impacts can be extremely severe in such unpredictable changes.

Agriculture is of particular importance in terms of climate change. The relationship between agriculture and climate change is two-sided. Agriculture is a major emitter of greenhouse gases. The conducted research shows that farms are responsible for approximately 16–27% of all anthropogenic emissions [14]. Emissions in agriculture take place at every stage of production, from seed preparation to harvesting and storage of finished products [15]. Agriculture is also the sector of the economy most affected by the ongoing processes, which requires large-scale adaptation measures [16]. For most areas of the world, climate change is a growing problem in ensuring an adequate level of food production for an ever-growing world population due to declining yields [17] and rising food prices [18–20]. This is evidenced by the value of the so-called transferable stocks of cereals (which are the main food product), determining the level of food security, which fell from 74 days in 2002 to 54 days in 2011 [21]. The amount of available food varies greatly between regions, and its shortages are particularly visible in the poorest regions of the world [22]. In terms of the energy value of food, 870 million people go hungry worldwide. The worst situation is in the sub-Saharan region, where almost 30% of the population does not have enough food, and, in South Asia, where this situation affects 300 million people [23]. The situation related to the climate crisis is exacerbated by the COVID-19 pandemic [24]. The reduction of agricultural production is directly caused by the fact that climate change causes:

- changing weather patterns, reducing rainfall in many regions of the world. Where rainfall is constant, its nature changes from long-term rainfall to long periods of drought, interrupted by storm rain,
- much more frequent occurrence of extreme phenomena, unfavourable for agriculture: storms, hail, frosts,
- the emergence of new species of pests, diseases that have not been encountered so far, do not have natural enemies [25],
- periods of extremely high temperatures, dangerous for crops and livestock. They also reduce the productivity of human labour, making it impossible at certain times.

Apart from these problems, activities to reduce GHG emissions turn out to be another risk factor for agriculture. The high emissivity of agriculture is becoming a subject of political and social discussion. This is related to a wider issue, such as achieving, by 2050, climate neutrality by the EU-zero net emissions [26].

Modern agriculture is dependent on external industrial energy sources. Fossil fuels and electricity have become an indispensable element of modern agricultural production. They are used directly to power machines and indirectly for their construction, extraction of mineral fertilizers, or the synthesis of nitrogen compounds.

The dominant role in this respect is played by non-renewable energy sources (fossil fuels), which contribute to the emission of greenhouse gases and, consequently, the degradation of the natural environment. Therefore, it becomes obvious to strive to improve the efficiency of energy use and to change the structure of its sources [27].

Taking into account the total dependence of agriculture on fossil fuels, which are a significant source of greenhouse gas emissions, research was undertaken on GHG emissions from energy inputs used in agricultural production. The main purpose of the study was therefore to assess the size and structure of greenhouse gas emissions from energy carriers used in farms of various production directions and then to indicate the possibility of reducing them.

2. Background

The main cause of climate change is the high consumption of energy produced by burning fossil fuels and the excessive development of transport. This sector is responsible for 75% of EU emissions. It is worth noting the evolution of views on the availability and use of fossil fuels. Fifty years ago, it was thought that the diminishing availability of fossil fuels would force a switch to renewable resources [28,29]. Currently, there is a clear trend in the development of renewable energy related to the fight against climate change. Thus, the availability of fossil fuels is less of a problem than predicted, while the question of their negative impact on the environment has turned out to be more serious.

Between 1950 and 1984, there was a "Green Revolution" which increased the grain yield by 250%. However, this increase required a multiple increase in energy inputs in agriculture, even 50 times [30]. Only rough calculations can be made to trace the increase in direct and indirect use of fossil fuels and electricity in modern agriculture. In the 20th century, when the world population increased 3.7 times and the inhabited area increased by about 40%, the energy input increased from 0.1 EJ to almost 13 EJ. As a result, in 2000, on average, about 90 times more energy was used per hectare of arable land than in 1900 [31]. This causes a decrease in the efficiency of energy use in farms [32]. The level of energy consumption and the efficiency of its use were the subject of research both in countries and in such sectors of agricultural production as beef production [33], milk [34] soy [35], or wheat [36,37]. The issues of energy consumption in agriculture are directly related to GHG emissions [38,39]. Some of the studies conducted indicate that the improvement of the energy efficiency of agriculture and the wider use of renewable energy sources is the best way to reduce GHG emissions [40,41].

Energy Consumption in Agriculture

Energy consumption in EU agriculture has had an upward trend since 2015, which is a clear change in the direction observed before 2015 (Figure 1).

Figure 1. Energy consumption by agriculture in EU in thousand tonnes of oil equivalent. Reproduced from [42], Eurostat: 2021.

In 2018, the amount of energy consumption in agriculture in the EU countries accounted for 3.2% of the final energy consumption in the EU (Table 1). In the years 2004–2018, the share of agriculture in the total final energy consumption did not change on average in the EU (it decreased to the greatest extent in Greece-by 3.9 pp). By far, the largest share of agriculture in total energy consumption among all EU countries was in the Netherlands (8.1%) and Poland (5.6%) [43].

Table 1. Share of energy consumption by agriculture in final energy consumption.

Countries	Energy Consumption by Agriculture in 2018	Change 2018/2004 (%)	Total Energy Consumption in 2018	Change 2018/2004 (%)	Share of Energy Consumption by Agriculture in Final Energy Consumption in 2018	Change 2018/2004 (pp.)
EU-28 *	25,166	−5.4	860,754	−5.4	3.2	0.0
Belgium	792	−3.0	33,111	−5.2	2.4	0.1
Bulgaria	185	−33.0	9750	6.5	1.9	−1.1
Czechia	619	11.2	24,180	−3.7	2.6	0.3
Denmark	596	−13.8	14,070	−3.9	4.2	−0.5
Estonia	124	18.4	2889	3.7	4.3	0.5
Ireland	223	−28.8	11,219	0.2	2.0	−0.8
Greece	264	−76.3	15,169	−23.0	1.7	−3.9
Spain	2458	−26.6	82,020	−9.4	3.0	−0.7
France	4089	−3.2	139,829	−7.7	2.9	0.1
Croatia	211	−0.7	6682	−3.6	3.2	0.1
Italy	2798	−5.5	114,422	−10.7	2.4	0.1
Cyprus	42	332.7	1581	3.8	2.7	2.0
Latvia	181	44.8	4025	4.3	4.5	1.3
Lithuania	108	2.3	5446	24.8	2.0	−0.4
Luxembourg	24	8.6	3737	−5.6	0.6	0.1
Hungary	641	9.3	17,865	4.8	3.6	0.1
Malta	5	–	515	50.5	0.9	0.9
Netherlands	3647	−3.3	44,933	−9.4	8.1	0.5
Austria	529	−3.5	26,036	3.7	2.0	−0.2
Poland	3918	−8.9	69,983	23.3	5.6	−2.0
Portugal	382	−28.6	16,201	−11.0	2.4	−0.6
Romania	566	144.0	23,445	−1.3	2.4	1.4
Slovenia	73	−1.3	4940	0.1	1.5	0.0
Slovakia	133	−18.0	9912	0.3	1.3	−0.3
Finland	688	−6.7	25,074	0.6	2.7	−0.2
Sweden	613	−19.7	31,777	−1.7	1.9	−0.4
United Kingdom	1257	46.2	121,944	−12.2	1.0	0.4

* Germany is not included as many data points are not available. Reproduced from [42], Eurostat: 2021.

In 2018, six EU countries with the highest energy consumption in agriculture accounted for almost 70% of energy consumption in agriculture in the entire EU, which proves a high level of concentration (Figure 2). The phenomenon meets the assumptions of the Pareto principle, and, in this case, 20% of the EU countries use 70% of energy in the agriculture of the Community.

In the EU, the greatest amount of energy used in agriculture came from gas oil and diesel oil, which in the analysed period accounted for over 50% of the structure of energy used (Figure 3). Electricity and natural gas were also important sources of energy. In the years 2004–2018, on average in the EU, the share of energy from renewable sources increased from 5 to 10%, although it seems that the pace of increasing the share of these sources is too slow. In the EU countries, the structure of energy consumption in agriculture varied considerably depending on the country. In almost all countries, gas oil and diesel oil were the most important, despite clear differences between countries (from about 90% in Slovenia to 9% in the Netherlands, which in this respect differed from other EU countries). In the Netherlands, like in no other country, more than 50% of the energy used in agriculture comes from natural gas. In Belgian agriculture, about 1/3 of the energy used came from natural gas. Natural gas was also important in Romania, Lithuania, and Hungary (20%, 19%, and 17%, respectively, in 2018). Poland, as the only country in the EU, to a large extent uses other bituminous coal (about 20%) as an energy source in agriculture. It is worth paying attention to Sweden and Austria, where over 30% of the energy used in agriculture came from renewable sources, which in the context of the current EU climate policy should be considered an example to be followed by other countries. Czechia, Slovakia, and Finland also stood out in this area, where renewable sources accounted for a quarter of the energy used for agriculture in 2018. For Germany, Malta and Cyprus, complete data for 2004 were not available. Therefore, data from the years 1998 (for Germany) and 2005 (Cyprus and Malta) were adopted for the study—these were the years closest to 2004 with complete data available.

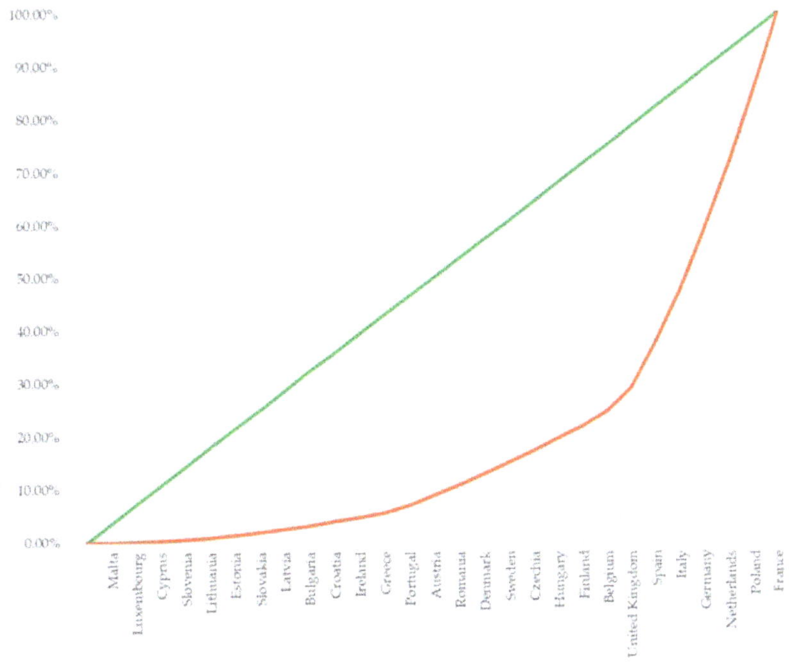

Figure 2. The concentration of energy consumption in agriculture in 2018. Reproduced from [42], Eurostat: 2021.

The Netherlands was characterized by the highest energy consumption in agriculture per hectare of arable land. In 2018, the Netherlands used nearly four times more energy per hectare of UAA (2052.93 kgoe) than in Belgium, second in the ranking, and over 15 times more than the average in all EU countries (Figure 4). This was due to very intensive agriculture and a high share of energy-intensive greenhouse production. The lowest final energy consumption per hectare of UAA was observed in Romania (33.5 kgoe/ha), Lithuania (35.3 kgoe/ha), and Bulgaria (36.8 kgoe/ha). In the case of Germany, the data for 2010 was used, as the data for 2008 were incomplete.

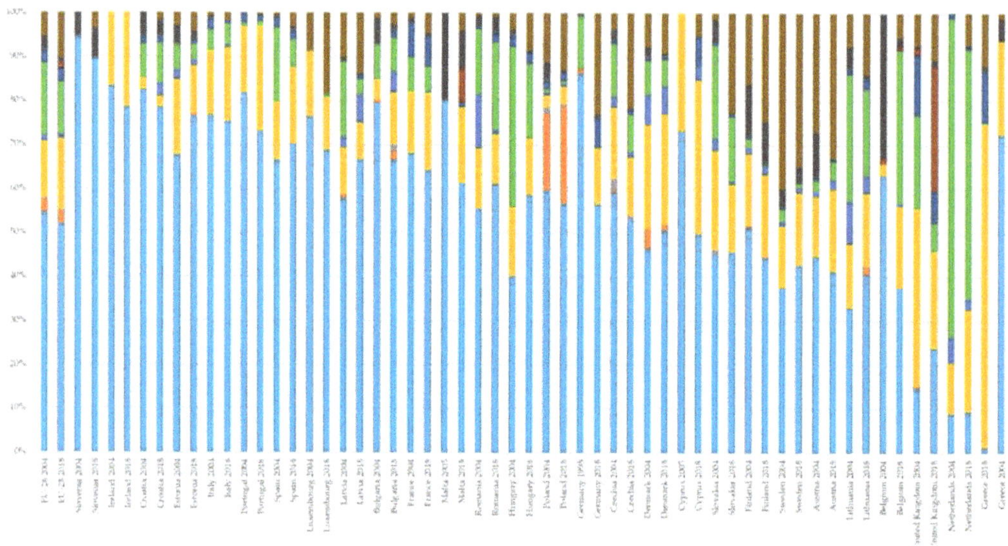

Figure 3. The structure of energy consumption in agriculture in the EU countries in 2004–2018. Reproduced from [44], Eurostat: 2021.

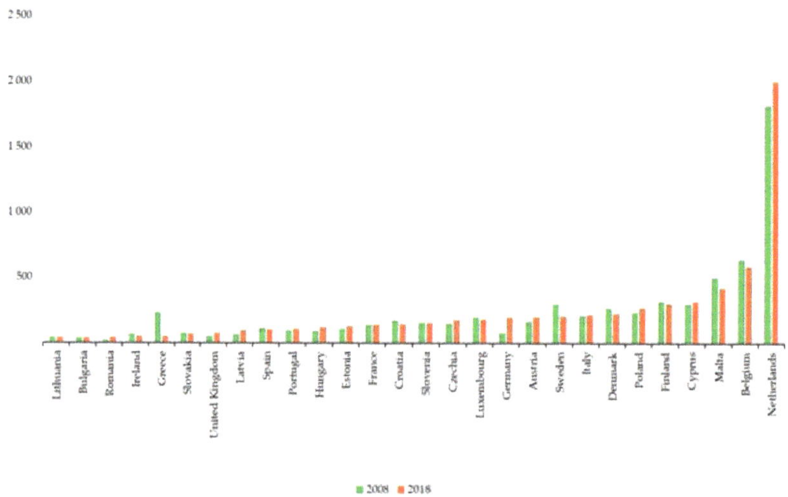

Figure 4. Energy consumption in agriculture per hectare of arable land in kgoe. Reproduced from [45], Eurostat: 2021.

3. Materials and Methods

3.1. Overview

In the research on the level and structure of emissions from Polish farms, data from the FADN (Farm Accountancy Data Network) from 2017 were used. The FADN operating in Poland is part of the European system, operating since 1965, based on Regulation of the Council of 15 June 1965 setting up a network for the collection of accountancy data on the incomes and business operation of agricultural holdings in the European Economic Community [46]. Data in FADN are collected in the management accounting convention. The FADN database is economic and organizational. It is now the most complete source of information on the situation of agricultural holdings. The identical principles of operation of the FADN system throughout the EU make the results comparable for all EU countries. The obtained data are used both for decision-making by EU bodies, monitoring the effects of these activities, and scientists dealing with the economics and organization of agriculture. Participation in the FADN system is voluntary. Farmers participating in the research write down every economic event that took place on their farm, in a special book, then agricultural advisors transfer them to the system.

The FADN observation field covers only commercial farms, i.e., farms supplying the market. In 2017, the results in Poland were calculated for 12,100 farms with an economic size greater than or equal to EUR 4000.

3.2. Types of Farms

The type of farm is defined based on the share of individual agricultural activities in the creation of the entire Standard Output of a farm. In the conducted research, grouping was made according to eight basic types. In practice, there were seven types because type 3-Vineyards does not occur in Poland (Table 2). Farms classified to a particular type are specialized in this type of agricultural production.

Table 2. Grouping of farms by type.

Symbol	Name	Description of the Type of Farm
1	Field crops	Specializing in the cultivation of cereals (including rice), oilseeds, and protein crops for seeds
2	Horticultural crops	Specializing in outdoor horticulture, under high cover, (vegetables, strawberries, flowers, and ornamental plants) and the cultivation of mushrooms and in nursery and horticulture
3	Vineyards	Specializing in viticulture
4	Permanent crops	Specializing in the cultivation of fruit trees and shrubs
5	Dairy cows	Specializing in dairy cattle farming
6	Herbivorous animals	Specializing in rearing cattle for slaughter (including breeding), sheep, goats, and other animals fed on roughage
7	Granivorous animals	Specializing in rearing pigs, poultry, and other animals fed with concentrated fodder
8	Mixed	Mixed-different crops, different animals

While there are some doubts about the use of the FADN for environmental issues [47], it is the most comprehensive source of information on farms in the EU. Basic organizational and economic information on the researched farms, grouped by type of farm, is presented in Table 3.

Table 3. Characteristics of the researched farms.

Description	Unit	Type of Farm							Average
		Field Crops	Horticultural Crops	Permanent Crops	Dairy Cows	Herbivorous Animals	Granivorous Animals	Mixed	
Sample size	pcs.	3922	304	445	2665	735	729	3313	–
Economic size	EUR	38,380	80,157	24,251	50,189	27,662	120,671	37,008	45,432
Labor inputs	AWU	1.73	3.48	2.28	1.99	1.59	2.08	1.74	1.87
Agricultural land area	ha	47.89	7.21	13.14	31.99	27.76	33.77	29.68	35.04
Total production value	EUR	47,111	79,738	35,891	65,427	23,821	144,360	40,855	54,272
The value of livestock production	EUR	1790	292	126	57,532	17,321	118,932	22,073	27,470

Table 3. *Cont.*

Description	Unit	Type of Farm							Average
		Field Crops	Horticultural Crops	Permanent Crops	Dairy Cows	Herbivorous Animals	Granivorous Animals	Mixed	
The value of plant production	EUR	44,931	79,257	35,311	7642	5973	24,974	18,519	26,462
Income from the farm	EUR	20,263	17,744	14,747	31,918	11,551	38,177	15,458	21,794
Income per full-time employee	EUR	13,844	10,033	9612	16,699	7393	22,036	9239	13,197

3.3. *Methodology of Estimating Emissions in Farms*

The problem of estimating the amount of greenhouse gas emissions in farms is difficult. GHG emission depends on numerous variables such as soil type, species, cultivation technology, breeding, the weather pattern in a given year, etc. Research carried out in one country does not have to be useful in other countries, and the obtained results are often very divergent [48].

The work attempts to link the internationally recognized methodology used by The National Centre for Emissions Management (KOBiZE) with data from the FADN database. The first attempts to calculate GHG emissions based on these data took place in Italy [49]. The authors of this study focused on a group of 695 farms in the Veneto region. They identified six emission sources, which were then calculated based on FADN data and national emission factors. Later, the research was extended to cover the entire FADN population [50]. In Poland, research combining FADN and greenhouse gas emissions is carried out at the Institute of Agricultural and Food Economics-National Research Institute [48,51,52]. Similar works are also carried out in other EU countries [53,54].

This study adopts its methodology for calculating GHG emissions, taking into account the latest Intergovernmental Panel on Climate Change (IPCC) guidelines. Contrary to Polish studies, GHG emissions at farms, emissions from fuel combustion (liquid, solid and gaseous), and electricity consumption were also taken into account. The main sources of emissions in agriculture, together with the data and indicators necessary for their estimation (in an IPCC-compliant format), are divided into three main categories: Energy (Sector 1), Agriculture (Sector 3), Land use (Sector 5) [55,56].

Within individual sectors, a total of 15 emission streams were identified (Table 4), each of which required a separate approach and determination of the GHG emission level based on the available FADN data and based on the guidelines contained in Guidelines for National Greenhouse Gas Inventories [56–58], modified in a way that allows the use of data collected in the FADN system. The amount of emissions in farms was calculated according to the formula:

$$Y = X_1 + X_2 + \ldots + X_{15} \quad (1)$$

Table 4. Calculation of GHG emissions in farms.

Emission Source	Emission Factor	Reference
X_1—Energy production for agriculture	Energy consumption [MWh] × Factors of the produced electricity for the end-user [1 MWh = 781 kg CO_2]	[59]
X_2—Combustion of fuels in agriculture	Fuel consumption x Emission factor for fuels [Diesel: 1 GJ = 74.1 t CO_2; Petrol: 1 GJ = 69.3 t CO_2]	
X_3—Intestinal fermentation	Number of animals of a certain species and age × Emission factor for species and age × 28 (Global Warming Potential-GWP) [Emission factor: from 5 kg CH_4/year for goats to 75.59 kg CH_4/year for bulls over 2 years of age]	[56,60]

Table 4. *Cont.*

Emission Source	Emission Factor	Reference
X_4—Methane emissions from livestock manure	Number of animals per species × Emission factor for species × 28 (GWP) [Emission factor for species: from 0.02 kg CH_4/year for broilers to 11.87 kg CH_4/year for dairy cows]	[56]
X_5—Direct emission of nitrous oxide from livestock manure	Number of animals of a certain species and age × Emission factor for species and age (Nex) × N_2O-N to N_2O conversion factor × 265 (GWP) [Nex: from 1 kg N_2O/year for turkeys to 83 kg N_2O/year for dairy cows; N_2O-N to N_2O conversion factor = 44/28]	[56]
X_6—Indirect emission of nitrous oxide from livestock manure	Composed of two processes: Indirect N_2O emissions due to volatilization of N from manure management and Indirect N_2O emissions due to leaching from manure management	[58] Equations: 10.27 and 10.29
X_7—Use of mineral fertilizers	Amount of mineral fertilizers applied × Fertilizer emission factor × 44/28 × 265 (GWP) [Fertilizer emission factor = 0.01 kg N_2O out of 1 kg of N]	[56,58]
X_8—Use of organic fertilizers	Amount of organic fertilizers applied × Fertilizer emission factor × 44/28 × 265 (GWP) [Fertilizer emission factor = 0.01 kg N_2O out of 1 kg of N]	[56,58]
X_9—Animal manure on pastures and grasslands	Number of animals of a certain species and age × Emission factor for species and age (Nex) × Pasture maintenance factor × Emission factor for manure from grazing animals × 265 (GWP) [Pasture maintenance factor-from 0.103 (dairy cows) to 0.44 (sheep); Emission factor for manure from grazing animals–0.2 for cattle and pigs and 0.01 for sheep, goats, and horses]	[56,58]
X_{10}—Plant residues	Annual harvest of a given crop × Dry matter share × Nitrogen content in biomass × (1—Share of burnt biomass—Share of biomass removed from the field)	[56,58]
X_{11}—Nitrogen deposition from the atmosphere (indirect emissions)	Annual amount of mineral fertilizers × Factor of nitrogen participation in fertilizers emitted in the form of NH_3 and NO_x + Annual amount of organic fertilizers + Annual amount of animal manure on pastures x Factor of the share of nitrogen from the manure emitted in the form of NH_3 and NO_x × 44/28 × 265 (GWP) [Factor of nitrogen participation in fertilizers emitted in the form of NH_3 and NO_x = 0.01; Factor of the share of nitrogen from the manure emitted in the form of NH_3 and NO_x = 0.2]	[56,58]
X_{12}—Leaching and oxidation of nitrogen from the ground (indirect emissions)	(Annual amount of mineral fertilizers + Annual amount of organic fertilizers + Annual amount of plant residues) × Factor of the share of nitrogen leached from the ground into the waters × Emission factor of leached nitrogen × 44/28 × 265 (GWP) [Factor of the share of nitrogen leached from the ground into the waters = 0.3; Emission factor of leached nitrogen = 0.0075]	[56,58,60]
X_{13}—Liming	Annual amount of calcium fertilizers $CaCO_3$ × $CaCO_3$ emission factor + Annual amount of calcium fertilizers $CaMg(CaCO_3)_2$ × $CaMg(CaCO_3)_2$ emission factor [$CaCO_3$ emission factor = 0.12; $CaMg(CaCO_3)_2$ emission factor = 0.13]	[56,58]
X_{14}—Burning crop residues	(Annual harvest of a given crop × Dry matter share × Nitrogen content in biomass × Share of burnt biomass × Combustion efficiency) × Carbon content in biomass = Total amount of carbon released	[56,58]
X_{15}—Urea fertilization	Amount of urea used during the year × Emission factor × Conversion factor [Emission factor = 0.2 kg C/kg N; Conversion factor of C in CO_2 = 44/12]	[56,58]

The Global Warming Potential (GWP) was used to calculate the emissions of individual GHGs, i.e., a conversion factor enabling the determination of individual GHG emissions as a CO_2 equivalent. GWP is a measure of how much energy the emissions of 1 kg of a gas will absorb over a given period of time, relative to the emissions of 1 kg of CO_2. The individual factors are presented in Table 5.

Table 5. Global warming potential of greenhouse gases.

Greenhouse Gas	Global Warming Potential (GWP)
CO_2	1
CH_4	28
N_2O	265
SF_6	23,500
NF_3	16,100

For example, the emission of 1 kg of methane for the climate equates to the emission of 28 kg of CO_2 [60]. This allows the emissions of all GHGs to be reduced to one value.

The amount of taxes/fees for GHG emissions was calculated based on the price of emission allowances, which was achieved at the auction on the European Energy Exchange (EEX) on 23 September 2020—27.31 EUR/t [61]. This method was used in other studies [62]; it is also similar to the calculations made by Richard Tol on the social costs of GHG emissions [63].

4. Results and Discussion

4.1. Total GHG Emissions from Agriculture

In 2018, the total EU GHG emissions amounted to 4.4 billion tonnes. In the years 1990–2018, the share of individual GHG emission sources in the EU did not change. In the case of Agriculture, the share fluctuated in the range of 1–14%, which is comparable to the Industry (Figure 5) [64–66]. In absolute terms, agriculture emitted an annual average of 436 million tonnes of greenhouse gases. In the context of the GHG emission reduction process, it should be noted that, since 1990, emissions in agriculture have been reduced by 23%. This was due to several factors. First of all, the livestock stock decreased and the consumption of nitrogen compounds was limited [67]. Except for Spain, each EU Member State has reduced GHG emissions between 1990 and 2018. The largest decreases were recorded in Germany, Romania, and Poland [66]. However, globally, the agricultural sector has increased GHG emissions by 1.1% [64].

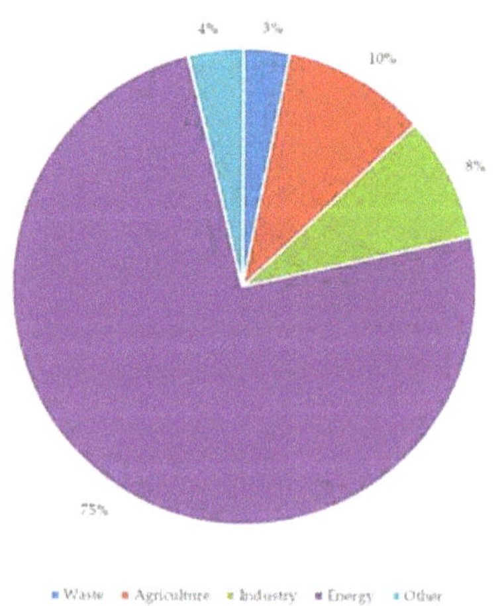

Figure 5. Structure of GHG emissions in the EU in 2018 by sector. Reproduced from [68], Eurostat: 2021.

Poland, with GHG emissions at the level of 416 million tons per year, ranks 5th in the EU. The sectoral structure of GHG emissions in Poland is slightly different than the EU average. The dominant sector is energy with a share of over 80% of the total emissions, while agriculture is responsible for 8% of the emissions in the country, recording a decrease in emissions by almost 1/3 in the years 1990–2018. This was due to a reduction in the number of livestock, the collapse of inefficient State Agricultural Farms, and more rational use of fertilizers based on the principles of a market economy or shaping the production structure [69,70].

4.2. GHG Emission from Energy Inputs in Agriculture

Energy consumption in EU agriculture increased in the years 2004–2018 by 3%, while the emissions accompanying this consumption increased by almost 6%, which proves that, on average, in the entire Community structure of energy sources, there were more sources with a higher greenhouse gas emission index (Table 6). Agricultural energy consumption was reduced most in Greece, Bulgaria, and Ireland by 76%, 33%, and 29%, respectively. However, Slovakia deserves special attention, as it has reduced energy consumption by 1/5 while reducing emissions from this energy consumption by almost 40%, which shows the replacement of high-emission energy carriers, e.g., with renewable energy. Slovakia, along with Czechia and Slovenia, had the lowest emissivity of energy inputs in agriculture, far below the average for the entire EU [71].

Table 6. GHG emission from energy inputs in agriculture in the EU countries in 2004–2018.

Countries	Energy Inputs 2018 (TJ)	Change 2018/2004 (%)	GHG Emissions 2018 (t)	Change 2018/2004 (%)	Emissivity of Energy Inputs 2004 (t GHG/TJ)	Emissivity of Energy Inputs 2018 (t GHG/TJ)	Change 2018/2004 (%)
EU-28	1,193,555	3.1	103,671,715	5.6	84.80	86.86	2.4
Slovenia	3059	−1.3	219,729	−4.9	74.54	71.82	−3.6
Ireland	9355	−28.8	972,756	−24.0	97.37	103.98	6.8
Croatia	8821	2.1	658,303	0.5	75.79	74.63	−1.5
Estonia	5199	18.4	447,061	7.2	95.00	85.98	−9.5
Italy	117,157	−5.5	11,117,330	−3.2	92.64	94.89	2.4
Portugal	15,992	−28.6	1,704,399	−19.7	94.83	106.58	12.4
Spain	102,896	−26.6	9,691,288	−20.7	87.24	94.19	8.0
Luxembourg	990	8.6	75,993	−5.6	88.35	76.76	−13.1
Latvia	7565	44.8	576,977	42.5	77.48	76.27	−1.6
Bulgaria	7757	−33.0	690,938	−23.7	78.26	89.07	13.8
France	171,192	−3.2	15,890,349	0.1	89.77	92.82	3.4
Malta	203	–	19,327	–	–	95.14	–
Romania	23,690	144.0	1,979,516	122.5	91.61	83.56	−8.8
Hungary	26,834	9.3	2,167,087	7.3	82.19	80.76	−1.7
Poland	164,050	−8.9	13,125,832	−6.6	78.07	80.01	2.5
Germany	139,904	1573.4	10,392,644	7231.5	16.96	74.28	338.1
Czechia	25,933	11.2	1,937,408	−9.7	91.98	74.71	−18.8
Denmark	24,938	−13.8	2,555,070	−13.9	102.56	102.46	−0.1
Cyprus	1776	332.7	210,908	142.8	211.65	118.76	−43.9
Slovakia	5555	−18.0	415,070	−38.2	99.20	74.72	−24.7
Finland	28,822	−6.7	2,379,596	−9.4	84.96	82.56	−2.8
Sweded	25,656	−19.7	1,831,157	−9.5	63.37	71.37	12.6
Austria	22,156	−3.5	1,664,652	−0.5	72.83	75.13	3.2
Lithuania	4507	2.3	373,509	2.5	82.66	82.87	0.2
Belgium	33,148	−3.0	2,841,637	6.3	78.23	85.73	9.6
United Kingdom	52,631	46.2	4,961,670	17.8	116.98	94.27	−19.4
Netherlands	152,697	−3.3	12,942,532	16.5	70.36	84.76	20.5
Greece	11,069	−76.3	1,828,980	−62.0	103.07	165.24	60.3

Reproduced from [44], Eurostat: 2021; Reproduced from [58], IPCC: 2006.

The amount of emissions from consumed energy directly depends on the amount of energy consumed and on the structure of energy carriers with different greenhouse gas emissivity. In the years 2004–2018, emissions in Poland, similarly to energy consumption, reached a minimum level of 11.18 million tonnes in 2015. It was followed by an increase, also visible in the rest of the Polish economy.

The emissions from energy sources in agriculture are dominated by diesel oil, which is constantly growing, accounting for half of the emissions in 2018. Two more energy carriers play an important role in the emission structure-bituminous coal 34% and electricity 11%. Searching for opportunities to reduce energy consumption and, at the same time, to reduce greenhouse gas emissions, in-depth research was carried out to find the answers to which farms emit greenhouse gases from energy carriers the most and where to look for opportunities to reduce energy consumption and thus greenhouse gas emissions in the first place. [31,72–74].

4.3. GHG Emissions from Energy Carriers Depending on the Type of Farm

As part of the research, the GHG emissions were calculated in individual production types of farms in the Polish FADN system. Calculations were made for all 15 emission streams. For the sake of legibility, they have been aggregated into categories related to Plant production, Animal production, and Fertilization. The Energy category has been presented broken down into Electricity and Fuels. Figure 6 shows the emission volumes for the subsequent emission categories.

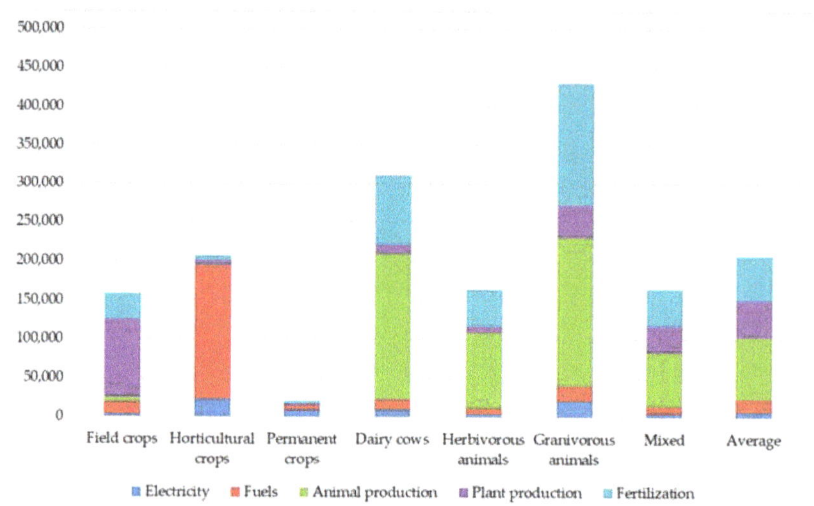

Figure 6. GHG emission in particular types of farms in kg. Source: own study.

The average level of all GHG emissions in Polish farms covered by the FADN system was over 207,000 kg per farm, including 24,000 kg from energy inputs, which accounted for 12% of all emissions. The highest total emission level was observed for two types of farms involved in livestock production: dairy cows and granivorous animals, respectively 311,000 kg and 430,000 kg of GHG per farm (Figure 6). This is confirmed by studies [75–77], that animal production is the main source of emissions. The lowest emission level was found on farms of the type of permanent crops, which in the Polish FADN system include fruit-growing farms. As already mentioned, one of the important sources of emissions in the surveyed farms were fuels and electricity, which together accounted for the average emission on the farm from 11,700 kg of GHG in the type of herbivorous animals to 194,500 kg of GHG in horticultural crops.

The share of Energy in the emission structure in the researched farms was very diversified and ranged from 7% for dairy cows and herbivorous animals to 84% for horticultural crops (Figure 7). The high share of energy is related to production technology. In general, vegetable growing is a type of production associated with extremely intensive use of production factors such as land, water, energy [78]. In the case of the horticultural crops type, especially for cultivation in greenhouses, high costs are incurred to ensure the appropriate temperature. This requires the combustion of fossil fuels, gas, coal, or the use of electricity. The situation is slightly different for permanent crops. These are fruit-growing farms, with the dominant role of apples. The high emissions in the Energy category are related to two issues. The production of fruit requires intensive protection and many operations performed by machines, which causes high consumption of fuels, especially diesel oil. During the season, even a dozen or so agrotechnical treatments are performed, such as sprinkled fertilization, foliar fertilization, disease and pest control, and weed control. Each of these treatments requires the use of agricultural tractors. After harvest, the apples are placed in various types of storage (with a normal, modified, or controlled atmosphere) [79]. Maintaining the assumed conditions, temperature, and atmosphere composition require the consumption of electricity, which directly translates into the structure of emissions in these farms [80]. The next stage is also important—packing and often distributing the fruit on the farm's own. It is worth noting that within the energy section, fuels were dominant, accounting for an average of 69% of emissions from energy inputs in the researched farms. The highest share of fuels was recorded in horticultural crops, 88%, while the lowest share of fuels among the researched farms was in the case of permanent crops and amounted to 41% of GHG emissions in the total emission from energy sources [81].

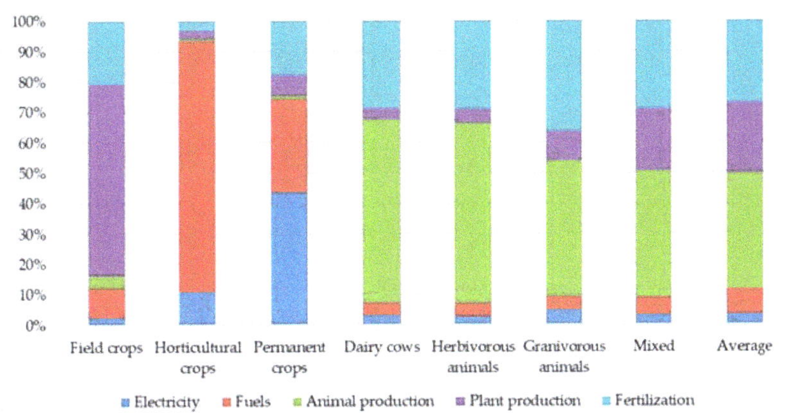

Figure 7. Structure of GHG emissions in types of farms. Source: own study.

Subsequently, the intensity of emissions from energy inputs in the researched farm types was determined by relating the emission level to the area of agricultural land (Table 7).

Table 7. Emission from energy inputs per 1 ha of agricultural land and production value per 1 kg of GHG from energy inputs.

Index	Field Crops	Horticultural Crops	Permanent Crops	Dairy Cows	Herbivorous Animals	Granivorous Animals	Mixed	Average
GHG from energy inputs (kg/ha)	407.42	26,976.23	1178.09	724.65	422.35	1196.15	511.28	689.40
Production value per kg GHG from energy inputs (EUR)	2.41	0.41	2.32	2.82	2.03	3.57	2.69	2.25

Source: own study.

The highest ratio was achieved by horticultural crops type-26,976.2 kg of GHG emissions/1 ha of UAA, and the lowest-field crops type, only 407.4 kg of GHG/1 ha of UAA. The issue of environmental efficiency is also important, as shown in Table 7 as the production value per 1 kg of GHG emissions from the energy used in the production process. Except for horticultural crops, 1 kg of GHG emissions from the Energy category allowed to generate production worth EUR 2–3, in the case of horticultural crops it was only EUR 0.41. By far the highest environmental efficiency in this respect was presented by farms of the granivorous type, where 1 kg of GHG from the energy used allowed to generate over EUR 3.5 of the production value.

4.4. Farm Income and GHG Emission Costs from Energy Inputs

Taking into account the economic aspect and social costs of GHG emissions, the impact of introducing charges/taxes on emissions on farm income was determined (Figure 8). Two variants were presented: introducing taxes/fees related only to energy inputs as well as to all GHG emissions in the farm.

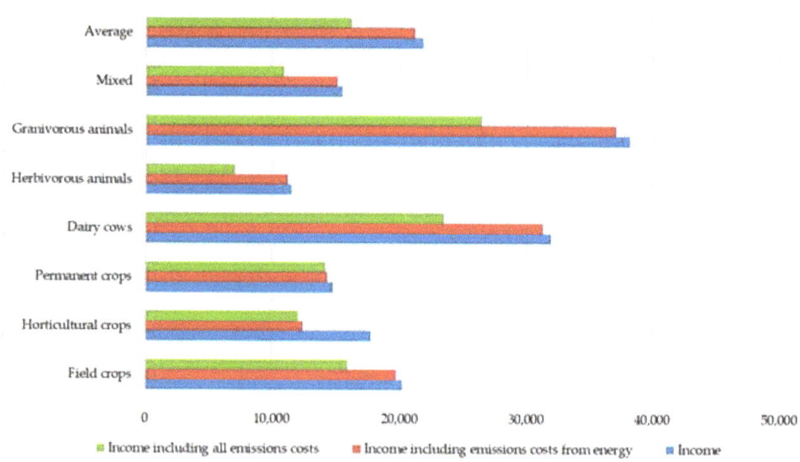

Figure 8. Farm income before and after taking into account GHG emission costs in EUR. Source: own study.

In the first case, if emission charges were introduced only from energy inputs, the impact of these solutions on farm income would not be large, except for horticultural crops, where the income would be reduced by about 30%. The decline in income for the remaining types is only 2–3%. The situation is completely different in the variant of taxation of all emissions on the farm. The income of the surveyed farms would drop from 21% for field crops to 40% for herbivorous animals. The low-emission farms in the type of permanent crops are a phenomenon here, where the decrease in income would amount to only 3.85%.

4.5. Outlook

The conducted research shows the types of production and the main types of emissions. For many years, research has been conducted on the possibility of reducing these emissions. The methods of reducing emissions can be divided into two groups: economic, influencing eating habits and related to production technology.

Various administrative and economic instruments are considered to encourage farmers to reduce emissions and the society to reduce the consumption of goods that require high emissions. This problem is particularly relevant to livestock production [82]. There are more and more calculations of the hidden environmental costs of this production, combined with the calculation of the benefits that can be achieved by switching to a vegan or vegetarian diet [24]. Research confirms that maintaining current eating habits will lead to high GHG emissions [83]. The European Parliament discussed the taxation of meat so that its price fully corresponds to environmental costs. A tax at a rate of 60 EUR/t CO_2 equivalent emissions would reduce total GHG emissions in the EU by 5% [84]. The research conducted in Denmark determined that the introduction of the burden at the level of 150–1730 DKK per 1 ton of CO_2 equivalent emissions results in a reduction of the emis−sion footprint from food production by 2.3–8.8% [85].

In addition to changes in food consumption, it is also postulated to introduce various technological changes aimed at reducing the level of GHG emissions. They relate to different emission areas [86]:

1. Reduction of emissions from nitrogen fertilizers:
   - limiting the consumption of mineral fertilizers,
   - selection of appropriate forms of nitrogen fertilizers,
   - use of inhibitors,
   - maintaining an appropriate soil pH [87].
2. Carbon retention in soil and biomass.
3. Breeding progress:
   - increasing the area of legume crops,
   - introducing more fats into the diet of ruminants [88],
4. Genetic improvement of animals.
5. The use of animal excrements for the production of biogas, which prevents the escape of nitrogen compounds into the atmosphere [89,90].
6. Increasing energy efficiency, the use of renewable energy and improved sources of nuclear energy [91].

Taking into account the research carried out (Figure 6), it seems that the application of GHG emission reduction methods should cover two directions. First, there is a need to change food habits, move away from ruminant animal products. The main role should be played by economic tools, taxes and fees. At the same time, production methods limiting GHG emissions, especially related to livestock production, should be introduced.

In the case of reducing emissions from energy carriers, the problem is extremely complex. Research shows that the intensification of energy consumption in agriculture has made it possible to feed a rapidly growing world population [27]. With the current level of production intensity and a large number of agricultural operations, the possibilities of reducing these emissions are small. However, a decrease in GHG emissions can be achieved in two ways:

1. Fossil fuel consumption reduction

In research and studies carried out all over the world, there are various examples of how to reduce fuel consumption. They are mainly related to changes in production technology:

- Cultivation without plowing (simplified cultivation)—although it is difficult to convince farmers to this type of cultivation, it causes even a threefold decrease in GHG levels [92].
- Precision agriculture and precision agriculture technologies (PAT) [93]. One of the main tasks of precision agriculture is to optimize the use of agricultural inputs, fertilisers, fuel. From the point of view of GHG emissions, techniques that reduce the consumption of nitrogen fertilizers and the number of activities seem to be crucial. This allows a reduction in fuel consumption [94].
- Electrically powered agricultural tractors. Despite the serious obstacle of low battery capacity, agricultural tractor manufacturers are trying to placing them on the market. Two versions of the machines are tested: with batteries and with a cable connection to the power source [94,95]. There are also ideas for introducing agricultural tractors with modern combustion engines, powering electric motors.

- Technical progress in the construction of traditional combustion engines. 2020 is a transition period for engines below 75hp and above 175hp due to the introduction of the Stage V standard. Until 30 June 2020, manufacturers could install transition engines on their machines and market them until 31 December 2020. For machines with a capacity of 75–175 hp, the transition year is 2021 [96].
- Appropriate use of existing agricultural tractors. Appropriate management of tires and weights, use of start-stop systems, longer work sequence, eco-driving, replacing agricultural tractors with more energy-efficient machines [97].

2. Renewable energy

The development of renewable energy in rural areas will be a key element in reducing GHG emissions from energy carriers. Different types of RES are possible: biomass, solar energy, wind farms. Agricultural biogas plants are particularly promising. In addition to solving the problem of CH_4 emissions from animal manure, they provide electricity and heat necessary for agricultural production. It is interesting to combine different types of technologies, where the farmer is both a producer and consumer of energy (prosumer). This makes it possible to combine renewable energy sources with electric vehicles charged from own sources. Another solution may be to combine livestock farming that supplies input to a biogas plant, which supplies electricity and provides heating for the farm.

5. Conclusions

The implementation of the ambitious vision of Europe by 2050 as a climate-neutral continent set out in the European Green Deal requires the intensification of efforts to reduce GHG emissions in all sectors. Such actions must also be taken in agriculture, which is responsible for about 10–14% of their emissions. From simulations by The National Centre for Emission Management (KOBiZE) [98] it results that in Poland if the current production technologies are continued to be used, achieving the ambitious targets for reducing emissions from the agricultural sector will be very difficult. Attempts to implement more ambitious reduction targets may lead not only to a decrease in farm income but also to a relatively high reduction in the level of production, which may increase food prices. This study does not take into account GHG emissions related to the consumption of energy carriers, as well as in the materials and databases of FAO, EPA, and other organizations. It is not included in the agriculture section but belongs to the general category of energy. This is the reason for difficulties in comprehensive assessments of the effectiveness of activities aimed at reducing GHG gas emissions in agriculture.

Looking for ways to reduce energy consumption, and at the same time to emit greenhouse gases, in-depth research was carried out to find out which farms emit greenhouse gases from energy carriers the most, and where to look for ways to reduce energy consumption and thus greenhouse gas emissions in the first place. The average GHG emission level in Polish farms covered by the FADN system was over 207 Mg per farm, of which 24 Mg came from energy inputs, which accounted for 12% of the total GHG emission. The lowest share, amounting to 7%, was characteristic for farms keeping dairy cows and herbivorous animals, and the highest (84%) for horticultural crops farms. The amount of GHG emission from the consumed energy was directly dependent on the amount of its consumption and the structure of the energy carriers used. Emission from diesel oil consumption (50%) dominated, followed by bituminous coal (34%) and electricity (11%).

Author Contributions: Conceptualization, P.G. (Piotr Gołasa) and M.W.; methodology, P.G. (Piotr Gołasa) and P.G. (Piotr Gradziuk); software, A.S.; validation, P.G. (Piotr Gołasa), M.W. and P.G. (Piotr Gradziuk); formal analysis, P.G. (Piotr Gradziuk); investigation, B.G. and W.B.-G.; resources, W.B.-G. and A.S.; data curation, B.G. and A.G.; writing—original draft preparation, M.W.; writing—review and editing, P.G. (Piotr Gradziuk); visualization, A.G. and M.G.; supervision, M.W.; project administration, P.G. (Piotr Gołasa); funding acquisition, M.W. and M.G. All authors have read and agreed to the published version of the manuscript.

Funding: This research received no external funding.

Institutional Review Board Statement: Not applicable.

Informed Consent Statement: Not applicable.

Data Availability Statement: Not applicable.

Conflicts of Interest: The authors declare no conflict of interest.

References

1. Intergovernmental Panel on Climate Change (IPCC). *Climate Change. The IPCC Scientific Assessment*; Cambridge University Press: Cambridge, MA, USA, 1990; p. 202.
2. Stern, N. The Structure of Economic Modelling of the Potential Impacts of Climate Change: Grafting Gross Underestimation of Risk onto Already Narrow Science Models. *J. Econ. Lit.* **2013**, *51*, 838–859. [CrossRef]
3. Zachos, J.C.; Dickens, G.R.; Zeebe, R.E. An early Cenozoic perspective on greenhouse warming and carbon-cycle dynamics. *Nature* **2008**, *451*, 279–283. [CrossRef] [PubMed]
4. Arrhenius, S. On the Influence of Carbonic Acid in the Air upon the Temperature of the Ground. *Philos. Mag. J. Sci.* **1896**, *41*, 268. [CrossRef]
5. Revelle, R.; Suess, H.E. Carbon dioxide exchange between atmosphere and ocean and the question of an increase of atmospheric CO_2 during the past decades. *Tellus* **1957**, *9*, 18–27. [CrossRef]
6. Freudenburg, W.R.; Muselli, V. Global warming estimates, media expectations, and the asymmetry of scientific challenge. *Glob. Environ. Chang.* **2010**, *20*, 483–491. [CrossRef]
7. Brysse, K.; Oreskes, N.; O'Reilly, J.; Oppenheimer, M. Climate change prediction: Erring on the side of least drama? *Glob. Environ. Chang.* **2013**, *23*, 327–337. [CrossRef]
8. Hansen, J.; Sato, M.; Hearty, P.; Ruedy, R.; Kelley, M.; Masson-Delmotte, V.; Russell, G.; Tselioudis, G.; Cao, J.; Rignot, E.; et al. Ice melt, sea level rise and superstorms: Evidence from paleoclimate data, climate modeling, and modern observations that 2 °C global warming is highly dangerous. *Atmos. Chem. Phys. Discuss.* **2016**, *16*, 3761–3812. [CrossRef]
9. Sherwood, S.C.; Huber, M. An adaptability limit to climate change due to heat stress. *Proc. Natl. Acad. Sci. USA* **2010**, *107*, 9552–9555. [CrossRef]
10. Slater, T.; Hogg, A.E.; Mottram, R. Ice-sheet losses track high-end sea-level rise projections. *Nat. Clim. Chang.* **2020**, *10*, 879–881. [CrossRef]
11. Kriegler, E.; Hall, J.W.; Held, H.; Dawson, R.; Schellnhuber, H.J. Imprecise probability assessment of tipping points in the climate system. *Proc. Natl. Acad. Sci. USA* **2009**, *106*, 5041–5046. [CrossRef]
12. Lenton, T.M.; Livina, V.N.; Dakos, V.; van Nes, E.H.; Scheffer, M. Early warning of climate tipping points from critical slowing down: Comparing methods to improve robustness. *Philos. Trans. R. Soc. A* **2012**, *370*, 1185–1204. [CrossRef] [PubMed]
13. Turetsky, M.R.; Abbott, B.W.; Jones, M.C.; Walter Anthony, K.M.; Olefeldt, D.; Schuur, E.A.G.; Grosse, G.; Kuhry, P.; Hugelius, G.; Koven, C.; et al. Carbon release through abrupt permafrost thaw. *Nat. Geosci.* **2020**, *13*, 138–143. [CrossRef]
14. Intergovernmental Panel on Climate Change (IPCC). *Climate Change and Land: An IPCC Special Report on Climate Change, Desertification, Land Degradation, Sustainable Land Management, Food Security, and Greenhouse Gas Fluxes in Terrestrial Ecosystems*; IPCC: Geneva, Switzerland, 2019.
15. Lal, R. Carbon emission from farm operations. *Environ. Int.* **2004**, *30*, 981–990. [CrossRef]
16. Leclère, D.; Jayet, P.A.; de Noblet-Ducoudré, N. Farm-level autonomous adaptation of European agricultural supply to climate change. *Ecol. Econ.* **2013**, *87*, 1–14. [CrossRef]
17. Brisson, N.; Gate, P.; Gouache, D.; Charmet, G.; Oury, F.X.; Huard, F. Why are wheat yields stagnating in Europe? A comprehensive data analysis for France. *Field Crops Res.* **2010**, *119*, 201–212. [CrossRef]
18. Lobell, D.B.; Schlenker, W.; Costa-Roberts, J. Climate Trends and Global Crop Production since 1980. *Science* **2011**, *333*, 616–620. [CrossRef] [PubMed]
19. Intergovernmental Panel on Climate Change (IPCC). *Climate Change 2014: Impacts, Adaptation, and Vulnerability. Part. A: Global and Sectoral Aspects. Contribution of Working Group II to the Fifth Assessment Report of the Intergovernmental Panel on Climate Change*; Cambridge University Press: Cambridge, MA, USA, 2014; p. 495.
20. Godfray, H.C.J.; Beddington, J.R.; Crute, I.R.; Haddad, L.; Lawrence, D.; Muir, J.F.; Pretty, J.; Robinson, S.; Thomas, S.M.; Toulmin, C. Food Security: The Challenge of Feeding 9 Billion People. *Science* **2010**, *327*, 812–818. [CrossRef] [PubMed]
21. Brown, L.R. *Full Planet, Empty Plates: The New Geopolitics of Food Scarcity*; W.W. Norton Company: New York, NY, USA; London, UK, 2012; p. 5.
22. Food and Agricultural Organization of the United Nations (FAO). *The State of the World's Biodiversity for Food and Agriculture*; FAO Commission on Genetic Resources for Food and Agriculture Assessment: Rome, Italy, 2019; Available online: http://www.fao.org/3/CA3129EN/CA3129EN.pdf (accessed on 5 May 2021).
23. Food and Agricultural Organization of the United Nations (FAO); International Fund for Agricultural Development (IFAD); World Food Programme (WFP). *The State of Food Insecurity in the World 2012. Economic Growth Is Necessary but Not Sufficient to Accelerate Reduction of Hunger and Malnutrition*; Food and Agriculture Organization of the United Nations: Rome, Italy, 2012; p. 9. Available online: http://www.fao.org/3/i3027e/i3027e.pdf (accessed on 5 May 2021).

24. Food and Agricultural Organization of the United Nations (FAO); International Fund for Agricultural Development (IFAD); United Nations Children's Fund (UNICEF); World Food Programme (WFP); World Health Organization (WHO). *The State of Food Security and Nutrition in the World 2020. Transforming Food Systems for Affordable Healthy Diets*; Food and Agriculture Organization of the United Nations: Rome, Italy, 2020. [CrossRef]
25. European Enviroment Agency (EEA). *Climate Change, Impacts and Vulnerability in Europe 2016—An Indicator-Based Report*; Publications Office of the European Union: Luxembourg, 2017; pp. 223–240. [CrossRef]
26. Communication from the Commission to the European Parliament; the European Council; the Council; the European Economic and Social Committee; the Committee of the Regions and the European Investment Bank. *A Clean Planet. for All—A European Strategic Long-Term Vision for a Prosperous, Modern, Competitive and Climate Neutral Economy*; COM: Brussels, Belgium, 2018.
27. Smil, V. *Energy and Civilization—A History*; MIT Press: Cambridge, MA, USA, 2017. [CrossRef]
28. Solow, R.M. The Economics of Resources or the Resources of Economics. *Am. Econ. Rev.* **1974**, *64*, 1–14.
29. Nordhaus, W.D. The Allocation of Energy Resources. *Brook. Pap. Econ. Act.* **1973**, *4*, 529–576. [CrossRef]
30. Pfeiffer, D.A. *Eating Fossil Fuels: Oil, Food, and the Coming Crisis in Agriculture*; New Society Publishers: Gabriola Island, BC, Canada, 2006.
31. Smil, V. *Energy in Nature and Society: General Energetics of Complex Systems*; MIT Press: Cambridge, MA, USA, 2008. [CrossRef]
32. Martinho, V.D. Energy consumption across European Union farms: Efficiency in terms of farming output and utilized agricultural area. *Energy* **2016**, *103*, 543–556. [CrossRef]
33. Veysset, P.; Lherm, M.; Bébin, D.; Roulenc, M.; Benoit, M. Variability in greenhouse gas emissions, fossil energy consumption and farm economics in suckler beef production in 59 French farms. *Agric. Ecosyst. Environ.* **2014**, *188*, 180–191. [CrossRef]
34. Sefeedpari, P.; Rafiee, S.; Akram, A.; Komleh, S.H.P. Modeling output energy based on fossil fuels and electricity energy consumption on dairy farms of Iran: Application of adaptive neural-fuzzy inference system technique. *Comput. Electron. Agric.* **2014**, *109*, 80–85. [CrossRef]
35. Ramedani, Z.; Rafiee, S.; Heidari, M. An investigation on energy consumption and sensitivity analysis of soybean production farms. *Energy* **2011**, *36*, 6340–6344. [CrossRef]
36. Ilahi, S.; Wu, Y.; Raza, M.A.A.; Wei, W.; Imran, M.; Bayasgalankhuu, L. Optimization Approach for Improving Energy Efficiency and Evaluation of Greenhouse Gas Emission of Wheat Crop using Data Envelopment Analysis. *Sustainability* **2019**, *11*, 3409. [CrossRef]
37. Khoshnevisan, B.; Rafiee, S.; Omid, M.; Mousazadeh, H. Applying data envelopment analysis approach to improve energy efficiency and reduce GHG (greenhouse gas) emission of wheat production. *Energy* **2013**, *58*, 588–593. [CrossRef]
38. Saldukaitė, L.; Šarauskis, E.; Lekavičienė, K.; Savickas, D. Predicting energy efficiency and greenhouse gases reduction potential under different tillage management and farm size scenarios for winter wheat production. *Sustain. Energy Technol. Assess.* **2020**, *42*. [CrossRef]
39. Yan, Q.; Yin, J.; Baležentis, T.; Makutėnienė, D.; Štreimikienė, D. Energy-related GHG emission in agriculture of the European countries: An application of the Generalized Divisia Index. *J. Clean. Prod.* **2017**, *164*, 686–694. [CrossRef]
40. Li, T.; Baležentis, T.; Makutėnienė, D.; Streimikiene, D.; Kriščiukaitienė, I. Energy-related CO_2 emission in European Union agriculture: Driving forces and possibilities for reduction. *Appl. Energy* **2016**, *180*, 682–694. [CrossRef]
41. Dyer, J.; Desjardins, R. Simulated Farm Fieldwork, Energy Consumption and Related Greenhouse Gas Emissions in Canada. *Biosyst. Eng.* **2003**, *85*, 503–513. [CrossRef]
42. Simplified Energy Balances. Available online: https://appsso.eurostat.ec.europa.eu/nui/show.do?dataset=nrg_bal_s&lang=en (accessed on 5 May 2021).
43. Agri-Environmental Indicator—Energy Use. Available online: https://ec.europa.eu/eurostat/statistics-explained/index.php/Agri-environmental_indicator_-_energy_use (accessed on 5 May 2021).
44. Complete Energy Balances. Available online: https://appsso.eurostat.ec.europa.eu/nui/show.do?dataset=nrg_bal_c (accessed on 5 May 2021).
45. Final Energy Consumption by Agriculture/Forestry per Hectare of Utilised Agricultural Area. Available online: http://appsso.eurostat.ec.europa.eu/nui/show.do?dataset=tai04&lang=en (accessed on 5 May 2021).
46. Regulation No 79/65/EEC of the Council of 15 June 1965 Setting Up a Network for the Collection of Accountancy Data on the Incomes and Business Operation of Agricultural Holdings in the European Economic Community. Available online: https://eur-lex.europa.eu/legal-content/EN/TXT/PDF/?uri=CELEX:31965R0079&from=EN (accessed on 23 June 2021).
47. Kelly, E.; Latruffe, L.; Desjeux, Y.; Ryan, M.; Uthes, S.; Diazabakana, A.; Dillon, E.; Finn, J. Sustainability indicators for improved assessment of the effects of agricultural policy across the EU: Is FADN the answer? *Ecol. Indic.* **2018**, *89*, 903–911. [CrossRef]
48. Zieliński, M. *Emisja Gazów Cieplarnianych a Wyniki Ekonomiczne Gospodarstw Specjalizujących się w Uprawach Polowych*; Institute of Agricultural and Food Economics National Research Institute, Studia i Monografie: Warsaw, Poland, 2016; p. 71.
49. Coderoni, S.; Bonati, G.; D'Angelo, L.; Longhitano, D.; Mambella, M.; Papaleo, A.; Vanino, S. Using FADN data to estimate agricultural GHG emissions at farm level. In *Pacioli 20—Complex Farms and Sustainability in Farm Level Data Collection*; Vrolijk, H., Ed.; LEIWageningen UR: The Hague, The Netherlands, 2013; pp. 86–102.
50. Coderoni, S.; Esposti, R. CAP payments and agricultural GHG emissions in Italy. A farm-level assessment. *Sci. Total Environ.* **2018**, *627*, 427–437. [CrossRef]

51. Syp, A.; Osuch, D. Assessing Greenhouse Gas Emissions from Conventional Farms Based on the Farm Accountancy Data Network. *Pol. J. Environ. Stud.* **2018**, *27*, 1261–1268. [CrossRef]
52. Klepacki, B.; Gołasa, P.; Wysokiński, M. Efektywność emisji gazów cieplarnianych w rolnictwie w UE. *Wieś i Rolnictwo* **2016**, *3*, 129–144. [CrossRef]
53. Dabkienė, V.; Baležentis, T.; Štreimikienė, D. Calculation of the carbon footprint for family farms using the Farm Accountancy Data Network: A case from Lithuania. *J. Clean. Prod.* **2020**, *262*. [CrossRef]
54. Baldoni, E.; Coderoni, S.; Esposti, R. The productivity and environment nexus with farm-level data. The Case of Carbon Footprint in Lombardy FADN farms. *Bio-Based Appl. Econ. J.* **2017**, *6*, 119–137. [CrossRef]
55. The National Centre for Emission Management (KOBiZE) at the Institute of Environmental Protection—National Research Institute. *Krajowy Raport Inwentaryzacyjny 2014. Inwentaryzacja Gazów Cieplarnianych w Polsce dla lat 1988–2012*; The National Centre for Emission Management: Warsaw, Poland, 2014.
56. The National Centre for Emission Management (KOBiZE) at the Institute of Environmental Protection—National Research Institute. *Poland's National Inventory Report 2017. Greenhouse Gas. Inventory for 1988–2015*; The National Centre for Emission Management: Warsaw, Poland, 2017.
57. Intergovernmental Panel on Climate Change (IPCC). *Revised 1996 IPCC Guidelines for National Greenhouse Gas. Inventories*; Intergovernmental Panel for Climate Change: Geneva, Switzerland, 1996.
58. Intergovernmental Panel on Climate Change (IPCC). *IPCC Guidelines for National Greenhouse Gas—Inventories 2006, Volume 4 Agriculture, Forestry and Other Land Use*; Institute for Global Environmental Strategies: Hayama, Japan, 2006.
59. The National Centre for Emission Management (KOBiZE) at the Institute of Environmental Protection—National Research Institute. *Wskaźniki emisyjności CO_2, SO_2, NO_x, CO i Pyłu Całkowitego dla Energii Elektrycznej na Podstawie Informacji Zawartych w Krajowej Bazie o Emisjach Gazów Cieplarnianych i Innych Substancji za 2018 rok*; The National Centre for Emission Management: Warsaw, Poland, 2019.
60. Intergovernmental Panel on Climate Change (IPCC). *Climate Change 2013: Working Group I: The Physical Science Basis*; Cambridge University Press: Cambridge, MA, USA, 2013.
61. Aukcje Polskich Uprawnień do Emisji. Available online: https://www.kobize.pl/pl/article/aktualnosci-2020/id/1717/aukcje-polskich-uprawnien-do-emisji (accessed on 5 May 2021).
62. Prandecki, K.; Gajos, E. Ekonomiczna Wycena Emisji Wybranych Substancji do Powietrza w Polsce ze Szczególnym Uwzględnieniem Rolnictwa. *Kwartalnik Naukowy Uczelni Vistula* **2017**, *3*, 189–207. Available online: http://cejsh.icm.edu.pl/cejsh/element/bwmeta1.element.desklight-537927e7-daaf-4405-bc01-637fcd3b30a4 (accessed on 5 May 2021).
63. Tol, R.S.J. On the Uncertainty About the Total Economic Impact of Climate Change. *Environ. Resour. Econ.* **2012**, *53*, 97–116. [CrossRef]
64. Tubiello, F.N.; Salvatore, M.; Rossi, S.; Ferrara, A.; Fitton, N.; Smith, P. The FAOSTAT database of greenhouse gas emissions from agriculture. *Environ. Res. Lett.* **2013**, *8*. [CrossRef]
65. Smith, P.; Martino, D.; Cai, Z.; Gwary, D.; Janzen, H.; Kumar, P.; McCarl, B.; Ogle, S.; O'Mara, F.; Rice, C.; et al. Policy and technological constraints to implementation of greenhouse gas mitigation options in agriculture. *Agric. Ecosyst. Environ.* **2007**, *118*, 6–28. [CrossRef]
66. Mohammed, S.; Alsafadi, K.; Takács, I.; Harsányi, E. Contemporary changes of greenhouse gases emission from the agricultural sector in the EU-27. *Geol. Ecol. Landsc.* **2020**, *4*, 282–287. [CrossRef]
67. Pérez Domínguez, I.; Fellmann, T.; Weiss, F.; Witzke, P.; Barreiro-Hurlé, J.; Himics, M.; Jansson, T.; Salputra, G.; Leip, A. *An Economic Assessment of GHG Mitigation Policy Options for EU Agriculture (EcAMPA 2)*; JRC Science for Policy Report: Seville, Spain, 2016. [CrossRef]
68. Greenhouse Gas Emissions by Source Sector. Available online: http://appsso.eurostat.ec.europa.eu/nui/show.do?dataset=env_air_gge&lang=en (accessed on 5 May 2021).
69. Piwowar, A. Low-Carbon Agriculture in Poland: Theoretical and Practical Challenges. *Pol. J. Environ. Stud.* **2019**, *28*, 2785–2792. [CrossRef]
70. The National Centre for Emission Management (KOBiZE) at the Institute of Environmental Protection—National Research Institute. *Poland's National Inventory Report 2019. Greenhouse Gas. Inventory for 1988–2017*; The National Centre for Emission Management: Warsaw, Poland, 2019.
71. Rokicki, T.; Koszela, G.; Ochnio, L.; Golonko, M.; Żak, A.; Szczepaniuk, E.K.; Szczepaniuk, H.; Perkowska, A. Greenhouse Gas Emissions by Agriculture in EU Countries. *Rocznik Ochrona Środowiska* **2020**, *22*, 809–824.
72. Balafoutis, A.; Beck, B.; Fountas, S.; Vangeyte, J.; Wal, T.V.d.; Soto, I.; Gómez-Barbero, M.; Barnes, A.; Eory, V. Precision Agriculture Technologies Positively Contributing to GHG Emissions Mitigation, Farm Productivity and Economics. *Sustainability* **2017**, *9*, 1339. [CrossRef]
73. Measures at Farm Level to Reduce Greenhouse Gas Emissions from EU Agriculture. Available online: http://www.europarl.europa.eu/RegData/etudes/note/join/2014/513997/IPOL-AGRI_NT(2014)513997_EN.pdf (accessed on 5 May 2021).
74. Gradziuk, P.; Gradziuk, B. Gospodarka niskoemisyjna—Nowe wyzwanie dla gmin wiejskich. *Wieś i Rolnictwo* **2016**, *1*, 105–126. [CrossRef]
75. Steinfeld, H.; Gerber, P.J.; Wassenaar, T.; Castel, V.; Rosales, M.; de Haan, C. *Livestock's Long Shadow: Environmental Issues and Options*; Food and Agriculture Organization of the United Nations: Rome, Italy, 2006.

76. Intergovernmental Panel on Climate Change (IPCC). *Climate Change 2014: Mitigation of Climate Change. Contribution of Working Group III to the Fifth Assessment Report of the Intergovernmental Panel on Climate Change*; Cambridge University Press: Cambridge, MA, USA, 2014.
77. MacLeod, M.J.; Vellinga, T.; Opio, C.; Falcucci, A.; Tempio, G.; Henderson, B.; Makkar, H.; Mottet, A.; Robinson, T.; Steinfeld, H.; et al. Invited review: A position on the Global Livestock Environmental Assessment Model (GLEAM). *Animal* **2018**, *12*, 383–397. [CrossRef] [PubMed]
78. Wainwright, H.; Jordan, C.; Day, H. Environmental impact of production horticulture. In *Horticulture: Plants for People and Places*; Dixon, G.R., Aldous, D.E., Eds.; Springer: Dordrecht, The Netherlands, 2014; Volume 1, pp. 503–522.
79. Gołasa, P.; Wysokiński, M.; Bieńkowska, W. Magazynowanie jabłek jako proces logistyczny w rolnictwie. *Logistyka* **2012**, *6*, 429–433.
80. Akdemir, S.; Akcaoz, H.; Kizilay, H. An analysis of energy use and input costs for apple production in Turkey. *J. Food Agric. Environ.* **2012**, *10*, 473–479. [CrossRef]
81. Król, E. Porównanie emisji zanieczyszczeń pojazdów z napędem elektrycznym i spalinowym. *Napędy Sterow.* **2017**, *7/8*, 140–143.
82. Hedenus, F.; Wirsenius, S.; Johansson, D.J.A. The importance of reduced meat and dairy consumption for meeting stringent climate change targets. *Clim. Chang.* **2014**, *124*, 79–91. [CrossRef]
83. World Health Organization (WHO). *Healthy Diet*; Fact. Sheet No. 394; WHO: Geneva, Switzerland, 2018; Available online: https://www.who.int/publications/m/item/healthy-diet-factsheet394 (accessed on 5 May 2021).
84. Wirsenius, S.; Hedenus, F.; Mohlin, K. Greenhouse gas taxes on animal food products: Rationale, tax scheme and climate mitigation effects. *Clim. Chang.* **2010**, *108*, 159–184. [CrossRef]
85. Edjabou, L.D.; Smed, S. The effect of using consumption taxes on foods to promote climate friendly diets—The case of Denmark. *Food Policy* **2013**, *39*, 84–96. [CrossRef]
86. Pellerin, S.; Bamière, L.; Angers, D.; Béline, F.; Benoit, M.; Butault, J.-P.; Chenu, C.; Colnenne-David, C.; De Cara, S.; Delame, N.; et al. Identifying cost-competitive greenhouse gas mitigation potential of French agriculture. *Environ. Sci. Policy* **2017**, *77*, 130–139. [CrossRef]
87. Sosulski, T.; Szymańska, M.; Szara, E. Assessment of various practices of the mitigation of N_2O emissions from the arable soils of Poland. *Soil Sci. Annu.* **2017**, *68*, 55–64. [CrossRef]
88. MacLeod, M.; Eory, V.; Gruère, G.; Lankoski, J. Cost-Effectiveness of Greenhouse Gas Mitigation Measures for Agriculture. *OECD Food Agric. Fish. Pap.* **2015**, *89*. [CrossRef]
89. Piechota, G.; Igliński, B. Biomethane in Poland—Current status, potential, perspective and development. *Energies* **2021**, *14*, 1517. [CrossRef]
90. Pilarski, G.; Kyncl, M.; Stegenta, S.; Piechota, G. Emission of biogas from sewage sludge in psychrophilic conditions. *Waste Biomass Valorization* **2020**, *11*, 3579–3592. [CrossRef]
91. Haneklaus, N.; Schnug, E.; Tulsidas, H.; Tyobeka, B. Using high temperature gas-cooled reactors for greenhouse gas reduction and energy neutral production of phosphate fertilizers. *Ann. Nucl. Energy* **2015**, *75*, 275–282. [CrossRef]
92. Syp, A.; Faber, A. Zastosowanie Modelu DNDC do Symulacji Plonów Roślin i Oceny Wpływu Zmian na Środowisko w Zmieniających Się Warunkach Klimatycznych i Różnych Systemach Uprawy. *Roczniki Naukowe Stowarzyszenia Ekonomistów Rolnictwa i Agrobiznesu* **2012**, *14*, 183–187. Available online: https://rnseria.com/resources/html/article/details?id=172128 (accessed on 9 June 2021).
93. Soto Embodas, I.; Barnes, A.; Balafoutis, A.; Beck, B.; Sanchez Fernandez, B.; Vangeyte, J.; Fountas, S.; Van Der Wal, T.; Eory, V.; Gomez Barbero, M. *The Contribution of Precision Agriculture Technologies to Farm Productivity and the Mitigation of Greenhouse Gas Emissions in the EU*; Publications Office of the European Union: Luxembourg, 2019. [CrossRef]
94. Future of Farming. Available online: https://www.deere.co.uk/en/agriculture/future-of-farming/ (accessed on 9 June 2021).
95. Monarch Tractor: The World's Smartest, Fully Electric, Autonomous Tractor. Available online: https://www.monarchtractor.com/monarch-blogs/#/news/1toEOumnkEksWakieoeC6M (accessed on 9 June 2021).
96. Regulation (EU) 2016/1628 of the European Parliament and of the Council of 14 September 2016 on Requirements Relating to Gaseous and Particulate Pollutant Emission Limits and Type-Approval for Internal Combustion Engines for Non-Road Mobile Machinery, Amending Regulations (EU) No 1024/2012 and (EU) No 167/2013, and Amending and Repealing Di-rective 97/68/EC. Available online: https://eur-lex.europa.eu/legal-content/EN/TXT/PDF/?uri=CELEX:32016R1628&from=EN (accessed on 23 June 2021).
97. Pellerin, S.; Bamière, L.; Angers, D.; Béline, F.; Benoît, M.; Butault, J.P.; Chenu, C.; Colnenne-David, C.; De Cara, S.; Delame, N.; et al. *How Can French Agriculture Contribute to Reducing Greenhouse Gas Emissions? Abatement Potential and Cost of Ten Technical Measures*; Synopsis of the Study Report; INRA: Paris, France, 2013; Available online: https://hal.inrae.fr/hal-02809908/document (accessed on 9 June 2021).
98. Ocena Wpływu Polityki Klimatycznej na Sektor Polskich Gospodarstw Rolnych. Available online: http://climatecake.pl/wp-content/uploads/2020/07/CAKE_Rolnictwo_ocena_wplywu_polityki_klimatycznej_Podsumowanie_final.pdf (accessed on 5 May 2021).

energies

Article

The Role of Agriculture and Rural Areas in the Development of Autonomous Energy Regions in Poland

Grzegorz Ślusarz [1], Barbara Gołębiewska [2,*], Marek Cierpiał-Wolan [1], Dariusz Twaróg [3], Jarosław Gołębiewski [2] and Sebastian Wójcik [4]

1 Institute of Economics and Finance, University of Rzeszów, M. Ćwiklińskiej Street 2, 35-601 Rzeszów, Poland; gslusarz@ur.edu.pl (G.Ś.); M.Cierpial-Wolan@stat.gov.pl (M.C.-W.)
2 Institute of Economics and Finance, Warsaw University of Life Sciences-SGGW, Nowoursynowska Street 166, 02-787 Warsaw, Poland; jaroslaw_golebiewski@sggw.edu.pl
3 Statistical Office in Rzeszów, Jana III Sobieskiego Street 10, 35-001 Rzeszów, Poland; D.Twarog@stat.gov.pl
4 Institute of Mathematics, University of Rzeszów, Pigonia Street, 35-959 Rzeszów, Poland; swojcik@ur.edu.pl
* Correspondence: barbara_golebiewska@sggw.edu.pl; Tel.: +48-22-5934228

Abstract: In many countries, energy security is treated as a priority for the coming decades, and at the same time energy production from the vast majority conventional energy sources does not meet environmental protection criteria. Hence, the need to use renewable energy sources (RES), which can largely satisfy energy needs. The aim of the study was to identify possibilities of creating autonomous energy regions (ARE) in Poland, based on renewable energy sources. Attention was paid to the role and significance of the potential of rural areas in this respect, taking into account the possibilities of increasing energy production from these sources in individual regions of Poland. The research was conducted on a regional level (division into voivodships) and on a local level (division into powiats, which form voivodships). When assessing the potential for constructing ARE based on RES, the following energy sources were taken into account: water, wind, sun, biogas and biomass. It was found that the highest RES potential versus energy consumption can be obtained in powiats where the share of arable land and forests exceeds 80%. The research showed that in most regions of Poland (powiats, voivodships), there is a large potential for obtaining additional energy from RES, which would cover over 73% of the country's demand for electricity. This could be the basis for building energy independence on a local scale. The results of the study indicated that as many as seven regions would become self-sufficient in terms of electricity demand.

Keywords: regional potential; rural areas; renewable energy sources; autonomous energy regions

Citation: Ślusarz, G.; Gołębiewska, B.; Cierpiał-Wolan, M.; Twaróg, D.; Gołębiewski, J.; Wójcik, S. The Role of Agriculture and Rural Areas in the Development of Autonomous Energy Regions in Poland. *Energies* **2021**, *14*, 4033. https://doi.org/10.3390/en14134033

Academic Editor: David Borge-Diez

Received: 31 May 2021
Accepted: 1 July 2021
Published: 4 July 2021

1. Introduction

Energy production, like food production, is of strategic importance to every country. At the current stage of development, there is an increasing demand for energy, which affects the socioeconomic development and quality of life of the population [1,2]. In the report *International Energy Outlook*, released by the Energy Information Administration (EIA), it was assumed that by the middle of this century energy consumption will increase by about 50% [3], and only from 2000 to 2018 global energy consumption increased by 48.2% [4]. At the same time, it was emphasized that energy production, in its vast majority, is based on conventional energy carriers [5,6], which poses an increasing threat to the environment [7] and human health, especially children's health [8]. Increased energy consumption determines the competitiveness and growth opportunities of businesses and the well-being of households [9], and as Gielen et al. [10] state, renewable energy can meet two-thirds of total global energy demand, but this will require new technologies and innovations.

Herington et al. [11] indicate that billions of people worldwide remain without access to modern energy services, most of whom live in rural areas [1,12]. To support the deployment of such services, local energy solutions must be taken into account [13]. Hence, there is a need to look for innovative solutions to meet growing energy demands [14–16], improving energy security while reducing the negative effects of its production on the environment [17–20], and thus on the health (quality of life) of people. A solution to this search is the development of Autonomous Energy Regions (ARE) based on RES, seen as the construction of energy self-sufficient regions (local development of RES potential, also sometimes referred to as Municipal Energy Centres) that fit into the socio-economic conditions of the development of a given area and their sustainable development [21,22]. As energy production based on RES has a spatial character, a special role in the creation of ARE is played by rural areas, which constitute the dominant part of Poland's area (93%) [23] and endogenous potential related to agriculture and the development of nonagricultural functions of these areas [24,25].

The idea of ARE is in line with the latest global trends of energy distribution based on locally available energy sources that contribute to increased efficiency of the energy system [4,7,26]. Undoubtedly, RES-related technologies are a part of smart technologies and, at the same time, part of the Directive (EU) 2018/2001 of the European Parliament and of the Council of 11 December 2018 on the promotion of the use of energy from renewable sources. The directive highlights the coupling of the particular importance of rural areas in the development of RES and, vice versa, the development of RES for rural development [27,28].

The need for RES development is widely recognised. The possibility of transmitting energy from remote areas is even indicated to move towards a sustainable, resource-efficient and low-carbon energy system [29]. Müller et al. [30] report that the use of renewable energy sources (RES) in small autonomous (decentralised) power systems can reduce emissions and increase the cost-effectiveness of energy supply. This is confirmed in a study by Marchenko and Solonim [31], which indicates the use of renewable energy sources in autonomous energy supply systems proves to be economically efficient. However, it also highlights that deployment of renewable energy sources is cost-intensive [32,33].

For some renewable sources, electricity production depends on environmental factors such as water resources (hydroelectric power plants) or weather (wind and solar power plants), and, as studies indicate, these resources vary regionally [34–36]. RES, in contrast to conventional power plants, are characterized by a much wider range of net efficiency: from 9% for photovoltaic panels to about 80% for hydroelectric power plants [37]. This lower efficiency (except for hydroelectric power plants) translates into higher costs of energy production in the micro account, which, in the case of RES, should be considered on a macro scale taking into account the theory of public goods. A special feature of public goods related to agriculture and rural areas is the fact that they may be an external effect of agricultural production [38,39], an intentional effect or a common resource held by society [40]. This is particularly important in relation to the creation of autonomous energy regions based on energy from RES.

It is worth emphasizing that in recent years there have been many studies focusing on the description of the renewable energy potential [41] or the dynamics of electricity production from renewable sources in Poland [42]. Some of the works usually focused on one of the renewable sources, such as the description of hydropower resources made by Kowalczyk and Cieśliński at the voivodship level [43] or the energy potential of straw and hay estimated by Jarosz [44] at the municipal level. Therefore, due to different levels of spatial aggregation and selective presentation of particular issues, a need arose to describe the entire potential of renewable energy in the era of the upcoming energy transformation. Research in this field is an important contribution to work on improving the energy policy supporting the development of renewable energy sources as an important factor in improving the quality of the environment and limiting the causes and effects of greenhouse effects.

The aim of the paper is to identify potential possibilities of constructing autonomous energy regions (ARE) in Poland based on renewable energy sources on the basis of diverse endogenous potential of regions. The main research problem relates to finding an answer to the fundamental question of what role do and can rural areas play in the construction of ARE and in the production of energy from renewable sources in Poland. In particular, we seek to find out:

- The current role of RES in balancing the total energy production in the regions;
- The possibilities for increasing energy production from these sources in individual regions of Poland;
- The current balance of energy production and consumption in each region;
- The status and potential for development of energy production: wind, hydro, biogas, biomass and photovoltaic;
- The potential for energy generation (energy density) from RES per hectare of nonurbanised land (including agricultural land and forests) and per capita.

The opportunities for the ARE development resulting from varied endogenous potential at regional and local levels are still poorly recognised in Poland. Such recognition is important from the point of view of the need to build the ARE development strategy, taking into account the specificity of regions resulting from their endogenous potential as well as from the point of view of the necessary measures supporting the ARE development.

2. Materials and Methods

The research was carried out on a regional basis in Poland. The spatial scope of the research covered all voivodships (regional level) of the country and their powiats (local level). In the projection of the assessment of potential for constructing RES-based ARE, the study was limited to the main sources, namely energy from: water, wind, sun, biogas and biomass. The potential of electricity production from these sources was calculated for each powiat. In the case of biogas and biomass, it was assumed that these raw materials are burned in cogeneration devices with an overall efficiency factor of 65% (this is the lower efficiency limit of devices currently available on the market) [45]. Thus, the obtained values of electricity production from the considered RES were compared with the data included in the balance report of energy carriers and heating infrastructure (G-02b) prepared by Statistics Poland in 2018 [46]. The choice of electricity for the assessment of RES potential and the construction of ARE relates to the fact that it is a form of energy that is significantly more difficult to produce than heat. Therefore, comparing the potential of electricity production from renewable sources against its consumption allows for a more detailed picture of RES potential for individual powiats.

Bearing in mind that some powiats, due to the location of large power plants, may have a considerable surplus of electricity, in further calculations only the consumption of electricity was taken into account excluding the energy industry and lignite mining (lignite-fired power plants and lignite mines are in fact the same complex). The above assumption made it possible to calculate the demand for energy in each of the powiats, disregarding the so-called parasite power from conventional utility power plants. The calculation of electricity production potential was carried out for each renewable energy source separately. Subsequently, their total potential was compared with the current electricity consumption.

2.1. Calculation of Small Hydropower Potential

In Poland, as in the rest of Europe, the construction of large hydropower plants (with a capacity of several hundred MW) is impossible, mainly due to the geographical and ecological conditions. Therefore, the study focuses on the possibility of hydropower development based on the expansion of MEW, which, according to the terminology used in Poland are facilities with an installed capacity of less than 5 MW. In contrast to large hydropower plants, small hydropower plants can be built on existing water stages and the electricity produced in these installations could be used primarily to meet local needs. Most MEW can operate on the basis of an Archimedes turbine using a low water drop of 1~10 m and low flow, so it can be easily seen that they fit perfectly into the hydrological conditions of the country. The potential of MEW was calculated based on Renewable

Energy Sources Transforming Our Regions (RESTOR) Hydro database, which shows that there are more than 6000 sites in Poland suitable for MEW construction [47]. In order to calculate the so-called technical potential, it was assumed that the installations operate with a net efficiency of 80% for 70% of hours in a year [48]. The calculations of the potential of small hydropower plants (MEW) were carried out based on the RESTOR Hydro database taking into account facilities for which the size of the minimum drop exceeds 1.6 m and the annual average flow is not less than 0.1 m^3/s [49]. The facilities that were found in the inventory materials as dams that once existed and were decommissioned were not taken into account. In this way, the potential of MEW power was achieved, which was feasible for economic reasons.

2.2. Calculation of Wind Energy Potential

An analysis of the wind energy zone map prepared by the Institute of Meteorology and Water Management—National Research Institute shows that about 60% of Polish territory has favourable conditions for wind energy development [50]. Wind has the largest share in energy production from renewable sources. Onshore wind power is currently the cheapest new energy generation technology in Poland [51]. The installed capacity of wind farms was over 6.7 GW in March 2021 [52]. However, it should be remembered that a large part of such areas is excluded from the possibility of their use by various forms of nature protection, e.g., national parks and their buffer zones, landscape parks and Natura 2000 areas [53]. Buildings or inaccessibility of the terrain (mountainous areas, swampy areas or dense forest complexes) are also a limitation. The factor of the so-called roughness of the terrain, which is determined mainly by the proximity of forest complexes and buildings, is indicated as very important [54]. In practice, the introduction of the so-called 10H rule [55], under which the permissible distance of a windmill from residential buildings is to be equal to or greater than ten times the height of the wind turbine measured from ground level to the highest point of the windmill, including the rotor with blades, has limited the possibility of building new wind turbines in Poland.

Furthermore, an important aspect recently raised in public discussion is the impact on human health. Negative factors include the stroboscopic effect [56] and noise caused by rotor blades in operation [57–59]. The above-mentioned factors make it necessary to locate wind turbines at a distance of at least 500 m from the nearest buildings or terrain obstacles. In our calculations, the condition of roughness of the terrain was taken into account, among other things, by eliminating all cities with powiat rights. An important limitation is also the accessibility to the energy network and the possibilities of connecting the power.

Applying the above restrictions, a list of 257 powiats with areas convenient from the point of view of wind energy was obtained, then only powiats where the wind power exceeds 750 $kWh/m^2/year$ were taken into account (i.e., areas ranging from extremely favourable to favourable wind energy zones in Poland). As a result, a list of 102 powiats was obtained in which the construction of wind turbines with the use of current technologies would be profitable (marked with a windmill symbol in Figure 1). Taking into account the average power of wind turbines built in Poland so far, ~2.2 MW, and an average effective onshore working time of up to 22% hours per year (according to PSE—Polish Power Grid Company), it was calculated that the total value of electricity production would reach 17,498 GWh, which would cover more than 10.3% of electricity production in 2018. Subsequently, the potential of wind electricity production was divided by electricity consumption (G-02b report, Statistics Poland), which allowed for a more accurate assessment of the importance of wind energy at the powiat level.

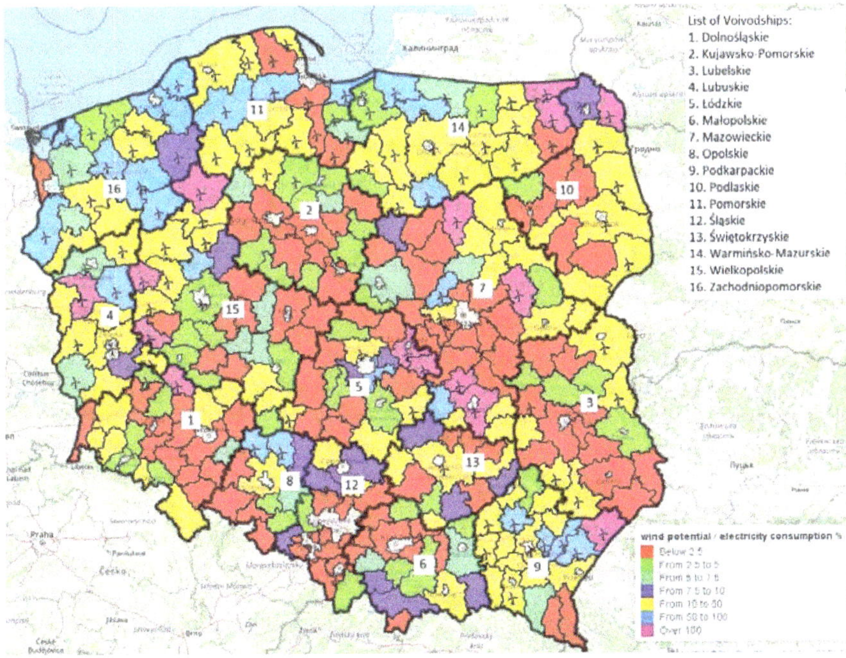

Figure 1. Wind energy potential relative to electricity consumption in 2018, powiats meeting the criteria necessary for wind energy development are marked with a windmill symbol. Source: own study.

2.3. Methods for Calculating Solar Energy Potential

In the case of our country, 80% of the total annual amount of insolation falls on the spring and summer term, from the beginning of April to the end of September. This means that the energy obtained from this source will not be evenly distributed over time, and the highest yield is achieved in the period from spring to autumn. This study focuses on the possibility of using solar energy to generate electricity (using photovoltaic cells). The most important value for the potential energy yield is insolation. Monthly sums of insolation for particular voivodships were used in the calculations, obtained with the help of the solar radiation potential distribution taken from the Atlas of the Republic of Poland [60]. On average, in Poland, each square meter of land (horizontal area) receives annually from 950 kWh to 1160 kWh in the form of solar radiation. In the case of direct radiation reaching directly from the solar disk, the optimal use of the absorbing surface would require that such surface be oriented perpendicular to the direction of the radiation. In practice, it is unrealistic in the latitudes of Poland. Therefore, it is necessary to select the installation inclination in the optimal angle to the incident radiation, which would allow for the highest energy yields and proper operation of the installation. For the purposes of the calculations, it was assumed that the inclination of the installation at an optimal angle in relation to the incident radiation is in the range of 25 to 40 degrees for the southern regions of the country and from 30 to 50 degrees in the north of Poland, such an arrangement ensures the most effective use of solar energy [61]. Next, the average cycle of the length of the day in the following months of the year was taken into account, based on the data of the Central Office of Measures [62] and the changes in insolation, which are influenced by meteorological conditions (Institute of Meteorology and Water Management). Moreover, the calculations take into account the influence of the orientation of photovoltaic modules in relation to the directions of the world (for the southern orientation, the average energy production reaches 3300 kWh/year, while for the northern orientation it is only 1900 kWh/year) [63]. When calculating the technical potential, it was assumed that electricity would be produced by monocrystalline silicon cells with an average standard efficiency of 15%. In real conditions, with the panels facing south at the optimal all-year angle, the average annual efficiency is in the range from 55% to 60% of the normative efficiency. For a monocrystalline silicon cell with a normative efficiency of 15%, this means an average annual efficiency of about 9%.

When calculating the potential of solar energy, the focus was on the possibility of generating solar energy on the premises of photovoltaic farms and investigating the potential related to the use of this type of installation in the case of single-family buildings (so-called microinstallations). Pursuant to the legal requirements, solar farms may be built only in areas with IV or lower valuation class, as well as on wastelands. In this study, it was assumed that farms may be located on wasteland, but excluding meadows and pastures, and in order not to compete with agricultural crops, they should not occupy more than one tenth of the wasteland [64]. Due to the terrain (e.g., slope inclination, shading, exposure direction) and the possibility of connecting to the electric network, the area of wastelands that could be used in each of the powiats was a subject to further restrictions. In the case of home photovoltaic installations it was assumed that the most frequently used place for investment is the roof of a building with an area of approx. 25 m^2, a slope of 35 degrees and southern exposure [63].

2.4. Calculation of Biogas Potential

The basis for the calculation of biogas potential was statistical data from Statistics Poland concerning agriculture and waste management [65]. When calculating the biogas potential, the possibility of its production from organic waste in landfills, animal and plant waste in farms and sludge in sewage treatment plants was taken into account. Under optimal conditions, about 400~500 m^3 of biogas can be produced from one tonne of municipal waste. However, in reality not all organic waste is fully decomposed and the fermentation process depends on many factors. Therefore, in calculations it is assumed that a maximum of 200 m^3 of biogas can be obtained from one tonne of waste [66].

Livestock farms generate significant amounts of waste that can be used for biogas production. The potential of agricultural biogas production is determined mainly by the amount of agricultural waste available [67]. In conducted calculations it was assumed that from 1 m^3 of liquid faeces it is possible to obtain 20 m^3 of biogas on average, and from 1 m^3 of manure, 30 m^3 of biogas [49]. The analysed data included the livestock by species (cattle, pigs, horses, sheep, poultry, goats) and performance groups and the average annual production volumes, natural fertilisers depending on the animal species, its age and performance and the housing system. In order to calculate biogas emission from manure originating from individual animal species, a study prepared by the Zootechnics Institute was used [68].

Sludge from sewage treatment plants is also used for biogas production. In order to calculate biogas emissions from the treatment plant, it was assumed that 100 m^3 of biogas can be obtained from 1000 m^3 of incoming sewage [69].

Another source that can be used for biogas production is agricultural crops [70,71]. In the calculations, only losses and wastage of agricultural crops were taken into account, such as: basic cereals with mixtures, potatoes, vegetables, fruit, maize, legumes and oil plants. Data on losses and wastage of these crops were taken from the Agricultural Statistical Yearbooks of Statistics Poland [72]. On the other hand, data taken from [73,74] were used to estimate the amount of biogas obtainable from plant biomass.

2.5. Calculation of Biomass Potential

For the estimation of biomass potential, it was assumed that it would come from crop production; including straw surpluses, hay surpluses, energy crops, orchards, forestry production as well as annual prefelling and tending cuts. In calculating the potential offered by the wood industry, the methodology proposed by Bujakowski [49] was used:

- One hectare of forest may yield 45 tonnes of wood, this amount is assumed for 1% of the forest area, furthermore it is assumed that 12 tonnes of wood may be harvested from one hectare of forest from pre-cutting and tending cuts, and this amount refers to 5% of the forest area.
- For every 100 m^3 of wood mass harvested in the forest, after deducting 36 m^3 of sawn wood for finished wood products, it is assumed that the remaining 64 m^3 of wood mass can be used for energy purposes.

The formula [75,76] was used to assess the surplus straw available for energy use:

$$N = P - (Z_s + Z_p + Z_n) \quad (1)$$

where: N—surplus straw for nonagricultural use (tonnes), P—volume of cereal straw production (tonnes), Z_s—straw demand for bedding (tonnes), Z_p—straw demand for fodder (tonnes), Z_n—straw demand for ploughing (tonnes).

In calculating the volume of straw production, demand for fodder, litter and ploughing, the methodology used in Hrynkiewicz's work [77] was used. In the case of straw shortages in individual powiats, it was assumed that they are supplemented by "imports" from neighbouring powiats having surpluses of this raw material. Then, the estimated straw surplus was converted into energy, assuming that 1 ton of straw with 15% moisture content has a calorific value of 13.1 GJ [44].

Despite being less popular, as indicated by Pudełko [78], surplus hay is also a significant biomass resource, which counts in potential use for energy purposes. In order to calculate the hay surplus, the statistical data of Statistics Poland covering the harvest of permanent meadows and permanent pastures, the population of ruminants (cattle, horses, sheep, goats) and their annual demand for fodder were used [79]. The problem of hay shortages in individual powiats was solved using the same assumption as for straw. The estimated hay surplus was converted into energy by assuming that 1 tonne of hay with 15% moisture content has a calorific value of 13.4 GJ [44].

An important role among the potential resources of solid biomass is also played by energy crops (willow, miscanthus, sida hermaphrodita, poplar). The analysis of the potential from energy plantations was carried out on the basis of statistical data from Statistics Poland and information obtained from the Agency for Restructuring and Modernisation of Agriculture (ARiMR) [80]. The potential of energy crops was calculated using the equation [44]:

$$P_w = [P_e + (P_g \cdot w_e)] \cdot Y_e, \quad (2)$$

where: P_w—potential of perennial energy crops (tonne), P_e—area of existing plantations of perennial energy crops (ha), P_g—sum of land of the soil quality class V, VI and VIz bonitation class (ha), w_e—land use coefficient for perennial energy crops (%), Y_e—average perennial yield (tonne/ha).

In the calculations, the w_e value equal to 1/10 was assumed as a safe limit eliminating competition between the production of raw materials and production for food purposes [81]. For the purpose of calculations, it was also assumed that in the case of marginal soils, the average yield will be at the level of 7.5 tonnes/ha [82], while the energy value was assumed at the average level of 17 MJ/kg [83]. When calculating the biomass potential, the biomass obtainable from orchard stand maintenance and replacement was also taken into account; for the purpose of the calculations, it was assumed that 3 tonnes of dry matter could be obtained from one hectare of orchard per year [84].

2.6. Ranking of Voivodships

At a further stage of the work, the level of utilisation of renewable energy sources was assessed using the TOPSIS method (Technique for Order Preference by Similarity to an Ideal Solution) in positional terms with the application of Weber's spatial median [85,86]. The TOPSIS method is based on the idea of constructing a synthetic pattern and antipattern of values of diagnostic features and enables synthetic evaluation of a phenomenon described by many features.

3. Results

3.1. Small Hydropower Potential

Water is essential in the production of electricity. According to Statistics Poland data, generation and supply of electricity, gas, steam and hot water consumed 6284.4 hm^3 of water in 2018, which accounted for more than 66.5% of the total water withdrawal in Poland. For comparison, the EU average in 2018 was 13.7% [87]. According to the Energy Regulatory Office, in 2018, in Poland there were 586 small hydropower plants (SHP) with an installed capacity of up to 300 kW, 96 hydropower plants with an installed capacity of up to 1 MW, 68 hydropower plants with an installed capacity of up to 5 MW, and 20 with a capacity greater than 5 MW. In 2018, the total electricity production of hydroelectric power stations was 2.387 TWh, which accounted for more than 1.4% of the total electricity production in Poland. According to some studies, this is only 20% of the energy potential of Polish rivers. Thus, the maximum use of the energy of falling water would cover about 7% of current electricity consumption [88]. Thus, hydropower in Poland can only play a supporting role, offering a stable (compared to, e.g., wind or sun) power source.

The results of calculations show that the activation of the SHP potential consisting of more than 6000 facilities would allow the production of 584 GWh of electricity, which would account for only 0.34% of energy production in 2018. On the other hand, the potential of SHP in covering electricity consumption for most powiats in Poland would be at the level of <1%. The situation was more favourable in the eastern and northern parts of the country. This was mainly due to low energy consumption in areas where there are no energy-intensive industries. Thus, as can be seen, the potential of small hydropower cannot solve Poland's energy problems, but the development of SHP and construction of small water reservoirs would have a beneficial effect on increasing water retention. This is especially important in the central part of Poland threatened by desertification/stagnation, as pointed out by Kowalczak [89] and Kudlicki [90]. The construction of MEW would allow, first of all, to rebuild the country's water resources (the so-called small retention), greatly needed both by agriculture and the "traditional" coal-burning power industry.

3.2. Wind Energy Potential

In the calculations of wind potential the most important factors limiting the development of onshore wind energy were taken into account. Figure 1 presents the differentiation of Poland's areas (broken down into powiats) with respect to wind energy production possibilities by establishing the wind potential in relation to electricity consumption. The largest wind energy potential is found in the powiats located in the western, north-western and north-eastern parts of the country. Powiats in Świętokrzyskie and Podkarpackie voivodships also belong to the group of areas with significant wind energy potential. It should be noted that the high share of wind energy in meeting the demand for electricity in individual powiats results not only from favourable natural conditions (high average annual wind power and low roughness coefficient), but primarily from low electricity consumption in these powiats.

When broken down by voivodships, the largest wind energy potential is found in Pomorskie voivodship (2258 GWh/year), while the smallest in Śląskie voivodship. It is worth noting that, according to PSE data, in the same year wind energy covered over 7.6% of the electricity demand, while in 2019 it was over 9.2%. Summing up, the results of our calculations show that the developmental potential of this branch of renewable energy in the inland part of the country is close to exhaustion, while taking into account the good wind conditions, it can be seen that the decisions on the location of this type of investment in the Baltic Sea are perfectly justified. Due to the emergence of new technologies allowing the storage of energy from renewable sources, the sense of building wind power plants acquires new importance, as indicated by Watson et al. [91]. A good example is the CO_2 methanation system for storing electricity through SNG (substitute natural gas) production [92]. It is indicated that this kind of solution would stabilise the operation of wind turbines through energy storage [93].

3.3. Solar Energy Potential

The obtained results indicate that, on a national scale, large-scale PV farms could produce 10,592 GWh of electricity per year), while the potential of domestic PV micro-installations reached 430.9 GWh per year. The total potential of Polish photovoltaic installations is estimated at 11,022.9 GWh per year, which corresponds to 6.48% of the electricity production in 2018 [23]. The results of the calculation by powiats are presented in Figure 2.

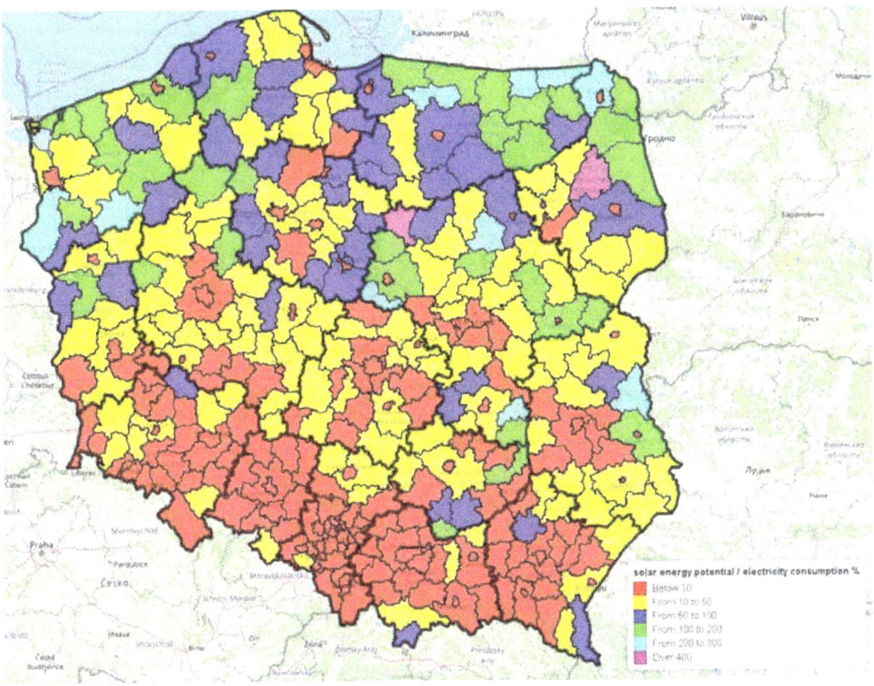

Figure 2. Solar energy potential relative to electricity consumption in 2018. Source: own study.

As can be seen in Figure 2 for most powiats the calculated solar energy potential did not exceed 10% of electricity consumption. In the scale of voivodships the leader in solar energy potential is still Zachodniopomorskie voivodship (1682 GWh/year), while the smallest potential is in Opolskie voivodship (101 GWh/year). However, the potential of solar energy is more significant in the northern and northeastern parts of Poland. This is partly due to the fact that, with the exception of urban centres, energy-intensive industries are mainly located in the central and southern parts of the country.

3.4. Biogas Potential

The biogas production potential was calculated according to the methodology presented in subchapter 2. Further, in order to calculate electricity production, it was assumed that 2.1 kWh of electricity can be produced in cogeneration from 1 m^3 of biogas [94]. The obtained results were compared with the data of Statistics Poland for 2018 included in the balance report on energy carriers and heating infrastructure (G-02b). They were further presented in graphical form in Figure 3.

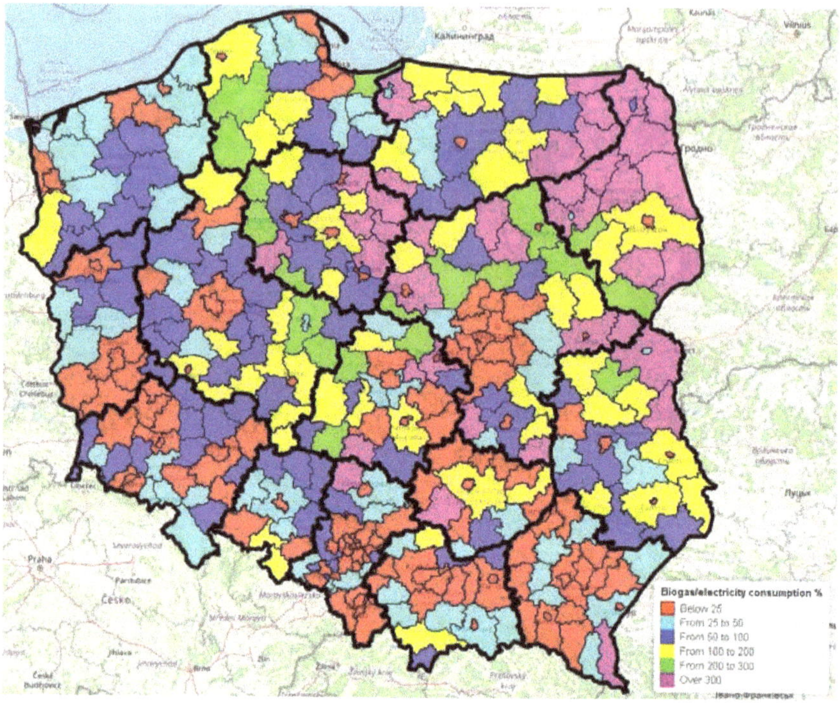

Figure 3. Electricity demand coverage through biogas electricity generation in 2018. Source: own study.

The results of Table 1 (Chapter 3.6) indicate that on a national scale biogas energy production could cover more than 11% of electricity consumption, while in the case of the voivodships, Podlaskie voivodship remains the leader, where biogas energy production could cover almost 53% of electricity consumption. This is mainly connected with intensive cattle breeding in that region, but also with the lack of energy-intensive industries. Warmińsko-Mazurskie (27.8%) and Wielkopolskie (21.4%) voivodships also have a high potential of biogas energy production, which is also connected with intensive agricultural production in those areas.

The smallest potential in terms of biogas production is shown by Śląskie voivodship (3.6%), whose biogas production potential is more than double that of Podlaskie voivodship, but the consumption of natural gas due to the heavy industry located there is more than 11 times higher. When analysing the structure of biogas production sources it can be seen that natural fertilisers, especially cattle manure, have the greatest potential in its creation, as also indicated by Sefeedpari et al. [95]. According to our calculations, on a national, scale cattle manure can cover more than 81% of biogas production from natural fertilizers. In the case of Podlaskie voivodship even 96%, which results from the highest cattle density in the country: 93.9 heads/ha [23]. The results of our calculations indicate that in the case of using biogas for electricity production, only 42 powiats with a developed agricultural economy could achieve energy independence.

Table 1. Relative potential to electricity consumption in 2018.

Voivodships	RES Potential [GWh/Year]				Share of RES Potential in Total Consumption [%]			
	WWS (Water + Wind + Sun)	Bg (Biogas)	Bm (Biomass)	WWSBgBm (Water + Wind + Sun + Biogas + Biomass)	WWS/Total Consumption	Bg/Total Consumption	Bm/Total Consumption	WWSBgBm Total Consumption
Dolnośląskie	358.75	660.31	3656.89	4675.94	3.45	6.35	35.16	44.95
Kujawsko-Pomorskie	1012.58	1011.20	2758.11	4781.89	15.78	15.76	42.98	74.52
Lubelskie	1412.84	705.21	3398.27	5516.32	26.67	13.31	64.16	104.15
Lubuskie	1373.73	295.40	2694.20	4363.33	46.63	10.03	91.44	148.10
Łódzkie	2027.25	1000.05	2126.65	5153.95	24.82	12.24	26.04	63.10
Małopolskie	340.43	745.80	1828.58	2914.80	3.41	7.46	18.29	29.16
Mazowieckie	2015.53	2298.34	5157.81	9471.68	8.01	9.14	20.51	37.66
Opolskie	405.93	362.03	1833.81	2601.76	11.65	10.39	52.62	74.65
Podkarpackie	4969.47	351.82	3585.95	8907.23	123.66	8.75	89.23	221.65
Podlaskie	2140.90	1328.49	2145.55	5614.95	85.37	52.97	85.55	223.90
Pomorskie	3259.70	642.14	5198.66	9100.50	48.79	9.61	77.80	136.20
Śląskie	389.06	711.06	1774.33	2874.45	1.99	3.64	9.07	14.70
Świętokrzyskie	240.87	335.64	1536.30	2112.81	5.90	8.22	37.63	51.75
Warmińsko-Mazurskie	3238.66	797.55	3405.22	7441.43	113.00	27.83	118.81	259.64
Wielkopolskie	2542.59	2147.49	4006.33	8696.40	25.40	21.45	40.01	86.86
Zachodniopomorskie	3353.21	493.99	4053.75	7900.95	80.61	11.88	97.45	189.93
Poland	29,081.50	13,886.49	49,160.41	92,128.40	23.12	11.04	39.10	73.27

Source: own study.

3.5. Solid Biomass Potential

The solid biomass potential was determined according to the methodology presented in Chapter 2. The results obtained by powiats are presented in graphic form in Figure 4.

As in the case of the renewable sources mentioned above, also in the case of solid biomass there is a clear tendency for powiats with the highest potential to be located in the eastern and northern parts of the country, i.e., regions where energy-intensive industries are found mainly in large urban centres. At the voivodship level, Pomorskie voivodship remains the leader in terms of solid biomass potential (5198 GWh/year), while Świętokrzyskie voivodship is characterised by the lowest potential (1536 GWh/year). More than 63% of solid biomass potential in the country is based on wood that is post-production waste, the result of cultivation operations or cutting for energy purposes. The second source, reaching 15% in the scale of the country, are energy crops (e.g., willow, miscanthus). In this case, according to the methodology used, their potential is related to the area of the weakest land—the soil quality class V, VI and VIz. The third source on the national scale is straw (~13.6%), in the case of Opolskie voivodship its share can reach over 45%, while for Podlaskie voivodship it is only 0.5%, which is due to, inter alia, differences in the intensity of keeping animals (mainly cattle) in these regions.

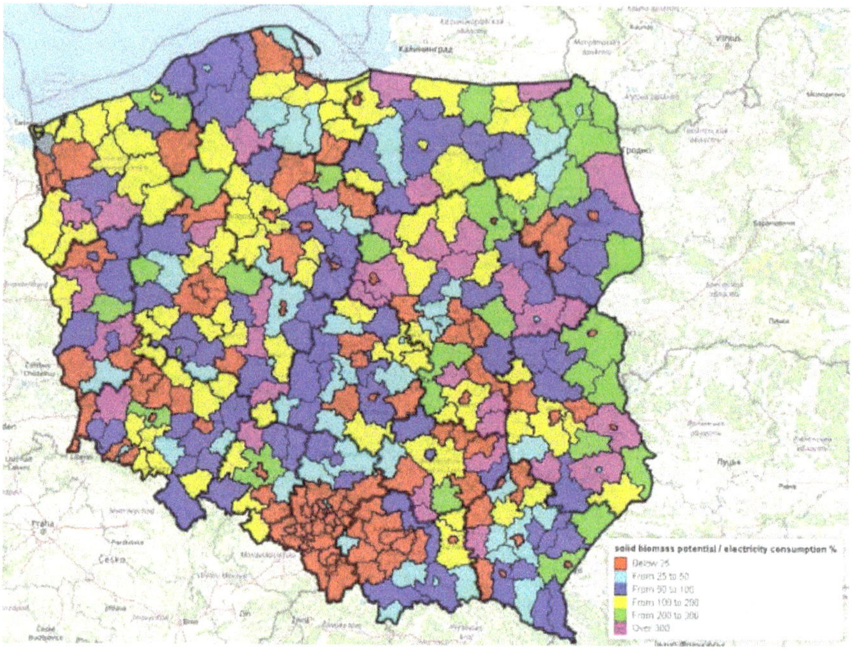

Figure 4. Electricity demand coverage through solid biomass electricity generation. Source: own study.

3.6. Total Potential of Renewable Sources

Our calculations showed that mobilising the potentials of wind, sun and water would cover over 23% of electricity demand, biogas combustion would cover over 11% of demand, while the use of solid biomass alone could cover over 39% of electricity consumption nationwide. In total, the renewable sources considered would cover more than 73% of electricity demand (Table 1).

The results presented in Table 1 indicate that, on a national scale, electricity production based on biomass resources could cover nearly 50% of demand. The importance of agriculture and forestry in building RES potential becomes more visible at the level of powiats. The data presented in Figure 5 imply that the biggest RES potential was found in powiats where the percentage share of agricultural land and forests in the total area of the powiat exceeded 80%. Out of the total number of 380 studied powiats, as many as 220 (57.9%) had RES potential exceeding their own demand for electricity. From this group as many as 214 powiats (56.3%) had a percentage share of agricultural land and forests exceeding 80%. It is worth noting that in the case of 25 powiats RES potential exceeded the demand for electricity more than 10 times. In Figure 5, powiats from Śląskie voivodship, characterized by the lowest RES potential (14.7%), are marked with green colour, while powiats from Warmińsko-Mazurskie voivodship, possessing a high surplus of RES potential (259.64%), are marked with blue colour.

Powiats from other voivodships are marked in red. As can be seen in Figure 5, only one powiat (city of Elbląg) from Warmińsko-Mazurskie voivodship would not be able to cover its total demand for electricity on the basis of RES potential. The situation is different in the case of Śląskie voivodship, where only one of the powiats shows RES potential exceeding the electricity consumption. It is worth noting that in the case of most of the powiats in Śląskie voivodship the share of arable land and forest land did not exceed 80%, which negatively influences the building of RES potential.

Spatial diversification of RES potential is also presented in Figure 6, where it can be seen that most of the powiats with low RES potential (<100%) are located in the southern and central part of the country (mainly Śląskie, Małopolskie and Opolskie voivodships). Moreover, the majority of cities with powiats status, regardless of the size of area, number of inhabitants or type of prevailing economic activity show RES potential below 50%. The powiats with the highest RES potential are mostly located in the northern and eastern part of the country.

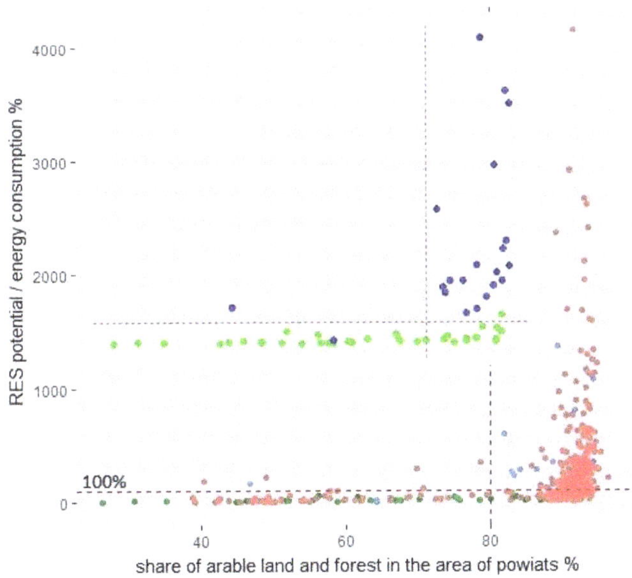

Figure 5. Dependence of RES potential on forest and agricultural land area at powiat level. In the upper left corner only powiats from the Warmińsko-Mazurskie and Śląskie voivodships. Source: own study.

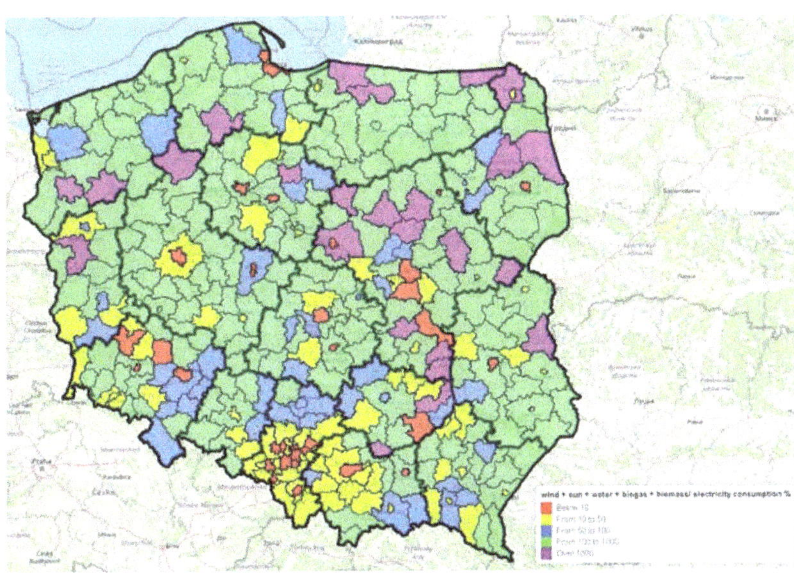

Figure 6. Total RES potential relative to electricity consumption in 2018. Source: own study.

Subsequently, a ranking of voivodships was prepared using the TOPSIS method. The results of the calculations allowed not only to establish the ranking of voivodships using renewable energy sources in 2018, but also allowed to compare them with RES potential available for use. When preparing the ranking of voivodships, the following indicators were taken into account: RES potential in relation to electricity consumption, RES potential in relation to arable land and forest area, and RES potential in relation to population. The obtained results are presented in Table 2.

Table 2. Ranking of voivodships according to TOPSIS with renewable energy production and production potential.

Voivodships	Year 2018			Unleashing the Full Potential			Current Utilisation of RES Potential [%]
	Energy Production from Renewable Energy Sources [GWh]	Topsis	Share of Renewable Energy in Electricity Consumption [%]	Sun, Wind, Water, Biogas, Biomass [GWh]	Topsis	Share of Renewable Energy in Electricity Consumption [%]	
Dolnośląskie	644.3	14	6.02	4675.9	13	44.9	13.8
Kujawsko-Pomorskie	3311.2	2	39.74	4781.9	11	74.5	69.2
Lubelskie	473.3	16	8.29	5516.3	9	104.2	8.6
Lubuskie	655.1	7	17.53	4363.3	5	148.1	15.0
Łódzkie	1466.1	6	17.91	5153.9	10	63.1	28.5
Małopolskie	413.1	15	3.25	2914.8	16	29.2	14.2
Mazowieckie	1450.2	10	5.87	9471.7	8	37.7	15.3
Opolskie	521.6	12	11.03	2601.8	12	74.7	20.1
Podkarpackie	568.8	13	10.43	8907.2	1	221.7	6.4
Podlaskie	717.5	11	23.49	5614.9	6	223.9	12.8
Pomorskie	2104.2	4	27.44	9100.5	2	136.2	23.1
Śląskie	803.1	9	3.96	2874.5	15	14.7	27.9
Świętokrzyskie	1822.1	3	41.32	2112.8	14	51.8	86.2
Warmińsko-Mazurskie	969.2	8	24.79	7441.4	4	259.6	13.0
Wielkopolskie	2092.6	5	18.57	8696.4	7	86.9	24.1
Zachodniopomorskie	3604.8	1	64.27	7900.9	3	189.9	45.6
Poland	21,617.2	x	15.39	92,128.4	x	73.3	23.5

Source: own study.

In 2018, the leaders in terms of electricity production from renewable sources were the following voivodships: Zachodniopomorskie, Kujawsko-Pomorskie, Świętokrzyskie and Pomorskie, respectively. On the other hand, in the case of "unleashing" RES potential, the following voivodships would become the leaders: Podkarpackie, Pomorskie, Zachodniopomorskie and Warmińsko-Mazurskie. In this situation, Świętokrzyskie and Kujawsko-Pomorskie voivodships would record the largest drops, from third to 14th place and from second to 11th place, respectively. Thus, the regions that in 2018 were characterised by the best use of RES potential would record the largest decrease, respectively: 86.24% (Świętokrzyskie voivodship) and 69.2% (Kujawsko-Pomorskie voivodship).

According to the ranking of voivodships, which takes into account the activation of the total RES potential, Podkarpackie voivodship would be the biggest beneficiary (promotion from 13th place to first place). In the case of this voivodship, RES could jointly cover over 221%, of which energy from wind, solar radiation and falling water could cover over 123% of electricity consumption. A further 98% of the surplus could come from biomass and biogas combustion. The situation is similar in the case of Warmińsko-Mazurskie voivodship, where wind, sun and water could cover more than 113% of electricity demand. In the case of the remaining voivodships: Lubelskie, Lubuskie, Podlaskie, Pomorskie and Zachodniopomorskie, achieving self-sufficiency would be possible only through the activation of all discussed RES.

3.7. Discussion

The issue of ARE development in Poland at the regional and local level is relatively poorly recognised. This was pointed out by Maśloch et al. [2] in their publications, that indicate the lack of proposals for regional development directions in terms of autonomous energy regions, but their research referred to only one region—Mazowieckie voivodship. A comprehensive assessment at the level of voivodships was made by a team of experts led by Wiśniewski [96], drawing the conclusion already in 2011 that regions in Poland have a huge, only slightly used technical and economic potential of renewable energy resources. The authors predicted that even taking into account environmental and spatial constraints, the existing potential makes it possible to cover at least 20% of the country's energy needs by 2020, and will ultimately make it possible to ensure 100% of the energy supply comes from renewable energy resources available in the country. Our research for 2018 indicates that the existing RES potential would cover more than 73% of the country's electricity needs, significantly more than the projections.

The spatial variation of RES potential has also been studied in other countries. In Romania, research in this area was conducted by Benedek et al. [27]. The results referred to the potential of three main renewable energy sources—solar, wind and biomass. The authors mapped the renewable energy potential indicating that the use of local RES resources is a development opportunity for farmers, wood producers, technology providers and small and medium-sized enterprises. The great importance of wood waste is indicated by Borzęcki et al. [97] taking into account its spatial distribution and determining the potential in the NUTS-2 regions of the EU-28. The total potential of this waste is about 7.85% of the estimated waste biomass and by-products of the European Union, which represents a significant fraction suitable for recycling or use in biofuel production.

On the other hand, Hartmann and coworkers [98] analysed the issue of wind energy in Hungary noting the need to find ways to maximise the use of locally available resources. The authors point out the great importance of local resources due to the emergence of more and more energy saving buildings and renewable energy resources. The use of local resources would enable the potential to be used more efficiently, but, as they point out, geospatial potential for wind energy is not an easy task to accomplish in the absence of accurate data in remote and extensive areas. The results of this research also show this problem.

Research on identifying the most favourable locations for wind generators was conducted by Potić et al. [99] in Serbia. The researchers considered the issue of how to select the best locations for renewable energy investments and minimise environmental impacts. Similar to our study, they showed that alternative energy sources represent a great potential as a solution to the energy crisis and their advantage over other energy sources is their wide availability.

As shown by Godlewska-Majkowska and Komor [100], access to energy and energy management, especially in rural areas, increase the locational advantages and reduce the economic risk of farming, which contributes to increasing the sustainability of agricultural production. This is of particular importance in rural areas in Poland, which in the past were dominated by state-owned agricultural and cooperative enterprises. The authors showed the significant influence of the energy factor on investment attractiveness at the local level.

An interesting study was conducted by Scaramuzzino and team [101] grouping the EU-28 regions and Switzerland into 17 clusters in terms of renewable energy potential. The results show a heterogeneous distribution of potential across countries, but there are cross-border similarities in the distribution of renewable energy potential. Poland, the Czech Republic and the Baltic States were included in cluster 2 forming the East European plain, with low renewable energy potential. However, our research indicates that the importance of renewable energy in the Polish economy is growing and the potential is significant, only the share of RES in energy production is still low. This is important because the use of renewable resources as energy sources is a factor that improves the security of energy supply, as indicated by Zhu et al. [102]. This is pointed out by Islam et al. [103] stating that renewable energy sources are becoming more common as more electricity generation is needed and renewable energy sources could provide half of the total energy demand by 2050.

In recent years, in Western Europe, mainly in Germany, a very rapid development of biogas plants has been observed (annual production reached 10 billion m^3). This is due to the consistently conducted energy transformation and the need to diversify supplies (the German economy consumes over 87 billion m^3 of natural gas annually) [104]. Poland, with an annual production of ~0.48 billion m^3 in 2018, is at the beginning of this road [105]. According to calculations, the use of such sources as natural fertilizers, plant production waste, landfills and sewage treatment plants could provide over 6.6 billion m^3 of biogas (i.e., approximately 4.3 billion m^3 of methane) annually. This would cover about 25% of the demand for natural gas (according to the Statistics Poland in 2018, natural gas consumption was 17.2 billion m^3). Therefore, it would be a very important support for the national economy in the era of energy transformation. The results of our calculations show that this type of energy production would cover over 11% of the electricity demand in the country (Table 1). According to Statistics Poland, in 2018 for Podlaskie voivodship, the share of renewable energy in energy production reached 68%, while for Warmińsko-Mazurskie voivodship, it was over 82%. These voivodships have become leaders in the development of renewable energy sources in the country, mainly due to the intensive development of wind energy and solid biomass combustion (Białystok power plant). Such a large share of renewable energy sources may suggest that they are close to achieving full energy transformation, however, taking into account the fact that biogas potential has not been released so far, one can risk a statement that they are only at the initial stage of transformation leading not only to energy independence (ARE on a voivodship scale), but also to become exporters of energy/gas. While investigating the potential of solid biomass, it was noticed that large power plants with units specialized in biomass combustion can exert a significant influence on the local energy policy aimed at the use of local resources of solid biomass. A good example is the Połaniec power plant (Świętokrzyskie voivodship), which burns ~1.4 million tonnes of biomass annually in its "green unit", of which 57% in 2016 was imported [106]. However, the remaining 43% of biomass came from the surrounding powiats. Taking into account the biomass resources we calculated for Świętokrzyskie voivodship (~0.53 million tonnes), it can be seen that such a large consumption by one entity may significantly affect local RES development plans, limiting the possibility of building local biomass-fired cogeneration installations in favour of the use of other renewable resources. A similar situation occurs in the case of the Białystok power plant (Podlaskie voivodship), which consumes over 0.4 million tonnes of biomass annually, while the potential of this energy carrier in the entire region is estimated at about 0.8 million tonnes.

The results presented in Table 1 indicate the dominant share of biomass in building RES potential. Both biogas and solid biomass are energy carriers strongly dependent on the broadly understood condition of agriculture and the condition of forest management. The amount of polonium depends not only on the degree of mechanization of agriculture, soil quality or the amount of mineral fertilizers used, but also on the water resources in a given area. Therefore, in the era of climate change, it would be worth following the changes that have occurred in the biomass potential in recent decades, with particular emphasis on drought (2015) and floods (2010) in Poland.

4. Conclusions

Autonomous energy regions (ARE) based on renewable energy sources appear to be one of the solutions which can ensure coverage of the growing energy needs while at the same time taking into account measures respecting the environment. In contrast to the traditional energy industry, RES can be distributed throughout the country. This is due not only to the "dispersed" character of renewable energies (wind, sun, water), but also to the fact that almost every powiat has an intensive agricultural and forestry economy. Therefore, each examined administrative unit in Poland (voivodship, powiat) may be treated as a source of biomass, which, supplemented with energy obtained from wind, water and sun, may constitute the basis for construction of ARE, which may improve not only local energy security but also reduce energy costs for local consumers, improve utilisation of endogenous potential of regions and, above all, reduce negative environmental impact caused by traditional methods of energy production.

The research showed that in Poland, on the one hand, the share of RES in energy production is still small, but at the same time, in the majority of regions (powiats, voivodships) there is a great potential to obtain additional energy from RES, which may constitute the basis for development of ARE. In this respect, there is a large spatial diversification both in relation to the status of RES energy production and its share in energy balance of particular regions and their constituent powiats as well as potential possibilities to increase the volume and structure of production from particular RES.

Activation of RES potential would make it possible to cover over 73% of the country's demand for electricity. Activation of RES potential could provide a basis for building energy independence on a local scale, where only large industrial and urban centres would require "support" from the traditional energy sector covering at least 30% of the national demand for electricity. In the proposed "scenario", as many as seven regions would become self-sufficient in terms of electricity demand. Such a large potential present in the powiats is a premise for the construction of ARE, but one should be aware that its launch is connected to the necessity of incurring huge costs. Obtaining significant financial assistance is a necessary condition for launching RES potential located in rural areas to boost their development.

The results of our calculations presented in Figure 5 indicate that the development of ARE could start first in the group of 220 powiats (57.9% of the total 380) with RES potential exceeding their own electricity consumption. Therefore, in addition to covering their own demand, these powiats could also "export" their surplus energy, thus making the whole project more economically stable.

The creation of RES-based ARE is an important undertaking related to the improvement of energy security, but at the same time it favours the improvement of environmental quality. Such regions may contribute to counteracting marginalisation of areas where undesirable socioeconomic phenomena intensify.

Author Contributions: Conceptualization, G.Ś. and B.G.; methodology, M.C.-W., D.T. and S.W.; software, S.W. and D.T.; validation, S.W. and D.T.; formal analysis, B.G., D.T., S.W. and G.Ś.; investigation, J.G., M.C.-W., D.T. and S.W.; resources, S.W., M.C.-W. and D.T.; data curation, S.W. and D.T.; writing—original draft preparation, G.Ś., B.G., D.T., S.W. and M.C.-W.; writing—review and editing, B.G. and J.G.; visualization, B.G. and S.W.; supervision, G.Ś.; project administration, G.Ś., B.G. and J.G.; funding acquisition, G.Ś. and B.G. All authors have read and agreed to the published version of the manuscript.

Funding: This research received no external funding.

Institutional Review Board Statement: Not applicable.

Informed Consent Statement: Not applicable.

Data Availability Statement: https://bdl.stat.gov.pl/BDL/start (accessed on 19 April 2021).

Conflicts of Interest: The authors declare no conflict of interest.

References

1. Owusu, P.A.; Asumadu-Sarkodie, S.A. review of renewable energy sources, sustainability issues and climate change mitigation. *Cogent Eng.* **2016**, 3. [CrossRef]
2. Maśloch, P.; Maśloch, G.; Kuźmiński, Ł.; Wojtaszek, H.; Miciuła, I. Autonomous energy regions as a proposed choice of selecting selected EU regions—Aspects of their creation and management. *Energies* **2020**, *13*, 6444. [CrossRef]

3. IEO. U.S. Energy Information Administration's International Energy Outlook 2020. Center for Strategic and International Studies. Independent Statistics & Analysis, U.S. Energy Information Administration. 2020. Available online: www.iea.gov (accessed on 12 March 2021).
4. Gradziuk, P.; Gradziuk, B. Renewable energy sources as a development opportunity for peripheral areas. *Econ. Reg. Stud.* **2020**, *13*, 184–196.
5. Zou, C.; Zhao, Q.; Zhang, G.; Xiong, B. Energy revolution: From a fossil energy era to a new energy era. *Nat. Gas Ind. B* **2016**, *3*, 1–11. [CrossRef]
6. Sonter, L.J.; Dade, M.C.; Watson, J.E.M.; Valenta, R.K. Renewable energy production will exacerbate mining threats to biodiversity. *Nat. Commun.* **2020**, *11*, 4174. [CrossRef]
7. Johnsson, F.; Kjärstad, J.; Rootzén, J. The threat to climate change mitigation posed by the abundance of fossil fuels. *Clim. Policy* **2019**, *19*, 258–274. [CrossRef]
8. Perera, F. Pollution from fossil-fuel combustion is the leading environmental threat to global pediatric health and equity: Solutions exist. *Int. J. Environ. Res. Public Health* **2017**, *15*, 16. [CrossRef]
9. Greenstone, M.; Reguant, M.; Ryan, N.; Dobermann, T. *Energy and Environment*; Evidence Paper; International Growth Centre, University of Oxford: Oxford, UK, 2019.
10. Gielen, D.; Boshell, F.; Saygin, D.; Bazilian, M.D.; Wagner, N.; Gorini, R. The role of renewable energy in the global energy transformation. *Energy Strategy Rev.* **2019**, *24*, 38–50. [CrossRef]
11. Herington, M.J.; van de Fliert, E.; Smart, S.; Greig, C.; Lant, P.A. Rural energy planning remains out-of-step with contemporary paradigms of energy access and development. *Renew. Sustain. Energy Rev.* **2017**, *67*, 1412–1419. [CrossRef]
12. Akhtar, S.; Zahedi, K.; Liu, H. *Regional Cooperation for Sustainable Energy in Asia and the Pacific*; ESCAP, United Nations: Bangkok, Thailand, 2017.
13. Ford, R.; Maidment, C.; Fell, M.; Vigurs, C.; Morris, R. *A Framework for Understanding and Conceptualising Smart Local Energy Systems. Energy REV*; University of Strathclyde Publishing: Strathclyde, UK, 2019; pp. 1–18.
14. Omer, A.M. Energy use and environmental impacts: A general review. *J. Renew. Sustain. Energy* **2009**, *1*, 053101. [CrossRef]
15. IRENA. *Innovation Landscape for a Renewable-Powered Future: Solutions to Integrate Variable Renewables*; International Renewable Energy Agency: Abu Dhabi, United Arab Emirates, 2019.
16. Plich, M.; Skrzypek, J. Trendy Energochłonności Polskiej Gospodarki. *Wiadomości Stat. Pol. Stat.* **2016**, 16–38. [CrossRef]
17. Clinch, P. Cost–benefit analysis applied to energy. *Encycl. Energy* **2004**, *1*, 715–725. [CrossRef]
18. Sharma, B.; Ingalls, R.G.; Jones, C.L.; Khanchi, A. Biomass supply chain design and analysis: Basis, overview, modeling, challenges, and future. *Renew. Sustain. Energy Rev.* **2013**, *24*, 608–627. [CrossRef]
19. Dampier, J.E.E.; Lemelin, R.H.; Shahi, C.; Luckai, N. Small town identity and history's contribution to a response in policy change: A case study of transition from coal to biomass energy conversion. *Energy Sustain. Societ.* **2014**, *4*, 1–14. [CrossRef]
20. Famielec, J.; Kijanka, A.; Żaba-Nieroda, R. Economic growth and carbon dioxide emissions. *Pol. Stat. Stat. Pract.* **2019**, *64*, 5–21. [CrossRef]
21. Gradziuk, P. Innowacje organizacyjne w bioenergetyce. In *Bioenergetyka Podkarpacka. Innowacje Technologiczne i Organizacyjne w Podkarpackiej Bioenergetyce*; Kościk, B., Ed.; Wydawnictwo Naukowe Państwowej Wyższej Szkoły Zawodowej im. ks. Bronisława Markiewicza w Jarosławiu: Jaroslaw, Poland, 2007; pp. 215–233.
22. Maśloch, G. Budowa autonomicznych regionów energetycznych w Polsce—Utopia czy konieczność? *Studia Prawno Ekon.* **2018**, *CVI*, 251–256.
23. Statistics Poland. *Rural Areas in Poland in 2018. Statistical Analyses*; Statistics Poland: Warszawa, Olsztyn, 2020.
24. Wójcik, M. Non-agricultural economic functions of rural areas in the Łódzkie Voivodship (1999–2009). *Reg. Barom. Anal. Progn.* **2013**, *2*, 43–50.
25. Sima, E. Non-agricultural activities in the economic development of the Romanian rural area. In Proceedings of the Agrarian Economy and Rural Development—Realities and Perspectives for Romania. In Proceedings of the 6th Edition of the International Symposium, ICEADR, Bucharest, Romania, 19 November 2015; pp. 390–395.
26. Kiciński, J. Autonomiczne Regiony Energetyczne (ARE)—Szansa Dla Polskiej Wsi, Instytut Maszyn Przepływowych PAN w Gdańsku, Uniwersytet Warmińsko-Mazurski w Olsztynie. Available online: https://docplayer.pl/13433401-Autonomiczne-regiony-energetyczne-are-szansa-dla-polskiej-wsi.html (accessed on 19 April 2021).
27. Benedek, J.; Sebestyén, T.-T.; Bartók, B. Evaluation of renewable energy sources in peripheral areas and renewable energy-based rural development. *Renew. Sustain. Energy Rev.* **2018**, *90*, 516–535. [CrossRef]
28. Directive (EU) 2018/2001 of the European Parliament and of the Council of 11 December 2018 on the Promotion of the Use of Energy from Renewable Sources. *Off. J. Eur. Union* **2018**, *L 328/82*, 1–128. Available online: https://eur-lex.europa.eu/legal-content/EN/TXT/PDF/?uri=CELEX:32018L2001&from=EN (accessed on 20 March 2021).
29. Benasla, M.; Hess, D.; Allaoui, T.; Brahami, M.; Denaï, M. The transition towards a sustainable energy system in Europe: What role can North Africa's solar resources play? *Energy Strategy Rev.* **2019**, *24*, 1–13. [CrossRef]
30. Müller, S.; Brown, A.; Ölz, S. *Renewable Energy Policy Considerations for Deploying Renewables*; Information Paper; OECD/IEA: Paris, France, 2011.
31. Marchenko, O.; Solomin, S. Economic efficiency of renewable energy sources in autonomous energy systems in Russia. *Int. J. Renew. Energy Res.* **2014**, *4*, 548–554.

32. Blazquez, J.; Fuentes-Bracamontes, R.; Bollino, C.A.; Nezamuddin, N. The renewable energy policy Paradox. *Renew. Sustain. Energy Rev.* **2018**, *82*, 1–5. [CrossRef]
33. Marks-Bielska, R.; Bielski, S.; Pik, K.; Kurowska, K. The Importance of renewable energy sources in Poland's energy mix. *Energies* **2020**, *13*, 4624. [CrossRef]
34. Matyka, M. *Rolnictwo a Odnawialne Źródła Energii—Szanse i Zagrożenia*; Zegar, J.S., Ed.; Z Badań Nad Rolnictwem Społecznie Zrównoważonym. No 11; Instytut Ekonomiki Rolnictwa i Gospodarki Żywnościowej: Warsaw, Poland, 2011; pp. 95–120.
35. Wang, Y.; Zhang, D.; Ji, Q.; Shi, X. Regional renewable energy development in China: A multidimensional assessment. *Renew. Sustain. Energy Rev.* **2020**, *124*, 109797. [CrossRef]
36. Ślusarz, G.; Gołębiewska, B.; Cierpiał-Wolan, M.; Gołębiewski, J.; Twaróg, D.; Wójcik, S. Regional diversification of potential, production and efficiency of use of biogas and biomass in Poland. *Energies* **2021**, *14*, 742. [CrossRef]
37. Jezierski, G. Elektrownia Jądrowa a Konwencjonalna. *Energetyka Ciepl. Zawodowa* **2009**, *10*, 1–12. Available online: https://www.cire.pl/pliki/2/jadrowa_a_konwencjonalna.pdf (accessed on 25 April 2021).
38. Maciejczak, M.; Zakharov, K. Public goods as a source of rural development. In Proceedings of the International Conference Development Prospects of Rural Areas Lagging Behind in the CEE Region, Szent Istvan University, Godollo, Hunagry, 24–26 May 2011.
39. Santos, J.L.; Madureirab, L.; Ferreirab, A.C.; Espinosac, M.; Gomez y Paloma, S. Building an empirically- based framework to value multiple public-goods of agriculture at broad supranational scales. *Land Use Policy* **2016**, *53*, 56–70. [CrossRef]
40. Plantinga, A.J.; Miller, D. Agricultural land values and future development. *Land Econ.* **2001**, *77*, 56–67. [CrossRef]
41. Koryś, K.A.; Latawiec, A.E.; Grotkiewicz, K.; Kuboń, M. The review of biomass potential for agricultural biogas production in Poland. *Sustainability* **2019**, *11*, 6515. [CrossRef]
42. Kukuła, K. Dynamika produkcji energii elektrycznej wykorzystującej źródła energii odnawialnej. *Wiadomości Stat. Pol. Stat.* **2015**, *12*, 58–69.
43. Kowalczyk, K.; Cieśliński, R. Analiza potencjału hydroenergetycznego oraz możliwości jego wykorzystania w województwie pomorskim. *Woda Sr. Obsz. Wiej.* **2018**, *18*, 69–86.
44. Jarosz, Z. Potencjał energetyczny biomasy roślinnej i możliwości wykorzystania do celów energetycznych. *Probl. Rol. Swiat. Probl. World Agric.* **2017**, *17*, 81–92. [CrossRef]
45. Cogeneration, European Biomass Industry Association. 2021. Available online: https://www.eubia.org/cms/wiki-biomass/combustion/cogeneration (accessed on 16 May 2021).
46. Sprawozdanie Bilansowe Nośników Energii i Infrastruktury Ciepłowniczej. Available online: http://form.stat.gov.pl/formularze/2018/passive/G-02b.pdf (accessed on 25 April 2021).
47. Available online: https://eref-europe.org/restor-hydro-database/ (accessed on 25 April 2021).
48. Blok, M.; Tomaszewska, B. Ocena potencjału technicznego wybranej MEW w powiecie rzeszowskim (Evaluation of the Technical Potential of a Selected Small Hydropower Plant in The Rzeszów District). *Geol. Explor. Technol. Geotherm. Energy Sustain. Dev.* **2015**, *2*, 61–78.
49. Bujakowski, W.; Barbacki, A.; Grzybek, A.; Hołojuch, G.; Pająk, L.; Skoczek, A.; Skrzypczak, M.; Skrzypczak, S. *Opracowanie Metody Programowania i Modelowania Systemów Wykorzystania Odnawialnych Źródeł Energii na Terenach Nieprzemysłowych Województwa Śląskiego, wraz z Programem Wykonawczym dla Wybranych Obszarów Województwa*; Część I: Metodyka Opracowania; Instytut Gospodarki Surowcami Mineralnymi i Energią PAN: Kraków, Poland; Katowice, Piland, 2005.
50. Filipowicz, M.; Żołądek, M.; Ochalik, J. Potencjał Rozwoju i Problemy Energetyki Wiatrowej (Growth Potential and Problems in Development of Wind Energy). *Rynek Instal.* **2020**, 1–2. Available online: http://www.rynekinstalacyjny.pl/artykul/id4944 (accessed on 25 April 2021).
51. PSEW. Analiza Krajowego Planu na Rzecz Energii i Klimatu na Lata 2021–2030 z Dnia 30 Grudnia 2019 r. Polskie Stowarzyszenie Energetyki Wiatrowej. 2019. Available online: http://psew.pl/wp-content/uploads/2020/06/Analiza-KPEiK_PSEW.pdf (accessed on 25 April 2021).
52. Energetyka Wiatrowa. 2021. Available online: https://globenergia.pl/ (accessed on 20 May 2021).
53. Stolińska, B. The factors of wind turbines location. *World Real Estate J.* **2014**, *88*, 27–31.
54. Dragusha, B.; Hoxha, B. Impact of field roughness and power losses, turbulence intensity on electricity production for an onshore wind farm. *Int. J. Power Electron. Drive Syst.* **2020**, *11*, 1519–1526. [CrossRef]
55. Dz.U. 2016 Poz. 961, Ustawa z Dnia 20 Maja 2016 r. o Inwestycjach w Zakresie Elektrowni Wiatrowych. Available online: https://isap.sejm.gov.pl/isap.nsf/download.xsp/WDU20160000961/U/D20160961Lj.pdf (accessed on 27 March 2021).
56. Voicescu, S.A.; Michaud, D.S.; Feder, K.; Marro, L.; Than, J.; Guay, M.; Denning, A.; Bower, T.; van den Berg, F.; Broner, N.; et al. Estimating annoyance to calculated wind turbine shadow flicker is improved when variables associated with wind turbine noise exposure are considered. *J. Acoust. Soc. Am.* **2016**, *139*, 1480–1492. [CrossRef]
57. Pantazopoulou, P. Wind turbine noise measurements and abatement methods. *WIT Trans. State Art Sci. Eng.* **2010**, *44*, 621–659. [CrossRef]
58. Botelho, A.; Arezes, P.; Bernardo, C.; Dias, H.; Pinto, L.M.C. Effect of wind farm noise on local residents' decision to adopt mitigation measures. *Int. J. Environ. Res Public Health* **2017**, *14*, 753. [CrossRef]
59. Díaz, K.M.A.; Velarde-Suárez, S.; Oro, J.M.F.; Perez, J.G. Simplified assessment on the wind farm noise impact of the E2O experimental offshore station in the Asturian coast. *Energies* **2020**, *13*, 5788. [CrossRef]

60. Kozłowska-Szczęsna, T. Promieniowanie, Temperatura Powietrza, Główny Geodeta Kraju, Warszawa 1994. In Atlas Rzeczypospolitej Polskiej. Cz.2. Warszawa: Główny Geodeta Kraju. 1994. Available online: https://www.igipz.pan.pl/mapy-all.html (accessed on 3 April 2021).
61. Tytko, R. *Urządzenia i Systemy Energetyki Odnawialnej, Wyd*; Towarzystwo Słowaków w Polsce: Krakow, Poland, 2013.
62. Available online: https://www.gum.gov.pl/pl/transfer-wiedzy/inne/mapy-godzin-wschodu-i-z/2652,Mapy-godzin-wschodu-i-zachodu-slonca-w-Polsce.html (accessed on 1 April 2021).
63. Mirowski, T.; Sornek, K. Potencjał energetyki prosumenckiej w Polsce na przykładzie mikroinstalacji fotowoltaicznych w budownictwie indywidualnym. *Energy Policy J.* **2015**, *18*, 73–85.
64. Thompson, E.P.; Bombelli, E.L.; Shubham, S.; Watson, H.; Everard, A.; D'Ardes, V.; Schievano, A.; Bocchi, S.; Zand, N.; Howe, C.J.; et al. Tinted semi-transparent solar panels allow concurrent production of crops and electricity on the same cropland. *Adv. Energy Mater.* **2020**, *10*, 1–9.
65. Available online: https://bdl.stat.gov.pl/BDL/start (accessed on 19 April 2021).
66. Wiśniewski, G. (Ed.) *Odnawialne Źródła Energii Jako Element Rozwoju Lokalnego*; Europejskie Centrum Energii Odnawialnej; EC BREC: Warszawa, Poland, 2003.
67. Wąs, A.; Sulewski, P.; Krupin, V.; Popadynets, N.; Malak-Rawlikowska, A.; Szymańska, M.; Wysokiński, M. The potential of agricultural biogas production in Ukraine—Impact on GHG emissions and energy production. *Energies* **2020**, *13*, 5755. [CrossRef]
68. Walczak, J.; Krawczyk, W.; Szewczyk, A.; Mazur, D.; Pająk, T.; Radecki, P. *Oszacowanie Wielkości Produkcji Oraz Jednostkowej Zawartości Azotu Nawozów Naturalnych, Powstałych w Różnych Systemach Utrzymania Zwierząt Gospodarskich w Polsce*; Instytut Zootechniki Państwowy Instytut Badawczy: Kraków, Poland, 2012; p. 17.
69. Kołodziejak, G. Możliwości wykorzystania potencjału energetycznego biogazu powstającego w trakcie procesu oczyszczania ścieków. Analiza opłacalności proponowanych rozwiązań. *Naft. Gaz* **2012**, *12*, 1036–1043.
70. Cherevko, G. Rapeseed growing for energy purposes in Ukraine. *Rocz. Nauk. Stowarzyszenia Ekon. Rol. Agrobiz.* **2016**, *18*, 29–34.
71. Gradziuk, P. Potencjał i prognozy wykorzystania biogazu rolniczego w Polsce. Potential and forecast for use of agricultural biogas in Poland. *Rocz. Nauk. Stowarzyszenia Ekon. Rol. Agrobiz.* **2017**, *19*, 64–70. [CrossRef]
72. Available online: https://stat.gov.pl/obszary-tematyczne/rolnictwo-lesnictwo/uprawy-rolne-i-ogrodnicze/produkcja-upraw-rolnych-i-ogrodniczych-w-2018-roku,9,17.html (accessed on 18 April 2021).
73. Martinát, S.; Dvořák, P.; Frantál, B.; Klusáček, P.; Kunc, J.; Kulla, M.; Mintálová, T.; Navrátil, J.; van der Horst, D. Spatial consequences of biogas production and agricultural changes in the Czech Republic after EU accession: Mutual symbiosis, coexistence or parasitism? *AUPO Geogr.* **2013**, *44*, 75–92.
74. Bharathirajaa, B.; Sudharsanaa, T.; Jayamuthunagaib, J.; Praveenkumarc, R.; Chozhavendhand, S.; Iyyappan, J. Biogas production—A review on composition, fuel properties, feed stock and principles of anaerobic digestion. *Renew. Sustain. Energy Rev.* **2018**, *90*, 570–582. [CrossRef]
75. Ludwicka, A.; Grzybek, A. Bilans masy rolnej (słomy) na potrzeby energetyki (Balance of the agricultural biomass (straw) for energy purposes). *Probl. Inż. Rol. Probl. Agric. Eng.* **2010**, *2*, 101–110.
76. Gradziuk, P. *Gospodarcze Znaczenie i Możliwości Wykorzystania słomy na Cele Energetyczne w Polsce*; Monografie i Rozprawy naukowe nr. 45, Wyd; Instytutu Uprawy, Nawożenia i Gleboznawstwa—Państwowy Instytut Badawczy w Puławach/Institute of Soil Science and Plant Cultivationstate Research Institute: Puławy, Poland, 2015; p. 176.
77. Hrynkiewicz, M.; Grzybek, A. Nadwyżka słomy dostępnej do wykorzystania na potrzeby energetyczne w 2016 r. *Probl. Inżynierii Rol.* **2017**, *3*, VII–IX.
78. Pudełko, R. Ocena potencjału biomasy ubocznej z rolnictwa w UE. *Nowa Energ.* **2013**, *4*, 44–47.
79. Koprowicz, A. *Planowanie Bilansu Paszowego w Gospodarstwach Mlecznych*; MPODR: Karniowice, Poland, 2016.
80. ARiMR-Agencja Restrukturyzacji i Modernizacji Rolnictwa. Available online: https://www.arimr.gov.pl/ (accessed on 4 April 2021).
81. Kuś, J.; Faber, A. Alternatywne kierunki produkcji rolniczej. *Studia Rap. IUNG-PIB Puławy* **2007**, *7*, 138–158.
82. Kołodziej, A.; Matyka, M. *Odnawialne Źródła Energii. Rolnicze Surowce Energetyczne*; PWRiL: Poznan, Poland, 2012; p. 594.
83. Han, S.H.; Shin, S.J.; Kim, B.R.; Aggangan, N.S.; Yun, C.W. Growth chemical composition and energy contents of seven clones of one-year-old Salix caprea L. as short rotation cappice. *Asia Life Sci.* **2013**, *22*, 413–426.
84. Lach, A.; Dyjakon, A.; Noszczyk, T. Sady jabłoniowe jako lokalne źródło biomasy do ogrzewania gospodarstw domowych w Województwie Mazowieckim (Apple orchards as a source of local biomass for households heating in the Mazowieckie Voivodeship). In *Rolnictwo XXI Wieku—Problemy i Wyzwania*; Łuczycka, D., Ed.; Idea Knowledge Future: Wroclaw, Poland, 2018; pp. 528–539.
85. Pavić, Z.; Novoselac, V. Notes on TOPSIS Method. *Int. J. Res. Eng. Sci.* **2013**, *1*, 5–12.
86. Wysocki, F. *Metody Taksonomiczne w Rozpoznawaniu Typów Ekonomicznych Rolnictwa i Obszarów Wiejskich*; Wydawnictwo Uniwersytetu Przyrodniczego w Poznaniu: Poznań, Poland, 2010.
87. EC 2021. Eurostat. Available online: https://ec.europa.eu/eurostat/data/database (accessed on 18 April 2021).
88. Igliński, B. Hydro energy in Poland: The history, current state, potential, SWOT analysis, environmental aspects. *Int. J. Energy Water Resour.* **2019**, *3*, 61–72. [CrossRef]
89. Kowalczak, P. *Hierarchia Potrzeb Obszarowych Małej Retencji w Dorzeczu Warty*; Instytut Meteorologii i Gospodarki Wodnej: Warsaw, Poland, 2001.

90. Kudlicki, Ł. Zagrożenie Pustynnieniem w Polsce. *Bezpieczeństwo Nar.* **2006**, *1*, 201–211. Available online: https://www.bbn.gov.pl (accessed on 21 April 2021).
91. Watson, S.; Moro, A.; Reis, V.; Baniotopoulos, C.; Barth, S.; Bartoli, G.; Wiser, R. Future emerging technologies in the wind power sector: A European perspective. *Renew. Sustain. Energy Rev.* **2019**, *113*, 109270. [CrossRef]
92. Available online: https://www.tauron-wytwarzanie.pl/innowacje/co2-sng (accessed on 23 April 2021).
93. Leahy, M.J.; Connolly, D.; Buckley, D.N. Wind energy storage technologies. *WIT Trans. State Art Sci. Eng.* **2010**, *44*, 621–714. [CrossRef]
94. Nsair, A.; Cinar, S.O.; Alassali, A.; Abu Qdais, H.; Kuchta, K. Operational parameters of biogas plants: A review and evaluation study. *Energies* **2020**, *13*, 3761. [CrossRef]
95. Sefeedpari, P.; Pudełko, R.; Jędrejek, A.; Kozak, M.; Borzęcka, M. To what extent is manure produced, distributed, and potentially available for bioenergy? A step toward stimulating circular bio-economy in Poland. *Energies* **2020**, *13*, 6266. [CrossRef]
96. Wiśniewski, G. (Ed.) *Określenie Potencjału Energetycznego Regionów Polski w Zakresie Odnawialnych Źródeł Energii—Wnioski dla Regionalnych Programów Operacyjnych na Okres Programowania 2014–2020*; Ministerstwo Rozwoju Regionalnego: Warsaw, Poland, 2011.
97. Borzęcki, K.; Pudełko, R.; Kozak, M.; Borzęcka, M.; Faber, A. Spatial distribution of wood waste in Europe. *Sylwan* **2018**, *162*, 563–571.
98. Hartmann, B.; Talamon, A.; Sugár, V. Renewable energy potentials in the administrative regions of Hungary. *Strateg. Plan. Energy Environ.* **2017**, *37*, 33–57. [CrossRef]
99. Potić, I.; Joksimović, T.; Milinčić, U.; Kićović, D.; Milinčić, M. Wind energy potential for the electricity production—Knjaževac Municipality case study (Serbia). *Energy Strategy Rev.* **2021**, *33*, 100589. [CrossRef]
100. Godlewska-Majkowska, H.; Komor, A. Energy as a factor of investment attractiveness of regions for agricultural enterprises. *Energies* **2021**, *14*, 2731. [CrossRef]
101. Scaramuzzino, C.; Garegnani, G.; Zambelli, P. Integrated approach for the identification of spatial patterns related to renewable energy potential in European territories. *Renew. Sustain. Energy Rev.* **2019**, *101*, 1–13. [CrossRef]
102. Zhu, D.; Mortazavi, S.M.; Maleki, A.; Aslani, A.; Yousefi, H. Analysis of the robustness of energy supply in Japan: Role of renewable energy. *Energy Rep.* **2020**, *6*, 378–391. [CrossRef]
103. Islam, M.A.; Hasanuzzaman, M.; Rahim, N.A.; Nahar, A.; Hosenuzzaman, M. Global renewable energy-based electricity generation and smart grid system for energy security. *Sci. World J.* **2014**, 1–13. [CrossRef] [PubMed]
104. Mardani, A.; Streimikiene, D.; Balezentis, T.; Saman, M.; Nor, K.M.; Khoshnava, S. Data envelopment analysis in energy and environmental economics: An overview of the state-of-the-art and recent development trends. *Energies* **2018**, *11*, 2002. [CrossRef]
105. Available online: file:///C:/Users/p138602/AppData/Local/Temp/2020-EurObserv_ER-biogas-barometer-GB-20201215.pdf (accessed on 16 April 2021).
106. Available online: https://powermeetings.eu/konferencja/forum-biomasy/wp-content/uploads/sites/19/2017/04/06-Rafal-Domaszewski-Enea.pdf (accessed on 8 April 2021).

energies

MDPI

Article

Factors Affecting the Adoption of Photovoltaic Systems in Rural Areas of Poland

Marek Angowski [1,*], Tomasz Kijek [1], Marcin Lipowski [2] and Ilona Bondos [3]

1 Department of Microeconomics and Applied Economics, Faculty of Economics, Maria Curie-Sklodowska University in Lublin, Pl. M. Curie-Skłodowskiej 5, 20-031 Lublin, Poland; tomasz.kijek@mail.umcs.pl
2 Department of Marketing, Faculty of Economics, Maria Curie-Sklodowska University in Lublin, Pl. M. Curie-Skłodowskiej 5, 20-031 Lublin, Poland; marcin.lipowski@mail.umcs.pl
3 Department of IT Systems and Logistics, Faculty of Economics, Maria Curie-Sklodowska University in Lublin, Pl. M. Curie-Skłodowskiej 5, 20-031 Lublin, Poland; ilona.bondos@mail.umcs.pl
* Correspondence: marek.angowski@mail.umcs.pl

Abstract: The paper aims to identify and explain the factors influencing the decision-making process on the behavioural intention to use home photovoltaic systems by Polish households and potential buyers. The survey was conducted in 2021 on a sample of 521 participants. The research used a random sample of households without PV systems located in the rural areas in Poland, where the adoption of innovative technologies related to obtaining energy from renewable sources is especially important. Structural equation modelling (SEM) was applied to measure structural relationships. The main finding indicates that consumer innovativeness has the strongest impact on the intention to purchase a photovoltaic installation. The perceived value also affects the intention to purchase a photovoltaic installation. The perceived value is affected by perceived economic benefits and indirectly by the subjective knowledge of PV. Surprisingly, environmental concerns negatively affect the intention to use PV installations.

Keywords: PV systems; renewable energy resources; rural areas; economic value; consumer behaviour; consumer innovativeness

Citation: Angowski, M.; Kijek, T.; Lipowski, M.; Bondos, I. Factors Affecting the Adoption of Photovoltaic Systems in Rural Areas of Poland. *Energies* **2021**, *14*, 5272. https://doi.org/10.3390/en14175272

Academic Editors: Carlo Renno and Anastasios Dounis

Received: 23 June 2021
Accepted: 24 August 2021
Published: 25 August 2021

1. Introduction

For several years there has been much interest in renewable energy, which also includes solar energy. Among these technologies is photovoltaics (PV), i.e., converting solar energy into electricity, considered one of the most promising and environmentally friendly energy sources. Photovoltaics guarantees energy obtained completely naturally. Any amount of energy produced from the sun reduces CO_2 emissions, which significantly impacts the environment [1]. A significant advantage of solar radiation is its availability and the daily and annual variability—an equally important disadvantage. The development of energy storage methods should minimise the limitations related to converting solar energy into energy useful for humans and its wider use [2,3]. In Poland, there are generally good natural and spatial conditions for the use of solar radiation energy. Satisfactory results can be achieved by adapting the type of systems and the properties of solar devices to the nature, structure and time distribution of the solar radiation. The basis for the development of photovoltaics in Poland is the introduction of appropriate legal regulations that guarantee the investor the profitability and predictability of the investment. Opportunities should also be seen in prosumer energy, where the main assumption should be the production of energy for one's own needs, and in the case of producing excess energy, the possibility of selling it at a favourable price [2,4]. The large increase in investments in photovoltaic systems in Poland is related primarily to the decrease in installation costs with the simultaneous increase in energy costs for end-users. Therefore, photovoltaics becomes a better alternative in terms of reducing energy costs, both in households and companies [3,4].

The issue of photovoltaics as a renewable energy source finds a place as a new area of activity both in the national energy policy and the EU energy policy. The EU policy supports the development of renewable energy to ensure energy security in the conditions of sustainable development and competitiveness [5]. Therefore, the EU provides significant financial resources for implementing investments related to the development of renewable energy sources. The selection of the appropriate source of financing depends on the type of beneficiary and the scale of the investment. RDP is the available form of financing for applicants located in rural areas. So far, investors of solar PV systems have benefited from support under the financial perspective for 2014–2020. This support will be continued in the coming years as part of the Act on renewable energy sources [2], which is to enter into force, and introduce new regulations, to a greater extent supporting the development of renewable energy sources in photovoltaic systems. In Poland, the development of photovoltaic systems in rural areas may also be favoured by the environmentally neutral nature of this type of investment, which, unlike wind farms or agricultural biogas plants, is more socially acceptable. What is important, is that rural areas encompass more than 93% of Poland and are inhabited by almost 40% of the country's population. The high consumers' demand for energy in rural areas combined with increased use by agriculture result in the need for energy security in these areas. Due to convenient conditions (geographical location and climate zone), the countryside has the real potential to increase the share of green energy in Poland's energy mix [6]. Owing to the enormous financial aid from the European Union for Polish agriculture, there is a great alternative in the form of installation of photovoltaics, which may contribute to the improvement of the economic situation of individual regions and households located there [3,7]. Therefore, it can be concluded that in the longer term of activities undertaken for the use of solar energy as an energy source, photovoltaics is not only a new direction in meeting energy needs but also a way of professional activation (activating entrepreneurship) of the population and rural development.

Despite the potential of photovoltaic systems, the dynamics of the development and implementation of these solutions are still limited. The literature identifies several barriers to the adoption of photovoltaics on the market. Key barriers include consumer passivity, high initial costs, the long payback period for investments, planning and installation pains, various information gaps, and customer concerns about the ability to use PV [1,8]. Additionally, it is possible to indicate the motivations and priorities shaping the behaviour of consumers, which differ and depend on their knowledge, personality, preferences, values and attitudes towards ecology or innovation [9]. For example, open-minded users and looking for novelty (innovation) are more likely to take risks despite the technical complexity and lack of short-term benefits of modern solutions such as PV than other consumers [10]. Among the factors influencing the development of photovoltaics in rural areas, the following limitations can be indicated [11,12]: public reluctance (with regard to the location of these investments in the vicinity); low public awareness of the role and importance of renewable energy sources; lack of stable support systems for this area of economic activity; the need to adapt the applied solutions to local conditions; problems with connecting new installations to energy and heating networks; lack of appropriate, time-stable legal solutions regarding the functioning of various groups of entities dealing with renewable energy on the energy market; poorly developed energy network in rural areas.

Despite the development barriers, the photovoltaic sector has enormous potential and is able to meet the growing global demand for energy in the near future. With the current trends in technology and market development, photovoltaics will significantly improve the natural environment, ensure energy security, and contribute to protecting the global economy from energy crises. As a result, this sector is increasingly becoming an area of interest for analysts and scientists from various fields. The research and publications conducted so far in the field of the photovoltaic sector have focused mainly on the analysis and assessment of technical parameters, R&D activity and the possibilities of increasing their efficiency [12,13]. The economic publications are dominated by items related to the analysis of costs and investment efficiency. Some empirical studies apply the technology acceptance model (TAM) to explain the adoption of PV solar technologies by consumers [14]). Unfortunately, using the TAM model in the context of PV adoption is questionable since this model focuses on the limited number of factors affecting consumers' decisions. Therefore, it seems important to explore the issues of using innovative photovoltaic technologies, emphasising the aspects of a buyers' behaviour in the purchasing process. In recent years, analyses of the issue of prosumerism have appeared more and more frequently. On the other hand, the issue of buyers' behaviour in relation to the process of selecting and purchasing photovoltaic installations seems to be still insufficiently explored.

In order to close this research gap; our contribution is twofold. First, to the best of our knowledge, this study is the first attempt to take into account the broad set of factors (i.e., environmental concerns, consumer innovativeness, subjective knowledge on PV, perceived economic benefits of PV, perceived risk of PV, the perceived value of PV) influencing the decision-making process on the selection and implementation of investments in home photovoltaic systems. Such an approach allows us to model and empirically verify the complex hypothetical relationships between the buyer's characteristics, beliefs, attitudes and knowledge, and their behaviour in the process of purchasing photovoltaic installations. Second, the paper considers the specificity of consumers living in rural areas, where the development of innovative technologies related to obtaining energy from renewable sources is important, because the energy sector in these regions, away from agglomeration and industrial centres, is often characterised by: isolation, high dispersion population centres, limited supply as well as the lagging behind in technical and business infrastructure.

We consider it also important that the research was carried out in a country with a specific energy culture—the coal culture. Poland is one of these countries where the coal market is particularly relevant as coal is the first choice to meet energy demand in this country. From 2019 to 2020, the percentage change of hard coal and lignite generation (bars) production was −8.0%, while the average decline for the EU was −18.0 [15]. In 2020, 83% of electricity came from fossil fuels (to samo źródło) in Poland. While the trend to move away from coal in the power sector is increasing, the Polish government aims to maintain coal as a dominant energy source until at least 2050. This position is in clear contradiction with the European and global climate and energy policy. [16,17]. What is more, according to the forecasts, the legislative changes planned for 2022 by the Ministry of Climate and Environment will reduce the profitability of PV installations in Poland. Therefore, it is worth analysing other factors influencing the intention to use PV than just those related to the perceived profitability of such a solution.

2. Theoretical Background

Beliefs and attitudes about the natural environment are translated into actions or behaviour. Some studies indicated the lack of strong impact of environmental beliefs on environmental behaviours; the main reason is that general opinions are not strong enough to pro-social acting—Gadenne et al. [18] called it the value-action gap. Nevertheless, it was also acknowledged that environmental concerns impact behavioural intention [19,20]. Additionally, Thi Khanh and Phong [21] noted that consumers' beliefs and awareness of the natural environment could create environmentally responsible behaviour.

Therefore, we proposed the hypotheses:

Hypothesis 1 (H1). *Environmental concerns positively impact intention to use PV;*

Hypothesis 2 (H2). *Environmental concerns positively impact the perceived value of PV.*

Consumer innovativeness is seen as an important part of personality. It refers to the tendency of purchasing and using new products more quickly and more often than other people [22]. The relation between consumer innovativeness and behavioural intention became the main research topic in reference to many offers such as robotic restaurants [23], a drone food delivery service [24], smart toys [25], smartwatches [26], autonomous cars [27].

Consumer innovativeness is vital to create a positive response towards new products and positively impacts willingness to pay [25].

Therefore, we proposed the hypotheses:

Hypothesis 3 (H3). *Consumer innovativeness positively impacts intention to use PV;*

Hypothesis 4 (H4). *Consumer innovativeness negatively impacts perceived risk.*

As stated by Buratti and Allwood [28], not only an individual's objective knowledge but also subjective knowledge (consumers assessment of the level of their knowledge) can influence consumers risk perception and actions. According to researchers, subjective knowledge plays a bigger role in predicting environmental behaviour than other knowledge types [29]. In our study, we concentrated on subjective (environmental) knowledge that is defined as people's perceptions of how much they know about a particular environmental issue [30]. Subjective knowledge can be identified as the result of highly objective knowledge and previous experience [31]. According to much research, people exhibit overconfidence when assessing their knowledge within many different kinds of domains. It is important to emphasise that the research on the impact of subjective knowledge on risk perception or risk behaviour provide mixed results—positive influence, negative influence or no influence [28]. Using the example of smart home technologies, Wilson et al. [32] noted that early adopters acquire greater knowledge. Their positive perceptions of benefits are strengthened, but greater knowledge does not significantly weaken early adopters' perceptions of risks. Zhu et al. [33] tried to prove that the greater the perceived knowledge, the lower the perceived risk. As a result of their research model, it turned out not be supported the research hypothesis. Additionally, Dursun et al. [34] could not find support to the assumption that high subjective environmental knowledge will decrease the tendency to deny the problem.

Therefore, we proposed the hypotheses:

Hypothesis 5 (H5). *Subjective knowledge of PV negatively impacts perceived risk;*

Hypothesis 6 (H6). *Subjective knowledge of PV positively impacts perceived benefits.*

The concept of 'perceived value' emerged as the defining business issue of the 1990s and has continued to receive extensive research interest in the 21st century [35]. El-Adly [36] noted that the definition of customer perceived value has changed over time. The perceived value has received much attention from academics and practitioners due to its close relationship with customer satisfaction and competitive advantages. However, the concept of perceived value was seen by Khalifa [37] as one of the most overused and misused concepts in the social sciences in general and in the management literature in particular. Among many definitions, one of the more commonly cited is that supplied by Zeithaml [38], who proposed a general perspective where perceived value is the consumer's overall assessment of the utility of a product based on what is obtained and provided. Her conceptualisation of perceived value was one of the first. Monroe's [39] research approach was similar; he defined perceived value as "a trade-off between the quality or benefits they perceive in the product relative to the sacrifice they perceive by paying the price". Simply, the perceived value is a difference between the benefits obtained and the sacrifices made; it is a trade-off between benefits acquired and perceived costs. In our study, the concept of perceived value was adapted in the context of PV installation offer. As it was stated by Slovic [40], a risk–benefit trade-off is used in order to evaluate specific technology to accept it or reject it. The most promising technologies generate both risk and benefits [41], and the PV system is no different. In our study, two separated variables were identified as main components of perceived value—perceived risk and perceived benefits.

Perceived benefits can be identified in the scientific literature in many different ways—functional, experiential, symbolic [42], functional benefits [43], utilitarian, emotional [44], hedonic. Although PV installation can provide its users with multiple benefits, only perceived economic benefits were included as an extremely strong incentive in our study. Santos [45] noted the importance of financial aspects for potential adopters of PV systems due to their deep interests in economic benefits when installing a photovoltaic system. We defined perceived economic benefits of PV installation by analogy to economic benefits from participating in sharing economy analysed by Lee et al. [46]. Kim [47] considered the general idea of perceived benefits and showed a positive impact of perceived benefit on the value.

Therefore, we proposed the hypothesis:

Hypothesis 7 (H7). *Perceived economic benefits positively impact the perceived value.*

By analogy to smart retail technology [48], the benefits offered by PV are not without potential risks, uncertainties and adverse consequences. In this study, we decided to resign from focusing on price perception aspects due to the vital financial support provided to potential PV buyers. Instead of analysing perceived costs in general, we put our research attention to risk perception. Impacts risk is seen as one antecedent of a perceived value [48]. Perceived risk was conceptualised by Raymond Bauer in 1960 in the context of consumer behaviour and is based on the notion that any buying activity includes risk. The perceived risk works when consumers are unsure whether the intended purchase will permit them to accomplish their buying purposes [49]. Scholars defined perceived risk as a kind of subjective, expected, possible loss when pursuing a prospective outcome [50]. It was usually defined as the subjective expectation of a loss [51] and the consequences of such a loss if it occurs [52]. What is important, this loss can have both monetary and non-monetary nature [53]. These negative consequences are expected to emerge from particular technology/innovation adoption or use [54]. In our study, by analogy to Chin et al.'s [55] definition, perceived risk referred to the consumer's subjective belief in the possibility of loss or harm emerging from the PV installation in a consumers' household. The information asymmetry to the sellers' advantage is the main reason most buyer–supplier relationships are characterised by risks [56]. Understanding elements that can decrease perceived risk is truly vital [57]. Different risk reducers can be useful depending on the specific type of perceived risk (risk factors) [58]. Perceived risk is well explained in literature; however, it is still an important avenue of research [59], such as PV installation as an innovative environmentally friendly technology in consumers' households.

Therefore, we proposed the hypothesis:

Hypothesis 8 (H8). *Perceived risk negatively impacts the perceived value of PV.*

Perceived value is highly relevant to marketers as it has been found to positively affect behavioural intention [60]. A series of studies confirmed that perceived value increases purchase intention. The impact of perceived value on behavioural intention has been studied in various interesting trendy areas such as online group buying [61], private labels [62], smart retail technology [48], green branding [63]. The positive impact on behavioural intention is connected with the psychological explanation of the perceived value that involves an attraction toward the outcome of the goal pursuit [64].

Purchasing intention indicates an individual's readiness to buy a product that one has preferred for oneself after some evaluations on the basis of personal experience, perception, attitude, subjective norm [65]. Green purchase intention refers to the probability and individual inclination to choose environmentally friendly energy products over the conventional products in their purchase decision [66]. In line with previous research findings, it is reasonable to predict that consumers' perceptions of PV installation value may increase their purchase intention towards PV installation.

Therefore, we proposed the hypothesis:

Hypothesis 9 (H9). *Perceived value positively impacts intention to use PV.*

3. Materials and Methods

Our research used a random sample consisting of households without PV systems located in the rural areas in Poland. A survey instrument was developed to understand the factors that shape households' use intention of PV technology in Poland. A total of 526 questionnaires were collected, of which five respondents reported that they used PV systems. These were removed from the sample, leading to a final sample size of 521 participants. Of the respondents, 53 per cent were male, and 47 per cent were female. Their ages varied from 18 to 81 years, with a mean of 40.5 years. The dominant household's size was four persons. In terms of the average monthly amount of a household's electricity bill, the largest group of respondents reported that they paid between EUR 22 and 44 (Table 1).

Table 1. Sample characteristics.

Characteristics		Number of Respondents	Percentage of Sample
Gender	Female	245	47
	Male	276	53
Age (years)	18–24	96	18.4
	25–34	133	25.5
	35–44	93	17.9
	45–54	101	19.4
	Over 55	98	18.8
Average monthly energy bills (euro)	Below 22.0	49	9.4
	22.0–44.0	241	46.3
	45.0–66.0	140	26.9
	67.0–89.0	58	11.1
	Over 90.0	33	6.3
Household size (number of persons)	1	29	5.6
	2	108	20.7
	3	136	26.1
	4	138	26.5
	5 or more	110	21.1

For the purpose of this study, the endogenous and exogenous constructs were adapted from prior research (Appendix A). Six constructs represented the influential factors (i.e., environmental concerns—EC, consumer innovativeness—CI, subjective knowledge on PV—SK, perceived economic benefits of PV—PEB, perceived risk of PV—PR, the perceived value of PV—PV). The remaining construct is related to the use intention of PV—UI. In total, 28 questions were applied to operationalise these constructs. All questions were measured on a seven-point Likert scale ranging from 1 (strongly disagree) to 7 (strongly agree).

The measurement model was verified via construct reliability and validity tests. To measure the scale's reliability, we applied Raykov's reliability coefficient—RRC. This coefficient is preferred over Cronbach's alpha since it relaxes the assumption of tau-equivalent measures. Other reliability tests used in our study include Loevinger's H coefficient and Ferguson's delta coefficient. The former provides information on scalability. The latter allows us to verify scales discrimination. Convergent validity was measured by average variance extracted (AVE). Finally, structural equation modelling (SEM) was applied to measure structural relationships [67]. SEM can be regarded as a combination of several traditional multivariate procedures such as factor analysis, regression analysis, discriminant analysis or canonical correlation.

SEM consists of the structural model and the measurement model [68]. The structural model with latent variables takes the form:

$$\eta_i = \alpha_y + \mathrm{B}\eta_i + \Gamma\xi_i + \zeta_i \tag{1}$$

where η_i is a vector of latent endogenous variables for unit i, α_y is a vector of constants for the equations, B is the matrix of coefficients showing the expected impacts of the latent endogenous variables (η) on each other, ξ_i is the vector of latent exogenous variables, Γ is the coefficient matrix showing the expected impacts of the latent exogenous variables (ξ) on the latent endogenous variables (η) and ζ_i is the vector of disturbances.

The measurement model has two equations:

$$y_i = \alpha_y + \Lambda_y \eta_i + \varepsilon_i \quad (2)$$

$$x_i = \alpha_x + \Lambda_x \xi_i + \delta_i \quad (3)$$

where y_i and x_i are vectors of the observed responses of η_i and ξ_i, accordingly, α_y and α_x are constant vectors, Λ_y and Λ_x are matrices of factor loadings showing the effect of the latent η_i and ξ on y_i and x_i, accordingly, and ε_i and δ_i are the unique factors of y_i and x_i.

Prior to the measurement model verification and the SEM analysis, the multivariate normality assumption was checked by the Doornik–Hansen test. Its outcome resulted in the application of Satorra–Bentler adjustment after maximum likelihood.

4. Results and Discussion

Table 2 shows the results of the reliability and validity tests. All of the RRCs are higher than the lowest acceptable level (0.7). Moreover, Loevinger's H coefficients for the seven scales are higher than 0.3, indicating good scalability. Regarding the generalised delta index of scale discrimination, it meets a threshold level (0.9) for all scales. The results of Confirmatory Factor Analysis show that all constructs can explain more than an average amount of 50% variance of its indicators. In other words, convergent validity is demonstrated since the average variance extracted (AVE) exceeds the cut-off value (0.5). There is no problem with discriminant validity, as AVE values of all constructs are larger than their squared correlations (SC) with other constructs in the model. All except one of the standardised factor loadings are above 0.5. This means that they are considered satisfactory items. Consequently, the item from the construct of subjective knowledge on PV with low loading was dropped in further analysis. CFA outcomes reveal that the entire measurement model indicates acceptable goodness of fit as RMSEA = 0.06 and CFI = 0.93 meet satisfactory thresholds [69].

Table 2. Reliability and validity of measurement model.

Construct	Items	Loadings Range	RRC	H	Delta	AVE
Environmental concerns—EC	4	0.75–0.92	0.93	0.67	0.94	0.76
Consumer innovativeness—CI	4	0.71–0.85	0.88	0.80	0.96	0.65
Subjective knowledge on PV—SK	3	0.63–0.95	0.88	0.74	0.97	0.74
Perceived economic benefits of PV—PEB	4	0.79–0.89	0.92	0.81	0.96	0.74
Perceived risk of PV—PR	4	0.62–0.91	0.88	0.76	0.96	0.65
Perceived value of PV—PV	4	0.88–0.93	0.96	0.73	0.95	0.82
Use intention of PV—UI	4	0.94–0.98	0.98	0.74	0.93	0.91

The standardised path coefficient between EC and UI did not confirm a positive relationship between the environmental concerns and the intention to use PV installations; thus, H1 has been rejected (Figure 1). Such a relation can be influenced by the increasing controversy over the possible environmental impact of environmentally friendly technologies, such as electric cars or wind farms [70]. Solar photovoltaic installations are also not free from this type of controversy, especially regarding the "end of life" panel and hazardous materials used in PV [71]. Apart from the factors related to recycling, the negative impact of PV installations on the environment may be associated with visual pollutions, especially in the case of large installations or the exclusion of the land from agricultural crops [72]. These effects can be especially felt in rural areas [73]. For this reason, respondents' environmental concerns do not have to positively affect the installation of voltaic panels, and even—as in the results of the conducted research—can have a negative effect on installing intention. A high level of environmental concerns can create a wider perspective of the effects of PV technology and its potential long-term impact on the household and its surroundings.

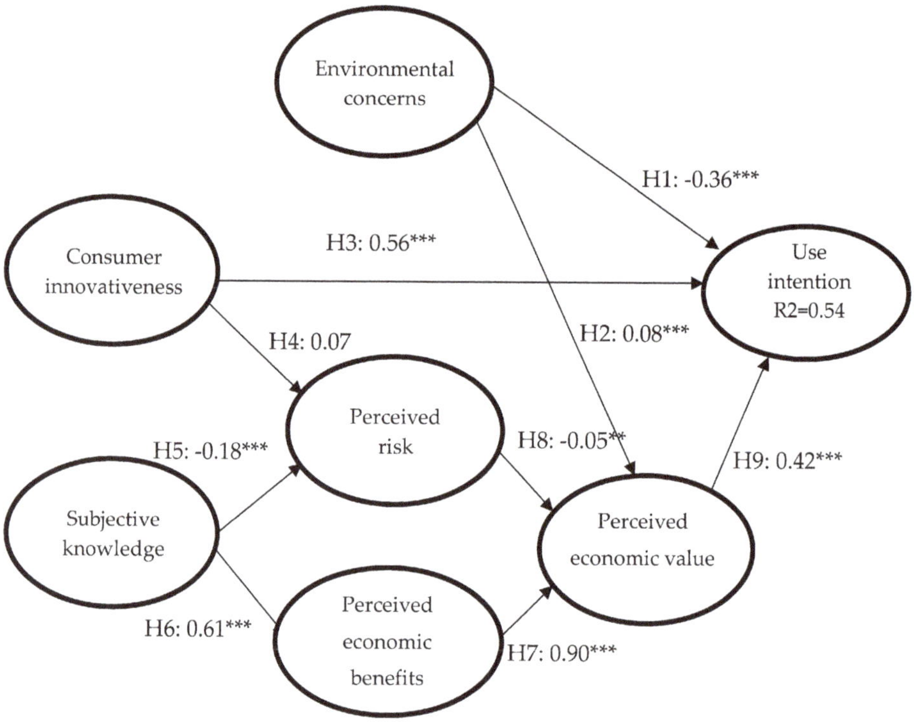

Figure 1. Results of structural model estimation. Note: *** $p < 0.01$, ** $p < 0.05$, RMSA = 0.08, CFI = 0.89.

The indirect and positive impact of the environmental concerns on the intention to use PV installation occurs via the perceived value (Table 3). Even though this dependence is weak, it has confirmed H2. The perceived value moderates the impact of environmental concerns on the intention to use PV installations. Thus, the perception of the photovoltaic installation value affects the intention to use this technology.

Table 3. Indirect effects of endogenous and exogenous variables on use intention.

Path	Coefficient
EC→UI	0.04 ***
PR→UI	−0.04 **
PEB→UI	0.48 ***
CI→UI	−0.003
SE→UI	0.28 ***

Note: *** $p < 0.01$, ** $p < 0.05$.

The hypothesis H3, indicating that consumer innovativeness has a positive effect on purchasing a photovoltaic installation, has been confirmed. Thus, the research results confirmed the conclusions provided by Zhang et al. [25]. In the analysed model, the impact of consumer innovativeness on the intention to purchase a photovoltaic installation is the strongest. Consumer innovation is a personal characteristic. The greater the consumer openness to innovative ecological technologies, the more likely they are to use PV installations. This consumer characteristic is even more important than the perceived value of the PV. According to the research model, the consumer's innovativeness does not affect the perceived risk of installing photovoltaic panels—the relationship is statistically insignificant (Figure 1). Therefore, hypothesis H4 has not been confirmed. The obtained results are consistent with the conclusions of Xue et al. [10]); in our research, consumer innovation does not affect the risk of the PV installation decision. The impact of consumer innovativeness on the perceived risk is irrelevant, probably due to the increasing usage of solar panels by households and the spread of this technology. As Sommerfeld et al. noted, PV technologies have already reached technological maturity, which is why they are perceived as low risky [74]

The subjective knowledge of PV installations reduces the perceived risk and increases the perceived economic benefits of PV panels (Figure 1). The results confirm the Alrashoud and Tokimatsu study [10], indicating that consumer knowledge impacts PV installation usage. Therefore, H5 and H6 have been verified. The impact of consumer knowledge on benefits is stronger than on the perceived risk. These results confirm the conclusions of Zhu et al. [33]. However, we agree that knowledge may have a different influence on the perceived risk depending on the type of technology and the stage of its acceptance, as shown by Wilson et al. [32].

The perceived value of PV in the research model depends on the perceived risk and the economic benefits (Figure 1). As Lewandowska et al. [75] noted, the use of renewable energy sources is innovative, and its implementation involves certain effects related not only to possible profits but also costs. The study confirmed that perceived risk lowers the perceived value while the economic benefits increase value perception. The positive impact of financial benefits on the perceived value in the case of PV installation is much stronger than the negative one on the value of the perceived risk. Another study showed that in the case of Polish households, the installation of PV brings some financial benefits [8]. Among some factors determining the profitability of PV investments in households, there can be indicated such as the level of electricity prices, the prices of electricity distribution, as well as the insolation condition. Despite the fact that energy prices for households in Poland are subsidised, they are constantly growing [8,76]. As a result, the homeowners who use more due to various systems (e.g., family farms) have greater economic benefits from using PV.

The research results have confirmed the hypotheses H7 and H8. Thus, both variables (perceived benefit and perceived risk) indirectly affect the intention of the photovoltaic installation. Importantly, the perceived economic benefits have the greatest positive impact on installation intention. Thus, the conclusions indicated by Santos [45] are confirmed by these findings. In the case of Poland, the value of PV installations is additionally supported by the support programs in the form of co-financing the PV installation from the European Union or domestic funds [4,77]. Some of these funds are dedicated to rural areas. The perceived value of a PV installation may also be increased by its declining cost and the possibility of obtaining income from selling surplus energy. This improves the ratio of expenditure to revenues.

The last of the hypotheses, H9, has also been confirmed (Figure 1). The perceived value affects the intention to purchase a photovoltaic installation. As expected, the perceived effect of using the installation determines its purchase, as are other studies on the impact of perceived value on the intention to purchase. This perceived value is greater for larger energy consumers, who often live in rural areas. They use the electricity for their own needs but also their family farm. They also have land for the location of slightly larger photovoltaic investments, which does not have to generate additional costs.

5. Conclusions

The development of PV installations is an issue that may combine the needs of individual consumers and the development factors of economies on a global, national or local scale. The impact of changes in the diversification of energy sources may contribute to changes in technical (development of technical infrastructure, innovation), economic (energy costs, new jobs) and improve the condition of the natural environment.

The objective of this study was to investigate the factors influencing the intention to install photovoltaics in the household. To achieve the set research goal, we collected data and used them to verify the proposed research model in this area. To the best of our knowledge, factors influencing the intention of PV installations in rural areas are not fully recognised. We believe that this issue deserves the interest of researchers also due to the largely untapped potential for PV installations in households as well as the European Union policy in the area of renewable energy sources.

The main finding shows that consumer innovativeness has the strongest impact on the intention to purchase a photovoltaic installation. The perceived value also affects the intention to purchase photovoltaic installation. The perceived value is affected by perceived economic benefits and indirectly by the subjective knowledge of PV. The surprising results of studies pointing to the negative impact of financial concerns on the intention to use a photovoltaic installation may be due to the greater sensitivity of rural residents to the possible negative ecological effects of PV installations. Currently, the issue of disposal of photovoltaic panels may be an important issue for potential customers. This issue seems vital because the lifetime of the panels is relatively short. The issue of disposal of PV panels can be treated as an important element of communication with customers. In rural areas, large solar farms can also be a concern by disturbing the natural landscape. The location of this type of farm should be regulated to a greater extent than just by the class of land on which farms are allowed to be located. It follows that consumers currently adopt a microeconomic perspective (the impact of installations on their immediate surroundings) and do not consider the macro perspective—improving the state of the natural environment in the country. This is important because Poland is one of the countries with very high environmental pollution and a dominant share of coal in energy production.

While encouraging the use of PV installations in rural areas, the economic benefits of such a solution should still be emphasised. The perceived value of PV installations increases along with the rising energy prices in the case of Polish households. Conclusions from the conducted research may indicate that communication may turn out to be a key element of the strategy for the development of photovoltaic systems in rural areas. Our results suggest that a well-disseminated information campaign must contain precise details on the economic benefits of PV (e.g., the sample calculation of energy cost reduction). Regarding environmental effects of PV, they should be presented in the context of the socio-economic framework since we found the indirect impact of environmental concern appeared to be positive and significant. What is more, to mitigate the perceived risk of PV evidence-based approach should be implemented in communication strategy. All government and local authority support/funding programs must be maintained as they increase the benefits of PV installations in rural areas.

Although the results of this study have useful implications, some limitations must be considered. These limitations offer three lines of further research. First, future research could extend empirical testing to these dependencies (the impact of environmental concerns as well as consumer innovativeness), which proved to be quite surprising and are not fully confirmed by the literature. The second limitation concerns the model specification. It seems necessary to incorporate other factors affecting the adoption of PV systems (e.g., PV system characteristics) into the research model. Finally, a limitation is the sample size and its composition. Although the sample consists of 521 respondents, there is a need to increase the sample size to improve the generalisability of the findings. It is worth noticing that the sample is limited to households located in rural areas. In order to conduct comparative analyses, it would be recommended to include respondents from the urban areas in future research.

Author Contributions: Research and paper concept, M.A., I.B., T.K. and M.L.; Abstract, M.L., 1. Introduction, M.A.; 2. Theoretical background, I.B.; 3. Materials and Methods, T.K. 4. Discussion, M.L.; 5. Conclusions, M.A., I.B., T.K. and M.L.; Collection and development of statistical material, M.A., I.B., T.K. and M.L.; Preparation of figures and text editing, M.A. and T.K. All authors have read and agreed to the published version of the manuscript.

Funding: This research received no external funding.

Institutional Review Board Statement: Not applicable.

Informed Consent Statement: Not applicable.

Data Availability Statement: Data will be available on request.

Conflicts of Interest: The authors declare no conflict of interest.

Appendix A

Latent Variables	Items
Environmental concern (EC) adapted from; Wung, Cao and Zhang [20]	EC1: I worry about air pollution. EC2: I am concerned about environmental problems. EC3: I think that environmental problems have become increasingly serious in recent years. EC4: I try to take care of the natural environment.
Consumer innovativeness (CI) adapted from Zhang et al. [25]	CI1: I know more about new ecological technologies than people around me. CI2: I eagerly reach for ecological products. CI3: I am interested in new energy-saving technologies. CI4: I believe that new green technologies are worth using.
Subjective knowledge on PV (SK) adapted from Dursun [34]	SK1: I know that photovoltaics is a good solution for obtaining energy from renewable sources (delated). SK2: I am interested in photovoltaics as a source of energy. SK3: I have much knowledge of photovoltaics. SK4: I have more knowledge about photovoltaics than the average person.
Perceived economic benefits of PV (PEB) adapted from Kim [47]	I'm able to save on energy expenses by installing photovoltaic panels. Nowadays, the installation of a photovoltaic installation is economically advantageous. Using photovoltaic installation could reduce my energy expenses. Taking into account financial support, the photovoltaic installation is financially beneficial.
Perceived risk (PR) adapted from Park et al. [78]	PR1: I believe that installing photovoltaics requires much effort. PR2: Using photovoltaic installation can cause problems. PR3: I perceive the installation of photovoltaics as risky. PR4: I am concerned that installing photovoltaics might be difficult.
Perceived value (PV) adapted from Oyedele et al. [79]	PV1: Installation of the photovoltaic system is cost-effective. PV2: Considering the benefits and costs, a photovoltaic installation is profitable. PV3: Production of energy from the photovoltaic system is more advantageous than buying it from the energy supplier. PV4: Considering the advantages and disadvantages, a photovoltaic installation is valuable.
Use intention of PV (UI) adapted from Li et al. [22]	UI1: I'm going to set up a photovoltaic installation in the near future. UI2: There is a high probability that I will install photovoltaics. UI3: Most likely, I will be installing a photovoltaic installation. UI4: I plan to use energy from a photovoltaic installation.

References

1. Strupeit, L.; Palm, A. Overcoming barriers to renewable energy diffusion: Business models for customer-sited solar photovoltaics in Japan, Germany and the United States. *J. Clean. Prod.* **2016**, *123*, 124–136. [CrossRef]
2. Wójcik, M. *Regionalny Wymiar Przemian Polskiej Wsi—Aspekty Społeczne i Środowiskowe*; Rural Studies XXXV, IGiPZ PAN; KOW PTG: Łódź, Poland, 2014; ISBN 978-83-62089-27-7.
3. Alipour, M.; Salim, H.; Stewart, R.A.; Sahin, O. Residential solar photovoltaic adoption behaviour: End-to-end review of theories, methods and approaches. *Renew. Energy* **2021**, *170*, 471–486. [CrossRef]
4. Kowalczyk-Juśko, A. *Rozwój Innowacyjnych Technologii Odnawialnych Źródeł Energii na Obszarach Wiejskich*; Centrum Doradztwa Rolniczego w Brwinowie: Radom, Poland, 2020; ISBN 978-83-66776-01-2.
5. Mirowski, T.; Sornek, K. Potencjał energetyki prosumenckiej w Polsce na przykładzie ikroinstalacji fotowoltaicznych w budownictwie indywidualnym, Potential of prosumer power engineering in Poland by example of micro PV installation in private construction. *Polityka Energetyczna Energy Policy J.* **2015**, *18*, 73–84.
6. Marks-Bielska, R.; Bielski, S.; Pik, K.; Kurowska, K. The Importance of Renewable Energy Sources in Poland's Energy Mix. *Energies* **2020**, *13*, 4624. [CrossRef]

7. Chmieliński, M. Profitability analysis of photovoltaic microinstallation in Poland based on electrical energy production for own use. *Econ. 21st Century* **2015**, *3*, 113–129. [CrossRef]
8. Reeves, D.C.; Rai, V.; Margolis, R. Evolution of consumer information preferences with market maturity in solar PV adoption. *Environ. Res. Lett.* **2017**, *12*, 1–14. [CrossRef]
9. Alrashoud, K.; Tokimatsu, K. Factors Influencing Social Perception of Residential Solar Photovoltaic Systems in Saudi Arabia. *Sustainability* **2019**, *11*, 5259. [CrossRef]
10. Xue, Y.; Lindkvist, C.M.; Temeljotov-Salaj, A. Barriers and potential solutions to the diffusion of solar photovoltaics from the public-private-people partnership perspective—Case study of Norway. *Renew. Sustain. Energy Rev.* **2021**, *137*, 110636. [CrossRef]
11. Nyholm, E.; Odenberger, M.; Johnsson, F. An economic assessment of distributed solar PV generation in Sweden from a consumer perspective—The impact of demand response. *Renew. Energy* **2017**, *108*, 169–178. [CrossRef]
12. Alipour, M.; Salim, H.; Stewart, R.A.; Sahin, O. Predictors, taxonomy of predictors, and correlations of predictors with the decision behaviour of residential solar photovoltaics adoption: A review. *Renew. Sustain. Energy Rev.* **2020**, *123*, 109749. [CrossRef]
13. Londo, M.; Matton, R.; Usmani, O.; van Klaveren, M.; Tigchelaar, C.; Brunsting, S. Alternatives for current net metering policy for solar PV in the Netherlands: A comparison of impacts on business case and purchasing behaviour of private homeowners, and on governmental costs. *Renew. Energy* **2020**, *147*, 903–915. [CrossRef]
14. Alam, S.S.; Ahmad, M.; Othman, A.S.; Shaari, Z.B.H.; Masukujjaman, M. Factors Affecting Photovoltaic Solar Technology Usage Intention among Households in Malaysia: Model Integration and Empirical Validation. *Sustainability* **2021**, *13*, 1773. [CrossRef]
15. Agora Energiewende and Ember. Available online: https://ember-climate.org/european-electricity-transition/#digin (accessed on 7 August 2021).
16. Kuchler, M.; Bridge, G. Down the black hole: Sustaining national socio-technical imaginaries of coal in Poland. *Energy Res. Soc. Sci.* **2018**, *41*, 136–147. [CrossRef]
17. LaBelle, M.C. Energy Cultures. In *Technology, Justice, and Geopolitics in Eastern Europe*; Edward Elgar Publishing: Northamton, MA, USA, 2020; ISBN 9781788975759.
18. Gadenne, D.; Sharma, B.; Kerr, D.; Smith, T. The influence of consumers' environmental beliefs and attitudes on energy saving behaviours. *Energy Policy* **2011**, *39*, 7684–7694. [CrossRef]
19. Chu, W.; Im, M.; Song, M.R.; Park, J. Psychological and behavioral factors affecting electric vehicle adoption and satisfaction: A comparative study of early adopters in China and Korea. *Transp. Res. Part D Transp. Environ.* **2019**, *76*, 1–18. [CrossRef]
20. Wung, X.-W.; Cao, Y.-M.; Zhang, N. The influences of incentive policy perceptions and consumer social attributes on battery electric vehicle purchase intentions. *Energy Policy* **2021**, *151*, 1–9. [CrossRef]
21. Thi Khanh, C.N.; Phong, L.T. Impact of environmental belief and nature-based destination image on ecotourism attitude. *J. Hosp. Tour. Insights* **2020**, *3*, 489–505. [CrossRef]
22. Li, L.; Wang, Z.; Li, Y.; Liao, A. Impacts of consumer innovativeness on the intention to purchase sustainable products. *Sustain. Prod. Consum.* **2021**, *27*, 774–786. [CrossRef]
23. Hwang, J.; Park, S.; Kim, I. Understanding motivated consumer innovativeness in the context of a robotic restaurant: The moderating role of product knowledge. *J. Hosp. Tour. Manag.* **2020**, *44*, 272–282. [CrossRef]
24. Hwang, J.; Kim, J.J.; Lee, K.-W. Investigating consumer innovativeness in the context of drone food delivery services: Its impact on attitude and behavioral intentions. *Technol. Forecast. Soc. Chang.* **2021**, *163*, 120433. [CrossRef]
25. Zhang, F.; Sun, S.; Lu, C.; Chang, V. Consumer innovativeness, product innovation and smart toys. *Electron. Commer. Res. Appl.* **2020**, *41*, 100974. [CrossRef]
26. Hong, J.-C.; Lin, P.-H.; Hsieh, P.-C. The effect of consumer innovativeness on perceived value and continuance intention to use smartwatch. *Comput. Hum. Behav.* **2017**, *67*, 264–272. [CrossRef]
27. Leicht, T.; Chtourou, A.; Youssef, K.B. Consumer innovativeness and intentioned autonomous car adoption. *J. High Technol. Manag. Res.* **2018**, *29*, 1–11. [CrossRef]
28. Buratti, S.; Allwood, C.M. The effect of knowledge and ignorance assessments on perceived risk. *J. Risk Res.* **2019**, *22*, 735–748. [CrossRef]
29. Rizkalla, N.; Erhan, T.P. Sustainable Consumption Behaviour in The Context of Millennials in Indonesia—Can Environmental Concern, Self-efficacy, Guilt and Subjective Knowledge Make a Difference? *Manag. J. Sustain. Bus. Manag. Solut. Emerg. Econ.* **2020**, *25*, 43–53. [CrossRef]
30. Park, C.W.; Mothersbaugh, D.L.; Feick, L. Consumer knowledge assessment. *J. Consum. Res.* **1994**, *21*, 71–82. [CrossRef]
31. Utkarsh; Sangwan, S.; Agarwal, P. Effect of consumer self-confidence on information search and dissemination: Mediating role of subjective knowledge. *Int. J. Consum. Stud.* **2018**, *43*, 46–57. [CrossRef]
32. Wilson, C.; Hargreaves, T.; Hauxwell-Baldwin, R. Benefits and risks of smart home technologies. *Energy Policy* **2017**, *103*, 72–83. [CrossRef]
33. Zhu, W.; Wei, J.; Zhao, D. Anti-Nuclear Behavioral Intentions: The Role of Perceived Knowledge, Information Processing and Risk Perception. *Energy Policy* **2016**, *88*, 168–177. [CrossRef]
34. Dursun, I.; Kabadayı, E.T.; Tuğer, A.T. Overcoming the psychological barriers to energy conservation behaviour: The influence of objective and subjective environmental knowledge. *Int. J. Consum. Stud.* **2019**, *43*, 402–416. [CrossRef]
35. Sánchez-Fernández, R.; Iniesta-Bonillo, M.A. The concept of perceived value: A systematic review of the research. *Mark. Theory* **2007**, *7*, 427–451. [CrossRef]

36. El-Adly, M.I. Modelling the relationship between hotel perceived value, customer satisfaction, and customer loyalty. *J. Retail. Consum. Serv.* **2019**, *50*, 322–332. [CrossRef]
37. Khalifa, A.S. Customer value: A review of recent literature and an integrative configuration. *Manag. Decis.* **2004**, *42*, 645–666. [CrossRef]
38. Zeithaml, V.A. Consumer Perceptions of Price, Quality, and Value: A Means-end Model and Synthesis of Evidence. *J. Mark.* **1988**, *52*, 2–22. [CrossRef]
39. Monroe, K.B. *Pricing: Making Profitable Decisions*; McGraw-Hill: New York, NY, USA, 1990.
40. Slovic, P. Perception of risk. *Science* **1987**, *236*, 280–285. [CrossRef]
41. Fischhoff, B.; Slovic, P.; Lichtenstein, S.; Read, S.; Combs, B. How safe is safe enough? A psychometric study of attitudes towards technological risks and benefits. *Policy Sci.* **1978**, *9*, 127–152. [CrossRef]
42. Ryu, S.; Park, J. The effects of benefit-driven commitment on usage of social media for shopping and positive word-of-mouth. *J. Retail. Consum. Serv.* **2020**, *55*, 102094. [CrossRef]
43. Anisimova, T.A. The effects of corporate brand attributes on attitudinal and behavioural consumer loyalty. *J. Consum. Mark.* **2007**, *27*, 395–405. [CrossRef]
44. Scholz, J.; Duffy, K. We are at home: How augmented reality reshapes mobile marketing and consumer-brand relationships. *J. Retail. Consum. Serv.* **2018**, *44*, 11–23. [CrossRef]
45. Santos, L.L.C.; Canha, L.N.; Bernardon, D.P. Projection of the diffusion of photovoltaic systems in residential low voltage consumers. *Renew. Energy* **2018**, *116*, 384–401. [CrossRef]
46. Lee, Z.; Chan, T.; Balaji, M.; Chong, A. Why people participate in the sharing economy: An empirical investigation of Uber. *Internet Res.* **2018**, *28*, 829–850. [CrossRef]
47. Kim, J. The influence of perceived costs and perceived benefits on AI-driven interactive recommendation agent value. *J. Glob. Sch. Mark. Sci.* **2020**, *30*, 319–333. [CrossRef]
48. Adapa, S.; Fazal-e-Hasan, S.M.; Makam, S.B.; Azeem, M.; Mortimer, G. Examining the antecedents and consequences of perceived shopping value through smart retail technology. *J. Retail. Consum. Serv.* **2020**, *52*, 101901. [CrossRef]
49. Guru, S.; Nenavani, J.; Patel, V.; Bhatt, N. Ranking of perceived risks in online shopping. *Decision* **2020**, *47*, 137–152. [CrossRef]
50. Wang, Y.; Min, Q.; Hin, S. Understanding the effects of trust and risk on individual behavior toward social media platforms: A meta-analysis of the empirical evidence. *Comput. Hum. Behav.* **2016**, *56*, 34–44. [CrossRef]
51. Nepomuceno, M.V.; Laroche, M.; Richard, M.-O. How to reduce perceived risk when buying online: The interactions between intangibility, product knowledge, brand familiarity, privacy and security concerns. *J. Retail. Consum. Serv.* **2014**, *21*, 619–629. [CrossRef]
52. Bonnin, G. The roles of perceived risk, attractiveness of the online store and familiarity with AR in the influence of AR on patronage intention. *J. Retail. Consum. Serv.* **2020**, *52*, 101938. [CrossRef]
53. Mueller, C.E. Examining the inter-relationships between procedural fairness, trust in actors, risk expectations, perceived benefits, and attitudes towards power grid expansion projects. *Energy Policy* **2020**, *141*, 111465. [CrossRef]
54. Featherman, M.S.; Pavlou, P.A. Predicting e-services adoption: A perceived risk facets perspective. *Int. J. Hum. Comput. Stud.* **2003**, *59*, 451–474. [CrossRef]
55. Chin, A.G.; Harris, M.A.; Brookshire, R. A bidirectional perspective of trust and risk in determining factors that influence mobile app installation. *Int. J. Inf. Manag.* **2018**, *39*, 49–59. [CrossRef]
56. Pavlou, P.A.; Gefen, D. Building Effective Online Marketplaces with Institution-Based Trust. *Inf. Syst. Res.* **2004**, *15*, 37–59. [CrossRef]
57. Beck, M.; Crié, D. I virtually try it ... I want it! Virtual Fitting Room: A tool to increase on-line and off-line exploratory behavior, patronage and purchase intentions. *J. Retail. Consum. Serv.* **2018**, *40*, 279–286. [CrossRef]
58. Rehman, Z.U.; Baharun, R.; Salleh, N.Z.M. Antecedents, consequences, and reducers of perceived risk in social media: A systematic literature review and directions for further research. *Psychol. Mark.* **2020**, *37*, 74–86. [CrossRef]
59. Pappas, N. Marketing Strategies, Perceived Risks, and Consumer Trust in Online Buying Behaviour. *J. Retail. Consum. Serv.* **2016**, *29*, 92–103. [CrossRef]
60. Mathmann, F.; Chylinski, F.; Higgins, E.T.; de Ruyterd, K. Every step counts: When physical movement affects perceived value. *J. Retail. Consum. Serv.* **2017**, *39*, 279–285. [CrossRef]
61. Sharma, V.M.; Klein, A. Consumer perceived value, involvement, trust, susceptibility to interpersonal influence, and intention to participate in online group buying. *J. Retail. Consum. Serv.* **2020**, *52*, 101946. [CrossRef]
62. Konuk, F.A. The role of store image, perceived quality, trust and perceived value in predicting consumers' purchase intentions towards organic private label food. *J. Retail. Consum. Serv.* **2018**, *43*, 304–310. [CrossRef]
63. Lin, J.; Lobo, A.; Leckie, C. The role of benefits and transparency in shaping consumers' green perceived value, self-brand connection and brand loyalty. *J. Retail. Consum. Serv.* **2017**, *35*, 133–141. [CrossRef]
64. Higgins, E.T. Value from hedonic experience and engagement. *Psychol. Rev.* **2006**, *113*, 439–460. [CrossRef]
65. Hai, M.A.; Moula, M.M.E.; Seppälä, U. Results of intention-behaviour gap for solar energy in regular residential buildings in Finland. *Int. J. Sustain. Built Environ.* **2017**, *6*, 317–329. [CrossRef]
66. Rashid, N.R.N.A. Awareness of eco-label in Malaysia's green marketing initiative. *Int. J. Bus. Manag.* **2009**, *4*, 132–149. [CrossRef]

67. Hair, J.F.; Hult, G.T.M.; Ringle, C.M.; Sarstedt, M. *A Primer on Partial Least Squares Structural Equation Modeling*; Sage: Thousand Oaks, CA, USA, 2014.
68. Bollen, K.A.; Noble, M.D. Structural equation models and the quantification of behavior. *Proc. Natl. Acad. Sci. USA* **2011**, *108*, 15639–15646. [CrossRef]
69. Kline, R.B. *Principles and Practice of Structural Equation Modeling*; The Guilford Press: New York, NY, USA, 2010.
70. Premalatha, M.; Abbasi, T.; Abbasi, S.A. Wind energy: Increasing deployment, rising environmental concerns. *Renew. Sustain. Energy Rev.* **2014**, *31*, 270–288. [CrossRef]
71. Delang, C.O.; Wai-Tung, C. Consumers' attitudes towards electric cars: A case study of Hong Kong. *Transp. Res. Part D* **2012**, *17*, 492–494. [CrossRef]
72. Tawalbeh, M.; Al-Othman, A.; Kafiah, F.; Abdelsalam, E.; Almomani, F.; Alkasrawi, M. Environmental impacts of solar photovoltaic systems: A critical review of recent progress and future outlook. *Sci. Total Environ.* **2021**, *759*, 143528. [CrossRef]
73. Dhar, A.; Naeth, M.A.; Dev Jennings, P.; El-Din, M.G. Perspectives on environmental impacts and a land reclamation strategy for solar and wind energy systems. *Sci. Total Environ.* **2020**, *718*, 134602. [CrossRef]
74. Del Carmen Torres-Sibille, A.; Cloquell-Ballester, V.A.; Cloquell-Ballester, V.A.; Artacho Ramírez, M.Á. Aesthetic impact assessment of solar power plants: An objective and a subjective approach. *Renew. Sustain. Energy Rev.* **2009**, *13*, 986–999. [CrossRef]
75. Lewandowska, A.; Chodkowska-Miszczuk, J.; Rogatka, K.; Starczewski, T. Smart Energy in a Smart City: Utopia or Reality? Evidence from Poland. *Energies* **2020**, *13*, 5795. [CrossRef]
76. Freund, C.L.; Wallich, C.I. The welfare effects of raising household energy prices in Poland. *Energy J.* **1996**, *17*, 53. [CrossRef]
77. Cieślik, S. Prosumer Micro-Installation in Poland—Benefits and Risks. *Sci. Pap. Fac. Electr. Control Eng. Gdańsk Univ. Technol.* **2015**, *45*, 29–34.
78. Park, J.; Gunn, F.; Han, S.-L. Multidimensional trust building in e-retailing: Cross-cultural differences in trust formation and implications for perceived risk. *J. Retail. Consum. Serv.* **2012**, *19*, 304–312. [CrossRef]
79. Oyedele, A.; Simpson, P.M. Streaming apps: What consumers value. *J. Retail. Consum. Serv.* **2018**, *41*, 296–304. [CrossRef]

energies

Article

Economic Implications of Agricultural Land Conversion to Solar Power Production

Yanay Farja [1] **and Mariusz Maciejczak** [2,*]

[1] Department of Economics and Management, Tel-Hai College, Tel Hai 1220800, Israel; yanay@telhai.ac.il
[2] Institute of Economics and Finance, Warsaw University of Life Sciences—SGGW, 02-787 Warsaw, Poland
* Correspondence: mariusz_maciejczak@sggw.edu.pl; Tel.: +48 22 5934235

Abstract: Meeting greenhouse gas (GHG) reduction targets will require a significant increase in electricity production from sustainable and renewable sources such as solar energy. Farmers have recognized this need as a chance to increase the profitability of their farms by allocating farmland to solar power production. However, the shift from agriculture to power production has many tradeoffs, arising primarily from alternative land uses and other means of production. This paper models the farmers' decision as a constrained profit maximization problem, subject to the amount of land owned by the farmers, who have to allocate it between agriculture and solar power fields, while considering factors affecting production costs. The farmers' problem is nested in the social welfare maximization problem, which includes additional factors such as ecological and aesthetical values of the competing land uses. Empirical analysis using data from a solar field operating in Israel shows that landowners will choose to have solar power production on their land unless agricultural production generates an unusually high net income. Adding the values of non-market services provided by agricultural land does not change this result. The consideration of the reduction in GHG emissions further increases the social welfare from solar fields.

Keywords: renewable energy production; agricultural land; profit maximization; social welfare; greenhouse gas emissions; landscape; biodiversity

Citation: Farja, Y.; Maciejczak, M. Economic Implications of Agricultural Land Conversion to Solar Power Production. *Energies* **2021**, *14*, 6063. https://doi.org/10.3390/en14196063

Academic Editor: Jesús Polo

Received: 3 August 2021
Accepted: 18 September 2021
Published: 23 September 2021

1. Introduction

Anthropogenic climate change due to greenhouse gas (GHG) emissions requires a shift in many areas of human activities. Keeping in mind that most of today's operations rely on power supply, one of the highest priority areas where change is needed is electric power production. This change requires a shift in production from fossil fuels to renewable sources, such as solar, wind, and hydroelectric power generation. Of all those alternatives, the sharpest observed rise is in the utilization of naturally available solar energy. Being a stable and consistently available source of clean energy, solar energy has the significant potential to cater to the ever-increasing world electricity requirements [1]. Keeping in mind the sustainability paradigm, this should be achieved in a technically feasible, cost-effective, socially acceptable, and environmentally reasonable way. The gradual transition to energy production from non-renewable sources to energy from renewable sources requires, in particular, attention to the appropriate dynamics of change of the energy mix of specific countries and its economic and environmental effects. According to Adebayo et al. [2], who investigated the case of the energy mix in Japan, the intensity of the transition to a larger share of renewable energy sources is crucial for reducing GHG emissions, which on the other hand, can influence future economic growth.

While taking into account renewable energy needs, the International Energy Agency (IEA) is calling for an energy revolution. The IEA released a roadmap in which it set up a goal for net-zero carbon dioxide (CO_2) emissions in the energy sector by 2050. The suggested pathway to obtain this challenging goal assumes, among other things, scaling up solar and wind energy production in this decade, reaching annual additions of

630 gigawatts (GW) of solar photovoltaics (PV) [3]. The International Renewable Energy Agency (IRENA) predicts that the share of renewable energy in the primary energy supply will grow from less than one-sixth today to nearly two-thirds in 2050. On the demand side, electricity is predicted to be a primary carrier of energy, with its final consumption being near a 50% share by 2050, and renewable power will be able to provide 86% of the power demand [4]. In this revolution, there is crucial importance placed on technology development and diffusion that allow the implementation of global policy requirements aimed at increasing the utilization of renewable sources of energy [5].

Solar power technologies for sustainable and clean electricity generation are considered one of the most promising alternatives for application on a global scale. As reported by IRENA, in 2010–2019, the cost of solar PV production dropped globally by 82%, mainly as a result of more efficient technologies [6]. The costs of local energy-storing systems also dropped [7]. This made solar energy production competitive with traditional power generation and ensured broad diffusion. However, among the constraints for its increased production is land area needed for it to be able to replace coal or natural gas-fueled power stations. It needs to be stressed that PV panels require large production areas, and these are not always available due to competing land uses, such as industrial, residential, agricultural, and environmental uses [8,9]. Additionally, as claimed by Oudes and Stremke [10], solar power plants transform the existing landscapes and, as further indicated by Picchi et al. [11], they also impact ecosystem services. Therefore, due to its large land consumption, PV energy production is challenging from economic, social, and environmental perspectives, as an activity with many tradeoffs [12,13].

One possible solution to this problem that researchers have discussed is to allocate marginal land for solar energy production. Milbrand et al. [14] defined marginal land as "*... areas with inherent disadvantages or lands that have been marginalized by natural and/or artificial forces. These lands are generally underused, difficult to cultivate, have low economic value, and varied developmental potential*". According to these authors, solar technologies present the best opportunity to capture value from marginal land and increase their development potential. Hoffacker et al. [15] identified different marginal land types for solar siting: Built environment, salt-affected land, contaminated land, water reservoirs, and others. They claim that each of these land types has the potential to create synergies between land and solar energy development. However, as stressed by some, e.g., Howard et al. [16] or Cialdea and Maccarone [17], marginal agricultural land, i.e., land with low agricultural productivity, has the highest potential for photovoltaic installations allocation and solar energy production. There is common agreement among scientists and policymakers that agriculture has a seemingly high potential for PV power generation due to different options to install PV on land or farm buildings [18].

At this point, it needs to be highlighted that although the transition to renewable energy will intensify the global competition for land, especially for agricultural purposes, the potential impacts of solar energy production seem not to be addressed enough in the literature. Existing studies focus mainly on the opportunities created by agrophotovoltaic solutions. Cho et al. [19] claim that solar energy installations and agricultural crop cultivation could simultaneously operate on the same land, both with economic justification. However, as investigated by Sacchelli et al. [13], there exist tradeoffs while making decisions about the sole utilization of land for energy or food/feed production. The constraints are related to landscape maintenance, morphological variables, specialization, and crop yields. Case studies confirmed that the coexistence of agriculture and renewable energy production is possible. They recognized roof-mounted or umbrella-shaped facilities using a photovoltaic system as new alternatives to conventional PV plants [20] with many positive direct and synergy effects, such as economic profitability, electricity production for self-consumption, or wildlife benefits [21].

Additionally, ground-mounted PV installations located on arable or grazing lands have been tested as another possible alternative. The results of such experiments are promising, especially for farmers, showing economic benefits [22]. Feasibility studies also highlight several risks connected to investment capacities, energy storage and grid infrastructure availability, biodiversity enhancement, or social limitations [23,24]. The studies also show that agrophotovoltaic installations change agricultural landscapes [25] and can potentially disrupt ecosystems [26] through reduced agricultural production [27]. Studies also identified several drivers that lead to farmers' decisions and show the role of policies that facilitate such changes in agricultural land utilization [18,28]. Policy interventions are crucial, concerning complex issues of climate change, agricultural land scarcity, and food security. According to Gomiero [29], new models need to be promoted to provide key social, economic, and environmental safety objectives. Pretty and Bharucha [30] suggest that sustainable intensification of agricultural production—thanks to which the land could be used optimally in a local dimension—provides food and energy-production opportunities.

However, only a few studies have investigated marginal agricultural land utilization for solar power production, through the sole allocation of PV installations. The knowledge obtained from these studies shows the importance of the different perspectives—energy-centric, agricultural-centric, or agricultural–energy-centric—in search of the benefits or constraints [31]. Leirpol et al. [32] indicate several constraints: Landscape, local, environmental, and socio-economic, in the search for optimal coexistence of agricultural production that is possible on marginal lands and solar energy production. As part of that, Milbrandt et al. [14] report the importance of the availability of PV technologies to facilitate the farmer's decisions, and Maye [33] pays attention to the environmental impact of the life cycle of PV infrastructure.

Bearing in mind the growing importance of solar energy production on marginal agricultural land, a key question arises regarding how to assess the efficiency of the decision to install such PV operations, in a way that will satisfy both private and public expectations.

Caputo et al. [34] present a nexus approach to decision-making in cases that affect food production, energy, water, and societal effects. Our study considers most of these aspects by analyzing the economic value of the different impacts in the case of solar PV field development. Spyridonidou et al. [35] present a planning framework for solar and wind power projects that incorporates many of the aspects mentioned above. However, their model does not consider the economic efficiency of the different choices from the private and social perspectives. Thus, in this study, we ask the following question: What are the conditions under which converting agricultural land to solar power production is economically efficient, both from the landowner's perspective and a social perspective? In our study, we aim at addressing that existing gap in the literature, and in doing so, help with better decision-making by both private and public entities. The analysis in the paper compares two scenarios: The status quo with fossil fuel electricity generation, resulting in GHG emissions, but also with more land in agricultural production, and the scenario with solar power generation on marginal agricultural land.

Our goal is to create a tool that will help farmers and policymakers forecast the economic efficiency of solar power installations on agricultural land and other open spaces. This will be achieved by looking at the decision to produce solar power on agricultural land at the margin, i.e., the profit or net benefit from the last hectare of the lowest-productivity land owned by the farmer. The paper brings into the existing body of knowledge a complex and systemic analysis of private and public perspectives of decision justification for installing solar installations on marginal agricultural land, along with empirical evidence from a representative case study, a field in Israel.

1.1. Climate Change Effects in Israel

Climate change is already affecting Israel, and its effects are expected to increase in the future. Mean annual temperatures have already increased by 1.5 °C in 2020 and are expected to increase by an additional 1 °C until 2050. By 2100, it is forecasted that the overall increase in temperature will be 3–5 °C, depending on the emissions scenario used (RCP4.5 or RCP8.5) [36]. The expected changes in precipitation are a significant decrease in the center and North parts of Israel, reaching up to a 40% decrease in autumn, fewer precipitation days, and more extreme weather events that could lead to floods [37]. The southern, more arid region could potentially experience an increase in precipitation.

These processes have an effect on the agricultural sector in Israel that will become more pronounced as the changes described above intensify. Haim et al. [38] show that some crops such as wheat, grown mainly in Southern Israel, might benefit from the expected changes. Other crops that rely on a more humid climate, such as cotton, will experience decreases in yield and net revenue. Zelingher et al. [39] forecast a partial abandonment of agricultural land, and a shift to production in controlled environments such as greenhouses.

Experts predict that climate change will also influence biodiversity in nearly all ecosystems, mainly due to the changes in temperature and precipitation, with some of these effects already evident in Israel [40]. The abandonment of agricultural lands and their potential conversion to built-up land poses an additional threat of habitat reduction and fragmentation for different species. Solar power production on (former) agricultural land could potentially aggravate the problem if these installations prevent the free movement of animals and the growth of native plant species. Hence, to determine the economic efficiency of solar power production on marginal agricultural land, we include the value of (potentially) lost biodiversity on that land.

1.2. Climate Change Policy in Israel

Israel has ratified the Paris agreement on Climate Change in November 2016. It has submitted its Intended National Determined Contribution (INDC) that promises to reduce per capita greenhouse gas (GHG) emissions to 26% below their 2005 level by 2030 [41]. Given Israel's relatively high rate of population, with a projected population growth of 36–51% between 2015 and 2035 [42], this does not necessarily mean a reduction in overall GHG emissions. In 2017, Israel's government decided on a goal of 10% energy production from renewable sources in 2020. This goal was not achieved, with only 6% of energy consumption in 2020 coming from renewable sources [43]. However, the high rate of growth of solar power and other renewable energy installations, 34% annually in 2012–2019, has led the Israeli government to decide, in 2020, on a more ambitious goal of 30% electricity from renewable sources by 2030. Weiss et al. [44] simulated and showed the feasibility of a 100% renewable energy scenario for Israel in 2030, acknowledging that this will need "radical" market designs.

Similarly, Solomon et al. [45] considered seven different energy transition scenarios in Israel, representing a larger class of Sun Belt countries. They show how the goal of net-zero emissions energy is possible by adopting an explicit pro-solar PV policy and/or using a GHG emissions price. Our study will include the gain to society from reducing GHG emissions as an essential component of the value created by solar power generation.

Israel's unique situation concerning its neighboring states has added two additional geopolitical goals to the renewable energy discourse and policymaking: Energy independence, since it cannot rely on energy supply from some of its hostile neighbors, and cooperation in energy production and supply, supposedly leading to increased economic growth through trade in renewable energy [46].

Additional renewable energy in Israel faces other challenges as well. One example is congestion in the electricity transmission network, because of the recent rise in solar installations [47]. The immediate solution to this problem is reducing energy production from conventional electricity sources, but this requires new agreements with the producers that own these sources.

1.3. The Response of the Agricultural Sector

In the past decades, the agricultural sector in Israel and other developed countries has been subject to processes that lead to rural households' diversification of income sources. These processes include a deterioration in terms of trade for agricultural products, with rising costs of inputs and a relative fall in the price of outputs; increased efficiency in the agricultural sector, leading to reduced demand for labor and food surpluses; and an overall decline in the importance of agriculture as a source of income [48]. The share of agriculture in Israel's GDP has been declining, and in 2020 it was 1.1%, compared to 4.8% in 1980 [49].

Diversification has led to an increase in non-agricultural land uses such as retail, storage and hospitality, and to the household members looking for employment off-farm [50]. Since 2002, the year the Israeli government decided on the first renewable energy production target of 2% by 2007, another source of income for farmers is solar power generation on rooftops and fields [51].

Agricultural land is converted to other uses, including solar power generation, according to its agricultural productivity. The most unproductive land—the marginal land—is converted to non-agricultural uses first. In this paper, we model the decision faced by agricultural landowners by including the opportunity cost of agricultural production on land converted to a PV installation. The opportunity cost is the value of the alternative use of the resource, in this case, agricultural production.

1.4. Renewable Energy Regulation in Israel

Agricultural fields are not the only option for large-scale solar power production in Israel. The current policy of planning authorities in Israel, whose permission is needed to build large solar projects, is that permits are not given while there is still a potential for rooftop solar power generation on large buildings owned by the landowner [52]. This decision is backed by research showing that in the long run, up to 32% of Israel's electricity consumption could be generated on available rooftop areas [53].

Planning authorities also prioritize building solar power facilities on land adjacent to land meant for buildings or other development; building these facilities on detached open space has a low priority. The guidelines also state that the committee will prefer "*plans that maintain the agricultural appearance and use and correspond to the rural texture in the district and the surroundings of the plan*" [52]. As a result, the land allocated to solar power installations needs to be of low agricultural value and have a low value for future residential or commercial development. The latter could be overcome if PV facilities do not require irreversible infrastructure changes to the land on which they are built.

2. Methods

2.1. Conceptual Model

The problem of deciding if and how much land to allocate from agricultural production to solar PV production is modeled with a constrained maximization setup, used in many microeconomic applications, e.g., [54,55]. Our model differs from other works that have looked at land allocation between agriculture and solar power, e.g., [56] by explicitly adding the amenity value of land and the value of biodiversity. We solve the maximization problem using the Lagrange multiplier method.

2.1.1. Private Profit Maximization

The decision to divert land from agricultural production to solar energy production will result from profit maximization by the landowner. Assuming that regulations allow the construction of such installations and that climate conditions are favorable, as they are in Israel, the landowner's problem can be written as:

$$\max \pi = P_{ag} \cdot Q_{ag}(L_{ag}, \theta, T) + P_{el} \cdot Q_{el}(L_{el}, T) - TC_{ag}(L_{ag}, \theta, T) - TC_{el}(L_{el}, d) \quad \text{s.t. } L_{ag} + L_{el} = \overline{L} \tag{1}$$

Table 1 explains the notation used in the preceding equation and throughout this section.

Table 1. Notation used in the conceptual model.

Symbol	Meaning
π	Profit of the landowner
P_{ag}	Price of agricultural product
Q_{ag}	Quantity of agricultural production
L_{ag}	Land in agricultural production (hectares)
θ	Agricultural productivity of land
T	Index of climate conditions (higher values are higher temperatures and lower precipitation)
P_{el}	Price of electricity (feed-in tariff, per kWh)
Q_{el}	Quantity of electricity produced (kWh)
L_{el}	Land with solar power production (hectares)
TC_{ag}	Total cost of agricultural production
TC_{el}	Total cost of solar power production
MC_{ag}	Marginal cost of agricultural production on an additional hectare
MC_{el}	Marginal cost of solar power production on an additional hectare
d	Distance of solar installation from electricity grid (km)
$\overline{L}$	Total amount of land owned (hectares)
SCC	Social cost of carbon (per kWh of electricity produced from natural gas)
α	Amenity value of hectare of land in agricultural production
γ	Loss per hectare due to fragmentation of ecosystem by solar power production

We assume that production increases with the quantity of land allocated to an activity, i.e., $Q_L{'}> 0$, and that costs also increase with land $TC_L{'} > 0$. The landowner maximizes her profit by choosing the values of L_{ag} and L_{el}, i.e., allocating her land between agricultural production and solar energy production. Maximization of the profit function with the Lagrange method with respect to L_{ag} and L_{el} and the multiplier λ leads to the following first-order conditions (FOCs):

$$P_{ag} \cdot MP_{ag,\,L} = MC_{ag} + \lambda \quad (2)$$

$$P_{el} \cdot MP_{el,\,L} = MC_{el} + \lambda \quad (3)$$

The FOCs show that the value of the marginal product of the last hectare of land in agricultural production is equal to the marginal cost of production on the last hectare of land and the shadow value of the land; similarly, the marginal product of the last hectare of land in solar energy production equals the marginal cost of energy production on the last hectare of land. The FOCs also show that the profit from allocating the last hectare of land to either agricultural or solar energy production must be equal; otherwise, profits can be increased by allocating that hectare to the higher net value activity.

Hence, we predict that if a landowner has the opportunity to allocate some of her land to solar production, she will do so by assigning her lowest-productivity agricultural land to that activity. The effect of climate change, either current or expected, is uncertain and depends on the kind of crops grown and its impact on markets through the price P_{ag}.

Land that is more distant from the electricity grid will be less profitable in solar energy production due to the additional costs that this entails.

2.1.2. Social Welfare Maximization

The landowner's profit is included when considering the social perspective of the land allocation problem. In addition, social welfare includes external costs and benefits that do not influence private decision-making. The sustainability of agricultural production is not necessarily a consideration for the landowner since inter-generational aspects are not always considered. However, they cannot be ignored when considering the welfare of the entire population, not only agricultural landowners. Society's net benefit maximization problem is:

$$\max NB = \pi + SCC \cdot Q_{el}(L_{el}, T) + \alpha \cdot L_{ag} - \gamma \cdot L_{el} \quad s.t.\ L_{ag} + L_{el} = \overline{L} \tag{4}$$

This extended problem includes the benefit to society from reducing GHG emissions resulting from the solar energy field, the value of amenities such as agricultural view resulting from agricultural production, and the damage to biodiversity when solar energy installations damage ecosystems and cause habitat fragmentation.

The first-order conditions of the extended problem, obtained using the Lagrange method, are:

$$P_{ag} \cdot MP_{ag,\,L} + \alpha = MC_{ag} + \lambda \tag{5}$$

$$(P_{el} + SCC) \cdot MP_{el,\,L} = MC_{el} + \gamma + \lambda \tag{6}$$

Including these considerations in the social perspective can potentially change the amount of land allocated between the two activities in the optimal case: A higher *SCC* will make solar energy production more valuable, but a higher amenity value of agricultural land can make that activity more worthwhile. In addition, the potential damage to ecosystems from habitat fragmentation and damage to ecosystems in the case of solar energy production means that in some cases, these considerations could lead to a lower amount of solar energy production.

After performing the single-period calculations shown above, we will also conduct a multi-period cost–benefit analysis with a time horizon corresponding to the project's expected life. The following formula will be used for the calculation of the net present value (*NPV*), the difference between the present value of benefits and the present value of costs:

$$NPV = \sum_{t=0}^{T} \left[(B_t - C_t) \cdot (1 + r)^{-t} \right] \tag{7}$$

In this formula, T is the time horizon in years; B_t is the benefit in year t; C_t is the cost in year t, and r is the interest rate used in the analysis. Higher values of r denote higher uncertainty and risk.

2.2. Empirical Methodology

In our analysis, we will estimate the necessary conditions for maximum profit for the landowner and a maximal net benefit for society derived in the conceptual model. This will be done using data from a solar power installation case study in Northern Israel, a representative example of a large-scale PV project (≥10 MW). Many more such projects are planned in the near future due to the Israeli government's decision to reduce carbon emissions from the electricity sector by 85% from 2015 levels by the year 2050 [57].

The values of the parameters in the model will be obtained from different sources: Actual cost and price data received from the owners and operators of the solar field, academic articles and reports for non-market values, and personal communications with stakeholders.

2.3. Data

The solar field we examine in this study was built in 2020 on 11 hectares, with a production capacity of 10 MW. The landowner in this case is a cooperative village, a kibbutz, that contracted with a renewable energy company. The company leased the land from its owners for 23 years, which is the maximum period allowed by Israel's land authority regulations for such contracts. The landowner does not assume any of the construction costs or other related expenses and risks, and is paid an annual fixed sum for the project's life. Figure 1 shows the solar field and its location between almond orchards, with the village that owns it in the background.

Figure 1. The solar power field in Northern Israel. Photo courtesy of Avihu Biali.

The renewable energy company that built and operates the solar power field is a publicly traded (in the Tel Aviv Stock Exchange) firm specializing in such installations. This fact enables us to obtain cost data and the price of electricity from the firm's announcements to the stock exchange.

Total agricultural land owned by the kibbutz that includes field crops and orchards is 330 hectares. Hence, the current solar installation takes slightly more than 3% of the total agricultural land area. The land where the facility was built was considered unsuitable for field crops and orchards because of severe drainage problems. These do not pose a problem for the solar installation but make the land unproductive for agriculture. Our analysis conservatively assumes that the land is not entirely unproductive, meaning that it could be used for almond orchards. We obtained data on agricultural costs and income from two sources: The manager of the farming operations of the village and the official input–output calculations published by the Israeli Ministry of Agriculture and Rural Development.

The location of the solar power field was selected from 8 possible alternatives in the area seen in Figure 1. The criteria for selection and approval of the installation site by the landowners and the authorities were low agricultural productivity, zoning restrictions, amenity value of the landscape, ecological significance as a wildlife corridor, and possible damage to archeological sites found in the area.

The consulting firm that provided its services to the landowners in the selection process did not perform an economic calculation of the different non-market impacts. Their assessment used a 4-color measure of the severity of the solar power field's impact in each potential location within each category: Green, yellow, orange, and red, from least impact to most impact, respectively.

In our analysis, we use data from reliable academic sources for the values of the non-market impacts of the solar field. The data on the landscape amenity value of farmland per hectare in Israel is from Fleischer and Tsur [58]. They used the contingent valuation method and obtained values between 208 € and 416 € per hectare/year.

The value of biodiversity per hectare comes from the Intergovernmental Science-Policy Platform on Biodiversity and Ecosystem Services (IPBES). The organization publishes assessment reports for different parts of the world. Israel is included in its report on Europe and Central Asia. The report contains the non-market values of many functions performed by nature. The median and mean values of habitat creation and maintenance, i.e., biodiversity, are 638 € and 1318 € [59] per hectare per year, respectively.

Since the primary motivation of the shift to renewable energy sources such as solar power is to reduce GHG emissions, we expect the value of the damages prevented to be relatively high. These damages are calculated with the social cost of carbon (*SCC*). Different researchers obtained several possible values of the *SCC*, and we will examine the sensitivity of our results to changes in this parameter. The range of values is between 42 € [60] and 354 € [61] per tonne of CO_2, with 105 € as a value in between the extremes [62].

Since solar power replaces fossil fuel power production, we use the amount of GHG emissions from natural gas electricity production in our calculations, since this represents Israel's most abundant fossil fuel, accounting for the largest share of power production. De Gouw et al. [63] estimate these emissions at 436–549 g CO_2/kWh.

3. Results

The values used in our calculations of both the private profit conditions and the social benefit are given in Table 2. As shown in the conceptual model (Equations (1) and (4)), all the parameters of the private profit maximization problem are also in the social benefit problem.

Table 2. Values used in the empirical application.

Description	Value (€/Hectare/Year)
Net income to landowner from agriculture	2000
Net income to landowner from solar power production	6800
Landscape amenity value	208 to 416
Biodiversity value of agricultural land	638 to 1318
Social cost of carbon (assuming 500 g CO_2/kWh)	23,360 to 238,950

3.1. *Profit Maximization*

The net income to landowners from the highest-value agricultural crop, currently almond orchards, is 2000 € hectare/year. This is a relatively high value. Other marginal lands could have no value in agricultural production, i.e., 0 € hectare/year. The rate at which the renewable energy firm sells the power to the grid is 0.05 €/kWh. This means an income of approximately 67,500 €/hectare/year. The firm pays the landowner 75,000 €/year for the project's life, i.e., a net income of approximately 6800 € hectare/year, or 4800 € hectare/year when considering the opportunity cost. It is clear what a landowner will be inclined to do as a profit maximizer when choosing between agricultural production and leasing the land to the renewable energy firm—the latter one.

When looking at a longer time horizon, that of the life of the project or the contract with the renewable energy firm, which in this case is 23 years, the *NPV* per hectare is 55,000–66,000 €, using discount rates of 5–7%. Lower discount rates, reflecting a lower risk of the project or lower capital costs, will result in even higher sums.

3.2. *Social Welfare Maximization*

Using the values shown in Table 2 for the landscape amenity value and biodiversity value per hectare/year, we see that the upper bound of the annual value of a hectare in agricultural production from society's point of view is 2000 + 416 + 1318 = 3734 €. This is still not high enough to justify giving up the higher value of the land in solar power production. Adding the savings in GHG emissions resulting from substituting natural gas power production with solar power tilts the inequality even more in favor of the solar field.

To find the *SCC* prevented by a hectare of solar power production, we multiply the amount of GHG emissions from natural gas electricity production, which is 412–549 g CO_2/kWh [63,64]. Assuming a social cost of carbon of 354 € per tonne of CO_2 [60,61,63], 1500 MWh produced per MW installed, and 0.9 MW per hectare, this translates to 23,360 € to 238,950 € hectare/year of avoided climate change damage, depending on the values used.

Thus, the annual net benefit of a hectare of solar power production is between 26,426 and 244,904 €. Longer time horizons, such as the project's life of 23 years, mean a social *NPV* of 297,879 € to more than 3.3 million € per hectare. The calculations performed are shown in Table 3. We also show the minimum and maximum social welfare values. The minimum values are obtained with the lowest benefits and the highest costs, and the maximum values are obtained with the highest benefits and lowest costs.

Table 3. Social welfare calculations with minimum and maximum values.

Description	Minimum Value (€/Hectare/Year)	Maximum Value (€/Hectare/Year)
Benefits		
Net income to landowner from solar power production	6800	6800
Social cost of carbon	23,360	238,950
Costs		
Net income to landowner from agriculture	0	2000
Landscape amenity value	208	416
Biodiversity value of agricultural land	638	1318
Social welfare		
Benefits—Costs	26,426	244,904
NPV—23 years—5% interest rate	356,449	3,303,406
NPV—23 years—7% interest rate	297,879	2,760,604

The profits and social welfare values obtained in our analysis are compared in Figure 2. It is evident that the conversion of marginal agricultural land is efficient when adopting the landowners' perspective, but even more so when adopting a social perspective. The minimum private profit value reached 4800 €/hectare/year, while the maximum was 42% higher and reached 6800 €/hectare/year. Concerning social welfare, the sensitivity analysis showed a difference between minimal and maximal values of more than nine times (819%) from 26,646 €/hectare/year to 244,904 €/hectare/year.

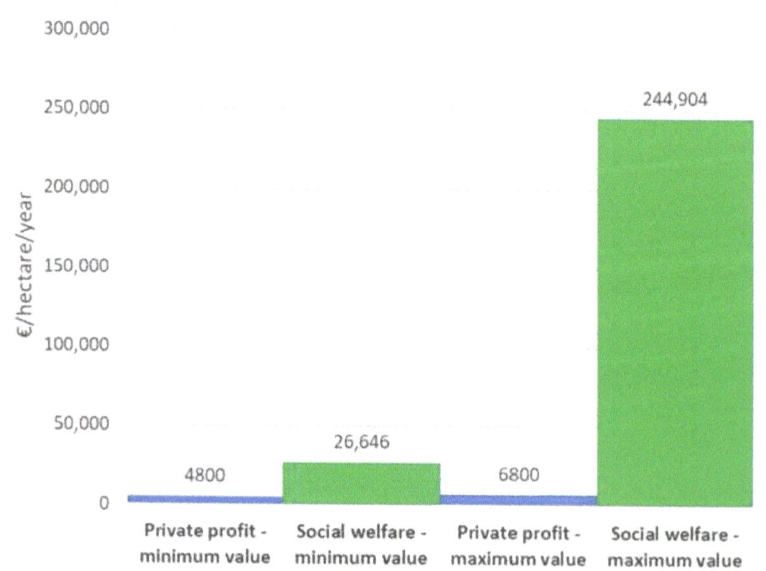

Figure 2. A comparison between annual private profit and annual social welfare of solar power production on one hectare of land.

4. Discussion

The results of our analysis show that as long as climate change is the leading global environmental and societal concern, substituting agricultural land of marginal productivity with solar fields is beneficial both to landowners and to society. This considers a case study of a small solar power field in Israel, where climate conditions make solar power the lowest-cost option for electricity generation. The conclusions might be different in other locations. Furthermore, Capellán-Pérez et al. [65] showed that land-use requirements and solar radiation impact the effectiveness of solar renewable energy production on marginal lands in different countries. Nevertheless, our methodology can be helpful also for those cases.

It is evident from our analysis that the conversion of marginal agricultural land is efficient when adopting the landowners' perspective, but even more so when adopting a social perspective. The latter is associated with the provisions of public goods. Zavalloni et al. [66] paid attention to the importance of the provision of socio-environmental public goods, showing the relationship between public goods provision and land use, as well as their societal value. Therefore, it is important to search for the welfare composition that considers private agricultural income and public good benefits.

The fact that converting agricultural land to a solar power production facility reduces the risk that farmers face in agricultural operations makes such an option, when available, preferable to many crops, certainly to agricultural commodities, as is also reported by other researchers. As part of economic risk reduction, Li et al. [67] indicate that the willingness to change and adoption behaviors of farmers depend on photovoltaic investment costs. However, Ghazeli and Di Corato [68] showed that solar installations reduced the uncertainty in agricultural production.

Changing the assumptions about the non-market value of land can change the results of the analysis. When considering the conversion of non-agricultural lands, such as wetlands or forests, to solar power production, the viability of solar power production might not be straightforward. In those cases, the importance of the land for carbon sequestration, maintenance of biodiversity, landscape, recreation, and other ecosystem services can tip the results in favor of maintaining the land in its current state. Sutherland et al. [69] claim that environmental motives play an important role in decision-making by farmers and are one of the critical factors for policymakers' decisions for supporting such actions.

On the other hand, although Amaducci et al. [70] showed that PV installations could be a valuable system for renewable energy production on farms without negatively affecting land productivity, one also needs to take into account growing food security concerns. Those concerns are rising, also due to possible disruptions to global markets, such as those experienced in 2020 [71]. In such a case, policymakers could become reluctant to give up agricultural production, even on relatively marginal land. It is also possible that in such cases, farmers would also not be willing to enter long-term commitments that could increase their uncertainty in farm profit generation.

5. Conclusions

Global actions towards more sustainable energy production from renewable sources form a movement that can already be recognized as an energy revolution. This revolution is significant from the scope of changes but relatively slow from the perspective of their implementation. Nonetheless, it is taking place. One of the most crucial changes is the use of solar radiation as a source of energy production. PV installations are also built in rural areas, where tradeoff questions arise regarding land allocation between agricultural and energy production. The farmer or the landowner is the final decision maker and needs to consider the short- and long-term effects of what they will decide. Such dilemmas are crucial, especially in marginal agricultural lands, where the costs of agricultural production and the economic gains from solar installations are uncertain. The farmers' dilemma should also be viewed as part of a social welfare problem that includes additional factors such as ecological and aesthetical values of the competing land uses.

The analysis presented in this paper regarded one PV installation on marginal agricultural land in Israel. The results show that the higher economic gains justified the landowners' decision to install a photovoltaic system. Furthermore, from the social point of view, regarding carbon sequestration, biodiversity enhancement, or land productivity, the analysis favors the investment in photovoltaics on marginal agricultural land. The analysis performed in this paper can be readily applied to future projects in Israel and elsewhere that involve land use conversion from agricultural use to energy production.

A possible direction for future research could be using life cycle analysis to further examine the costs associated with the different land use options, both solar power production and agricultural production. As technological knowledge in climate change mitigation and renewable power generation and storage advances, options such as carbon storage and sequestration and energy storage could further increase the attractiveness of solar power generation.

The problem investigated in this paper should also be considered in a much broader perspective that takes into account the correlation between the use of resources such as land, water, and energy, and food production. The nexus approach requires special attention to marginal land allocation as lands of this type become valuable resources with rising significance in sustainable and resilient growth.

Author Contributions: Conceptualization, Y.F. and M.M.; formal analysis, Y.F. and M.M.; investigation, Y.F. and M.M.; methodology, Y.F. and M.M.; writing—original draft, Y.F. and M.M.; writing—review and editing, Y.F. and M.M. All authors have read and agreed to the published version of the manuscript.

Funding: This research received no external funding.

Institutional Review Board Statement: Not applicable.

Informed Consent Statement: Not applicable.

Data Availability Statement: The data presented in this study are available on request from the corresponding author. The data are not publicly available due to privacy restrictions.

Conflicts of Interest: The authors declare no conflict of interest.

References

1. Khan, J.; Arsalan, M.H. Solar power technologies for sustainable electricity generation—A review. *Renew. Sustain. Energy Rev.* **2016**, *55*, 414–425. [CrossRef]
2. Adebayo, T.; Awosusi, A.; Oladipupo, S.; Agyekum, E.; Jayakumar, A.; Kumar, N. Dominance of Fossil Fuels in Japan's National Energy Mix and Implications for Environmental Sustainability. *Int. J. Environ. Res. Public Health* **2021**, *18*, 7347. [CrossRef] [PubMed]
3. International Energy Agency. *Net Zero by 2050: A Roadmap for the Global Energy System*, 3rd ed.; International Energy Agency: Paris, France, 2021.
4. International Renewable Energy Agency. *Global Energy Transformation: A Roadmap to 2050*, 2019th ed.; International Renewable Energy Agency: Abu Dhabi, United Arab Emirates, 2019.
5. Bilgili, M.; Ozbek, A.; Sahin, B.; Kahraman, A. An overview of renewable electric power capacity and progress in new technologies in the world. *Renew. Sustain. Energy Rev.* **2015**, *49*, 323–334. [CrossRef]
6. International Renewable Energy Agency. *Renewable Power Generation Costs in 2019*; International Renewable Energy Agency: Abu Dhabi, United Arab Emirates, 2020.
7. Jäger-Waldau, A. Snapshot of Photovoltaics—February. *Energies* **2020**, *13*, 930. [CrossRef]
8. Chivelet, N.M. Photovoltaic potential and land-use estimation methodology. *Energy* **2016**, *94*, 233–242. [CrossRef]
9. Hernandez, R.R.; Armstrong, A.; Burney, J.; Ryan, G.; Moore-O'Leary, K.; Diédhiou, I.; Grodsky, S.M.; Saul-Gershenz, L.; Davis, R.; Macknick, J.; et al. Techno–Ecological synergies of solar energy for global sustainability. *Nat. Sustain.* **2019**, *2*, 560–568. [CrossRef]
10. Oudes, D.; Stremke, S. Next generation solar power plants? A comparative analysis of frontrunner solar landscapes in Europe. *Renew. Sustain. Energy Rev.* **2021**, *145*, 111101. [CrossRef]
11. Picchi, P.; van Lierop, M.; Geneletti, D.; Stremke, S. Advancing the relationship between renewable energy and ecosystem services for landscape planning and design: A literature review. *Ecosyst. Serv.* **2019**, *35*, 241–259. [CrossRef]
12. Fthenakis, V.; Kim, H.C. Land use and electricity generation: A life-cycle analysis. *Renew. Sustain. Energy Rev.* **2009**, *13*, 1465–1474. [CrossRef]

13. Sacchelli, S.; Garegnani, G.; Geri, F.; Grilli, G.; Paletto, A.; Zambelli, P.; Ciolli, M.; Vettorato, D. Trade-off between photovoltaic systems installation and agricultural practices on arable lands: An environmental and socio-economic impact analysis for Italy. *Land Use Policy* **2016**, *56*, 90–99. [CrossRef]
14. Milbrandt, A.R.; Heimiller, D.M.; Perry, A.D.; Field, C.B. Renewable energy potential on marginal lands in the United States. *Renew. Sustain. Energy Rev.* **2014**, *29*, 473–481. [CrossRef]
15. Hoffacker, M.K.; Allen, M.F.; Hernandez, R.R. Land-Sparing Opportunities for Solar Energy Development in Agricultural Landscapes: A Case Study of the Great Central Valley, CA, United States. *Environ. Sci. Technol.* **2017**, *51*, 14472–14482. [CrossRef] [PubMed]
16. Howard, D.C.; Wadsworth, R.A.; Whitaker, J.W.; Hughes, N.; Bunce, R.G. The impact of sustainable energy production on land use in Britain through to 2050. *Land Use Policy* **2009**, *26*, S284–S292. [CrossRef]
17. Cialdea, D.; Maccarone, A. The Energy Networks Landscape. Impacts on Rural Land in the Molise Region. *TeMA-J. Land Use Mobil. Environ.* **2014**, 223–234. [CrossRef]
18. Lane, A.-L.; Boork, M.; Thollander, P. Barriers, Driving Forces and Non-Energy Benefits for Battery Storage in Photovoltaic (PV) Systems in Modern Agriculture. *Energies* **2019**, *12*, 3568. [CrossRef]
19. Cho, J.; Park, S.M.; Park, A.R.; Lee, O.C.; Nam, G.; Ra, I.-H. Application of Photovoltaic Systems for Agriculture: A Study on the Relationship between Power Generation and Farming for the Improvement of Photovoltaic Applications in Agriculture. *Energies* **2020**, *13*, 4815. [CrossRef]
20. Dupraz, C.; Marrou, H.; Talbot, G.; Dufour, L.; Nogier, A.; Ferard, Y. Combining solar photovoltaic panels and food crops for optimising land use: Towards new agrivoltaic schemes. *Renew. Energy* **2011**, *36*, 2725–2732. [CrossRef]
21. Beckman, J.; Xiarchos, I.M. Why are Californian farmers adopting more (and larger) renewable energy operations? *Renew. Energy* **2013**, *55*, 322–330. [CrossRef]
22. Roy, S.; Ghosh, B. Land utilization performance of ground mounted photovoltaic power plants: A case study. *Renew. Energy* **2017**, *114*, 1238–1246. [CrossRef]
23. Van Campen, B.; Guidi, D.; Best, G. *Solar Photovoltaics for Sustainable Agriculture and Rural Development*; FAO: Rome, Italy, 2000. [CrossRef]
24. Weselek, A.; Ehmann, A.; Zikeli, S.; Lewandowski, I.; Schindele, S.; Högy, P. Agrophotovoltaic systems: Applications, challenges, and opportunities. A review. *Agron. Sustain. Dev.* **2019**, *39*, 35. [CrossRef]
25. Chiabrando, R.; Fabrizio, E.; Garnero, G. The territorial and landscape impacts of photovoltaic systems: Definition of impacts and assessment of the glare risk. *Renew. Sustain. Energy Rev.* **2009**, *13*, 2441–2451. [CrossRef]
26. Pascaris, A.S.; Schelly, C.; Pearce, J.M. A First Investigation of Agriculture Sector Perspectives on the Opportunities and Barriers for Agrivoltaics. *Agronomy* **2020**, *10*, 1885. [CrossRef]
27. Chel, A.; Kaushik, G. Renewable energy for sustainable agriculture. *Agron. Sustain. Dev.* **2011**, *31*, 91–118. [CrossRef]
28. Brudermann, T.; Reinsberger, K.; Orthofer, A.; Kislinger, M.; Posch, A. Photovoltaics in agriculture: A case study on decision making of farmers. *Energy Policy* **2013**, *61*, 96–103. [CrossRef]
29. Gomiero, T. Soil Degradation, Land Scarcity and Food Security: Reviewing a Complex Challenge. *Sustainability* **2016**, *8*, 281. [CrossRef]
30. Pretty, J.; Bharucha, Z.P. Sustainable intensification in agricultural systems. *Ann. Bot.* **2014**, *114*, 1571–1596. [CrossRef] [PubMed]
31. Abidin, M.Z.; Mahyuddin, M.; Zainuri, M.M. Solar Photovoltaic Architecture and Agronomic Management in Agrivoltaic System: A Review. *Sustainability* **2021**, *13*, 7846. [CrossRef]
32. Leirpoll, M.E.; Næss, J.S.; Cavalett, O.; Dorber, M.; Hu, X.; Cherubini, F. Optimal combination of bioenergy and solar photovoltaic for renewable energy production on abandoned cropland. *Renew. Energy* **2021**, *168*, 45–56. [CrossRef]
33. Maye, D. The New Food Insecurity. In *Routledge Handbook of Landscape and Food*; Zeunert, J., Waterman, T., Eds.; Routledge: London, UK, 2014; pp. 380–390.
34. Caputo, S.; Schoen, V.; Specht, K.; Grard, B.; Blythe, C.; Cohen, N.; Fox-Kämper, R.; Hawes, J.; Newell, J.; Poniży, L. Applying the food-energy-water nexus approach to urban agriculture: From FEW to FEWP (Food-Energy-Water-People). *Urban For. Urban Green.* **2021**, *58*, 126934. [CrossRef]
35. Spyridonidou, S.; Sismani, G.; Loukogeorgaki, E.; Vagiona, D.G.; Ulanovsky, H.; Madar, D. Sustainable Spatial Energy Planning of Large-Scale Wind and PV Farms in Israel: A Collaborative and Participatory Planning Approach. *Energies* **2021**, *14*, 551. [CrossRef]
36. Yosef, Y.; Baharad, A.; Uzan, L.; Furshpan, A.; Levi, Y. *Israel Temperature Projections by Research Report No. 4000-0802-2020-0000044*; Israel Meteorological Service: Jerusalem, Israel, 2020. (In Hebrew)
37. Hochman, A.; Mercogliano, P.; Alpert, P.; Saaroni, H.; Bucchignani, E. High-resolution projection of climate change and extremity over Israel using COSMO-CLM. *Int. J. Clim.* **2018**, *38*, 5095–5106. [CrossRef]
38. Haim, D.; Shechter, M.; Berliner, P. Assessing the impact of climate change on representative field crops in Israeli agriculture: A case study of wheat and cotton. *Clim. Chang.* **2007**, *86*, 425–440. [CrossRef]
39. Zelingher, R.; Ghermandi, A.; De Cian, E.; Mistry, M.; Kan, I. Economic Impacts of Climate Change on Vegetative Agriculture Markets in Israel. *Environ. Resour. Econ.* **2019**, *74*, 679–696. [CrossRef]

40. Sternberg, M.; Gabay, O.; Angel, D.; Barneah, O.; Gafny, S.; Gasith, A.; Grünzweig, J.M.; Hershkovitz, Y.; Israel, A.; Milstein, D.; et al. Impacts of climate change on biodiversity in Israel: An expert assessment approach. *Reg. Environ. Chang.* **2015**, *15*, 895–906. [CrossRef]
41. Tal, A. Will We Always Have Paris? Israel's Tepid Climate Change Strategy. *Isr. J. Foreign Aff.* **2016**, *10*, 405–421. [CrossRef]
42. Central Bureau of Statistics. *Population Forecasts for 2025, 2035, 2045, 2055 and 2065, by Population Group*; Central Bureau of Statistics: Jerusalem, Israel, 2018. (In Hebrew)
43. Shahak, M. *Renewable Energy in Israel—Background and Issues for Discussion*; Research and Information Center: The Knesset, Jerusalem, Israel, 2020. (In Hebrew)
44. Weiss, O.; Bogdanov, D.; Salovaara, K.; Honkapuro, S. Market designs for a 100% renewable energy system: Case isolated power system of Israel. *Energy* **2017**, *119*, 266–277. [CrossRef]
45. Solomon, A.; Bogdanov, D.; Breyer, C. Solar driven net zero emission electricity supply with negligible carbon cost: Israel as a case study for Sun Belt countries. *Energy* **2018**, *155*, 87–104. [CrossRef]
46. Fischhendler, I.; Nathan, D.; Boymel, D. Marketing Renewable Energy through Geopolitics: Solar Farms in Israel. *Glob. Environ. Polit.* **2015**, *15*, 98–120. [CrossRef]
47. Navon, A.; Kulbekov, P.; Dolev, S.; Yehuda, G.; Levron, Y. Integration of distributed renewable energy sources in Israel: Transmission congestion challenges and policy recommendations. *Energy Policy* **2020**, *140*, 111412. [CrossRef]
48. Sofer, M. The future of family farming in Israel: The second generation in the Moshav. *Geogr. J.* **2005**, *171*, 357–368. [CrossRef]
49. Central Bureau of Statistics. *Gross Domestic Product and Gross Domestic Business Product, by Industry*; Central Bureau of Statistics: Jerusalem, Israel, 2021.
50. Kimhi, A. Family Composition and Off-Farm Participation Decisions in Israeli Farm Households. *Am. J. Agric. Econ.* **2004**, *86*, 502–512. [CrossRef]
51. Fischhendler, I.; Boymel, D.; Boykoff, M.T. How Competing Securitized Discourses over Land Appropriation Are Constructed: The Promotion of Solar Energy in the Israeli Desert. *Environ. Commun.* **2016**, *10*, 147–168. [CrossRef]
52. Planning Administration. *North District Committee for Planning and Construction: Decisions from Meeting 2020019—Updated Principles for Photo-Voltaic Facilities in the District*; Planning Administration: Jerusalem, Israel, 2020. (In Hebrew)
53. Vardimon, R. Assessment of the potential for distributed photovoltaic electricity production in Israel. *Renew. Energy* **2011**, *36*, 591–594. [CrossRef]
54. Balezentis, T.; Chen, X.; Galnaityte, A.; Namiotko, V. Optimizing crop mix with respect to economic and environmental constraints: An integrated MCDM approach. *Sci. Total. Environ.* **2020**, *705*, 135896. [CrossRef] [PubMed]
55. Zilberman, D.; Lu, L.; Reardon, T. Innovation-induced food supply chain design. *Food Policy* **2019**, *83*, 289–297. [CrossRef]
56. Amigues, J.-P.; Moreaux, M. Competing land uses and fossil fuel, and optimal energy conversion rates during the transition toward a green economy under a pollution stock constraint. *J. Environ. Econ. Manag.* **2019**, *97*, 92–115. [CrossRef]
57. Israel Ministry of Energy. The Israeli Government Set to Approve an Unprecedented Decision Mandating That by the Year 2050 Israel Will Move to a Low Carbon Emissions Economy, While Dealing with the Climate Crisis That Threatens All of Humanity. Available online: https://www.gov.il/en/departments/news/press_250721 (accessed on 12 September 2021).
58. Fleischer, A.; Tsur, Y. The Amenity Value of Agricultural Landscape and Rural-Urban Land Allocation. *J. Agric. Econ.* **2009**, *60*, 132–153. [CrossRef]
59. Rounsevell, M.; Fischer, M.; Torre-Marin, A.; Mader., A.; Bonn, G. (Eds.) Intergovernmental Science-Policy Platform on Biodiversity and Ecosystem Services, The IPBES Regional Assessment Report on Biodiversity and Ecosystem Services for Europe and Central Asia. 2018. Available online: https://www.ipbes.net/assessment-reports/eca (accessed on 20 September 2021).
60. Wagner, G.; Anthoff, D.; Cropper, M.; Dietz, S.; Gillingham, K.T.; Groom, B.; Kelleher, J.P.; Moore, F.C.; Stock, J.H. Eight priorities for calculating the social cost of carbon. *Nat. Cell Biol.* **2021**, *590*, 548–550. [CrossRef]
61. Ricke, K.; Drouet, L.; Caldeira, K.; Tavoni, M. Country-level social cost of carbon. *Nat. Clim. Chang.* **2018**, *8*, 895–900. [CrossRef]
62. Carleton, T.; Greenstone, M. *Updating the United States Government's Social Cost of Carbon. Becker Friedman Institute for Economics Working Paper No. 2021-04*; University of Chicago: Chicago, IL, USA, 2021.
63. De Gouw, J.A.; Parrish, D.D.; Frost, G.J.; Trainer, M. Reduced emissions of CO_2, NOx and SO_2 from U.S. power plants owing to switch from coal to natural gas with combined cycle technology. *Earths Future* **2014**, *2*, 75–82. [CrossRef]
64. US Energy Information Administration. How Much Carbon Dioxide Is Produced Per Kilowatthour of US Electricity Generation? 2020. Available online: https://www.eia.gov/tools/faqs/faq.php?id=74 (accessed on 12 June 2021).
65. Capellán-Pérez, I.; de Castro, C.; Arto, I. Assessing vulnerabilities and limits in the transition to renewable energies: Land requirements under 100% solar energy scenarios. *Renew. Sustain. Energy Rev.* **2017**, *77*, 760–782. [CrossRef]
66. Zavalloni, M.; D'Alberto, R.; Raggi, M.; Viaggi, D. Farmland abandonment, public goods and the CAP in a marginal area of Italy. *Land Use Policy* **2021**, *107*, 104365. [CrossRef]
67. Li, B.; Ding, J.; Wang, J.; Zhang, B.; Zhang, L. Key factors affecting the adoption willingness, behavior, and willingness-behavior consistency of farmers regarding photovoltaic agriculture in China. *Energy Policy* **2021**, *149*, 112101. [CrossRef]
68. Gazheli, A.; Di Corato, L. Land-use change and solar energy production: A real option approach. *Agric. Finance Rev.* **2013**, *73*, 507–525. [CrossRef]
69. Sutherland, L.-A.; Toma, L.; Barnes, A.; Matthews, K.B.; Hopkins, J. Agri-environmental diversification: Linking environmental, forestry and renewable energy engagement on Scottish farms. *J. Rural Stud.* **2016**, *47*, 10–20. [CrossRef]

70. Amaducci, S.; Yin, X.; Colauzzi, M. Agrivoltaic systems to optimise land use for electric energy production. *Appl. Energy* **2018**, *220*, 545–561. [CrossRef]
71. Mogues, T. *Food Markets During COVID-Fiscal Affairs*; Special Series on COVID; International Monetary Fund: Washington, DC, USA, 2020.

energies

MDPI

Article

GHGs Emission from the Agricultural Sector within EU-28: A Multivariate Analysis Approach

Endre Harsányi [1], Bashar Bashir [2], Gafar Almhamad [3], Omar Hijazi [4], Mona Maze [5], Ahmed Elbeltagi [6], Abdullah Alsalman [2], Glory O. Enaruvbe [7], Safwan Mohammed [1,*] and Szilárd Szabó [8]

1 Faculty of Agricultural and Food Sciences and Environmental Management, Institute of Land Utilization, Technology and Regional Development, University of Debrecen, Böszörményi Str. 138, 4032 Debrecen, Hungary; harsanyie@agr.unideb.hu
2 Department of Civil Engineering, College of Engineering, King Saud University, P.O. Box 800, Riyadh 11421, Saudi Arabia; bbashir@ksu.edu.sa (B.B.); asalman@ksu.edu.sa (A.A.)
3 Institute of Management and Organization Sciences, Debrecen University, 4032 Debrecen, Hungary; Almhamad.gafar@econ.unideb.hu
4 Chair of Wood Science, Technical University of Munich, 85354 Freising, Germany; hijazi@hfm.tum.de
5 Central Laboratory for Agricultural Climate (CLAC), Agricultural Research Center, Giza 12613, Egypt; mona_maze@yahoo.de
6 Agricultural Engineering Department, Faculty of Agriculture, Mansoura University, Mansoura 35516, Egypt; ahmedelbeltagy81@mans.edu.eg
7 African Regional Institute for Geospatial Information Science and Technology, Obafemi Awolowo University, Ile-Ife 220282, Nigeria; enaruvbe@gmail.com
8 Department of Physical Geography and Geoinformatics, University of Debrecen, Egyetem tér 1, 4032 Debrecen, Hungary; szabo.szilard@science.unideb.hu
* Correspondence: safwan@agr.unideb.hu

Citation: Harsányi, E.; Bashir, B.; Almhamad, G.; Hijazi, O.; Maze, M.; Elbeltagi, A.; Alsalman, A.; Enaruvbe, G.O.; Mohammed, S.; Szabó, S. GHGs Emission from the Agricultural Sector within EU-28: A Multivariate Analysis Approach. *Energies* **2021**, *14*, 6495. https://doi.org/10.3390/en14206495

Academic Editor: Piotr Gradziuk

Received: 7 September 2021
Accepted: 6 October 2021
Published: 11 October 2021

Abstract: Climate mitigation and adaptation planning (CMAP) has recently been implemented across the EU-28 to reduce GHG emissions (CO_2, CH_4, N_2O). Thus, the aim of this study was to provide an overview of GHG emissions from the agricultural sector in the EU-28 from 1990 to 2019, and cluster the EU-28 countries regarding their total GHG emissions. The results emphasize the positive impact of CMAP through a negative trend of the total GHG emissions (−2653.01 thousand tons/year, $p < 0.05$). Despite the positive and not significant trend of the total CO_2 emissions, both CH_4 and N_2O exhibited a negative and significant trend. At the country scale, Italy, the United Kingdom, and the Netherlands showed the highest reduction in total GHG emissions, by −282.61thousand tons/year ($p < 0.05$), −266.40 thousand tons/year ($p < 0.05$), and −262.91 thousand tons/year ($p < 0.05$), respectively. The output of the multivariate analysis approach indicates changes in the pattern of GHG emissions between 1990 and 2019, where CO_2 emissions decreased in the case of Poland and Czechia. The output of this study highlights the positive impact of CMAP, adopted by EU countries, in minimizing GHG emissions. Despite some fluctuations in CO_2 emissions, strategies for attaining carbon neutrality in the agricultural sector, across the European Union, should be pursued.

Keywords: climate policy; GHGs emissions; PCA; IPCC; CSA

1. Introduction

Rapid population growth and the concomitant increase in anthropogenic activities have resulted in climate change-induced challenges, and pose major threats to the sustainability of natural resources and the stability of the Earth's biosphere, especially in the recent past [1]. These challenges are leading to uncontrolled accumulation of greenhouses gases in the Earth's atmosphere [2]. The global concentration of greenhouse gas emissions (GHGs) has been accelerating particularly rapidly since the beginning of the industrial era because of various anthropogenic activities [3]; for instance, although the concentration of CO_2 in the 1760s was 280 ppm, the current estimate is 410 ppm, and is expected to reach

590 ppm by the end of 2100 [2]. The global tracking of greenhouse gas emissions provides a framework for assessing the contribution of individual countries to the climate change challenge. Climate change indicators define the emissions of the most significant GHGs from human activity, atmospheric concentrations, and how emissions and concentrations have evolved over time [4,5]. These indicators employ the concept of "global warming potential" for comparing the emissions of gases, in order to convert the quantities of other gases into CO_2 equivalents. The emissions of GHGs from human activities are rising and exacerbating climate change. This increasing level of GHGs is resulting in many more climate-related changes at the local to global scale [6].

Since the beginning of the industrial age, CO_2 and other GHGs in the atmosphere have been on the rise, primarily because of human activity. The net emissions of greenhouse gasses from human activities worldwide increased to 43% between 1990 and 2015. During this period, carbon dioxide emissions, representing approximately 35% of the total emissions, have grown by 51% [7]. The industrial and agricultural sectors accounted for 31.6% and 13.8%, respectively, and were considered to be major sources of GHG emissions, while 12.2% of the emissions came from land use changes [8]. Unpredictably, GHG emissions from the agricultural sector increased by 1.1% between 2000 and 2010 [9]. Many factors, such as agricultural expansion and/or intensification, deforestation, land clearing, fertilization, livestock production, and traditional soil management and cultivation, alter the global geochemical cycle and enhance GHG emissions from the agricultural sector, especially in developing countries [10]. Interestingly, Tian et al. [11] reported that 87% of the total N_2O emissions originated from the agricultural sector (71% agricultural + 16% N-fertilization), with cropland farming accounting for roughly 5% of all anthropogenic GHG emissions [12]. Therefore, many databases and analyses were developed to address the current and future contribution of the agricultural sector to climate change, and formulate adaptation and mitigation strategies. Politically, the assessment of GHG emissions from the agricultural sector, along with related sectors, such as forestry and other land use (i.e., AFOLU), will support the discussion about the role of agriculture in climate mitigation within the United Nations Climate Change Conference (COP26).

Recently, the GHG emissions from the European Union's agricultural sector were estimated to be 10% of the total GHG emissions [13]. Although the total amount of EU-27 GHG emissions in 2019 was 4.065 $MtCO_{2e}$ [14], the annual estimate of GHG emissions by agriculture was 436 million tons [15]. This suggests that energy is one of the main inputs in the agricultural system [16], whereas the energy from creating fossil fuels is mainly utilized by agriculture and multiple other actions, including forestry, which form 2.78% of the European Union correlated activities [17]. The Netherlands has the highest share of agricultural energy usage, with 8.1 percent, followed by Poland, which has 5.6 percent. In contrast, Romania accounts for the lowest percentage overall [18]. The growing quantity of energy is due to the neoteric agricultural activities, which are partly responsible for the persistent increase in GHG emissions [19]. About half of the energy used in the agricultural sector is derived from diesel and gas oil, which make up the highest share of energy utilized in the agricultural sector in the EU [15]. Regardless of the size and variation in the contribution of the agricultural sector in the national GDP in each member country of the EU, the EU has achieved a 23% reduction in GHG emissions in the last two decades [13].

Since its foundation, the EU has adopted many strategies, plans, and programs for environmental sustainability, with emphasis on energy management and the reduction in GHG emissions [20,21]. The common agricultural policy (CAP) is a policy created by the European Union, with the aim of implementing activities to integrate climate change reduction procedures into its policies [22]. During 2014–2020, over one hundred billion Euros, accounting for about 25% of the CAP budget, was the contribution of the commission to reduce, alleviate, and adjust to climate change. The European Green Deal strategy has recently been adopted, which is designed to promote climate neutral actions and resource-efficient consumption [20,23]. Many studies were carried out to assess low-carbon economy (LCE) within the agricultural sector. In view of this, Piwowar et al. [24]

stress the importance of raising the awareness of farmers about LCE practice in rural areas of Poland. In Spain, Baccour et al. [25] suggested that a combination of measures could help reduce GHG emissions from the agricultural sector by 75%. Interestingly, Bajan et al. [20] proved that the usage of renewable energy in food production is approaching the expected strategic goals within V4 countries (Czechia, Slovakia, Hungary, Poland), resulting in some of the successes recorded in the reduction in GHG emissions from the agricultural sector in some EU countries (1990–2018). However, EU polices in agricultural sector led to a reduction of GHG emissions [26].

Promoting efforts towards minimizing GHG emissions at the field scale are required to attain the aspiration of GHG reduction in Europe by 2050 [27]. In addition to an efficient energy toolkit, water and carbon footprints for agriculture output are being established by the European Union. This is aimed at reducing water shortage, enhancing energy efficiency, and excluding gas emissions by 2050 [28]. The objectives laid out under the EU effort sharing law may vary slightly among member states, and there are exceptions, even though most EU member states do not have agricultural targets. The Netherlands, for instance, has established an emission reduction target of 3.5 $MtCO_2eq\ yr^{-1}$ by 2030, which should be reached by the co-funding of mitigation measures, and governmental and business cooperation in their National Agreement on Climate Change (NACC) [29]. Other member states have set carbon budgets in their national low-carbon strategy. France, for example, projected a cut in GHG emissions of 8% by 2023, 13% by 2028, and 20% by 2033, based on a benchmark of the 2015 levels [30]. The UK has also created carbon budgets that have strategic sector objectives, including a 20% reduction in agriculture, forestry, and other land use emissions from 2016 to 2030 [31]. In its 2050 climate action plan, Germany has more aggressive targets of reducing agriculture emissions by 31–34% in 2030, using a 1990 benchmark [32]. The climate action plan of Ireland provides a de-carbonization route to 2030, consistent with the adoption, by 2050, of net zero emission objectives [33]. There are some measures for reducing GHGs from the agricultural sector, such as cost-effectiveness analysis [34], reducing water consumption in different agricultural systems [35], no or reduced tillage (NT/RT) combined with crop rotations (i.e., legumes and cover crops) [36], and others [37,38]. On this basis, the objective of this study was to (1) evaluate the changes in GHG emissions from the agricultural sector of the EU-28 from 1990 to 2019, and (2) cluster the EU-28 countries regarding their total GHG emissions.

2. Materials and Methods

2.1. Data Collection

Data for the EU-28 countries between 1990 and 2019 were collected from the European Environment Agency (EEA) [39] (Table 1). These data were checked and updated in June 2021. All countries and their abbreviations are listed in Table 2.

Table 1. Type of collected data.

Data Type	Unit	Time Frequency
Total GHGs emission from agricultural sector *	Thousand tons	Annual
CO_2	Thousand tons	Annual
CH_4 (CO_2 equivalent)	Thousand tons	Annual
N_2O (CO_2 equivalent)	Thousand tons	Annual

* Greenhouse gases (CO_2, N_2O in CO_2 equivalent, CH_4 in CO_2 equivalent, HFC in CO_2 equivalent, PFC in CO_2 equivalent, SF6 in CO_2 equivalent, NF_3 in CO_2 equivalent).

2.2. Trend Analysis

Trend analysis could be conducted using parametric and non-parametric methods. Despite the effectiveness of parametric methods, they require independent and normally distributed data. In contrast, non-parametric methods simply require independent data [40]. In this research, the Mann–Kendall (MK) test [41,42] was used for detecting trend of GHGs across EU-28 countries. The MK is a well-known rank-based non-parametric test used to detect decrease (−) or increase (+) for studied variables through time. The advantage of using the MK test is that data do not have to be normally distributed, and they are not affected by outliers. The null hypothesis (H_0) in MK test states that there is no trend over time. For time series X = (X_1, X_2, ... X_n), the static test of MK (S) could be interoperated as shown in Equations (1) and (2) as follow [41,42]:

$$S = \sum_{i=1}^{N-1} \sum_{j=i+1}^{N} \nabla_{ij} \quad (1)$$

$$\nabla_{ij} = \text{sign}(x_j - x_i) = \begin{cases} 1 & x_j > x_i \\ 0 & x_j = x_i \\ -1 & x_j < x_i \end{cases} \quad (2)$$

where N: length of the data, x_i and x_j: observations.

The variance in S is denoted as shown in Equation (3), as follows:

$$\text{Var}\ (S) = \frac{n(n-1)(2n+5) - \sum_{i=1}^{P} t_i(t_i - 1)(2t_i + 5)}{18} \quad (3)$$

where P: tied group, t_i: number of data. Then Z standard can be calculated. More details about MK calculations could be found in [40].

In this study the M_k was adopted for detecting the trend of GHG emissions to overcome the presence of outliers and skewed data [43]. Also, the Sen slope (ρ) [44] was used to determine the amount of GHG changes per time. The ρ is a non-parametric method that captures the slope of the trend in a dataset (N pairs) as depicted in Equations (4) and (5), as follows:

$$\rho = \frac{x_j - x_i}{j - k}\ (j > k) \quad (4)$$

where x_j, x_i are values of the data. Then the median of ρ is computed as follows:

$$\rho_{median} = \begin{cases} \rho_{\frac{(N+1)}{2}} & \text{if N is odd} \\ \frac{\rho_{(\frac{N}{2})} + \rho_{\frac{(N+2)}{2}}}{2} & \text{if N is even} \end{cases} \quad (5)$$

2.3. Multivariate Analysis

Principal component analysis (PCA) is a multivariate method for reducing a large number of inter-correlated quantitative data (dependent variables) to a smaller number of representative variables, known as principal components (PCs), by employing complex underlying mathematical functions [45,46]. In this study, the similarities and differences in GHG emissions (CO_2, CH_4 and NO_2) across EU-28 were determined using principal component analysis (PCA). The PCA was performed with the standardized approach using the correlation matrix to reveal the pattern of GHG emissions (CO_2, CH_4 and NO_2) in the ordination space defined by the principal components (PCs).

To show the differences between GHG emissions in 1990 and 2019, we conducted two PCAs by using biplots. Biplots can depict the cases considering the three dimensions with the correlations. We tested the model fit with the root mean square residual (RMSR), where values <0.1 are considered good and <0.05 indicated very good [47].

The EU-28 countries were divided into the following 6 groups: western (w) (Belgium, France, Luxemburg, the Netherlands, Ireland, the United Kingdom); northern (n) (Denmark, Finland, Sweden, Iceland); middle (m) (Austria, Germany); southern (s) (Greece, Spain, Italy, Cyprus, Portugal); post-socialist (Bulgaria, Estonia, Croatia, Latvia, Lithuania, Romania, Slovenia); and Visegrad 4s (v4) (Czechia, Slovakia, Hungary, Poland) based on their location and historical basis (having the heritage of communism on the economy). Using these groups, we tested the following hypothesis: H_0 that each group was from the same statistical population with the same medians using Kruskal–Wallis test. We performed the test with PCs of the PCA for the dataset of 1990 and 2019.

We also conducted a cluster analysis on the GHG emissions with the change, i.e., the ratio of 2019 and 1990 as percentages. We applied the Ward's method and the output was visualized with a hierarchical dendrogram and in boxplot diagram.

3. Empirical Findings

3.1. Trend Analysis of GHGs Emissions between 1990 and 2019

3.1.1. Total GHGs Emissions between 1990 and 2019

The result of the MK test indicated that there was a significant decline in GHG emissions from the agricultural sector in the majority of the EU-28 countries (Figure 1, Table 2). Table 2 shows that 20 European countries, Belgium (BE), Czechia (CZ), Denmark (DK), Germany (DE), Greece (EL), France (FR), Croatia (HR), Italy (IT), Latvia (LV), Malta (MT), the Netherlands (NL), Austria (AT), Poland (PL), Portugal (PT), Romania (RO), Slovenia (SI), Slovakia (SK), Finland (FI), Sweden (SE), and the United Kingdom (UK), had significant ($p < 0.05$) negative trends. The GHG emission trends were negative, but not significant, in Bulgaria (BG), Cyprus (CY), Lithuania (LT), Luxembourg (LU), and Iceland (IS). In contrast, positive, but not significant, trends were recorded in Estonia (EE), Ireland (IE), Spain (ES), and Hungary (HU).

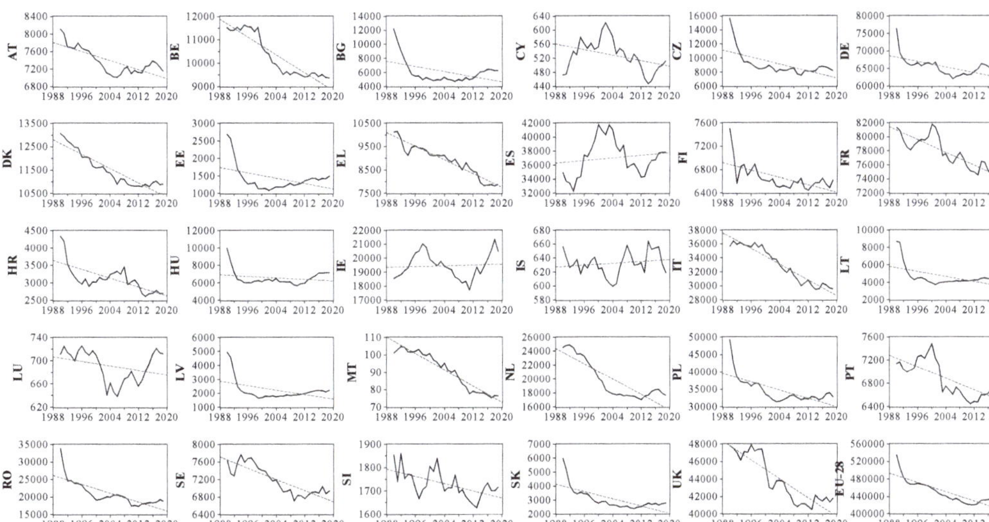

Figure 1. Time evolution of total GHG emissions between 1990 and 2019 from agricultural sector within EU-28.

The highest reduction in GHG emissions was recorded in IT (−282.61 thousand tons/year, $p < 0.05$), followed by the UK (−266.40 thousand tons/year, $p < 0.05$), and NL (−262.91 thousand tons/year, $p < 0.05$). The lowest reduction was recorded in MT (−1.18 thousand tons/year, $p < 0.05$), and SI (−3.55 thousand tons/year, $p < 0.05$). Nonetheless, the total emissions from the EU-28 depicted a significant reduction by −2653.01 thousand tons/year.

3.1.2. CO_2 Emissions between 1990 and 2019

The majority of the EU-28 countries exhibited a positive CO_2 emissions trend from the agricultural sector between 1990 and 2019 (Figure 2, Table 2). Only a few countries showed a negative significant trend, e.g., DK (−8.98 thousand tons/year, $p < 0.05$), EL (−1.03 thousand tons/year, $p < 0.05$), IT (−5.26 thousand tons/year, $p < 0.05$), CY

(−0.05 thousand tons/year, $p < 0.05$), NL (−0.99 thousand tons/year, $p < 0.05$), PL (−33.94 thousand tons/year, $p < 0.05$), SI (−0.81 thousand tons/year, $p < 0.05$), FI (−11.44 thousand tons/year, $p < 0.05$), and SE (−1.05 thousand tons/year, $p < 0.05$). In contrast, some EU countries exapted a positive significant trend, such as BG (+0.57 thousand tons/year, $p < 0.05$), CZ (+5.41 thousand tons/year, $p < 0.05$), DE (+16.33 thousand tons/year, $p < 0.05$), FR (+9.17 thousand tons/year, $p < 0.05$), HR (+1.16 thousand tons/year, $p < 0.05$), LU (+0.17 thousand tons/year, $p < 0.05$), AT (+2.7 thousand tons/year, $p < 0.05$), and IS (+0.13 thousand tons/year, $p < 0.05$). Nonetheless, the rest of the EU countries showed positive, but not significant, trends (Table 2).

Table 2. Trend analysis of GHG emissions and its component across EU-28 countries between 1990 and 2019.

EU-28 Countries		**Total GHGs Emissions**		CO_2		CH_4		N_2O	
		ρ	MK	ρ	MK	ρ	MK	ρ	MK
Total *	EU-28	<0.0001	−2653.01	0.42	−9.61	<0.0001	−1675.72	<0.0001	−916.06
Belgium	BE	<0.0001	−92.22	0.97	0.01	<0.0001	−31.48	<0.0001	−61.12
Bulgaria	BG	0.38	−22.62	0.00	0.57	<0.0001	−42.93	0.24	23.42
Czechia	CZ	<0.0001	−65.77	0.00	5.41	<0.0001	−65.89	0.70	−3.03
Denmark	DK	<0.0001	−78.51	<0.0001	−8.98	0.00	−7.76	<0.0001	−55.91
Germany	DE	0.00	−166.74	0.02	16.33	<0.0001	−177.53	0.83	5.06
Estonia	EE	0.72	3.00	0.40	0.14	0.83	−1.22	0.30	3.55
Ireland	IE	0.86	3.90	0.94	0.15	0.06	27.96	0.02	−22.92
Greece	EL	<0.0001	−73.27	<0.0001	−1.03	0.09	−6.02	<0.0001	−58.88
Spain	ES	0.34	66.99	0.32	1.69	0.92	9.10	0.05	52.44
France	FR	<0.0001	−227.20	<0.0001	9.17	<0.0001	−122.76	<0.0001	−109.34
Croatia	HR	0.00	−24.17	0.00	1.16	<0.0001	−16.16	0.00	−11.35
Italy	IT	<0.0001	−282.61	0.00	−5.26	<0.0001	−136.56	<0.0001	−143.50
Cyprus	CY	0.02	−2.37	<0.0001	−0.05	0.75	−0.11	0.00	−2.16
Latvia	LV	0.57	5.62	0.01	0.74	0.36	−4.73	0.04	7.97
Lithuania	LT	0.34	−7.94	0.34	0.19	<0.0001	−34.91	0.00	24.14
Luxembourg	LU	0.17	−0.91	<0.0001	0.17	0.34	0.51	<0.0001	−1.61
Hungary	HU	1.00	0.07	0.18	2.44	<0.0001	−33.08	<0.0001	37.97
Malta	MT	<0.0001	−1.18	-	-	<0.0001	−0.82	<0.0001	−0.37
Netherlands	NL	<0.0001	−262.91	0.01	−0.99	0.00	−92.96	<0.0001	−173.15
Austria	AT	0.00	−27.57	<0.0001	2.70	<0.0001	−21.41	0.00	−7.01
Poland	PL	0.00	−216.02	<0.0001	−33.94	<0.0001	−137.11	0.03	−36.94
Portugal	PT	0.00	−24.21	0.83	0.03	0.00	−14.47	0.01	−10.32
Romania	RO	<0.0001	−257.30	0.13	0.77	<0.0001	−187.77	0.02	−72.44
Slovenia	SI	0.00	−3.55	<0.0001	−0.81	0.18	−1.04	0.00	−1.56
Slovakia	SK	<0.0001	−43.52	0.42	0.54	<0.0001	−40.52	0.48	−3.10
Finland	FI	0.00	−11.64	<0.0001	−11.44	0.01	−3.27	0.14	2.70
Sweden	SE	<0.0001	−32.85	0.02	−1.05	<0.0001	−18.21	<0.0001	−15.46
Iceland	IS	0.46	0.27	0.00	0.13	0.13	−0.45	0.02	0.49
United Kingdom	UK	<0.0001	−266.40	0.17	8.47	<0.0001	−161.26	<0.0001	−115.55

* (2013–2020).

3.1.3. CH_4 Emissions between 1990 and 2019

The total emissions of CH_4 from the agricultural sector decreased significantly across the EU-28 (Table 2, Figure 3). The highest significant reduction was recorded in RO (−187.77 thousand tons/year, $p < 0.05$), DK (−177.53 thousand tons/year, $p < 0.05$), the UK (−166 thousand tons/year, $p < 0.05$), and PL (−137.11 thousand tons/year, $p < 0.05$). Despite the negative significant changes in CH_4, some countries, e.g., EE, CY, LV, SI, and IS, exhibited a negative, but not significant, trend. Notably, apart from IE (+27.96 thousand tons/year, $p > 0.05$), ES (9.10 thousand tons/year, $p > 0.05$), and LU (0.51 thousand tons/year, most of the EU-28 countries witnessed a negative trend of CH_4 emissions ($p > 0.05$) (Table 2, Figure 3).

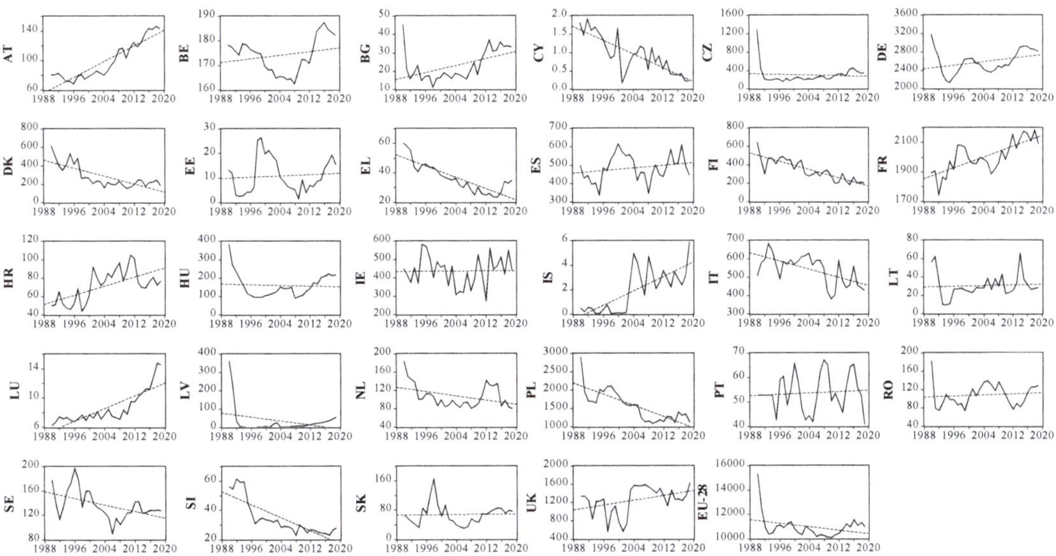

Figure 2. Time evolution of CO_2 emissions between 1990 and 2019 within EU-28 (agricultural sector).

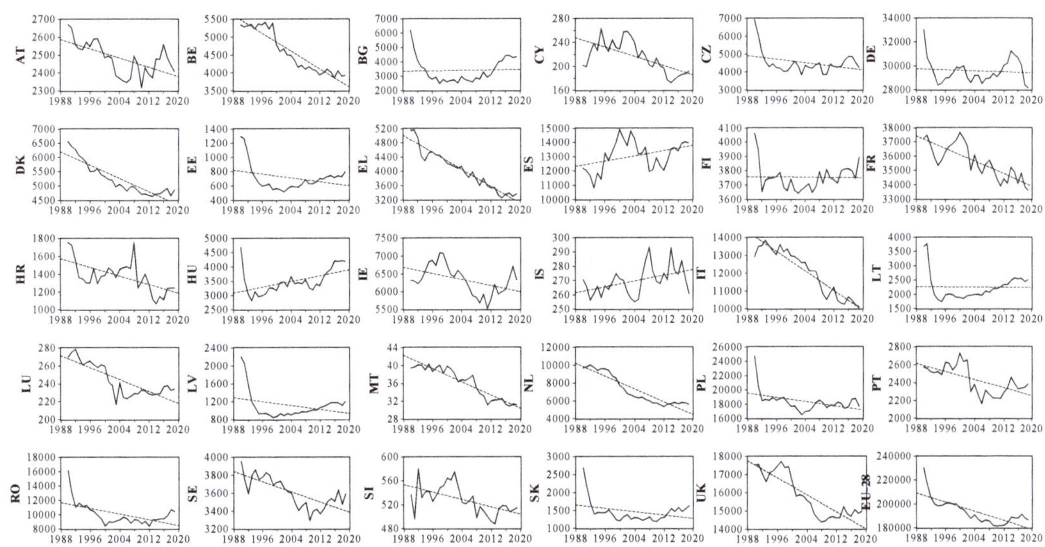

Figure 3. Time evolution of CH_4 emissions between 1990 and 2019 within EU-28 (agricultural sector).

3.1.4. N_2O Emissions between 1990–2019

Similarly to CH_4 emissions, N_2O emissions exhibited a negative trend from most of the EU-28 countries (Table 2, Figure 4). The total reduction in N_2O emissions from all the EU-28 countries was −916 thousand tons/year ($p < 0.05$) (Table 2). However, some countries showed a significant positive trend of N_2O emissions; for instance, LV (+7.79 thousand tons/year, $p < 0.05$), LT (+24.14 thousand tons/year, $p < 0.05$), HU (+37.97 thousand tons/year, $p < 0.05$), and IS (0.49 thousand tons/year, $p < 0.05$). On the other hand, some other countries, such as BG (+23.42 thousand tons/year, $p > 0.05$),

DE (+ 5.06 thousand tons/year, $p > 0.05$), EE (+3.5 thousand tons/year, $p > 0.05$), ES (+52.44 thousand tons/year, $p > 0.05$), and FI (2.7 thousand tons/year, $p > 0.05$), showed a positive, but not significant, trend in N_2O emissions. Interestingly, the other EU countries showed a significant negative trend of N_2O emissions, whereas the highest value of reduction was recorded in NL (−143.5 thousand tons/year, $p < 0.05$), followed by IT (−143.5 thousand tons/year, $p < 0.05$), then the UK (−115.55 thousand tons/year, $p < 0.05$).

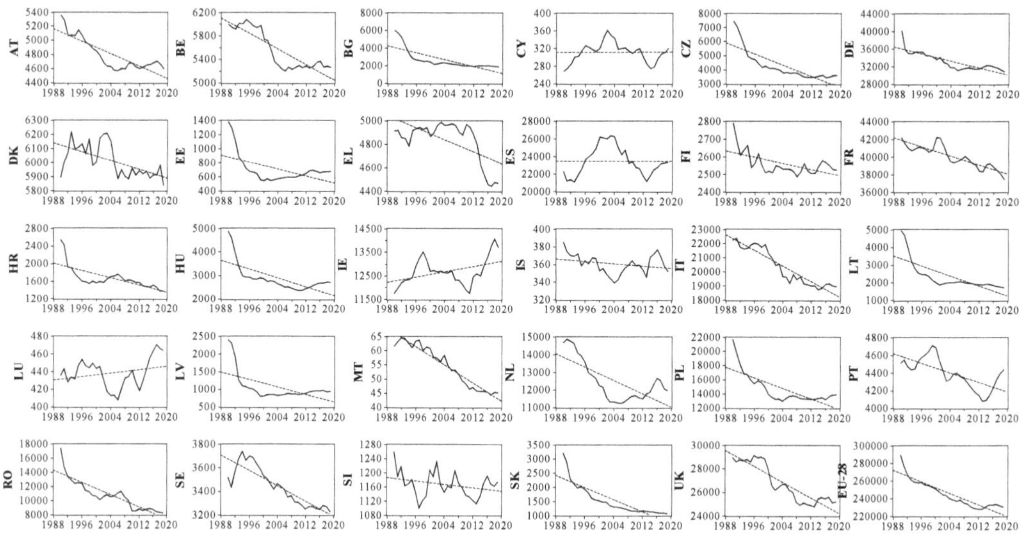

Figure 4. Time evolution of N_2O emissions between 1990 and 2019 within EU-28 (agricultural sector).

3.2. *Multivariate Analysis of GHGs Emissions in 1990 and 2019*

The total variance explained was 99% for the data of 1990 and 98% for 2019. Goodness-of-fit analysis also indicated a very good fit, with RMSRs of 0.01 for both years. Two PCs were confirmed by the RMSRs; PC1 correlated with the N_2O and CH_4 emissions, explaining 58% (in 1990) and 51% (in 2019) of the variance, while PC2 correlated with CO_2, and explained 41% (in 1990) and 46% (in 2019) of the variance. Consequently, PC2 involved only one variable, CO_2, while PC1 followed the state of N_2O and CH_4 on the same axis.

Most countries formed a compact group in both years in the lower section of the diagram, regarding N_2O and CH_4, and with a larger variance in CO_2. There were also outlier countries (having scores larger than two) regarding the lower or higher GHG emissions (Figure 5). In 1990, DE, FR, the UK, and PL represented the highest emissions of N_2O and CH_4, and CZ, PL, and DE had larger emissions of CO_2. The lowest CO_2 belonged to RO, FR, IT, and ES. The scores were mainly between −1 and 1 regarding the CO_2; thus, the variance was smaller than what was observed in PC2 (N_2O and CH_4). There were changes in the pattern in 2019, but the most important ones were observed in the lowest and largest values. The positions regarding PC1 (i.e., N_2O and CH_4) did not change, but the CO_2 emission decreased in PL and CZ, and increased in the UK, DE, and FR. The lowest CO_2-emitting countries were ES, IT, and NL.

There were no significant differences regarding PC1 in 1990 (Kruskal–Wallis H: 8.334, $p = 0.138$) and in 2019 (Kruskal–Wallis H: 6.654, $p = 0.254$). Similarly, there was no significant difference in CO_2 emissions in 1990 (Kruskal–Wallis H: 9.973, $p = 0.076$) and in 2019 (Kruskal–Wallis H: 10.59, $p = 0.052$). Accordingly, the spatial distribution of the countries did not discriminate the emissions.

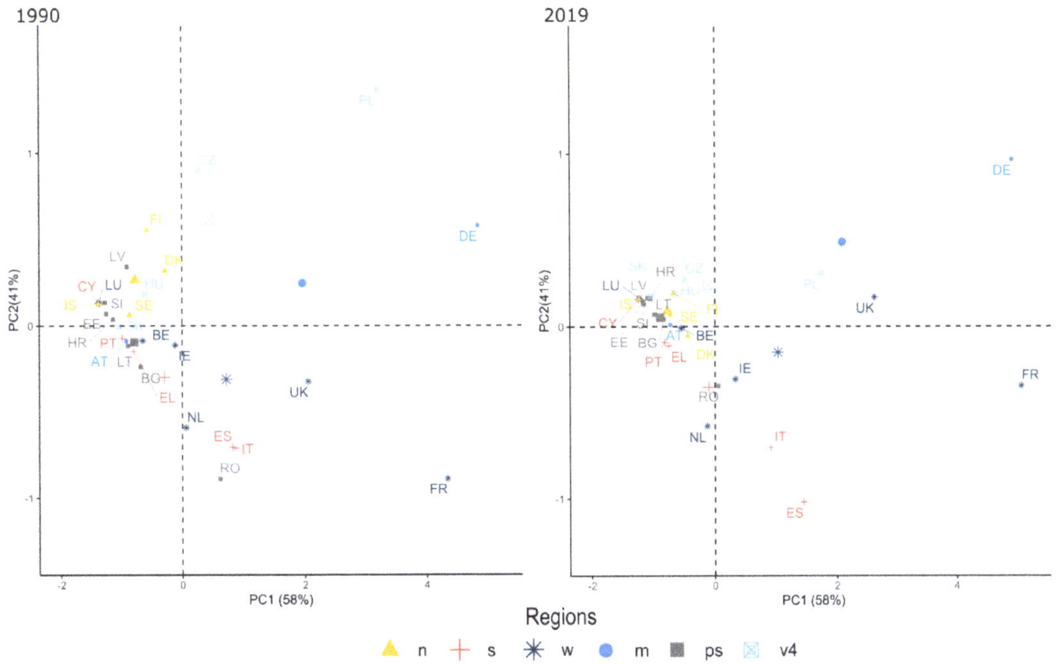

Figure 5. Ordination diagram of GHG emissions (N_2O, CH_4, CO_2) in 1990 and 2019 within EU-28 (agricultural sector; n: Northern Europe, s: Southern Europe, w: Western Europe, m: Middle Europe, ps: post-socialist countries, v4: Visegrad 4 countries; large symbols: group centroids).

Cluster analysis, focusing on the positive and negative changes, revealed that the difference was the most discriminative in the case of Iceland, due to its high increase in CO_2 emissions. Other clusters were only partly formed by their location (e.g., southern or western), and the differences were relatively smaller than those observed in Iceland. As Iceland formed a unique cluster in itself, we did not involve it in the statistical evaluation. Cluster 1 (C1) was formed by purely post-socialist countries, but all the other clusters were a mixture of different locations and historic heritage (Figure 6). This approach maximized the variance among the countries; thus, the clusters reflected similarity in the changes (Figure 7).

The C1 cluster contained the countries that had the largest negative change, i.e., these countries made the largest progress in reducing GHG emissions. The countries of the C2 cluster gained relevant results on decreasing CO_2 and CH_4, and the N_2O emissions also decreased, but by a smaller measure. In the C3 cluster, the countries only reached a small decrease in each GHG, and a limited increase was observed for CO_2. In both the C4 and C5 clusters, the CO_2 emissions increased, while the N_2O and CH_4 emissions decreased; the difference between the two clusters was that in C5, the decrease was smaller.

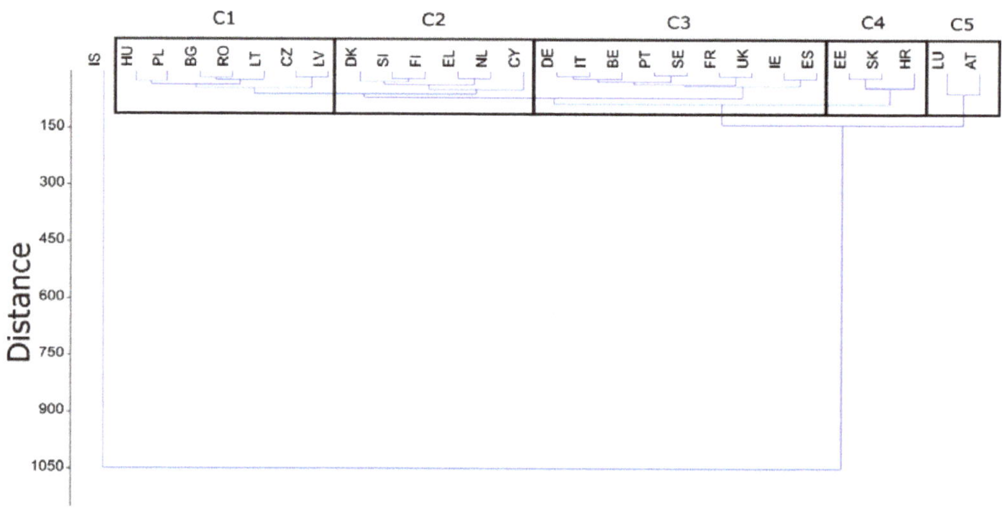

Figure 6. Clusters of GHG emission change (input: emissions of CO_2, CH_4 and N_2O in 2019 were divided by the emission of 1990, expressed in %, C1–C5: clusters).

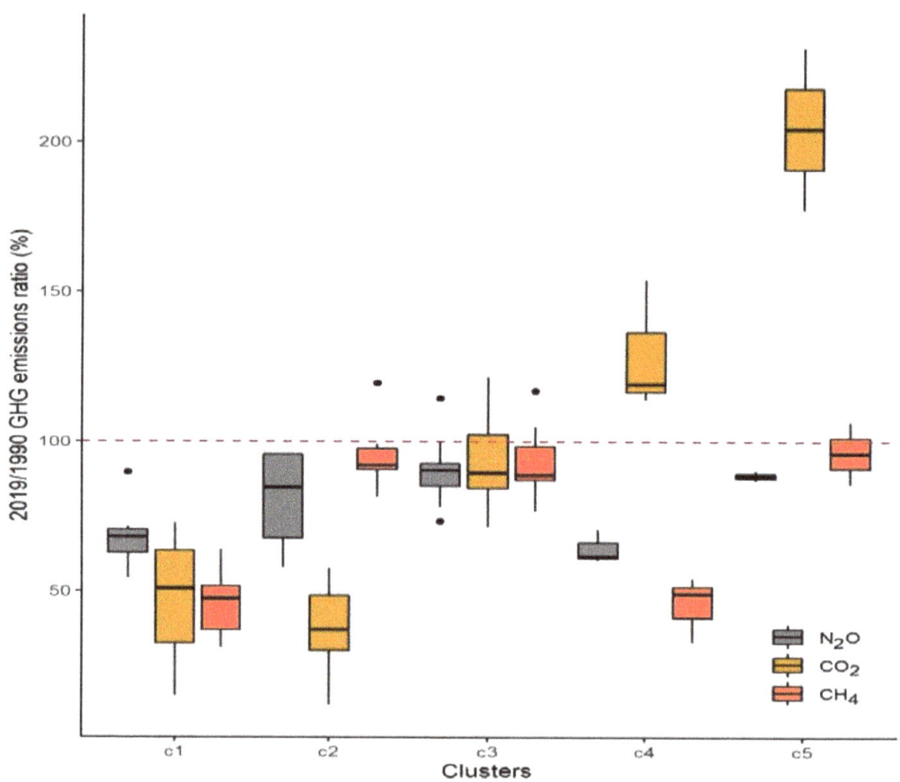

Figure 7. Ratio of GHG emissions 2019/1990 (dashed line indicates no change, below the line changes were negative, above the line they were positive; C1–C5: clusters of countries indicated by Figure 6.

4. Discussion

In general, the trend of GHG emissions from the agricultural sector in EU countries was negative during 1990–2019, except for some countries, such as Estonia, Ireland, and Latvia, which displayed a slight increase from 3–5.6 thousand tons/year. Iceland and Hungary both exhibited insignificant increases. A significant increase in GHG emissions was exhibited by Spain (Figure 1), with a significant increase in N_2O emissions (Figure 4). In fact, more than 72% of the Spanish land was used for agricultural practices and forestry, while 19% was used for meadows, which accelerated the GHG emissions from this sector [48,49]. Nonetheless, 11% of the total emissions in Spain originated from the agricultural sector [50]. The increase in GHG emissions in Spain, between 1990 and 2019, could be attributed to the lack of clear national strategies for minimizing and mitigating GHG emissions from the Spanish agricultural sector [51] (Table 2). This was exacerbated by the highly intensive agricultural production per capital and technological advancements in the agricultural sector in Spain in the recent decades [52]. The significant positive trend of GHG emissions was mainly dependent on the increase in N_2O emissions from 1990 to 2019, with an overall trend of 52.44 thousand tons/year. This trend may be because of the mismanagement of soil fertilization, agrochemicals. Livestock manure was the main cause of N_2O emissions, eutrophication of water courses, and atmosphere acidification. Similar conclusions have been reported by Albiac et al. [52] and other workers [53,54]. Magrama [55] noted that the overdose of N fertilizer, along with the neglect of livestock manure, added more than 780.000 tN of fertilizer to the soil, leading to severe environmental pollution. Nevertheless, the mean annual increase in CH_4 and CO_2 emissions during the period of this study showed much lower values of 9.10 and 1.69 thousand tons/year, respectively. Other countries, such as Italy, the Netherlands, the UK, among others, manifested a significant decline in the total GHG emissions.

In Italy, only 7.1% of the GHG emissions originated from the agriculture sector in 2016. The decline in GHG emissions can be attributed to a decrease in the number of animals, especially the dairy cattle heads from 1990 to 2016, which resulted in a decline of about 40% [56]; this may have contributed to the negative trend of CH_4 emissions (Table 2 and Figure 3). Also, the CH_4 and N_2O from manure management decreased with the decline in the number of animals during the studied period. The more efficient manure management system may have also contributed to the reduction in N_2O [56]. The CH_4 emissions from rice cultivation have also decreased, according to the revised CH_4 daily EF measurements in Italian rice fields [57,58], considering the different agronomic practices between the different cultivars [59,60], and the different irrigation regimes [61]. The N_2O emissions from managed soils declined from 29.72 Gg (80.6% of N_2O emissions for the agriculture sector) in 1999 to 23.99 Gg (78.2%) in 2016, where this decline agreed with Table 2 and Figure 4. The decline in the N_2O emissions from managed soils may be because of a reduction in the use of inorganic and organic fertilizers, which was about 25% from 1990 until 2016 [56]. Romano et al. [56] reported that the application of carbonate for decreasing soil acidity is one of the main sources for CO_2 emissions. In this context, the liming process in Italy was responsible for 2.3% (2016) of the total CO_2 emissions from the agricultural sector [56].

The agricultural sector in the UK accounted for 10% (2018) of the GHG emissions [62], where livestock and manure accounted for 56% of the emissions, synthetic fertilizers accounted for 31%, and fuel and machinery accounted for 12% [62]. There was a significant decline in CH_4 and N_2O emissions in the UK during the studied period, with values of −166 thousand tons/year, $p < 0.05$, and −115.55 thousand tons/year, $p < 0.05$, respectively; this resulted in a decline in the total GHG emissions, with a trend of −266.40 thousand tons/year, $p < 0.05$ (Table 2). Similarly, Nair et al. [63] and NFS [64] reported that, overall, the GHG emissions from agriculture in the UK have decreased by 16% from 1990 to 2018. Ortiz et al. [62] mentioned three factors that led to a significant GHG emission decrease from the agricultural sector. The factors are as follows: (1) adaptation of new technology in the agricultural sector, (2) national policies, and (3) changing the incentives model, which reduced the number of ruminants to meet the EU-CAP (Common Agricultural Policy). The UK have launched a national framework for evaluating the annual reduction in GHG emissions since 2012. This framework covers ten indictors, including mitigation and adaptation methods, farmer knowledge and behavior, and emission per product [65]. It is good to mention here that a large amount of research in the UK was focused on the improvement in the agricultural GHG inventory [65]. The livestock population in the UK reduced by 19.8% from 1990 to 2018, while only the dairy cattle category also decreased by 33.6%. The application of N fertilizer had been dropped by permanent grasslands, which represented almost half of the area of the UK's major crop area, with 55.6% from 1990 to 2018, while the other crops have been fluctuating between declining, such as by grass leys, oilseed rape, and potatoes, with 41%, 19%, and 27.2%, respectively; stabilizing, such as by wheat; and increasing in N fertilizer, such as by Spring and Winter barely, with 15.6% and 10%, respectively [66].

Studies indicate that liming is a major contributor to CO_2 emissions in the agricultural sector. The contribution of the agricultural sector to GHG emissions in the Netherlands was about 9.7% in 2019. There was a reduction in the application of lime, which also caused a reduction in CO_2 emissions during 1990–2008 and 2016–2019 in the Netherlands, although there was a slight increase in emissions in 2009. The reduction in liming resulted in the decline of CO_2 emissions, by 80.9%, from 1990 to 2019 (0.18–0.03 Tg CO_2 eq), while the CO_2 emissions from urea application increased from 0.002 to 0.045 Tg CO_2 eq in the same period [67]. This behavior explained the results of the CO_2 emissions of NL (Figures 2 and 3, and Table 2). The trend of methane during the studied period decreased because of a reduction in the application of mature dairy cattle, where the CH_4 emissions of enteric fermentation and manure management decreased from 1990 to 2005, increased from 2007 to 2016, then start to decrease again [67]. The significant negative trend of N_2O emissions in NL (Figure 4) is explained by the decrease in organic and inorganic N fertilizer application, the decrease in animal numbers, and the decrease in animal production on pasture, from 1990 to 2010, and, after 2010, the decline in N_2O emissions was stabilized at 44.8% (8.7–4.8 Tg CO_2 eq). This is similar to the decrease in N_2O from the agriculture soil reported by Ruyssenaars et al. [67].

For Romania, a number of measures have contributed to the reduction in GHG emissions, such as minimizing the amount of synthetic nitrogen fertilizer, decreasing the number of livestock, and reducing the area under rice cultivation [68]. The Romanian agricultural sector contributed 18.98% (2015) to the total GHG emissions [68]. Compared to 1989, the reduction in GHG emissions reached 78.93% by liming, 72.98% by rice cultivation, 61.86% by manure management, 49.20% by enteric fermentation, 48.48% by agricultural soils, 46.34% by urea application, and 12.84% by the field burning of agricultural residues [68]. These findings agreed with the negative trend of Romania in Table 2 and Figures 1–4. Table 2 shows that the negative trend of CH_4 was the highest, −187.77 thousand tons/year, followed by N_2O emissions with a negative trend, −72.44 thousand tons/year.

The total share of GHG emissions from agriculture in 2017 in Poland was 7.16% [69]. However, Poland showed a significant decline in the agricultural GHG emissions from 1990 to 2019, with a trend of −216.02 thousand tons/year ($p < 0.05$), which was categorized into −33.94 thousand tons/year ($p < 0.05$), −137.11 thousand tons/year ($p < 0.05$), and −36.94 thousand tons/year ($p < 0.05$) for the CO_2, CH_4, and N_2O emissions (Table 2), respectively. Poland's National Inventory Report [69] stated that the decline in CH_4 emissions was due to the dramatic decrease in the livestock population after 1989, especially for the dairy cattle population that decreased by almost 50% from 1990 to 2017. This decline in the livestock population decreased the CH_4 emissions for enteric fermentation and manure management. As well as this, the N_2O emissions from manure management dropped by 31% from 1988 to 2017, also depending on the diminishing livestock population. N_2O emissions mainly come from the agriculture soil, which was significantly decreased from 1988 to 2017, by 21%. However, nitrogen fertilization accounted for 47% of direct N_2O emission (2017). Piwowar [70] explained that the liming process and carbonate usage were balanced from 2000 to 2004, and were relatively low later on. The CO_2 emissions from lime, dolomite and urea utilization were significantly decreased from 1988 to 2017 [69].

French agriculture GHG emissions contributed 16.8% of the total GHG emissions in 2019 [71]. France accounts for 25% of the livestock in Europe, and 40% of the agricultural land in France is used for cereal production, making France the largest producer of cereal in the EU [72]. The large proportion of livestock and cereal production in France also implies that France will account for a large part of the GHG emissions, and will have a difficult challenge in reducing GHGs. However, there was a significant decline in CH_4 emissions resulting from livestock (Table 2). The National Inventory Report for France [71] noted that a 34% decline in the dairy cattle population resulted in a 13% decrease in enteric fermentation emission from 1990 to 2019. An increase in the number of pig herds has been linked to an increase in manure management and a 7% increase in CH_4 emissions over the period 1990–2019. However, other parameters, such as the increase in manure management systems in the form of slurry, are contributing inversely to this trend. Rice cultivation is also a major contributor to CH_4 emissions. The area under rice cultivation in 1990 was 22,458 ha. This increased to 34,405 ha in 1994, but has declined to 15,100 ha in 2019. A decrease in the area under rice cultivation results in a decrease in rice cultivation-induced CH_4 emissions. Table 2 shows that the N_2O emissions also decreased significantly during 1990–2019. The National Inventory Report for France [71] indicated a decrease in N_2O emissions by minimizing the use of mineral nitrogen fertilizers, which reduced the N_2O emissions by −16% (1990–2019), and decreasing the cattle herds, resulting in a reduction in both the nitrogen excreted in the pasture and the organic nitrogen to be applied, leading to a 12% decline in N_2O emissions from 1990 to 2019. The total N_2O emissions from agricultural soils decreased by −9% over the period 1990–2019 (Table 2).

In Germany, the agriculture sector was responsible for 7.6% of the total GHG emissions in 2019 [73]. The total GHG emissions from the agricultural sector decreased by 19.2% in the period 1990–2021 [74]. This decline is consistent with the negative trend in Table 2 (−166.74 thousand tons/year, $p < 0.05$), which mainly depends on the CH_4 emissions trend that revealed −177.53 thousand tons/year ($p < 0.05$) (Table 2). As mentioned before, the CH_4 emissions come from enteric fermentation and manure management, and both rely the most on the population of animals, especially the dairy cattle, and pig for manure management. The German NIR [75] reported that, from 1990 to 2019, the decline in the animal population was almost 37%, 42%, 18.5%, and 41.6% for the dairy cattle, swine, sheep, and goats, respectively. This notable decline in the animal population leads to a decrease in CH_4 emissions, by 27.7% and 21.3% for enteric fermentation and manure management, respectively [74] (Table 2 and Figure 3). The N_2O emissions include manure management, energy crops (from digester and storage of digestate from the anaerobic digestion of energy crops, and include both CH_4 and N_2O emissions), and agricultural soil. In regards to the emissions from energy crops (CH_4 and N_2O), which presented the smallest share of the total agricultural GHG emissions (2.5% in 2019), they increased from zero in 1990 to 1573 Tg CO_2 eq in 2019, with a gradual utilization of energy crops since 1991. The N_2O emissions decreased by 18.9% for manure management and 15.3% for agricultural soil. The smaller dwindling of N_2O emissions from 1990 to 2019 may be attributed to variation in its components' behavior (decrease in mineral fertilizer N quantities by 35%, decrease in manure N quantities, including energy crops, by 18.6%, increase in crop residues N quantities by 16%, and the relatively unchanged indirect soil emissions) [74,75]. In contrast, our results indicate a positive, but not significant, trend of N_2O (Table 2). The reasonable cause of that could be the increase in the applied N fertilization quantities between 2014 and 2016. The CO_2 emissions trend exhibited an increase, with a positive trend (16.66 thousand tons/year ($p > 0.05$)) (Table 2, Figure 2). Similar results were reported by the German NIR [75], which highlighted a 10.68% and 8.8% increase in the application of limestone and urea, respectively, and a decrease in the application of dolomite and calcium ammonium nitrate (84.5% and 61.6%, respectively). This resulted in a total increase in CO_2 emission from 1990 to 2019, by 11.6% [74].

The highest share of GHG emissions in the agricultural sector is presented by the CH_4 and N_2O emissions, accounting for more than 90% of the total GHG emissions. This majority share explained the clear proximity of the CH_4 and N_2O emission values of the contributing countries to the PC1 (Figure 5), where CH_4 and N_2O emissions, which are the closest components to PC1, best approximate the total GHG emissions. Nevertheless, the CO_2 emission values with a lower share of GHG emissions, and sometimes with opposite behavior to the total GHG emission values, were close to PC2. However, analysis of the spatial pattern, or even the historical heritage (post-socialist shared heritage on the agriculture, with the transformation from large agricultural co-operations to private farming), did not reveal any result. There were exceptions in each group; thus, the differences were not significant in the change between 1990 and 2019. The biplots pointed out countries that had increasing or decreasing changes during the 29 years. Furthermore, cluster analysis was the best method to show the country groups of similar changes, but, in this case, we involved the ratio of the change in the GHG emissions between the earliest and latest dates, dividing the data of 2019/1990. This approach was an efficient tool to identify the countries that have similar gains in GHG emission reduction, or in pointing out the ones that have issues in reaching the goals.

Iceland formed a sole cluster, but we have to highlight that the cause of this was the 1128% surplus in the CO_2 emissions from agricultural sources. This seems to be a large increase, but in 1990, the initial emissions were only 0.52 thousand tons, the smallest in the EU countries; even Luxembourg and Cyprus could approach it (with 6.3 and 1.8 thousand tons). In 2019, the CO_2 emissions of Iceland increased to 5.87 thousand tons, which was the second lowest in the EU, and Cyprus was the first, with 0.22 thousand tons in this year. In 2021, the world's first CO_2 removal plant started operating in Iceland, which will remove 4000 t of CO_2 a year [76]. Austria and Luxembourg formed the cluster, both having the worst performance, but the increase should also be evaluated carefully; Luxembourg is still third in the ranks of CO_2 emissions in the EU. Accordingly, the countries that had the smallest emissions in 1990 can appear as inefficient ones in CO_2 reduction, but there are lot of components of these numbers. Besides local food production, transportation and even food import can also count, and can have direct consequences on the emissions too [77]. Although the population did not increase in the European countries, globalization can generate demands and, therefore, food or agricultural product import and export. In the case of Iceland, the food product import was USD 268,000 in 1990 and USD 1,266,638 in 2019; therefore, the increase was almost five times as large [78].

Climate change mitigation, for finding more efficient farming, is one of the global challenges in the EU. The utilization of optimal agricultural practice management, provided by convenient technologies, assists by not only reducing the GHG emissions, but also promoting agricultural productivity and income [79]. Precision agriculture can achieve this, where precision agriculture based on utilizing digital techniques can aid in monitoring and optimizing agriculture production processes at different field scales [80]. Precision agriculture supports the optimization of field management based on the actual crop needs; for example, using sensors to identify the specific area in field that needs a particular treatment, such as irrigation, fertilizers, insecticides, and herbicides [81]. However, promoting precision agriculture in Europe could be one of the solutions for mitigating climate change across the EU-28.

In this research, the trend of GHG emissions from agricultural sectors across the EU-28 was analyzed between 1990 and 2019, accompanied by multivariate analysis. The results only highlighted the GHG trend, with no further information about the GHGs origins (soil, fertilization, livestock, food production), which is one of the drawbacks of this research. On the other hand, this study did not investigate the relationship between GHGs and GDP, where GDP can play an important role in GHG emissions, and could help in discriminating and categorizing European countries regarding their emissions. However, future steps will employ the environmental Kuznets curve (EKC) for exploring the relationship between economic growth and environmental degradation. Previous studies also reveal that, in the long run, economic growth could lead to the reduction in atmospheric pollution [2,82].

5. Conclusions

Since the release of the Paris Agreement in 2015, the EU have strived to reduce GHG emissions from all sectors to achieve carbon neutrality (zero emission). Thus, a package of policies and strategies was released in order to achieve this aim. To reach this ambitious goal by 2050, GHG emissions need to be evaluated on a sectorial level (i.e., industry, agriculture, energy) to provide decision makers with a full overview of the changes, and the efficiency of mitigation and adaptation strategies.

In this research, the GHG emissions from the agricultural sector within the EU-28, from 1990 to 2019, was analyzed by using the MK test and multivariate approach. The analysis revealed that most of the EU-28 countries exhibited a significant reduction in GHG emissions (1990–2019). The highest reduction in the total GHG emissions was recorded in Italy (−282.61 thousand tons/year, $p < 0.05$), followed by the United Kingdom (−266.40 thousand tons/year, $p < 0.05$), and the Netherlands (−262.91 thousand tons/year, $p < 0.05$). Similarly, the CH_4 and N_2O emissions exhibited a negative emission trend from most of the EU−28 countries. However, a positive CO_2 emissions trend from the agricultural sector, between 1990 and 2019, was recorded. Nonetheless, the accumulation of CO_2 emissions from all the EU-28 countries depicted a non-significant negative trend (−9.61 thousand tons/year). Interestingly, the multivariate analysis approach indicates changes in the pattern of GHG emissions between 1990 and 2019. In 1990, DE, FR, the UK, and PL represented the highest emissions of N_2O and CH_4; where CZ, PL, and DE had larger emissions of CO_2. In 2019, the patterns were changed, in terms of the lowest and largest values.

The findings of this study highlight the need for policy makers in the European Union to evaluate the strategies for mitigating CO_2 emissions, and underline the need to formulate new policies for reducing CO_2 emissions from the agricultural sector. However, future studies should focus on analyzing the relationship between GHG emissions from the agricultural sector and environmental degradation, through the application of environmental Kuznets curve hypothesis.

Author Contributions: Conceptualization, S.M. and E.H.; methodology, S.M. and S.S; writing—original draft preparation, S.M., A.E., G.A., O.H. and M.M.; writing—review and editing, B.B., A.A., and G.O.E.; supervision, S.S. and E.H.; funding acquisition, B.B., and A.A. All authors have read and agreed to the published version of the manuscript.

Funding: This research was supported by RESEARCHERS SUPPORTING PROJECT, grant number RSP-2021/296, King Saud University, Riyadh, Saudi Arabia. Szabo S. was supported by the Project TKP2020-IKA-04 (National Research, Development and Innovation Fund of Hungary, financed under the 2020-4.1.1-TKP2020) funding scheme.

Institutional Review Board Statement: Not applicable.

Informed Consent Statement: Not applicable.

Data Availability Statement: Original data available on the EEA website.

Acknowledgments: The authors are grateful to EEA team for providing data used in this study. Authors would like to thank Debrecen University (Hungary) and King Saud University for their unlimited support.

Conflicts of Interest: The authors declare no conflict of interest.

References

1. Dale, A.; Robinson, J.; King, L.; Burch, S.; Newell, R.; Shaw, A.; Jost, F. Meeting the climate change challenge: Local government climate action in British Columbia, Canada. *Clim. Policy* **2020**, *20*, 866–880. [CrossRef]
2. Mohammed, S.; Gill, A.R.; Alsafadi, K.; Hijazi, O.; Yadav, K.K.; Khan, A.H.; Harsanyi, E. An overview of greenhouse gases emissions in Hungary. *J. Clean. Prod.* **2021**, 127865. [CrossRef]

3. Shakoor, A.; Ashraf, F.; Shakoor, S.; Mustafa, A.; Rehman, A.; Altaf, M.M. Biogeochemical transformation of greenhouse gas emissions from terrestrial to atmospheric environment and potential feedback to climate forcing. *Environ. Sci. Pollut. Res.* **2020**, *27*, 38513–38536. [CrossRef]
4. IPCC (Intergovernmental Panel on Climate Change). *Climate Change 2014: Mitigation of Climate Change. Working Group III Contribution to The IPCC Fifth Assessment Report*; Cambridge University Press: Cambridge, UK, 2014; Available online: www.ipcc.ch/report/ar5/wg3 (accessed on 3 March 2021).
5. WRI (World Resources Institute). Climate Watch Historical GHGE. Available online: www.climatewatchdata.org/ghg-emissions. (accessed on 3 March 2021).
6. FAO (Food and Agriculture Organization). FAOSTAT Emissions Database—Land Use Indicators. Available online: www.fao.org/faostat/en/#data/EL. (accessed on 24 March 2021).
7. IPCC (Intergovernmental Panel on Climate Change). *Climate Change: The Physical Science Basis. Working Group I Contribution to the IPCC Fifth Assessment Report*; Cambridge University Press: Cambridge, UK, 2013; Available online: www.ipcc.ch/report/ar5/wg1 (accessed on 3 March 2021).
8. Herzog, T. World Greenhouse Gas Emissions in 2005. World Resources Institute 2009, 1–7. Available online: https://www.wri.org/research/world-greenhouse-gas-emissions-2005 (accessed on 5 October 2021).
9. Tubiello, F.N.; Salvatore, M.; Rossi, S.; Ferrara, A.; Fitton, N.; Smith, P. The FAOSTAT database of greenhouse gas emissions from agriculture. *Environ. Res. Lett.* **2013**, *8*, 015009. [CrossRef]
10. Tongwane, M.I.; Moeletsi, M.E. A review of greenhouse gas emissions from the agriculture sector in Africa. *Agric. Syst.* **2018**, *166*, 124–134. [CrossRef]
11. Tian, H.; Xu, R.; Canadell, J.G.; Thompson, R.L.; Winiwarter, W.; Suntharalingam, P.; Davidson, E.A.; Ciais, P.; Jackson, R.B.; Janssens-Maenhout, G.; et al. A comprehensive quantification of global nitrous oxide sources and sinks. *Nature* **2020**, *586*, 248–256. [CrossRef]
12. Folberth, C.; Khabarov, N.; Balkovič, J.; Skalský, R.; Visconti, P.; Ciais, P.; Janssens, I.A.; Peñuelas, J.; Obersteiner, M. The global cropland-sparing potential of high-yield farming. *Nat. Sustain.* **2020**, *3*, 281–289. [CrossRef]
13. Zafeiriou, E.; Mallidis, I.; Galanopoulos, K.; Arabatzis, G. Greenhouse Gas Emissions and Economic Performance in EU Agriculture: An Empirical Study in a Non-Linear Framework. *Sustainability* **2018**, *10*, 3837. [CrossRef]
14. Statista: Emissions in the EU—Statistics & Facts. Available online: https://www.statista.com/topics/4958/emissions-in-the-european-union/ (accessed on 25 September 2021).
15. Gołasa, P.; Wysokiński, M.; Bieńkowska-Gołasa, W.; Gradziuk, P.; Golonko, M.; Gradziuk, B.; Siedlecka, A.; Gromada, A. Sources of Greenhouse Gas Emissions in Agriculture, with Particular Emphasis on Emissions from Energy Used. *Energies* **2021**, *14*, 3784. [CrossRef]
16. Rokicki, T.; Perkowska, A.; Klepacki, B.; Bórawski, P.; Bełdycka-Bórawska, A.; Michalski, K. Changes in Energy Consumption in Agriculture in the EU Countries. *Energies* **2021**, *14*, 1570. [CrossRef]
17. Eurostat. Final Energy Consumption by Sector and Fuel. 2014. Available online: https://www.eea.europa.eu/data-and-maps/indicators/final-energy-consumption-by-sector-9/assessment-1#tab-related-briefings (accessed on 30 June 2021).
18. Agri-Environmental Indicator—Energy Use. Available online: https://ec.europa.eu/eurostat/statistics-explained/index.php/Agri-environmental_indicator_-_energy_use (accessed on 30 June 2021).
19. Li, T.; Baležentis, T.; Makutėnienė, D.; Streimikiene, D.; Kriščiukaitienė, I. Energy-related CO_2 emission in European Union agriculture: Driving forces and possibilities for reduction. *Appl. Energy* **2016**, *180*, 682–694. [CrossRef]
20. Bajan, B.; Łukasiewicz, J.; Mrówczyńska-Kamińska, A. Energy Consumption and Its Structures in Food Production Systems of the Visegrad Group Countries Compared with EU-15 Countries. *Energies* **2021**, *14*, 3945. [CrossRef]
21. Masi, M.; Vecchio, Y.; Pauselli, G.; Di Pasquale, J.; Adinolfi, F. A Typological Classification for Assessing Farm Sustainability in the Italian Bovine Dairy Sector. *Sustainability* **2021**, *13*, 7097. [CrossRef]
22. European Commission. The Post-2020 Common Agricultural Policy: Environmental Benefits and Simplification. 2019. Available online: https://ec.europa.eu/info/sites/info/files/food-farming-fisheries/key_policies/documents/cap-post-2020-environ-benefits-simplification_en.pdf (accessed on 18 June 2020).
23. Malinovský, V. System Macromodel of Agricultural Building with Aim to Energy Consumption Minimization. *Agris on-line Pap. Econ. Informatics* **2018**, *10*, 25–35. [CrossRef]
24. Piwowar, A. Challenges associated with environmental protection in rural areas of Poland: Empirical studies' results. *Econ. Sociol.* **2020**, *13*, 217–229. [CrossRef] [PubMed]
25. Baccour, S.; Albiac, J.; Kahil, T. Cost-Effective Mitigation of Greenhouse Gas Emissions in the Agriculture of Aragon, Spain. *Int. J. Environ. Res. Public Health* **2021**, *18*, 1084. [CrossRef] [PubMed]
26. Solazzo, R.; Donati, M.; Tomasi, L.; Arfini, F. How effective is greening policy in reducing GHG emissions from agriculture? Evidence from Italy. *Sci. Total Environ.* **2016**, *573*, 1115–1124. [CrossRef]
27. Pleßmann, G.; Blechinger, P. How to meet EU GHG emission reduction targets? A model based decarbonization pathway for Europe's electricity supply system until 2050. *Energy Strategy Rev.* **2017**, *15*, 19–32. [CrossRef]
28. European Environment Agency (EEA). *Climate Change Adaptation in the Agriculture Sector in Europe*; Publications Office of the European Union: Luxembourg, 2019; Available online: https://www.euroseeds.eu/app/uploads/2019/09/Climate-change-adaptation-in-the-agriculture-sector-in-Europe.pdf (accessed on 30 June 2021).

29. Klimaatakkoord. Proposal for Key Points of the Climate Agreement. 2018. Available online: https://www.klimaatakkoord.nl/documenten/publicaties/2018/09/19/proposal-for-key-points-of-the-climate-agreement (accessed on 3 November 2019).
30. Ministère de l'Ecologie, du D.D. Stratégie Nationale Bas-Carbone: Summary for Decision-Makers. 2015. Available online: https://unfccc.int/files/focus/long-term_strategies/application/pdf/national_low_carbon_strategy_en.pdf (accessed on 30 June 2021).
31. Climate Change Committee (CCC): Carbon Budgets: How We Monitor Emissions Targets. 2019. Available online: http://www.theccc.org.uk/tackling-climate-change/reducing-carbon-emissions/carbon-budgets-and-targets/ (accessed on 30 November 2020).
32. Federal Ministry for the Environment, Nature Conservation, Building and Nuclear Safety (BMUB): *Principles and Goals of the German Government's Climate Policy*; Federal Ministry for the Environment, Nature Conservation, Building and Nuclear Safety (BMUB) Public Relations Division: Berlin, Germany, 2016.
33. DCCAE. Climate Action Plan 2019: To Tackle Climate Breakdown. Available online: https://assets.gov.ie/10206/d042e174c1654c6ca14f39242fb07d22.pdf (accessed on 15 July 2021).
34. Bakam, I.; Balana, B.B.; Matthews, R. Cost-effectiveness analysis of policy instruments for greenhouse gas emission mitigation in the agricultural sector. *J. Environ. Manag.* **2012**, *112*, 33–44. [CrossRef]
35. Aguilera, E.; Guzmán, G.; Alonso, A. Greenhouse gas emissions from conventional and organic cropping systems in Spain. I. Herbaceous crops. *Agron. Sustain. Dev.* **2015**, *35*, 713–724. [CrossRef]
36. Zhang, L.; Zheng, J.; Chen, L.; Shen, M.; Zhang, X.; Zhang, M.; Bian, X.; Zhang, J.; Zhang, W. Integrative effects of soil tillage and straw management on crop yields and greenhouse gas emissions in a rice-wheat cropping system. *Eur. J. Agron.* **2015**, *63*, 47–54. [CrossRef]
37. Sáez-Almendros, S.; Obrador, B.; Bach-Faig, A.; Serra-Majem, L. Environmental footprints of Mediterranean versus Western dietary patterns: Beyond the health benefits of the Mediterranean diet. *Env. Health* **2013**, *12*, 1–8. [CrossRef] [PubMed]
38. Lassaletta, L.; Billen, G.; Grizzetti, B.; Garnier, J.; Leach, A.M.; Galloway, J.N. Food and feed trade as a driver in the global nitrogen cycle: 50-year trends. *Biogeochemistry* **2014**, *118*, 225–241. [CrossRef]
39. European Environment Agency (EEA). Available online: https://ec.europa.eu/eurostat/databrowser/view/ENV_AIR_GGE__custom_1118502/default/table?lang=en (accessed on 30 June 2021).
40. Hamed, K.H.; Rao, A.R. A modified Mann-Kendall trend test for autocorrelated data. *J. Hydrol.* **1998**, *204*, 182–196. [CrossRef]
41. Mann, H.B. Nonparametric tests against trend. *Econom. J. Econom. Soc.* **1945**, *13*, 245–259. [CrossRef]
42. Kendall, M.G. *Rank Correlation Methods*; Griffin: London, UK, 1975.
43. Yue, S.; Wang, C.Y. Applicability of prewhitening to eliminate the influence of serial correlation on the Mann-Kendall test. *Water Resour. Res.* **2002**, *38*, 4-1–4-7. [CrossRef]
44. Sen, P.K. Estimates of the regression coefficient based on Kendall's tau. *J. Am. Stat. Assoc.* **1968**, *63*, 1379–1389. [CrossRef]
45. Abdi, H.; Williams, L.J. Principal component analysis. *Wiley Interdiscip. Rev. Comput. Stat.* **2010**, *2*, 433–459. [CrossRef]
46. Richardson, M. Principal Component Analysis. Available online: http://people.maths.ox.ac.uk/richardsonm/SignalProcPCA.pdf (accessed on 20 June 2021).
47. Basto, M.; Pereira, J.M. An SPSS R-Menu for Ordinal Factor Analysis. *J. Stat. Softw.* **2012**, *46*, 1–29. [CrossRef]
48. MAGRAMA. *Anuario de Estadística 2014*; MAGRAMA: Madrid, Spain, 2015.
49. MAGRAMA. *Encuesta Sobre Superficies y Rendimientos De Cultivos (ESYRCE): Resultados Nacionales y Autonómicos*; MAGRAMA: Madrid, Spain, 2015.
50. MAGRAMA. *Inventario de Emisiones de Gases de Efecto Invernadero de España. Serie 1990–2014*; Informe Resumen, Secretaría de Estado de Medio Ambiente, Dirección General de Calidad y Evaluación Ambiental, MAGRAMA: Madrid, Spain, 2016.
51. Álvaro-Fuentes, J.; Plaza-Bonilla, D.; Arrúe, J.L.; Lampurlanés, J.; Cantero-Martínez, C. Soil organic carbon storage in a no-tillage chronosequence under Mediterranean conditions. *Plant Soil.* **2014**, *376*, 31–41. [CrossRef]
52. Albiac, J.; Kahil, T.; Notivol, E.; Calvo, E. Agriculture and climate change: Potential for mitigation in Spain. *Sci. Total Environ.* **2017**, *592*, 495–502. [CrossRef] [PubMed]
53. Hijazi, O.; Berg, W.; Moussa, S.; Ammon, C.; von Bobrutzki, K.; Brunsch, R. Comparing methane emissions from different sheep-keeping systems in semiarid regions: A case study of Syria. *J. Saudi Soc. Agric. Sci.* **2014**, *13*, 139–147. [CrossRef]
54. Hijazi, O.; Berg, W.; Moussa, S.; Ammon, C.; Von Bobrutzki, K.; Brunsch, R. Awassi sheep keeping in the Arabic steppe in relation to nitrous oxide emission from soil. *J. Assoc. Arab. Univ. Basic Appl. Sci.* **2014**, *16*, 46–54. [CrossRef]
55. MAGRAMA. *Balance Del Nitrógeno En La Agricultura Española 2012*; MAGRAMA: Madrid, Spain, 2014.
56. Romano, D.; Arcarese, C.; Bernetti, A.; Caputo, A.; Cóndor, R.D.; Contaldi, M.; De Lauretis, R.; Di Cristofaro, E.; Federici, S.; Gagna, A.; et al. *Italian Greenhouse Gas Inventory 1990–2007. National Inventory Report*; ISPRA: Rome, Italy, 2009.
57. Marik, T.; Fischer, H.; Conen, F.; Smith, K. Seasonal variations in stable carbon and hydrogen isotope ratios in methane from rice fields. *Glob. Biogeochem. Cycles* **2002**, *16*, 41. [CrossRef]
58. Leip, A.; Bidoglio, G.; Smith, K.A.; Conen, F.; Russo, S. Rice cultivation by direct drilling and delayed flooding reduces methane emissions. InNon-CO2 greenhouse gases: Scientific understanding, control options and policy aspects. In Proceedings of the Third International Symposium, Maastricht, The Netherlands, 21–23 January 2002; pp. 457–458.
59. Schütz, H.; Holzapfel-Pschorn, A.; Conrad, R.; Rennenberg, H.; Seiler, W. A 3-year continuous record on the influence of daytime, season and fertilizer treatment on methane emission rates from an Italian rice padd. *J. Geophys. Res.* **1989**, *94*, 16405–16415. [CrossRef]

60. Schütz, H.; Seiler, W.; Conrad, R. Processes involved in formation and emission of methane in rice paddies. *Biogeochemistry* **1989**, *7*, 33–53. [CrossRef]
61. Yan, X.; Yagi, K.; Akiyama, H.; Akimoto, H. Statistical analysis of the major variables controlling methane emission from rice fields. *Glob. Chang. Biol.* **2005**, *11*, 1131–1141. [CrossRef]
62. Ortiz, M.; Baldock, D.; Willan, C.; Dalin, C. *Towards Net Zero in UK Agriculture Key Information, Perspectives and Practical Guidance*; University College London, HSBC UK: London, UK, 2021; p. 55.
63. Nair, P.K.R.; Kumar, B.M.; Nair, V.D. Agroforestry as a strategy for carbon sequestration. *J. Plant Nutr. Soil Sci.* **2009**, *172*, 10–23. [CrossRef]
64. National Food Strategy (NFS). 2020. Available online: https://www.nationalfoodstrategy.org/ (accessed on 1 August 2021).
65. Department for Environment Food & Rural Affairs (Defra). 2021. Available online: https://www.gov.uk/government/statistical-data-sets/greenhouse-gas-emissions-from-agriculture-indicators (accessed on 1 August 2021).
66. Brown, P.; Cardenas, L.; Choudrie, S.; Jones, L.; Karagianni, E.; MacCarthy, J.; Passant, N.; Richmond, B.; Smith, H.; Thistlethwaite, G.; et al. *UK Greenhouse Gas Inventory, 1990 to 2018: Annual Report for Submission under the Framework Convention on Climate Change*; Department for Business, Energy & Industrial Strategy: Oxfordshire, UK, 2020.
67. Ruyssenaars, P.G.; Coenen, P.W.H.G.; Rienstra, J.D.; Zijlema, P.J.; Arets, E.J.M.M.; Baas, K.; Dröge, R.; Geilenkirchen, G.; t Hoen, M.; Honig, E.; et al. *Greenhouse gas emissions in the Netherlands 1990–2019*; National Inventory Report 2021; National Institute for Public Health and the Environment: Bilthoven, The Netherlands, 2021.
68. Ministry of Environment collaborated with Institute for Studies and Power Engineering. *Third Biennial Report of Romania. Romania's Third Biennial Report under the UNFCCC*; Ministry of Environment collaborated with Institute for Studies and Power Engineering: Bucharest, Romania, 2017.
69. Poland's National Inventory Report (NIR). Greenhouse Gas Inventory for 1988–2017. Submission under the UN Framework Convention on Climate Change and its Kyoto Protocol; Reporting entity: National Centre for Emission Management (KOBiZE). Institute of Environmental Protection—National Research Institute: Warsaw, Poland, 2019.
70. Piwowar, A. Low-Carbon Agriculture in Poland: Theoretical and Practical Challenges. *Pol. J. Environ. Stud.* **2018**, *28*, 2785–2792. [CrossRef]
71. National Inventory Report for France. *Inventory of Greenhouse Gas Emissions in France from 1990 to 2019. Under the United Nations Framework Convention on Climate Change and the Kyoto Protocol*; Citepa: Paris, France, 2021.
72. Garnier, J.; Le Noë, J.; Marescaux, A.; Sanz-Cobena, A.; Lassaletta, L.; Silvestre, M.; Thieu, V.; Billen, G. Long-term changes in greenhouse gas emissions from French agriculture and livestock (1852–2014): From traditional agriculture to conventional intensive systems. *Sci. Total Environ.* **2019**, *660*, 1486–1501. [CrossRef]
73. Fuß, R.; Vos, C.; Rösemann, C.; Greenhouse Gas Emissions from Agriculture. Thünen Institute (TI). Federal Research Institute for Rural Areas, Forestry and Fisheries. 2021. Available online: https://www.thuenen.de/en/topics/climate-and-air/emission-inventories-accounting-for-climate-protection/greenhouse-gas-emissions-from-agriculture/ (accessed on 27 September 2021).
74. Rösemann, C.; Haenel, H.; Vos, C.; Dämmgen, U.; Döring, U.; Wulf, S.; Eurich-Menden, B.; Freibauer, A.; Döhler, H.; Schreiner, C.; et al. *Calculations of Gaseous and Particulate Emissions from German Agriculture 1990–2019. Report on Methods and Data (RMD) Submission 2021*; Johann Heinrich von Thünen-Institute: Braunschweig, Germany, 2021.
75. German National Inventory Report (NIR). National Inventory Report for the German Greenhouse Gas Inventory from 1990–2019. Submission under the United Nations Framework Convention on Climate Change and the Kyoto Protocol 2021. Federal Environment Agency. Thünen Institute (TI): Braunschweig, Germany, 2021.
76. Graham, F. Daily briefing: Ground-Breaking Carbon Capture Plant Starts Up. Available online: https://www.nature.com/articles/d41586-021-02473-y (accessed on 1 August 2021).
77. Peek, M. What Is the Taste of CO2? The Sustainability of Icelandic Food Systems in the Face of Climate Change. Independent Study Project (ISP) Collection. 2016. Available online: https://digitalcollections.sit.edu/isp_collection/2452 (accessed on 1 August 2021).
78. WITS. Available online: https://wits.worldbank.org/CountryProfile/en/Country/ISL/Year/1990/TradeFlow/Import/Partner/all/Product/16-24_FoodProd# (accessed on 27 September 2021).
79. Soto, I.; Barnes, A.; Balafoutis, A.; Beck, B.; Sanchez, B.; Vangeyte, J.; Fountas, S.; Wal, T.V.D.; Eory, V.; Gómez-Barbero, M. *The Contribution of Precision Agriculture Technologies to Farm. Productivity and the Mitigation of Greenhouse Gas. Emissions in the EU.*; EU Science Hub. Publications Office of the European Union: Brussels, Belgium, 2019; ISBN 978-92-79-92834-5. ISSN 1831-9424.
80. Kritikos, M. *Precision Agriculture in Europe: Legal and Ethical Reflections for Law-Makers. European Parliament's Science and Technology Options Assessment (STOA)*; European Parliamentary Research Service (EPRS): Brussels, Belgium, 2017.
81. Euractiv. Europe Entering the Era of 'Precision Agriculture'. 2016. Available online: https://www.euractiv.com/section/science-policymaking/news/europe-entering-the-eraof-precision-agriculture/ (accessed on 27 September 2021).
82. Sadik-Zada, E.R.; Ferrari, M. Environmental Policy Stringency, Technical Progress and Pollution Haven Hypothesis. *Sustainability* **2020**, *12*, 3880. [CrossRef]

energies

MDPI

Article

An Economic Assessment of the Impact on Agriculture of the Proposed Changes in EU Biofuel Policy Mechanisms

Piotr Gradziuk [1,*], Krzysztof Jończyk [2], Barbara Gradziuk [3], Adrianna Wojciechowska [1], Anna Trocewicz [4] and Marcin Wysokiński [5]

1 Economic Modelling Department, Institute of Rural and Agricultural Development, Polish Academy of Sciences, Nowy Świat 72, 00-330 Warsaw, Poland; awojciechowska@irwirpan.waw.pl
2 Institute of Soil Science and Plant Cultivation—State Research Institute, Czartoryskich 8, 24-100 Puławy, Poland; kjonczyk@iung.pulawy.pl
3 Department of Management and Marketing, Faculty of Agrobioengineering, University of Life Sciences in Lublin, Akademicka 13, 20-950 Lublin, Poland; barbara.gradziuk@up.lublin.pl
4 Department of Economics, Faculty of Economic Sciences, Pope John Paul II State School of Higher Education, Sidorska 95/97, 21-500 Biała Podlaska, Poland; usanna@op.pl
5 Institute of Economics and Finance, Warsaw University of Life Sciences—SGGW, 02-787 Warsaw, Poland; marcin_wysokinski@sggw.edu.pl
* Correspondence: pgradziuk@irwirpan.waw.pl

Abstract: In Poland, rapeseed production has been the fastest growing branch of plant production since 2000. Rapeseed yields have increased 2.5 times in the last 20 years. The main reason for this trend was the implementation of obligations resulting from legal acts by Member States relating to increasing the share of RES in the structure of primary energy production, and in particular relating to the share of biofuels in fuels used in transport. In Poland in the years 2010–2020, about 1.0–1.6 million tonnes of rape seeds were used for this purpose annually. Due to the fact that biofuel production competes for raw materials with the food economy, at the end of the first decade of the 21st century, many representatives of various circles intensified their voices, calling for withdrawal from the policy supporting the biofuel sector, which may have resulted in a decrease in oilseed plant cultivation areas. As a result of the research conducted here, it was determined that the place of oilseed rape in the sowing structure will be taken by rye, triticale and, on good soils, by wheat. Compared to rape, their production is characterised by lower income per 1 ha; in the years 2013–2019, these differences amounted to: wheat—8 EUR, triticale—102.3 EUR, and rye—168 EUR. This situation will deteriorate the value cereal cultivation sites and will result in a decrease in their yields. On the basis of the conducted research, the estimated value of rape as a forecrop for wheat, triticale, and rye was, respectively: 103.7; 64.6 and 46.7 EUR. An additional advantage of oilseed rape is that it is an excellent bee resource and is classified as a commodity crop, i.e., one from which significant amounts of honey can be obtained, with a net value of EUR 55 per hectare. In addition, in many agricultural holdings, as a result of forecasted changes in plant production, there will be an accumulation of field work during the harvest period, which will also affect the worse use of machinery and storage areas. The consequence of increasing the area under which cereal crops and their supply can grow may be the decline in production profitability and thus the income situation of farms, but this will be assessed at the next stage of research.

Keywords: renewable energy; biofuels; biodiesel; legal sources on renewable energy; oilseeds and rape; profitability of production; crop rotation; beekeeping

Citation: Gradziuk, P.; Jończyk, K.; Gradziuk, B.; Wojciechowska, A.; Trocewicz, A.; Wysokiński, M. An Economic Assessment of the Impact on Agriculture of the Proposed Changes in EU Biofuel Policy Mechanisms. *Energies* **2021**, *14*, 6982. https://doi.org/10.3390/en14216982

Academic Editor: David Borge-Diez

Received: 22 September 2021
Accepted: 19 October 2021
Published: 25 October 2021

1. Introduction

The first attempts to utilise biofuels to power engines were made by the end of the 19th century [1–3]. The self-ignition engine constructed in 1893 by Rudolf Diesel could be fuelled with both petroleum-derivative fuels and oils of both vegetable and animal origin [4–6].

Similarly, ethanol has a long history as an engine fuel [1,2,7–9]. It had already been used at the end of the 19th century in engines designed by Samuel Morey, Nicholas Otto, and Henry Ford. In the 1920s in the USA, about the fuel sold 20% accounted for ethanol [10,11]. Due to much higher production costs in relation to fuels produced from crude oil, their importance was marginalized [5,12–14]. In the 20th century, a downward price trend prevailed, with periods of growth occurring sporadically and most often as a result of political or economic crises, especially after 1974 (OPEC embargo, Iranian revolution, Gulf War). After their subsidence, prices declined, and some economic crises, such as the recession of the early 1980s, the Asian crisis (1997), the financial market crisis (2008) and the coronavirus pandemic crisis (2020), caused significant price decreases. The concept of Rudolf Diesel or Henry Ford, based on the use of vegetable oils to run engines and produce other biofuels, was reintroduced in the 1980s. This was mainly due to four reasons:

The first of them can be expressed in the shortest terms by quoting Alvin Toffler [15]: "The condition for the existence of any civilisation—old or new—is energy". According to the International Energy Outlook 2002 report prepared by the Energy Information Administration (EIA), which is a part of the US Department of Energy, between 2000 and 2020, a significant increase in energy consumption in the world was projected, which was to reach 60% and increase from 382 to 612 quadrillion Btu [16]. These predictions turned out to be accurate because by 2018, the world's energy consumption increased by 48.2% compared to 2000. However, in a report published in 2020 by the International Energy Agency (IEA), it was assumed that by the middle of this century, energy consumption would increase by nearly 50% compared to 2018 [17]. Thus, energy was, is, and will be the main determinant of economic activity and the development of civilization because every management process must be powered by energy [18–21].

The second consideration is that resources are limited, which may reduce business efficiency or even hinder economic development. The energy barrier to economic development as a result of dwindling coal reserves and rising mining costs was already presented by William Stanley Jevons [22] in the mid-19th century, who stated that technological progress and other energy sources would not change this situation. This view was rejected by neoclassical economists, who did not acknowledge the possibility of limiting growth in the long term. Indeed, they considered that under optimal market conditions, there is a harmonised adjustment of individual resources, ensuring balance while fully exploiting production potential [23–25]. The basis of such a judgment is the assumption that prices increase with the depletion of resources. This triggers incentives to increase the supply of these resources or their substitutes or to introduce more efficient use methods, which reduces demand for them [26,27]. An example of this is the market for energy resources. In all of the scenarios considered, the share of conventional fuels—oil, gas and coal—is projected to decrease more rapidly after 2020, which is in line with the depletion of their resources and the associated increase in energy prices. Conventional resources will be replaced by renewable energy sources (RES) [17,28–31].

Another aspect is concern for the environment, as the intensive use and processing of traditional energy resources has a negative impact on natural resources. Since the last three decades of the 20th century, the most serious threat is related to climate change caused by the anthropogenic heating of the atmosphere as a result of increasing concentrations of greenhouse gases (GHG), especially CO_2 [32]. There is a well-founded concern that this phenomenon may pose a threat to life for most of humanity and even to the whole of civilisation [33]. Hence, many circles and international organisations have made initiatives to take action to reduce GHG emissions. The first World Climate Conference held in 1979 established the World Climate Research Programme (WCRP), and its management was entrusted to the World Meteorological Organisation (WMO), the United Nations Environment Programme (UNEP), and the International Council for Scientific Union (ICSU). These organisations established the Intergovernmental Panel on Climate Change (IPCC) in 1988. The first significant effect of the IPCC activities was the preparation of the United Nations Framework Convention on Climate Change (UNFCCC), which was presented in 1992 at the United Nations Conference on Environment and Development (UNCED) in Rio de Janeiro. The main body of the convention became the Conference of the Parties (COP), which, since its first meeting in Berlin in 1995 (COP 1), regularly assesses the scale and course of climate change and its effects and develops strategies to respond to these changes [34]. The first significant effect of these activities was the signing of the Kyoto Protocol during COP 3, in which the 38 most industrialised countries and the European Union committed to reduce GHG, which was expressed in carbon dioxide equivalent, by at least 5% below 1990 levels between 2008 and 2012 [35]. Due to protracted negotiations on a new global "climate agreement", COP 18 extended its validity until 31 December 2020 [36]. Although the Kyoto Protocol was a first significant step towards reducing greenhouse gas emissions, it did not solve the problem of global warming. This did not occur until climate policy re-prioritisation (among other things, under the influence of the financial crisis), which began to be seen as a factor for economic growth through "the development of clean or low-carbon technologies, the creation of new markets, industries and jobs" [37]. The latter led to the acceleration of negotiations and agreement on the content of a global climate agreement at COP 21 in Paris in December 2015 (the Paris Agreement) [38]. The European Union (EU) plays a very important role in reducing greenhouse gas emissions. The actions taken by the EU go far beyond the obligations arising from global climate agreements [39]. The European Green Deal has set out a clear vision of how to achieve climate neutrality by 2050 [40].

The fourth reason is the stagnation in demand for agricultural raw materials and food products, which is becoming a barrier to agricultural development. In countries with a developed economy, surpluses of agricultural raw materials have started to occur, which has led to a deterioration in the profitability of production and reduced incomes for farming families. One way of managing these surpluses is to use them for non-food purposes. The idea of "Chemurgy" was already promoted in the 1920s as a strategy for industries and governments who were interested in reviving the agricultural economy [41]. The USA reverted to this concept in the early 1980s. As part of the Growing Industrial Materiale programme, more than two thousand plant species have been tested for the raw material content sought by industries, of which several dozen have been selected and recommended for cultivation [42]. In Europe, the intensification of research into the cultivation of plants for industrial purposes dates back to 1982, when the European Commission recommended cooperation between agriculture and industry. This research resulted in a very long list of arable crops that can be used in several industries and branches of industry [43–45]. However, the direction of bioenergy has become dominant, which is mainly due to the growing interest in obtaining inexhaustible and ecologically clean energy sources [46–48]. The records in the White Paper "Energy for the Future: renewable sources of energy" prepared by the European Commission in 1997 showed that by 2010, the production of first-generation biofuels, mainly comprising biodiesel produced from rapeseed, will increase the most [49].

As a result of these factors, between 1997 and 2017, biodiesel production in the EU 28 increased from 332.9 ktoe to 12,239.4 ktoe. First-generation biofuels are very controversial. Their use is questioned for ethical [50–53], economic [54–56], and environmental [57,58] reasons. The production of biofuels has thus become the subject of numerous discussions, polemics, comments, and contradictory judgments: from extreme disapproval and objections [59,60] to equally extreme affirmation and approval [61,62]. Hence, the frequent changes in legal regulations governing this market [63–66].

The interest of scientific environment concerning the issue of biofuels was mainly stimulated by the discussion in the context of climate change, energy and food security, and the legitimacy of support for the development of their use on both national and EU (European Union) levels. On the other hand, there is a lack of comprehensive assessments relating to the agricultural sector on a micro-scale, which is the key supplier of raw materials for their production. This primarily results from the problem of identification, and especially quantification, of a wide range of effects that are the result of changes in the structure of plant production. The main motivation of our study has been to assess the impact of the EU biofuel policy on the agricultural sector. In this paper, we illustrate this for biodiesel in Poland, which is the largest producer of biodiesel produced from domestic feedstock in the EU.

2. Background

2.1. What Are Biofuels?

In the RES literature, the term "biofuel" is defined very differently. It is most often used to refer to fuels produced from biomass, which can take solid, liquid, or gaseous forms [2,67,68]. However, since it started to be widely used as motor fuel, the term is dedicated to any type of liquid or gas produced from biomass that can be used as a substitute for fossil fuels [69,70]. According to the International Energy Agency [71], biofuels are "liquid and gaseous fuels produced from biomass—organic matter derived from plants or animals". Biofuels are usually classified according to two categories: type of biomass and production technologies.

Biomass sources are defined in Directive 2018/2001 of the European Parliament and of the European Council (EU) as "the biodegradable fraction of products, waste and residues from biological origin from agriculture, including vegetal and animal substances, from forestry and related industries, including fisheries and aquaculture, as well as the biodegradable fraction of waste, including industrial and municipal waste of biological origin". Biomass fuels refer to gaseous and solid fuels produced from biomass and, biofuels refer to liquid fuel for transport produced that is from biomass [66].

Due to the diverse composition and suitability for the various conversion methods that are used, the following biomass categories can be distinguished [72–74]:

- Raw materials containing significant amounts of sugar and starch (sugar beet, cereals, potatoes);
- Lignocellulosic biomass (wood and its waste, targeted wood crops, straw);
- Oilseeds and animal fats;
- Organic waste (organic fertilisers and food and municipal waste);
- Algal biomass.

Depending on the type of biomass that is used, the following biofuel generations can be distinguished [75]:

- First generation (edible feedstocks);
- Second generation (non-edible biomass sources);
- Third generation (microalgae biomass);
- Fourth generation (genetic modification of the microalgae).

In 2017, the main raw materials that were used in the production of bioethanol were sugar cane, maize grain (Brazil, USA), biodiesel soybean and palm oils, animal fats, used cooking oils, and rapeseed, which was mainly used in the EU [47,74]. Biofuels are commonly referred to as first-generation fuels, which is mainly due to the fact that they use conventional technologies during their production: alcoholic fermentation, mechanical pressing, and transesterification (hydrogenation) of oils and anaerobic digestion of organic biodegradable wastes to produce biogas [71]. Due to the controversy arising from the significant quantities of agricultural raw materials used to produce biofuels [50–60], research on the production of second, third, and fourth generation biofuels, known as advanced biofuels, has intensified since the beginning of the 21st century. The main substrates for their production are waste and residues of biological origin from agriculture, forestry and related industries, fisheries, aquaculture, and municipal and industrial waste of biological origin. The prospective development of next-generation biofuel production is [76]:

- Microbial conversion of lignocellulosic biomass (e.g., stalks, corn stover) into bioethanol or biobutanol;
- Transesterification of sustainably sourced FAME (i.e., biodiesel);
- Hydrotreatment of sustainably sourced vegetable oils or animal fats followed by alkane isomerisation and cracking to produce drop-in fuels (HVO/HEFA);

- Thermochemical pathways starting with pyrolysis to produce biocrude or gasification of biomass for syngas.

Based on experience to date, it can be concluded that apart from HVO technology, the production of other advanced biofuels is still under intensive development and work on optimising production efficiency, minimising production costs, and seeking non-commercial sources of financing is being undetaken [76–78].

2.2. Legal Conditions

The growing interest in opportunities to increase energy production from renewable sources in the EU began after the first oil crisis. However, the energy obtained in this way was more expensive than conventional energy in most applications. Therefore, the EU and individual countries have taken political, legal, administrative, and financial measures to achieve this objective as efficiently as possible. The first regulations concerning the support for renewable energy sources were included in Council Regulation (EEC) No 1302/78 of 12 June 1978, which discussed the granting of financial support for projects to exploit alternative energy sources [79]. In contrast, the Council resolution of 9 June 1980 concerning Community energy policy objectives for 1990 and the convergence of the policies of the Member States required the Commission to integrate RES into the framework of community energy policies [80]. Further actions include an assessment of the potential, the state of the technology, economic conditions, and barriers related to increasing the use of RES [81,82]. Research and development work has also been intensified, among other initiatives, within the framework of the programmes Valoren, Altener, Coopener, Intelligent Energy-Europe Programme, Joule-thermie, Save, Steer, and Synergy of subsequent European Framework Programme for Research and Innovation.

A milestone on the way to increase the importance of RES in the EU was the publication of The Green Paper [83] and White Paper [49] between 1996 and 1997, which were entitled "Energy for the future: renewable sources of energy". At that time, these were key documents that were political and strategic in character, setting directions for long-term policy, with quantitative targets in the form of doubling the share of RES in the structure of primary energy production from 6 to 12% between 1998 and 2010. They indicated that biomass would be the most important among renewable energy sources. Its share in the production of liquid fuels was predicted to increase (40–60 times) compared to electricity (ten times) and thermal energy (two times). These documents also formulate the need to introduce appropriate legal regulations and to secure sources of funding to achieve these ambitious goals [84].

In 2000, the Commission proposed the first two EU directives for RES, the promotion of renewable electricity and the development of biofuels in transport. The first was adopted in 2001 (2001/77/EC), and the second objective pertaining to the development of biofuels was adopted in 2003 (2003/30/EC). The biofuels directive obliged Member States to set national indicative targets to set reference values of 2% share for biofuel consumption in transport by 31 December 2005 and obliged them to increase those shares to 5.75% in 31 December 2010 [85]. To meet these requirements Member States used two main tools: tax exemptions and biofuels obligations. Additionally, they introduced a special "energy crop payment" of EUR 45 per hectare (a maximum guaranteed area of 1.5 million hectares). These measures were complemented by the extension of offers for preferential loans, guaranteed lending, and loans to small businesses for renewable energy investments by financial institutions such as the European Investment Bank (EIB) and the European Bank for Reconstruction and Development (EBRD). Despite the instruments used, the market share of biofuels in 2005 was only 1.4% [86].

Although in those first years, there were problems with the implementation of Directive 2003/30/EC in some countries, as there were intense discussions in the EU regarding increasing the market share of biofuels [87,88]. In 2009, the European Parliament and the Council adopted a climate policy package in which the European Union committed to reducing greenhouse gas emissions expressed as a CO_2 equivalent by 20% by 2020 (if other developed countries made similar commitments, then the reduction could be as high as 30%). In the same period, the EU should have also increased the share of renewable energy in terms of total energy production from 8.5 to 20% to 10%—the share of biofuel in transport fuel—and reduce energy consumption by 20%. The biofuel sector was mainly covered by two directives:

- Directive 2009/28/EC (RED) of the European Parliament and by the Council meeting of 23 April 2009 on the promotion of the use of energy from renewable sources and the amendment and subsequent appeal of Directives 2001/77/EC and 2003/30/EC (Text with EEA relevance) [89];
- Directive 2009/30/EC of the European Parliament and by the Council meeting of 23 April 2009 that amended Directive 98/70/EC regarding the specification of petrol, diesel, and gas-oil and introduced a mechanism to monitor and reduce greenhouse gas emissions and amended Council Directive 1999/32/EC regarding the specification of fuel used by inland waterway vessels and repealed Directive 93/12/EEC (Text with EEA relevance) [90].

The results of the researched obtained within the European Framework Programme—Horizon 2020 have shown low efficiency in reducing CO_2 emissions through the use of traditional biofuels, the so-called first generation, hence the proposals to reform the biofuel directives [91]. As a result of the discussions and analyses that have been conducted, the current solutions were modified and were included in Directive 2015/1513 of the European Parliament and during the Council meeting on 9 September 2015 [92]. One of the most important changes introduced by this Directive was to set a limit for the level of first-generation biofuels, with the Directive stating that their maximum quantity in 2020 could not exceed 7%. Moreover, the condition for including such biofuels as renewable energy was to prove that the raw materials obtained for their production did not come from areas with high biodiversity value and high carbon intensity, and that their production complied with environmental requirements, which are regulated by the Code of Good Agricultural Practice in Poland [92]. The remaining part (at least 3%) was to be produced from algae, by-products (e.g., straw, manure, seed hulls, etc.), or waste. A detailed list is provided in Annex IX of Directive 2015/1513 [65].

The necessity of meeting the EU's obligations arising from the Paris Agreement was the main determinant of the adoption of a new directive on the promotion of the use of energy from renewable sources (EU) 2018/2001 (RED II). In this document, the Member States agreed that the share of energy from renewable sources in gross final energy consumption in 2030 will be at least 32%. After 2023, a proposal to increase this target will be considered if its production costs are significantly reduced or due to the EU's international commitments. This Directive also contains many significant changes relating to the issue of biofuels [66]. The most important are:

- A 14% share of renewable energy in final energy consumption in the transport sector by 2030 at least;
- Renewable energy used in the transport sector should also comprise renewable liquid and gaseous transport fuels of non-biological origin (e.g., hydrogen) and recycled carbon fuels (e.g., derived from plastic waste, rubber);
- First-generation biofuels should be divided into two categories: low (certification required) and high-risk Indirect Land Use Change -ILUC (cannot be higher than 2019 consumption levels—reduction from 31 December 2023 to 0% by 31 December 2030);
- Input of advanced biofuels and biogas produced from raw materials listed in Annex IX:
    - Part A—min 0.2% in 2022, min 1% in 2025 and min 3.5% in 2030;
    - Part B—maximum 1.7%.
- New methodology for calculating GHG emissions.

2.3. Development of Biofuel Production in the UE

Between 1996 and 1997, when the Green Paper [83] and White Paper [49] "Energy for the future: renewable sources of energy" were presented, the assumptions they made regarding the development of biofuel production in the European Union were considered unrealistic by most experts dealing with the issue [84]. However, the systematic implementation of the provisions contained in both documents and Directive 2003/30/EC contributed to the development of this economic sector. Between 1996 and 2010, the production of biodiesel in the EU increased by more than thirty times, and the production of bioethanol increased by nearly fifty times. This growth dynamic, which was mainly due to the continuation of the current EU policy on RES (Directives 2009/28/EC and 2015/1513/EC), continued. In 2018, bioethanol and biodiesel production increased by 60 and 50 times in relation to their production in 1996, respectively. In the considered period (1996–2018), the share of biofuels in the RES production structure also increased significantly, from 0.36% to 7.06% (Figure 1).

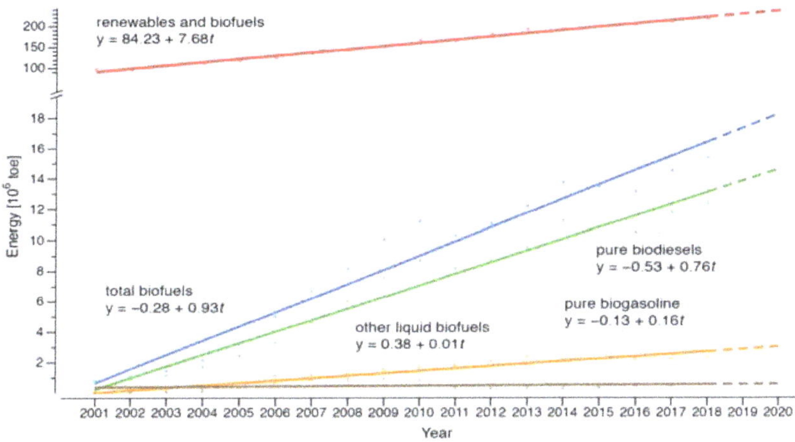

Figure 1. Share of biofuels in the energy production structure from renewable sources in the EU between 1996 and 2018.

In the EU, the predominant role among biofuels is played by biodiesel, the use of which increased from 85.8% in 1996 to 81.0% in 2018. On an energy basis, biodiesel represents about 75 percent of the total transport biofuel market [93]. Globally, the share of biodiesel in the production of biofuels in 2018 was only 28.1%, with bioethanol accounting for over 70% [94]. The term biodiesel (pure) includes traditional biodiesel fatty acid methyl ester (FAME) and hydrotreated vegetable oil (HVO). The main factors that determined greater interest in the production of biodiesel in the EU rather than bioethanol were:

- The Blair House Agreement (provisions on the production of oilseeds under the Common Agricultural Policy) [93,95];
- Higher margin income in the production of oilseeds, which are the primary feedstock in the production of biodiesel, than cereals [96,97];
- The possibility of using by-products for feed purposes and thus reducing protein feed imports [93,96];
- Ensuring that oilseeds had a higher pre-crop value than cereals [96,98–103];
- Beekeeping profits [96,104–106];
- Protection of the natural environment through the reduction of NOx emissions and a closed CO_2 cycle [47,61,98,107];
- Increasing the number of jobs mainly in rural areas [61,62,70,96,98,108];
- Improvements in energy security [55,70,98,107].

The largest biodiesel producers in the EU are Germany, France, the Netherlands, Spain, Poland, and Italy (Table 1).

Table 1. Biodiesel production and oilseed areas in selected EU countries.

Specification	Pure Biodiesels (Ktoe)							Oilseed Areas (Kha)													
								Total						Including Rape and Turnip Rape						
	1993	2000	2005	2010	2015	2018	2019	1993	2000	2005	2010	2015	2018	2020	1993	2000	2005	2010	2015	2018	2020
Germany	0	222	1323	2736	2765	2960	3176	1088	1104	1371	1486	1304	1272	1018	1007	1078	1344	1461	1286	1228	957
France	0	265	542	1788	2170	2435	1868	1443	1993	1935	2207	2239	2323	2082	559	1186	1232	1465	1499	1617	1122
Netherlands	0	0	0	338	1440	1625	1739	2	1	2	3	2	3	2	2	1	2	3	2	2	1
Spain	0	0	147	764	1011	1561	1804	2155	871	523	704	811	771	725	13	29	5	21	71	79	73
Poland	0	0	59	348	695	784	849	349	438	555	949	955	856	876	349	437	550	946	947	845	864
Italy	0	0	177	706	510	664	771	311	506	286	280	436	445	397	6	36	4	20	12	14	17
Portugal	0	0	0	280	317	321	347	95	52	7	14	20	9	8	0	0	0	0	0	0	0
Finland	0	0	0	297	431	281	339	69	53	77	158	55	53	30	69	53	77	158	55	53	30
Sweden	0	0	7	111	124	259	323	169	48	82	110	95	97	99	169	48	82	110	95	97	99
Belgium	0	0	0	309	223	227	229	4	5	6	11	11	11	11	4	5	6	11	11	11	11
Austria	0	17	37	237	303	206	255	148	90	87	114	113	130	124	59	52	35	54	38	41	32
Others	0	60	155	763	889	1099	1236	2107	2927	3522	4639	4852	5329	5179	447	819	964	2215	1799	2331	2005
UE 27	0	564	2447	8677	10,878	12,422	12,936	7940	8088	8453	10,675	10,893	11,299	10,551	2684	3744	4301	6464	5815	6318	5211

Source: Own study based on data from Eurostat (Oilseed areas—EU_Cereal_Balance—europa.eu (accessed on 18 October 2021); Pure biodiesels—https://ec.europa.eu/eurostat/web/energy/data/energy-balances (accessed on 18 October 2021).

Rapeseed remains the dominant raw material used for the production of biofuels (France, Germany, Poland), but its share is systematically decreasing. In 2008, it was 72%, and in 2019, it was only 43%. This is the result of the growing use of used cooking oil (UCO) and palm oil. In 2019, the share of UCO was 21%, and it was mainly used in the Netherlands, Portugal, and Austria. The high biodiesel production in the Netherlands, Portugal and Belgium is based on imports. The incentive for its application is provided by Annex IX, point B of the RED and RED II Directives. In determining the contribution of biofuels to the final energy consumption of the transport sector, the use of UCO can be considered equivalent to twice the energy value of biofuels products from UCO. Palm oil, which had a share of 16% in 2019, has been used on a large scale in Spain, Italy, France, and the Netherlands. It has been used on a smaller scale in Finland, Germany, and Portugal. In the EU, biodiesel is also produced from sunflower seeds (Greece, Bulgaria, Hungary, Lithuania, France, Romania, Austria), animal fats (Denmark, Finland, France, The Netherlands), tall oil (Finland, Sweden), and cottonseed oil (Greece). The volume of biodiesel production supplies about 80% of the demand for this biofuel, hence the need for imports. The EU mostly imports biodiesel from Argentine, Malaysia, China, and Indonesia.

2.4. Biodiesel Production and Changes in the Area under Basic Crops

In the EU, the main raw material used for the production of first-generation biodiesel are oilseeds, so as demand for this type of biofuel increases, so does the area under cultivation. Based on tests that were performed independently—using the Pearson correlation—it was found that both the sown areas of oilseed plants (Y1), rapeseed and colza seed (Y2), soybean (Y3), and sunflower (Y4) were significantly correlated to biodiesel production (x). As expected, these correlations were positive, but their strength was characterised by significant differentiation. The characteristics of the estimated parameters of the models are summarised in Table 2. The model expressing the relationship between rapeseed and colza seeding areas (Y2) and biodiesel production (x), followed by Y1(x), Y3(x), and Y4(x), turned out to be the best suited to empirical data (R2 = 0.909). Among the EU countries, the production of biodiesel to the greatest extent produced determined the sown area of oilseed crops in Poland (R2 = 0.803).

Table 2. Basic statistic relationships between oilseeds (Y_1), rape areas (Y_2), soybean (Y_3), sunflower (Y_4), and biodiesel production (x) in UE, Germany, France, and Poland.

Dependent Variable	b_0				b_1				b_2				b_3				Error Variance	R^2
	Estimate	Stan. Error	t-Stat.	*p*-Value	Estimate	Stan. Error	t-Stat.	*p*-Value	Estimate	Stan. Error	t-Stat.	*p*-Value	Estimate	Stan. Error	t-Stat.	*p*-Value		
UE																		
Y_1	8162.552	112.46	72.580	0.000	-	-	-	-	0.00007	0.000	9.169	0.000	0.000	0.000	−6.892	0.000	168205	0.904
Y_2	3151.389	122.87	25.648	0.000	0.647	0.062	10.456	0.000	−0.00004	0.000	−7.153	0.000	-	-	-	-	140047	**0.909**
Y_3	516.030	33.412	15.444	0.000	−0.034	0.011	−3.158	0.004	-	-	-	-	0.000	0.000	5.861	0.000	11564	0.711
Y_4	4404.115	112.99	38.977	0.000	−0.474	0.123	−3.862	0.001	0.000082	0.000	3.550	0.002	0.000	0.000	−3.133	0.005	83821	0.359
Germany																		
Y_1	995.231	44.249	22.492	0.000	0.375	0.065	5.776	0.000	-	-	-	-			−4.682	0.000	16693	0.587
Y_2	995.231	44.249	22.492	0.000	0.375	0.065	5.776	0.000	-	-	-	-			−4.682	0.000	16693	0.587
France																		
Y_1	1764.149	64.072	27.534	0.000	0.603	0.187	3.225	0,003	−0.00019	0.000	−2.403	0.024	-	-	-	-	22705	0.515
Y_2	623.984	86.412	7.221	0.000	2.057	0.402	5.118	0.000	−0.00134	0.000	−3.600	0.001	0.000	0.000	2.650	0.014	20948	0.752
Poland																		
Y_1	461.034	26.846	17.173	0.000	1.547	0.242	6.392	0.000	−0.00127	0.000	−4.206	0.000	-	-	-	-	10013	**0.803**
Y_2	459.682	26.706	17.213	0.000	1.549	0.241	6.435	0.000	−0.00129	0.000	−4.281	0.000	-	-	-	-	9909	0.800

Source: Own study based on data from Eurostat (EU_Cereal_Balance—europa.eu (accessed on 18 October 2021); Pure biodiesels—https://ec.europa.eu/eurostat/web/energy/data/energy-balances (accessed on 18 October 2021).

These relationships are reflected in changes in the sown areas of basic crops (Figure 2). The sowing area of oilseeded crops, with the exception of sunflower, increased, and the sowing area of cereals decreased (except for triticale). Trend models for the sowing of basic crops and their statistical characteristics are presented in Table 3. The estimated trend models for the sown area of oilseed crops, including oilseed rape, cereals (except wheat and maize for grain), and biofuels (except other liquid biofuels), are very well fitted to the characterised phenomena (R^2 for the mentioned variables ranges from 0.754 to 0.916).

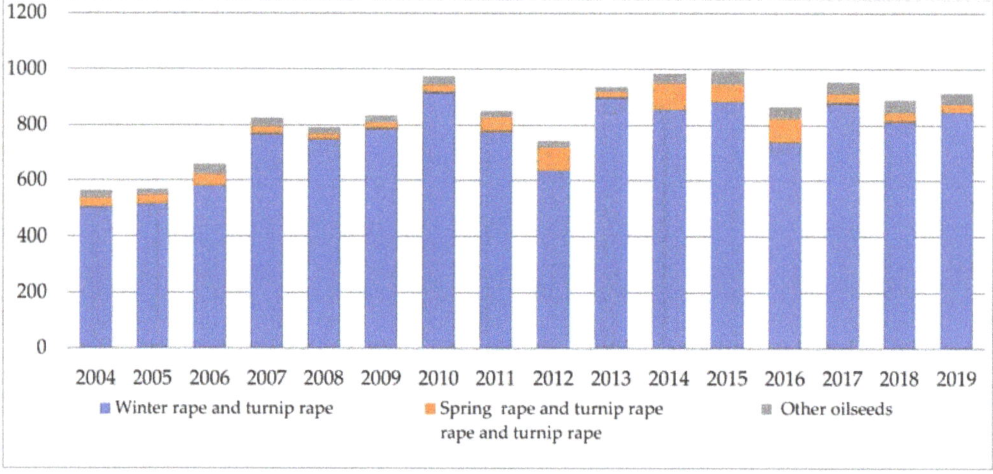

Figure 2. Sown area of oilseed crops in Poland in 2004–2019 (per thousand hectares).

Table 3. Linear trend models for the sowing of major crops and selected biofuels and their statistical characteristics.

Scheme 1	b_1 Parameter Characteristic				Error Variance	Adjusted R-Squared
	Estimate	Stan. Error	t-Stat.	*p*-Value		
Total oilseeds	0.144	0.014	10.358	0.000	0.354	0.797
Rape and turnip rape	0.131	0.014	9.148	0.000	0.377	0.754
Soybean	0.015	0.004	3.798	0.001	0.027	0.332
Sunflower	−0.002	0.009	−0.194	0.847	0.136	−0.037
Total cereals	−0.233	0.029	−8.063	0.000	1,526	0.703
Wheat	0.021	0.017	1.202	0.240	0.544	0.016
Barley	−0.147	0.011	−13.370	0.000	0.222	0.868
Rye	−0.093	0.008	−11.944	0.000	0.111	0.840
Triticale	0.060	0.006	10.427	0.000	0.061	0.800
Oats	−0.031	0.003	−11.756	0.000	0.013	0.836
Grain maize	−0.025	0.012	−2.033	0.052	0.284	0.104
Other cereals	−0.017	0.004	−4.389	0.000	0.027	0.403
Renewable energy sources	6.042	0.276	21.932	0.000	111.011	0.950
Total biofuels	0.715	0.043	16.506	0.000	2.744	0.916
Pure biodiesels	0.570	0.035	16.082	0.000	1.834	0.912
Pure biogasoline	0.119	0.008	14.053	0.000	0.105	0.887
Other liquid biofuels	0.026	0.008	3.155	0.004	0.102	0.264

Source: own study based on data from Eurostat (EU_Cereal_Balance—europa.eu (accessed on 18 October 2021) Pure biodiesels—https://ec.europa.eu/eurostat/web/energy/data/energy-balances (accessed on 18 October 2021).

Linear models turned out to be the most fitted, except for in the case of the sown area of total oilseed crops and oilseed rape and colza. For the total oilseed crops and for oilseed rape and colza, these were quadratic trends (Table 4). These trends were characterised by a very high coefficient of determination ($R^2 = 0.926$), which may indicate that the used model is correct.

Table 4. Quadratic trend models for total oilseed crops and oilseed rape and colza area and their statistical characteristics.

Specification	b_1 Parameter Characteristic				b_2 Parameter Characteristic				Error Variance	Adjusted R-Squared
	Estimate	Stan. Error			Estimate	Stan. Error	t-Stat.	*p*-Value		
Total oilseeds	0.507	0.054	9.467	0.000	−0.016	0.002	−6.296	0.000	0.108	0.926
Rape and turnip rape	0.490	0.065	7.518	0.000	−0.018	0.003	−5.995	0.000	0.159	0.817

Source: own study based on data from Eurostat (EU_Cereal_Balance—europa.eu (accessed on 18 October 2021); Pure biodiesels—https://ec.europa.eu/eurostat/web/energy/data/energy-balances (accessed on 18 October 2021).

3. Materials and Methods

The analyses in Section 2.4 show that the implementation of the EU biofuel policy has contributed to a significant increase in oilseed sowing. In Poland, the average acreage occupied by these crops in 2017–2019 was more than 58% higher than the 2004–2006 average (Table 5). Hence, it was first necessary to identify the crop species that were abandoned in favour of oilseed crops. To this end, statistical relations between the areas sown to oilseed crops (Y1) and the areas taken up by other crops (xn) were evaluated. In the next stages, on the basis of research conducted at the Institute of Plant Cultivation, Fertilisation and Soil Science National Research Institute in Puławy (IUNG PIB), the Institute of Agricultural and Food Economics National Research Institute in Warsaw (IERiGŻ PIB), literature on the subject, and data from the Central Statistical Office, the following five factors were identified and quantified, and it was on the basis of this that a synthetic assessment of the economic benefits of increasing the area of oilseed crop cultivation in Poland was made:

- The sown area of oilseeds;
- The area of sown crops replaced by oilseeds;
- The direct surplus for the above-mentioned crops;
- The value of oilseeds as a forecrop in relation to the crops that were replaced;
- The profits of beekeeping;
- The possibilities of using by-products for feed purposes and thereby reducing protein feed imports.

Table 5. Changes in the sown area of main crops in Poland in 2004–2019 (per thousand hectares).

Specification	Average		Difference
	2004–2006	2017–2019	
Wheat (winter)	1848.2	1972.8	124.6
Wheat (spring)	386.7	467.3	80.6
Barley(winter)	143.0	205.7	62.7
Triticale (winter)	1032.7	1137.2	104.5
Triticale (spring)	114.9	181.0	66.1
Maize for grain	353.5	624.1	270.6
Maize for feed	317.4	599.2	281.8

Table 5. *Cont.*

Specification	Average		Difference
	2004–2006	2017–2019	
Rape and turnip rape (winter)	534.6	846.6	312.0
Total	4731.0	6033.9	1302.9
Rye	1427.6	890.3	−537.3
Barley (spring)	972.9	762.4	−210.5
Cereal mixed for grain (spring)	1412.6	861.8	−550.8
Potatoes	632.9	311.7	−321.2
Sugar beets	286.9	237.1	−49.8
Total	4732.9	3063.3	−1669.6

Source: Own study based on data from Eurostat (EU_Cereal_Balance—europa.eu (accessed on 18 October 2021) Pure biodiesels—https://ec.europa.eu/eurostat/web/energy/data/energy-balances (accessed on 18 October 2021).

4. Results and Discussion

4.1. Land Use Change

In the years 2004–2019, the sown area of oilseed crops in Poland increased from 564.8 thousand ha to 915.9 thousand ha. Rapeseed and winter oilseed rape accounted for the largest share in the structure of these crops, from 85.2% in 2016 to 95.9% in 2013, with the average for the whole period under study being 91.0% (Figure 2).

In the same period, the total sown area decreased from 11,285.4 thousand ha to 10,897.7 thousand ha. Apart from the decrease in the sown area, there were significant changes in its structure. Apart from rapeseed and colza, maize, wheat, and triticale areas increased to the greatest extent. These plant species were mainly introduced in place of spring cereal mixtures, rye, potatoes, and spring barley (Table 5). Similar trends were observed in most EU countries [100,109–111]. The main reason or this was the profitability of production [96–98,100,112].

In order to illustrate these changes in relation to oilseed rape and colza seed, cause-and-effect models were built and subjected to detailed verification, where the dependent variable was the area sown with oilseed rape and colza seed, and the independent variables were the areas of other crops, and these models were constructed using the following procedure:

- The model was estimated with all of the independent variables and then statistically insignificant and non-coincident variables were removed by a posteriori elimination method;
- The model was estimated using all of the independent variables as potential variables using the stepwise regression algorithm (assuming that the variable left in the model must be statistically significant at least at the level of $p < 0.05$) and following the rule of coincidence;
- The model with independent variables negatively correlated with the dependent variable was estimated, and then statistically insignificant and non-correlated variables were removed by a posteriori elimination method;
- The model was estimated by using only independent variables as the potential variables for winter crops, which were negatively correlated with the dependent variables, using the stepwise regression algorithm (assuming that the variable left in the model must be statistically significant at least at the level of $p < 0.05$) and following the rule of coincidence;
- The dependence model of the sown area of winter rape and colza (Y) and rye (X) was estimated with the use of an additional artificial zero-one variable (with value 1 for the periods when the variable Y had significantly lower values than those resulting

from the linear model; and 0—in the remaining periods). Both variables in the model were statistically significant at the $p < 0.0001$ level.

The obtained econometric models, whose parameters were estimated with the use of the Classic Least Squares Method, were subjected to further verification to assess their quality and the validity of their specification (e.g., tests of non-linearity, RESET specification, stability of QUSUM parameters, distribution of residuals). Finally, the selected models were characterised by the best values of the corrected coefficient of determination and the Akaike information index. No significant residual autocorrelation was found in the approximated models (LM test for autocorrelation of order 1). Due to a small number of observations, testing integration and the cointegration of the examined time series was abandoned. The only variable fulfilling these conditions was the winter rye sown area. The trend of decreasing the share of this crop in the sowing structure has persisted since the second half of the 1960s. Between 1965 and 2015 in Poland, the share of rye in the cereal sowing structure decreased from 52.8% to 9.7%. Initially, its place was taken by wheat and rye, and since 2004, its share has also been replaced by oilseed rape [113]. The introduction of oilseed rape to crops was regionally differentiated and depended on the share of good soil and the structural area of farms [114]. Stable oilseed rape yields can only be obtained in good and very good soils, which constitute about 50% of the arable land in Poland. Moreover, only larger farms can apply the correct technology needed for the production of the seeds of this plant. At present, over 70% of rape crops are grown in farm with over 50 ha of arable land.

4.2. Revenues of Operations

As a principle, the activities of agricultural producers aim to obtain the highest possible income from their activities. This is true outside of Poland as well, with the key factor being based on which farmers made decisions to increase the production of winter oilseed rape and colza due to its higher profitability in relation to most cereal crops, especially winter ones [96–98,100,101,112]. Table 6 compares the average incomes obtained from the production of winter oilseed rape and rapeseed as well as rye, triticale, and winter wheat in 2013–2019. These values were determined within the framework of the AGROKOSZTY and Polish FADN agricultural product data collection system conducted at the Institute of Agricultural Economics and Food Economics—National Research Institute in Warsaw, in cooperation with agricultural advisory centres. Over the entirety of the analysed period, the income obtained from the production of winter oilseed rape and colza was significantly higher than that of winter rye (by 59.7%) and triticale (by 29.4%) and was comparable to winter wheat [96–99,115,116].

Table 6. Income from production of winter rape and colza, rye, triticale, and wheat in 2013–2019 (PLN/ha).

Specification	2013	2014	2015	2016	2017	2018	2019	Average 2013–2019
Rape and turnip rape (winter)	2147	2225	2077	1508	1793	1918	2050	1960
Wheat (winter)	2177	2247	1982	1409	1945	1975	1739	1925
Rye (winter)	1273	1480	1252	1186	1446	1166	1137	1227
Triticale (winter)	1624	1724	1387	1411	1568	1452	1427	1513

Source: Own study on the basis of statistic belongings to the Institute of Agricultural and Food Economics—National Research Institute in Warsaw, Poland.

4.3. Pre-Crop Value

Apart from the financial benefits, oilseed rape cultivation is distinguished by a whole range of other favourable characteristics that are important for farms. The most important of these is its value as a forecrop, especially on farms specialising in cereal production. The cultivation of oilseed rape enables the effective interruption of the natural development cycle of cereal plant diseases and prevents the spread of weeds and pests. This makes the management of successive cereal crops easier, which helps to increase yields and reduce cultivation costs [100,101,103]. In addition, soil cover for 11 months of the year and a deep and extensive root system counteract erosion, improve soil aeration, and reduce nitrate leaching. The large amounts of biomass produced by oilseed rape both above and below the soil surface also contribute to the build-up of fertile humus [102].

In Poland, oilseed rape is mainly grown in simplified crop rotations (3-field rotations) after cereals, and it is most often the only crop that interrupts the succession of cereals. If oilseed rape is excluded from crop rotation, its place will be taken by cereals with greater economic value, mainly winter wheat, and that can be grown in weaker soils—triticale or rye. This situation will cause a deterioration in the value of the site for cereal cultivation and will generally result in lower yields. It is assumed that wheat yields are 15–20% higher in the stand after rape compared to pre-crops. Many years of research indicate that a negative stand for cereal cultivation cannot be fully compensated by increased fertilisation or higher doses of plant protection products [117]. The effect of lower yield of cereals under the conditions of the increased cereal shares in the sowing structure should be associated with the deterioration of phytosanitary conditions (increased intensity of diseases of the stem base and root system), weed infestation in the field (including possible compensation of noxious weeds), and the accumulation of toxic phenolic compounds in the soil [99–104].

The expanding cultivation area is also a factor stimulating the yield level of wheat, which both in Poland and in the world is traditionally sown in the position after rape [118]. The important significance of oilseed rape as a forecrop for cereals also results from its favourable effect on the soil environment under cultivation conditions, especially in terms of long monoculture sequences of monocotyledonous vegetation [119]. The attractiveness of winter oilseed rape as a forecrop is not only due to the rapid decomposition of crop residues (narrow C:N ratio) but is also due to their biofumigant effects [120]. Manfred Schoepe [96] estimated the value of a post-rape stand at 130 EUR/ha. In the presented paper, these values for wheat, triticale, and rye were set being equivalent to 11% of the yield (Table 7). Such an assumption was based on the results of research conducted in IUNG PIB [102,117,121,122] and in the literature [96,100,101,103,104].

Table 7. Average yields and prices of grain wheat, triticale, rye, and pre-crop value after rape in 2013–2019.

Specification	Yield Mg/ha	Prices		Pre-Crop Value	
		PLN/Mg	EUR/Mg	PLN	EUR
Wheat (winter)	6.3	654.2	149.7	453.4	103.7
Rye (winter)	3.5	530.3	121.4	204.2	46.7
Triticale (winter)	4.8	534.4	122.3	282.2	64.6

Source: Own study on the basis of statistic belongings of Institute of Agricultural and Food Economics—National Research Institute in Warsaw, Poland.

4.4. *Profits from Beekeeping*

Beekeeping is a very important part of the bioeconomy. However, the literature is dominated by studies on the ecosystem services provided by pollinators. According to estimates made by Launtenbach and associates [123], the global value of pollinator services in 2009 was EUR 265 billion. In Europe, the largest benefits were obtained in Italy, Greece, Spain, France, the UK, Germany, the Netherlands, Switzerland, Austria, Poland, Romania, and Hungary. The latest published estimates on the value of the ecosystem service provided to the human economy by pollinators, mainly by honeybees, puts this work at between USD 235 and 577 billion. These values may vary depending on the assessment method used and the inflation levels that are assumed. It is worth noting that successive evaluations of pollination benefits to the food economy become higher and higher [124,125]. In Poland, the economic value of bees as pollinators of crop plants alone was estimated to be around EUR 2.0 billion in 2015 [126].

Agriculture, however, mainly through so-called melliferous plants, can contribute to the development of apiary management. The beekeeping value of a given plant species is mainly determined by the time and abundance of flowering as well as by the abundance of nectar and pollen. Winter oilseed rape is an excellent source of honey in the first two decades of May and is classified as a commodity crop, i.e., one from which significant quantities of honey can be obtained. The flowering period of oilseed rape lasts, depending on weather conditions, from 15 to 20 days, during which flowering plants provide insects with approximately 90–120 kg of sugars and 115–160 kg of pollen from 1 hectare of crops [127–129]. The high beekeeping value of oilseed rape is evidenced by the intensity of its flight by pollinating insects, reaching up to 5–6 individuals per 1 m^2 of the flowering canopy in the peak insect flight hours in good weather, among which the honeybee constitutes approximately 90% of all of the insects found on flowers [130]. The value of net profit of beekeeping (the calculation as food fields for apiculture) from one hectare of oilseed rape cultivation was determined on the basis of research conducted at the Apiculture Division in Pulawy of to The National Institute of Horticultural Research, at the level of 55 EUR/ha. This amount is similar to that estimated by the Institute for Economic Research at the University Munich [96].

5. Conclusions

In Poland, rapeseed production has been the fastest growing branch of plant production since the year 2000. Rapeseed yields have increased 2.5 times in the last 20 years. The main reason for this trend was the implementation of obligations resulting from legal acts by Member States relating to increasing the share of RES in the structure of primary energy production and to the share of biofuels in fuels used in transport in particular. In the White Paper, which was entitled "Energy for the Future: renewable sources of energy", prepared by the European Commission in 1997, it was indicated that the fulfilment of these intentions would take place through the increased production of first-generation biofuels, mainly biodiesel produced from rapeseed.

In Poland, in the years 2010–2020, about 1.0–1.6 million tons of rapeseed was used for this purpose annually. Such utilization had an impact on the increase in agricultural incomes, contributed to the decrease in income disparity, and increased the chances of gaining equal—with respect to urban residents—access to goods and services. Moreover, an increase in the demand for agricultural raw materials for biofuel production created an opportunity to abolish the demand barrier that hampers the development of agriculture. Another important benefit connected to the development of the liquid biofuel sector is the processing of oilseed, thanks to which the country obtains considerable quantities of high-protein post-extraction meal, which is an important component of feedstuffs. This makes it possible to limit imports of high-protein feedstuffs, mainly soya meal, including that produced from genetically modified seeds.

Due to the fact that biofuel production competes for raw materials with the food economy, at the end of the first decade of the 21st century, many called for withdrawal from the policy supporting the biofuel sector. Its implementation was to lead, inter alia, to changes in land use, mainly in the reduction of the area comprising forests and land with natural values. The research conducted here shows that in Poland in the period 2000–2020, the opposite trend occurred. The area of forest land increased from 9.1 to 9.6 million hectares, including increases in the area taken up by forests from 8.9 to 9.3 million hectares, and the sown area decreased from 12.4 to 10.8 million hectares despite a significant increase in rape sowing from 437 to 864 thousand hectares. The introduction of changes in the present EU biofuel policy may result in a significant reduction in the area where oilseed rape is sown and thus in a reduction in the income generated from its production.

Taking into account the factors determining the cultivation of oilseed rape: soil quality, the share in the sowing structure of farms, the area structure of farms, and regionalisation related to the risk of crop freezing, it can be assumed that the growth and production of oilseed rape will be abandoned first by farms that produce the crop on land that is less suitable for oilseed rape production, e.g., medium soils (complex 5) and some good soils (complexes 8, 11), as well as smaller farms. Only in good and very good soils, which in Poland constitute around 50%, and on larger farms (over 50 ha) can a smaller reduction in rape growing area be expected. Rape will be replaced in the sowing structure by rye, triticale, and, in good soils, wheat. Compared to oilseed rape, their production is characterised by lower income per hectare; in 2013–2019 these differences amounted to EUR 8 for wheat, EUR 102.3 for triticale, and EUR 168 for rye.

The expanding area of rape cultivation is a factor stimulating the yield level of other plants, mainly wheat, which both in Poland and worldwide is traditionally sown in the position after rape. The significant importance of oilseed rape as a forecrop for cereal crops results from its favourable impact on the soil environment in terms of cultivation conditions and long monoculture sequences of monocotyledonous vegetation. At present, oilseed rape is mainly grown in simplified rotations (3-field) after cereals, and it is usually the only plant that is able to interpret the succession of cereals. If oilseed rape is removed from the rotation, cereals will take its place. This situation will cause the value of the growing area used to grow cereals to decrease and thus a decrease in the yield of those cereals. On the basis of the conducted research, the estimated value of oilseed rape as a fore crop for wheat, triticale, and rye was EUR 103.7, 64.6, and 46.7, respectively. An additional advantage of oilseed rape is that it is an excellent bee resource and is classified as a commodity crop, i.e., one from which significant amounts of honey can be obtained, with a net value of EUR 55 per hectare.

In addition, in many agricultural holdings, as a result of the forecasted changes in crop production, there will be an accumulation in field work during the harvest period, which will also affect the worse use of machinery and storage areas. The consequence of increasing the acreage of cereal cultivation and its supply may be worse production profitability and thus the income situation of farms, but this will be assessed at the next stage of research.

Author Contributions: Conceptualization, P.G.; methodology, P.G. and K.J.; software, P.G. and K.J.; validation, P.G. and B.G.; formal analysis, M.W. and P.G.; investigation, P.G. and A.T.; resources, P.G.; data duration, P.G. and B.G.; writing—original draft, P.G., K.J., M.W., A.W., B.G. and A.T.; writing—review and editing, K.J., M.W., B.G., P.G., A.T. and A.W.; supervision, P.G., K.J. and M.W.; funding acquisition, P.G. and A.W. All authors have read and agreed to the published version of the manuscript.

Funding: This research received no external funding.

Institutional Review Board Statement: Not applicable.

Informed Consent Statement: Not applicable.

Data Availability Statement: Not applicable.

Conflicts of Interest: The authors declare no conflict of interest.

References

1. Songstad, D.D.; Lakshmanan, P.; Chen, J.; Gibbons, W.; Hughes, S.; Nelson, R. Historical perspective of biofuels: Learning from the past to rediscover the future. *In Vitro Cell. Dev. Biol. Plant* **2009**, *45*, 189–192. [CrossRef]
2. Ayadi, M.; Sarma, S.J.; Pachapur, V.L.; Brar, S.K.; Cheikh, R.B. History and Global Policy of Biofuels. In *Green Fuels Technology*; Soccol, C.R., Brar, S.K., Faulds, C., Ramos, L.P., Eds.; Springer International Publishing: Cham, Switzerland, 2016; pp. 1–14. [CrossRef]
3. Singh, R.S.; Pandey, A.; Gnansounou, E. *Biofuels: Production and Future Perspectives*; CRC Press, Taylor & Francis Group: Boca Raton, FL, USA, 2016; ISBN 978-149-872-359-6.
4. Knothe, G. The History of Vegetable Oil-Based Diesel Fuels. In *The Biodiesel Handbook*, 2nd ed.; Knothe, G., Krahl, J., Gerpen, J.V., Eds.; AOCS Press: Urbana, IL, USA, 2010; pp. 5–19. [CrossRef]

5. Ma, F.; Hanna, M.A. Biodiesel production: A review. *Bioresour. Technol.* **1999**, *70*, 1–15. [CrossRef]
6. Ghobadian, B.; Rahimi, H.; Nikbakht, A.M.; Najafi, G.; Yusaf, T.F. Diesel engine performance and exhaust emission analysis using waste cooking biodiesel fuel with an artificial neural network. *Renew. Energy* **2009**, *34*, 976–982. [CrossRef]
7. Carolan, M.S. A Sociological Look at Biofuels: Ethanol in the Early Decades of the Twentieth Century and Lessons for Today. *Rural. Sociol.* **2009**, *74*, 86–112. [CrossRef]
8. Solomon, B.D.; Barnes, J.R.; Halvorsen, K.E. Grain and cellulosic ethanol: History, economics, and energy Policy. *Biomass Bioenergy* **2007**, *31*, 416–425. [CrossRef]
9. Balat, M.; Balat, H.; Öz, C. Progress in bioethanol processing. *Prog. Energy Combust. Sci.* **2008**, *34*, 551–573. [CrossRef]
10. Rosillo-Calle, F.; Walter, A. Global market for bioethanol: Historical trends and future prospects. *Energy Sustain. Dev.* **2006**, *10*, 20–32. [CrossRef]
11. Kovarik, B. Henry Ford, Charles, F. Kettering and the Fuel of the Future. *Automot. Hist. Rev. Spring* **1998**, *32*, 7–27. Available online: https://environmentalhistory.org/people/henry-ford-charles-kettering-and-the-fuel-of-the-future/ (accessed on 14 November 2020).
12. Dimitri, C.; Effland, A. Fueling the automobile: An economic exploration of early adoption of gasoline over ethanol. *J. Agric. Food Ind. Organ.* **2007**, *5*, 1–21. [CrossRef]
13. Yergin, D. *The Prize: The Epic Quest for Oil Money and Power*; Simon and Schuster: New York, NY, USA, 1991; ISBN 0-671-50248-4.
14. BP. *Statistical Review of World Energy*, 69th ed.; 2020. Available online: https://www.bp.com/content/dam/bp/business-sites/en/global/corporate/pdfs/energy-economics/statistical-review/bp-stats-review-2020-full-report.pdf (accessed on 16 November 2020).
15. Toffler, A. *The Third Wave*; Bantam Books: New York, NY, USA, 1980; ISBN 0-553-24698-4.
16. Energy Information Agency, U.S. Department of Energy, Washington. International Energy Outlook 2003. Available online: https://www.hsdl.org/?view&did=15902 (accessed on 14 November 2020).
17. International Energy Agency. Statistics Report. Key World Energy Statistics 2020. Available online: https://webstore.iea.org/download/direct/4093?fileName=Key_World_Energy_Statistics_2020.pdf (accessed on 18 November 2020).
18. Soddy, F. *Wealth, Virtual Wealth and Debt. The Solution of the Economic Paradox*; Britons Publishing Company: London, UK, 1933; Available online: https://www.fadedpage.com/link.php?file=20140873-a5.pdf (accessed on 16 November 2020).
19. Trenn, T.J. The central role of energy in soddy's holistic and critical approach to nuclear science, economics, and social responsibility. *Br. J. Hist. Sci.* **1979**, *12*, 261–276. [CrossRef]
20. Csereklyei, Z.; Rubio-Varas, M.d.M.; Stern, D.I. Energy and economic growth: The stylized facts. *Energy J.* **2016**, *37*, 223–255. [CrossRef]
21. Stern, D.I.; Kander, A. The role of energy in the industrial revolution and modern economic growth. *Energy J.* **2012**, *33*, 125–152. [CrossRef]
22. Jevons, W.S. *The Coal Question: An Inquiry Concerning the Progress of the Nation, and the Probable Exhaustion of Our Coal-Mines*; Macmillan & Co.: London, UK, 1865. Available online: https://books.google.pl/books?id=gAAKAAAAIAAJ&dq=editions:AAotKDT6KKcC&pg=PR3&redir_esc=y&hl=pl#v=onepage&q&f=false (accessed on 20 November 2020).
23. Dasgupta, P.; Heal, G. The optimal depletion of exhaustible resources. *Rev. Econ. Stud.* **1974**, *41*, 3–28. [CrossRef]
24. Solow, R.M. Intergenerational equity and exhaustible resources. *Rev. Econ. Stud.* **1974**, *41*, 29–45. [CrossRef]
25. Stiglitz, J. Growth with exhaustible natural resources: Efficient and optimal growth paths. *Rev. Econ. Stud.* **1974**, *41*, 123–137. [CrossRef]
26. Slade, M. Trends in natural-resource commodity prices: An analysis of the time domain. *J. Environ. Econ. Manag.* **1982**, *9*, 122–137. [CrossRef]
27. Stern, D.I. Limits to substitution and irreversibility in production and consumption: A neoclassical interpretation of ecological economics. *Ecol. Econ.* **1997**, *21*, 197–215. [CrossRef]
28. Ghasemian, S.; Faridzad, A.; Abbaszadeh, P.; Taklif, A.; Ghasemi, A.; Hafezi, R. An overview of global energy scenarios by 2040: Identifying the driving forces using cross-impact analysis method. *Int. J. Environ. Sci. Technol.* **2020**, 1–24. [CrossRef]
29. U.S. Energy Information Agency. International Energy Outlook 2020 with Projections to 2050. 2020. Available online: https://www.eia.gov/outlooks/aeo/pdf/aeo2020.pdf (accessed on 19 November 2020).
30. International Renewable Energy Agency. Global Energy Transformation. A Roadmap to 2050. 2018. Available online: https://www.irena.org/-/media/Files/IRENA/Agency/Publication/2018/Apr/IRENA_Report_GET_2018.pdf (accessed on 18 November 2020).
31. Marks-Bielska, R.; Bielski, S.; Pik, K.; Kurowska, K. The importance of renewable energy sources in Poland's energy mix. *Energies* **2020**, *13*, 4624. [CrossRef]
32. Hansen, J.; Sato, M.; Ruedy, R.; Lo, K.; Lea, D.W.; Medina-Elizade, M. Global temperature change. *Proc. Natl. Acad. Sci. USA* **2006**, *103*, 14288–14293. [CrossRef] [PubMed]
33. Cambridge University Press. Stern Review: The Economics of Climate Change. Available online: http://mudancasclimaticas.cptec.inpe.br/~{}rmclima/pdfs/destaques/sternreview_report_complete.pdf (accessed on 10 November 2020).
34. Schipper, L. Conceptual history of adaptation in the UNFCCC process. *Rev. Eur. Community Int. Environ. Law* **2006**, *15*, 82–92. [CrossRef]

35. United Nations. Kyoto Protocol to the United Nations Framework Convention on Climate Change. Available online: https://unfccc.int/resource/docs/convkp/kpeng.pdf (accessed on 18 November 2020).
36. Report of the Conference of the Parties Serving as the Meeting of the Parties to the Kyoto Protocol on Its Eighth Session, Held in Doha from 26 November to 8 December 2012. Available online: https://unfccc.int/resource/docs/2012/cmp8/eng/13a01.pdf (accessed on 18 November 2020).
37. Blusz, K.; Hinc, A.; Brodzikowski, J. *W Kierunku Niskoemisyjnej Strategii Gospodarczej dla Polski. Energia i Klimat Pomiędzy Keynesem i Hayekiem?* DemosEUROPA—Centrum Strategii Europejskiej: Warszawa, Poland, 2011.
38. United Nations. Adoption of the Paris Agreement. Available online: https://unfccc.int/resource/docs/2015/cop21/eng/l09r01.pdf (accessed on 18 November 2020).
39. Kulovesi, K.; Oberthür, S. Assessing the EU's 2030 Climate and Energy Policy Framework: Incremental change toward radical transformation? *Rev. Eur. Comp. Int. Environ. Law* **2020**, *29*, 151–166. [CrossRef]
40. European Commission. Communication from the Commission to the European Parliament, the European Council, the Council, the European Economic and Social Committee and the Committee of the Regions: The European Green Deal. Available online: https://ec.europa.eu/info/sites/info/files/european-green-deal-communication_en.pdf (accessed on 10 November 2020).
41. Bertram, R.B. New crops and the international agricultural research centers. In *New Crops*; Janick, J., Simon, J.E., Eds.; Wiley: New York, NY, USA, 1993; pp. 11–22. Available online: https://hort.purdue.edu/newcrop/proceedings1993/V2-011.html (accessed on 10 November 2020).
42. Rexen, F.; Munck, L.; Andersen, F. *Cereal Crops for Industrial Use in Europe. Report Prepared for: The Commission of the European Communities*; EEC and the Carlsberg Research Laboratory: Copenhagen, Denmark, 1984.
43. Mangan, C.; Kerckow, B. *Practical Information and Programmes-Eclair-Agro-Industrial Research-Mid-Term Progress Reports-1988–1993*; Commission of the European Communities Directorate-General XII Science-Research-Development, European Commission: Luxembourg, 1994.
44. European Parliament; Directorate General Internal Policies of the Union; Policy Department Structural and Cohesion Policies, Agriculture and Rural Development. The Promotion of Non-Food Crops. 2005. Available online: https://www.europarl.europa.eu/meetdocs/2004_2009/documents/dv/studynon-foodcrops_/studynon-foodcrops_%20en.pdf (accessed on 12 November 2020).
45. Hall, D.O.; Overend, R.P. *Biomass: Regenerable Energy*; Wiley and Sons: Hoboken, NJ, USA, 1987; ISBN 0-471-90919-X.
46. Gradziuk, P.; Wojtaszek, Z. Alternatywne wykorzystanie gruntów rolniczych na cele niezwiązane z produkcją żywności. In *Procesy Dostosowawcze Produkcji Roślinnej w Polsce w Kontekście Integracji z Unią Europejską*; Klepacki, B., Ed.; SGGW: Warsaw, Poland, 2002; pp. 213–228. ISBN 83-88368-19-2.
47. Su, Y.; Zhang, P.; Su, Y. An overview of biofuels policies and industrialization in the major biofuel producing countries. *Renew. Sustain. Energy Rev.* **2015**, *50*, 991–1003. [CrossRef]
48. Finlay, M.R. Old efforts at new uses: A brief history of Chemurgy and the American search for biobased materials. *J. Ind. Ecol.* **2003**, *7*, 33–46. [CrossRef]
49. European Commission. Communication from the Commission. Energy for the Future: Renewable Sources of Energy. White Paper for a Community Strategy and Action Plan, COM(97)599 Final, Brussels. 1997. Available online: https://europa.eu/documents/comm/white_papers/pdf/com97_599_en.pdf (accessed on 10 May 2021).
50. Buyx, A.; Tait, J. Biofuels: Ethics and policy-making. *Biofuels Bioprod. Biorefin.* **2011**, *5*, 631–639. [CrossRef]
51. Thompson, P.B. The agricultural ethics of biofuels: The food vs. fuel debate. *Agriculture* **2012**, *2*, 339–358. [CrossRef]
52. Gomiero, T.; Paoletti, M.; Pimentel, D. Biofuels: Efficiency, ethics, and limits to human appropriation of ecosystem services. *J. Agric. Environ. Ethics* **2009**, *23*, 403–434. [CrossRef]
53. Mohr, A.; Raman, S. Lessons from first generation biofuels and implications for the sustainability appraisal of second generation biofuels. *Energy Policy* **2013**, *63*, 114–122. [CrossRef]
54. Huang, J.; Yang, J.; Msangi, S.; Rozelle, S.; Weersink, A. Global biofuel production and poverty in China. *Appl. Energy* **2012**, *98*, 246–255. [CrossRef]
55. Timilsina, G.; Shrestha, A. An overview of global markets and policies. In *The Impacts of Biofuels on the Economy, Environment and Poverty*; Springer Science and Business Media LLC: Berlin/Heidelberg, Germany, 2014; pp. 1–14. [CrossRef]
56. Gomiero, T. Are biofuels an effective and viable energy strategy for industrialized societies? A reasoned overview of potentials and limits. *Sustainability* **2015**, *7*, 8491–8521. [CrossRef]
57. Mol, A.P. Environmental authorities and biofuel controversies. *Environ. Polit.* **2010**, *19*, 61–79. [CrossRef]
58. Selbmann, K.; Pforte, L. Evaluation of Ecological Criteria of Biofuel Certification in Germany. *Sustainability* **2016**, *8*, 936. [CrossRef]
59. Lane, J. Letter Sent to Dutch Cabinet by 178 Dutch Scientists Urges End to Food-Based Biofuels. Available online: http://www.biofuelsdigest.com/bdigest/2017/12/04/letter-sent-to-dutch-cabinet-by-178-dutch-scientists-urges-end-to-food-based-biofuels/ (accessed on 23 November 2020).
60. Cao, Y.; Pawlowski, L. Effect of Biofuels on Environment and Sustainable Development. *Ecol. Chem. Eng.* **2013**, *20*, 799–804. [CrossRef]
61. Copa Cogeca. Phasing Out First Generation Biofuels: What Is at Stake? Available online: http://copa-cogeca.eu/Download.ashx?ID=1590445 (accessed on 22 November 2020).

62. Falck-Zepeda, J.; Msangi, S.; Sulser, T.; Zambrano, P.; Falconi, C. *Biofuels and Rural Economic Development in Latin America and the Caribbean*; Cooperation Programme FAO/Inter-American Development Bank Latin America and the Caribbean Service Investment Centre Division: Washington, DC, USA, 2010. Available online: https://publications.iadb.org/publications/english/document/Biofuels-and-Rural-EconomicDevelopment-in-Latin-America-and-the-Caribbean.pdf (accessed on 20 November 2020).
63. USDA Foreign Agricultural Service; Global Agricultural Information Network. EU Biofuels Annual 2017. GAIN Report No. NL7015. Available online: https://apps.fas.usda.gov/newgainapi/api/report/downloadreportbyfilename?filename=Biofuels%20Annual_The%20Hague_EU-28_6-19-2017.pdf (accessed on 24 November 2020).
64. Stattman, S.L.; Gupta, A.; Partzsch, L.; Oosterveer, P. Toward sustainable biofuels in the European Union? Lessons from a decade of hybrid biofuel governance. *Sustainability* **2018**, *10*, 4111. [CrossRef]
65. Directive (EU) 2015/1513 of the European Parliament and of the Council of 9 September 2015 Amending Directive 98/70/EC Relating to the Quality of Petrol and Diesel Fuels and Amending Directive 2009/28/EC on the Promotion of the Use of Energy from Renewable Sources. 2015. Available online: https://eur-lex.europa.eu/legal-content/EN/TXT/PDF/?uri=CELEX:32015L1513&from=EN (accessed on 20 November 2020).
66. Directive (EU) 2018/2001 of the European Parliament and of the Council of 11 December 2018 on the Promotion of the Use of Energy from Renewable Sources (Recast). 2018. Available online: https://eur-lex.europa.eu/legal-content/EN/TXT/PDF/?uri=CELEX:32018L2001&from=EN (accessed on 21 November 2020).
67. Gradziuk, P.; Grzybek, A.; Kowalczyk, K.; Kościk, B. *Biopaliwa*; Wydawnictwo Wieś Jutra: Warsaw, Poland, 2002; ISBN 83-7160-217-0.
68. Guo, M.; Song, W.; Buhain, J. Bioenergy and biofuels: History, status, and perspective. *Renew. Sustain. Energy Rev.* **2015**, *42*, 712–725. [CrossRef]
69. Giampietro, M.; Ulgiati, S.; Pimentel, D. Feasibility of large-scale biofuel production—Does an enlargement of scale change the picture? *BioScience* **1997**, *47*, 587–600. [CrossRef]
70. Demirbas, A. Biofuels sources, biofuel policy, biofuel economy and global biofuel projections. *Energy Convers. Manag.* **2008**, *49*, 2106–2116. [CrossRef]
71. International Energy Agency (IEA). Technology Roadmap: Biofuels for Transport. Available online: https://webstore.iea.org/download/direct/632 (accessed on 27 November 2020).
72. United Nations Conference on Trade and Development. Biofuel Production Technologies: Status, Prospects and Implications for Trade and Development. Available online: https://unctad.org/system/files/official-document/ditcted200710_en.pdf (accessed on 21 November 2020).
73. Stafford, W.; Lotter, A.; Brent, A.; von Maltitz, G. WIDER Working Paper 2017/87. The United Nations University World Institute for Development Economics Research. Biofuels Technology. Available online: https://www.wider.unu.edu/sites/default/files/wp2017-87.pdf (accessed on 21 November 2020).
74. UFOP Union zur Förderung von Oel- und Proteinpflanzen E.V. UFOP Report on Global Market Supply 2018/2019. Available online: https://www.ufop.de/files/4815/4695/8891/WEB_UFOP_Report_on_Global_Market_Supply_18-19.pdf (accessed on 27 November 2020).
75. Aron, N.S.M.; Khoo, K.S.; Chew, K.W.; Show, P.L.; Chen, W.-H.; Nguyen, T.H.P. Sustainability of the four generations of biofuels—A review. *Int. J. Energy Res.* **2020**, *44*, 1–17. [CrossRef]
76. International Renewable Energy Agency (IRENA). Advanced Biofuels. What Holds Them Back? Available online: https://www.irena.org/-/media/Files/IRENA/Agency/Publication/2019/Nov/IRENA_Advanced-biofuels_2019.pdf (accessed on 21 November 2020).
77. Oh, Y.-K.; Hwang, K.-R.; Kim, C.; Kim, J.R.; Lee, J.-S. Recent developments and key barriers to advanced biofuels: A short review. *Bioresour. Technol.* **2018**, *257*, 320–333. [CrossRef]
78. IEA Bioenergy; Brown, A.; Waldheim, L.; Landälv, I.; Saddler, J.; Ebadian, M.; McMillan, J.D.; Bonomi, A.; Klein, B. Advanced Biofuels—Potential for Cost Reduction. Available online: http://task39.sites.olt.ubc.ca/files/2020/02/Advanced-Biofuels-Potential-for-Cost-Reduction-Final-Draft.pdf (accessed on 26 November 2020).
79. Commission of the European Communities. Proposal for a Council Regulation (EEC) on the Implementation in the Solar-Energy Sector of Council Regulation (EEC) No 1302/78 of 12 June 1978 on the Granting of Financial Support for Projects to Exploit Alternative Energy Sources. Available online: http://aei.pitt.edu/32677/1/COM_(78)_511_final_1.pdf (accessed on 26 November 2020).
80. Council Resolution of 9 June 1980 Concerning Community Energy Policy Objectives for 1990 and Convergence of the Policies of the Member States. Available online: https://eur-lex.europa.eu/legal-content/EN/TXT/PDF/?uri=CELEX:31980Y0618(01)&from=EN (accessed on 4 June 2021).
81. Opinion on a Community Orientation to Develop New and Renewable Energy Sources (') (86/C 207/05). Available online: https://eur-lex.europa.eu/legal-content/EN/TXT/PDF/?uri=CELEX:51986AC0504&from=EN (accessed on 26 November 2020).
82. Council Resolution of 16 September 1986 Concerning New Community Energy Policy Objectives for 1995 and Convergence of the Policies of the Member States. Available online: https://eur-lex.europa.eu/legal-content/EN/TXT/PDF/?uri=CELEX:31986Y0925(01)&qid=1606834484404&from=EN (accessed on 26 November 2020).

83. Commission of the European Communities. Communication from the Commission. Energy for the Future: Renewable Sources of Energy. Green Paper for a Community Strategy and Action Plan. Available online: http://aei.pitt.edu/1280/1/renewalbe_energy_gp_COM_96_576.pdf (accessed on 29 November 2020).
84. Solorio, I.; Jörgens, H. (Eds.) *A Guide to EU Renewable Energy Policy: Comparing Europeanization and Domestic Policy Change*; Edvard Elgar: Cheltenham, UK, 2017; ISBN 978 1 78347 155 3.
85. Directive 2003/30/EC of the European Parliament and of the Council of 8 May 2003 on the Promotion of the Use of Biofuels or Other Renewable Fuels for Transport. Available online: https://eur-lex.europa.eu/legal-content/EN/TXT/PDF/?uri=CELEX:32003L0030&from=en (accessed on 29 November 2020).
86. Communication from the Commission-Biomass Action Plan {SEC(2005) 1573}. Available online: https://eur-lex.europa.eu/legal-content/EN/TXT/PDF/?uri=CELEX:52005DC0628&from=EN (accessed on 29 November 2020).
87. Commission of the European Communities. Communication from the Commission An EU Strategy for Biofuels {SEC(2006) aaaa}. Available online: http://www.cres.gr/biodiesel/downloads/EU%20Strategy%20for%20biofuels.pdf (accessed on 29 November 2020).
88. Commission of the European Communities. Communication from the Commission to the Council and the European Parliament. Renewable Energy Road Map Renewable Energies in the 21st Century: Building a More Sustainable Future. Available online: https://eur-lex.europa.eu/LexUriServ/LexUriServ.do?uri=COM:2006:0848:FIN:EN:PDF (accessed on 29 November 2020).
89. Directive 2009/28/EC of the European Parliament and of the Council of 23 April 2009 on the Promotion of the Use of Energy from Renewable Sources and Amending and Subsequently Repealing Directives 2001/77/EC and 2003/30/EC (Text with EEA Relevance). Available online: https://eur-lex.europa.eu/legal-content/EN/TXT/PDF/?uri=CELEX:32009L0028&from=EN (accessed on 29 November 2020).
90. Directive 2009/30/EC of the European Parliament and of the Council of 23 April 2009 Amending Directive 98/70/EC as Regards the Specification of Petrol, Diesel and Gas-Oil and Introducing a Mechanism to Monitor and Reduce Greenhouse Gas Emissions and Amending Council Directive 1999/32/EC as Regards the Specification of Fuel Used by Inland Waterway Vessels and Repealing Directive 93/12/EEC (Text with EEA Relevance). Available online: https://eur-lex.europa.eu/legal-content/EN/TXT/PDF/?uri=CELEX:32009L0030&from=EN (accessed on 29 November 2020).
91. Lonza, L.; Deix, S.; Maas, H.; Caiado, C.; Hamje, H.D.C.; Reid, A. *EU Renewable Energy Targets in 2020: 2015 Legislative Update*; Publications Office of the European Union: Luxembourg, 2016. [CrossRef]
92. Ministry of Agriculture and Rural Development of Poland; Ministry of Environment of Poland. *Kodeks Dobrej Praktyki Rolniczej*; Foundation of Agricultural Assistance Programs: Warsaw, Poland, 2004; p. 98. Available online: http://iung.pl/dpr_eng/publikacje/kodeks_dobrej_praktyki_rolniczej.pdf (accessed on 30 November 2020).
93. USDA Foreign Agricultural Service; Global Agricultural Information Network. EU Biofuels Annual 2020. GAIN Report No. E42020-0032. Available online: https://apps.fas.usda.gov/newgainapi/api/Report/DownloadReportByFileName?fileName=Biofuels%20Annual_The%20Hague_European%20Union_06-29-2020 (accessed on 30 November 2020).
94. International Energy Agency. Renewables 2019 Analysis and Forecast to 2024. Available online: https://webstore.iea.org/download/direct/2854?fileName=Renewables_2019.pdf (accessed on 30 November 2020).
95. Official Journal of the European Communities. Court of Auditors Special Report 6/2002. Available online: https://eur-lex.europa.eu/LexUriServ/LexUriServ.do?uri=OJ:C:2002:254:0001:0022:EN:PDF (accessed on 30 November 2020).
96. Schoepe, M. Economic Effects of the Production of Biodiesel for Use as Fuel. Ifo Institute for Economic Research at the University Munich, Final Report Ifo Schnelldienst. Manuscript 59(17). 2006, pp. 1–10. Available online: http://www.globalbioenergy.org/uploads/media/0611_IFO_-_Economic_effects_of_the_production_of_biodiesel_for_use_as_fuel.pdf (accessed on 30 November 2020).
97. Jabłoński, K.; Skarżyńska, A.; Abramczuk, Ł. Determinants of income from wheat and rape production in projection for 2020 in Poland. *Žemės Ūkio Moksl.* **2015**, *22*, 229–239. [CrossRef]
98. Blanco Fonseca, M.; Burrell, A.; Gay, S.H.; Henseler, M.; Kavallari, A.; M'Barek, R.; Pérez Domínguez, I.; Tonini, A. *Impacts of the EU Biofuel Target on Agricultural Markets and Land Use: A Comparative Modelling Assessment*; European Commission, Joint Research Centre, Institute for Prospective Technological Studies: Luxembourg, 2010; ISBN 978-92-79-16310-4. [CrossRef]
99. Mazzilli, S.R.; Ernst, O.R. Rapeseed-to-wheat yield ratio in different production environments and effects on subsequent summer crops yields. *Agrosyst. Geosci. Environ.* **2019**, *2*, 190017. [CrossRef]
100. Petersen, J.; Haastrup, M.; Knudsen, L.; Olesen, J.E. *Causes of Yield Stagnation in Winter Wheat in Denmark. DJF Report PLANT SCIENCE NO. 147*; Faculty of Agricultural Sciences Aarhus University: Tjele, Denmark, 2010; ISBN 87-91949-46-7. Available online: https://dcapub.au.dk/djfpublikation/djfpdf/DJF-Rap.Mark147-internet.pdf (accessed on 29 November 2020).
101. Sieling, K.; Christen, O. Crop rotation effects on yield of oilseed rape, wheat and barley and residual effects on the subsequent wheat. *Arch. Agron. Soil Sci.* **2015**, *61*, 1531–1549. [CrossRef]
102. Kuś, J. Znaczenie płodozmianu we współczesnym rolnictwie. *Studia Rap. IUNG-PIB* **2015**, *43*, 65–87. [CrossRef]
103. Angus, J.F.; Kirkegaard, J.A.; Hunt, J.R.; Ryan, M.H.; Ohlander, L.; Peoples, M.B. Break crops and rotations for wheat. *Crop Pasture Sci.* **2015**, *66*, 523–552. [CrossRef]
104. Darguza, M.; Gaile, Z. Yield and quality of winter wheat, depending on crop rotation and soil tillage. *Res. Rural. Dev.* **2019**, *2*, 29–35. [CrossRef]
105. Kovács, A. The role of beekeeping in production of oil crops. *Appl. Stud. Agribus. Commer.* **2013**, *7*, 77–82. [CrossRef]

106. Perrot, T.; Gaba, S.; Roncoroni, M.; Gautier, J.-L.; Bretagnolle, V. Bees increase oilseed rape yield under real field conditions. *Agric. Ecosyst. Environ.* **2018**, *266*, 39–48. [CrossRef]
107. Ramos, M.; Soares Dias, A.P.; Puna, J.F.; Gomes, J.; Bordado, J.C. Biodiesel Production Processes and Sustainable Raw Materials. *Energies* **2019**, *12*, 4408. [CrossRef]
108. Gradziuk, P.; Gradziuk, B. Employment impacts of renewable energy in UE. Proceedings of the International Scientific Conference. *Econ. Sci. Agribus. Rural Econ.* **2018**, *1*, 259–267. [CrossRef]
109. Donfouet, H.P.P.; Barczak, A.; Détang-Dessendre, C.; Maigné, E. Crop production and crop diversity in France: A Spatial analysis. *Ecol. Econ.* **2017**, *134*, 29–39. [CrossRef]
110. Schils, R.; Olesen, J.E.; Kersebaum, K.-C.; Rijka, B.; Oberforsterd, M.; Kalyadae, V.; Khitrykaue, M.; Gobinf, A.; Kirchevg, H.; Manolovag, V.; et al. Cereal yield gaps across Europe. *Eur. J. Agron.* **2018**, *101*, 109–120. [CrossRef]
111. Nowak, A. Differentiation in cereal production among Member States of the European Union. *Pol. J. Agron.* **2020**, *40*, 7–15. [CrossRef]
112. Stuczyński, T.; Łopatka, A.; Faber, A.; Czaban, P.; Kowalik, M.; Koza, P.; Korzeniowska-Pucułek, R.; Siebielec, G. Prognoza wykorzystania przestrzeni rolniczej dla produkcji roślin na cele energetyczne. *Studia Rap. IUNG-PIB* **2008**, *11*, 25–42. [CrossRef]
113. Rachoń, L.; Kawczyńska, M. Changes in the structure of sowing area, yields and harvests of cereal crops in Poland in the years 1965–2015. *Agron. Sci.* **2018**, *73*, 5–13. [CrossRef]
114. Jarosz, Z.; Faber, A.; Syp, A. Influence of raw materials used for biofuels production on the structure sown. *Rocz. Nauk. Stow. Ekon. Rol. Agrobiz.* **2013**, *15*, 4.
115. Augustyńska, I. Opłacalność Produkcji Wybranych Produktów Rolniczych w Polsce w Latach 2013–2017. Instytut Ekonomiki Rolnictwa i Gospodarki Żywnościowej—Państwowy Instytut Badawczy w Warszawie. 2018. Available online: http://agro.icm.edu.pl/agro/element/bwmeta1.element.agro-b3d3b197-950c-4a08-9f8f-b3e727b25c18/c/152.pdf (accessed on 18 October 2021).
116. Skarżyńska, A.; Augustyńska, I.; Czułowska, M.; Abramczuk, Ł. *Produkcja, Koszty i Dochody z Wybranych Produktów Rolniczych w Latach 2018–2019*; Instytut Ekonomiki Rolnictwa i Gospodarki Żywnościowej—Państwowy Instytut Badawczy w Warszawie: Warszawa, Poland, 2020; ISBN 978-83-7658-840-7.
117. Jończyk, K. *Oddziaływanie Siedmiu Czynników Agrotechnicznych na Plonowanie Pszenicy Ozimej w Doświadczeniach z Replikacją Połówkową*; Seria R; Instytut Uprawy, Nawożenia i Gleboznastwa w Puławach: Puławy, Poland, 1997; p. 346.
118. Rondanini, D.P.; Gomez, N.V.; Agosti, M.B.; Miralles, D.J. Global trends of rapeseed grain yield stability and rapeseed-to-wheat yield ratio in the last four decades. *Eur. J. Agron.* **2012**, *37*, 56–65. [CrossRef]
119. Budzyński, W.; Ojczyk, T. *Rzepak–Produkcja Surowca Olejarskiego*; Wyd. ART: Olsztyn, Poland, 1996.
120. Oleszek, W.; Ascard, J.; Johansson, H. Brassicae jako rośliny alternatywne umożliwiające kontrole zachwaszczenia w rolnictwie zachowawczym. *Fragm. Agron.* **1994**, *11*, 5–19.
121. Kuś, J.; Jończyk, K.; Kawalec, A. Czynniki ograniczające plonowanie pszenicy ozimej w różnych systemach gospodarowania. *Acta Agrophys.* **2007**, *10*, 407–417.
122. Kuś, J.; Jończyk, K. Produkcyjne i środowiskowe Skutki Stosowania różnych Systemów Gospodarowania w Osinach. In *Eksperymenty Wieloletnie w Badaniach Rolniczych w Polsce*; Marks, M., Ed.; Wydawnictwo Uniwersytetu Warmińsko-Mazurskiego: Olsztyn, Poland, 2018; pp. 133–156. ISBN 978-83-8100-132-8.
123. Lautenbach, S.; Seppelt, R.; Liebscher, J.; Dormann, C. 2012: Spatial and temporal trends of global pollination benefit. *PLoS ONE* **2012**, *7*, e35954. [CrossRef] [PubMed]
124. Potts, S.G.; Imperatriz-Fonseca, V.; Ngo, H.T.; Aizen, M.A.; Biesmeijer, J.C.; Breeze, T.D.; Dicks, L.V.; Garibaldi, L.A.; Hill, R.; Settele, J.; et al. Safeguarding pollinators and their values to human well-being. *Nature* **2016**, *540*, 220–229. [CrossRef] [PubMed]
125. Porto, R.G.; de Almeida, R.F.; Cruz-Neto, O.; Tabarelli, M.; Viana, B.F.; Peres, C.A.; Lopes, A.V. Pollination ecosystem services: A comprehensive review of economic values, research funding and policy actions. *Food Sec.* **2020**, *12*, 1425–1442. [CrossRef]
126. Majewski, J. Determinanty opłacalności produkcji pszczelarskiej w Polsce. *Rocz. Nauk. Stowarzyszenia Ekon. Rol. Agrobiz.* **2018**, *20*, 105–111. [CrossRef]
127. Kołtowski, Z. Beekeeping value and pollination requirements of double-improved cultivars of spring rapeseed (*Brassica napus* L. var. oleifera Metzger f. annua Thell.). *J. Apic. Sci.* **2001**, *45*, 69–84.
128. Kołtowski, Z. Beekeeping value of recently cultivated winter rapeseed cultivars. *J. Apic. Sci.* **2002**, *46*, 23–34.
129. Lipiński, M. *Pożytki Pszczele, Zapylanie i Miododajność Roślin*; Kołtowski, Z., Ed.; PWRiL, Wydawnictwo Sądecki Bartnik-Stroże: Warszawa, Poland, 2010; p. 320. ISBN 978-83-09-99024-6.
130. Kołtowski, Z. The effect of pollinating insects on the yield of winter rapeseed (*Brassica napus* L. var. napus f. biennis) cultivars. *J. Apic. Sci.* **2005**, *49*, 29–41.

energies

MDPI

Article

Legal Framework for the Sustainable Production of Short Rotation Coppice Biomass for Bioeconomy and Bioenergy

Jakub Jan Zięty [1,*], Ewelina Olba-Zięty [2], Mariusz Jerzy Stolarski [2], Michał Krzykowski [1] and Michał Krzyżaniak [2]

[1] Department of Economic Law and Commercial Law, Faculty of Law and Administration, Centre for Bioeconomy and Renewable Energies, University of Warmia and Mazury in Olsztyn, 10-724 Olsztyn, Poland; michal.krzykowski@uwm.edu.pl

[2] Department of Genetics, Plant Breeding and Bioresource Engineering, Faculty of Agriculture and Forestry, Centre for Bioeconomy and Renewable Energies, University of Warmia and Mazury in Olsztyn, 10-724 Olsztyn, Poland; e.olba-ziety@uwm.edu.pl (E.O.-Z.); mariusz.stolarski@uwm.edu.pl (M.J.S.); michal.krzyzaniak@uwm.edu.pl (M.K.)

* Correspondence: jakub.ziety@uwm.edu.pl

Abstract: The production of lignocellulose biomass on dedicated plantations is an option that facilitates the implementation of sustainable development policy. The novelty of this type of research is that it involves the use of various types of methods—economic and legal analyses within a particular subject. This makes it possible to obtain a holistic view of the chosen case study. The purpose of this study was to determine whether setting up a Short Rotation Coppice (SRC) plantation of willow (*Salix* spp.) and poplar (*Populus* spp.) was economically profitable and if the legal regulations supported this type of production. The economic data are based on an experimental case study. The economic profitability of the plantations was assessed through an analysis of discounted cash flows, net present value (NPV), internal rate of return (IRR), and profitability index (PI). The legal analysis of the relevant EU and Polish legal solutions was based on a dogmatic approach. The study demonstrated that SRC cultivation was mostly hindered by economic factors, including the low selling price of biomass as an energy resource and the high costs of harvest. Meanwhile, in the analysed period, i.e., from 2015 to 2020, there were no additional legal or economic forms of support for this type of production that was addressed directly to lignocellulose biomass producers, with the exception of the standard support under the Common Agricultural Policy framework. The results of this study provide information for decisionmakers about the opportunities and challenges experienced during the development of SRC plantations.

Keywords: short rotation coppice of willow/poplar; marginal land use; agricultural law; agricultural policy; economic profitability

Citation: Zięty, J.J.; Olba-Zięty, E.; Stolarski, M.J.; Krzykowski, M.; Krzyżaniak, M. Legal Framework for the Sustainable Production of Short Rotation Coppice Biomass for Bioeconomy and Bioenergy. *Energies* **2022**, *15*, 1370. https://doi.org/10.3390/en15041370

Academic Editors: Ignacio Mauleón and Adam Smoliński

Received: 30 November 2021
Accepted: 9 February 2022
Published: 14 February 2022

1. Introduction

Willow and poplar are grown for biomass in both European countries (including Poland, Lithuania and Germany) and in the United States of America and Canada. The biomass resulting from short rotation coppice plants can be used for the cogeneration of renewable energy or for the manufacture of bioproducts. The generation of energy from SRC is efficient, and SRC plantations improve the diversity of production and soil use [1,2]. The costs of biomass production on SRC plantations compared to the cost of acquiring waste and by-products from forest production are high, which necessitates support for this type of production both while setting up a plantation and during the SRC cultivation cycle. Energy production from biomass (e.g., by its incineration) is an element of a changing strategy that is directed towards energy generation from renewable sources, including biomass [3]. In Poland, the main obstacle to the replacement of fossil fuels, mainly hard coal, in commercial energy generation is the relatively high cost of SRC biomass production. For this reason, although biomass contains less compounds that are harmful to human health and the environment, it has not become as popular as it might appear in the adopted strategies. An example is the recently made decision to build a heat and power plant fired with coal and then with natural gas rather than using renewable energy sources, such as biomass or alternative fuels, i.e., RDF. The production of SRC biomass as a raw material for the generation of energy needs to be supported at both stages: the generation of electric power and biomass cultivation. The generation of electricity has been supported by various kinds of subsidies, e.g., direct payments to energy producers, redeemed certificates, or by maintaining regulated prices for electric power and the obligation to receive the electricity generated from RES. Conversely, between 2015 and 2020, the production of SRC biomass was only subsidised by the state to a small extent, even though in previous years, the state had directly supported the production of energy crops and the establishment of energy crop plantations [4].

In terms of simplified accounting, the generation of energy from renewable energy sources (RES) is more expensive than producing it from conventional fuels. However, environmental changes and the growing social awareness of these changes and causes necessitate a change in our attitude towards this problem. An aware consumer prefers to buy products that have been made sustainably. The strategies that are being developed nowadays show the need to shift towards a closed-loop economy. Changes in legal regulations also proceed in this direction. A closed-loop economy and consumer attitudes can be supported by the generation of renewable energies from lignocellulose biomass grown in the SRC system. For years, energy crops have been the subject of interest among farmers and authorities. SRC plants are characterised by the rapid production of biomass, low fertiliser and herbicide consumption, and high elasticity of production management. Furthermore, SRCs ensure benefits to the environment compared to the crops they compete with and the move towards RES. However, the combination of high production costs and uncertain policies as well as the prices of the products discourage farmers from growing SRC plants. As of today, not many political instruments have been implemented to help increase the production of biomass in this systemin EU countries. Some examples of support solutions in the EU can be found in Germany and Lithuania, although it has been indicated that these solutions have not been very successful in these two countries [5].

The area under SRC plantations in Poland did not increase in the period of 2015–2020. The situation was similar in other EU countries. This is worrying because biomass, including SRC, can provide a source of renewable energy for fuel industry as well as for power generation and bioeconomy. Significant factors are at play here: the relatively high costs of starting a plantation, the long time needed to recover inputs, the immature market, and low profitability compared traditional crops and the lack of contract farming that would ensure the profitability of production over a time horizon corresponding to the longevity of a plantation (20–25 years). Considering the above as well as the fact that biomass in Poland is expected to become a major RES in 2030, a more effective support system for biomass production is needed. If the social and economic goal is to increase the volumes of produced biomass, then greater financial support is necessary, not only subsidies to energy generation, but also direct support to SRC plantations. Support does not only have to rely on subsidies but also on reducing legal barriers to starting manufacturing activities, e.g., establishing the precise legal requirements of such activity, including obtaining environmental assessments or the certification of leading production [6].

This article demonstrates the influence of support on the economic profitability of biomass production in the SRC system and suggests some legal solutions that may help SRC development. The novelty of the current research is that it involves various types of approaches—both economic and legal—from an environmental point of view. This makes it possible to obtain a holistic view of the chosen case study. Hence, this paper contains an analysis of the production costs of biomass that can be used as a feedstock for energy generation. It is followed by an analysis of different support mechanisms for SRC biomass production. It is worth underlining that biomass production is a link in the supply chain that is involved in the generation of energy from renewable sources. Thus, the costs of biomass production will have a direct impact on its use in a circular bioeconomy. The analysis is limited to an evaluation of the profitability of biomass production and the effect of its potential support mechanisms on biomass production. The analyses indicate how to account for the economic support for biomass production from the perspective of the EU's support for manufacturing activities. Research conducted in Poland has confirmed that society is convinced of the generation of energy from renewable sources as a means to increase environmental protection and to strengthen energy security. However, the production costs, particularly those related to the construction of infrastructure or to the distribution to the end-user, pose a significant barrier. Among the above-mentioned methods of reducing these costs, reducing the tax burden for renewable energy producers is also postulated [6].

This study aimed to determine whether setting up an SRC plantation of willow (*Salix* spp.) and poplar (*Populus* spp.) was economically profitable and if the legal regulations supported this type of production. The economic data are based on an experimental case study. The economic profitability of the plantations was assessed through an analysis of discounted cash flows, net present value, internal rate of return, and a profitability index.

2. Materials and Methods

2.1. Field Experiment

The analysis was based on two species of energy crops grown in an SRC system, i.e., willow and poplar. The data used for an economic evaluation of willowand poplar cultivation originated from a field experiment situated in north-eastern Poland (53°59′ N, 21°04′ E) and had been set up on a poor soil field owned by the University of Warmia and Mazury in Olsztyn. Details of the soil properties, weather conditions, design, and performance of the experiment as well as an economic assessment of the production were presented in [7,8]. Two of the most popular SRC species were taken for the analysis: willow (*Salix viminalis*, Żubr variety) and poplar (*Populusnigra* x P. Maximowiczii Henry cv. Max-5), both of which were grown under two fertilisation regimes. The first option was the control option, which was denoted with the letter C, in which no fertilisation was applied. The second option consisted of mineral fertilisation, which was applied in the second year after the plantation had been started and in each subsequent year after harvest. Both species were grown at a density of 11.11 thousand pieces per ha^{-1}. The cost breakdown took account the following treatments into account: winter ploughing, disking (2×), harrowing (2×), marking planting spots, manual planting, mechanical weeding (3×), mineral fertilisation, harvesting, field and road transport, and liquidation of the plantation. The data regarding the costs of harvest and transport were estimated based on earlier studies conducted on a commodity plantation [8,9]. Details of the analysed production technology and the equipment used in the field and other operations are presented in [8].

2.2. Economic Analysis

2.2.1. The Cost of Cultivation

The analysis of the economic efficiency of the cultivation and production of chips from two SRC species was based on the yield of the biomass dry matter obtained in the first four-year cycle of the cultivation system. The total direct costs that were incurred were divided between stages. The first stage was the establishment of the plantation, the second was its exploitation, and the third stage consisted of the liquidation of the plantation (Table 1). The values of the fresh chips of the analysed SRC species were assessed according to the market price during the analysed period (EUR 5.24 GJ^{-1}). Consequently, the monetary value of the chips was calculated based on the calorific value of fresh the SRC wood chips obtained from the soil fertilisation variants and the price of 1 GJ energy they contained. Both the income and costs were converted to 2020 values using the inflation rate. However, the tax was calculated in accordance with the binding taxation rate in 2020. The prices of the purchased materials and selling prices of the chips determined as of 2013 can be found in [6]. These prices were then converted to 2020 values using the inflation rates and were then expressed in euros based on the average currency rate in 2020.

Table 1. Data on SCR production and costs used for analyses in 2020.

Item	Unit	Value
Plantation life span	Years	20.00
Harvest cycle	Years	4.00
Number of harvests	-	5.00
Planting density	cuttings ha^{-1}	11.11
Interest rate	%	5.00
Price of chips	EUR GJ^{-1}	5.24
Single area payment	EUR ha^{-1}	96.52
Payment for greening	EUR ha^{-1}	71.24
Additional (redistribution) payment	EUR ha^{-1}	40.04
Payment for young farmers	EUR ha^{-1}	56.45
Area with natural constraints, sphere I	EUR ha^{-1}	39.37
N fertiliser	EUR kg^{-1}	0.96
P fertiliser	EUR kg^{-1}	0.84
K fertiliser	EUR kg^{-1}	0.65
Cuttings	EUR ha^{-1}	436.39 a; 1163.71 b
Workforce	EUR ha^{-1}	6.08
Application of lignin	EUR ha^{-1}	93.66
Ploughing	EUR ha^{-1}	64.51
Disking	EUR ha^{-1}	56.63
Harrowing	EUR ha^{-1}	40.84
Marking planting spots	EUR ha^{-1}	81.98
Manual planting	EUR ha^{-1}	134.98
Mechanical weeding	EUR ha^{-1}	103.10
Mineral fertilisation	EUR ha^{-1}	59.61
Land tax	EUR ha^{-1}	24.88
Liquidation of plantation	EUR ha^{-1}	295.20
Harvesting	EUR t^{-1} d.m.	26.19 a; 29.30 b
Field transport	EUR t^{-1} d.m.	11.45 a; 12.83 b

a for willow; b for poplar.

2.2.2. Profitability Calculation

The economic analysis comprised the following steps: An analysis of discounted cash flows was carried out to determine the discounted payback period (DPBP) for setting up, cultivating, and then terminating the plantation of both SRC species in all of the fertilisation regimes. In the discounted cash flow method, the approach of changing the value of money over time was used. All future cash flows were estimated and discounted by the discount rate in order to determine their present value. Annual cash flows were identified as the difference between the annual income and annual cost, and the value that was thus calculated was discounted for each year. To compare the profitability of SRC production in the analysed variants, the net present value (NPV), internal rate of return (IRR), and the profitability index (PI) were determined. A similar approach can be found in Hauk, S.; Knoke T.; and Wittkopf S. [10] and in Stolarski, M.J. [9]. In addition, the revenue was determined as an NPV annual annuity. The model assumptions enabled a comparative analysis of the production costs and production profitability through the use of the net present value and annuity methods. The discount rate taken for all calculations was 5%. The costs and revenues from the SRC plantations were spread over the entire cultivation period. To compare the results obtained from SRC with the results achieved from the annual plant cultivation, the net present value (NPV) approach was assumed, similar to other studies [11–13] in which the costs and incomes were converted to annual flows. The analysis was supported by the Invest for Excel 3.9 software programme.

In addition, the calculated revenue values from the production of willow and polar were corrected by the adding values of the payments to their production in the following scenarios. The first scenario, denoted as (I), did not include any direct area payments offered to farmers under the Common Agricultural Policy, production subsidies, or costs of land purchase. The second scenario, denoted as (II), included a single area payment as well as payments for agricultural practices, creating climate and environmental benefits (so-called payment for greening, i.e., EFA—ecological focus area) and additional (redistribution) payments. The third scenario, denoted as (III), also contained (apart from the payments mentioned in scenario II) payments for young farmers and payments to areas with natural constraints (ANC) allocated to sphere I, representing areas with natural constraints.

As part of the research, a one-way sensitivity analysis was performed. The impact of changing two independent variables (the discount rate and revenue) on the dependent variable (the NPV value) in all of the analysed scenarios was examined. The scope of the analysed independent variables ranged from −20% to +20%.

2.3. Methodology of Legal Research

The basic methods used in the scope of the legal analyses included the dogmatic legal method and the comparative method. The former takes the results of the linguistic (grammatical), systemic, and teleological interpretation into consideration. It was applied to analyse the legal acts relevant to the research problem. The linguistic interpretation was mostly carried out in respect to regulations introducing programmes involved in payment distribution under the Common Agricultural Policy (CAP). The EU regulations were a component of the legal systems of the Member States and only partially required additional implementation. The outcome of the analysis supported by the dogmatic method served as a starting point for the next stage of the research, where the legal comparative method was employed. Additionally, the historical legal method was used because many programmes connected to EU payments were implemented for specific periods of time. Using the historical method enabled an evaluation of the consequences in terms of the motivation to achieve the objectives defined by the EU with respect to lignocellulose biomass production.

3. Results and Discussion

3.1. The Legal Ground of the CAP in the EU Law

The origin of the Common Agricultural Policy is thought to be as early as the moment that the Treaties of Rome were signed. The basis for the implementation of the contemporary CAP is Article 38 and subsequent articles of the consolidated version of the Treaty on the Functioning of the European Union (Official Journal C 326, 26/10/2012 P. 0001–0390). The objectives of the CAP are (a) to increase agricultural productivity by promoting technical progress and by ensuring the rational development of agricultural production and the optimum utilisation of the factors of production, in particular, labour; (b) to ensure a fair standard of living for the agricultural community, particularly by increasing the individual earnings of people engaged in agriculture; (c) to stabilise markets; (d) to assure the availability of supplies; and (e) to ensure that supplies reach consumers at reasonable prices. These purposes, listed in Art. 39 of the TFEU, are not the only ones associated with the CAP. Other regulations envisage include supporting high employment levels (Art. 9) and protecting the natural environment in order to support sustainable development (Art. 11). Importantly, not all of these goals have to be pursued at the same time. Currently, it becomes increasingly more evident that the focus of the strategies adopted in the EU lies in sustainable development. Support for energy crop production should be viewed in the context of implementing the EU's Common Agricultural Policy for 2015–2020. The principles of the CAP were regulated in Regulation (EU) No 1307/2013 of the European Parliament and of the Council of 17 December 2013, which established rules for direct payments to farmers under support schemes within the framework of the Common Agricultural Policy and repealing Council Regulation (EC) No 637/2008 and Council Regulation (EC) No 73/2009 (Official Journal of the European Union L 2013.347.608). An example of the implementation of this regulation is the direct payment scheme and the single area payment scheme; the question of national ceilings; definition of 'the active farmer'; granting the rights to the Member States to make payments in amounts of less than EUR 100 or to an agricultural holding with an eligible area of less than 1 ha; or the reduction of payments. A more detailed interpretation of the above regulation was made by the Commission of the European Union, who delegated this task in the mentioned regulation and published it in the Commission Delegated Regulation (EU) No 639/2014 of 11 March 2014, supplementing Regulation (EU) No 1307/2013 of the European Parliament and of the Council, establishing rules for direct payments to farmers under support schemes within the framework of the common agricultural policy and amending Annex X to that Regulation (Official Journal of the European Union L 2014.181.1). This legal act brought to force regulations supplementing some of the provisions of the Regulation EU No 1307/2013 other than significant ones, including(a) general provisions on direct payments, (b) the basic payment system, (c) single area payment systems, (d) payments for farmers implementing agricultural practices beneficial for the climate and the natural environment; (e) payments for young farmers starting their agricultural activity; (f) voluntary production-coupled support; (g) crop-specific support for cotton cultivation; and (h) notification obligations of the Member States. Further details of the above regulation consisted in the issuing, in line with Regulation No 1307/2013, of the regulations of the EU Commission. Among these documents, something that is noteworthy is the Commission Implementing Regulation (EU) No 641/2014 of 16 June 2014, which lays down rules for the application of Regulation (EU) No 1307/2013 of the European Parliament and for the Council establishing rules for direct payments to farmers under support schemes within the framework of the Common Agricultural Policy (Official Journal of the EU L 2014.181.74). It specifies such issues as (a) general provisions on direct payments; (b) the basic payment scheme; (c) the payment for farmers implementing agricultural practices beneficial for the climate and the environment; (d) voluntary production coupled support; (e) the crop-specific payment for cotton; and (f) obligations for the Member States to make notifications. With respect to the funding of the CAP, significant problems are governed by Regulation (EU) No1306/2013 of the European Parliament and of the Council of 17 December 2013 on the financing, management, and monitoring of the common agricultural policy and repealing Council Regulations (EEC) No352/78, (EC) No165/94, (EC) No2799/98, (EC) No814/2000, (EC) No1290/2005, and (EC) No485/2008 (Official Journal of the European Union L 2013.347.549). This regulation specifies the following issues: (a) the financing of the expenses under the Common Agricultural Policy (CAP), including funds allocated to the development of rural areas; (b) the agricultural advisory system; (c) the management and quality systems to be instituted by the Member States; (d) the cross-compliance system; and (e) the clearing of accounts. Control matters are specified in the Commission Delegated Regulation (EU) No 640/2014, of 11 March 2014, supplementing Regulation (EU) No 1306/2013 of the European Parliament and of the Council with regard to the integrated administration and control system and conditions for refusal or the withdrawal of payments and administrative penalties applicable to direct payments, rural development support, and cross-compliance (Official Journal of the EU L 2014.181.48).

3.2. The CAP Pillars

Under the CAP, financing rests on direct payments (first pillar); funds allocated to the development of rural areas in the CAP; and additional means for the development of rural areas under the framework of the Next Generation EU (a temporary instrument to help repair damage caused by COVID) (second pillar). The direct payment mechanisms were altered by abandoning the idea of 'decoupling payments and production' in favour of 'targeting' payments, which has led to a system of seven functional payments, each corresponding to a specific goal: (1) a basic payment per hectare, the level of which should be harmonised in accordance with national or regional economic and administrative criteria and that is to undergo the process of convergence (so-called internal convergence);

(2) payments for greening in the form of an additional support dedicated to compensating the costs of producing environmental benefits that are not paid for by the market; (3) additional payments for young farmers; (4) 'redistributive payments', which enable increasing the support allocated to the first hectares on a farm; (5) additional aid to incomes earned in areas with natural constraints; (6) production-coupled payments to specific areas or branches of agricultural production; (7) a voluntary simplified system for small farms that receive less than EUR 1250 in payments a year. The first three components are mandatory for the EU Member States, while the other four components are elective. The Member States must allocate 30% of their national direct payment funds to greening payments. The remaining 70% are allocated to direct payments, having deducted all sums for obligatory national reserves (obligatory, up to 3% of the national envelope) and for additional redistributive payments (up to 30%), payments for young farmers (to 2%), payments to areas with natural constraints (to 5%), and production-coupled payments (to 15%). New payments to 1 ha will only be awarded to active farmers. Moreover, since 2019, they have been subject to partial convergence (external convergence) between the Member States. The basic payment system will receive approximately 70% of each Member State's budget for direct payments.

Regarding 'internal convergence', the Member States who maintained the payments established on historical references for payment entitlements in 2013 were told that they must gradually shift towards more uniform amounts of payments per ha. To this aim, these states can choose from a few options: they can adopt a national or regional approach, which will enable them to achieve a national or regional flat-rate level until 2019, or they must ensure that the farms receiving less than 90% of the country's national or regional payment will gradually receive increasing payments on the condition that every farmer will receive a payment that corresponds to at least 60% of the national or regional average payment no later than in 2019. The sums paid to farmers receiving payments below the regional or national average are proportionally corrected, and the Member States are allowed to reduce possible 'losses' in support to 30%. The Member States will also have the right to award a redistributive payment, i.e., for the first 30 ha or for an area corresponding to the area of an average farm in the country provided that it is no more than 30 ha at most. Another possibility the application of a maximum payment per ha. In addition, Member States are allowed to provide payments to young farmers (less than 40 years of age) who have commenced agricultural activity in the past five years. The young farmer payment system is mandatory for all Member States.

Another mandatory solution is the greening payment system. A farm can receive an additional payment per ha for using agricultural practices that are beneficial for the climate and nature. The Member States are obliged to allocate 30% of the national envelope to this payment. The three measures envisaged under this umbrella are (a) the diversification of crops: a farmer must grow two main crops if he has more than 10 ha of arable land and three crops if he has more than 30 ha of arable land; the main crop may cover no more than 75% of the arable land, and the two main crops may cover no more than 95% of the arable land; (b) a farmer must maintain permanent grassland; and (c) a farmer must maintain 'an ecological focus area' covering at least 5% of the arable land of a farm that is more than 15 ha of the arable land (excluding permanent grassland and perennial crops), for example, the edges of fields, hedges, trees, fallow land, landscape features, biotopes, buffer strips, afforested areas, or nitrogen-fixing crops. There will be severe penalty fees for failing to abide by these 'greening' rules. In order to avoid punishing farmers who have already implemented eco-friendly solutions, the regulation establishes a 'green equivalency system', which affirms pro-environmental practices that are already in place and that are deemed to meet these basic requirements. For instance, organic farmers are not obliged to meet any additional requirements because their agricultural activity brings about evident ecological benefits. The new regulation contains a list of practices considered to be equivalent.

In 2014, the Member States had to make fundamental choices in the face of a variety of rules regulating the implementation of the new direct payment system and to create room to manoeuvre the system they were asked to work within. Most Member States, except one (Germany), had the option of using coupled payments of highly varied rates. Regarding greening payments, some Member States allowed farmers to meet some of the requirements by using equivalent practices. With respect to the second pillar, the Commission approved 118 rural area development programmes. Twenty Member States decided to implement just one national programme, while eight opted for more than one programme (which, for example, allows them to take better account of the country's geographical or administrative structure).

As the legislative procedures concerning the CAP reform after 2020 had not been completed by 1 January 2021, the co-legislators passed Regulation (EU) 2020/2220 of the European Parliament and the Council of 23 December 2020, extending the currently binding regulations by two years (until 31 December 2022). The European Union's policy concerning the development of rural areas was established as the second pillar of the CAP during the Agenda 2000 reform. It was co-financed by the European Agricultural Fund for Rural Development (EAFRD) and by regional or national funds. The Commission determined the three overriding priorities in the rural development policy: increased agricultural competitiveness; ensuring the sustainable management of natural resources; and counteracting climate change by attaining the balanced territorial development of rural economies and communities as well as creating and maintaining jobs.

These three principal goals were reflected in the six priorities of the EU regarding the policy for the development of rural areas in 2014–2020: supporting the transfer of knowledge in agriculture and forestry; improved competitiveness of all branches of the agricultural economy and improved economic viability of farms; promoting food chain organisation and risk management in agriculture; restoring, protecting, and supporting ecosystems dependent on agriculture and forestry; promoting resource efficiency and supporting the conversion to low-carbon economy resistant to climate change in the agricultural, food, and forestry sectors; supporting social inclusion, poverty reduction, and economic growth in rural areas. Lowering production costs by changing (simplifying) technologies or obtaining additional public subsidies is also noted in other areas of research [14,15].

The rural development policy was implemented on the basis of programmes prepared by the Member States (or their regions) for the development of rural areas. Under these programmes, which cover several years, individualised strategies are executed, which respond to specific needs of the Member States (or regions) and that account for at least four of the six priorities mentioned above. These programmes rest on several financial means and have been selected from the set of EU funds, which are laid out in greater detail in the Regulation on the support of the development of rural areas (EU Regulation No 1305/2013) and co-financed from AEFRD funds (cf. see below for more specific data). The level of co-funding varies depending on the region and on the funds it engages. The programmes must be approved by the European Commission, and they must contain a financial plan as well as a set of indicators to evaluate the results. In the current programme perspective (for years 2014–2020), special focus is placed on coordination between the EAFRD and other EU structural and investment funds, such as funds dedicated to the cohesion policy (Cohesion Fund, European Regional Development Fund ERDF, and European Social Fund ESF) and the European Maritime, Fisheries, and Aquaculture Fund, EMFFA.

3.3. Proposed Directions in the Development of the CAP

The European Commission has formulated some proposals for new legal regulations concerning the CAP after 2021. Examples include the European Green Deal, proposed in November 2019, and the field-to-table strategy as well as the EU strategy for biodiversity 2030, issued by the Commission in May 2020. They all attest to the increasingly broader scope of issues related to agriculture and food. Furthermore, in the context of market opening and globalisation, Article 207 of the TFEU determined new guidelines for the common commercial policy of the EU, which will now be more applicable to the trade of agricultural products. The key assumptions of the EU's new agricultural policy are that (1) the focus on climate and environment are stronger than before; (2) the annual report of a Member State is based on achieved results rather than on compliance with the EU regulations; (3) the first pillar to be included in the programme; (4) changes in the so-called green architecture; (5) new options of sector interventions (promoting team activities from the first pillar; and(6) strengthening the role of technological progress and innovativeness, including the growing role of knowledge extension and science. Of the regulations binding to this day, the European Commission wishes to maintain the reduction in the share of the second pillar; create further though smaller reductions in the differences in direct payments; establish a simplified area payment system; make redistributive payments targeting small and medium farms; and make coupled payments. There are still three shared goals of the Common Agricultural Policy, and they continue to relate to the following issues: (1) economic, with more stress placed on the resilience of agriculture and smart development; (2) connected with the environment and climate; and (3) dealing with the development of rural areas.

Instead of the six priorities specified previously, the Commission proposed nine specific objectives: ensuring a fair income to farmers and supporting resilience of farms in the entire Union in order to improve food safety; a stronger focus on research, technology, and digitalisation; a stronger position for farmers in the food chain; contributing to climate action, including the use of sustainable energy; supporting the sustainable development and efficient management of resources, such as water, soil, and air; contributing to the preservation of biodiversity, strengthening ecosystem services, and protecting habitats and the landscape in rural areas; attracting young farmers and helping to start business activity in rural areas; promoting the employment, growth, cohesion, and social inclusion as well as local development, including bioeconomy and sustainable forestry in rural areas; and a better response of the EU to social needs regarding health and food, including healthy, nutritious, and sustainable food, preventing food waste, and ensuring animal welfare. The information on the support for the development of rural areas suggests that up to 40% of the total funds allocated to agricultural policy (at least 30% of the European Fund for Agriculture and Rural Development) is to be allocated to attaining environmental and climatic goals. At least 5% of the II pillar's budget is to be dedicated to the implementation of the community-led local development mechanism.

3.4. *Systems of Support to Agricultural Production for Energy Purposes in Poland until 2020*

The support for crop production for energy purposes in Polish law relies on the previously mentioned CAP mechanisms. Thus, a chance to acquire funds should be sought in terms of both area and greening payments. Detailed regulations, indicated in Section 2 of the EU regulations, entered into force in Poland's Act on Payments in the Direct Payments Scheme of 5 February 2015. These regulations pertain to both area payments and other types of payments, such as those for young farmers, green payments, or additional payments. Polish legislators have followed the rules of referring national definitions to terms introduced in regulations of the EU Council, for example, a farmer, farm, or greening. Incidentally, the law did not make any provisions for payments for the production of biomass for energy purposes available in 2015–2020. Therefore, any possible additional SRC payments are only possible by applying for payments for greening. The payment rates are established by the Minister for Agriculture every year. The rates for the years 2015–2020 are given in Table 2.

Table 2. Rates of payments for SRC production in Poland between 2015–2020.

Year	Type of Payment	Rate (PLN ha^{-1})	Rate (EUR ha^{-1})	Source
2020	Additional payment	182.02	40.04	[16]
	Payment for young farmers	256.62	56.45	[17]
	Payment for greening	323.85	71.24	[18]
	Area payment	483.79	106.42	[19]
2019	Additional payment	184.98	40.69	[20]
	Payment for young farmers	165.1	36.32	[21]
	Payment for greening	316.54	69.63	[22]
	Area payment	471.64	103.74	[23]
2018	Additional payment	178.01	39.16	[24]
	Payment for young farmers	175.62	38.63	[25]
	Payment for greening	308.18	67.79	[26]
	Area payment	459.19	101.01	[27]
2017	Additional payment	177.02	38.94	[28]
	Payment for young farmers	214.82	47.25	[29]
	Payment for greening	309.77	68.14	[30]
	Area payment	461.55	101.52	[31]
2016	Additional payment	172.79	38.01	[32]
	Payment for young farmers	231.97	51.03	[33]
	Payment for greening	310.1	68.21	[34]
	Area payment	462.05	101.63	[35]
2015	Additional payment	171.73	37.77	[36]
	Reduction of the additional payment rate	1.62	0.36	[37]
	Payment for young farmers	258.97	56.96	[38]
	Payment for greening	304.31	66.94	[39]
	Area payment	453.7	99.80	[40]

In the previous CAP perspective, it was possible to obtain subsidies for the establishment of energy crop plantations. In the years 2007–2009, under the CAP, Poland participated in the financial support programme by addressing the energy crop sector. To be eligible for such a payment, one had to meet the following conditions: the minimum area under a plantation—1 ha; not set up on permanent grassland; payment application submitted in the year when the plantation was started or in the consecutive year; the plantation had to be set up in accordance with agrotechnical requirements—location (at least 1.5 m from the border of a land parcel where a similar energy crop plantation is grown or a land plot used as forest; at least 3 m from the border of an adjacent land plot if it is used differently than mentioned; not set up on areas with nature protection: if the relevant documentation did not provide a possibility of starting such a plantation; on drained land if it was a poplar or willow plantation; on other types of land for which an payment applications for energy crop cultivation had already been submitted). In 2009, the payments for starting a willow plantation on 1 ha of land equalled 50% of the flat-rate costs. Assuming that those costs at that time were PLN 8600 per 1 ha^{-1} (about EUR 1892 ha^{-1}), then a payment was PLN 4300 (about EUR 946). This type of support attracted some interest, as the data on the numbers of filed applications suggest that applications for payments for energy willow plantations covered a total of 553.58 ha in 2008 and 779.29 ha in 2009, indicating growing interest. The provision ensuring the possibility of granting these payments was contained in the Council Regulation (EC) No 73/2009 of 19 January 2009, establishing common rules for direct support schemes for farmers under the common agricultural policy and establishing certain support schemes for farmers, amending Regulations (EC) No1290/2005, (EC) No 247/2006, and (EC) No 378/2007 and repealing Regulation (EC) No 1782/2003 (Official Journal of EU L 30/16 of 31 January 2009, with amendments). It contains transitional regulations that allow the provision of support for energy crops until the year 2009, as stipulated in Title IV, Chapter 5 of the Council Regulation (EC) No 1782/2003 of 29 September 2003, establishing common rules for direct support schemes under the common agricultural policy and establishing certain support schemes for farmers and amending Regulations (EEC) No 2019/93, (EC) No 1452/2001, (EC) No 1453/2001, (EC) No 1454/2001, (EC) 1868/94, (EC) No 1251/1999, (EC) No 1254/1999, (EC) No 1673/2000, (EEC) No 2358/71, and (EC) No 2529/2001 (Official Journal of EU L 270/1 of 21 October 2003 with amendments). In line with Article 90 of Regulation 1782/2003, support was only granted to areas producing energy crops for which a farmer had a contract with the processing industry, except when processing was undertaken by a farmer on the farm. This payment was financed from the EU budget, and it was distributed in the years 2007–2009. In turn, the aid for setting up permanent crop plantations for energy purposes was financed from national budgets. The provisions specifying the eligibility conditions were putinto force in Poland by the Act on amending the Act on payments to agricultural land and on sugar tax and the Act on the stamp duty, of 29 February 2008 (Dz. U No 44 item 262). In 2008–2009, farmers could therefore obtain support to permanent plantations by pursuing Article 29a Section 1 of the Act on payments from direct payments systems, of 26 January 2007, which stated that a farmer was eligible for support in an amount equal to 50% of the flat-rate costs of starting permanent plant plantations per 1 ha of a plantation. Currently, the only support for energy crop cultivation could be obtained from the single area payment system. It should be noted that a single area payment also applies to plantations of trees provided that compose short-rotation woods (they can be trees used for energy purposes). In conclusion, since 2010, there has been no specific support dedicated to energy crop cultivation under the direct payment schemes. However, willow plantations can constitute SRC plantations within the meaning of Regulation No 1120/2009. A single area payment can be obtained for the area covered with such plantations, provided that the conditions concerning the minimum cultivation area and maximum crop harvest period are satisfied. Moreover, beneficiaries of the direct payment system are obliged to maintain good agricultural practices on their land in accordance with the principles of environmental protection, which, in terms of short rotation woods, entails the requirement of keeping specific distances from the borders of adjacent land plots. Since 2010, there have not been direct payments for the establishment of energy crop plantations or directly for the production of energy crops, which were previously permitted by some EU regulations [41,42].

3.5. *Economic Analysis of the Profitability of Production*

Table 3 contains the results of the economic assessment of willow and poplar chip production. In all of the analysed variants, chip production was profitable, although the values of all of the indicators varied considerably depending on the SRC species, fertilisation regime, and, above all, on the payment scenario. When comparing the NPV, this index for willow plantations was evaluated from EUR 1540 to 5641, and it was EUR 861 higher for the control plantation and EUR 1172 higher for mineral fertilisation than the net present values (NPV) for poplar for the parallel fertilisation variants and payment scenarios. The revenue values for willow ranged from EUR 121 year^{-1} to EUR 568 year^{-1} and were EUR 68 higher than they were for the control and EUR 92 higher for mineral fertilisation than in the parallel fertilisation regimes and payment scenarios for poplar plantations.

Table 3. Profitability indices for willow and poplar chip production depending on the soil fertilisation option and subsidy scenarios.

SRC	Fertilisation Option	Subsidy Scenario	NPV (EUR ha^{-1})	IRR (%)	PI	DPBP (Years)	Revenue (EUR ha^{-1} year^{-1})
Willow	C	I	1 540	22	2.40	6.20	121
		II	4 389	57	4.99	2.27	344
		III	5 641	81	6.12	2.00	443
	F	I	3 135	31	3.85	2.74	246
		II	5 984	63	6.44	2.28	469
		III	7 236	83	7.57	2.10	568
Poplar	C	I	680	10	1.39	10.53	53
		II	3 528	29	3.00	2.99	277
		III	4 780	39	3.71	2.64	375
	F	I	1 964	17	2.12	6.59	154
		II	4 812	34	3.73	2.74	378
		III	6 064	43	4.44	2.53	476

The IRR values were approximately twice as high for willow as they were for poplar in the parallel fertilisation regimes and payment scenarios. However, the profitability index PI was about 40% higher for poplar than it was for willow. The possibility to obtain payments in scenario II, which comprised a single area payment, additional payment, and ecological focus area payment, which equalled EUR 207.79 per ha altogether in 2020, raised the revenue by EUR 233 and the NPV by EUR 2848. Including an additional payment for young farmers and a payment to an area with natural constraints in sphere I (i.e., considering scenario III), the revenue increased by EUR 322, and the NPV rose by EUR 4101 in each analysed case. The increase in the NPV due to the possibility of obtaining payments was very large. For SRC willow plantations with no fertilisation (option C), the NPV increased 2.8- and 3.7-fold (for scenario II and III, respectively), whereas on poplar plantations, this index was 5.2- and 7-fold higher (for II and III scenarios, respectively). The application of mineral fertilisers, despite the input costs, improved the yields, and the revenue from selling wood chips was therefore higher. The contribution of the profits earned from selling chips was large in these variants (willow F and poplar F) and, therefore, the effect of payments on the NPV was weaker, less by 1.9- and 2.3-fold for willow in scenario II and III and by 2.5- and 3.1-fold for poplar in scenario II and III, respectively.

A quite long DPBP, approximately six years, was obtained for willow CI and poplar FI, whereas the DPBP calculated for the other variants was much shorter, 2–3 years.

Figure 1 illustrates discounted cash flows for willow and poplar chip production depending on the soil fertilisation option and subsidy scenario. The highest DCF values were calculated for willow F III (EUR 7236), followed by poplar F III (EUR 6064), and then willow F II (EUR 5984) and willow C III (EUR 5641), confirming that a chance to receive payments has a considerable influence on the profitability of production, as reflected by from the willow F variant achieving over double revenue values. The strongest effect of payments on revenues, leading to an over 7-fold increase for poplar and a 3.6-fold increase for willow, was demonstrated in the C variants (without fertilisation). The above results are particularly important in terms of persuading farmers to consider low-input SRC cultivation, which has a much less intensive impact on the environment, preventing the adverse and high influence of fertilisation on the natural environment, particularly in terms of causing freshwater eutrophication by poplar plantations [43] or freshwater toxicity caused by willow plantations [44], although higher external costs incurred by fertilisation (as much as 23% more for willow production [45] and 20% for poplar production [46]) should not be neglected. The highest DCF values were also recorded on the willow F III and poplar F IIII plantations and were over 2- and 3-fold higher than they were for the willow F I and poplar F I variants.

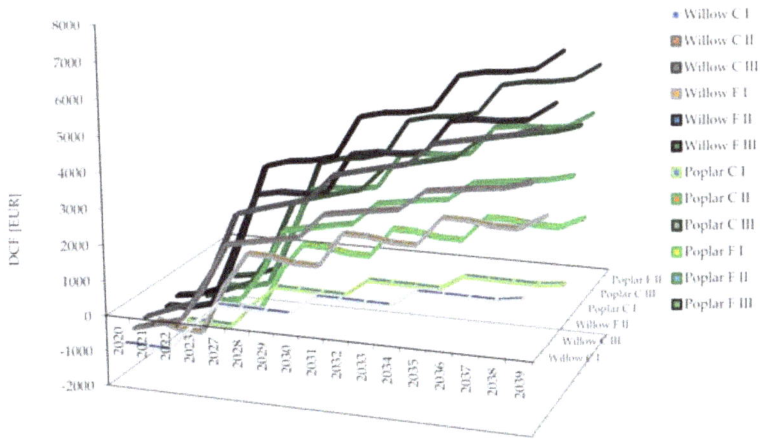

Figure 1. Discounted cash flows for willow and poplar chip production depending on the soil fertilisation option and subsidy scenario.

The sensitivity analysis determined the effect of the changes in the discount rate on NPV for willow and poplar chip production depending on the soil fertilisation option and payment scenario (Figure 2) and the effect of the changes on the income on the NPV for willow and poplar chip production depending on the soil fertilisation option and payment scenario (Figure 3).

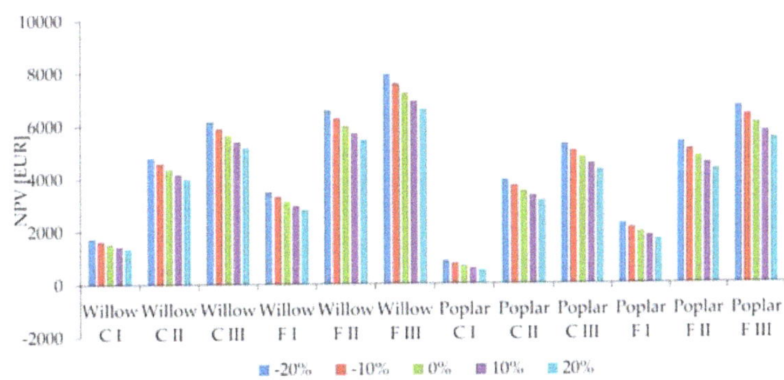

Figure 2. The effect of changes in the discounted rate on the NPV for willow and poplar chip production depending on the soil fertilisation option and payment scenarios.

A change in the discounted rate applied to the NPV at rates that were even as high as ±20% did not affect this index considerably. A change in the NPV was most distinctly seen in the variants comprising fertilisation, as it amounted to EUR 1000 ha^{-1} for both willow and poplar compared to around EUR 700 ha^{-1} for unfertilised willow and poplar plantations. Poplar plantations were more sensitive to a modification in the discounted rate; the differences in the NPV for both variants (willow C and willow F) reached 20% on average, while the differences in the NPV values for analogous poplar variants averaged 17%. Chip production was profitable in all cases. The current analysis of the effect of a change in the revenue on the NPV, likewise in a range of ±20%, showed that when the revenue decreased by 20% due to a lower harvest or lower selling price for poplar chips, in the unfertilised variant without payments (poplar C I), the production became unprofitable (NPV EUR-388 ha^{-1}), while poplar chip production in the same variant, with no fertilisation or payments (poplar FI), was on the brink of profitability (NPV EUR

176 ha^{-1}). A change in the revenue value by +20%, stemming from higher yields, higher prices, or higher payments (in scenario II and III) increased the NPV to over EUR 7000 for willow F II, willow F III, poplar F II, poplar F III, and willow C III.

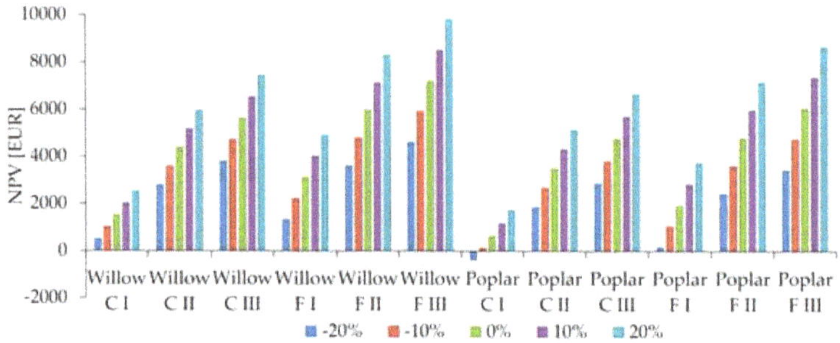

Figure 3. The effect of changes in revenue on the NPV for willow and poplar chip production depending on the soil fertilisation options and payment scenarios.

The analysis of willow production costs and coupled payments in Poland conducted in 2006 allowed us to estimate the revenue value at EUR 236 ha^{-1}, and when the available payments were considered, the revenue improved by EUR 17 ha^{-1}, reaching 220 EUR ha^{-1} from the establishment subsidy [13].

When comparing the economic situation in Poland in 2013–2015 [9] with the situation in 2020, the experimental data demonstrated a decline in the NPV: for willow C I, a decline from EUR 1653 to EUR 1540 was experienced, and for willow F I, a decline from EUR 3298 to EUR3135 was experienced; for poplar F I, there was a decline from EUR 2111 to EUR 1964, which represented a decline of 5%, and for poplar C I, there was a decline from EUR 844 to EUR 680, which was nearly 20%.

Fradj et al. [47] discussed the prospect of potential SRC willow plantation integration with the cultivation of other crops in Poland. Those researchers analysed the payment amounts and their effect on the total acreage cultivated with willow in Poland. They concluded that willow, which could make a large contribution to the Polish economy, could be produced sustainably and efficiently and could provide farmers with additional income.

In Lithuania, willow production has also been determined to be profitable, regardless of whether it was supported with payments or not [48]. The results of a cash flow analysis showed that at a 6% discounted rate and without EU payments, the net present value of willow cultivation was EUR 458. If the EU payments were granted, the net present value of a willow plantation in the 22nd year was EUR 1800. The DPBP without payments was 17 years, which was shortened to 9 years when payments were available.

The analysis of the policy for perennial energy crop production, which was based on poplar production in Germany for over 24 years [49], comprised four types of payments at 3 and 4 levels. One of the scenarios presumed a guaranteed price of EUR 50, 55, and 60 Mg d.m. I, and it was only the highest price that resulted in a positive effect on income related to the NPV at EUR 2826.29; this value is higher than the value of EUR 885 obtained in this paper, although the assumed market price was approximately EUR 20 Mg dm. higher. Another scenario assumed a one-off subsidy for starting a plantation, which was 500 EUR ha^{-1} and affected the NPV, which then reached EUR 3758.82 ha^{-1}; a similar value (NPV EUR 3528) was achieved for Poplar C II, in which the total value of annual payments was about EUR 200 ha^{-1}.

Faasch et al. also made an assessment of the SRC production profitability in Germany [50]. Their results confirmed that appropriate economic and political conditions, such as high subsidies, low costs, and higher prices of wood chips, could lead to SRC plantations achieving higher profitability than the production of conventional crops. The most favourable variant accounted for the subsidy in which farmers were reimbursed 30%

of the initial investment inputs and an area payment of EUR 200 per hectare per annum, which allowed the generation of the NPV of EUR 8660 per ha.

The Swedish experience in SRC production for energy purposes suggests a need to develop financial models that are orientated towards diminishing the risk connected with SRC cultivation [51], thus confirming the earlier assertion that a stable policy and long-term contracts between different subjects may reduce the uncertainty raised by SRC cultivation [52].

The latest studies on the trends and location of rapidly growing energy crops show that the total area covered by SRC plantations in Sweden has been declining for years and that willow has been planted increasingly on more productive farmland, and poplar plantations have been set up on less productive soils than previously [53].

4. Conclusions

The ongoing work on the implementation of the new CAP perspective, especially in terms of energy crops, should lead to the successful achievement of the sustainable development goals. The above study, which concerns SRC production and using willow and poplar as model species, shows that SRC production incurs high costs. The CAP payments are insufficient to offset these costs so as to make production competitive towards other sources of biomass. The market price that SRC biomass producers could obtain is not competitive in relation to other biomass sources, particularly production waste in the forestry industry. Stimulation mechanisms, such as subsidies or certificates, are not addressed to an SRC biomass producer but rather to companies using biomass. The only subsidies that a biomass producer can obtain are a single area payment, payments for young farmers, payments for greening, or payments to areas with natural constraints. It is noteworthy that the problem of insufficient support to SRC development has been raised for years. The analyses conducted in this research show that the current support to SRC production is too small for such plantations to be a serious alternative to other biomass sources. Moreover, this aid is now weaker than it was before 2014, as it does not comprise subsidies for the establishment of plantations or for the production of energy crops. The economic results that are achieved nowadays do not encourage farmers to set up SRC plantations, even though they might be a stable source of high-quality dendromass, which would facilitate the gradual phasing out of fossil fuels, especially hard coal, in individual households as well as in whole regions or countries. It should also be added that SRC plantations should be set up on land that is of little or no value for the production of food or fodder plants, i.e., mainly on marginal land, fallow land, contaminated soils, or wasteland. This approach would be extremely important for the economy, as it could activate the use of many areas left unused and that do not generate any profit but that could become a source of dendromass. Furthermore, such areas could serve as sites for the utilisation of sewage sludge and other organic residues, e.g., ash from the burning of biomass, which would improve the soil fertility, and this, in turn, would have a positive effect on plant yields while limiting the consumption of fertilisers. Such an integrated solution (i.e., using residues for the enrichment of marginal or degraded land, production of biomass for energy and industrial purposes, and returning processing waste and by-products to circulation) fully agrees with the idea of a closed-loop bioeconomy. It appears that the mechanisms of direct support to SRC producers should be intensified in order to launch this process and to suggest a new approach in this field, including possible pathways for the development of bioenergy and bioeconomy. In this context, perhaps the simplest thing to do is to return to direct payments to new plantations, e.g., return 50% of the flat-rate costs. In addition, tax incentives could be created, e.g., reducing or foregoing some taxes, a refund of excise duty on the materials used, or income or indirect tax relief. In addition, it might be helpful to restore the obligation of the power industry to purchase SRC-produced biomass by the power industry. The promotion of such solutions could increase the amounts of non-forest biomass used in heat, combined heat and power, or power plants for energy generation and to replace hard coal with the solid biofuel produced on dedicated SRC plantations, which would be in agreement with the concept of bioeconomy.

Author Contributions: Conceptualization, J.J.Z., E.O.-Z. and M.J.S.; methodology, J.J.Z., E.O.-Z. and M.J.S.; validation, J.J.Z., E.O.-Z., M.J.S., M.K. (Michał Krzykowski) and M.K. (Michał Krzyżaniak); formal analysis, J.J.Z., E.O.-Z. and M.J.S.; investigation, J.J.Z., E.O.-Z. and M.J.S.; resources, J.J.Z., E.O.-Z. and M.J.S.; data curation, J.J.Z., E.O.-Z. and M.J.S.; writing—original draft preparation, J.J.Z., E.O.-Z. and M.J.S.; writing—review and editing, J.J.Z., E.O.-Z. and M.J.S.; M.K. (Michał Krzykowski) and M.K. (Michał Krzyżaniak); visualization, J.J.Z. and E.O.-Z.; supervision, J.J.Z., E.O.-Z. and M.J.S.; project administration, J.J.Z. All authors have read and agreed to the published version of the manuscript.

Funding: The results presented in this paper were obtained as part of a comprehensive study financed by the University of Warmia and Mazury in Olsztyn, 24.610.043—110 Faculty of Agriculture and Forestry, Department of Genetics, Plant Breeding and Bioresource Engineering (grant No. 30.610.007-110) and has been co-financed by the Interreg Baltic Sea Region Programme 2014–2020 co-funded by the European Regional Development Fund under the project "Unlocking the Potential of Bio-based Value Chains in the Baltic Sea Region" (BalticBiomass4Value-BB4V), No. #R095 and co-financed from the funds of the Ministry of Science and Higher Education under the program "PMW" in the years 2019–2021; No. 5047/INTERREG BSR/2019/2.

Acknowledgments: We would also like to thank the staff of the Centre for Bioeconomy and Renewable Energies for their administrative support during the study.

Conflicts of Interest: The authors declare no conflict of interest.

References

1. Djomo, S.N.; El Kasmioui, O.; Ceulemans, R. Energy and greenhouse gas balance of bioenergy production from poplar and willow: A review. *Glob. Chang. Biol. Bioenergy* **2011**, *3*, 181–197. [CrossRef]
2. Parra-Lopez, C.; Holley, M.; Lindegaard, K.; Sayadi, S.; Esteban-Lopez, G.; Duran-Zuazo, V.H.; Knauer, C.; von Engelbrechten, H.G.; Winterber, R.; Henriksson, A.; et al. Strengthening the development of the short-rotation plantations bioenergy sector: Policy insights from six European countries. *Renew. Energy* **2017**, *114*, 781–793. [CrossRef]
3. Lopez, P.; Lindegaard, K.N.; Adams, P.W.R.; Holley, M.; Lamley, A.; Henriksson, A.; Larsson, S.; von Engelbrechten, H.G.; Esteban Lopez, G.; Pisarek, M. Short rotation plantations policy history in Europe: Lessons from the past and recommendations for the future. *Food Energy Secur.* **2016**, *5*, 125–152.
4. Grodecki, A. Dopłaty do upraw roślin energetycznych. *Kwart. Nauk. Uczel. Vistula* **2014**, *3*, 54–69.
5. Spiegel, A.; Britz, W.; Djanibekov, U.; Finger, R. Policy analysis of perennial energy crops cultivation at the farm level: The case of short rotation coppice (SRC) in Germany. *Agric. Resour. Econ. Discuss. Pap.* **2017**, *3*, 1–33. [CrossRef]
6. Marks-Bielska, R.; Bielski, S.; Pik, K.; Kurowska, K. The Importance of Renewable Energy Sources in Poland's Energy Mix. *Energies* **2020**, *13*, 4624. [CrossRef]
7. Stolarski, M.J.; Krzyżaniak, M.; Szczukowski, S.; Tworkowski, J.; Załuski, D.; Bieniek, A.; Gołaszewski, J. Effect of increased soil fertility on the yield and energy value of short-rotation woody crops. *BioEnergy* **2015**, *8*, 1136e1147. [CrossRef]
8. Stolarski, M.J.; Krzyżaniak, M.; Tworkowski, J.; Szczukowski, S.; Niksa, D. Analysis of the energy efficiency of short rotation woody crops biomass as affected by different methods of soil enrichment. *Energy* **2016**, *113*, 748e761. [CrossRef]
9. Stolarski, M.J.; Olba-Ziety, E.; Rosenqvist, H.; Krzyżaniak, M. Economic efficiency of willow, poplar and black locust production using different soil amendments. *Biomass Bioenergy* **2017**, *106*, 74–82. [CrossRef]
10. Hauk, S.; Knoke, T.; Wittkopf, S. Economic evaluation of short rotation coppice systems for energy from biomass—A review. *Renew. Sustain. Energy Rev.* **2014**, *29*, 435–448. [CrossRef]
11. Rosenqvist, H.; Dawson, M. Economics of willow growing in Northern Ireland. *Biomass Bioenergy* **2005**, *28*, 7–14. [CrossRef]
12. Ericsson, K.; Rosenqvist, H.; Ganko, E.; Pisarek, M.; Nilsson, L. An agro-economic analysis of willow cultivation in Poland. *Biomass Bioenergy* **2006**, *30*, 16–27. [CrossRef]
13. Ericsson, K.; Rosenqvist, H.; Nilsson, L.J. Energy crop production costs in the EU. *Biomass Bioenergy* **2009**, *33*, 1577–1586. [CrossRef]
14. Shao, H.C.; Zhang, Y.Y.; Hussain, S.; Liu, X.C.; Zhao, L.J.; Zhang, X.Z.; Liu, G.W.; Qiao, G.J. Effect of Preform Structure on the Performance of Carbon and Carbon Composites. *Sci. Adv. Mater.* **2019**, *11*, 945–953. [CrossRef]
15. Hou, H.G.; Hussain, S.; Shao, H.C.; Liu, G.W.; Wang, M.S.; Qiao, G.J.; Shaheen, A. Experimental Insights on Factors Influencing Sensitivity of Thin Film Narrow Band-Pass Filters. *Sci. Adv. Mater.* **2019**, *14*, 1548–1554. [CrossRef]
16. Rozporządzenie Ministra Rolnictwa I Rozwoju WSI z dnia 14 października 2020 r. w sprawie stawki płatności dodatkowej (Regulation of the Minister of Agriculture and Rural Development on the Rate of Additional Payment)za 2020 r., O.J. 2020 pos. 1812. Available online: https://isap.sejm.gov.pl/isap.nsf/download.xsp/WDU20200001812/O/D20201812.pdf (accessed on 12 October 2021).

17. Rozporządzenie Ministra Rolnictwa I Rozwoju Wsi z dnia 14 października 2020 r. w sprawie stawki płatności dla młodego rolnika (Regulation of the Minister of Agriculture and Rural Development on the Payment Rate for Young Farmer) za 2020 r., O.J. 2020 pos. 1811. Available online: https://isap.sejm.gov.pl/isap.nsf/download.xsp/WDU20200001811/O/D20201811.pdf (accessed on 12 October 2021).
18. Rozporządzenie Ministra Rolnictwa I Rozwoju Wsi z dnia 14 października 2020 r. w sprawie stawki płatności za zazielenienie (Regulation of the Minister of Agriculture and Rural Development on the Payment Rate for Greening) za 2020 r., O.J. 2020 pos. 1806. Available online: https://isap.sejm.gov.pl/isap.nsf/download.xsp/WDU20200001806/O/D20201806.pdf (accessed on 12 October 2021).
19. Rozporządzenie Ministra Rolnictwa i Rozwoju Wsi z dnia 14 października 2020 r. w sprawie stawki jednolitej płatności obszarowej (Regulation of the Minister of agriculture and Rural Development on the Single Area Payment Rate) za 2020 r., O.J. 2020 pos. 1798. Available online: https://isap.sejm.gov.pl/isap.nsf/download.xsp/WDU20200001798/O/D20201798.pdf (accessed on 12 October 2021).
20. Rozporządzenie Ministra Rolnictwa i Rozwoju Wsi z dnia 7 października 2019 r. w sprawie stawki płatności dodatkowej (Regulation of the Minister of Agriculture and Rural Development on the Rate of Additional Payment) za 2019 r., O.J. 2019 pos. 1958. Available online: http://isap.sejm.gov.pl/isap.nsf/download.xsp/WDU20190001958/O/D20191958.pdf (accessed on 12 October 2021).
21. Rozporządzenie Ministra Rolnictwa i Rozwoju Wsi z dnia 7 października 2019 r. w sprawie stawki płatności dla młodego rolnika (Regulation of the Minister of Agriculture and Rural Development on the Payment Rate for Young Farmer) za 2019 r., O.J. 2019 pos. 1957. Available online: http://isap.sejm.gov.pl/isap.nsf/download.xsp/WDU20190001957/O/D20191957.pdf (accessed on 12 October 2021).
22. Rozporządzenie Ministra Rolnictwa I Rozwoju Wsi z dnia 7 października 2019 r. w sprawie stawki płatności za zazielenienie (Regulation of the Minister of Agriculture and Rural Development on the Payment Rate for Greening) za 2019 r., O.J. 2019 pos. 1964. Available online: http://isap.sejm.gov.pl/isap.nsf/download.xsp/WDU20190001964/O/D20191964.pdf (accessed on 12 October 2021).
23. Rozporządzenie Ministra Rolnictwa I Rozwoju Wsi z dnia 7 października 2019 r. w sprawie stawki jednolitej płatności obszarowej (Regulation of the Minister of Agriculture And Rural Development on the Rate of Additional Payment) za 2019 r., O.J. 2019 pos. 1963. Available online: http://isap.sejm.gov.pl/isap.nsf/download.xsp/WDU20190001963/O/D20191963.pdf (accessed on 12 October 2021).
24. Rozporządzenie Ministra Rolnictwa I Rozwoju Wsi z dnia 11 października 2018 r. w sprawie stawki płatności dodatkowej (Regulation of the Minister of Agriculture and Rural Developmenton the Rate of Additional Payment) za 2018 r., O.J. 2018 pos. 1961. Available online: http://isap.sejm.gov.pl/isap.nsf/download.xsp/WDU20180001961/O/D20181961.pdf (accessed on 12 October 2021).
25. Rozporządzenie Ministra Rolnictwa I Rozwoju Wsi z dnia 11 października 2018 r. w sprawie stawki płatności dla młodego rolnika (Regulation of the Minister of Agriculture and Rural Development on the Payment Rate for Young Farmer) za 2018 r., O.J. 2018 pos. 1964. Available online: http://isap.sejm.gov.pl/isap.nsf/download.xsp/WDU20180001964/O/D20181964.pdf (accessed on 12 October 2021).
26. Rozporządzenie Ministra Rolnictwa I Rozwoju Wsi z dnia 11 października 2018 r. w sprawie stawki płatności za zazielenienie (Regulation of the Minister of Agriculture and Rural Development on the Payment Rate for Greening) za 2018 r., O.J. 2018 pos.1960. Available online: http://isap.sejm.gov.pl/isap.nsf/download.xsp/WDU20180001960/O/D20181960.pdf (accessed on 12 October 2021).
27. Rozporządzenie Ministra Rolnictwa I Rozwoju Wsi z dnia 11 października 2018 r. w sprawie stawki jednolitej płatności obszarowej (Regulation of the Minister of Agriculture and Rural Development on the Single Area Payment Rate) za 2018 r., O.J 2018 pos. 1966. Available online: http://isap.sejm.gov.pl/isap.nsf/download.xsp/WDU20180001966/O/D20181966.pdf (accessed on 12 October 2021).
28. Rozporządzenie Ministra Rolnictwa I Rozwoju Wsi z dnia 9 października 2017 r. w sprawie stawki płatności dodatkowej (Regulation of the Minister of Agriculture and Rural Development on the Rate of Additional Payment) za 2017 r., O.J. 2017 pos. 1895. Available online: http://isap.sejm.gov.pl/isap.nsf/download.xsp/WDU20170001895/O/D20171895.pdf (accessed on 12 October 2021).
29. Rozporządzenie Ministra Rolnictwa I Rozwoju Wsi z dnia 9 października 2017 r. w sprawie stawki płatności dla młodego rolnika (Regulation of the Minister of Agriculture and Rural Development on the Payment Rate for Young Farmer) za 2017 r., O.J. 2017 pos. 1890. Available online: http://isap.sejm.gov.pl/isap.nsf/download.xsp/WDU20170001890/O/D20171890.pdf (accessed on 12 October 2021).
30. Rozporządzenie Ministra Rolnictwa I Rozwoju Wsi z dnia 9 października 2017 r. w sprawie stawki płatności za zazielenienie (Regulation of the Minister of Agriculture and Rural Development on the Payment Rate for Greening) za 2017 r., O.J. 2017 pos. 1894. Available online: http://isap.sejm.gov.pl/isap.nsf/download.xsp/WDU20170001894/O/D20171894.pdf (accessed on 12 October 2021).
31. Rozporządzenie Ministra Rolnictwa I Rozwoju Wsi z dnia 9 października 2017 r. w sprawie stawki jednolitej płatności obszarowej (Regulation of the Minister of Agriculture and Rural Development on the Single Area Payment Rate)za 2017 r., O.J. 2017 pos.

1887. Available online: http://isap.sejm.gov.pl/isap.nsf/download.xsp/WDU20170001887/O/D20171887.pdf (accessed on 12 October 2021).
32. Rozporządzenie Ministra Rolnictwa I Rozwoju Wsi z dnia 13 października 2016 r. w sprawie stawki płatności dodatkowej (Regulation of the Minister of Agriculture and Rural Development on the Rate of Additional Payment)za 2016 r., O.J. 2016 pos. 1708. Available online: http://isap.sejm.gov.pl/isap.nsf/download.xsp/WDU20160001708/O/D20161708.pdf (accessed on 12 October 2021).
33. Rozporządzenie Ministra Rolnictwa I Rozwoju Wsi z dnia 13 października 2016 r. w sprawie stawki płatności dla młodego rolnika (Regulation of the Minister of Agriculture and Rural Development on the payment rate for young farmer) za 2016 r., O.J. 2016 pos. 1701. Available online: http://isap.sejm.gov.pl/isap.nsf/download.xsp/WDU20160001701/O/D20161701.pdf (accessed on 12 October 2021).
34. Rozporządzenie Ministra Rolnictwa I Rozwoju Wsi z dnia 13 października 2016 r. w sprawie stawki płatności za zazielenienie (Regulation of the Minister of Agriculture and Rural Development on the Payment Rate for Greening)za 2016 r., O.J. 2016 pos. 1707. Available online: http://isap.sejm.gov.pl/isap.nsf/download.xsp/WDU20160001707/O/D20161707.pdf (accessed on 12 October 2021).
35. Rozporządzenie Ministra Rolnictwa I Rozwoju Wsi z dnia 13 października 2016 r. w sprawie stawki jednolitej płatności obszarowej (Regulation of the Minister of Agriculture and Rural Development on the Single Area Payment Rate) za 2016 r., O.J. 2016 pos. 1702. Available online: http://isap.sejm.gov.pl/isap.nsf/download.xsp/WDU20160001702/O/D20161702.pdf (accessed on 12 October 2021).
36. Rozporządzenie Ministra Rolnictwa I Rozwoju Wsi z dnia 9 października 2015 r. w sprawie stawki płatności dodatkowej (Regulation of the Minister of Agriculture and Rural Development on the Rate of Additional Payment) za 2015 r., O.J. 2015 pos. 1622. Available online: http://isap.sejm.gov.pl/isap.nsf/download.xsp/WDU20150001622/O/D20151622.pdf (accessed on 12 October 2021).
37. Rozporządzenie Ministra Rolnictwa I Rozwoju Wsi z dnia 31 grudnia 2015 r. w sprawie wielkości zmniejszenia (redukcji) stawki płatności dodatkowej (Regulation of the Minister of Agriculture and Rural Development on the Rate of Additional Payment) za 2015 r., O.J. 2015 pos. 5. Available online: https://isap.sejm.gov.pl/isap.nsf/download.xsp/WDU20160000005/O/D20160005.pdf (accessed on 12 October 2021).
38. Rozporządzenie Ministra Rolnictwa I Rozwoju Wsi z dnia 9 października 2015 r. w sprawie stawki płatności dla młodego rolnika (Regulation of the Minister of Agriculture and Rural Development on the Payment Rate for Young Farmer) za 2015 r., O.J. 2015 pos. 1621. Available online: http://isap.sejm.gov.pl/isap.nsf/download.xsp/WDU20150001621/O/D20151621.pdf (accessed on 12 October 2021).
39. Rozporządzenie Ministra Rolnictwa I Rozwoju Wsi z dnia 9 października 2015 r. w sprawie stawki płatności za zazielenienie (Regulation of the Minister of Agriculture and Rural Development on the Payment Rate for Greening) za 2015 r. Available online: http://isap.sejm.gov.pl/isap.nsf/download.xsp/WDU20150001620/O/D20151620.pdf (accessed on 12 October 2021).
40. Rozporządzenie Ministra Rolnictwa I Rozwoju Wsi z dnia 9 października 2015 r. w sprawie stawki jednolitej płatności obszarowej (Regulation of the Minister of Agriculture and Rural Development on the Single Area Payment Rate) za 2015 r. Available online: http://isap.sejm.gov.pl/isap.nsf/download.xsp/WDU20150001619/O/D20151619.pdf (accessed on 12 October 2021).
41. Council of the European Union. *Council Regulation (EC) No 1782/ 2003 of 29 September 2003 Establishing Common Rules for Direct Support Schemes under the Common Agricultural Policy and Establishing Certain Support Schemes for Farmers and Amending Regulations (EEC) No 2019/93, (EC) No 1452/2001, (EC) No 1453/2001, (EC) No 1454/2001, (EC) 1868/94, (EC) No 1251/1999, (EC) No 1254/1999, (EC) No 1673/2000, (EEC) No 2358/71 and (EC) No 2529/2001*; Council of the European Union: Brussels, Belgium, 2003; Available online: https://eur-lex.europa.eu/legal-content/EN/TXT/PDF/?uri=CELEX:32003R1782&from=EN (accessed on 12 October 2021).
42. Council of the European Union. *Council Decision of 22 March 2004 Adapting the Act Concerning the Conditions of Accession of the Czech Republic, the Republic of Estonia, the Republic of Cyprus, the Republic of Latvia, the Republic of Lithuania, the Republic of Hungary, the Republic of Malta, the Republic of Poland, the Republic of Slovenia, the Slovak Republic and the Adjustments to the Treaties on Which the European Union Is Founded, Following the Reform of the Common Agricultural Policy*; Council of the European Union: Brussels, Belgium, 2004; Available online: https://eur-lex.europa.eu/LexUriServ/LexUriServ.do?uri=OJ:L:2004:093:0001:0017:EN:PDF (accessed on 12 October 2021).
43. Krzyżaniak, M.; Stolarski, M.J.; Warmiński, K. Life cycle assessment of poplar production: Environmental impact of different soil enrichment methods. *J. Clean. Prod.* **2019**, *206*, 785e796. [CrossRef]
44. Krzyżaniak, M.; Stolarski, M.J.; Szczukowski, S.; Tworkowski, J. Life Cycle Assessment of New Willow Cultivars Grown as Feedstock for Integrated Biorefineries. *Bioenergy Res.* **2016**, *9*, 224–238. [CrossRef]
45. Olba-Zięty, E.; Stolarski, M.J.; Krzyżaniak, M.; Gołaszewski, J. Environmental external cost of poplar wood chips sustainable production. *J. Clean. Prod.* **2020**, *252*, 119854. [CrossRef]
46. Olba-Zięty, E.; Stolarski, M.J.; Krzyżaniak, M.; Warmiński, K. Willow Cultivation as Feedstock for Bioenergy-External Production Cost. *Energies* **2020**, *13*, 4799. [CrossRef]
47. Ben Fradj, N.; Jayet, P.A.; Rozakis, S.; Georganta, E.; Jędrejek, A. Contribution of agricultural systems to the bioeconomy in Poland: Integration of willow in the context of a stylised CAP diversification. *Land Use Policy* **2020**, *99*, 104797. [CrossRef]

48. Konstantinavičienė, J.; Varnagirytė-Kabašinskienė, I.; Škėma, M.; Aleinikovas, M. Assessment of the socio-economic factors affecting the development of willow energy plantations in Lithuania. *Balt. For.* **2020**, *26*, 1–8. [CrossRef]
49. Spiegel, A.; Britz, W.; Djanibekov, U.; Finger, R. Policy analysis of perennial energy crop cultivation at the farm level: Short rotation coppice (SRC) in Germany. *Biomass Bioenergy* **2018**, *110*, 41–56. [CrossRef]
50. Faasch, R.J.; Patenaude, G. The economics of short rotation coppice in Germany. *Biomass Bioenergy* **2012**, *45*, 27–40. [CrossRef]
51. Mola-Yudego, B.; Dimitriou, I.; Gonzalez-Garcia, S.; Gritten, D.; Aronsson, P. A conceptual framework for the introduction of energy crops. *Renew. Energy* **2014**, *72*, 29–38. [CrossRef]
52. Helby, P.; Rosenqvist, H.; Roos, A. Retreat from Salix—Swedish experience with energy crops in the 1990s. *Biomass Bioenergy* **2006**, *30*, 422–427. [CrossRef]
53. Xu, X.; Mola-Yudego, B. Where and when are plantations established? Land-use replacement patterns of fast-growing plantations on agricultural land. *Biomass Bioenergy* **2021**, *144*, 105921. [CrossRef]

MDPI
St. Alban-Anlage 66
4052 Basel
Switzerland
Tel. +41 61 683 77 34
Fax +41 61 302 89 18
www.mdpi.com

Energies Editorial Office
E-mail: energies@mdpi.com
www.mdpi.com/journal/energies

www.ingramcontent.com/pod-product-compliance
Lightning Source LLC
Chambersburg PA
CBHW061453220926
51850CB00004B/78
9783036551753